Advances in Computational and Theoretical Chemistry:

From Molecules to Materials

Nadia Faska
https://orcid.org/0000-0001-6203-6904
Ibn Zohr University, Morocco

IGI Global
Scientific Publishing
Publishing Tomorrow's Research Today

Vice President of Editorial	Melissa Wagner
Director of Acquisitions	Mikaela Felty
Director of Book Development	Jocelynn Hessler
Production Manager	Mike Brehm
Cover Design	Jose Rosado

Published in the United States of America by
IGI Global Scientific Publishing
701 East Chocolate Avenue
Hershey, PA, 17033, USA
Tel: 717-533-8845 | Fax: 717-533-7115
Website: https://www.igi-global.com E-mail: cust@igi-global.com

Library of Congress Cataloging-in-Publication Data

LCCN: 2025044761 (CIP Data Pending)
ISBN13: 9798337360584
Isbn13Softcover: 9798337360591
EISBN13: 9798337360607

British Cataloguing in Publication Data
A Cataloguing in Publication record for this book is available from the British Library.

Table of Contents

Detailed Table of Contents

Mobarek Dib, Faculty of Sciences Agadir, University Ibn Zohr, Morocco
Essediq Youssef El-Yakoubi, Faculty of Sciences Agadir, University Ibn Zohr, Morocco

In a dynamic field along the magnetization hard and Easy axis, Permalloy nanoparticles (Ni81Fe19)1-x(Al2O3)x were deposited by RF sputtering on Corning glass substrates. With argon pressure ranges of 4.10-3 and 60.10-3, the samples were exposed to a magnetic field, resulting in Al2O3 substitution rates of 24, 34, and 44%. The permeability spectra calculated from the Gilbert Lifshitz Landau model are consistent with the data and damping coefficient inferred from this change. The damping ratio values are in agreement with the results from the earlier studies following this change. Similar to pressure, the damping coefficient augments linearly, and the Full Width at Half-Maximum values of the imaginary part of the permeability-dependent frequency exhibits a linear correlation with the depth coefficient. Fits used a convolution of Lorentzian (TMS, magnon-electron term) and Gaussian (inhomogeneity) linewidths. Py film FMR showed (4πM_s) increased with H_a, nearing theory. This work clarifies Gilbert damping in magnetic films for spintronic devices, sensor and photovoltaic applications.

Himadri Sekhar Das, Haldia Institute of Technology, Haldia, India
Heranmoy Maity, Ideal Institute of Engineering, Kalyani, India
Sudipta Banerjee, Symbiosis Institute of Technology, Symbiosis International University, Pune, India
Maitrayee Chakrabarty, Narula Institute of Technology, Agarpara, India
Gaurav Sharma, Saveetha School of Engineering, Saveetha Institute of Medical and Technical Sciences, Thandalam, India

The electronic structure of semiconductors and topological materials determines their electrical and optical properties. Semiconductors possess a tunable band gap crucial for electronic and optoelectronic devices, while topological materials exhibit unique surface states protected by symmetry and strong spin–orbit coupling. These states enable robust, dissipationless transport. Understanding their band topology and electronic interactions is key to designing next-generation quantum, spintronic, and energy-efficient devices.

The pursuit of sustainable, lead-free perovskites has fueled the growing interest in halide double perovskites. This study utilizes first-principles calculations to explore the structural, elastic, electronic, and optical properties of Rb2NaInX6 (X=Br, I) and to evaluate their photovoltaic potential. The structural stability and mechanical strength of Rb2NaInBr6 have been substantiated through the application of Goldschmidt's tolerance factors, elastic constants, and bulk moduli. These findings collectively indicate that Rb2NaInBr6 exhibits superior mechanical performance. Electronic analysis indicates direct band gaps of 2.879 eV and 1.35 eV (TB-mBJ) for $Rb_2NaInBr_6$ and Rb_2NaInI_6, respectively, thereby enabling robust absorption across the ultraviolet (UV) to visible (Vis) spectrum. The simulated solar cell architectures (FTO/TiO2/Rb2NaInX6/Cu2O/Au) demonstrate efficiencies of 4.81% for the bromide and 26.6% for the iodide compound, with an absorber thickness of 0.8 μm. These findings identify Rb2NaInI6 as a promising eco-friendly material for high-efficiency, lead-free solar cells.

Accurate computational prediction of NMR chemical shifts requires proper treatment of solvent effects, which influence nuclear shielding through electrostatic polarization, hydrogen bonding, and dispersion interactions. This review examines computational methodologies from continuum solvation models (PCM, COSMO) to explicit approaches incorporating ab initio molecular dynamics and QM/MM schemes. We evaluate method performance and provide practical guidelines for methodology selection based on solute-solvent interaction types and system size. Emerging machine learning approaches are discussed. While significant progress has been achieved, quantitative accuracy requires careful integration of quantum mechanical theory, statistical mechanics, and validation against experimental data.

Molecular docking and virtual screening are computational methods for drug discovery that can help predict the interaction between ligands and targets at high speed. This chapter describes recent developments with

techniques using classical methods, as these models cannot understand the relationships between the different parameters that affect the process. For instance, the adsorption process is a commonly used process for removing contaminants from wastewater. The adsorption technique requires two elements: the adsorbent material and the pollutant; however, the complexity of molecule-surface interactions makes the simulation challenging. The utilization of artificial neural networks in the adsorption process is crucial. Considering that ANN provides close prediction results for adsorption performance and materials efficiency, which helps to design and improve the next generation of adsorbent materials.

Widad Ezhani, Faculty of Sciences, Ibn Tofail University, Kenitra, Morocco
Mohamed El Bakkali, Instituts Supérieurs des Professions Infirmières et Techniques de Santé, Morocco
Said Bouchefra, Higher Institute of Nursing and Health Techniques, Morocco
Wafaa Benkayba, Chemistry, Laboratory of Materials, Nanotechnology, and Environment, Faculty of Sciences, University of Mohammed V, Rabat, Morocco
Yassine Mouniane, Faculty of Sciences, Ibn Tofail University, Kenitra, Morocco
Mohamed Rektouti, Faculty of Sciences, Ibn Tofail University, Kenitra, Morocco
Brahim El Ouardi, Faculty of Sciences, Ibn Tofail University, Kenitra, Morocco
Imane Aitouhanni, ENSIAS, SSLAB, Mohammed V University, Rabat, Morocco
Nour Eddine Amenzou, DEC, Division of Water and Climate, Direction of Studies and Scientific Research, National Center for Nuclear Energy, Sciences and Techniques, Kenitra, Morocco
Yassine Kadmi, LASIRE, Equipe Physico-Chimie de l'Environnement, CNRS UMR 8516, Université Lille, France
Amar Habsaoui, Faculty of Sciences, Ibn Tofail University, Kenitra, Morocco

The adulteration of sesame oil poses significant challenges for public health, food safety, and environmental sustainability, requiring advanced analytical and decision-making tools for detection. This study presents an integrated framework combining Fourier Transform Infrared (FTIR) spectroscopy, $\delta^{13}C$ isotope ratio mass spectrometry, and artificial intelligence within an environmental engineering perspective. FTIR spectra were processed using an unsupervised deep autoencoder, producing a Spectral Deviation Index (SDI) with excellent discriminatory power (AUC = 1.00). In parallel, a Box-Behnken design quantified $\delta^{13}C$ isotopic responses, while machine learning models, including Random Forest ($R^2 = 0.93$) and symbolic regression, accurately predicted adulteration patterns even at low levels. Symbolic regression provided interpretable equations supporting traceability and quantitative decision-making. The hybrid methodology is rapid, non-destructive, and automatable, offering a scalable pathway for sustainable food quality control and environmental monitoring.

Preface

Advances in Computational and Theoretical Chemistry: From Molecules to Materials represents a comprehensive exploration of the transformative role that computational methodologies now play across the chemical and materials sciences. This edited volume is conceived to bridge the gap between foundational theory and cutting-edge application, demonstrating how the integration of quantum mechanics, molecular simulation, and artificial intelligence is reshaping discovery, design, and optimization in fields ranging from quantum materials and renewable energy to environmental remediation and pharmaceutical development.

The increasing sophistication and accessibility of computational tools have fundamentally altered the scientific workflow. What was once a domain of specialized theoretical analysis has expanded into a generative engine for innovation, enabling the *in-silico* prediction of properties, the virtual screening of vast chemical libraries, and the multiscale modeling of complex dynamical processes. This book captures this paradigm shift, curating contributions that illustrate the transition from computation as a tool for explanation to its role as a primary driver of hypothesis generation and materials engineering.

Organized to reflect the logical progression from fundamental principles to interdisciplinary applications, this volume serves both as an academic reference and a practical guide. It is intended for a broad audience, including graduate students and early-career researchers seeking to understand the application landscape, established scientists aiming to integrate new computational approaches into their work, and industry professionals navigating the transition toward data-driven and AI-enhanced research and development.

ORGANIZATION OF THE BOOK

Chapter 1. Quantum Chemical Methods and Pressure's Effect on Permalloy Permeability: Spintronic Devices and Sensor Applications

This opening chapter establishes a critical link between nanofabrication conditions and functional performance. Utilizing micromagnetic models informed by quantum chemical insights, it demonstrates how processing parameters like argon pressure govern magnetic damping in Permalloy nanocomposites. This work provides a foundational case study in tailoring nanomagnetic properties for high-speed spintronic memory and ultrasensitive sensor technologies.

Chapter 2. Electronic Structure of Semiconductors and Topological Materials

Providing the essential theoretical bedrock, this chapter elucidates the quantum mechanical principles of band theory, spin-orbit coupling, and topological invariants that define the behavior of modern electronic materials. It frames the understanding necessary for the rational design of next-generation semiconductors and topological insulators for quantum computing and energy-efficient electronics.

Chapter 3. First-Principles Insights into the Structure and Physical Properties of Perovskite Materials

Focusing on a pivotal material family, this chapter reviews the central role of Density Functional Theory (DFT) in perovskite research. It details how first-principles calculations predict and interpret structural, electronic, and optical properties, establishing the computational foundation for their exceptional performance in photovoltaics and optoelectronics.

Chapter 4. Computational Design of Efficient Lead-Free Double Perovskites for Photovoltaic Applications

Showcasing a proactive design paradigm, this chapter applies a complete computational workflow combining DFT stability analysis with device physics simulations to identify and validate environmentally benign, lead-free double perovskites as high-potential absorbers for sustainable solar energy conversion.

Chapter 5. Computational Engineering of Direct-Bandgap Double Perovskites for Solar Cells

Building directly on the previous chapter, this work performs targeted computational engineering on a related chemical series. Through systematic evaluation, it identifies a specific composition (Rb_2NaInI_6) with a near-ideal electronic structure and predicts superior photovoltaic efficiency, offering a clear synthetic target for experimental validation.

Chapter 6. Theoretical and Computational Advances in Solvent Effect on NMR Spectroscopy: A Comprehensive Review

Addressing a central challenge in molecular characterization, this chapter surveys advanced quantum chemical and solvation models for accurately predicting NMR spectra in solution. This capability is indispensable for bridging computation with experiment in structural elucidation and mechanistic studies across synthetic and medicinal chemistry.

Chapter 7. Molecular Docking and Virtual Screening: Advances, Applications, and Integration with AI in Modern Drug Discovery

This chapter details the computational revolution in pharmaceuticals, illustrating how molecular docking and virtual screening increasingly integrated with machine learning accelerate the identification of novel drug candidates by predicting and optimizing ligand-target interactions with high speed and precision.

Chapter 8. Multiscale Simulation of Reactive Polymeric Systems with RMD and SPH Transport Modeling

Tackling a core challenge in chemical engineering, this chapter presents a framework for coupling atomistic reaction dynamics with continuum transport modeling. This integrated multiscale approach is critical for simulating complex processes like reactive extrusion, enabling the predictive design of polymers with tailored properties.

Chapter 9. Green Synthesis, Characterization and Applications of Tin Nanoparticles

Bridging nanotechnology with sustainable chemistry, this chapter reviews eco-friendly, plant-mediated synthesis routes for tin oxide nanoparticles. It highlights how green synthesis principles can yield nanomaterials with valuable therapeutic and catalytic properties, aligning advanced material production with environmental consciousness.

Chapter 10. Predictive Analytics in Phytochemistry: Modeling Extraction Yields from Opuntia ficus-indica via Solvent Polarity and Regional Variation

Introducing data science into natural product research, this chapter employs machine learning models to predict phytochemical extraction yields. It demonstrates how predictive analytics can optimize solvent selection and experimental design, reducing resource consumption and accelerating the discovery of bioactive compounds.

Chapter 11. Modeling of Adsorption by Artificial Neural Networks: Predicting the Performance of Adsorbent Materials

Focusing on environmental remediation, this chapter advocates for Artificial Neural Networks (ANNs) to model the complex, non-linear adsorption processes central to wastewater treatment. ANNs provide superior predictive capability for contaminant removal, guiding the development of more efficient adsorbent materials.

Chapter 12. Artificial Intelligence-Enhanced Detection of Sesame Oil Adulteration Using FTIR, $\delta^{13}C$ Analysis, and Symbolic Regression

Presenting a sophisticated AI-driven framework for analytical chemistry, this chapter combines spectroscopic and isotopic data with deep learning and interpretable machine learning (symbolic regression) to create a rapid, accurate system for detecting food adulteration, showcasing AI's transformative potential in quality control and consumer safety.

CONCLUSION

The contributions compiled in *Advances in Computational and Theoretical Chemistry: From Molecules to Materials* collectively underscore a decisive shift in scientific inquiry. Computational science has evolved into a primary, predictive discipline that not only interprets the world but also intelligently designs its future constituents. From the quantum spin states in a nanomagnet to the global dynamics of environmental systems, the methodologies explored herein demonstrate an unparalleled capacity to accelerate discovery, enhance precision, and promote sustainability.

A unifying theme across this volume is the synergistic fusion of physical first-principles with data-driven artificial intelligence. This convergence is creating a new paradigm for research one that is more iterative, predictive, and efficient. By presenting a trajectory from core theory to broad application, this book aims to serve as both an inspiration and a practical resource. It is our belief that the insights contained within these chapters will contribute meaningfully to the ongoing evolution of the field, empowering researchers and engineers to tackle the complex, interdisciplinary challenges of the 21st century with increasingly powerful and insightful computational tools.

Acknowledgment

We extend our sincere gratitude to all the authors and contributors who devoted their time and expertise to this volume, Advances in Computational and Theoretical Chemistry: From Molecules to Materials. Their collective efforts have been indispensable in shaping a work that synthesizes cutting-edge research and insights within the field.

Our particular thanks go to the distinguished authors from France, India, Iran, Thailand, and Morocco. Their commitment to advancing knowledge through innovative research was vital to the project's success, and the diversity of their scholarly perspectives has significantly enriched this truly international collaboration.

We are profoundly grateful to the peer reviewers for their rigorous and meticulous evaluation of the manuscripts. Their insightful feedback and constructive critiques were essential in upholding the highest standards of scholarly quality and rigor throughout the publication process. This book's excellence is a direct result of their expertise and dedication.

Finally, we thank IGI Global for providing the platform to disseminate this significant scholarly work to the global academic and professional community.

To all, we express our profound appreciation for your invaluable contributions.

Acknowledgment

Chapter 1

Quantum Chemical Methods and Pressure's Effect on Permalloy Permeability:

Spintronic Devices and Sensor Applications

Mobarek Dib
https://orcid.org/0000-0002-8562-1839
Faculty of Sciences Agadir, University Ibn Zohr, Morocco

Essediq Youssef El-Yakoubi
https://orcid.org/0009-0005-9096-3402
Faculty of Sciences Agadir, University Ibn Zohr, Morocco

ABSTRACT

In a dynamic field along the magnetization hard and Easy axis, Permalloy nanoparticles (Ni81Fe19)1-x(Al2O3)x were deposited by RF sputtering on Corning glass substrates. With argon pressure ranges of 4.10-3 and 60.10-3, the samples were exposed to a magnetic field, resulting in Al2O3 substitution rates of 24, 34, and 44%. The permeability spectra calculated from the Gilbert Lifshitz Landau model are consistent with the data and damping coefficient inferred from this change. The damping ratio values are in agreement with the results from the earlier studies following this change. Similar to pressure, the damping coefficient augments linearly, and the Full Width at Half-Maximum values of the imaginary part of the permeability-dependent frequency exhibits a linear correlation with the depth coefficient. Fits used a convolution of Lorentzian (TMS, magnon-electron term) and Gaussian (inhomogeneity) linewidths. Py film FMR showed (4πM_s) increased with H_a, nearing theory. This work clarifies Gilbert damping in magnetic films for spintronic devices, sensor and photovoltaic applications.

1. INTRODUCTION

The combination of iron and nickel forms a ferromagnetic alloy recognized as Permalloy, which is characterized by a large anisotropic magnetoresistance, low coercivity, almost zero magnetostriction, and high permeability. In this instance, Permalloy, which is constituted principally of 81% nickel and 19%

DOI: 10.4018/979-8-3373-6058-4.ch001

iron ($Ni_{0.81}Fe_{0.19}$), includes a variety of nickel-iron combinations. Permalloy is a dependable spin injector in nonlocal spin-valve experiments due to its high itinerant d-electron spin density asymmetry (Bocklage et al., 2007; Jedema et al., 2001; Miura et al., 2006; van Staa et al., 2008; Vogel et al., 2009; T. Yang et al., 2008). The most common methods for creating Permalloy microstructures and nanostructures are heat evaporation and electron-beam lithography (Bocklage et al., 2009). Both strategies are thoroughly examined, and the Permalloy films are examined in terms of magnetic field associated with damping, coercivity, saturation magnetization with using the GLL model, Lorentzian and Gaussian fitting. Because of their exceptional mechanical, electrical, and magnetic properties, permalloy soft magnetic materials play a crucial in current magnetic materials research (D. C. Jiles, 2003; Koohkan et al., 2008; Lu et al., 2007; Shokrollahi & Janghorban, 2006). Many studies have been conducted in recent decades on the use of permalloy and other soft magnetic materials in magnetic devices, including electric motors, sensors, transformers, and inductive devices (Bas et al., 2003; Hamler et al., 2006; Ripka, 2008) . Thin soft magnetic films have been the subject of much research in the recent past because of their potential use in high-frequency applications, including electromagnetic shielding (J. H. Park et al., 2019) patch antennas (Huitema et al., 2013; T. Wang et al., 2017), tunable microwave filters (Rahman et al., 2015; Sohn et al., 2006), magnetoelastic sensors (Fergen et al., 2002; Ludwig et al., 2002), micro-transformers (Bekker et al., 2006), magnetic inductors (Bekker et al., 2006; Fergen et al., 2002; Ludwig et al., 2002; Rahman et al., 2015; Ripka, 2008; Seemann et al., 2006; Sohn et al., 2006; T. Wang et al., 2016; Yamaguchi et al., 2000), and more. High microwave permeability values are necessary in these applications. A material's microwave permeability can be estimated using its and static permeability (μ_S) and frequency of ferromagnetic resonance (f_{res}).

Thick films of permalloy, which have strong soft magnetic properties, have been employed for electric devices like magnetic sensors (H. S. Park et al., 2004; Ripka, 1992, 2003; Yanai et al., 2018).

Permalloy, a soft magnetic alloy distinguished by its exceptionally low coercivity and high permeability (Tang et al., 2021; W. Xu et al., 2024), finds significant utility in photovoltaic applications (Bose & Wang, 2019; W. Peng et al., 2022), making it a promising for efficient energy transmission and magnetic shielding in high-frequency magnetic links and power-electronic converters in advanced grid components.

Permalloy nanoparticles are studied for magnetic recording. Thin films with high saturation magnetization and permeability are key for 5G, electronics, and miniaturization.

This is primarily attributed to its promising potential in the development and implementation of spintronic and magnetic devices. The unique properties of permalloy facilitate its interaction with a range of materials, including semiconductors and topological quantum materials (Duan et al., 2024).

This interaction is what makes permalloy particularly valuable for the manipulation of quantum properties at the atomic and subatomic levels. The capacity to manipulate these quantum properties underscores permalloy's critical role in the rapidly evolving field of quantum engineering. In turn, the parameter is clearing an area for the creation of the most recent type of electronics, electronic structure of semiconductors and topological materials

The study shows hysteresis loops in the $(Py)_{1-x}(Al_2O_3)_x$ thin film, revealing exchange bias and magnetic anisotropy. Permeability spectra show no resonance at 24% concentration, but decreases with pressure. Thin films have adjustable resonance, suitable for microwave use. Anisotropy and demagnetization cause a reduction of argon pressure.

The lowest α occurs with excellent soft magnetic characteristics, as indicated by the minimal increase in H_{ce} with argon pressure. A high α indicates inhomogeneity and extrinsic damping from spin waves (Manago et al., 2015).

The study explores Gilbert coefficient physical source in thin film magnetic, using radiofrequency sputtering to deposit Permalloy nanoparticles on Corning glass substrates developing in spintronic devices, sensor applications, advanced quantum computing technologies.

These advancements promise not only enhanced performance capabilities but also significant improvements in energy efficiency across various applications. Therefore, conducting thorough research into the utilization of permalloy within photovoltaic systems holds the key to unlocking potential performance improvements and exploring a diverse array of applications that can benefit from its unique characteristics.

Micromagnetic simulations serve as a valuable tool for modeling the intricate magnetization dynamics within Permalloy nanodots when subjected to Ferromagnetic Resonance (FMR) (De Moraes et al., 2025). These simulations allow researchers to predict the generation of spin currents and understand the resonance behavior exhibited by these nanodots and multiscale modeling. Furthermore, micromagnetic simulations play a key element in enhancing our recognizing of various magnetic concepts, including the dynamics of domain walls, the mechanisms behind magnetization switching processes, and the propagation characteristics of spin waves.

The simulations also contribute significantly to the exploration of more complex magnetic structures. Permalloy has unique properties, which make it useful for many different programs in advanced technologies such as the design of functional materials. These applications include its use in Magneto-resistive Random-Access Memory (MRAM) devices (Dempsey et al., 2025), the development of topological transistors, the creation of spin filters, the construction of highly sensitive quantum sensors, and the advancement of neuromorphic computing architectures.

The ability to accurately simulate magnetization dynamics in Permalloy is therefore essential for optimizing the performance and functionality of these emerging technologies. By using micromagnetic simulations, we can gain deeper insights into the fundamental magnetic processes that govern the behavior of Permalloy nanostructures and unlock their full potential for future technological innovations.

2. FOCUS OF THE CHAPTER

The main objectives of this chapter are:

- Introduce the fundamentals of Permalloy thin film, their Permeability Magnetic, Hysteresis Loops, and the ferromagnetic resonance of their magnetic properties.
- Present the theoretical framework of Landau-Lifshitz-Gilbert Model (GLL) as a powerful computational approach for investigating Permalloy nanoparticles. Describe practical components of a GLL calculation choice of exchange–correlation functional. Explain how GLL yields total energies, damping coefficient, and Magnetization alimentation information (coercivity, damping, magnetic field), making it suitable for predicting many derived properties of permalloy.
- Describe how to execute Gaussian and Lorentzian fitting to examine magnetic properties, information electronics, ferromagnetic resonance, and low damping of permeability with the best agreement between simulation and experiment. Describe workflows and explain the hysteresis loops with hard, easy axis: two magnon scattering, surface roughness, eddy current (Zezyulina et al., 2016).
- Present case studies on the Hard and Easy axis planes, hysteresis loops, and magnetic permeability, which emphasize the impact of magnetic fields and doping or substitutions on the various

argon pressures. Provide illustrative examples of hysteresis loops, such as easy and hard axis, emphasizing the low damping and coercivity anticipated by GLL, magnetization computations, and various pressures. Describe how the various parameters, magnetic characteristics, and low damping may be accurately predicted using GLL, Lorentzian, and Gaussian fitting.

- Discuss the strengths and limitations of GLL and utilized the Lorentzian and Gaussian fitting, particularly in relation to the challenges of modeling complex permeability of the permalloy nanoparticle. Summarize and make connections between these models and the influence on predicting the nature of the permalloy: Agreement of the models for the surface roughness and the eddy currant, relatively modest computational design of functional materials compared with higher-level methods, and adaptability to several types of magnetic characteristics and families of nanomaterials.
- Explore future perspectives, such as the integration of GLL with high-volume screening and computational intelligence, for accelerating the discovery and design of next-generation for the computational design of functional materials

3. FUNDAMENTALS OF PERMALLOYS NANOPARTICLES

3.1. Landau-Lifshitz Equation of Motion

The response of a magnetic moment to an external magnetic field can be treated analogously to the response of a spinning top with angular momentum$J$ to an external force F (Gurevich & Melkov, 2020). For such a spinning top, the formula of motion is as given:

$$\frac{\partial J}{\partial t} = \gamma J \times F \tag{1}$$

An electron's orbital in an atom produces a magnetic moment M, which is proportional to the angular moment through the gyromagnetic γ ratio:

$$M = -\gamma J \tag{2}$$

Consequently, the magnetization's equation of motion is expressed as follows when taking into take into consideration an outside factor, such as a magnetic field H_{eff}:

$$\frac{\partial M}{\partial t} = -\gamma \mu_0 M \times H_{eff} \tag{3}$$

The permeability of a vacuum is represented as μ_0. The famous Landau-Lifshitz equation (Eq.3) (Landau & Lifshitz, 1935) explains a precessional motion of magnetization. Here, the effective magnetic field is denoted by H_{eff}:

$$H_{eff} = H_0 + H_d + H_{ex} + H_K \tag{4}$$

which is the total of all the fields impacting the magnetization: the exchange field H_{ex}, which contains both symmetric (Heisenberg) and antisymmetric (Dzyaloshinskii-Moriya) exchange interactions, the dipolar field H_d, the applied magnetic field H_0, and the anisotropy field H_K, incorporating surface and magnetocrystalline magnetic anisotropies. The conservation of the magnetization vector's magnitude is a significant aspect of this equation.

$$\frac{\partial m^2}{\partial t} = 0 \tag{5}$$

This indicates that the magnetization vector is moving on a sphere's surface.

3.2. A model of Susceptibility

We shall discuss the Landau-Lifshitz (LL) equation for a ferromagnetic system in the presence of a magnetic field, which has a low-amplitude moving part h(t) in the *xy* plane and a static part H_{eq} oriented with the z-axis. The equilibrium magnetization M_{eq} is aligned with the static field, and the dynamic component m(t) oscillates at the equilibrium location (Figure.1(b)). Thus, the total magnetic field and total magnetization of (Eq. 3) is defined as:

$$H_{eff} = H_{eq} + h(t) \tag{6}$$

$$M_{eff} = M_{eq} + m(t) \tag{7}$$

With

$$h \ll H_{eff}\ ,\ m \ll M_{eff} \tag{8}$$

Equations (6)– (7) can be substituted into the equation of motion (Eq.3) to provide:

$$\frac{\partial m}{\partial t} = -\gamma\ \mu_0\left[M_{eq} \times H_{eq} + M_{eq} \times h + m \times H_{eq} + m \times h\right] \tag{9}$$

The first term vanishes because the equilibrium magnetization in that situation is parallel to the static field, and the last term can also be ignored as both h and m are minor compared to the static parts' magnitude and the result is of second order in tiny amounts. It in the simplified LL formula:

$$\frac{\partial m}{\partial t} = -\gamma\ \mu_0\left[M_{eq} \times h + m \times H_{eq}\right] \tag{1}$$

supposing that h and m have a sinusoidal based on time modulation, we solve (Eq.10), $h(t) = h e^{i\omega t}$ and $m(t) = m e^{i\omega t}$

$$i\omega m = -\gamma\ \mu_0\left[M_{eq} z \times h + m \times H_{eq} z\right] = -z \times \left[\omega_M h - \omega_M m\right] \tag{11}$$

Where $\omega_H = \gamma \mu_0 H_{eq}$ and $\omega_M = \gamma \mu_0 H_S$. The following is obtained by projecting a coordinate system defined in Figure.1 onto its axes:

$$\begin{cases} i\omega m_x = \omega_M h_y - \omega_H m_y \\ i\omega m_y = -\omega_M h_x + \omega_H m_x \\ i\omega m_z = 0 \end{cases} \tag{12}$$

Where the magnetic field's constituent parts have the following solutions:

$$\begin{pmatrix} h_x \\ h_y \end{pmatrix} = \frac{1}{\omega_M} \begin{pmatrix} \omega_H & -i\omega \\ -\omega & \omega_H \end{pmatrix} \begin{pmatrix} m_x \\ m_y \end{pmatrix} \tag{13}$$

Equation (7) can be expressed as $m = \overline{\chi}_p h$, where $\overline{\chi}_p$ is the Polder susceptibility tensor defined as:

$$\overline{\chi}_p = \begin{bmatrix} \chi & i\kappa \\ -i\kappa & \chi \end{bmatrix} \tag{14}$$

With $\chi = \frac{\omega_H \omega_M}{\omega_H^2 - \omega^2}$ and $\kappa = \frac{\omega \omega_M}{\omega_H^2 - \omega^2}$

The nonsymmetric, non-diagonal polar matrix reveals the gyrotropic qualities of the medium. With an angle shift of $\frac{\pi}{2}$, it exhibits how longitudinal magnetic field parts collaborate with both their parallel and perpendicular magnetization components. The magnetization vector's precessional motion in response to microwave field stimulation is emphasized by this characteristic. Also, the components of the Polar matrix become more resonant as the frequency gets nearer to the pole.

$$\omega \equiv \omega\omega_M = \gamma\mu_0 H_{eq} \tag{15}$$

This is equivalent to the bulk isotropic ferromagnet's uniform resonance condition. Since the dynamic effective field additionally includes the exchange field, the magneto-crystalline anisotropy field, and the demagnetizing field specified by the sample geometry, the resonance condition in a finite sample and/or non-uniform mode is different from that defined by (Eq.15). These topics will be covered in the sections that follow.

3.3. Ferromagnetic Resonance in Various Sample Configurations

Kittel demonstrated in 1947 that the resonance frequency is highly dependent on the sample shape, and that the demagnetizing variables can be used to explain this phenomenon (Kittel, 1948). Let's examine the resonance requirement for a uniform magnetization oscillation in an ellipsoid with principal axes parallel to the x, y, and z axes, as illustrated in Figure. 1(b). A small amplitude external oscillating field h_1 is applied in the x, yplane, while a static external field H_0 is applied along z. Thus, the system's effective field is described as:

Kittel demonstrated in 1947 that the resonance frequency is highly dependent on the sample shape, and that the demagnetizing variables can be used to explain this phenomenon (Kittel, 1948). Let's examine the resonance requirement for a uniform magnetization oscillation with principal axis parallel with x and, y axis, as illustrated on Figure. 1(b). A static external field H_0 is applied along z, and a small amplitude external oscillation field h_1 is adapted in the x, y plane. Thus, the effective field of the system is defined as:

$$H_{eff} = H_0 + H_{dem} \tag{16}$$

where the demagnetizing tensor N provides the demagnetizing field H_{dem} as:

$$H_{dem} = -\overline{N}M = \begin{pmatrix} N_x & 0 & 0 \\ 0 & N_y & 0 \\ 0 & 0 & N_z \end{pmatrix} \begin{pmatrix} m_x \\ m_y \\ M_S \end{pmatrix} \tag{17}$$

N_x, N_y, and N_z are the demagnetizing coefficients. Thus, effective field's static and dynamic parts can be described as:

$$\begin{cases} h_x = h_{1x} - N_x m_x \\ h_y = h_{1y} - N_y m_y \\ h_{eq,z} = H_0 - N_z M_S \end{cases} \tag{18}$$

The Kittel susceptibility tensor can be obtained by solving the linearized LL (Eq.3) for h_1.

$$h_1 = \overline{\chi}_\kappa^{-1} \mathrm{m} = \begin{pmatrix} h_{1x} \\ h_{1y} \end{pmatrix} = \frac{1}{\omega_M} \begin{pmatrix} \omega_H & -i\omega \\ -\omega & \omega_H \end{pmatrix} \begin{pmatrix} m_x \\ m_y \end{pmatrix} \tag{19}$$

where $\omega_0 = \gamma\mu_0(H_0 - N_z M_S), \omega_x = \omega_H - N_x\omega_M$, and $\omega_y = \omega_H - N_y\omega_M$. As the frequency ω gets in proximity to the pole, the components of the tensor diverge $(det(\overline{\chi}_\kappa^{-1}) = 0)$.

$$\omega = \sqrt{\omega_x \omega_y} \tag{20}$$

The resonance frequency for the previously defined ω_x and ω_y can be determined by applying the Kittel formula:

$$\omega^2_{r,\alpha=0} = \gamma^2\left(H + 4\pi[N_y - N_z]M\right)\left(H + 4\pi[N_x - N_z]M\right) \tag{21}$$

Figure 1. (a) Represented the Precession of magnetization with damping, (b) adding the damping around the applied bias field.

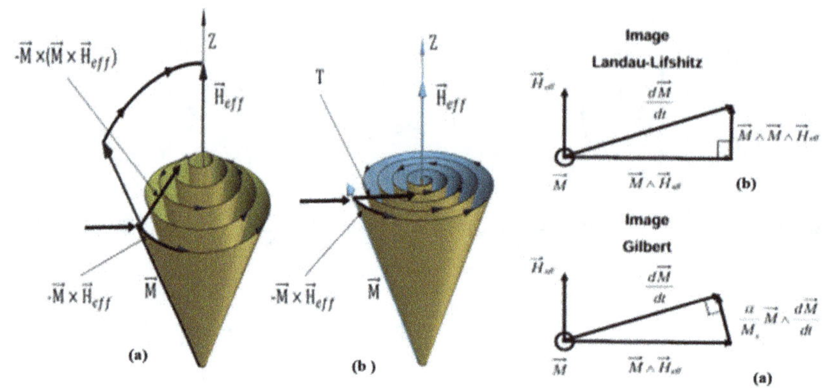

3.4. Resonance Frequency and Uniaxial Magnetic Anisotropy

Resonance state is influenced by the magnetic anisotropy energy, which changes the total effective field. The anisotropy energy density's first order term in a uniaxial crystal is summarized as follows:

$$H_{an} = -K_u cos^2\theta = -K_u \left(\frac{M_i}{M_S}\right)^2 \tag{22}$$

where K_u is the anisotropy constant, M_i is the part of M along this axis, θ is the angle associated with the magnetization and the main axis. The embedded effective field of this energy writes:

$$H_{an} = -\frac{1}{\mu_0}\frac{\partial U_{an}}{\partial M} \tag{23}$$

This can be represented as follows for the uniaxial crystal:

$$H_{an} = -\frac{2}{\mu_0}\frac{K_u M_i}{M_S^2} \tag{24}$$

We can analyze the generic ellipsoid situation when the uniaxial anisotropy axis e_i is across y and H_0 is at z. By expressing the anisotropy field similarly to the demagnetizing field, the effective field can be described as the sum of a form and a uniaxial anisotropy contribution: $H_{an} = -\overline{N}_{an} M$

$$H_{dem} = -(N + \overline{N}_{an})M = -\begin{pmatrix} N_x & 0 & 0 \\ 0 & N_y + \frac{2K_u}{\mu_0 M_s^2} & 0 \\ 0 & 0 & N_z \end{pmatrix}\begin{pmatrix} m_x \\ m_y \\ M_S \end{pmatrix} \tag{25}$$

In the event that the nanoparticle is magnetized in-plane, with $N_x = N_y = 0$ but $N_z = 1$, the resonance frequency defined by:

$$\omega_{res} = \gamma\mu_0\sqrt{H_0\left(H_0 + M_S - 2\frac{K_u}{\mu_0 M_S}\right)} = \gamma\mu_0\sqrt{H_0\left(H_0 + M_{eff}\right)} \tag{26}$$

where the effective magnetization is defined as $M_{eff} = M_S - 2\frac{K_u}{\mu_0 M_S}$ Symmetry breaking at the film's surface can cause magnetic anisotropy in thin magnetic films (Bruno & Renard, 1989). The anisotropy constant in this case may be physically divided into two sections; a second aspect symbolizes the surface contribution, while the start is related to the volume contribution: $K_u = K_V + \frac{\sum K_S}{t}$ (where $\sum K_S$ denotes the respective contributions of the two interfaces). The uniaxial surface anisotropy constant K_S is responsible for the thickness attachment of K_u and thus, the thickness reliance of the effective magnetization:

$$M_{eff} = M_S - 2\sum\frac{K_S}{\mu_0 M_S t} \tag{27}$$

Figure 2. Dispersion relation is altered when a uniaxial anisotropy is present.

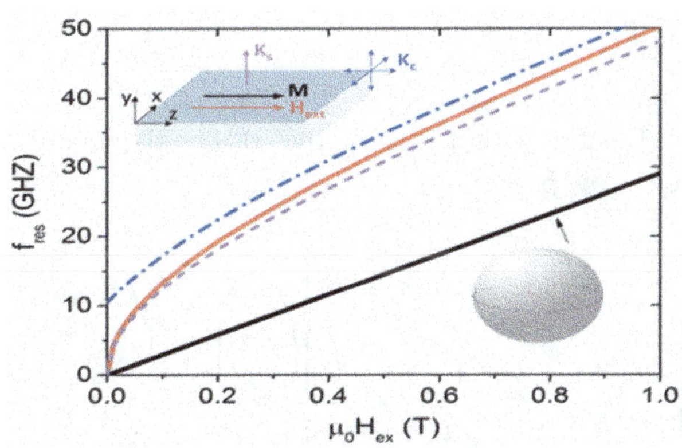

Figure 2 depicts the dispersion relations for a sphere (black line) and an in-plane magnetic film (red line) with parameters $\mu_0 M_S = 2T$ and $\frac{\gamma}{2\pi} = 29\left(\frac{GHz}{T}\right)$. Solid and dash- lines, respectively, represent the thin film's distribution characteristics by a uniaxial and cubic anisotropy with $K_u = 50\frac{kJ}{m^3}$ and $K_u = 100kJ/m^3$ at out-of-plane.

3.5. Cubic magnetic Isotropy's Impact on the Resonance Frequency

When a cubic crystal is involved, the lowest-order magneto-crystalline anisotropy energy is expressed as:

$$U_c = K_c\left(\alpha_1^2\alpha_2^2 + \alpha_2^2\alpha_3^2 + \alpha_3^2\alpha_1^2\right) \tag{28}$$

The anisotropy field is provided by:

$$H_c = -\frac{2}{\mu_0}\frac{K_u}{M_S^4}\begin{pmatrix} M_1(M_2^2 + M_3^2) \\ M_2(M_3^2 + M_1^2) \\ M_3(M_1^2 + M_2^2) \end{pmatrix} \tag{29}$$

Similar to the demagnetizing field, the anisotropy field can be characterized with a little time-varying part $m(t)$ and a static part, M_0 (Walker, 1967):

$$H_c = -\frac{2}{\mu_0}\frac{K_u}{M_S^4}\begin{pmatrix} M_{01}(M_{02}^2 + M_{03}^2) \\ M_{02}(M_{03}^2 + M_{01}^2) \\ M_{03}(M_{01}^2 + M_{02}^2) \end{pmatrix} + \overline{N}_C M \tag{30}$$

where the anisotropic tensor's off-diagonals $\overline{N}_C$:

$$N_C^{ij} = -\frac{4}{\mu_0}\frac{K_u}{M_S^4} M_{0i} M_{0j} (i \neq j) \tag{31}$$

And the perpendicular components are:

$$N_C^{ij} = -\frac{2}{\mu_0}\frac{K_u}{M_S^4}\left(M_S^2 - M_{0i}^2\right) \tag{32}$$

Pretend for the moment M that direction is along the $\langle 001 \rangle$ axis, where $Kc > 0$ (easyaxis). Consequently, the effective field can now be written as:

$$H'_{\text{dem}} = -\left(N + N_c\right) \mathrm{m} = -\begin{pmatrix} N_x + \frac{2K_C}{\mu_0 M_s^2} & 0 & 0 \\ 0 & N_y + \frac{2K_C}{\mu_0 M_s^2} & 0 \\ 0 & 0 & N_z \end{pmatrix}\begin{pmatrix} m_x \\ m_y \\ M_S \end{pmatrix} \tag{33}$$

Hence the dispersion relation for the magnetic film on the planet is as follows:

$$\omega_{res} = \gamma \mu_0 \sqrt{H_{eff}\left(M_S + M_{eff}\right)} \tag{34}$$

Where $H_{eff} = H_0 + 2\frac{K_C}{\mu_0 M_s^2}$A dashed line in Figure. (1) represents the dispersion that takes cubic anisotropy into account.

3.6. Landau-Lifshitz-Gilbert (LLG) Model

A formula for distribution can be obtained for single spin movement using quantum mechanics. This equation is called the Landau–Lifshitz (LL) (Landau & Lifshitz, 1935) equation which is described as follows:

$$\frac{d}{dt}\langle S \rangle = \frac{g\mu_B}{\hbar} \langle S \times B \rangle \quad (35)$$

3.6.1. View of Phenomenology

The gyromagnetic rapport is the global term, with B and S respectively the magnetic field and the spin:

$$\gamma = \frac{g\mu_B}{\hbar}$$

According to the macroscopic model, the dispersion of uniform spin defines the magnetization vector, which is stated as follows:

$$M = -\frac{g\mu_B}{\hbar}\langle S \rangle \quad (36)$$

Thus, under the influence of a magnetic outside field, the magnetization formula of motion is expressed by:

$$\frac{dM}{dt} = -\frac{g\mu_B}{\hbar}\langle M \times H \rangle \quad (37)$$

Equation (3) can be extended to the Landau–Lifshitz (LL) formula by substituting H_{eff} for H.

Based to the Landau–Lifshitz model, the total angular momentum and the total magnetic moment are associated, resulting in a cumulative torque and produces a processional trajectory.

As seen in Figure.1(a), It indicates the point of the magnetization vector spins in a circular orbit around the effective magnetic field for an indefinite amount of time with an angular frequency of $\omega = \gamma H_{eff}$. In normal situations, magnetization's precession amplitude diminishes over time, causing a spiral pattern. Gilbert (Gilbert, 1955, 2004) proposed a "viscous" damping model to dampen magnetization precession, allowing for practical systems by studying a phenomenologically described dissipation concept. The Landau-Lifshitz-Gilbert equation is commonly used in conjunction with the Landau-Lifshitz equation to accurately predict macro-spin dynamics, despite the latter's importance in understanding magnetization dynamics in ferromagnetic materials (Landau & Lifshitz, 1935), which Gilbert (Gilbert, 2004) subsequently reformulated as:

$$\frac{d\vec{M}}{dt} = -\mu_0\gamma\left(\vec{M} \times \vec{H}_{eff}\right) + \frac{\alpha_{eff}}{M_S}\left(\vec{M} \times \frac{d\vec{M}}{dt}\right) \quad (38)$$

In this case, $\vec{H}$, $\vec{M}$, and α_{eff} denote for the effective field, magnetization, and Gilbert damping, respectively. A key concept in explaining the behavior of charged particles with spin in a magnetic field is the gyromagnetic ratio, which is calculated using the Lande g-factor formula (Salahun et al., 2001): $\gamma = \frac{ge}{2m}$.

The effective Magnetic Fields: $H_{eff} = H_a + H_{dem} + H_{ext} + H_{exc}$

It combines the exchange field (H_{exc}), externally applied magnetic field (H_{ext}), demagnetizing field (H_{dem}), and anisotropy fields (H_a). Crucially, the (LLG) equation also conserves the magnetization's magnitude, just like the LL model does. To clarify, The Gilbert coefficient term provides an accurate representation of the experimental process. Numerous magnetic systems' magnetization dynamics have been investigated using the (LLG) equation. Anisotropy, demagnetization, and the applied external field add up to the effective field. It is reasonable to assume that the magnetization will be constant during the movements if the size of the magnetic system is appropriate. For more data points and in the case of inhomogeneous fluctuations, the magnetic moment then relies on location in environment. The effective magnetic field is influenced by exchange interactions, which allows for non-homogeneous magnetic excitations (Bloch, 1930). Numerous lattice sites are involved in these excitations, which are known as spin waves. Existing literature provides more information on these topics (Barman & Sinha, 2018; Demokritov et al., 2001). It is commonly believed that intrinsic Gilbert coefficient originates in the link of spin and orbit, as this process doesn't retain spin; nevertheless, at any time has a suggestion formed a basis of logic. Gilbert coefficient is ameliorated in ferromagnetic nanostructures and thin films due to the coupling of disorder broadening and non-local spin relaxation movement to the spin dynamics. Among the renowned techniques for validating the type of damping is to determine the ferromagnetic resonance linewidth, is often calculated by rephrasing it in terms of a relaxation time and taking use of the dominant relaxation processes.

That's also interesting that effective field H_{eff} takes into account internal fields, particularly isotropy, which can play an important role and can be incorporated into the resulting final equation. Several key points and considerations are proposed to deepen our understanding of these phenomena by rigorously accounting for demagnetization factors as well as the domains used in this specific geometry.

$$H_{eff} = H_a - 4\pi N_{ij} M_j = \begin{bmatrix} H_x - 4\pi N_x m_x \\ H_y - 4\pi N_y m_y \\ H - 4\pi N_z M \end{bmatrix} \tag{39}$$

Using this expression, we can examine the right side of equation (Eq 36) while leaving out the damping.

$$-\gamma M H_{eff} = -\gamma \begin{bmatrix} i & j & k \\ m_x & m_y & M \\ H_x & H_y & H_z \end{bmatrix} = -\gamma \begin{bmatrix} \left(H + 4\pi [N_y - N_z] M\right) m_y - M h_y \\ -\left(H + 4\pi [N_x - N_z] M\right) m_x - M h_x \\ O(m^2) \end{bmatrix} \tag{40}$$

This condition will be useful because it is of the order O (m^2), and it can be expressed by separating the above expression and using (Eq 36).

$$\frac{dM}{dt} = \begin{bmatrix} 0 & -\gamma D_y \\ \gamma D_y & 0 \end{bmatrix} m + \begin{bmatrix} 0 & \gamma M \\ -\gamma M & 0 \end{bmatrix} h \tag{41}$$

With $\begin{cases} D_x = H + [N_y - N_z] 4\pi M \\ D_y = -H - [N_x - N_z] 4\pi M \end{cases}$

To make things easier, we limit our analysis to two dimensions because magnetization's duration is nearly constant, implying that its time derivative is 0. The main magnetization, which is parallel to the applied magnetic field M, remains constant over time. The small magnetization m evolves due to a magnetic field perturbation, which we can express as $h = h_0 e^{-i\omega t}$. This gives us the equation $\frac{dM}{dt} = imh$. Thus, we can reformulate our expression for susceptibility $m = \chi h$ by adjusting our (Eq.39) accordingly.

$$\begin{bmatrix} i\omega & \gamma D_x \\ \gamma D_y & i\omega \end{bmatrix} m = \begin{bmatrix} 0 & \gamma M \\ -\gamma M & 0 \end{bmatrix} h \tag{42}$$

The expression now becomes:

$$m = \frac{1}{-\omega^2 - \gamma^2 D_x D_y}\begin{bmatrix} \gamma^2 D_x M & i\omega\gamma M \\ -i\omega\gamma M & -\gamma^2 D_y M \end{bmatrix} h \tag{43}$$

The resonance of the undamped oscillation occurs when a critical element, namely the denominator's determinant, approaches zero. This phenomenon then causes a resonance peak, resulting in a significant increase in oscillations. Resonance can be defined as follow:

$$\omega^2_{r,\alpha=0} = \gamma^2\left(H + 4\pi[N_y - N_z]M\right)\left(H + 4\pi[N_x - N_z]M\right) \tag{44}$$

With using the $H_{ani} = \frac{2K_u}{M}$ and simplification we find:

$$\omega^2_{r,\alpha=0} = \gamma^2\left(H + \frac{2K_u}{M} + 4\pi M\right)\left(H + \frac{2K_u}{M}\right) \tag{45}$$

The Gilbert damping term, which plays an important role in our analysis, significantly complicates the search for a solution. To better understand this complication, we will investigate this specific term in depth, with the goal of precisely determining the resulting susceptibility:

$$\frac{\alpha}{M} M \times \frac{dM}{dt} = \begin{bmatrix} i & j & k \\ M_x & M_y & M_z \\ \frac{dM_x}{dt} & \frac{dM_y}{dt} & \frac{dM_z}{dt} \end{bmatrix} \tag{46}$$

Using the information provided above, we have:

$$M_x = m_x,\ M_y = m_y, M_z = M,\ \frac{dMx}{dt} = \frac{dmx}{dt}, \frac{dMy}{dt} = \frac{dmy}{dt} \text{ and } \frac{dM_z}{dt} = 0.$$

The damping term is thus:

$$\frac{\alpha}{M}\begin{bmatrix} -M\frac{dm_y}{dt} \\ M\frac{dm_x}{dt} \\ O(m^2) \end{bmatrix} = \alpha\begin{bmatrix} 0 & -1 & 0 \\ 1 & 0 & 0 \\ 0 & 0 & 0 \end{bmatrix}\begin{bmatrix} \frac{dm_x}{dt} \\ \frac{dm_y}{dt} \\ 0 \end{bmatrix} \tag{47}$$

Ignoring the order O(m^2). This expression can be added to (Eq. 39) to give:

$$\frac{dM}{dt} = \begin{bmatrix} 0 & -\gamma D_x \\ \gamma D_y & 0 \end{bmatrix} m + \begin{bmatrix} 0 & \gamma M \\ -\gamma M & 0 \end{bmatrix} h + i\omega\alpha\begin{bmatrix} 0 & -1 \\ 1 & 0 \end{bmatrix} m \tag{48}$$

According to the theory, the characteristics of films with uniaxial anisotropy in plane and the resolution of a particular equation (Eq. 45) using the macro-spin approximation are the main topics. The expressions for μ' and μ" are as follows (Ge et al., 2007; Kuanr et al., 2003; Spenato et al., 2007):

$$\begin{cases} \mu' = 1 + 4\pi M_S \gamma^2 \dfrac{(H_k^{dyn} + 4\pi M_S)(\omega_r^2 - \omega^2) + \alpha_{eff}^2 \omega^2 H_k^{dyn}}{\left(\omega_r^2 - (1+\alpha_{eff}^2)\omega^2\right)^2 + \left(\omega\gamma\alpha_{eff}(2H_k^{dyn} + 4\pi M_S)\right)^2} \\ \mu'' = 4\pi M_S \omega\gamma\alpha_{eff} \dfrac{(H_k^{dyn} + 4\pi M_S)^2\gamma^2 + (1+\alpha_{eff}^2)\omega^2}{\left(\omega_r^2 - (1+\alpha_{eff}^2)\omega^2\right)^2 + \left(\omega\gamma\alpha_{eff}(2H_k^{dyn} + 4\pi M_S)\right)^2} \end{cases} \quad (49)$$

Here, $4\pi M_S$ represents saturation magnetization, $H_k^{dyn}$the dynamic magnetic anisotropy field, and $\omega_r = 2\pi f_{FMR}$. Fitting the Landau-Lifshitz-Gilbert (LLG) model to Eq. (46) and minimizing the difference between the resulting permeability spectra and experimental data refines parameter values.

We can further simplify the numerator of each expression by noting that in practice we always have $f_M \gg f_0$.We therefore have the approximate expressions for susceptibility: We finally obtain a Lorentzian resonance profile, The resonance frequency of such a resonance profile isf_r The linewidth f is proportional to the damping factor, In excellent accordance with the data accomplished in the time domain:

$$f = \alpha\left(2H_{eff} + M_S\right) = \frac{2}{\tau} \quad (50)$$

Remember that in most cases, we have $f_0 \ll f_M$. the measured line width is therefore:

$$f \approx \alpha\gamma_0 M_S \quad (51)$$

In an inductive technique measurement, the linewidth is constant throughout the entire studied frequency range as long as the damping factor α is constant. This contrasts with the linewidth in the field H evolving continuously as a function of frequency, in FMR measurements, given by:

$$\frac{\gamma_0}{2\pi} H = 1.16\alpha f \quad (52)$$

It is interesting to calculate the maximum amplitude of μ"", obtained for $f = f_r$:

$$\mu''_{max} = \frac{\gamma_0 M_S}{2\pi\alpha f_r} \quad (53)$$

3.6.2. Ferromagnetism for Hyperfrequency.

As shown in Figure.3, the permeability spectrum is generally divided into three application zones, which are described below. Key characteristic quantities, regularly cited throughout this manuscript, are noted first. (H_{eff}), the material's effective magnetic field, and (M_S), the saturation magnetization, will have a direct impact on the three primary permeability-describing quantities: μ_s : the quasi-static permeability, which is a property of the permeability level of the material and is expressed as:

$$\mu_s = 1 + \frac{M_S}{H_{eff}} \tag{54}$$

f_r: the ferromagnetic resonance frequency, which is expressed as follows and corresponds to the maximum losses:

$$f_r = \frac{\gamma_0}{2\pi}\sqrt{\left(H_{eff} + M_S\right)H_{eff}} \tag{55}$$

Figure 3. μ'' the permeability's real and imaginary parts, namely along the hard axis are shown for typical Permalloy parameters

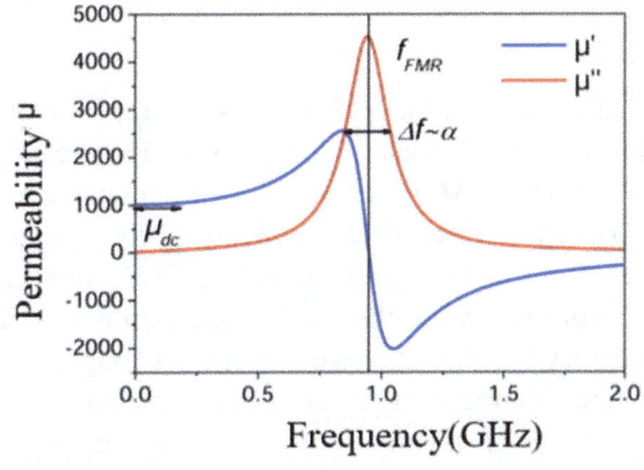

$\frac{\gamma_0}{2\pi} = 30GHz.T^{-1}$, $\mu_0 M_S = 1T$, $H_k = 5(Oe)$, $H_{dc} = 5(Oe)$ and $\alpha = 0.007$.

We have represented the width of the resonance line, both by the peak-to-peak width of the real part (μ'), as well as by the imaginary part the width at half of its greatest extent (μ''). We represented the resonance frequency and maximum of the imaginary part (μ''), but also the annulation point and the inflection point of the real part (μ').

3.7. Magnetic Properties

3.7.1. Magnetization

The density of magnetic dipole moments, whether induced or permanent, within a substance is possible symbolized by a vector field magnetization. Both the intrinsic spin of electrons or atomic nuclei and tiny electric currents brought on by electron mobility in atoms are the sources of these dipole moments. The total magnetization of a material is determined by its internal magnetic moments as well as how it adapts to an external magnetic field, as ferromagnetic materials demonstrate. A material's magnetization, which varies from point to point inside the material, regulates how the material interacts with and modifies magnetic fields, making it possible to identify the forces resulting from these interactions. The magnetic moment per unit mass or bulk of specimen is known as magnetization (M) (D. C. Jiles, 2017).

$$M = \frac{m}{Volume} \tag{56}$$

3.7.1.1. Saturation Magnetization

If a material has n atomic magnetic dipoles per unit volume, each with a magnetic moment of m, then the saturation magnetization, which is the largest magnetic moment per unit volume when all dipoles align properly, becomes n × m. There are two types: total saturation, which only happens at higher fields, and technical saturation (Ms), which is attained when the material turns into a single magnetic domain. Forced magnetization, which is a progressive strengthening of the unplanned magnetization in a single field, causes the magnetization to grow slightly even after technical saturation.

3.7.1.2. Magnetic Hysteresis

Hysteresis, which is crucial for devices like electric motors, transformers, and magnetic recording, is significant for the behavior of magnetic materials. Hysteresis loops can have many different forms, and parameters constituted by $B(H)$ or $M(H)$ capture important characteristics. The formula for the relationship is $B = \mu_0(H + M)$. In soft materials, B and $M(H)$ differ very little, while in hard materials, H and M are similar and their loops differ significantly. The $B(H)$ loop more correctly characterizes behavior under operational conditions, while the $M(H)$ loop exhibits intrinsic magnet qualities. The definitions of coercivity fields vary based on the zero induction or magnetization point. Coercivity field (H_C)and remanent magnetization (H_r)are key parameters in this context

Remanence: This displays the magnetization that occurs when a specimen is exposed to a strong field and subsequently taken out. The ability of a ferromagnet to become magnetized on its own, even in the absence of external pressures, is illustrated by this natural quantity. While being of the same order of magnitude as the spontaneous magnetization M_S, $H_r$may be much lower than $M_S$due to a variety of structural or geometrical factors.

Coercivity Field (H_C): This field is required to reduce the magnetization from the residual value to zero. The (H_C) quantifies the order of magnitude of the fields required to reverse the magnetization of a material. In opposite of remanance, the coercive field ranges from less than $1A\,m^{-1}$ to more than $106A$ m^{-1}, an incredibly broad range. Handling The magnetic hysteresis cycle, which is represented by the magnetization curves $M = f(H)$ or $B = f(H)$, shows how the magnetic material reacts to the applied field H. The applied field (H) and the mobility of the Bloch walls determine the $B = f(H)$ diagram's shape (Robert, 1999). First, the magnetization process in a magnetic substance subjected to a magnetic field will be examined. Let's investigate a virtually flawless iron single crystal. Figure. 4(a) depicts the existence of magnetic domains. Figure. 4(c) illustrates that when an excitation magnetic field H is applied, the energy of the magnetic moments increases in domains 2 and 3 but decreases in domain 5. The volume of these domains is lowered by respecting the sample's minimal energy constraint. The domain structure will gradually take on the appearances seen in figure. 4(b-c) as the value of the field H rises. At last, when the magnetic field is sufficiently increased, just one domain oriented in accordance with H remains (Bertotti, 1998; Chen, 2013; Cullity & Graham, 2011; Robert, 1999).

Figure 4. A ferromagnetic single crystal's Weiss domain structure changes as the applied magnetic field increases (Cullity & Graham, 2011).

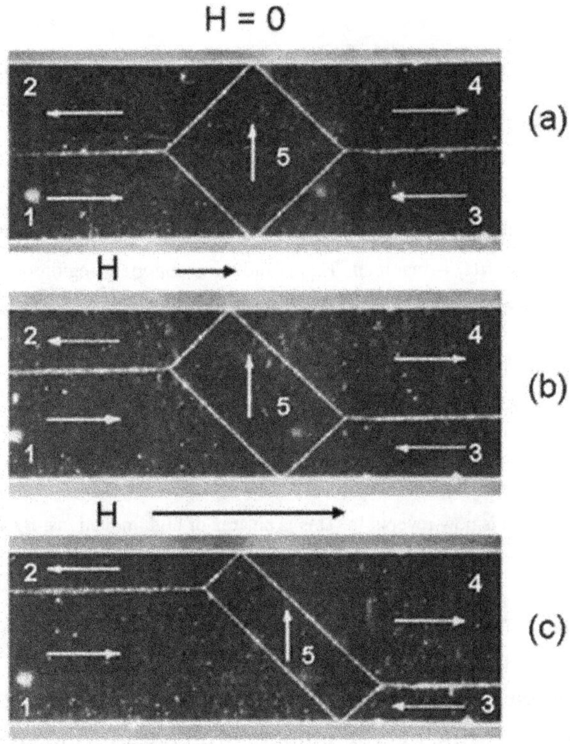

Bloch walls are completely free to move in this perfect example. This indicates that the process is reversible and that the domain structure returns to its starting state by shutting off the magnetic field. The $B = f(H)$ diagram's shape is significantly influenced by structural flaws. Table 1 shows how point defects affect Bloch wall mobility inside a single crystal. Each grain in a polycrystalline media (such as powders made by mechanical grinding) goes through the same procedure as that outlined in Table. 1 Additionally, the structure of the grains is crucial, and the magnetostatic and magneto-strictive interactions between nearby grains complicate the situation (Robert, 1999).

Table 1. Point defects' effects on Bloch wall motion in a single crystal (Robert, 1999).

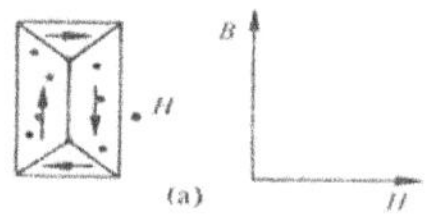	Due of the two flaws supporting them, the Bloch walls on this typical domain structure seem rectilinear.
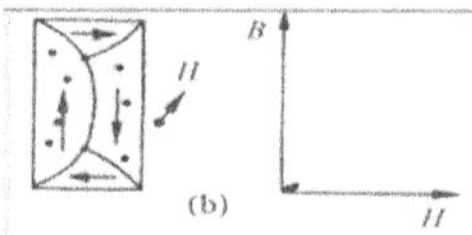	Similar to elastic membranes, the walls undergo reversible deformation when exposed to a weak field. There is no change in the anchor points. It is the area where inductions are weak: B/Bmax is less than 0.05.
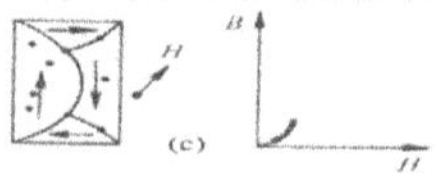 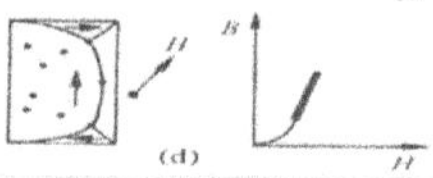	In the average field, the anchoring points give way due to wall pressure. The walls ascend in a sequence of movements as noted by the Barkhausen experiment. These wall movements cannot be reversed. The slope of B = f(H) is maximal. This is the most important region of intermediate inductions, defined by the irreversible displacement of Bloch walls over distances on the order of Weiss domain widths.
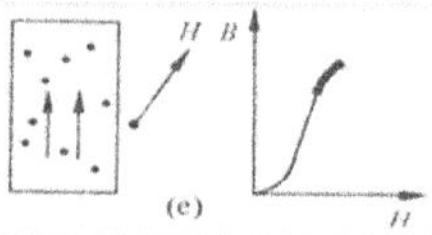	The polarization is kept parallel to the direction of easy magnetization nearest to H. Through irreversible wall displacements, polarization has fully increased.
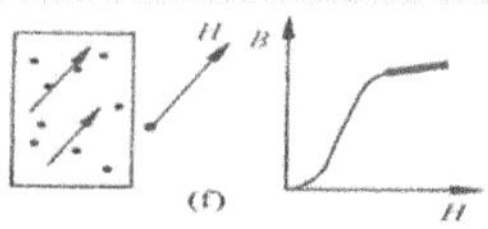	As H increases, the polarization deviates from the easiest magnetization direction to align with H. This phenomenon is now reversible. This is an area of high inductions. $B/Bmax$ exceeds 0.95.

When a material is demagnetized, its macroscopic magnetism is zero. When an excitation field is applied, its magnetization increases in accordance with the first magnetization curve (Figure 2) until it reaches saturation magnetization, which is the point at which all of the atomic magnetic moments are fully aligned in the direction of the field (Couderchon, 1998; Getzlaff, 2007; Moliton, 2007). The induction no longer follows the same path and keeps a non-zero value for $H = 0$ when the excitation field is decreased from saturation induction. This is referred to as the remanent induction. Apply the coercive field (H_C), an opposite field, to cancel it. This phenomenon is a hallmark of hysteresis. The hysteresis loop (Figure. 5) occurs when the field fluctuates between $B = f(H)$ and $\mp H_m$ (Couderchon, 1998)

Figure 5. Variation of induction B with applied field H in a ferromagnetic material (Couderchon, 1998)

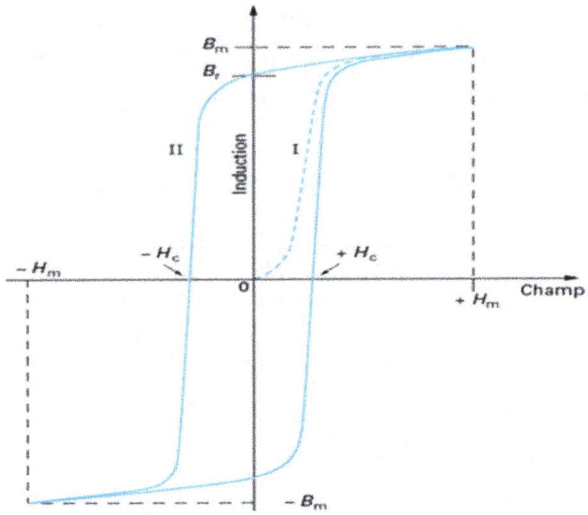

3.8. Permalloy' Structure

3.8.1. Permalloy's Matrix

Phase Distribution and Microstructure Al_2O_3 matrix with isolated Py between 25 at 40% and more than 40% Al_2O_3 dominating depending on thickness and pressure; Py particles are orientated spontaneously in a matrix(Goldman & Celinski, 2023; H. Wang et al., 2025). Many factors, including as shape anisotropy, magneto-crystalline anisotropy, magnetoelastic anisotropy, and the contact between the nanoparticles, affect the magnetic behavior of a collection of nanoparticles. The shape anisotropy orients the aimantation down the length of the films, whilst the dipolar interaction promotes alignment of the aimantation perpendicular to the nanoparticles. Depending on crystal structure and mechanical limitations, magneto-crystalline and magneto-elastic anisotropy also influence aim direction. Depending on the length of the nanoparticles and the porosity of the alumina matrix, the aimantation curve may exhibit distinct characteristics in terms of field and remanence at saturation. Phase transitions from metastable to stable states occur during sintering in common catalyst support materials like Al_2O_3, or Si_2O_3 (Dib et al., 2025; Fnidiki et al., 2010; Goldman & Celinski, 2023; H. Wang et al., 2025).

3.8.2. Alloy With Approximately 30% Nickel (invar)

These compositions include the well-known alloys with 36% nickel (INVAR). These alloys exhibit numerous properties, such as very low expansibility (Figure 6) and significant electrical resistivity (Couderchon & Tiers, 1982), for temperatures ranging from cryogenic to ambient, ensuring their versatilit (Cherrad Ounissa, 2023; Couderchon, 1994). However, it was not until 1921 that a 78% nickel alloy (Permalloy 78) was used as a magnetic material in telephony. Their structure is then face-centered cubic, and they do not undergo any phase transformations in the solid state (except for ordering near Ni_3Fe) (Dupeux, 2004).

Depending on the nickel content, high-permeability NiFe alloys have a number of unusual properties that can be used in conjunction with magnetic properties. However, the physical properties of the used iron-nickel alloys are primarily magnetic. Aside from that, they are primarily classified into two families:

Figure 6. The thermal expansion coefficient α varies with nickel content (Béranger et al., 1996)

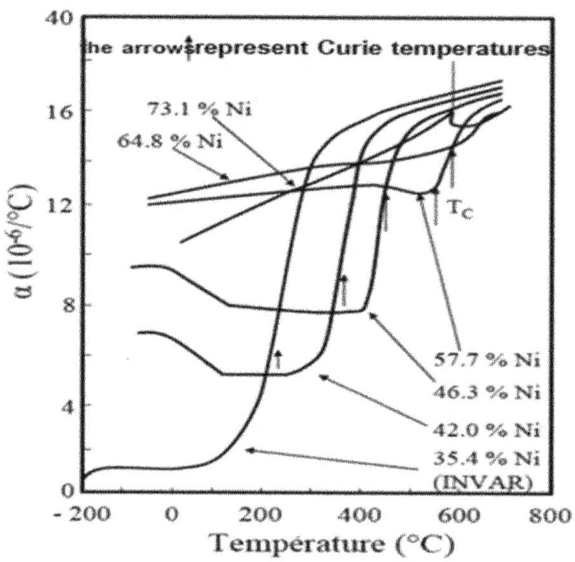

3.8.3. Alloy With Approximately 50% Nickel

High saturation magnetization, high saturation polarization (1.6 T), and intriguing permeabilities are present in alloys with 45–50% Ni (Mazaleyrat, 2006). Solid versions are used in safety devices, wires, bars, and relays. These alloys are used in a variety of applications, including stators and rotors in small motors and thin sheets (0.1 to 0.3 mm) in wound circuits. Stacks of sheets are used for telephone receivers and headphones that operate in a low field (Couderchon, 1994). The 56% Ni alloy is utilized for sealing and has isotropic magnetostriction and expansibility comparable to glass or certain ceramics (Huffer, 1977; Pfeifer & Radeloff, 1980).

3.8.4. Alloy With Approximately 80% Nickel (Permalloy)

Permalloy (Py), discovered in 1928 by H. D. Arnold and G. W. Elmen at Bell Laboratories, is a metallic alloy made up of roughly 80% Ni and 20% Fe. Although there are others, Permalloy has been extensively studied for its remarkable magnetic properties, which offer a very good compromise between zero magnetocrystalline anisotropy and very low (almost zero) crystalline magnetostriction (Guechi Nacir, 2017; D. Jiles, 2015), resulting in very high magnetic permeabilities that make the alloy useful in a wide range of applications where higher magnetic permeability is required. The variations in permeability as a function of Ni content in FeNi binary alloys at different cooling rates are given in Figure 7. Permalloy is a ferromagnetic material with a saturation magnetization of 9.30 ×105 A/m, low polarization (0.8T), and extremely low coercivity. Permalloy's Curie temperature is Tc = 853 K (Ranchal et al., 2008).

Figure 7. The permeabilities of FeNi alloys as a function of nickel concentration after rapid cooling (double treatment) and slow cooling (furnace cooling)(Bozorth & Ferromagnetism, 1951)

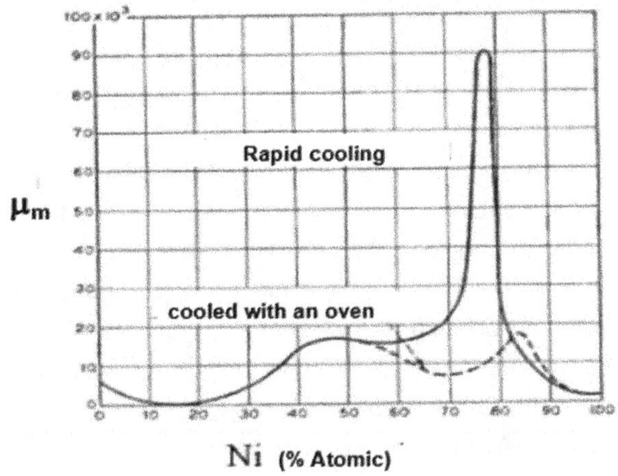

4. METHODS

Employing a radiofrequency sputtering device (f = 13.56 MHz), the composite samples $(Py)_{1-x}$ $(Al_2O_3)_x$ were created on Corning Glass substrates with different Py concentrations (24%, 34%, and 44%). The films were placed on 9 mm square glass substrates, which were placed on the anode in a 300 Oe magnetic field to generate magnetization easy. Film thicknesses of around 300 nm were achieved by deposition about $1.9.10^{-3}$ to 6.10^{-2} mbar of argon pressure. Magnetization loops were used to evaluate magnetic characteristics such as saturation magnetization and coercive field, and a simplified micro-strip transmission line perturbation approach was used to analyze permeability spectra for frequencies at 250 MHz to 10 GHz.

5. DATA AND ANALYSIS

5.1. Hysteresis Loops:

The static magnetic characteristics can be evaluated via measurements of magnetic hysteresis loops at room temperature of samples using vibration sample magnetometers (VSMs), a popular and efficient method in materials science and physics (Phuoc et al., 2012, 2013; Phuoc & Ong, 2013, 2014). Figure. 8 depicts the magnetic hysteresis loops of several samples. At P = 10.10^{-3} (mbar), the uniaxial magnetic anisotropy seen in Figure.8 (b) becomes more noticeable, indicating that pressure can affect the magnetic properties under study. Figure.8 (b) plots static parameters from hysteresis loops, incorporating the static anisotropy field (H_a) derived from static measurements (Figure. 8). They handle coercivity for the easy and hard axis respectively (H_{ce}) and (H_{ch}). The anisotropy field (H_k^{Stat}), this is identified as the difference in magnetic fields essential for saturation magnetization along the hard and easy axis, is offered to quantitatively describe the change in magnetic anisotropy, with the formula: $H_k^{Stat} = H_{HA}^{Stat} + H_{EA}^{Stat}$ (Camarero et al., 2005; Kipgen et al., 2012; Zhang et al., 2015) and the rotational-like magnetization curve's incline is predicted from the HA (Lamy & Viala, 2006; Liu et al., 2021). The hysteresis loops' easy and hard axes are presented in the Permalloy thin film in Figure. 8 (a). The hard axis in deposition is inside the sample plane but perpendicular to the easy axis, while the easy axis is the direction of the magnetic field (Jiang et al., 2011). On the magnetization curve, a tilted loop in the hard and easy axis represents magnetic anisotropy and exchange bias, respectively (Jiang et al., 2011; Phuoc et al., 2009; Spenato et al., 2007). Static magnetic anisotropy (H_{EA}^{Stat}) can be extracted thanks to the hysteresis loop (Phuoc et al., 2009; Spenato et al., 2007).

Low pressure may result in more surface characteristics and roughness because of increased deformation, encourages uniform stress distribution and reduces stress concentrations generate coarser grains with fewer nucleation sites. On the other hand, Higher pressure produces finer, homogeneous grains by recrystallization and dynamic recovery, produces smoother surfaces by improving polishing and minimizing microcracking, pressure raises internal and residual stresses that affect mechanical qualities and facilitates the incorporation of Al_2O_3 into the material matrix and enhances dynamic recovery and recrystallization, leading to finer, more uniform grain sizes.

Figure 8. Under differing pressures, (a) Illustrates the M/Ms-H loop components quantitated within to -10 at 10 Oe field interval, in parallel to the easy and hard axis at Permalloy. (b) Correlations with the static, dynamic anisotropy (H_k^{Stat}, H_k^{dyn})and coercivities $\left(H_{ce}, H_{ch}\right)$ of the diverse constituents.

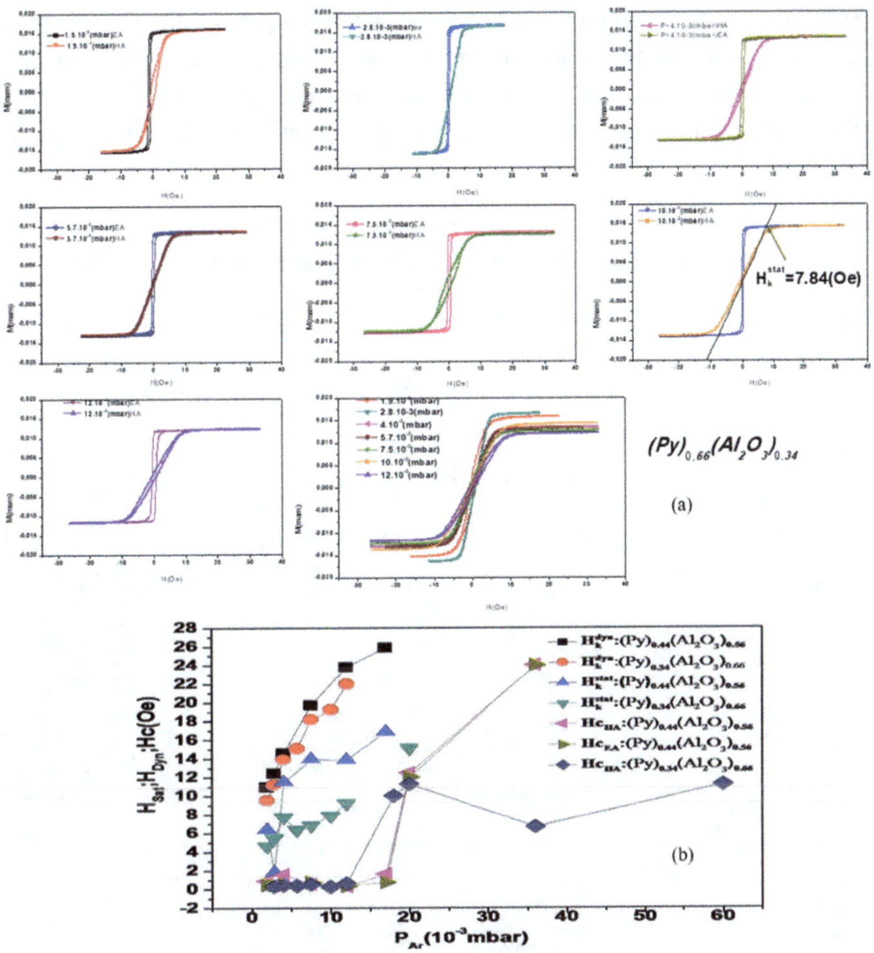

5.2. Microwave and High-Frequency Properties

5.2.1. Permeability Spectra

Microwave frequencies from 250 MHz to 1 GHz, no resonance was detected at any argon pressures with a 24% concentration. Figure. 9 (d-f), depicts comparable analyses of theoretical and experimental spectra, which indicate how an increase in pressure causes the measured permeability levels to minimize simultaneously as well. The theoretical and experimental results are in agreement, as illustrated by the diagrams in Figure. 9 (c-d). At P_{Ar}= 4(10^{-3} mbar) we identified that the real part of permeability attains a maximum of 700 (emu) as the pressure decreases. It suggests that an excellent resonance frequency is present a minimal pressure in the permeability variation with argon pressure (P_{Ar}). The real part illustrates a change in sign as the imagined part is at the greatest. Both the width of resonance and the damping factor spectra influence the quantity of degradation μ" and the location of resonance frequency. As pressure rises, both the resonance frequency and the initial permeability diminish. Our findings are adequate in comparison to Samuel Queste's (Spenato et al., 2007). The dissipative parameter, (α_{eff}), is less than 0.088 and the resonance frequency disappears when the pressure is superior to 17.10^{-3} (mbar). With a narrow linewidth, the lower absorption resonance μ" is significant. For low field levels, these curves demonstrate a notable extension of the magnetic loss area (Ledieu et al., 2003). The damping values and experimental outcomes are investigated from computations in theory using the GLL equation.

Figure 9. (a-f) describes how, at room temperature and constant thickness, theoretical and data of permeability varies with frequency at different pressures for concentrations (24%, 34%, and 44%).

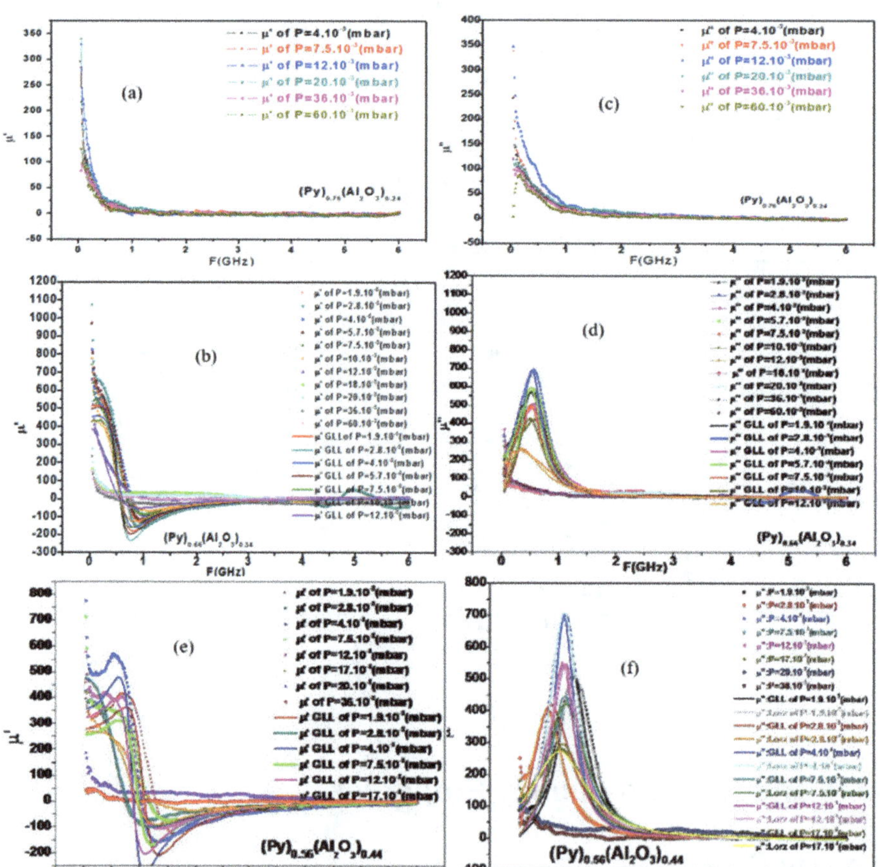

As Observed in Figure. 8, pressure and saturation magnetizations at varying concentrations appear to have a significant impact. For all three rates, we find that the magnetization varies in a similar way. With increasing pressure, the magnetization disappears. The best Permalloy deposit has high magnetization and is located below 17.10^{-3} (mbar) of pressure. For the P_{Ar} flow amounts were 17.10^{-3} (mbar) or less, the easy axis was perpendicular to the applied external magnetic field, which may have been caused by changes in the Al_2O_3 concentration gradient. It is widely acknowledged that the decrease in magnetization saturation with increasing pressure is due to damping coefficient, which is too little and causes a possible regarding the recent equilibrium point, damped precession. Figure. 8 displays the (H_k^{Stat}) and coercivity (H_c) values derived from the VSM measurement. (H_c) appears to be impacted by both reversal processes of the magnetization reaction and magnetic anisotropy. However, similar to our situation, while the coercivity essentially remained constant, the magnetic anisotropy depicted in Figure. 8 underwent a substantial modification. Pressure lead to increase both (H_k^{Stat}) and (H_k^{dyn}), but (H_k^{dyn}) value was higher than (H_k^{Stat}), this variance is explained by rotational magnetic anisotropy, which has already been observed in many other magnetic thin films (Phuoc & Ong, 2011; Rantschler & Alexander Jr, 2003). In both direction and magnetization, rotatable magnetic anisotropy is constantly aligned.

As demonstrated by the M-H loops, the anisotropy direction likewise varies in proportion to the reversal of magnetization (Fan et al., 2009; Olamit & Liu, 2007; P. Y. Yang et al., 2009). The permeability data illustrate that rotatable magnetic anisotropy occurs for reasons the magnetization changes directions at microwave frequencies and there is not enough small induced magnetic field to reverse it (McCord et al., 2004; McMichael et al., 2000; Phuoc & Ong, 2011; X. Yang et al., 2012).

An anisotropy caused by ripples or local disorder can be thought of as having an energy minimum parallel to a direction of FM magnetization (Gilmore et al., 2010; McCord et al., 2004; H. Wang et al., 2025). Figure. 8 suggests the magnetic hysteresis loops of the $(Py)_{1-x}(Al_2O_3)_x$ thin film created at changing Argon pressures. The easy axis coercivity (H_{ce}) was 3.58, 9.31, and 42.93 (Oe) for nanoparticles deposited at Argon pressures of 1.9, 4, and 60 (10^{-3} mbar), respectively. This indicates that (H_{ce}) rises as the Argon pressure increases. Interestingly, coercivity increased in the hard axis's orientation (H_{ch}). According to the results, higher argon pressure can result in higher (H_c). One factor contributing to the advantageous soft magnetic properties of low argon pressure films is the presence of a certain pressure (Bai et al., 2004). The exchange coupling model (Rudeichuk et al., 2024) can clarify the rise in coercivity and anisotropy field fluctuation and effectively characterizes the soft magnetic properties of granular films, nano-crystalline materials, and amorphous films (Goldman & Celinski, 2023; H. Wang et al., 2025). The calculated real and imaginary permeability data, exhibited in Figure. 9 (c-d-e-f), are fitted using formulas (Eq.46) using the previously mentioned parameters. Approximately 0.032 and 0.0885 are the Gilbert damping that fit $(Py)_{1-x}(Al_2O_3)_x$ films at 1.9 and 60 (10^{-3} mbar) argon pressures. The phenomenological quantity called damping is used to identify many origins of damping, including spin-orbit interactions, magnon interactions, and the spin pumping effect (Ben Youssef et al., 2004; Gerrits, 2006; Tserkovnyak et al., 2002; Zakeri et al., 2007). The good fit with Eqs.(46) (Ben Youssef et al., 2004) shows that our magnetic permeability data still shows a significant within its framework, Gilbert coefficient can be included or quantified.

5.2.2. Resonance Ferromagnetic

5.2.2.1. Computational Modeling: The LLG Approach

Reducing the pressure in this thin film would induce a deviation based on the perfect magnetic characteristics. This constitutes tenable explanation for the slight increase in (H_{ce})seen in Figure. 8 (b) when the argon pressure is progressively raised. The best static soft magnetic characteristics are usually found when the α is at its lowest. The large value of α in this case indicates extrinsic damping effects, including inhomogeneity (Bonin et al., 2005) and even spin-wave contributions (Counil et al., 2004), as it clearly exceeds the thin film normal range (Tserkovnyak et al., 2002). However, the exact origin of these extrinsic damping effects is still up for debate. Many damping mechanisms have been discussed in detail in relation to soft magnetic thin films (Gilbert, 2004; Kalarickal et al., 2008; Kuanr et al., 2005; Lenz et al., 2006; Maksymov & Kostylev, 2015; Mills & Arias, 2006; Zakeri et al., 2007). In fact, few extrinsic contributions, particularly Arias-Mill's TMS, can be sufficiently characterized by the phenomenological (LLG) formula (Kalarickal et al., 2008; Lenz et al., 2006; Tserkovnyak et al., 2002; Zakeri et al., 2007). As argon pressure increases Figure. 10 (a) shows how the frequency linewidth (Δf)and effective damping factor (α_{eff}) vary. Both are found to be insensitive to pressure changes, and P_{Ar} increases slightly as pressure rises ((α_{eff}) augment at 0.03 to 0.0885 and at 1.27 to 1.49), completely satisfying the requirement for independent component control by pressure fluctuation across a large frequency band. The frequency linewidth Δf may be obtained using the formula below (Phuoc & Ong, 2011):

$$\Delta f = \frac{\gamma \alpha_{eff}\left(2H_k^{dyn} + 4\pi M_s\right)}{2\pi} \propto \alpha \qquad (57)$$

5.2.2.2. Lorentzian and Gaussian fitting

Lorentzian (Two magnon scattering (TMS), magnon-electron term) and Gaussian (inhomogeneity) linewidths were convolutionally employed in the fits. Utilizing Lorentzian fits to calculate linewidth from experimental data. When compared to the declining trend of (μ''_{max}), (Δf) and $\left(\alpha_{eff}\right)$ show similar increasing tendencies. Figure. 10 (a-b) shows that the damping and linewidth are compatible when the pressure is less than 20.10^{-3}(mbar); conversely, when the pressure approaches P=20.10^{-3}(mbar), there is no resonance. The crystallization dynamic is revealed by plotting the FWHM of the Permeability mode as a function of frequency. Aside from the degree of disorder (Balaji et al., 2006) (FWHM rises with disorder), non-stoichiometry effects (Richter et al., 1981), the presence of oxygen vacancies (W F Zhang et al., 2000), and phonon confinement (Ager et al., 1991), the trend of FWHM can actually give general information about the degree of crystallization (the wider the peak, the lower the degree of crystallization). The linewidth (Δf) and peak position are obtained from the Lorentzian fit and then modified by the instrumental broadening, which is expressed as (Pakhomov et al., 2025; Serrano et al., 2003). Magnetic anisotropy dispersion, surface roughness, non-uniform spin wave excitation, two magnon scattering, and eddy currents can all contribute to extrinsic linewidth.

Figure 10. (a) Link with a 34% ratio between linewidth and damping. (b) relation of Line-width, damping and Permeability (μ''), (μ'),(μ_{Max}) with the gas pressure (Argon) at different concentration 34%. (c) Progression of the line-width with on the resonance field of Py-Al_2O_3 for different Pressures in $\theta°= 90$. (d) Frequency reliance on the gas pressure (Argon). (e) Variability of line-width, frequency and damping as a result of the gas pressure (Argon) at different substitution 34%.

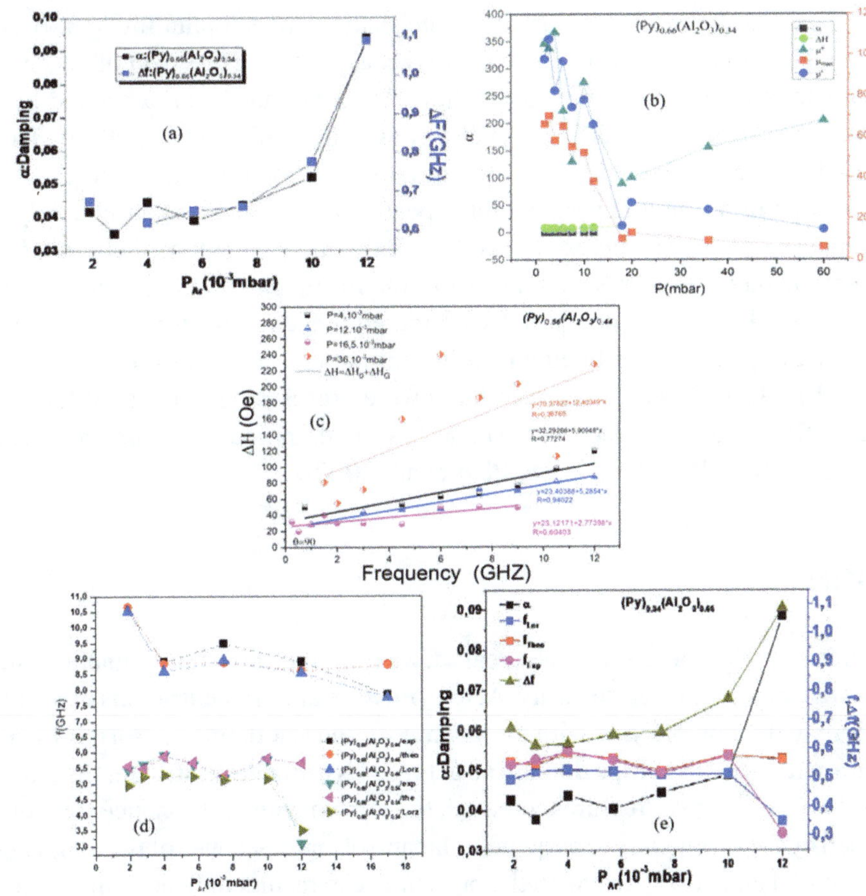

Concluded from Figure. 9 that, independent of pressure, the frequency never appears at the charge rate of 24%. On the opposed hand, the frequency rises with concentration in the 34 and 44% examples, suggesting the presence of a resonance frequency. The rise in the real component of permeability (from 60 to 800) is shown in Figure. 10 (b). Accordingly, the resonance frequency drops at 934 to 517 MHz with increasing argon pressure. Our samples are of excellent quality, evidenced by the higher resonance frequency of 44% as opposed to 34%. In the pressure range of 1.9 to 60 (10^{-3} mbar), experimental data (f_{FMR}) confirms with the (f_{LOR}) and (f_{FMR})hypotheses.

Figure 10 (a) demonstrates how damping and linewidth (Δf) grows at higher argon pressure for thin films, indicating excellent tenability. Consequently, under pressure, the imaginary permeability rises until it achieves saturation. When the softest magnetic properties are discovered, the damping value is lowest, demonstrated due to the damping change trend being nearly identical to (H_c). As the pressure of the argon gas elevates, it is clear that the damping increases considerably. The Pressure affects microstructure by reducing interparticle distances and adding strain, which may result in interface defects. Additionally, it enhances spin-phonon coupling by affecting lattice vibrations that interact with spins. Also, two-magnon scattering may be enhanced by pressure because of inhomogeneities that scatter magnons. Usually, the perfect static soft magnetic properties appear to generate a minimal quantity of damping. The study shows that the intrinsic Gilbert damping α scales inversely with different pressures, with the supplied argon pressure ranging from 4.10^{-3} to 6010^{-3} mbar (Srinivas et al., 2014; X. Xu et al., 2021). Damping as a phenomenological feature implies a variety of physical sources, including spin-orbit interactions and magnon-orbit interactions (Lenz et al., 2006; Zakeri et al., 2007). The effect of pressure on the static soft magnetic characteristics in a response to the thin-film stress has been explained by a roundabout method (Dib et al., 2025; Ellis et al., 2020; B. Peng et al., 2013; Zhou et al., 2020). This approach is inaccurate for broad frequency line-width, but it can be used to approximate the FMR frequency by determining the peak point during the permeability spectrum experiment. Consequently, In Figure. 10 (e), The rising propensity of (α_{eff}) is represented although the increasing tendencies of (f_{Lor}), (f_{FMR}), and (Δf) are identical among themselves, respectively, show the fitting parameters, which include (H_k^{dyn}), f_{FMR} ,Damping and frequency line-width. compared with the equivalent anisotropy dynamic values (H_k^{dyn}), which were somewhat higher and exhibited an equivalent pattern to the anisotropy static values (H_k^{stat}) shown in Figure. 8 (b). The differences between them were presented by impact of the magnetic ripple fields (Hoffmann, 1979; Takeda & Naoe, 2018) and have been noted in a number of research papers (Buznikov & Rozanov, 2005; Pan et al., 2018; Platt et al., 2000).

6. CONCLUSION

As the Argon pressure of permalloy films increased, we examined how the magnetic structure changed based on the measured microwave permeability. At low pressure and concentrations of 34% and 44%, high permeability exhibits and accompanying with low coercivity and low damping. An out-of-plane magnetic anisotropy for films with a regular stripe domain structure was utilized to interpret the experimental data (Lordan et al., 2021). As the argon Pressure increased, the ferromagnetic resonance frequency decreased, which was explained by the emergence of perpendicular anisotropy. Because of the magnetoelastic effect, both vertical and parallel anisotropy rise with the pressure. The results indicated an excellent agreement between the data and those deducted from the Gilbert model, which took into account a dynamic field along the hard axis of magnetization. Using the equivalent technique, the permeability and resonance frequency varied with argon pressure. A greater permeability was the result of the effective field rising with increasing pressure. In narrow linewidth, resonance absorption was significant when the dissipative parameter (α_{eff}) was less. It has been demonstrated that measurements in the frequency domain, however, provide greater validity assurances. Approximately 0.032 and 0.0885 are the Gilbert coefficient that fit Permalloy films at pressures varying from 1.9 to 60 (10^{-3}mbar). The phenomenological quantity called damping is used to identify many sources of damping, including two magnon scattering, spin orbit interactions, and the spin pumping effect meaningful in solvation Modeling strategies.

Future Research Directions

- The global study can be focused in the next steps:
- Magnon relaxation using eddy currents

- Simulate the Ab-initio model and search to determine the surface magnetic
- Density Functional Theory is a majority successful and extensively employed ab initio method.
- Data Representing and transformation.
- Utilizing the algorithm and exploitation the results.
- Model testing and optimization

Next steps for other work and questions? For example:

- Quantify energy dissipation in Permalloy-based magnetic components and the prediction study model edge resonance in Permalloy structures?
- Model magnetization dynamics and estimate switching energy in a device using Dzyaloshinskii–Moriya interaction?

REFERENCES

Ager, J. W., Veirs, D. K., & Rosenblatt, G. M. (1991). Spatially resolved Raman studies of diamond films grown by chemical vapor deposition. *Physical Review B: Condensed Matter*, *43*(8), 6491–6499. DOI: 10.1103/PhysRevB.43.6491

Bai, D. Z., Zhu, J.-G., Yu, W., & Bain, J. A. (2004). Micromagnetic simulation of effect of stress-induced anisotropy in soft magnetic thin films. *Journal of Applied Physics*, *95*(11), 6864–6866. DOI: 10.1063/1.1667445

Balaji, S., Djaoued, Y., & Robichaud, J. (2006). Phonon confinement studies in nanocrystalline anatase-TiO 2 thin films by micro Raman spectroscopy. *Journal of Raman Spectroscopy : JRS*, *37*(12), 1416–1422. DOI: 10.1002/jrs.1566

Barman, A., & Sinha, J. (2018). *Spin Dynamics and Damping in Ferromagnetic Thin Films and Nano-structures*. Springer International Publishing., DOI: 10.1007/978-3-319-66296-1

Bas, J. A., Calero, J. A., & Dougan, M. J. (2003). Sintered soft magnetic materials. Properties and applications. *Journal of Magnetism and Magnetic Materials*, *254–255*(255), 391–398. DOI: 10.1016/S0304-8853(02)00934-4

Bekker, V., Seemann, K., & Leiste, H. (2006). Development and optimisation of thin soft ferromagnetic Fe–Co–Ta–N and Fe–Co–Al–N films with in-plane uniaxial anisotropy for HF applications. *Journal of Magnetism and Magnetic Materials*, *296*(1), 37–45. DOI: 10.1016/j.jmmm.2005.01.025

Ben Youssef, J., Vukadinovic, N., Billet, D., & Labrune, M. (2004). Thickness-dependent magnetic excitations in Permalloy films with nonuniform magnetization. *Physical Review B: Condensed Matter and Materials Physics*, *69*(17), 174402. DOI: 10.1103/PhysRevB.69.174402

Béranger, G., Duffaut, F., Morlet, J., & Tiers, J.-F. (1996). A Hundred Years After the Discovery of Invar... the Iron-nickel Alloys. Londres.

Bertotti, G. (1998). *Hysteresis in magnetism: for physicists, materials scientists, and engineers*. Academic press.

Bloch, F. (1930). Zur theorie des ferromagnetismus. *The European Physical Journal A*, *61*(3), 206–219. DOI: 10.1007/BF01339661

Bocklage, L., Krüger, B., Matsuyama, T., Bolte, M., Merkt, U., Pfannkuche, D., & Meier, G. (2009). Dependence of Magnetic Domain-Wall Motion on a Fast Changing Current. *Physical Review Letters*, *103*(19), 197204. DOI: 10.1103/PhysRevLett.103.197204

Bocklage, L., Scholtyssek, J. M., Merkt, U., & Meier, G. (2007). Spin polarization of Ni2MnIn and Ni80Fe20 determined by point-contact Andreev spectroscopy. Journal of Applied Physics, 101(9), 09J512. DOI: 10.1063/1.2710238

Bonin, R., Schneider, M. L., Silva, T. J., & Nibarger, J. P. (2005). Dependence of magnetization dynamics on magnetostriction in NiFe alloys. *Journal of Applied Physics*, *98*(12), 123904. Advance online publication. DOI: 10.1063/1.2143121

Bose, B. K., & Wang, F. (Fred). (2019). ENERGY, ENVIRONMENT, POWER ELECTRONICS, RENEWABLE ENERGY SYSTEMS, AND SMART GRID. In Power Electronics in Renewable Energy Systems and Smart Grid (pp. 1–83). Wiley. DOI: 10.1002/9781119515661.ch1

Bozorth, R. M., & Ferromagnetism, D. (1951). Van Nostrand Company. Inc., New York, 867.

Bruno, P., & Renard, J.-P. (1989). Magnetic surface anisotropy of transition metal ultrathin films. *Applied Physics. A, Solids and Surfaces*, *49*(5), 499–506. DOI: 10.1007/BF00617016

Buznikov, N. A., & Rozanov, K. N. (2005). The effect of stripe domain structure on dynamic permeability of thin ferromagnetic films with out-of-plane uniaxial anisotropy. *Journal of Magnetism and Magnetic Materials*, *285*(3), 314–326. DOI: 10.1016/j.jmmm.2004.07.050

Camarero, J., Sort, J., Hoffmann, A., García-Martín, J. M., Dieny, B., Miranda, R., & Nogués, J. (2005). Origin of the Asymmetric Magnetization Reversal Behavior in Exchange-Biased Systems: Competing Anisotropies. *Physical Review Letters*, *95*(5), 057204. DOI: 10.1103/PhysRevLett.95.057204

Chen, C.-W. (2013). *Magnetism and metallurgy of soft magnetic materials*. Courier Corporation.

Cherrad Ounissa. (2023). *Élaboration sous vide et investigations de propriétés physiques de films minces ferromagnétiques de Permalloy*. Dépôt Institutionnel de l'Université Ferhat ABBAS.

Couderchon, G. (1994). Alliages fer-nickel et fer-cobalt Propriétés magnétiques. Techniques de l'ingénieur. Matériaux Fonctionnels, 2(D2130).

Couderchon, G. (1998). *Alliages magnétiques doux* (Ingénieur, T., Ed.).

Couderchon, G., & Tiers, J. F. (1982). Some aspects of magnetic properties of Ni-Fe and Co-Fe alloys. *Journal of Magnetism and Magnetic Materials*, *26*(1–3), 196–214. DOI: 10.1016/0304-8853(82)90152-4

Counil, G., Kim, J.-V., Devolder, T., Chappert, C., Shigeto, K., & Otani, Y. (2004). Spin wave contributions to the high-frequency magnetic response of thin films obtained with inductive methods. *Journal of Applied Physics*, *95*(10), 5646–5652. DOI: 10.1063/1.1697641

Cullity, B. D., & Graham, C. D. (2011). *Introduction to magnetic materials*. John Wiley & Sons.

de Moraes, A., Bohn, F., Gamino, M., da Silva, E. F., de Oliveira, D. F., Lopes, C., Ferreira, A., Vaz, F., & Correa, M. A. (2025). Asymmetrical magnetoimpedance on Permalloy/Ag multilayer for high-frequency sensor applications. *Surface and Coatings Technology*, *495*, 131580. DOI: 10.1016/j.surfcoat.2024.131580

Demokritov, S. O., Hillebrands, B., & Slavin, A. N. (2001). Brillouin light scattering studies of confined spin waves: Linear and nonlinear confinement. *Physics Reports*, *348*(6), 441–489. DOI: 10.1016/S0370-1573(00)00116-2

Dempsey, J. L., Li, F., O'Mullane, A. P., & Unwin, P. R. (2025). Welcome to ACS Electrochemistry. *ACS Electrochemistry*, *1*(1), 1–2. DOI: 10.1021/acselectrochem.4c00205

Dib, M., El-Yakoubi, E. Y., ElHarfaoui, M., & Oublal, M. (2025). The study analyzes the impacts of pressure the permeability of permalloy $(Ni81Fe19)1-x(Al2O3)x$ thin films. *Applied Physics. A, Materials Science & Processing*, *131*(9), 707. DOI: 10.1007/s00339-025-08844-w

Duan, J., Hu, S., Wang, P., Zhang, D., & Jiang, Y. (2024). Topological quantum materials for spintronics. *MetalMat, 1*(2), e24. Advance online publication. DOI: 10.1002/metm.24

Ellis, E. A. I., Chmielus, M., Han, S., & Baker, S. P. (2020). Effect of sputter pressure on microstructure and properties of β-Ta thin films. *Acta Materialia, 183*, 504–513. DOI: 10.1016/j.actamat.2019.10.056

Fan, B., Zeng, F., Chen, C., Yang, Y. C., Yang, P. Y., & Pan, F. (2009). Influence of strain and grain boundary variations on magnetism of Cr-doped AlN films. *Journal of Applied Physics, 106*(7), 073907. Advance online publication. DOI: 10.1063/1.3234377

Fergen, I., Seemann, K., Weth, A., & Schüppen, A. (2002). Soft ferromagnetic thin films for high frequency applications. *Journal of Magnetism and Magnetic Materials, 242–245*, 146–151. DOI: 10.1016/S0304-8853(01)01185-4

Fnidiki, A., Lemarchand, D., & Ben Youssef, J. (2010). Structural and magnetic properties of granular (Ni0.84Fe0.16)54(alumina)46 sputtered thin films. *Physica B, Condensed Matter, 405*(12), 2632–2639. DOI: 10.1016/j.physb.2010.03.037

Ge, S., Yao, D., Yamaguchi, M., Yang, X., Zuo, H., Ishii, T., Zhou, D., & Li, F. (2007). Microstructure and magnetism of FeCo–SiO 2 nano-granular films for high frequency application. *Journal of Physics. D, Applied Physics, 40*(12), 3660–3664. DOI: 10.1088/0022-3727/40/12/016

Gerrits, M. L. (2006). Schneider, and TJ Silva. *Journal of Applied Physics, 99*, 023901. DOI: 10.1063/1.2159076

Getzlaff, M. (2007). *Fundamentals of magnetism*. Springer Science & Business Media.

Gilbert, T. L. (1955). A Lagrangian formulation of the gyromagnetic equation of the magnetization field. *Physical Review, 100*, 1243.

Gilbert, T. L. (2004). Classics in Magnetics A Phenomenological Theory of Damping in Ferromagnetic Materials. *IEEE Transactions on Magnetics, 40*(6), 3443–3449. DOI: 10.1109/TMAG.2004.836740

Gilmore, K., Stiles, M. D., Seib, J., Steiauf, D., & Fähnle, M. (2010). Anisotropic damping of the magnetization dynamics in Ni, Co, and Fe. *Physical Review B: Condensed Matter and Materials Physics, 81*(17), 174414. DOI: 10.1103/PhysRevB.81.174414

Goldman, S., & Celinski, Z. (2023). Magnetic properties of (Ni0.81Fe0.19/SiO2)n multilayers for high frequency on-wafer inductor applications. *Journal of Magnetism and Magnetic Materials, 569*, 170440. DOI: 10.1016/j.jmmm.2023.170440

Guechi Nacir. (2017). *Élaboration et caractérisations structurale, électrique et magnétique des couches minces de FexNi100-x déposées sur Si (100)*. Dépôt Institutionnel de l'Université Ferhat ABBAS.

Gurevich, A. G., & Melkov, G. A. (2020). *Magnetization oscillations and waves*. CRC press. DOI: 10.1201/9780138748487

Hamler, A., Goričan, V., Šuštaršič, B., & Sirc, A. (2006). The use of soft magnetic composite materials in synchronous electric motor. *Journal of Magnetism and Magnetic Materials, 304*(2), e816–e819. DOI: 10.1016/j.jmmm.2006.03.003

Hoffmann, H. (1979). Magnetic properties of thin ferromagnetic films in relation to their structure. *Thin Solid Films*, *58*(2), 223–233. DOI: 10.1016/0040-6090(79)90241-4

Huffer, J. M. (1977). Special aspects of primary care. *Postgraduate Medicine*, *62*(5), 223–228. DOI: 10.1080/00325481.1977.11714687

Huitema, L., Reveyrand, T., Mattei, J.-L., Arnaud, E., Decroze, C., & Monediere, T. (2013). Frequency Tunable Antenna Using a Magneto-Dielectric Material for DVB-H Application. *IEEE Transactions on Antennas and Propagation*, *61*(9), 4456–4466. DOI: 10.1109/TAP.2013.2269474

Jedema, F. J., Filip, A. T., & van Wees, B. J. (2001). Electrical spin injection and accumulation at room temperature in an all-metal mesoscopic spin valve. *Nature*, *410*(6826), 345–348. DOI: 10.1038/35066533

Jiang, C., Xue, D., & Sui, W. (2011). Broadband microwave absorption in [NiFe/FeMn]n exchange-coupled multilayer films. *Thin Solid Films*, *519*(8), 2527–2530. DOI: 10.1016/j.tsf.2010.11.030

Jiles, D. (2015). *Introduction to magnetism and magnetic materials*. CRC press. DOI: 10.1201/b18948

Jiles, D. C. (2003). Recent advances and future directions in magnetic materials. *Acta Materialia*, *51*(19), 5907–5939. DOI: 10.1016/j.actamat.2003.08.011

Jiles, D. C. (2017). *Introduction to the electronic properties of materials*. CRC Press.

Kalarickal, S. S., Krivosik, P., Das, J., Kim, K. S., & Patton, C. E. (2008). Microwave damping in polycrystalline Fe-Ti-N films: Physical mechanisms and correlations with composition and structure. *Physical Review B: Condensed Matter and Materials Physics*, *77*(5), 054427. DOI: 10.1103/PhysRevB.77.054427

Kipgen, L., Fulara, H., Raju, M., & Chaudhary, S. (2012). In-plane magnetic anisotropy and coercive field dependence upon thickness of CoFeB. *Journal of Magnetism and Magnetic Materials*, *324*(19), 3118–3121. DOI: 10.1016/j.jmmm.2012.05.012

Kittel, C. (1948). On the Theory of Ferromagnetic Resonance Absorption. *Physical Review*, *73*(2), 155–161. DOI: 10.1103/PhysRev.73.155

Koohkan, R., Sharafi, S., Shokrollahi, H., & Janghorban, K. (2008). Preparation of nanocrystalline Fe–Ni powders by mechanical alloying used in soft magnetic composites. *Journal of Magnetism and Magnetic Materials*, *320*(6), 1089–1094. DOI: 10.1016/j.jmmm.2007.10.033

Kuanr, B. K., Camley, R. E., & Celinski, Z. (2003). Exchange bias of NiO/NiFe: Linewidth broadening and anomalous spin-wave damping. *Journal of Applied Physics*, *93*(10), 7723–7725. DOI: 10.1063/1.1557964

Kuanr, B. K., Camley, R. E., & Celinski, Z. (2005). Extrinsic contribution to Gilbert damping in sputtered NiFe films by ferromagnetic resonance. *Journal of Magnetism and Magnetic Materials*, *286*(SPEC. ISS.), 276–281. DOI: 10.1016/j.jmmm.2004.09.080

Lamy, Y., & Viala, B. (2006). NiMn, IrMn, and NiO Exchange Coupled CoFe Multilayers for Microwave Applications. *IEEE Transactions on Magnetics*, *42*(10), 3332–3334. DOI: 10.1109/TMAG.2006.878871

Landau, L., & Lifshitz, E. (1935). On the theory of the dispersion of magnetic permeability in ferromagnetic bodies. *Phys. Z. Sowjetunion*, *8*(153), 101–114.

Ledieu, M., Schoenstein, F., Le Gallou, J.-H., Valls, O., Queste, S., Duverger, F., & Acher, O. (2003). Microwave permeability spectra of ferromagnetic thin films over a wide range of temperatures. *Journal of Applied Physics*, *93*(10), 7202–7204. DOI: 10.1063/1.1555902

Lenz, K., Wende, H., Kuch, W., Baberschke, K., Nagy, K., & Jánossy, A. (2006). Two-magnon scattering and viscous Gilbert damping in ultrathin ferromagnets. *Physical Review B: Condensed Matter and Materials Physics*, *73*(14), 144424. DOI: 10.1103/PhysRevB.73.144424

Liu, M., Du, S., Wang, F., Adam, R., Li, Q., Ma, X., Guo, X., Chen, X., Yu, J., Song, Y., Xu, J., Li, S., & Cao, D. (2021). Influence of surface pinning in the domain on the magnetization dynamics in permalloy striped domain films. *Journal of Alloys and Compounds*, *869*, 159327. DOI: 10.1016/j.jallcom.2021.159327

Lordan, D., Wei, G., McCloskey, P., O'Mathuna, C., & Masood, A. (2021). Origin of perpendicular magnetic anisotropy in amorphous thin films. *Scientific Reports*, *11*(1), 3734. DOI: 10.1038/s41598-020-78950-7

Lu, A. H., Salabas, E. L., & Schüth, F. (2007). Magnetic nanoparticles: Synthesis, protection, functionalization, and application. *Angewandte Chemie International Edition*, *46*(8), 1222–1244. DOI: 10.1002/anie.200602866

Ludwig, A., Tewes, M., Glasmachers, S., Löhndorf, M., & Quandt, E. (2002). High-frequency magnetoelastic materials for remote-interrogated stress sensors. *Journal of Magnetism and Magnetic Materials*, *242–245*(PART II), 1126–1131. DOI: 10.1016/S0304-8853(01)00979-9

Maksymov, I. S., & Kostylev, M. (2015). Broadband stripline ferromagnetic resonance spectroscopy of ferromagnetic films, multilayers and nanostructures. *Physica E, Low-Dimensional Systems and Nanostructures*, *69*, 253–293. DOI: 10.1016/j.physe.2014.12.027

Manago, T., Yamanoi, K., Kasai, S., & Mitani, S. (2015). Damping factor estimation using spin wave attenuation in permalloy film. *Journal of Applied Physics*, *117*(17), 17D121. Advance online publication. DOI: 10.1063/1.4914538

Mazaleyrat, F. (2006). *Introduction à la physique des matériaux électromagnétiques*. Inst. Univ. Form. Des Maîtres Crétl.

McCord, J., Mattheis, R., & Elefant, D. (2004). Dynamic magnetic anisotropy at the onset of exchange bias: The NiFe/IrMn ferromagnet/antiferromagnet system. *Physical Review. B*, *70*(9), 094420. DOI: 10.1103/PhysRevB.70.094420

McMichael, R. D., Lee, C. G., Bonevich, J. E., Chen, P. J., Miller, W., & Egelhoff, W. F.Jr. (2000). Strong anisotropy in thin magnetic films deposited on obliquely sputtered Ta underlayers. *Journal of Applied Physics*, *88*(9), 5296–5299. DOI: 10.1063/1.1323436

Mills, D. L., & Arias, R. (2006). The damping of spin motions in ultrathin films: Is the Landau–Lifschitz–Gilbert phenomenology applicable? *Physica B, Condensed Matter*, *384*(1–2), 147–151. DOI: 10.1016/j.physb.2006.05.209

Miura, K., Kasai, S., Kobayashi, K., & Ono, T. (2006). Non Local Spin Detection in Ferromagnet/Superconductor/Ferromagnet Spin-Valve Device with Double-Tunnel Junctions. *Japanese Journal of Applied Physics*, *45*(4R), 2888. DOI: 10.1143/JJAP.45.2888

Moliton, A. (2007). *Basic electromagnetism and materials*. Springer.

Olamit, J., & Liu, K. (2007). Rotational hysteresis of the exchange anisotropy direction in Co/FeMn thin films. *Journal of Applied Physics*, *101*(9), 09E508. Advance online publication. DOI: 10.1063/1.2694378

Pakhomov, A. S., Skirdkov, P. N., Yurlov, V. V., Kichin, G. A., Hai, N. T., Wu, J. C., Huang, J. C. A., Su, S. H., Chernov, A. I., & Zvezdin, K. A. (2025). Temperature resolved spin-pumping in permalloy-bismuth selenide heterostructure. *Journal of Applied Physics*, *137*(20), 203903. Advance online publication. DOI: 10.1063/5.0259659

Pan, L., Xie, H., Cheng, X., Zhao, C., Feng, H., Cao, D., Wang, J., & Liu, Q. (2018). Tuning the ferromagnetic resonance frequency of soft magnetic film by patterned permalloy micro-stripes with stripe-domain. *Journal of Magnetism and Magnetic Materials*, *457*, 46–51. DOI: 10.1016/j.jmmm.2018.02.060

Park, H. S., Hwang, J. S., Choi, W. Y., Shim, D. S., Na, K. W., & Choi, S. O. (2004). Development of micro-fluxgate sensors with electroplated magnetic cores for electronic compass. *Sensors and Actuators. A, Physical*, *114*(2–3), 224–229. DOI: 10.1016/j.sna.2003.11.037

Park, J. H., Lee, J. W., Choi, H. J., Jang, W. G., Kim, T. S., Suh, D. S., Jeong, H. Y., Chang, S. Y., Roh, J. C., Yoo, C. S., Kim, K. H., Park, C., & Suh, S. J. (2019). Electromagnetic interference shielding effectiveness of sputtered NiFe/Cu multi-layer thin film at high frequencies. *Thin Solid Films*, *677*, 130–136. DOI: 10.1016/j.tsf.2019.03.015

Peng, B., Zhang, W. L., Xie, Q. Y., Zhang, W. X., & Jiang, H. C. (2013). Effect of sputtering pressure on microstructure and magnetic properties of amorphous FeCoSiB films. *Journal of Non-Crystalline Solids*, *365*(1), 59–62. DOI: 10.1016/j.jnoncrysol.2013.01.028

Peng, W., Wang, L., Li, Y., Du, Y., He, Z., Wang, C., Zhao, Y., Jiang, Z., Zhou, Z., & Liu, M. (2022). Photoelectron-induced quantitative regulation of ferromagnetism in Permalloy at room temperature for photovoltaic flexible spintronics. *Journal of Alloys and Compounds*, *910*, 164903. DOI: 10.1016/j.jallcom.2022.164903

Pfeifer, F., & Radeloff, C. (1980). Soft magnetic Ni-Fe and Co-Fe alloys - some physical and metallurgical aspects. *Journal of Magnetism and Magnetic Materials*, *19*(1–3), 190–207. DOI: 10.1016/0304-8853(80)90592-2

Phuoc, N. N., Chai, G., & Ong, C. K. (2012). Temperature-dependent dynamic magnetization of FeCoHf thin films fabricated by oblique deposition. *Journal of Applied Physics*, *112*(8), 083925. Advance online publication. DOI: 10.1063/1.4763361

Phuoc, N. N., Chapon, P., Acher, O., & Ong, C. K. (2013). Large magneto-elastic anisotropy enhancement with temperature in composition-graded FeCoTa thin films. *Journal of Applied Physics*, *114*(15), 153903. Advance online publication. DOI: 10.1063/1.4825225

Phuoc, N. N., & Ong, C. K. (2011). Influence of ferromagnetic thickness on dynamic anisotropy in exchange-biased MnIr/FeCo multilayered thin films. *Physica B, Condensed Matter*, *406*(18), 3514–3518. DOI: 10.1016/j.physb.2011.05.063

Phuoc, N. N., & Ong, C. K. (2013). Anomalous Temperature Dependence of Magnetic Anisotropy in Gradient-Composition Sputterred Thin Films. *Advanced Materials*, *25*(7), 980–984. DOI: 10.1002/adma.201203995

Phuoc, N. N., & Ong, C. K. (2014). Gradient-Composition Sputtering: An Approach to Fabricate Magnetic Thin Films With Magnetic Anisotropy Increased With Temperature. *IEEE Transactions on Magnetics*, *50*(7), 1–6. DOI: 10.1109/TMAG.2013.2296936

Phuoc, N. N., Xu, F., Ma, Y., & Ong, C. K. (2009). Permalloy–FeMn exchange-biased multilayers grown on flexible substrates for microwave applications. *Journal of Magnetism and Magnetic Materials*, *321*(17), 2685–2690. DOI: 10.1016/j.jmmm.2009.03.073

Platt, C. L., Berkowitz, A. E., Smith, D. J., & McCartney, M. R. (2000). Correlation of coercivity and microstructure of thin CoFe films. *Journal of Applied Physics*, *88*(4), 2058–2062. DOI: 10.1063/1.1305833

Rahman, B. M. F., Divan, R., Rosenmann, D., Wang, T., Peng, Y., & Wang, G. (2015). Application of sub-micrometer patterned permalloy thin film in tunable radio frequency inductors. Journal of Applied Physics, 117(17), 17C121. DOI: 10.1063/1.4918766

Ranchal, R., Aroca, C., & López, E. (2008). Domain walls and exchange-interaction in Permalloy/Gd films. *New Journal of Physics*, *10*(1), 013013. DOI: 10.1088/1367-2630/10/1/013013

Rantschler, J. O., & Alexander, C.Jr. (2003). Ripple field effect on high-frequency measurements of FeTiN films. *Journal of Applied Physics*, *93*(10), 6665–6667. DOI: 10.1063/1.1556097

Richter, H., Wang, Z. P., & Ley, L. (1981). The one phonon Raman spectrum in microcrystalline silicon. *Solid State Communications*, *39*(5), 625–629. DOI: 10.1016/0038-1098(81)90337-9

Ripka, P. (1992). Review of fluxgate sensors. *Sensors and Actuators. A, Physical*, *33*(3), 129–141. DOI: 10.1016/0924-4247(92)80159-Z

Ripka, P. (2003). Advances in fluxgate sensors. *Sensors and Actuators. A, Physical*, *106*(1–3), 8–14. DOI: 10.1016/S0924-4247(03)00094-3

Ripka, P. (2008). Sensors based on bulk soft magnetic materials: Advances and challenges. *Journal of Magnetism and Magnetic Materials*, *320*(20), 2466–2473. DOI: 10.1016/j.jmmm.2008.04.079

Robert, P. (1999). *Matériaux de l'électronique* (Vol. 2). EPFL Press.

Rudeichuk, T., Olekšáková, D., Maciaszek, R., Matysiak, W., & Kollár, P. (2024). Exploring the Impact of Different Milling Parameters of Fe/SiO2 Composites on Their Structural and Magnetic Properties. *Materials (Basel)*, *17*(4), 862. DOI: 10.3390/ma17040862

Salahun, E., Tanné, G., Quéffélec, P., Lefloc'h, M., Adenot, A., & Acher, O. (2001). Application of ferromagnetic composite in different planar tunable microwave devices. *Microwave and Optical Technology Letters*, *30*(4), 272–276. DOI: 10.1002/mop.1288

Seemann, K., Leiste, H., & Bekker, V. (2006). A new generation of CMOS-compatible high frequency micro-inductors with ferromagnetic cores: Theory, fabrication and characterisation. *Journal of Magnetism and Magnetic Materials*, *302*(2), 321–326. DOI: 10.1016/j.jmmm.2005.05.042

Serrano, J., Manjón, F. J., Romero, A. H., Widulle, F., Lauck, R., & Cardona, M. (2003). Dispersive Phonon Linewidths: The E_2 Phonons of ZnO. *Physical Review Letters*, *90*(5), 055510. DOI: 10.1103/PhysRevLett.90.055510

Shokrollahi, H., & Janghorban, K. (2006). The effect of compaction parameters and particle size on magnetic properties of iron-based alloys used in soft magnetic composites. *Materials Science and Engineering B*, *134*(1), 41–43. DOI: 10.1016/j.mseb.2006.07.015

Sohn, J., Han, S. H., Yamaguchi, M., & Lim, S. H. (2006). Tunable electromagnetic noise suppressor integrated with a magnetic thin film. *Applied Physics Letters*, *89*(10), 103501. DOI: 10.1063/1.2335389

Spenato, D., Pogossian, S. P., Dekadjevi, D. T., & Ben Youssef, J. (2007). Study of dynamic properties and magnetic anisotropies of NiFe/MnPt in the critical thickness range. *Journal of Physics. D, Applied Physics*, *40*(11), 3306–3313. DOI: 10.1088/0022-3727/40/11/008

Srinivas, K., Manivel Raja, M., Sridhara Rao, D. V., & Kamat, S. V. (2014). Effect of sputtering pressure and power on composition, surface roughness, microstructure and magnetic properties of as-deposited Co2FeSi thin films. *Thin Solid Films*, *558*, 349–355. DOI: 10.1016/j.tsf.2014.02.052

Takeda, S., & Naoe, M. (2018). Size optimization for complex permeability measurement of magnetic thin films using a short-circuited microstrip line up to 30 GHz. *Journal of Magnetism and Magnetic Materials*, *449*, 530–537. DOI: 10.1016/j.jmmm.2017.10.025

Tang, G., Ghosez, P., & Hong, J. (2021). Band-Edge Orbital Engineering of Perovskite Semiconductors for Optoelectronic Applications. *The Journal of Physical Chemistry Letters*, *12*(17), 4227–4239. DOI: 10.1021/acs.jpclett.0c03816

Tserkovnyak, Y., Brataas, A., & Bauer, G. E. W. (2002). Enhanced Gilbert Damping in Thin Ferromagnetic Films. *Physical Review Letters*, *88*(11), 117601. DOI: 10.1103/PhysRevLett.88.117601

van Staa, A., Wulfhorst, J., Vogel, A., Merkt, U., & Meier, G. (2008). Spin precession in lateral all-metal spin valves: Experimental observation and theoretical description. *Physical Review B: Condensed Matter and Materials Physics*, *77*(21), 214416. DOI: 10.1103/PhysRevB.77.214416

Vogel, A., Wulfhorst, J., & Meier, G. (2009). Enhanced spin injection and detection in spin valves with integrated tunnel barriers. *Applied Physics Letters*, *94*(12), 122510. DOI: 10.1063/1.3109787

Walker, C. T. (1967). Solid State Physics over 13 Years: Introduction to Solid State Physics. Charles Kittel. Wiley, New York, ed. 3, 1966. 662 pp., illus. $12.50. Science, 155(3765), 991. DOI: 10.1126/science.155.3765.991.b

Wang, H., Zhang, Y., Yang, D., Liu, J., Huang, W., Dai, B., Ren, Y., Li, X., & Chen, M. (2025). Preparation and properties of high magnetic permeability [Fe20Ni80-Al2O3]n magnetic thin films. *Applied Physics. A, Materials Science & Processing*, *131*(4), 254. DOI: 10.1007/s00339-025-08416-y

Wang, T., Peng, Y., Jiang, W., Huang, Y. M., Rahman, B. M. F., Divan, R., Rosenmann, D., & Wang, G. (2016). Integrating nanopatterned ferromagnetic and ferroelectric thin films for electrically tunable RF applications. *IEEE Transactions on Microwave Theory and Techniques*, *65*(2), 504–512. DOI: 10.1109/TMTT.2016.2616869

Wang, T., Peng, Y., Jiang, W., Huang, Y. M., Rahman, B. M. F., Divan, R., Rosenmann, D., & Wang, G. (2017). Integrating Nanopatterned Ferromagnetic and Ferroelectric Thin Films for Electrically Tunable RF Applications. *IEEE Transactions on Microwave Theory and Techniques*, *65*(2), 504–512. DOI: 10.1109/TMTT.2016.2616869

Xu, W., Hou, X., Lu, T., Jiao, Z., Xu, X., & Dai, J. (2024). Research on the High Permeability Permalloy and its Application to the Hydrogen Maser. 2024 European Frequency and Time Forum (EFTF), 300–303. DOI: 10.1109/EFTF61992.2024.10722257

Xu, X., Jin, L., Wen, T., Liao, Y., Tang, X., Zhang, H., & Zhong, Z. (2021). Effects of substrate annealing on uniaxial magnetic anisotropy and ferromagnetic resonance frequency of Ni80Fe20 films deposited on self-organized periodically rippled sapphire substrates. *Vacuum*, *186*, 110047. DOI: 10.1016/j.vacuum.2021.110047

Yamaguchi, M., Suezawa, K., Takahashi, Y., Arai, K. I., Kikuchi, S., Shimada, Y., Tanabe, S., & Ito, K. (2000). Magnetic thin-film inductors for RF-integrated circuits. *Journal of Magnetism and Magnetic Materials*, *215–216*, 807–810. DOI: 10.1016/S0304-8853(00)00293-6

Yanai, T., Koda, K., Kaji, J., Aramaki, H., Eguchi, K., Takashima, K., Nakano, M., & Fukunaga, H. (2018). Electroplated Fe-Co-Ni films prepared in ammonium-chloride-based plating baths. *AIP Advances*, *8*(5), 056127. DOI: 10.1063/1.5007782

Yang, P. Y., Song, C., Fan, B., Zeng, F., & Pan, F. (2009). The role of rotatable anisotropy in the asymmetric magnetization reversal of exchange biased NiO/Ni bilayers. *Journal of Applied Physics*, *106*(1), 013902. Advance online publication. DOI: 10.1063/1.3157176

Yang, T., Kimura, T., & Otani, Y. (2008). Giant spin-accumulation signal and pure spin-current-induced reversible magnetization switching. *Nature Physics*, *4*(11), 851–854. DOI: 10.1038/nphys1095

Yang, X., Wei, J. Q., Li, X. H., Gong, L. Q., Wang, T., & Li, F. S. (2012). Thickness dependence of microwave magnetic properties in electrodeposited FeCo soft magnetic films with in-plane anisotropy. *Physica B, Condensed Matter*, *407*(3), 555–559. DOI: 10.1016/j.physb.2011.11.049

Zakeri, Kh., Lindner, J., Barsukov, I., Meckenstock, R., Farle, M., von Hörsten, U., Wende, H., Keune, W., Rocker, J., Kalarickal, S. S., Lenz, K., Kuch, W., Baberschke, K., & Frait, Z. (2007). Spin dynamics in ferromagnets: Gilbert damping and two-magnon scattering. *Physical Review B: Condensed Matter and Materials Physics*, *76*(10), 104416. DOI: 10.1103/PhysRevB.76.104416

Zezyulina, P. A., Iakubov, I. T., Lagarkov, A. N., Maklakov, S. A., Maklakov, S. S., Naboko, A. S., Osipov, A. V., Petrov, D. A., Rozanov, K. N., & Ryzhikov, I. A. (2016). Effect of Perpendicular Anisotropy and Eddy Currents on the Microwave Performance of Single-Layer and Multi-Layer Permalloy Films. *IEEE Magnetics Letters*, *7*, 1–4. DOI: 10.1109/LMAG.2016.2597230

Zhang, H., Li, Y.-Y., Yang, M.-Y., Zhang, B., Yang, G., Wang, S.-G., & Wang, K.-Y. (2015). Tuning the magnetic anisotropy of CoFeB grown on flexible substrates. *Chinese Physics B, 24*(7), 077501. DOI: 10.1088/1674-1056/24/7/077501

Zhang, W. F., He, Y. L., Zhang, M. S., Yin, Z., & Chen, Q. (2000). Raman scattering study on anatase TiO 2 nanocrystals. *Journal of Physics. D, Applied Physics, 33*(8), 912–916. DOI: 10.1088/0022-3727/33/8/305

Zhou, C., Li, T., Wei, X., & Yan, B. (2020). Effect of the sputtering power on the structure, morphology and magnetic properties of Fe films. *Metals, 10*(7), 896. DOI: 10.3390/met10070896

Chapter 2
Electronic Structure of Semiconductors and Topological Materials

Himadri Sekhar Das
https://orcid.org/0000-0002-3509-3388
Haldia Institute of Technology, Haldia, India

Heranmoy Maity
Ideal Institute of Engineering, Kalyani, India

Sudipta Banerjee
https://orcid.org/0000-0003-0150-6794
Symbiosis Institute of Technology, Symbiosis International University, Pune, India

Maitrayee Chakrabarty
Narula Institute of Technology, Agarpara, India

Gaurav Sharma
Saveetha School of Engineering, Saveetha Institute of Medical and Technical Sciences, Thandalam, India

ABSTRACT

The electronic structure of semiconductors and topological materials determines their electrical and optical properties. Semiconductors possess a tunable band gap crucial for electronic and optoelectronic devices, while topological materials exhibit unique surface states protected by symmetry and strong spin–orbit coupling. These states enable robust, dissipationless transport. Understanding their band topology and electronic interactions is key to designing next-generation quantum, spintronic, and energy-efficient devices.

DOI: 10.4018/979-8-3373-6058-4.ch002

1. INTRODUCTION

The study of electronic structure lies at the heart of modern materials science, serving as the essential foundation for understanding and predicting the behavior of matter across length scales-from molecules to crystalline solids (Beran, 2016). Among the vast array of materials explored by chemists and physicists, semiconductors and topological materials occupy a central position because of their unique combination of technological relevance and fundamental scientific importance. Their electronic structures not only dictate their functional properties but also embody profound concepts that connect chemistry, condensed matter physics, and computational science.

Semiconductors as the backbone of materials chemistry: For over half a century, semiconductors have been the cornerstone of electronics, enabling the transformation of society through transistors, diodes, integrated circuits, solar cells, and optoelectronic devices (Das et al., 2024a). Their importance lies in the tunability of their band gaps, which determine whether a material can absorb visible light, transport charge efficiently, or emit photons under excitation (Banerjee et al., 2024). The development of silicon-based technology, III–V compound semiconductors (such as GaAs, InP), and organic semiconductors demonstrates how diverse chemical compositions and structures can be engineered to optimize performance for electronics, energy harvesting, photonics, and sensors. Semiconductors also bridge fundamental research and application: from understanding electron–hole recombination and exciton binding energies to designing devices such as photodetectors, thermoelectrics, and high-power switches. In essence, semiconductors embody the interplay of chemistry, structure, and function, representing a paradigm where electronic structure theory directly impacts real-world technologies.

The rise of topological materials: Over the past two decades, a new class of quantum materials has revolutionized our understanding of electronic phases: topological materials. These include topological insulators, Dirac and Weyl semimetals, nodal-line semimetals, and topological superconductors. Unlike conventional semiconductors or metals, their properties are dictated not only by local bonding or band energetics but by global topological characteristics of their electronic wavefunctions. Topological insulators, for example, are insulating in the bulk but host robust metallic surface states protected by time-reversal symmetry. These edge or surface states are resistant to scattering and defects, making them promising for spintronics and quantum information. Dirac and Weyl semimetals exhibit relativistic fermion-like dispersions near their band crossings, enabling unusual transport phenomena such as the chiral anomaly, ultrahigh mobility, and large magnetoresistance. Topological superconductors, although still largely hypothetical, are anticipated to harbor Majorana fermions, which may underpin fault-tolerant quantum computing. Importantly, these exotic phases emerge from chemical and structural motifs such as heavy elements with strong spin-orbit coupling, band inversion between atomic orbitals, and nontrivial crystalline symmetries emphasizing the need to unify chemical intuition with advanced electronic structure theory (Liu & Zhang, 2023).

The convergence of chemistry and topology: The discovery of topological materials highlights how fundamental chemistry concepts, such as orbital hybridization and symmetry, directly contribute to quantum states protected by topology. For instance, the band inversion between Bi 6p and Se 4p orbitals underpins the topological insulator character of Bi_2Se_3. Similarly, the presence of d–p orbital mixing and spin–orbit coupling in transition-metal compounds can stabilize Weyl semimetal states. Thus, chemists are uniquely positioned to guide the design of new topological phases by understanding how composition and structure create the conditions necessary for topological order. At the same time, computational methods provide the quantitative framework to predict and verify these properties, linking local bonding with global band topology (Kumar et al., 2020). Schematic comparing molecular orbitals, small cluster, and periodic band structure (Figure.1).

Figure 1. Schematic comparing molecular orbitals, small cluster, and periodic band structure.

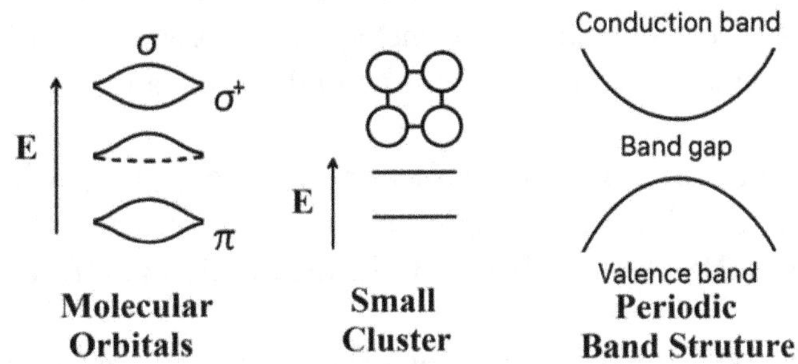

Molecular orbitals describe the quantum states formed by combining atomic orbitals in molecules, representing bonding, anti-bonding, and nonbonding interactions. In small clusters, as the number of atoms increases, discrete molecular orbitals begin to overlap and broaden, forming closely spaced energy levels that approximate the behavior of solids. When extended to an infinite periodic lattice, these overlapping orbitals evolve into continuous energy bands valence and conduction bands separated by a band gap that defines the material's electronic properties. This progression from molecular orbitals to cluster states to periodic bands illustrates the fundamental link between atomic bonding and bulk electronic behavior.

This chapter situates semiconductors and topological materials within the broader evolution of computational and theoretical chemistry, highlighting how electronic structure theory has expanded from molecules to extended materials. The scope can be divided into three interrelated themes:

Bridging Molecular Electronic Structure and Solid-State Band Theory

Molecular orbital theory, developed to describe discrete states in finite systems, provides the foundation for understanding how electronic states evolve in crystals. The transition from molecules to materials through clusters, nanostructures, and periodic solids illustrates how localized molecular orbitals form delocalized energy bands. We will explore the conceptual continuity between molecular orbital methods (Hartree–Fock, configuration interaction, coupled cluster) and solid-state approaches (tight-binding, Bloch wavefunctions, k·p perturbation), emphasizing how chemists and physicists describe the same underlying phenomena in different languages. This bridge also reveals how finite-size effects, defects, and interfaces link molecular-level descriptions with bulk band theory (Westermayr & Marquetand, 2020).

Recent Methodological Advances

Modern computational chemistry has witnessed remarkable progress in accurately describing electronic structure, particularly for semiconductors and topological materials. The chapter will review key methods and their significance:

Density Functional Theory (DFT): The workhorse of electronic structure, with generalized gradient approximations (GGAs), hybrid functionals, and meta-GGA approaches. Although standard DFT underestimates band gaps, it remains invaluable for structural optimization and qualitative band analysis (Blum et al., 2024).

Many-body perturbation theory (GW approximation): Corrects quasiparticle energies and provides accurate band structures, essential for predicting semiconducting gaps and validating topological order.

Bethe–Salpeter Equation (BSE): Extends GW to include excitonic effects, bridging electronic and optical properties.

Symmetry and topology tools: Berry phase, Wannier charge centers, and topological invariants (Chern numbers, Z_2 indices) allow rigorous identification of nontrivial topological phases.

Real-space vs reciprocal-space viewpoints: Real-space analysis (electron density topology, orbital chemistry) complements reciprocal-space band structure analysis, offering a dual perspective on materials.

Integrating Molecules, Semiconductors, and Topological Phases

A central theme is the continuity across length scales: from molecular systems, where electronic states are discrete and localized, to extended materials, where periodicity creates delocalized bands and new topological features. This perspective emphasizes:

- How finite clusters and molecular crystals serve as stepping stones to understanding bulk semiconductors.
- How chemical composition, orbital hybridization, and symmetry control topological behavior.
- How computational tools adapt across these scales, from highly correlated molecular methods to efficient band-structure codes for extended solids.

By placing semiconductors and topological materials in this continuum, the chapter seeks to show that advances in computational and theoretical chemistry are not only about better numerical predictions, but also about unifying chemical intuition with quantum topological principles.

The integration of semiconductors and topological materials into the framework of computational chemistry represents one of the most exciting frontiers in modern science. Semiconductors remain indispensable to technological development, while topological phases introduce entirely new paradigms of transport, protection, and symmetry-driven behavior. Both domains benefit from methodological advances in quantum theory, enabling accurate predictions and deeper understanding (Islam et al., 2025). By bridging molecules to materials, and chemistry to topology, this chapter aims to present a coherent narrative of how electronic structure theory continues to evolve, shaping the future of materials discovery and design.

The electronic properties of crystalline solids are fundamentally governed by their electronic band structure, which emerges from the interaction between electrons and the periodic potential of the lattice. Both conventional semiconductors and topological materials can be understood within this same theoretical framework, namely Bloch's theorem, band theory, and symmetry analysis. Despite their apparent differences in transport and surface behavior, these material classes share common physical origins, particularly in systems with highly symmetric morphologies such as bulk single crystals, thin films, nanowires, and layered heterostructures.

In symmetric morphology systems, where translational, rotational, and time-reversal symmetries are preserved, electronic interactions are strongly constrained by symmetry operations. This provides an ideal platform to compare how electronic states evolve in conventional semiconductors and how subtle changes-primarily due to spin–orbit coupling (SOC) and band inversion-give rise to topological phases. Understanding both classes within a single conceptual framework allows readers to appreciate how topology emerges as an extension of conventional semiconductor physics rather than a completely separate phenomenon.

Electronic Structure of Semiconductors in Symmetric Systems

Band Formation and Orbital Contributions

In conventional semiconductors such as silicon, germanium, and III–V compounds (GaAs, InP), the electronic structure arises from the hybridization of atomic orbitals under a periodic crystal potential. The valence band is primarily composed of bonding p-orbitals, while the conduction band originates from antibonding s or s–p hybridized states. The energy separation between these bands defines the band gap, which is central to semiconductor functionality.

In highly symmetric crystal structures (e.g., diamond or zinc-blende lattices), the symmetry of the lattice leads to well-defined degeneracies at high-symmetry points in the Brillouin zone. These degeneracies are lifted or preserved depending on the presence of inversion symmetry and spin–orbit interactions.

Role of Symmetry and Spin–Orbit Coupling

In symmetric semiconductor systems, time-reversal symmetry ensures Kramers degeneracy, while spatial symmetries determine band dispersion and effective masses. Spin–orbit coupling is typically weak to moderate in conventional semiconductors, leading to effects such as:

- Valence band splitting into heavy-hole, light-hole, and split-off bands
- Spin splitting in non-centrosymmetric structures (Rashba and Dresselhaus effects)

However, these SOC-induced modifications do not qualitatively alter the topological nature of the bands. The band gap remains topologically trivial, meaning that the electronic states can be continuously transformed into isolated atomic orbitals without closing the gap.

Electronic Interactions and Transport

Electronic transport in semiconductors is dominated by bulk states, with carrier dynamics influenced by phonons, impurities, and structural defects. Surface or interface states, if present, are highly sensitive to disorder and symmetry breaking. In symmetric morphologies such as quantum wells or nanowires, quantum confinement modifies band dispersion and increases the effective band gap but does not introduce protected conducting channels.

Electronic Structure of Topological Materials

Band Inversion and Its Origin

Topological materials including topological insulators and topological semimetals share the same band-theory foundation as semiconductors but differ in a crucial aspect: band inversion. In these materials, strong spin–orbit coupling causes the conventional ordering of conduction and valence bands to reverse at specific high-symmetry points in the Brillouin zone. For example, states with p-orbital character may shift above s-orbital states, leading to an inverted band structure. Importantly, this inversion occurs without breaking crystal symmetry, making symmetric morphology systems ideal for observing topological phases.

Topological Invariants and Symmetry Protection

Unlike conventional semiconductors, topological materials are characterized by global properties of the electronic wavefunctions, described by topological invariants such as the Z_2 index or Chern number. These invariants depend on:

- Time-reversal symmetry
- Inversion or crystalline symmetries
- Berry curvature distribution in momentum space

In symmetric systems, these symmetries protect the topological phase by preventing backscattering and localization of electronic states. Crucially, the topological nature of the band structure cannot change unless the band gap closes, making these phases robust against moderate perturbations.

Surface and Edge States in Symmetric Morphologies

A defining feature of topological materials is the emergence of gapless surface or edge states, a phenomenon explained by the principle of bulk–boundary correspondence. In symmetric morphologies:

- Bulk states remain gapped
- Surface states exhibit linear, Dirac-like dispersion
- Spin and momentum are locked, suppressing backscattering

These boundary states persist even in thin films and nanostructures, provided that the underlying symmetries are preserved. This sharply contrasts with semiconductor surface states, which lack symmetry protection.

Comparative Analysis in Symmetric Morphology Systems

When examined in identical symmetric systems, the similarities and differences between semiconductors and topological materials become particularly clear.

Shared Features

Both classes exhibit:

- Periodic band structures governed by crystal symmetry
- Energy gaps separating occupied and unoccupied states
- Modifications due to confinement and dimensional reduction

From a computational perspective, both can be analyzed using density functional theory, k•p perturbation theory, and tight-binding models.

Fundamental Distinctions

The key distinction lies in the topology of the wave functions rather than the existence of a band gap itself. In semiconductors, wave functions are locally defined and sensitive to perturbations. In topological materials, wave functions possess a non-trivial geometric phase, leading to robust boundary conduction. In symmetric morphologies, confinement in semiconductors primarily alters energy scales, whereas in topological materials it reshapes-but does not eliminate-topological features.

Interaction Perspective: Geometry of Electronic States

A unifying interpretation of electronic interactions in these systems arises from the geometry of electronic wave functions. In semiconductors, electron–lattice interactions dominate, and phase information plays a minimal role. In topological materials, the Berry phase accumulated by electrons moving through momentum space directly influences transport, leading to unusual responses such as suppressed backscattering and spin-polarized currents.

This geometric distinction explains why topological materials retain conducting channels even in highly symmetric and confined morphologies, while semiconductors do not.

2. ELECTRONIC STRUCTURE BASICS

The concept of electronic structure provides the fundamental language by which chemists and physicists describe how electrons behave in molecules and solids (Das et al., 2024b). Whether one is concerned with a simple diatomic molecule, a crystalline semiconductor, or a complex quantum material, the unifying thread is that electrons occupy quantized states determined by atomic orbitals, interatomic bonding, and the symmetry of the system. This section provides a comprehensive overview of electronic structure principles, beginning with the evolution from molecules to crystals, before addressing semiconductors, charge carriers, and the computational tools developed to understand them (Manni et al., 2023).

2.1 From Molecule to Crystal

Molecular Orbitals → Band Formation

In isolated molecules, the electronic structure is described by molecular orbital (MO) theory, where atomic orbitals combine to form bonding, anti-bonding, and non-bonding states. Each orbital is discrete, characterized by specific energies and symmetry. For small systems, molecular orbital diagrams give a clear picture of how electrons fill orbitals according to the Pauli principle and Hund's rule.

However, when many identical units come together to form a solid, the discrete energy levels of individual molecules or atoms overlap and broaden into bands. Consider a chain of hydrogen atoms: each atom contributes a 1s orbital. In a dimer, the orbitals split into bonding and anti-bonding levels. In a crystal of N atoms, the splitting produces N closely spaced levels that form a nearly continuous energy band. Thus, the band picture of solids emerges naturally as the collective extension of molecular orbital concepts.

Bloch Theorem and Wave Functions

In periodic crystals, the motion of electrons is governed by the translational symmetry of the lattice. Bloch's theorem states that the wave functions can be expressed as:

Bloch's Theorem and Wave Functions

In periodic crystals, the motion of electrons is governed by the translational symmetry of the lattice.

If the potential energy satisfies: $V(r + R) = V(r)$,

for every lattice translation vector R, then according to Bloch's theorem, the electron wavefunctions can be expressed as:

$$\psi_{nk}(r) = e^{ik\cdot r}u_{nk}(r)$$

where: k is the wave vector (or crystal momentum),

n is the band index,

$u_{nk}(r)$ is a periodic function that has the same periodicity as the crystal lattice:

$$u_{nk}(r+R) = u_{nk}(r)$$

Implication

Under a lattice translation R,

$$\psi_{nk}(r+R) = e^{ikR}\cdot \psi_{nk}(r)$$

Showing that the wavefunction changes only by a phase factor upon translation by a lattice vector.

Physical Meaning

e^{ikR}: plane-wave part → describes free-electron–like propagation.

$u_{nk}(r)$: periodic part → captures how the lattice potential modifies the electron's motion.

Together, they describe Bloch states, which form the basis for the band theory of solids.

Reciprocal lattice and Brillouin zone: The reciprocal lattice is a mathematical construct where wave vectors are defined. Its primitive vectors are related to those of the real-space lattice, and it provides the natural space for describing diffraction, phonons, and electron states. The first Brillouin zone-the Wigner-Seitz cell of the reciprocal lattice-is the unique domain where all distinct electronic states can be represented. The shape and size of the Brillouin zone reflect the underlying crystal symmetry, and critical points within it (Γ, X, L, k, etc.) often correspond to extrema in electronic band dispersions.

Energy bands, density of states, and effective mass: The set of allowed electron energies as a function of wavevector constitutes the band structure. While band dispersions provide microscopic details, a more global measure is the density of states (DOS), which quantifies how many states exist at a given energy. For semiconductors, the DOS near the band edges is critical for determining carrier concentrations and optical absorption. Figure .2 shows the workflow from molecules to crystal and topological invariant.

Figure 2. Workflow schematic: from molecule crystal topological invariant computation.

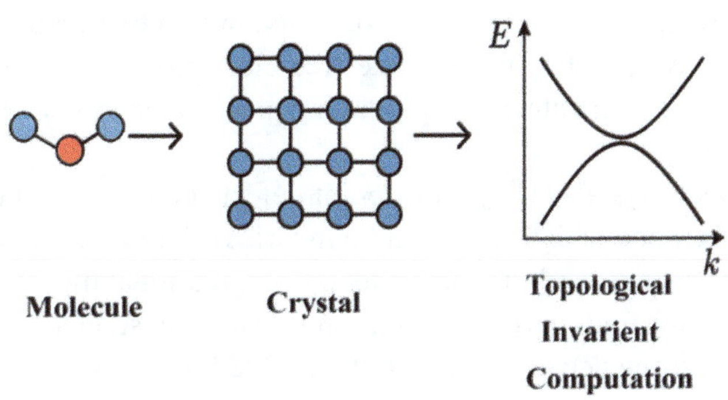

The workflow from molecule to crystal to topological invariant computation represents the hierarchical approach to understanding material properties from fundamental principles. At the molecular level, electronic orbitals and bonding characteristics are analyzed using quantum chemical methods such as Density Functional Theory (DFT). These molecular units serve as the building blocks for the crystal structure, where periodicity and atomic arrangement determine the overall electronic band structure. Using periodic boundary conditions, calculations reveal how overlapping molecular orbitals evolve into continuous energy bands, distinguishing metals, semiconductors, and insulators. The next stage involves topological invariant computation, where advanced techniques such as Berry curvature analysis, parity eigenvalue evaluation, or Wilson loop methods are applied to the crystal's electronic structure to classify its topological phase. These invariants (e.g., Chern number) characterize whether the material hosts protected surface or edge states, as in topological insulators or Weyl semimetals. This workflow bridges chemical bonding and quantum topology, showing how local molecular symmetries and spin–orbit coupling propagates into macroscopic properties. Ultimately, it highlights a computational framework that connects microscopic electronic structure to global topological behavior, enabling prediction and design of materials with novel quantum functionalities.

2.2 Semiconductors

Semiconductors are distinguished by the presence of an energy gap between the valence and conduction bands. If the conduction band minimum (CBM) and valence band maximum (VBM) occur at the same wavevector, the semiconductor is direct-gap, enabling efficient optical absorption and emission (e.g., GaAs, CdTe). In contrast, when the CBM and VBM are at different k-points, the semiconductor is indirect-gap, as in Si and Ge. Indirect-gap semiconductors require phonon assistance for optical transitions, making them less efficient for light-emitting devices but excellent for microelectronics (Benjamin et al., (2017).

The size of the band gap determines application domains:

– Narrow gaps (0.5–1 eV): infrared detectors, thermoelectrics.
– Intermediate gaps (1–2 eV): photovoltaics, LEDs.
– Wide gaps (>3 eV): power electronics, UV photodetectors.

Intrinsic vs extrinsic semiconductors: An intrinsic semiconductor is perfectly pure, where carrier concentrations arise only from thermal excitation across the band gap. However, practical semiconductors are often extrinsic, with intentional impurities (dopants) or unintentional defects altering electronic properties. Donor dopants (e.g., P in Si) contribute free electrons, producing n-type behavior, while acceptor dopants (e.g., B in Si) generate holes, producing p-type behavior. Doping control is the key to modern device functionality.

Defects and doping: Real crystals always host point defects (vacancies, interstitials, antisites), line defects (dislocations), and extended defects (grain boundaries). These introduce localized states in the band gap, trap carriers, or act as recombination centers. Computational modeling of defects often using super-cells with corrections for finite-size effects-remains central in semiconductor research, as defects often dominate material performance (Zaiba Zafar et al., (2022).

Charge Carriers: Electrons, Holes, Excitons

Electrons in the conduction band and holes in the valence band act as mobile charge carriers. Under optical excitation, an electron-hole pair can remain bound by Coulomb interaction, forming an exciton. Excitons are categorized into two limiting cases:

Wannier excitons: delocalized over many unit cells, weakly bound, common in inorganic semiconductors.

Frenkel excitons: tightly bound and localized, typical of organic molecular crystals.

Excitonic effects are crucial for interpreting optical spectra, understanding energy transfer, and designing optoelectronic materials. Table.1 shows the Computational Methods for Electronic Structure of Semiconductors and Topological Materials.

Table 1. Computational Methods for Electronic Structure of Semiconductors and Topological Materials.

Method	Principle / Approach	Key Applications	Advantages	Limitations
Density Functional Theory (DFT)	Solves many-body electron systems using electron density; based on Kohn–Sham equations.	Band structure, density of states, charge distribution in semiconductors and topological insulators.	Computationally efficient; good balance of accuracy and cost.	Limited accuracy for band gap prediction; inadequate for strong correlation effects.
GW Approximation	Many-body perturbation theory using Green's function (G) and screened Coulomb interaction (W).	Accurate quasiparticle energy calculations for semiconductors and topological materials.	Provides correct band gaps; includes electron self-energy effects.	Computationally demanding; often requires DFT as a starting point.
Hybrid Functionals (e.g., HSE06, PBE0)	Mixes a portion of exact Hartree–Fock exchange with DFT exchange-correlation.	Improved band gap estimation and defect-level studies in semiconductors.	Corrects DFT band gap underestimation; better electronic structure accuracy.	Higher computational cost; less suitable for very large systems.
Tight-Binding (TB) Method	Uses parameterized Hamiltonians with atomic orbitals; semi-empirical approach.	Modeling topological edge states, nanostructures, and large-scale semiconductors.	Efficient for large systems; intuitive for topological band structure analysis.	Accuracy depends on parameter fitting; lacks predictive capability.
k·p Perturbation Theory	Expands band structure near high-symmetry points using momentum operators.	Effective mass approximation; analysis of topological transitions.	Simple and analytical; connects to symmetry properties.	Valid only near band extrema; not suitable for full Brillouin zone.
Hartree–Fock (HF) Method	Solves Schrödinger equation using single Slater determinant; mean-field approximation.	Benchmark for small semiconductor clusters; study of exchange interactions.	Incorporates exchange exactly; foundation for post-HF methods.	Neglects correlation effects; expensive for large systems.
Quantum Monte Carlo (QMC)	Uses stochastic sampling of wavefunctions for electron correlation.	Highly accurate band gap and excitonic property calculations.	Excellent treatment of electron correlation; high accuracy.	Extremely computationally intensive; limited to small systems.
Dynamical Mean-Field Theory (DMFT)	Treats local electron correlations dynamically using impurity solvers.	Strongly correlated topological insulators and Mott semiconductors.	Captures local correlation effects; useful for correlated topological phases.	Neglects non-local correlations; complex implementation.
Topological Invariant Calculations (Z_2, Chern Number)	Computes Berry curvature, parity eigenvalues, or Wilson loops.	Identification of topological phases (e.g., Z_2 insulators, Weyl semimetals).	Direct link between band topology and observable properties.	Requires accurate electronic structure; sensitive to computational precision.
Ab Initio Molecular Dynamics (AIMD)	Simulates atomic motion using forces derived from electronic structure.	Thermal and structural effects on semiconductor and topological materials.	Realistic finite-temperature behavior; combines structure and electronics.	Computationally heavy; limited time and length scales.

2.3 Key Computational Tools

The accurate prediction of electronic structure requires sophisticated computational frameworks. Different methods are applicable to different systems and length scales, and each comes with its advantages and limitations.

Hartree–Fock (HF) provides a mean-field description of electrons, solving the many-electron problem approximately by representing the wavefunction as a Slater determinant of single-particle orbitals. Exchange is treated exactly, but correlation is neglected. HF is widely used for molecules and clusters but performs poorly for extended solids, especially in predicting band gaps (tending to overestimate them). Nonetheless, HF forms the foundation for more advanced post-HF methods (Matveeva et al., 2023).

Post-Hartree–Fock Methods

Configuration Interaction (CI): systematically includes electron correlation by mixing excited configurations. However, it scales exponentially with system size, limiting it to small molecules.

Coupled Cluster (CCSD, CCSD(T)): widely regarded as the "gold standard" for molecular systems, providing accurate correlation energies. Yet its steep computational cost makes it impractical for periodic solids, though local or embedding schemes attempt to extend its reach.

For semiconductors and materials, these methods are typically restricted to small clusters or benchmarking purposes.

Density Functional Theory (DFT) revolutionized electronic structure by recasting the many-body wavefunction problem into a problem of electron density. Practical implementations use approximate exchange–correlation functionals.

Local Density Approximation (LDA): simple, efficient, but underestimates gaps.

Generalized Gradient Approximation (GGA): includes gradient corrections, improves structural properties but still struggles with gaps.

Hybrid functionals (e.g., HSE06, PBE0): mix HF exchange with DFT, significantly improving gap predictions at higher cost.

Meta-GGAs (e.g., SCAN): offer balance of accuracy and efficiency, especially for structural properties.

Despite its success, DFT has known limitations: self-interaction errors, underestimation of quasiparticle gaps, and difficulties in capturing excitonic effects.

Many-Body Perturbation Theory

To go beyond DFT, many-body approaches explicitly consider electron correlation:

GW approximation: corrects single-particle energies by approximating the electron self-energy as the product of the Green's function (G) and the screened Coulomb interaction (W). GW is now the standard for quasiparticle band structures, yielding accurate band gaps for semiconductors.

Bethe–Salpeter Equation (BSE): built upon GW, BSE accounts for electron–hole interactions, allowing quantitative prediction of optical absorption spectra, including excitonic features.

Together, GW+BSE provide the most reliable picture of electronic and optical structure in semiconductors.

Tight-binding models: The tight-binding (TB) method represents electronic states as linear combinations of localized atomic orbitals, with parameters fitted or derived from first principles. TB is invaluable for intuitive insights, large systems, and topological materials, where model Hamiltonians reveal band inversion and Dirac points. For example, the Kane–Mele model describes quantum spin Hall effects in graphene-like systems (Pham *et al.,* 2025).

k·p perturbation theory: The k·p method is a semi-analytical approach where the Hamiltonian near a band extremum is expanded around a known point (e.g., Γ point). It provides effective models for low-energy carrier dynamics, optical transitions, and anisotropies. k·p theory underlies understanding of effective masses, band-edge symmetries, and multiband effects in semiconductors such as GaAs (Urban et al., 2011).

2.4 Integration of Concepts

The progression from molecular orbitals to band structures highlights the continuity of electronic structure theory. Chemists view bonding through localized orbitals, while physicists view electronic states through delocalized Bloch functions. Both perspectives are essential, and modern computational chemistry increasingly unifies them.

Semiconductors exemplify how electronic structure governs real-world applications: the band gap, carrier dynamics, and excitonic effects are direct manifestations of fundamental quantum mechanics. Computational tools from Hartree–Fock to GW provide different levels of accuracy and efficiency, allowing researchers to tackle problems ranging from small clusters to bulk solids. The balance of physical insight, computational feasibility, and predictive accuracy continues to drive the development of electronic structure methods, shaping the design of next-generation materials (Li et al., 2024).

This section has outlined the conceptual and methodological basics of electronic structure. We have seen how molecules evolve into crystals, how semiconductors derive their unique behavior from their band gaps and carriers, and how computational methods span from wave function-based quantum chemistry to many-body perturbation techniques for solids. The unifying message is that electronic structure is not only a theoretical abstraction but a practical framework guiding the discovery, design, and engineering of materials. In the subsequent sections, these principles will be applied to advanced topics such as topological phases and the role of symmetry, providing a broader perspective on how computational chemistry connects molecules to quantum materials.

3. SEMICONDUCTORS: COMPUTATIONAL ADVANCES

Understanding semiconductors at the quantum level has been one of the central achievements of computational chemistry and condensed matter physics. From predicting accurate band gaps to modeling charge transport and defect states, recent advances in computational methods have transformed semiconductor theory from a descriptive science into a predictive one. This section focuses on how modern electronic-structure techniques and theoretical frameworks illuminate the complexities of semiconductor materials, with applications spanning inorganic crystals, molecular solids, and hybrid systems.

3.1 Electronic Structure Prediction

Hybrid and advanced functional: Standard DFT with LDA or GGA functionals notoriously underestimates semiconductor band gaps. Hybrid functionals (HSE06, PBE0), which incorporate a fraction of exact exchange, have emerged as highly reliable for predicting realistic gaps and dielectric properties. Meta-GGAs (e.g., SCAN) and range-separated hybrids further refine the balance between accuracy and computational efficiency. These functionals are now widely used in computational materials discovery, enabling predictions of photovoltaic and power electronic materials (Jain et al., 2011).

GW-based quasi particle methods: To address quasi particle excitations, GW methods represent a gold standard. Starting from a DFT reference, the G_0W_0 approximation applies a perturbative correction, while self-consistent GW (scGW) iteratively refines both wave functions and energies. GW systematically improves conduction band energies and gap sizes, bringing them in line with experiment. Coupled with the Bethe–Salpeter Equation (BSE), one can predict excitonic absorption spectra, central for optoelectronic materials.

Spin–orbit coupling (SOC) effects: For semiconductors containing heavy elements (e.g., Pb, Bi, Sn compounds), spin–orbit coupling plays a decisive role. SOC not only modifies the splitting of conduction and valence bands but can invert them, giving rise to topological phases. In perovskites, chalcogenides, and bismuth-based semiconductors, accurate treatment of SOC is essential for predicting carrier lifetimes, Rashba splitting, and spintronic functionalities.

3.2 Carrier Transport, Phonons, and Defects

First-principles phonons and electron–phonon coupling: Carrier mobility is governed by electron–phonon interactions. Density Functional Perturbation Theory (DFPT) enables first-principles calculation of phonon spectra, vibrational lifetimes, and coupling to electronic states. From these, scattering rates and carrier mobilities can be estimated with no adjustable parameters. This approach has clarified why polar semiconductors (e.g., GaN, ZnO) exhibit strong polar optical phonon scattering, limiting mobilities despite wide gaps (Xiao et al., 2025).

Defects and dopants: Defect physics is at the heart of semiconductor performance. Supercell-based first-principles defect calculations, corrected for finite-size effects and spurious electrostatics, now allow quantitative predictions of defect formation energies, charge transition levels, and dopant solubilities. Such calculations explain deep-level traps in GaN LEDs, carrier compensation in ZnO, and self-healing in halide perovskites. Computational defect thermodynamics provides a pathway to defect engineering, where synthesis conditions are tuned for desired defect equilibria.

Complexities in real materials: Transport in semiconductors is rarely simple. Anisotropic effective masses in layered semiconductors (MoS_2, black phosphorus) lead to direction-dependent mobilities. Interface scattering becomes crucial in heterojunctions and nanostructures. Polar and piezoelectric fields in wurtzite III-nitrides strongly modulate optical transitions. Modern computational methods, often combining atomistic simulations with Boltzmann transport theory, are necessary to capture these coupled complexities.

3.3 Organic Semiconductors and Molecular Solids

Molecular vs solid-state frameworks: Organic semiconductors lie at the interface of molecular and solid-state theory. Their electronic structure arises from molecular orbitals of π-conjugated backbones, modified by intermolecular packing and long-range Coulomb interactions. Unlike inorganic crystals, disorder and polaronic effects play dominant roles. Computational methods must therefore bridge cluster quantum chemistry with periodic boundary approaches (Kawabata et al., 2024).

Advanced treatments

- Periodic boundary condition methods enable electronic band-structure calculations of organic crystals.
- Cluster embedding and QM/MM schemes capture local excitations while including environmental polarization.
- Non-adiabatic molecular dynamics track exciton and polaron transport under realistic thermal disorder.
- These approaches have explained why acenes (e.g., pentacene, tetracene) exhibit high mobilities, how thiophene derivatives form efficient charge-transport channels, and why disorder dominates exciton diffusion in amorphous organics.

Case Studies

Acenes (pentacene, tetracene): Modeling reveals that intermolecular packing controls π–π overlap and charge transport. Optical gaps are highly sensitive to environment, requiring dielectric embedding corrections.

Thiophenes (polythiophene, P3HT): Structural disorder strongly impacts electronic states. Simulations have shown how regioregularity enhances band-like transport.

Perovskite–organic hybrids: Emerging cases where inorganic band-edge physics couples with molecular excitons, requiring multiscale modeling.

Computational chemistry now provides a predictive toolkit for semiconductor physics. Hybrid functionals and GW+BSE allow quantitative band gap and optical predictions; DFPT provides phonon and mobility estimates; defect modeling explains dopant behavior; and multiscale methods capture organic and hybrid semiconductors. These tools are indispensable for guiding experimental synthesis and device design.

The next frontier lies in beyond-DFT approaches (quantum Monte Carlo, embedding theories, machine-learning potentials), enabling accurate yet scalable treatment of complex semiconductor phenomena. Together, these advances demonstrate how computational theory connects fundamental electronic structure with real-world semiconductor performance, bridging the molecular and material scales.

4. TOPOLOGICAL MATERIALS: THEORY, CHEMISTRY, AND COMPUTATION

Topological materials represent one of the most revolutionary advances in condensed matter physics and materials chemistry of the past two decades. Unlike conventional phases, which are described by broken symmetries or local order parameters, topological phases are distinguished by global invariants of the electronic wavefunctions. Their discovery has unified abstract mathematical ideas with real materials, leading to unprecedented properties such as robust metallic surface states immune to backscattering, Dirac and Weyl fermions in crystals, and potential applications in quantum computing and spintronics (Kumar et al., 2020).

This section surveys the theoretical foundations, chemistry, and computational strategies for understanding and predicting topological materials.

4.1 Topological Phases: Basic Concepts

Band Inversion and Topology Invariants

The hallmark of a topological phase is band inversion, where the normal order of conduction and valence bands is reversed due to strong interactions or relativistic effects. Such inversion creates a non-trivial topology in momentum space. Topological invariants classify these states:

Chern number (integer): characterizes the quantum Hall effect and Chern insulators, indicating the number of protected edge states.

Z_2 invariants: govern time-reversal-invariant topological insulators, distinguishing trivial and non-trivial insulating phases.

Symmetry and Protection

Symmetries play a central role in determining topology:

Time-reversal symmetry (TRS): ensures Kramers degeneracy and protects Z_2 topological insulators.

Inversion symmetry: simplifies invariant calculations through parity eigenvalues at high-symmetry k-points.

Crystalline symmetries (mirror, rotational, glide): give rise to topological crystalline insulators (TCIs), where boundary states are symmetry-protected.

Spin–orbit coupling (SOC): SOC is often the driving force for topological behavior. In systems with heavy elements, SOC can lift degeneracies, invert bands, and induce new topological phases. Materials such as Bi_2Se_3, Bi_2Te_3, and SnTe owe their topological nature to strong SOC.

4.2 Materials Chemistry Perspective

Chemical Composition and Orbital Character

Topological behavior is intimately linked to elemental composition and orbital hybridization:

- Heavy elements (Bi, Sb, Te, W, Ta) provide strong SOC needed for band inversion.
- Orbital competition (s–p, p–d, d–f) determines the type of inversion. For example, s–p inversion underlies Bi_2Se_3, while d–p inversion is seen in transition-metal oxides and perovskite analogues.

Crystal Structure and Symmetry

- The lattice symmetry defines allowed band crossings and protections:
- Layered structures (Bi chalcogenides) naturally host Dirac-like surface states.
- Non-centrosymmetric lattices facilitate Weyl points by breaking inversion symmetry.
- Hexagonal and tetragonal lattices host nodal lines stabilized by mirror or glide symmetries.

Defects, stoichiometry, and growth issues: Real materials depart from ideal models. Defects and stoichiometric deviations can dope carriers into bulk states, masking surface contributions. For instance, Se vacancies in Bi_2Se_3 make the bulk metallic, complicating observation of topological surface states.

Growth conditions and substrate choices can stabilize or destroy the delicate balance of band inversion. Thus, chemistry is as crucial as physics in realizing topological materials.

4.3 Computational Approaches

DFT with SOC:

Density Functional Theory (DFT) incorporating spin–orbit coupling is the starting point for predicting topological behavior. Band structure calculations can reveal inverted gaps and Dirac-like dispersions.

Wannier functions and tight-binding models

From DFT wavefunctions, maximally localized Wannier functions (MLWFs) can be constructed to build effective tight-binding Hamiltonians. These allow interpolation of band structures, calculation of Berry curvatures, and evaluation of topological invariants with high efficiency (Qiao et al., 2023).

Topological invariant calculations

Parity analysis: for inversion-symmetric materials, Z_2 invariants are obtained from parity eigenvalues at time-reversal invariant momenta (TRIM).

Wilson loop / Wannier charge centers: a general approach to extract Z_2 invariants and Chern numbers by tracking Berry phase evolution.

Berry curvature integration: gives Chern numbers directly for systems without TRS.

Challenges and limitations

- Accuracy of band ordering: standard DFT may incorrectly predict band alignment; hybrid functionals or GW corrections may be required.
- Correlation effects: in d- and f-electron systems, strong correlations complicate predictions, necessitating DFT+U, dynamical mean-field theory (DMFT), or beyond.
- Many-body effects: topological superconductivity and fractional topological states require approaches beyond single-particle band theory.

4.4 Case Studies and Examples

Topological insulators (TIs)

- Bi_2Se_3, Bi_2Te_3, Sb_2Te_3: canonical 3D TIs with single Dirac cone surface states. Computational screening first predicted their band inversions, later confirmed by ARPES experiments.
- SnTe: a topological crystalline insulator (TCI), where mirror symmetry protects surface states.

Dirac and Weyl semimetals

- Na_3Bi, Cd_3As_2: Dirac semimetals with linear band crossings at the Fermi level.
- TaAs, NbAs, TaP, NbP: Weyl semimetals lacking inversion symmetry, hosting Fermi arcs observable in ARPES (Angle-Resolved Photoemission Spectroscopy).

Nodal-line semimetals

- Systems such as ZrSiS exhibit line degeneracies in momentum space, stabilized by mirror symmetry. Their flat "drumhead" surface states offer new routes for correlation-driven phases.

Computationally discovered materials: High-throughput DFT databases and machine learning methods have screened thousands of compounds, revealing previously unrecognized topological candidates across oxides, half-Heuslers, and layered systems. Computational predictions increasingly drive experimental synthesis (Xu et al., 2023).

Hybrid Organic–Inorganic Systems

Though less explored, hybrid perovskites and metal–organic frameworks (MOFs) provide platforms where strong SOC from heavy metal centers could combine with tunable organic components to engineer topological phases.

Topological materials exemplify the synergy between theoretical physics, computational chemistry, and experimental synthesis. Topological invariants, rooted in mathematical physics, find material realization through chemical composition, orbital hybridization, and lattice symmetry. Computational tools, from DFT+SOC to Wannier-based models, allow predictive design, though challenges remain in treating correlations and disorder.

Future opportunities include:

- Topological superconductors for quantum computing platforms.
- Magnetic topological phases enabling dissipationless spintronics.
- Correlated topological matter where electron–electron interactions give rise to exotic quasiparticles.
- Machine-learning-assisted discovery to accelerate identification of novel compounds.

As computational chemistry increasingly integrates many-body and topological concepts, the exploration of these phases will expand from conventional inorganic systems to hybrid molecular and nanoscale architectures, completing the bridge from molecules to materials in the topological domain.

Why Magnetic Topology Depends on Periodic Structure

Magnetic topology describes global, non-trivial arrangements of spins that cannot be removed by smooth deformation without crossing an energy barrier. These magnetic textures do not arise arbitrarily; instead, they are strongly constrained and enabled by the periodic structure of the material. The crystal lattice defines the allowed symmetries, exchange pathways, and spin–orbit interactions, which together determine what kinds of magnetic orders and topological textures can exist.

In periodic solids, the relationship between lattice symmetry, magnetic ordering, and electronic structure is hierarchical. The periodic structure sets the symmetry rules, magnetism selectively breaks or preserves these symmetries, and magnetic topology emerges when the resulting spin configuration acquires a non-trivial global character. Understanding this hierarchy is essential for categorizing magnetic materials and designing systems with desired topological properties.

Categorization of Periodic Structures Relevant to Magnetism

Periodic structures can be categorized in several complementary ways. Each categorization directly influences the magnetic topology that can emerge.

Classification by Crystal Symmetry

(a) Centrosymmetric Periodic Structures

Examples:

- Simple cubic (SC)
- Body-centered cubic (BCC)
- Face-centered cubic (FCC)
- Perovskite oxides (ideal form)

Key features:

- Inversion symmetry present
- Dzyaloshinskii–Moriya interaction (DMI) is forbidden in the bulk

Magnetic consequence:

- Magnetism tends to be collinear (ferromagnetic or antiferromagnetic)
- Magnetic topology is usually trivial unless time-reversal symmetry is broken strongly

Typical materials:

- Fe (BCC ferromagnet)
- NiO (antiferromagnet)

(b) Non-Centrosymmetric Periodic Structures

Examples:

- B20 structure (MnSi, FeGe)
- Wurtzite lattices
- Polar tetragonal lattices

Key features:

- Broken inversion symmetry
- Strong spin–orbit coupling effects

Magnetic consequence:

- Allows Dzyaloshinskii–Moriya interaction
- Stabilizes chiral and non-collinear magnetic textures

These structures are central to real-space magnetic topology.

Classification by Dimensionality

Dimensionality Examples Dominant Magnetic Topologies

3D bulk Perovskites, Heuslers Weyl fermions, magnetic monopoles
2D layers van der Waals magnets Chern insulators, skyrmions

1D chains Spin chains Solitons, domain walls

Lower dimensionality enhances topological effects by reducing symmetry and increasing interaction anisotropy.

What Is Magnetic Topology?

Magnetic topology refers to spin configurations characterized by topological invariants, such as winding numbers or skyrmion numbers. These invariants describe how spins wrap around a mathematical space (usually a sphere).

Trivial Magnetic Orders

- Ferromagnetic (all spins aligned)
- Antiferromagnetic (alternating spins)
- Ferrimagnetic

These orders repeat periodically and can be continuously deformed into a uniform state. They are topologically trivial.

Non-Trivial Magnetic Topologies

These possess a global structure protected by topology:

- Skyrmions
- Magnetic vortices
- Hopfions
- Emergent magnetic monopoles

Such textures arise from competing interactions and symmetry constraints imposed by the lattice.

How Periodic Structures Enable Magnetic Topology

The connection between lattice structure and magnetic topology can be understood through key interactions.

Exchange Interaction

- Favors parallel or antiparallel spin alignment
- Dominant in centrosymmetric lattices

Dzyaloshinskii–Moriya Interaction (DMI)

Arises when:

- Inversion symmetry is broken

- Spin–orbit coupling is present

Mathematically: Dij·(Si×Sj)

Effect:
- Favors spin canting
- Essential for chiral magnetic textures

Magnetic Anisotropy
- Depends on lattice symmetry
- Stabilizes certain spin orientations
- Crucial for maintaining topological textures at finite temperatures

Examples Linking Periodic Structures to Magnetic Topologies

Example 1: Skyrmions in B20-Type Periodic Structures
Material: MnSi
Crystal structure: Non-centrosymmetric cubic
Magnetic topology: Skyrmion lattice

Why skyrmions form:

- Broken inversion symmetry → strong DMI
- Competing exchange interactions
- Applied magnetic field stabilizes skyrmion phase

Topological invariant:

- Skyrmion number = ±1 per unit cell

Schematic (top view of a skyrmion):

↑ ↑ ↑ ↑ ↑
↑ ↗ → ↘ ↑
↑ ← ↓ → ↑
↑ ↖ ← ↗ ↑
↑ ↑ ↑ ↑ ↑

Here, spins wrap the entire sphere once, giving a non-trivial topology.

Example 2: Magnetic Weyl Semimetals in Periodic Lattices
Material: $Co_3Sn_2S_2$
Structure: Layered kagome lattice
Magnetic order: Ferromagnetic
Topology: Weyl nodes in momentum space

Role of periodic structure:

- Kagome lattice produces Dirac-like band crossings
- Ferromagnetism breaks time-reversal symmetry

Outcome:

- Weyl nodes act as sources and sinks of Berry curvature
- Correspond to magnetic monopoles in momentum space

This is a momentum-space magnetic topology, directly linked to the lattice symmetry.

Example 3: Chern Insulators in Layered Magnetic Systems
Material: $MnBi_2Te_4$ (few layers)
Structure: Layered periodic lattice
Magnetic order: Antiferromagnetic
Topology: Chern insulator state
Key factors:

- Strong spin–orbit coupling
- Layer-dependent magnetic symmetry
- Periodicity along the stacking direction

Result:

- Quantized Hall conductance without external magnetic field
- Topology protected by magnetic symmetry rather than time-reversal symmetry

Real-Space vs Momentum-Space Magnetic Topology

A crucial concept for projects is the distinction between two types of topology:

Real-Space Magnetic Topology

- Skyrmions
- Vortices
- Domain walls

Defined by:

- Spatial spin textures
- Real-space winding numbers

Momentum-Space Magnetic Topology

- Weyl fermions
- Chern bands
- Berry curvature monopoles

Defined by:

- Electronic band structure
- Berry phase in reciprocal space

Unified schematic:

Real Space Momentum Space
----------- ---------------
Skyrmion Weyl Node
Spin wrapping Berry flux source

Both arise from the same periodic lattice but manifest differently.

Schematic Framework for Project Presentation

Flow chart: Structure–Magnetism–Topology Flow

Periodic Lattice
↓
Symmetry Constraints
↓
Magnetic Interactions
↓
Magnetic Order
↓
Magnetic Topology

Structure-Topology Map flow shown below:

Centrosymmetric → Collinear magnetism → Trivial Non-centrosymmetric → Chiral magnetism → Skyrmions Layered + SOC → Magnetic gaps → Chern phases.

5. FROM MOLECULES TO MATERIALS: THE SCALE-BRIDGING

One of the central challenges of modern computational and theoretical chemistry lies in connecting molecular-level understanding with extended solid-state behavior. Molecules, clusters, nanostructures, and bulk materials represent different regimes of electronic structure. Yet, advances in computational tools now make it possible to bridge these scales, allowing us to understand how fundamental chemical principles manifest in collective quantum phenomena such as band formation, excitonic states, and topological phases.

This section focuses on the methodologies and conceptual frameworks that link molecular electronic structure to solid-state band theory, with an emphasis on quantum confinement, embedding methods, multiscale modeling, and the duality of real-space and reciprocal-space descriptors.

5.1 Clusters, Nanostructures, and 2D Materials

The transition from molecular chemistry to materials science often proceeds through finite clusters and nanostructures. These intermediate systems retain aspects of both discrete and extended states, revealing how molecular orbitals evolve into bands.

Size quantization: As cluster size increases, molecular levels begin to merge into quasi-continuous energy bands. The quantization of electronic states leads to quantum confinement effects, observed prominently in semiconductor nanocrystals (quantum dots), where the band gap scales inversely with particle size (Giri et al., 2023).

Surface effects: Unlike bulk crystals, nanostructures possess a high surface-to-volume ratio. Surface states, dangling bonds, and termination chemistry strongly influence the electronic structure, often dominating optical and catalytic properties.

2D materials: Graphene, transition-metal dichalcogenides (TMDs), and phosphorene exemplify reduced-dimensional systems where confinement in one direction leads to unique band dispersions, valley degrees of freedom, and strong excitonic effects (Kreminska, 2025).

Together, these cases demonstrate how discrete molecular orbitals transition into extended Bloch states, making nanostructures a natural bridge between molecules and materials.

5.2 Quantum Confinement and the Transition to Band Structure

Quantum confinement arises when the material dimension approaches the exciton Bohr radius, leading to significant changes in energy levels and wavefunction localization.

From molecular orbitals to bands: In small molecules, electronic states are discrete. As molecular units aggregate, orbital overlap leads to splitting into bands, described by Bloch's theorem. In the limit of infinite periodicity, bands and gaps emerge.

Electronic and optical consequences: Confinement increases the effective band gap (blue-shifted absorption), modifies effective masses, and enhances Coulomb interactions, producing tightly bound excitons.

Chemical tunability: The interplay between size, shape, and composition allows chemists to engineer desired properties e.g., CdSe nanocrystals tuned across the visible spectrum, or MoS_2 monolayers with direct gaps distinct from bulk (Li et al., 2024).

Thus, quantum confinement connects chemistry (orbitals) with physics (bands) in a seamless spectrum.

5.3 Embedding Methods and Multiscale Approaches

Real materials often demand methods that balance accuracy and scale. Pure molecular quantum chemistry is too expensive for extended solids, while pure band theory may miss local chemistry. Embedding methods offer a compromise.

QM/MM (Quantum Mechanics/Molecular Mechanics): Widely used in chemistry, QM/MM partitions a system into a quantum core and a classical environment. In materials, QM/MM can describe defects or dopants in a periodic host.

Embedded fragment approaches: Cluster embedding methods treat a small fragment quantum-mechanically (e.g., a defect center or catalytic site) while embedding it into the electrostatic field of the host lattice (Li et al., 2025).

Periodic vs cluster models: Periodic boundary conditions capture long-range periodicity, but clusters are useful for localized chemistry. Multiscale approaches can combine both to capture defects, surfaces, and interfaces accurately.

Machine learning and high-throughput methods: Large-scale exploration of chemical/material space is enabled by surrogate models (e.g., neural networks trained on DFT data) and automated high-throughput DFT screenings. These methods allow prediction of stability, band topology, and defect energetics at scale.

Such methods create a hierarchy of modeling approaches, ensuring accuracy where necessary and scalability where possible.

5.4 Real-Space vs Reciprocal-Space Descriptors

Bridging molecules to materials also requires reconciling two complementary languages: real space, familiar to chemists, and reciprocal space, central to solid-state physics.

Real-Space Tools

Quantum Theory of Atoms in Molecules (QTAIM): Analyzes the topology of electron density, identifying bonds, critical points, and delocalization.

Chemical bond analysis: Tools like Natural Bond Orbital (NBO) analysis and Electron Localization Functions (ELF) reveal bonding patterns even in extended materials (Morales-Bayuelo & Sánchez-Márquez, 2020).

Charge density distributions: Allow visualization of bonding motifs, defects, and surface chemistry.

Reciprocal-Space Descriptors

Berry phase and Berry curvature: Central to polarization, anomalous Hall effects, and topological classification.

Wannier charge centers: Enable interpolation of band structures, mapping of topological invariants, and efficient construction of tight-binding models.

Surface states in topological materials: Reciprocal-space analysis reveals protected edge states and band crossings.

Thus, real-space and reciprocal-space frameworks provide complementary perspectives: chemistry in terms of bonds and density, physics in terms of bands and topology.

5.5 Toward Scale-Integrated Computational Chemistry

The ultimate goal of bridging molecules to materials is a unified predictive framework:

– Start from accurate molecular electronic structure (Hartree–Fock, coupled-cluster, high-level DFT).
– Scale up through clusters and nanostructures, incorporating confinement, disorder, and surface effects.
– Transition to periodic band-structure calculations, including many-body and topological corrections.

Integrate with data-driven machine learning to explore vast chemical/material spaces efficiently.

Such integration allows not only rational design of semiconductors and topological materials but also the discovery of hybrid systems organic–inorganic perovskites, covalent organic frameworks, and functionalized 2D materials where molecular chemistry meets extended quantum phases.

This section emphasizes the scale-bridging nature of electronic structure, from molecular orbitals to bands, from real-space bonding to reciprocal-space topology. The tools of computational chemistry-ranging from embedding methods to high-throughput DFT and machine learning are now capable of connecting these scales (Jiang et al., 2025).

In doing so, they provide a pathway toward understanding and designing advanced semiconductors and topological materials that embody the continuum from molecules to materials, a theme central to modern chemistry and condensed matter physics.

6. CHALLENGES AND FUTURE DIRECTIONS

The study of the electronic structure of semiconductors and topological materials at the interface of computational and theoretical chemistry has achieved remarkable advances. However, bridging the gap between molecular-scale chemistry and extended material physics introduces a set of unresolved challenges that require novel methods, conceptual frameworks, and computational strategies. In this section, we outline key difficulties and sketch pathways toward future solutions, focusing on accuracy and cost, disorder and many-body effects, the chemistry–topology nexus, and the integration of emerging methods with applications across technology (Olatomiwa et al., 2023).

6.1 Accuracy and Computational Cost

One of the fundamental challenges is balancing accuracy with computational efficiency.

Scaling Limitations of Many-Body Methods

Many-body perturbation techniques such as GW and Bethe–Salpeter Equation (BSE) are benchmarks for quasiparticle energies and optical excitations. However, their computational cost scales poorly with system size, making them prohibitive for large supercells or complex materials. For strongly correlated systems, dynamical mean-field theory (DMFT) and beyond are required, further amplifying the cost.

Limitations of density functional theory (DFT): Standard generalized-gradient approximation (GGA) or local-density approximation (LDA) functionals often underestimate band gaps and fail to capture long-range correlation or dispersion. For topological systems, accuracy in predicting spin–orbit coupling effects is crucial, yet subtle. Hybrid functionals (e.g., HSE06), meta-GGA, and range-separated functionals improve accuracy but at significantly higher computational demand.

Finite size and surface effects: Modeling nanoscale systems introduces spurious size effects, surface terminations, and artificial periodic interactions. Accurate embedding methods and extrapolation schemes are essential to correct such artifacts.

Temperature and disorder effects: Real materials exist at finite temperature, often with lattice vibrations (phonons), defects, and dynamic disorder. Electron–phonon coupling, polaron formation, and vibrational renormalization of band gaps require advanced methods such as ab initio molecular dynamics, self-consistent phonon theories, or stochastic approaches.

A critical frontier is developing computationally affordable yet accurate approximations for example, localized orbital implementations of GW, stochastic many-body approaches, and machine-learned potentials trained on high-level data.

6.2 Disorder, Temperature, and Many-Body Effects

Beyond the idealized perfect crystal picture, real-world materials exhibit structural disorder, defect chemistry, and temperature-driven fluctuations.

Disorder: Topological insulators and semiconductors often rely on alloying, doping, or off-stoichiometric compositions. Disorder alters band topology, localizes states, and complicates predictions. Capturing this requires large supercell calculations or methods such as the coherent potential approximation (CPA) (Liu et al., 2025; Van Veen et al., 2025).

Temperature effects: Phonons can close gaps, induce topological phase transitions, or renormalize transport properties. For example, lead-halide perovskites show strong electron–phonon coupling leading to dynamic band fluctuations, while in thermoelectrics, phonon scattering governs efficiency (Kumar et al., 2020).

Strong correlations: Materials with d- or f-electrons often host competing orders (magnetism, Mott localization, superconductivity). In these, topology interplays with correlation, requiring advanced approaches such as GW+DMFT or tensor-network methods.

Excited-state and finite-temperature many-body physics: Real-world applications often rely on non-equilibrium processes (e.g., light absorption in photovoltaics, ultrafast spin switching in spintronics). Real-time many-body approaches, time-dependent DFT (TDDFT), and reduced-density-matrix methods are promising but still under development.

The complexity of these effects highlights the need for multiscale modeling frameworks that combine quantum accuracy with statistical sampling.

6.3 The Interplay of Chemistry and Topology

Perhaps the most intellectually exciting frontier is the interplay between chemical design and topological features.

Topological classification vs chemical intuition: Topological phases are often described abstractly in terms of band inversions, symmetry eigenvalues, and invariants. Yet, chemists think in terms of orbital hybridization, electronegativity, and coordination. Bridging these languages remains a challenge: How do bond chemistry and orbital composition predispose a material to topological phases?

Systematic design of topological features: Can chemists craft materials where spin–orbit coupling, orbital overlap, or lattice symmetry are tuned to engineer topological phases? Strategies include alloying heavy elements (Bi, Sb, Te, Se), stabilizing non-centrosymmetric lattices, and building artificial heterostructures with designed band inversions.

Magnetism and topology: Magnetic topological insulators, Weyl semimetals, and axion insulators embody the deep connection between spin order and topology. The coexistence of magnetic exchange interactions and relativistic spin–orbit coupling requires chemical design strategies that balance local moments with itinerant carriers.

Correlated topological matter: The frontier of topological Mott insulators or fractional topological phases lies in designing strongly correlated lattices where electron interactions generate topological order.

Chemists can contribute by crafting molecular frameworks, low-dimensional lattices, and organometallic systems with correlated topological bands. Thus, the question shifts from identifying topological materials to designing them through chemistry.

6.4 Applications and Technological Relevance

The drive to understand and predict electronic structure is motivated by applications across multiple technological domains.

Spintronics: Topological insulators with spin-momentum–locked surface states enable low-power spin-based devices. Magnetic topological phases can be harnessed for spin current generation and manipulation.

Quantum computing: Majorana zero modes in topological superconductors are candidates for fault-tolerant qubits. Designing robust platforms for topological quantum computation remains a holy grail, requiring materials with clean, controllable band inversions and superconducting proximity (Jia et al., 2025).

Thermoelectrics: Narrow-gap semiconductors and topological semimetals exhibit high thermoelectric efficiencies due to large Seebeck coefficients and reduced thermal conductivity. Computational screening aids discovery of high-performance candidates.

Photovoltaics and optoelectronics: Band-gap engineering in nanostructured semiconductors, excitonic effects in 2D materials, and topological protection of edge states contribute to next-generation solar cells and light-emitting devices.

Catalysis: Electronic structure at surfaces determines catalytic activity. Topological surface states and nontrivial band dispersions offer new routes for enhancing catalytic selectivity and activity.

These applications emphasize that computational chemistry and theory are not only explanatory but also prescriptive tools in materials design.

6.5 Emerging Computational Methods

To address these challenges, new generations of computational techniques are under active development.

More efficient GW/BSE methods: Localized orbital GW, stochastic GW, and linear-scaling algorithms are enabling quasiparticle calculations for thousands of atoms. Efficient BSE implementations are being developed to study excitons in complex nanostructures.

Real-time electronic-structure methods: Real-time TDDFT and non-equilibrium Green's function (NEGF) methods allow simulation of ultrafast dynamics, light-matter interactions, and transport beyond linear response.

Integration with machine learning: Machine learning force fields, neural network wavefunction approximations, and data-driven screening pipelines are rapidly transforming computational chemistry. Materials informatics enables high-throughput discovery of semiconductors, topological insulators, and quantum materials (Wilhelm et al., 2017).

Quantum computing for electronic structure: Although still in early stages, quantum algorithms (VQE, QPE) hold promises for tackling correlated electronic structure intractable on classical machines. Hybrid quantum–classical methods may eventually be indispensable in exploring strongly correlated topological matter.

Multiscale embedding frameworks: Progress in embedding theories (e.g., density matrix embedding, dynamical cluster embedding) provides systematic ways to capture both local chemical accuracy and global band topology.

These methods signal a future where accuracy, efficiency, and scale are no longer mutually exclusive. The overarching challenge is to move from post-facto explanation of materials to predictive and design-driven frameworks. Achieving this will allow systematic crafting of functional materials for spintronics, quantum information, energy conversion, and catalysis. In short, the next decade will see theoretical and computational chemistry evolve from interpreting existing materials to actively shaping the materials landscape, guided by a scale-bridging, data-informed, and topologically aware understanding of electronic structure. Examples of materials: chemistry, structure, computed/inferred topological character shown in Table.2 to Table.7.

Table 2 presents a classification of three-dimensional (3D) topological insulators, which are quantum materials characterized by an insulating bulk and metallic surface states protected by time-reversal symmetry. These materials are distinguished from conventional insulators by a non-trivial Z_2 topological invariant, arising from strong spin–orbit coupling and band inversion at time-reversal invariant momenta in the Brillouin zone. In 3D Z_2 topological insulators, the bulk electronic structure features an inverted ordering of conduction and valence bands, while the surface hosts gapless Dirac-like states with spin–momentum locking. This helical spin texture suppresses backscattering from non-magnetic impurities, ensuring robust surface transport. The topological classification is often expressed in terms of four Z_2 indices ($\nu0;\nu1\nu2\nu3$), where $\nu0$ denotes the strong topological invariant and νi represent weak invariants associated with crystal lattice directions. Representative materials include Bi_2Se_3, Bi_2Te_3, and Sb_2Te_3, which have been extensively studied due to their simple surface-state spectra and relatively large bulk band gaps. These 3D topological insulators provide a foundational platform for exploring novel quantum phenomena, including topological superconductivity, magnetic topological phases, and spintronic device applications.

Table 2. 3D Topological Insulators (Z_2 type)

Material	Chemistry / Composition	Crystal Structure	Topological Character	Notes
Bi_2Se_3	Bismuth selenide	Rhombohedral (space group R-3m)	Strong Z_2 topological insulator ($\nu_0 = 1$)	Single Dirac cone on (111) surface; prototype TI.
Bi_2Te_3	Bismuth telluride	Rhombohedral (R-3m)	Z_2 TI ($\nu_0 = 1$)	Spin–orbit coupling induces band inversion at Γ.
Sb_2Te_3	Antimony telluride	Rhombohedral (R-3m)	Z_2 TI	Similar to Bi_2Te_3, tunable with alloying.
SnTe	Tin telluride	Rock-salt (Fm-3m)	Topological crystalline insulator (TCI)	Protected by mirror symmetry; mirror Chern number = 2.

Table 3 summarizes the principal classes of topological semimetals, which are quantum materials characterized by band crossings near the Fermi level protected by symmetry and topology. Unlike topological insulators, these systems do not possess a full bulk band gap; instead, conduction and valence bands touch at discrete points or along lines in momentum space, giving rise to exotic quasiparticles and unusual transport properties. Dirac semimetals host fourfold-degenerate Dirac points protected by crystal symmetry and time-reversal symmetry, as observed in materials such as Na_3Bi and Cd_3As_2. When either inversion or time-reversal symmetry is broken, Dirac points split into pairs of Weyl nodes with opposite chirality, leading to Weyl semimetals like TaAs and $Co_3Sn_2S_2$. These systems exhibit surface Fermi arcs and strong anomalous Hall responses. A third category, nodal-line semimetals, features band crossings along one-dimensional loops or lines in the Brillouin zone, stabilized by mirror or glide symmetries. Examples include ZrSiS and Ca_3P_2.Together, these classes illustrate how lattice symmetry, spin–orbit coupling, and symmetry breaking determine the nature of topological band crossings, making topological semimetals a bridge between conventional metals and fully gapped topological phases.

Table 3. Topological Semimetals

Material	Chemistry	Crystal Structure	Topological Character	Notes
Na_3Bi	Sodium bismuthide	Hexagonal ($P6_3$/mmc)	Dirac semimetal	Dirac nodes along Γ–A; protected by rotation symmetry.
Cd_3As_2	Cadmium arsenide	Tetragonal ($I4_1$/acd)	Dirac semimetal	High electron mobility; experimentally confirmed Dirac dispersion.
TaAs	Tantalum arsenide	Tetragonal ($I4_1md$)	Weyl semimetal	24 Weyl points ; Fermi-arc surface states.
NbAs / TaP / NbP			Weyl semimetals	Same structural family as TaAs; non-centrosymmetric.

Table.4 summarize the topological insulators are materials that behave as electrical insulators in the bulk while hosting robust, conducting states at their boundaries. The quantum spin Hall (QSH) insulator represents a two-dimensional realization of this concept, protected by time-reversal symmetry. In QSH systems, strong spin–orbit coupling leads to band inversion, where the conventional ordering of valence and conduction bands is reversed at specific points in momentum space. A defining feature of quantum spin Hall insulators is the presence of helical edge states, in which electrons with opposite spins propagate in opposite directions along the edges. These states are protected against backscattering from non-magnetic impurities due to time-reversal symmetry, resulting in dissipation less edge transport. The topological nature of the phase is characterized by a Z_2 topological invariant, which distinguishes it from ordinary insulating states.

The first experimental realization of the QSH (Query String Hash) effect was demonstrated in HgTe/CdTe quantum wells, followed by observations in other two-dimensional materials such as InAs/GaSb heterostructures and certain van der Waals systems. Quantum spin Hall insulators provide an important platform for exploring spin-based electronics and serve as building blocks for more complex topological phases, including topological superconductivity and higher-order topological states.

Table 4. 2D Topological Insulators (Quantum Spin Hall insulators)

Material	Chemistry	Structure	Topological Character	Notes
HgTe quantum wells	Mercury telluride	Zincblende-derived	2D $Z_2 = 1$ (Quantum Spin Hall)	Band inversion between s-like and p-like orbitals.
1T′–WTe_2	Tungsten ditelluride	Distorted octahedral ($P2_1$/m)	2D TI (monolayer)	Experimentally observed edge states.
Bismuthene on SiC	2D bismuth honeycomb	Planar hexagonal	2D Z_2 TI	Large topological gap (~0.8 eV).

Table.5 shows the magnetic insulators are materials in which long-range magnetic order coexists with an insulating bulk electronic structure. When magnetism breaks time-reversal symmetry, it can fundamentally alter the topology of electronic bands, giving rise to Chern insulators a paradigmatic class of topological phases. Unlike conventional topological insulators, which rely on time-reversal symmetry, Chern insulators exhibit a quantized Hall conductance in the absence of an external magnetic field, a phenomenon known as the quantum anomalous Hall effect. The defining characteristic of a Chern insulator is a nonzero Chern number, a topological invariant determined by the integral of Berry curvature over the Brillouin zone. This invariant guarantees the existence of chiral, unidirectional edge states that are immune to backscattering from non-magnetic disorder. These edge channels provide dissipation less transport and are a direct manifestation of bulk–boundary correspondence. Chern insulating behaviour can arise intrinsically in magnetic materials with strong spin–orbit coupling or be engineered in layered heterostructures combining magnetism and topological band structures. Notable examples include magnetically doped topological insulators and van der Waals magnetic systems. Due to their robust edge conduction and symmetry-controlled topology, magnetic and Chern insulators are promising candidates for low-power electronics, spintronic applications, and quantum metrology.

Table 5. Magnetic & Chern Insulators

Material	Chemistry	Structure	Topological Character	Notes
$MnBi_2Te_4$	Layered manganese bismuth telluride	Rhombohedral (R-3m)	Antiferromagnetic TI / Axion insulator	Alternating magnetic layers; quantized magnetoelectric effect predicted.
$Co_3Sn_2S_2$	Cobalt shandite	Hexagonal (R-3m)	Weyl semimetal with large anomalous Hall effect	Ferromagnetic Weyl semimetal; intrinsic AHE observed.
Fe_3GeTe_2	Iron-based vdW material	Hexagonal ($P6_3$/mmc)	Magnetic Weyl semimetal candidate	Tunable magnetism in few-layer form.

Table.6- Topological crystalline insulators (TCIs) are a class of quantum materials in which crystal symmetries, rather than time-reversal symmetry alone, protect topological electronic states. Unlike conventional topological insulators, where surface states are guaranteed on all boundaries, TCIs host symmetry-protected metallic states only on surfaces or edges that preserve specific crystalline symmetries, such as mirror, rotational, or glide symmetries. A prototypical example is SnTe, where mirror symmetry protects Dirac-like surface states on selected crystal facets.

Higher-order topological phases extend this concept by redefining the bulk–boundary correspondence. In these systems, an nnn-dimensional bulk does not host gapless states on its (n−1)-dimensional boundaries; instead, topologically protected modes appear on boundaries of even lower dimension, such as edges or corners. For example, a two-dimensional higher-order topological insulator can exhibit zero-energy corner states, while a three-dimensional system may support hinge states.

Both TCIs and higher-order phases are governed by crystal symmetry indicators and topological invariants, which encode the global structure of electronic wavefunctions. Their symmetry-dependent boundary states offer tunable platforms for robust electronic transport and hold promise for applications in nanoelectronics, spintronics, and quantum devices.

Table 6. Topological Crystalline Insulators & Higher-Order Phases.

Material	Chemistry	Structure	Topological Character	Notes
$Pb_{1-X}Sn_XSe$	Lead–tin selenide alloys	Rock-salt	TCI	Mirror-symmetry-protected Dirac cones on (001).
Bi_4Br_4	Bismuth bromide	Quasi-1D layered	1D higher-order TI candidate	Edge states at ribbon ends.
Bismuth (pure element)	Elemental Bi	Rhombohedral	Higher-order topological insulator	Topological hinge states predicted/ observed.

Table.7 shows the Superconducting topological phases represent a class of quantum states in which superconductivity coexists with non-trivial band topology, leading to exotic and highly robust quantum phenomena. In conventional superconductors, Cooper pairing opens an energy gap at the Fermi level without altering the topological nature of electronic states. In contrast, topological superconductors host a superconducting gap in the bulk while supporting gapless, topologically protected boundary states.

The defining feature of these phases is the emergence of Majorana zero modes at edges, surfaces, or vortex cores. These quasiparticles are their own antiparticles and arise due to the combination of superconducting pairing, strong spin–orbit coupling, and symmetry constraints such as time-reversal or particle–hole symmetry. From a topological perspective, these systems are characterized by invariants such as Chern numbers or Z_2 indices, which remain unchanged unless the superconducting gap closes. Superconducting topological phases can be realized intrinsically in certain materials or engineered through hybrid structures, such as semiconductor nanowires or topological insulators proximitized by conventional superconductors. Their robustness against local perturbations makes them promising platforms for fault-tolerant quantum computation, particularly through non-Abelian statistics of Majorana modes. Consequently, topological superconductivity lies at the intersection of condensed matter physics, materials science, and quantum technology.

Table 7. Superconducting Topological Phases

Material	Chemistry	Structure	Topological Character	Notes
$Cu_XBi_2Se_3$	Copper-doped Bi_2Se_3	Rhombohedral	Topological superconductor	Unconventional pairing; Majorana modes proposed.
$FeSe_{0.5}Te_{0.5}$	Iron-based chalcogenide	Tetragonal	Topological Dirac surface + s-wave superconductivity	Candidate for Majorana bound states.

Density Functional Theory (DFT) is commonly chosen as the primary electronic-structure method because it provides a robust first-principles description of ground-state properties at a reasonable computational cost. Within DFT, the generalized gradient approximation (GGA) is often preferred over the local density approximation (LDA) due to its improved treatment of inhomogeneous electron densities, which is important for layered and low-dimensional topological materials. However, GGA is known to underestimate band gaps, a limitation that can affect the quantitative prediction of topological phase transitions. This trade-off is addressed by focusing on band ordering and symmetry characteristics, which are generally more reliable than absolute gap values. To accurately capture relativistic effects crucial for topological phenomena, spin–orbit coupling (SOC) is explicitly included. Although SOC increases computational cost, its inclusion is essential, as neglecting it can eliminate band inversion and lead to incorrect topological classification. For magnetic systems, non-collinear spin calculations are employed to represent complex magnetic textures, at the expense of higher computational demands.

For improved band-gap accuracy and validation, hybrid functionals or GW approximations may be selectively applied. While these methods offer better quantitative precision, their high computational cost limits their routine use to benchmark calculations. Similarly, tight-binding and Wannier-based models are adopted to compute topological invariants and Berry curvature efficiently, trading some ab initio rigor for computational efficiency and physical transparency. Overall, the chosen methodologies reflect a strategic compromise: DFT-based approaches ensure material realism, while model Hamiltonians and selective higher-level corrections provide efficiency and deeper insight into the topological nature of the systems under investigation.

7. SUMMARY AND OUTLOOK

The electronic structure of semiconductors and topological materials represents one of the most intellectually rich frontiers at the intersection of computational chemistry, condensed matter physics, and materials science. In this chapter, we have traced the conceptual and methodological evolution from molecules to materials, highlighting how advances in theory and computation have enabled new insights and discoveries.

Key Takeaways

Methods and accuracy: The development of advanced density functional approximations (hybrids, meta-GGAs, range-separated functionals), many-body techniques (GW, BSE), and embedding frameworks has dramatically improved our ability to predict band structures, band gaps, and excitonic phenomena.

Semiconductors: Computational strategies now account for electron–phonon coupling, defect physics, anisotropic carrier masses, and the challenges of organic semiconductors, providing a much more complete picture of real-world materials.

Topological materials: By integrating spin–orbit coupling, symmetry analysis, and topological invariants into computational pipelines, theory has revealed the chemical origins of topology and enabled systematic identification of topological insulators, semimetals, and correlated quantum phases.

Scale-bridging: From clusters and nanostructures to 2D systems and bulk solids, the molecular-to-material continuum is now accessible, with quantum confinement and surface effects serving as critical connecting principles. Real-space chemical analysis and reciprocal-space topological descriptors complement each other, offering dual perspectives.

Applications: Predictive computational design has already informed the search for materials in spintronics, photovoltaics, quantum computing, thermoelectrics, and catalysis.

Outlook: Roadmap for the Next 5–10 Years

Looking ahead, the convergence of chemistry, physics, and data science will shape the next era of discovery.

Toward predictive accuracy at scale: Development of affordable many-body techniques, stochastic approaches, and machine-learned functionals will allow quasiparticle and excitonic predictions for complex, disordered, or finite-temperature systems.

Integration of chemical design and topology: Chemists will increasingly guide the discovery of topological phases by leveraging orbital chemistry, crystal engineering, and heterostructure design. The synthesis–theory feedback loop will become tighter and more predictive.

Multiscale and real-time modeling: Embedding strategies, QM/MM for solids, and real-time TDDFT will expand our ability to simulate ultrafast processes, quantum coherence, and nonequilibrium dynamics relevant to devices.

Data-driven exploration: High-throughput screening combined with materials informatics will systematically map chemical and structural spaces for semiconductors and topological matter, uncovering new candidates beyond serendipity.

Quantum computing and correlated matter: As quantum algorithms mature, they will offer transformative capabilities for tackling strongly correlated electronic structures and fractional topological phases inaccessible to classical methods. Table.9 shows the roadmaps of electronics structure coming 5-10 years.

Table 8. Roadmap for the Next 5–10 Years in Electronic Structure Research.

Time Frame	Research Focus	Key Goals / Milestones	Technological Applications
2025–2027 (Short Term)	Advanced characterization of electronic structures in 2D semiconductors and topological insulators	• High-resolution ARPES-(Angle-Resolved Photoemission Spectroscopy) and STM studies • Improved DFT + GW and hybrid functional models • Discovery of new narrow-gap topological materials	• High-speed transistors • Low-power optoelectronic devices • Surface-conduction-based sensors
2027–2030 (Mid Term)	Integration of semiconductor and topological functionalities	• Engineering heterostructures combining trivial and non-trivial bands • Band alignment control for quantum interfaces • Demonstration of topological field-effect transistors	• Spintronics and quantum logic devices • Room-temperature quantum Hall systems
2030–2033 (Mid to Long Term)	Quantum transport and spin–orbit coupling control	• Manipulation of Berry curvature and spin textures • Robust spin-polarized edge states • Experimental realization of Majorana modes in hybrid systems	• Quantum computing components • Topological quantum sensors
2033–2035 (Long Term)	Scalable and sustainable device implementation	• Low-cost synthesis of high-quality topological crystals • Integration with CMOS platforms • Development of standards for topological metrology	• Commercial quantum and spintronic technologies • Next-generation energy-efficient electronics

8. CONCLUSION

The study of the electronic structure of semiconductors and topological materials provides a fundamental understanding of how atomic arrangements and quantum mechanical interactions govern material properties. In semiconductors, the formation of valence and conduction bands, along with a tunable band gap, determines their electrical and optical behaviors, making them essential for modern electronics, photonics, and energy devices.

In contrast, topological materials represent a new quantum phase of matter where non-trivial band topology gives rise to protected surface or edge states that are robust against disorder and scattering. These states, derived from the interplay between spin–orbit coupling and crystal symmetry, extend conventional band theory and open pathways to novel phenomena such as quantum spin Hall effects, topological superconductivity, and dissipation less transport.

Overall, understanding and controlling the electronic structure across both semiconductors and topological systems not only deepens our theoretical insight into solid-state physics but also drives innovation in next-generation technologies, including spintronics, quantum computing, and advanced sensing devices.

REFERENCES

Banerjee, S., Das, H. S., & Maity, H. (2024). *An introduction to optoelectronics*. LAP Lambert Academic Publishing.

Beran, G. J. O. (2016). Modeling Polymorphic Molecular Crystals with Electronic Structure Theory. *Chemical Reviews*, *116*(9), 5567–5613. DOI: 10.1021/acs.chemrev.5b00648 PMID: 27008426

Blum, V., Asahi, R., Autschbach, J., Bannwarth, C., Bihlmayer, G., Blügel, S., Burns, L. A., Crawford, T. D., Dawson, W., De Jong, W. A., Draxl, C., Filippi, C., Genovese, L., Giannozzi, P., Govind, N., Hammes-Schiffer, S., Hammond, J. R., Hourahine, B., Jain, A., & Windus, T. L. (2024). Roadmap on methods and software for electronic structure-based simulations in chemistry and materials. *Electronic Structure*, *6*(4), 042501. DOI: 10.1088/2516-1075/ad48ec

Das, H. S., Basak, A., & Maity, S. (2024a). Materials Science and nanotechnology. In *Advances in chemical and materials engineering book series* (pp. 175–206). DOI: 10.4018/979-8-3693-3398-3.ch008

Das, H. S., Mishra, S., & Roymahapatra, G. (2024b). Advanced Nano-Structured materials for energy storage devices. In *Advances in chemical and materials engineering book series* (pp. 1–34). DOI: 10.4018/979-8-3693-5320-2.ch001

Giri, A., Park, G., & Jeong, U. (2023). Layer-Structured Anisotropic Metal Chalcogenides: Recent advances in synthesis, modulation, and applications. *Chemical Reviews*, *123*(7), 3329–3442. DOI: 10.1021/acs.chemrev.2c00455 PMID: 36719999

Islam, M. A., Khan, S., Mim, J. J., Rahman, S. M. M., Patwary, M. I., Islam, M. S., & Hossain, N. (2025). Recent advances of 2D materials in semiconductor application: A review. *Advanced Sensor and Energy Materials*, *4*(4), 100161. DOI: 10.1016/j.asems.2025.100161

Jain, M., Chelikowsky, J. R., & Louie, S. G. (2011). Reliability of hybrid functionals in predicting band gaps. *Physical Review Letters*, *107*(21), 216806. DOI: 10.1103/PhysRevLett.107.216806 PMID: 22181911

Jia, Z., Zhao, M., Chen, Q., Tian, Y., Liu, L., Zhang, F., Zhang, D., Ji, Y., Camargo, B., Ye, K., Sun, R., Wang, Z., & Jiang, Y. (2025). Spintronic Devices upon 2D Magnetic Materials and Heterojunctions. *ACS Nano*, *19*(10), 9452–9483. DOI: 10.1021/acsnano.4c14168 PMID: 40053908

Jiang, D., Tan, V. G. W., Gong, Y., Shao, H., Mu, X., Luo, Z., & He, S. (2025). Semiconducting covalent organic frameworks. *Chemical Reviews*, *125*(13), 6203–6308. DOI: 10.1021/acs.chemrev.4c00950 PMID: 40366230

Kawabata, K., Mashimo, K., & Takimiya, K. (2024). Core- versus End-Alkylation: Tailoring Solid-State Structures and Properties of Near-Infrared-Absorbing Organic Semiconductors Based on Naphthodithiophenediones. *Chemistry of Materials*, *36*(24), 11920–11933. DOI: 10.1021/acs.chemmater.4c02436

Kreminska, J. (2025). *Thickness Dependent Fermiology in 2D materials*. DOI: 10.33612/diss.1328567056

Kumar, N., Guin, S. N., Manna, K., Shekhar, C., & Felser, C. (2020). Topological Quantum Materials from the Viewpoint of Chemistry. *Chemical Reviews*, *121*(5), 2780–2815. DOI: 10.1021/acs.chemrev.0c00732 PMID: 33151662

Li, Q., Wu, K., Zhu, H., Yang, Y., He, S., & Lian, T. (2024). Charge Transfer from Quantum-Confined 0D, 1D, and 2D Nanocrystals. *Chemical Reviews, 124*(9), 5695–5763. DOI: 10.1021/acs.chemrev.3c00742 PMID: 38629390

Li, Q., Wu, K., Zhu, H., Yang, Y., He, S., & Lian, T. (2024b). Charge Transfer from Quantum-Confined 0D, 1D, and 2D Nanocrystals. *Chemical Reviews, 124*(9), 5695–5763. DOI: 10.1021/acs.chemrev.3c00742 PMID: 38629390

Liu, B., & Zhang, W. (2023). Research Progress of Topological Quantum Materials: From First-Order to Higher-Order. *Symmetry, 15*(9), 1651. DOI: 10.3390/sym15091651

Liu, X., Cai, J., Zhang, J., Zhao, P., Wang, G., Bi, S., & Hou, Z. (2025). Surface-state-constrained topological insulator Bi2Te3 nanorods for electromagnetic wave trapping and conversion into electricity. *Journal of Materials Science and Technology, 244*, 149–155. DOI: 10.1016/j.jmst.2025.03.102

Manni, G. L., Galván, I. F., Alavi, A., Aleotti, F., Aquilante, F., Autschbach, J., Avagliano, D., Baiardi, A., Bao, J. J., Battaglia, S., Birnoschi, L., Blanco-González, A., Bokarev, S. I., Broer, R., Cacciari, R., Calio, P. B., Carlson, R. K., Couto, R. C., Cerdán, L., & Lindh, R. (2023). The OpenMoLCAS Web: A Community-Driven Approach to Advancing Computational Chemistry. *Journal of Chemical Theory and Computation, 19*(20), 6933–6991. DOI: 10.1021/acs.jctc.3c00182 PMID: 37216210

Matveeva, R., Folkestad, S. D., & Høyvik, I. (2023). Particle-Breaking Hartree–Fock Theory for open Molecular Systems. *The Journal of Physical Chemistry A, 127*(5), 1329–1341. DOI: 10.1021/acs.jpca.2c07686 PMID: 36720055

Morales-Bayuelo, A., & Sánchez-Márquez, J. (2020). Discerning the thermal cyclotrimerizations of fluoro- and chloroacetylenes through ELF, NBO descriptors and QTAIM analysis: Pseudodiradical character. *Heliyon, 6*(7), e04441. DOI: 10.1016/j.heliyon.2020.e04441 PMID: 32715128

Olatomiwa, A., Adam, T., Edet, C., Adewale, A., Chik, A., Mohammed, M., Gopinath, S. C., & Hashim, U. (2023). Recent advances in density functional theory approach for optoelectronics properties of graphene. *Heliyon, 9*(3), e14279. DOI: 10.1016/j.heliyon.2023.e14279 PMID: 36950613

Pham, L. V. D., Hinsche, N. F., & Mertig, I. (2025). Quantum spin Hall phase in the truncated trihexagonal lattice: a topological archimedean structure. *2D Materials, 13*(1), 015021. DOI: 10.1088/2053-1583/ae23e1

Qiao, J., Pizzi, G., & Marzari, N. (2023). Automated mixing of maximally localized Wannier functions into target manifolds. *npj Computational Materials, 9*(1), 206. Advance online publication. DOI: 10.1038/s41524-023-01147-9 PMID: 38666055

Urban, A., Reese, M., Mrovec, M., Elsässer, C., & Meyer, B. (2011). Parameterization of tight-binding models from density functional theory calculations. *Physical Review B: Condensed Matter and Materials Physics, 84*(15), 155119. Advance online publication. DOI: 10.1103/PhysRevB.84.155119

Van Veen, F., Kölling, S., De Wit, S. R., Metsch, R., Rosenbach, D., Li, C., & Brinkman, A. (2025). Observation of the surface hybridization gap in the electrical transport properties of the ultrathin topological insulator (Bi1−xSbx)2Te3. *Physical Review. B. Physical Review. B, 112*(4). Advance online publication. DOI: 10.1103/d5xc-bvmx

Westermayr, J., & Marquetand, P. (2020). Machine learning for electronically excited states of molecules. *Chemical Reviews*, *121*(16), 9873–9926. DOI: 10.1021/acs.chemrev.0c00749 PMID: 33211478

Wilhelm, J., Golze, D., Talirz, L., Hutter, J., & Pignedoli, C. A. (2017). Toward GW calculations on thousands of atoms. *The Journal of Physical Chemistry Letters*, *9*(2), 306–312. DOI: 10.1021/acs.jpclett.7b02740 PMID: 29280376

Xiao, W., Hu, Y., Yan, K., Tang, L., Chen, X., D'Agosta, R., & Yang, K. (2025). Phonon-limited carrier mobility modeling of two-dimensional semiconductors based on first principles. *Journal of Physics Condensed Matter*, *37*(26), 263001. DOI: 10.1088/1361-648X/ade18e PMID: 40472870

Xu, D., Zhang, Q., Huo, X., Wang, Y., & Yang, M. (2023). Advances in data-assisted high-throughput computations for material design. *Materials Genome Engineering Advances*, *1*(1), e11. Advance online publication. DOI: 10.1002/mgea.11

Zafar, Z., Yi, S., Li, J., Li, C., Zhu, Y., Zada, A., Yao, W., Liu, Z., & Yue, X. (2021). Recent development in defects Engineered Photocatalysts: An overview of the experimental and theoretical strategies. *Energy & Environmental Materials*, *5*(1), 68–114. DOI: 10.1002/eem2.12171

Chapter 3
First-Principles Insights Into the Structure and Physical Properties of Perovskite Materials

Hicham Kacimi-Naciri
https://orcid.org/0009-0005-9576-818X
Laboratory of Ceramic Materials and Mathematics CERAMATHS, University Polytechnique Hauts-de-France, France

Assia Mabrouk
https://orcid.org/0000-0001-6399-0644
Faculté des Sciences Appliquées - Ait Melloul, Université Ibnou Zohr, Agadir, Morocco

Rachid Amrousse
https://orcid.org/0000-0002-2905-602X
Faculté des Sciences, Université de Chouaib Doukkali, El Jadida, Morocco

Mohamed Rguiti
Laboratory of Ceramic Materials and Mathematics CERAMATHS, University Polytechnique Hauts-de-France, France

Ahmed Bachar
Faculté des Sciences Appliquées - Ait Melloul, Université Ibnou Zohr, Agadir, Morocco

ABSTRACT

Perovskite-structured materials form a highly promising family whose electronic, optical, thermoelectric, mechanical, and structural properties are outstanding, making them valuable for photovoltaics, optoelectronics, sensors, and energy storage. This chapter reviews perovskites and summarizes density functional theory (DFT), the primary computational tool used to predict and interpret their structural, electronic, vibrational, mechanical, thermoelectric, and optical properties. Case studies KNbO3 system, BaTiO3 with Sr substitution illustrate how DFT captures doping or substitution induced changes in phase stability, band gaps, electronic and structural characteristics, and elastic constants, while noting

DOI: 10.4018/979-8-3373-6058-4.ch003

limitations from exchange correlation approximations and computational costs. The chapter concludes with perspectives on accelerating perovskite discovery using high throughput calculations and machine learning to streamline the screening and design of novel materials.

INTRODUCTION

Perovskite materials have been the subject of intensive research and have been considered as one of the most fascinating families of materials in recent years. Their general formula is ABO_3 for oxide perovskites and their halide and hybrid variants are derivatives of this formula (Arul & Nithya, 2020). These materials are recognized by their adaptable crystal lattice that allows virtually any chemical substitution. The ability of the lattice to accommodate various chemical substitutions results in the materials exhibiting a wide range of physical properties, such as ferroelectricity (Fan et al., 2018), piezoelectricity (Dai et al., 2020), high temperature superconductivity, ionic conductivity, and impressive optoelectronic properties(Chakraborty et al., 2023). This extraordinary combination of properties has led perovskites to be the most promising candidates for an extensive range of applications, including solar cells (Chakraborty et al., 2023) and light-emitting devices, as well as sensors, catalysts, and energy storage technologies(A. Sharma et al., 2009).

Among the various computational methods available, Density Functional Theory (DFT) (Schwarz, 2003) has emerged as the most powerful and widely used tool in the study of perovskite systems. Rooted in quantum mechanics, DFT allows researchers to describe the ground-state properties of matter by treating the electron density as the central variable. This approach provides access to critical information such as lattice stability, electronic band structure, density of states, vibrational spectra, dielectric constants, and optical absorption, all of which are crucial for understanding the multifunctional behavior of perovskites (Hao et al., 2019).

Over the last few decades, Density Functional Theory (DFT) based studies have elucidated the nature of ferroelectric transitions in oxide perovskites, the impact of doping in the improvement of piezoelectric and dielectric properties, and the electron pathways that explain the high performance of halide perovskites in solar cells. DFT, however, still has some issues along with the standard band gap underestimation and the strong electron correlation problem in transition metal perovskites. These drawbacks have led to the introduction of sophisticated computational methods such as hybrid functionals, DFT+U, and many-body perturbation theory.

The primary aim of this chapter is to give a detailed account of the theoretical modeling of perovskite materials by DFT. Initially, we present the basic structural characteristics of perovskites and give an outline of the theoretical basis of DFT. Subsequently, we describe the DFT methods to be used for the study of perovskite structural, electronic, vibrational, mechanical, and optical properties with some case studies as examples. At the end, we recollect the pros and cons of DFT in this area and glance at the new frontiers such as machine-learning-assisted simulations and high-throughput computational screening that can speed up the exploration of next-generation perovskite materials.

1 OBJECTIVES OF THE CHAPTER

The main objectives of this chapter are:

- Introduce the fundamentals of perovskite materials their crystal structure, chemical flexibility, and the diversity of their physical properties.
- Present the theoretical framework of Density Functional Theory (DFT) as a powerful computational approach for investigating perovskite systems. Describe practical components of a DFT calculation choice of exchange–correlation functional (LDA, GGA, hybrid functionals, meta-GGA). Explain how DFT yields total energies, forces, and electronic structure information (band structure, density of states, charge densities), making it suitable for predicting ground-state geometries, energetics, and many derived properties of perovskites. Mention commonly used software packages and emphasize the importance of benchmarking and validation against experiment.
- Demonstrate how DFT is applied to study the structural, electronic, vibrational, mechanical, and optical properties of perovskites. Describe workflows for structural optimization and phase stability. Explain electronic structure analyses: band structure, density of states, effective masses, band alignment. Present mechanical property calculations (elastic constants, bulk/shear moduli, and derived mechanical stability criteria) and how they are computed from stress–strain relations. Describe optical property predictions using DFT.
- Provide illustrative case studies on oxide or oxide perovskites doped, including the effects of chemical substitutions or doping on the different properties. Offer representative oxide perovskite examples: ferroelectric $BaTiO_3$ and $PbTiO_3$, highlighting DFT-predicted structural phase transitions, polarization calculations, and strain effects. Show how DFT can leads to predict with precision the different parameters and properties of the structure pure or doped.
- Discuss the strengths and limitations of DFT, particularly in relation to the challenges of modeling complex perovskite systems. Summarize strengths: predictive power for ground-state structures and trends, relatively modest computational cost compared with higher-level methods, and adaptability to a wide range of properties and materials classes. Then enumerate limitations.
- Explore future perspectives, such as the integration of DFT with high-throughput screening and machine learning, for accelerating the discovery and design of next-generation perovskite materials.

2 FUNDAMENTALS OF PEROVSKITE MATERIALS

2.1 Crystal Structure ABO_3

The term perovskite initially referred to the mineral $CaTiO_3$, discovered in 1839 (Glazer, 2002). It is now used to describe a family of materials with a similar crystal structure and the general formula ABX_3. In this structure, a large A cation (valence 1-3) is located at the corners of a cubic unit cell, a smaller B cation (valence 3-6) occupies the center, and X anions (often O^{-2}) are at the center of the faces, forming octahedra around the B ion. This arrangement allows the A cations to be housed within cavities created by these octahedra, giving these materials their distinctive structure.

Figure 1. Crystal structure of perovskite ABO_3

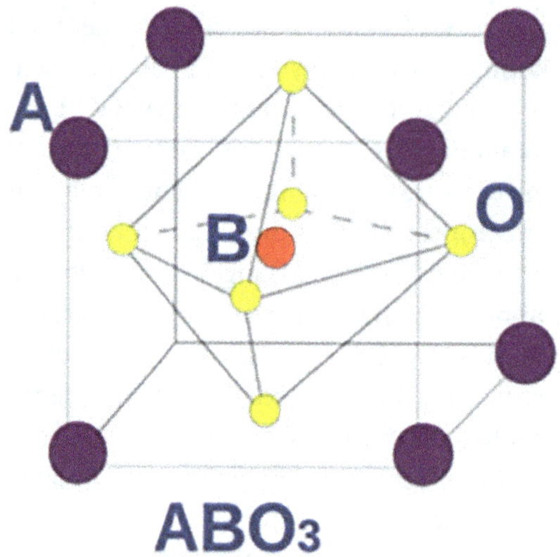

2.2 Classification Oxide Perovskites Case

Two main types of perovskites can be distinguished (Johnsson & Lemmens, 2007):

– Perovskites with a formula of ABO_3 are those in which a single cation is found in both the A and B sites, as in the cases of $BaTiO_3$, $CaTiO_3$, and $KNbO_3$...
– Perovskites complexes with the formula $AA'BB'O_3$ are such that different cations occupy the same site, for example, $(Na_{1/2}Bi_{1/2})\ TiO_3$ or $K_{1/2}Na_{1/2}NbO_3$ and $(Ba_{1-x}Ca_x)\ (Ti_{1-y}Zr_y)\ O_3$...

In an ideal scenario, the ions are tangents, which results in perfect alignment of the BO_6 octahedra. This system then crystallizes into a cubic, three-dimensional network without deformation. Perovskites with the ideal structure crystallize in the space group $Pm\bar{3}m$. $BaTiO_3$ can be used to characterize the cubic perovskites' archetype (Eglitis et al., 2004). The substitution or doping with different ions or element leads to a change in the structure and decrease the symmetry of the structure. The electronegativity and the atoms size inside of the perovskite structure determines or evaluate the stability of the structures

The Goldschmidt tolerance factor (t) is a measurement that defines the favorability of a perovskite structure based on the elements A, B, and O. This factor is represented as:

$$t = \frac{R_A + R_O}{\sqrt{2}\,(R_B + R_O)} \tag{1}$$

When the Goldschmidt factor is less than 1, it indicates that the crystal mesh is distorted. A wide range of stable perovskites can be obtained since perovskites are stable materials when $0,88 < t < 1,05$. The crystalline deforms in a more or less significant way depending on the atoms used.

- When t > 1, it indicates that the cation A has entirely occupied the cubic cavity space. Given that the cation B is smaller than the space it occupies, the cation B leaves the center of the octahedra cavity along one of the main axes of symmetry to shorten some of the B-O distances and hence to reduce the cation B's coherence. Consequently, a polar and ferroelectric structure was obtained.
- The cation is smaller than the theoretical value when (t) is less than 1. At that moment, the cuboctahedra cavity is not fully filled. This cavity deforms, causing the octahedra to rotate. This rotation has the effect of reducing The coordination of the cation A and the distances A-O (Glazer, 1972).

This type of material has great potential for a variety of functional properties, particularly ferroelectricity and piezoelectricity, because of its capacity to conduct or to convert mechanical energy into electrical energy and vice versa. According to (Ali et al., 2026), such materials show strong promise for next-generation solar energy technologies, where efficient energy conversion and novel device architectures are needed. They are also important for a range of optoelectronic devices, as discussed by (R. Wang et al., 2019), where their electronic and optical properties can be exploited to improve performance (Shaukat et al., 2023). In addition, the piezoelectric characteristics make these materials suitable for transducer and sonar applications (Ahmed, 2021), a point emphasized by (Thi et al., 2014), where reliable mechanical-to-electrical conversion and sensitivity to pressure or acoustic waves are essential.

2.3 Key Physical Properties (Ferroelectric, Piezoelectric, Thermoelectric)

The interaction between the lattice, electronic, and ionic degrees of freedom in perovskites is what gives them their multifunctional nature. Among the most important properties are:

2.3.1 Ferroelectricity

A lot of properties are related to the crystal structure of material that are linked by the space group. The following fig. 2, present the relation between the structure and the space group.

Figure 2. Different classes of crystals structures

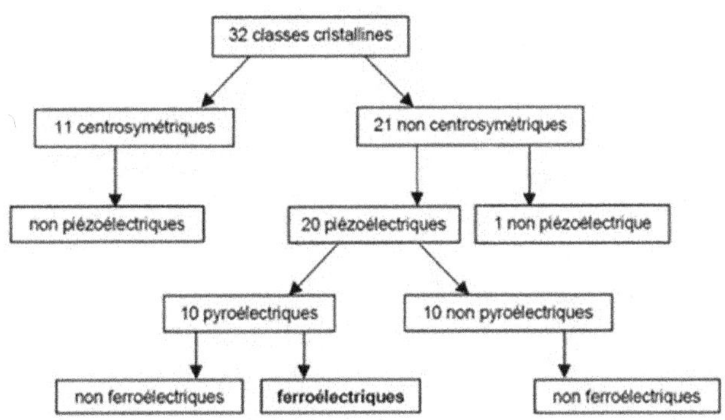

Out of the 32 crystal classes, 11 have an inversion center. They're called centrosymmetric. The remaining 21 classes lack a center of inversion, making them non centrosymmetric classes. Of these 21 classes,

20 allow for the piezoelectric effect. Some of these 20 punctual groups have ferroelectric properties, and 10 of them are pyroelectric. Figure 3 below illustrates the distinction between a centrosymmetric and non-centrosymmetric material following the application of stress. As is evident, applying stress causes a charge differential to form on the material's surface; this phenomenon is known as polarization.

Figure 3. The difference between a centrosymmetric and non-centrosymmetric material after a mechanical stress applying.

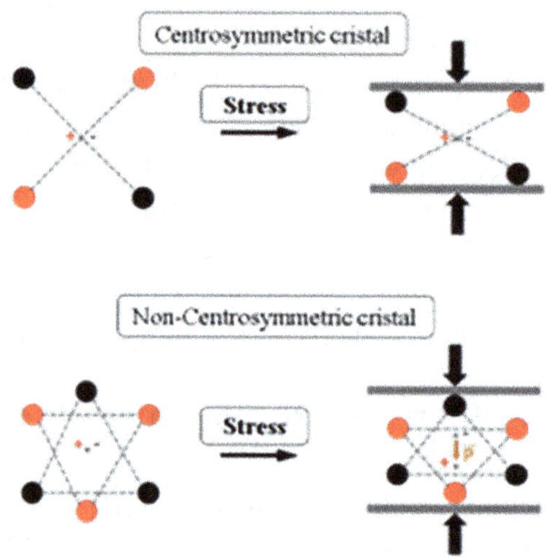

Ferroelectricity is a property where a material possesses a spontaneous remanent polarization (P_r) without any external electric field. The direction of this polarization can be reversed by applying an electric field stronger than a certain threshold, known as the coercive field (E_c). This behavior is a defining characteristic of ferroelectric materials and results in a distinctive dielectric hysteresis loop. The following fig. 4 presented the hysteresis loop.

Figure 4. Ferroelectric crystal polarization–electric field hysteresis loop with coercive field E_c and remanent polarization P_r (Acosta et al., 2017)

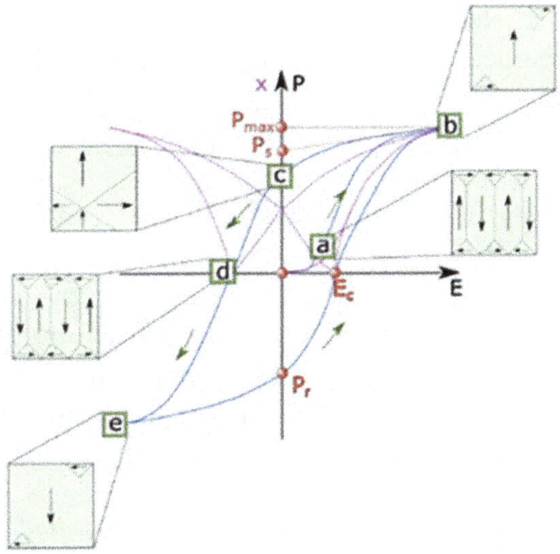

2.3.2 Piezoelectricity

Piezoelectricity is among the most valuable properties of perovskite materials, particularly those exhibiting ferroelectric phenomena. Piezoelectricity is the ability of a crystal material to charge electrically when subjected to a mechanical force is known as piezoelectricity (Bechmann, 1956). Certain natural or artificial crystals, such as quartz, exhibit this spontaneous behavior within the crystal's mesh. Positive and negative charges are moved to the scale of the mesh when a mechanical constraint is applied. The amount of charge generated by the material when it undergoes a deformation along axis z, which is parallel to the polarization axis, is translated by the piezoelectric constant d_{33}. The inverse phenomena of that are to measure the material's reaction after it undergoes an electrical charge; in our case, the material undergoes polarization in the form of an electrical field as showing in the fig 5 below.

Figure 5. The direct and inverse piezoelectric effect.

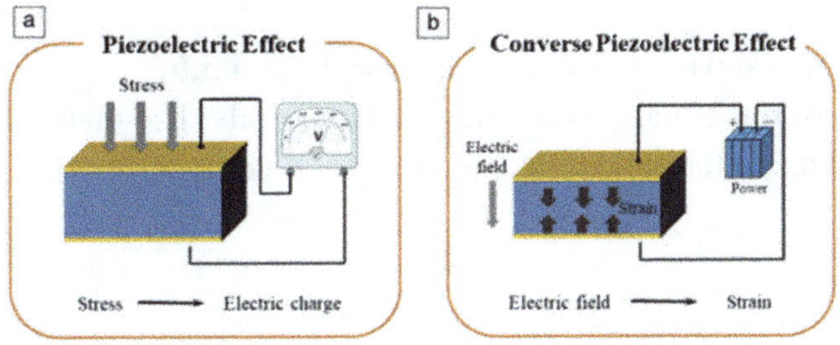

In theory, these kinds of properties can be reproduced by various software such as WIEN2k, VASP, ABINIT, etc. Actually, this chapter treats the method of using the WIEN2k software to simulate the piezoelectrical properties. The theoretical part the computation of the piezoelectric coefficient e_{33} is implemented by introducing an electric charge (polarization) which should considerably disturb the crystal, change the electronic structure of the material due to charge displacement or $d\varepsilon_{jk}$ deformation. The formula for the general case is given below:

$$e_{ijk}(C/m^2) = \frac{dP_i}{d\varepsilon_{jk}} \tag{2}$$

Using the method of the Berry phase (Ahmed et al., 2013) integrated into WIEN2k, the polarization vector, Pi, is determined. The applied deformation interval ranges from -0.01 to 0.01. To represent elastic (deformation) and piezoelectric tensive, respectively, the notation $d\varepsilon_{jk}$ and e_{ij} are used (with i = 1, 2, 3 and j = 1, 2, 3, 4, 5, 6). Or 1, 2, and 3 correspond to the Cartesian directions x, y, and z, respectively.

The structural matrix for the tetragonal structure is:

$$e_{ijk} = \begin{pmatrix} 0 & 0 & 0 & 0 & e_{15} & 0 \\ 0 & 0 & 0 & e_{15} & 0 & 0 \\ e_{31} & e_{31} & e_{33} & 0 & 0 & 0 \end{pmatrix} \tag{3}$$

The structural matrix for the rhombohedra structure is:

$$e_{ijk} = \begin{pmatrix} 0 & 0 & 0 & 0 & e_{15} & -e_{22} \\ -e_{22} & e_{22} & 0 & e_{15} & 0 & 0 \\ e_{31} & e_{31} & e_{33} & 0 & 0 & 0 \end{pmatrix} \tag{4}$$

For the constant piezoelectric d_{ijk}. The two essential parameters, the coefficient e_{ijk} (C/m^2) and the elastic constants of the material C_{ijkl} (GPa), are inextricably linked to the piezoelectric d_{ijk} constant (pC/N). The following is a detailed and simplified explanation. Theoretically, using the following equation, we can calculate the constant of piezoelectric d_{ij} (pC/N) from the coefficient of piezoelectric e_{ik} (C/m^2) and the constant of elastic C_{jk} (GPa) according to the following equation:

$$e_{ij} = \sum_{k=1}^{6} d_{ik} \times C_{jk} \tag{5}$$

with C_{jk}=(I,j=1,2,3,4,5,6) and e_{ij}, d_{ik}= (i=1,2,3; j, k= 1,2,3,4,5,6).

following the solution of this equation for tetragonal materials. The many constant characteristics of the material can be found using the following equations:

$$d_{31} = \frac{C_{33}e_{31} - C_{13}e_{33}}{(C_{11} + C_{12})C_{33} - 2C_{13}^2} \tag{6}$$

$$d_{33} = \frac{(C_{11} + C_{12})e_{33} - C_{13}e_{31}}{(C_{11} + C_{12})C_{33} - 2C_{13}^2} \tag{7}$$

$$d_{33} = \frac{e_{15}}{C_{44}} \tag{8}$$

The constants for the rhombohedra structure become as follows:

$$d_{31} = \frac{C_{33}e_{31} - C_{13}e_{33}}{(C_{11} + C_{12})C_{33} - 2C_{13}^2} \tag{9}$$

$$d_{33} = \frac{(C_{11} + C_{12})e_{33} - 2C_{13}e_{31}}{(C_{11} + C_{12})C_{33} - 2C_{13}^2} \tag{10}$$

$$d_{15} = \frac{(C_{11} - C_{12})e_{15} - C_{14}e_{22}}{(C_{11} - C_{12})C_{44} - C_{14}^2} \tag{11}$$

$$d_{22} = \frac{C_{44}e_{22} + C_{14}e_{15}}{(C_{11} - C_{12})C_{44} - C_{14}^2} \tag{12}$$

The theory of perturbation by density function (DFPT) is used to calculate the various piezoelectric coefficients (e_{ij}, d_{jk}).

Preliminary computations employ density functional perturbation theory (DFPT) (Kohn & Sham, 1965) to predict the piezoelectric properties of materials. This method accurately calculates elasticity, dielectric constants, piezoelectric constants, and phononic dispersion relations. Using $LiNbO_3$ as a case study (Nakamura et al., 2018), the authors illustrate how DFPT effectively predicts the influence of crystal phases and phase transitions, along with analyzing atomic contributions to piezoelectricity. They propose enhancing these properties through structural modifications and element substitution, demonstrating a strong correlation between stability, piezoelectric performance, and crystallographic deformation. This approach serves as a robust method for the development of novel, efficient piezoelectric materials.

2.3.3 Thermoelectricity

Thermoelectrical properties are the features of materials that allow them to change the heat to electricity directly and the other way round, mainly through the Seebeck and Peltier effects. These features are quantified by such parameters as the Seebeck coefficient, electrical conductivity, and thermal conductivity. The effectiveness of a thermoelectric material is generally assessed using the factor without dimension ZT, which is defined as follows:

$$ZT = \frac{S^2.\sigma.T}{\kappa} \tag{13}$$

With

S: is the Seebeck coefficient. (V/K)

σ: is electrical conductivity. (S/m),

T: Is this the absolute temperature (K),

κ: is the overall thermal conductivity (W/m. K), which is made up of electronic (κ_E) and phononic (κ) contributions.

The essential thermoelectrical properties are covered in further detail in the following paragraphs, where each key characteristic is examined more thoroughly, relevant mechanisms are explained, and their implications for practical applications and device performance are discussed.

2.4 Relevance in Modern Technologies (Solar Cells, Sensors, Catalysis, Optoelectronics)

Perovskites are at the forefront of developing technologies because of their adaptability and remarkable range of tunable properties. Solid oxide materials for solar cells, high-performance ferroelectric memory elements (Shaukat et al., 2023), and precision piezoelectric devices all depend on oxide perovskites for their unique structural and electronic characteristics. Their significance in both basic research and technological advancement is further highlighted by their capacity to integrate several functions into a single material platform, enabling multifunctional devices that combine electrical, optical, and mechanical responses. This multifunctionality, together with chemical versatility and the possibility of engineering at the atomic scale, makes perovskites indispensable in efforts to design next-generation energy conversion systems, data storage technologies, and sensing applications (Mishra et al., 2015).

3 BASICS OF DENSITY FUNCTIONAL THEORY (DFT)

3.1 Theoretical Foundation of DFT

Density Functional Theory (DFT) (« The Fundamental Equations of Quantum Mechanics », 1925), is an influential and computationally challenging technic; however, it is successfully used in physics and chemistry to figure out the electronic structure of atoms, molecules, and materials. DFT concentrates on a system's total electron density rather than dealing with each individual electron, which is a major factor for achieving computational efficiency. As a result of this simplified method, scientists can easily calculate the basic properties like energy, structure, and reactivity, which turn DFT into a crucial and largely utilized instrument for understanding complex systems, interdisciplinary research, and beyond. Density Functional Theory (DFT) is predicated upon two fundamental theorems articulated by Hohenberg and Kohn in 1964 (A. Zhou, 2012):

1-Hohenberg–Kohn Theorem I: The ground-state energy of a many-electron system is uniquely determined by the electron density $\rho(r)$.

$$E[\rho] = T[\rho] + V_{ee}[\rho] + V_{ext}[\rho] \tag{14}$$

This function E [ρ] comprises the chemical energy T [ρ], the electron-to-electron interaction V_{ee} [ρ], and the energy originating from the external potential. The density must meet the normalization constraint:

$$\int \rho(r)dr = N \tag{15}$$

Where N is the total number of electrons.

2-Hohenberg–Kohn Theorem II: The electron density that minimizes the energy functional corresponds to the true ground state of the system.

$$E_0 = \min_n E\left[n\left(\vec{r}\right)\right] \tag{16}$$

These principles provide a significant simplification to the quantum many-body problem. Instead of necessitating the solution of a 3N-dimensional wavefunction (where (N) denotes the number of electrons), DFT solely requires the three-dimensional electron density.

In practical applications, the Kohn–Sham formalism, introduced in 1965, posits a system of non-interacting electrons that produces an electron density identical to that of the actual interacting system. The resultant Kohn–Sham equations are solved self-consistently, thereby facilitating the computation of total energies, forces, and electronic structures. For more mathematical aspect about the resolution of this theorem. I recommended to consult the article of Zhou et al (A. Zhou, 2019).

In conclusion, DFT is a useful technique for comprehending materials at the molecular and atomic level. It is widely used in many fields, such as materials research and quantum chemistry.

3.2 Exchange-Correlation Functionals (LDA, GGA, Hybrid)

The hardest part about Density Functional Theory (DFT) is handling the exchange-correlation (XC) energy that involves all the electron-electron interactions in the system, except the ones that can be described by the usual Coulomb term. There are several approximations for that matter:

3.2.1 Local Density Approximation (LDA)

A key approximation in Density Functional Theory (DFT) for calculating a system's exchange-correlation energy is the Local Density Approximation (LDA). This approximation is based on the assumption that the exchange-correlation energy density at a specific point in space is equivalent to that of a uniform electron gas at the same point (Koch & Holthausen, 2001).

We need to establish an algebraic form that effectively captures the relationships between the motions of different electrons. The initial approximation discussed follows the method proposed by Kohn and Sham, which involves creating a reference system for which we have the most accurate expression available for the functional in question. The local density approximation treats the exchange-correlation potential as a local quantity defined at a specific point r, only slightly influenced by the density fluctuations in its vicinity. Therefore, the local density approximation (LDA) should replicate the physical properties of the spherical average of the exchange-correlation hole, rather than the hole itself. This approximation serves as the foundation for all contemporary exchange-correlation functionals and can be defined as follows

$$E_{XC}^{LDA}[\rho] = \int \rho(r)\, \varepsilon_{XC}(\rho(r))dr \tag{17}$$

The specific form of this function is almost known. One tries to approximate the non-interacting functional of the exchange and correlation (E_{xc}) starting from the uniform electron gas model, where the exchange-correlation energy unit (the energy of the exchange-correlation interaction in a uniform electron gas of density ρ) is denoted by the symbol ε_{xc} (ρ(r)). Besides, ε_{xc} (ρ(r)) may be considered as the sum of those parts which come from correlation and exchange.

$$\varepsilon_{XC}(\rho(r)) = \varepsilon_{X}(\rho(r)) + \varepsilon_{C}(\rho(r)) \quad (18)$$

3.2.2 Generalized Gradient Approximation (GGA)

Two essential methods within Density Functional Theory (DFT) for approximating a system's exchange-correlation energy are the Local Density Approximation (LDA) and the Generalized Gradient Approximation (GGA) (Perdew et al., 1996). LDA treats the system as a homogeneous electron gas and depends only on the electron density at a specific location. Despite being computationally efficient, LDA frequently underestimates correlation energy and overestimates exchange energy. Although these mistakes can somewhat compensate for one another, they nonetheless reduce the precision of computations for intricate systems. By adding details about the electron density gradient, GGA improves LDA. Compared to LDA, this extra information enables a more detailed description of the electronic structure, improving predictions of molecule geometries, reaction energy, and other characteristics.

By adding a degree of complexity that more accurately takes into account the fluctuations in electron density inside a system, GGA essentially expands on LDA. GGA improves the overall predictive capacity of DFT calculations by providing a more accurate depiction of the exchange-correlation energy by taking into account the gradient of the electron density (Koch & Holthausen, 2001).

$$E_{XC}^{GGA}[\rho; \nabla\rho] = \int \varepsilon_{xc}^{GGA}\, r\, f(\rho(r), \nabla\rho(r))dr \quad (19)$$

where the energy density of exchange correlation is denoted by ε_{xc}. The exchange-correlation energy per unit volume is represented by her. The electronic density at location r is denoted by ρ(r). At this moment, she provides the number of electrons per unit volume. The electronic density gradient at location r is denoted by ∇ρ(r). It calculates the variation in the electron density throughout space.

3.2.3 GGA-PBEsol

The PBEsol method (Perdew-Burke-Ernzerhof revised for solids)(Terentjev et al., 2018) is a generalized gradient approximation (GGA) for the exchange-correlation energy in density functional theory (DFT) that aims to provide a better description of the properties of close-packed solids and their surfaces. PBEsol captures short-range electronic correlation more accurately and gives better structural predictions than standard PBE for solids, however, as with most GGA functionals, it lacks the nonlocal (long-range) VDW interactions by default.

The exchange-correlation energy in PBEsol is expressed as:

$$Exc(n) = \int n(r)\, \varepsilon_{xc}(n(r), \nabla n(r))dr \quad (20)$$

where n(r) is the electron density and ε_{xc} is the exchange-correlation energy per particle, as a function of local density and its gradient.

PBEsol may also be coalesced with dispersion corrections to account the van der Waals interactions. The formula for the total exchange-correlation energy of the combined functional is:

$$E_{xc}^{total} = E_{xc}^{PBEsol} + E_{c}^{rVV10} \tag{21}$$

Where E_c^{VV10} is the nonlocal correlation energy accounting for long-range interactions, see the references for more details (Terentjev et al., 2018).

When compared to normal GGA functionals, PBEsol yields better results for lattice constants, bulk moduli, and surface characteristics, making it particularly helpful for electronic-structure computations of solids, layered materials, and metal surfaces.

3.2.4 DFT+U

The DFT+U method (Tolba et al., 2018) is an extension of density functional theory (DFT) designed to improve the treatment of strongly correlated electrons, especially in localized d or f orbitals of transition metals and rare earth elements. It adds a Hubbard-like (DFT+U) correction term U that compensates for the self-interaction error and electron over-delocalization common in standard local or semi-local exchange-correlation functionals. This correction effectively introduces a penalty for fractional occupation of localized orbitals, thereby improving the description of electron localization and related properties such as band gaps and magnetism.

$$E^{DFT+U} = E^{DFT} + E_U \tag{22}$$

Where DFT+U is the total energy equation and the Hubbard correction E_U is written following the Dudarev's approach (Tolba et al., 2018) as

$$E_U = \frac{U_{eff}}{2} \sum_{(m,\sigma)} \left(n_m - n_{m\sigma}^2\right) \tag{23}$$

Here U_{eff} =U-J. is the effective Hubbard parameter (on-site Coulomb interaction minus the exchange interaction), and $n_{m\sigma}$ is the occupation number of the localized orbital m with spin σ.

3.2.5 Hybrid Functionals (B3LYP, HSE06...)

Hybrid functionals (Garza & Scuseria, 2016) of density functional theory (DFT) are exchange-correlation functionals that mix a fraction of the exact exchange energy of Hartree-Fock theory with the rest from the standard DFT exchange-correlation sources. Such a combination is able to achieve better results than either one of the methods alone and thus can be used for a variety of molecular properties like atomization energies, bond lengths, and vibrational frequencies, that is to say, the issues raised by some pure DFT functionals can be partially solved by this kind of functionals.

For instance, B3LYP and HSE06 are typical hybrid functionals that scientists commonly use in density functional theory (DFT). These methods essentially blend one part of the so-called exact exchange (Hartree-Fock) and several density functional approximations to get better accuracy when doing electronic structure calculations.

B3LYP (Becke, 3-parameter, Lee–Yang–Parr) (Paier et al., 2007) incorporates a certain amount of pure Hartree–Fock exchange with local and gradient-corrected exchange and correlation functionals. The energy of its exchange-correlation is represented as follows:

$$E_{xc}^{B3LYP} = (1 - a)E_x^{LSDA} + a\,E_x^{HF} + b\,E_x^{B88} + (1 - c)E_c^{LSDA} + c\,E_c^{LYP} \quad (24)$$

Where a, b, and c are typical parameters with values 0.20, 0.72, and 0.81, respectively, and LYP denotes the Lee–Yang–Parr correlation.

HSE06 (Heyd–Scuseria–Ernzerhof 2006) (Janesko et al., 2009) is a range-separated hybrid functional. In order to achieve this, the functional breaks down the Coulomb operator into short-range and long-range parts and combines only the short-range exact exchange with the rest, which makes it very efficient from the computational point of view for solids and surfaces:

$$E_{xc}^{HSE06} = \alpha\,E_x^{HF,\,SR}(\omega) + (1 - \alpha)E_x^{GGA,\,SR}(\omega) + E_x^{GGA,\,LR}(\omega) + E_c^{GGA} \quad (25)$$

with α=0.25 (25% of the exact exchange in the short range) and ω controlling the delimitation between the short and the long range. These steps lead to better predictions of band gaps and the electronic properties of solids.

3.2.6 Basis Sets and Pseudopotentials

To be used effectively, DFT requires a representation of the electron wavefunctions: the Kohn–Sham orbitals and associated quantities must be expressed using a finite, calculable basis set or representation. Typical options include

Plane-wave basis sets: This is a type of basis that is widely used in solid-state physics (VASP, CASTEP, Quantum ESPRESSO). It allows one to achieve convergence in a system with a periodic structure, as well as to have an efficient and systematic work with periodic systems.

Linearized Augmented Plane Wave (LAPW): This is a method used in all-electron calculations, such as WIEN2k (Blaha et al., 2020), that allows one to obtain very high accuracy without using pseudopotentials.

Pseudopotentials and PAW (Projector Augmented-Wave): These are potentials that represent core electrons with effective potentials that allow one to save computational resources while still maintaining the accuracy of valence electrons.

The table 1 that follows provides an overview of the various approaches that we discussed in our chapter.

Table 1. Summary of the different methods, including their advantages, limitations, and applications.

Function	Description	Advantages	Limitations	Application in perovskites
LDA	Considers only the local electron density for XC energy	Simple, computationally efficient; good for systems with slowly varying electron density	Underestimates lattice constants; poor for strongly correlated systems; significant band-gap underestimation	Early studies of oxide perovskites (SrTiO3, BaTiO3) for structural optimization (Jayalakshmi & Nivetha, 2023; Parmar et al., 2023)
GGA-PBE	Includes gradient of electron density; most widely used	Better accuracy for lattice parameters and structural stability than LDA	Still underestimates band gaps; sometimes overestimates cell volumes	Structural studies, electronic DOS, ferroelectric transitions in BaTiO3, NaNbO3 (Machado et al., 2012; Sakhya et al., 2015)
GGA-PBESol	Modified PBE optimized for solids	Improved prediction of lattice constants and bulk moduli	Same band-gap underestimation problem as standard GGA	Accurate structural optimization of oxide perovskites
DFT+U	Adds Hubbard U correction for localized d/f electrons	Corrects for strong correlation; improves electronic structure of transition-metal oxides	Requires empirical U parameter; results depend on chosen U value	Mn-, Fe-, Co-Ni based perovskites; defect and doping studies (S. Sharma & Khan, 2023)
Hybrid fuction	Mix exact Hartree–Fock exchange with GGA	More accurate band gaps and electronic structures; good for optical property prediction	Computationally expensive; less practical for large systems	Band-gap engineering in halide perovskites ($MAPbI_3$, $CsPbBr_3$) (Lv et al., 2025)

3.3 Advantages and Limitations of DFT

The DFT method's limitations and greatest benefits are also shown in the following table 2.

Table 2. Advantages and limitations of the DFT method.

Advantages	Limitations
Equilibrium configurations and parameters of the lattice.	Band-gap underestimation: Conventional functionals usually give a band-gap value that is less than that obtained from experiments.
Electron band diagrams and the density of states.	Strongly correlated systems: Materials such as transition-metal oxides and f-electron systems need additional treatment (DFT+U, hybrid, GW) for accurate description.
Phonon spectra and vibrational characteristics.	Temperature effects: DFT is mainly designed for ground-state (0 K) properties, thus methods like molecular dynamics or equianharmonic approximation are needed for finite-temperature investigations.
Elastic, dielectric, and optical properties.	Neglect of Temperature and Dynamic Effects

4 COMPUTATIONAL MODELING OF PEROVSKITES

One of the ways to guide theory regarding perovskites is through density functional theory (DFT). The process is the entire workflow the authors mention starting from the initial arrangement of structures, and ending with the determination of physical properties. For this, every single step is essential in terms of providing dependable outcomes and logical understanding, thus wrapping up the factors contributing to perovskite anomalies such as lattice deformations, phase changes, and strong electronic interactions.

4.1 Setting up the Structural Model

The first step in using DFT for calculations is setting up the structure, which involves defining the initial atomic arrangement of the perovskite based on experimental crystallographic data. This data is typically available in databases like COD Crystallographic or Materials Project. Generally, for perovskite materials, the crystal structure is determined based on the space group and lattice parameters found in the database. For doped or defect-containing systems, supercells are generated to accommodate substitutions, vacancies, or interstitials. The following example of $KNbO_3$ structure created using WIEN2k package (Schwarz, 2003) and illustrated using XCrySDen program(Kokalj, 1999).

Figure 6. $KNbO_3$ structure created in WIEN2k.

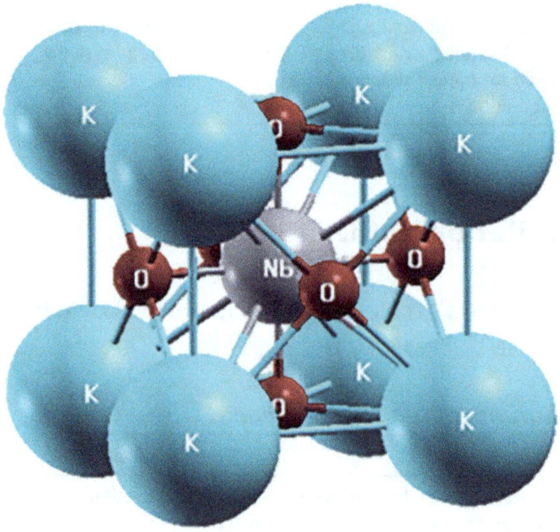

We typically employ a supercell structure, which involves multiplying the unit cell in accordance with the X, Y, and Z axes, to clearly demonstrate the doping or replacement in the crystal structure. Based on a $BaTiO_3$ unit cell, the following example of a supercell 2*2*2 of the $Ba_XCa_{1-X}TiO_3$ structure had 40 atoms overall. Following the use of a 2*2*2 supercell, the structure ABO3's A-site or B-site doping must be at least 0.125 (12.5%) percent. Using the following formula, $Ba_{0.875}Ca_{0.125}TiO_3$, to demonstrate doping with that percentage of 0.125 Ca atoms. When x equals 0.125, one Ba atom is replaced by one Ca atom. A Ca atom could be substituted for any one of the eight Ba atoms in the supercell in order to get the x=0.125 composition. As seen in fig. 7, in this example, we substituted Ca for the Ba atom (1/2, 1/2, 1/2) (Z. Hasan et al., s. d.).

Figure 7. The supercell of $BaTiO_3$ doped by 0.125 of Ca atom.

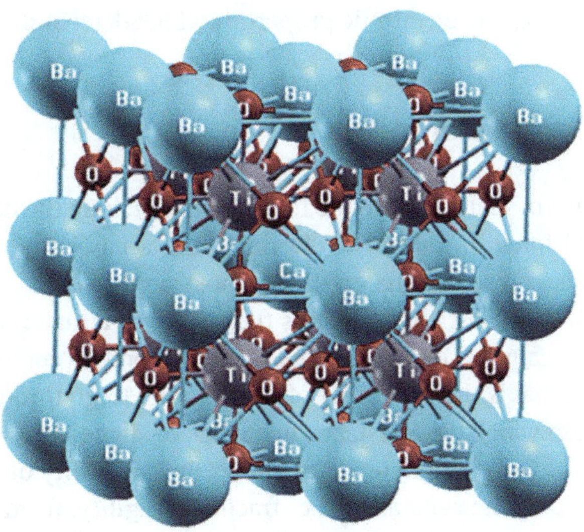

To illustrate structures with a small minimum of doping of element implicate to use supercell bigger than 2*2*2. For example, when we use the supercell of 3*3*3. The minimum percentage of doping will be in this case 0.0370. It means 3.7%.

4.2 Geometry Optimization and Convergence Criteria

DFT-based modeling of perovskite materials requires geometry optimization as the most basic procedure. The total energy of the system is minimized by the process of relaxation of the lattice parameters and the atomic positions until the structure reaches the most stable one. In perovskites, this phase is the most crucial because any slight distortion of the oxygen octahedra or shift of the cations can drastically change ferroelectricity, polarization, and the electronic structure. The application of convergence criteria is the measure of accuracy in the results during this process of optimization which is usually done by setting thresholds for the maximum force acting on atoms, the stress on the unit cell, and the change in total energy between successive iterations. Strict convergence guarantees that no artificial instabilities are present and also ensures that the predictions made for the structural and functional properties are reliable.

4.3 Calculation of electronic Band Structure and Density of States (DOS)

The electronic band structure and density of states (DOS) are definitely the most valuable signals that one can get from their DFT calculations. The band structure gives us a detailed view of the electronic states depending on their momentum, which in turn allows us to detect direct or indirect band gaps, effective masses, and the variation in the energy of conduction and valence bands. Along with this, the DOS also tells us which are the most contributing atomic orbitals and the elements they belong to, for the newly formed electronic states, thus making it easier to understand which are the cations and anions taking part in the charge transport. Moreover, the band structure in the case of oxide perovskites shows that the transition-metal d-orbitals and oxygen p-orbitals are strongly hybridized, and this is the reason for their ferroelectric and dielectric behavior. As a whole, the band structure and DOS constitute the main keys to the understanding of the electronic properties which determine the functional performance of perovskites.

4.4 Mechanical and Elastic Constants

The mechanical properties of materials are the features that include the description of their behavior when they apply an external force. In this case, elasticity has a major function. The other mechanical properties of materials are also present in materials, the most significant of which are mentioned below. Besides, these properties include elasticity, which is the ability of a material to change its shape when a force acts upon it and to regain the original shape once the force is removed if the elastic limit has not been exceeded; ductility, that is the capability of the material to go through permanent deformation without rupture; rigidity, from which the resistance to change of shape is derived; durability, which indicates resistance to radiation or penetration; tenacity, that is the capability of an object to take in energy, at least temporarily, before complete separation, i.e. fracture; fragility, from which the characteristic of breaking under low pressure. Elasticity is a very important property of materials as it is the basis for the others. A proper understanding of these properties is the key to choosing the right materials for different mechanical needs and specific application requirements.

Mechanical properties describe a materials reaction to mechanical stress and by that they decide the material's capacity or ability to resist (or even break) when treated with external forces these are the features that are quite necessary to define the perfector ideal material for a particular use. Elastic constants for tetragonal, cubic, and orthorhombic morphologies can be calculated from the first principles by the density functional theory (DFT) method with the help of the WIEN2k package in conjunction with the IRELAST module (Jamal et al., 2018). The stress-strain method for elastic constants involves a single or several deformations of the crystal lattice which are characterized by the change of volume and the total energy (E in eV) as well as the elastic constants that are obtained by the energy-strain profile fitting; the basic idea is the second derivative of total energy with respect to strain which is closest to the symmetry of the crystal under consideration.

$$C_{ijkl} = \frac{\partial^2 E}{\partial \delta_{ij} \partial \delta_{kl}} \qquad (26)$$

Where the E, $\partial\delta_{ij}$ and $\partial\delta_{kl}$ are the energy the strain and the stress respectively.

Six independent elastic constants (C_{11}, C_{12}, C_{13}, C_{14}, C_{33}, and C_{44}) and a tetragonal (P4bm) structure comprised the rhomboid crystal structures (R3c). Nevertheless, the number of independent elastic constants (C_{11}, C_{12}, and C_{44}) in cubic (Pm3m) is different due to the haut symmetry, rotation in axes 4 and 3, as well as mirrors and inversion. The C_{ijkl} matrix is given:

$$C_{ij}(R3c) = \begin{pmatrix} C_{11} & C_{12} & C_{13} & C_{14} & 0 & 0 \\ C_{12} & C_{11} & C_{13} & -C_{14} & 0 & 0 \\ C_{13} & C_{13} & C_{33} & 0 & 0 & 0 \\ C_{14} & -C_{14} & 0 & C_{44} & 0 & 0 \\ 0 & 0 & 0 & 0 & C_{44} & C_{14} \\ 0 & 0 & 0 & 0 & C_{14} & C_{66} \end{pmatrix} \quad (27)$$

$$C_{ij}(P4bm) = \begin{pmatrix} C_{11} & C_{12} & C_{13} & 0 & 0 & 0 \\ C_{12} & C_{11} & C_{13} & 0 & 0 & 0 \\ C_{13} & C_{13} & C_{33} & 0 & 0 & 0 \\ C_{14} & -C_{14} & 0 & C_{44} & 0 & 0 \\ 0 & 0 & 0 & 0 & C_{44} & 0 \\ 0 & 0 & 0 & 0 & 0 & C_{66} \end{pmatrix} \quad (28)$$

$$C_{ij}(Pm3m) = \begin{pmatrix} C_{11} & C_{12} & C_{12} & 0 & 0 & 0 \\ C_{12} & C_{11} & C_{12} & 0 & 0 & 0 \\ C_{12} & C_{12} & C_{11} & 0 & 0 & 0 \\ 0 & 0 & 0 & C_{44} & 0 & 0 \\ 0 & 0 & 0 & 0 & C_{44} & 0 \\ 0 & 0 & 0 & 0 & 0 & C_{44} \end{pmatrix} \quad (29)$$

The following table summarized the literature results of the elastic constant using GGA-PBE method According to Mubarak et al 2016 (Mubarak, 2016).

Table 3. The elastic constant simulated for the $ATiO_3$ with A= Ca, Sr, Ba.

Structures	**Elastic constant C_{ij} (GPa)**		
	C_{11}	**C_{12}**	**C_{44}**
$CaTiO_3$	286.91 (Mubarak, 2016) 342 (Sakhya et al., 2015)	112.82 (Mubarak, 2016) 97 (Sakhya et al., 2015) 93.27 (Tariq et al., 2015)	74.98(Mubarak, 2016) 91 (Sakhya et al., 2015) 94.47 (Tariq et al., 2015)
$SrTiO_3$	256.16 (Mubarak, 2016) 344 (Sakhya et al., 2015) 317 (Berlincourt & Jaffe, 1958)	118.56 (Mubarak, 2016) 96 (Sakhya et al., 2015) 102.5(Berlincourt & Jaffe, 1958)	84.13 (Mubarak, 2016) 108 (Sakhya et al., 2015) 123.5 (Berlincourt & Jaffe, 1958)
$BaTiO_3$	263.22 (Mubarak, 2016) 295 (Sakhya et al., 2015) 222 (Khalal et al., 1999)	112.3 (Mubarak, 2016) 100 (Sakhya et al., 2015) 134 (Khalal et al., 1999)	107.67 (Mubarak, 2016) 116 (Sakhya et al., 2015) 61 (Khalal et al., 1999)

References: (Sakhya et al., 2015), (Mubarak, 2016), (Tariq et al., 2015), (Berlincourt & Jaffe, 1958), (Khalal et al., 1999).

This section presents the elastic and mechanical properties of the $XTiO_3$ compounds. The elastic constants, calculated at absolute zero temperature (T = 0 K) and compiled in Table 3, provide essential insights into the materials' stability and stiffness. Discrepancies observed between the calculated elastic constants C_{ij} and previously reported values are attributed to differences in the exchange-correlation functional and computational approximations employed. The constant C_{11} characterizes uniaxial deformation along the crystallographic direction, C_{12} corresponds to pure shear stress on the (110) plane along the $[\bar{1}10]$ direction, and C_{44} represents pure shear deformation on the (100) plane. The compounds studied satisfy the mechanical stability criteria for cubic crystals, specifically $(C_{11}+2C_{12})/3>0$, $C_{44}>0$, $(C_{11}-C_{12})/2>0$, and $C_{12}<B<C_{11}$, thus confirming their elastic stability. Notably, the values of C_{11} exceed those of C_{44} by 73.9%, 67.2%, and 59.1% for $CaTiO_3$, $SrTiO_3$, and $BaTiO_3$, respectively, indicating a comparatively weaker resistance to shear deformation than to uniaxial compression. Furthermore, the higher C_{44} value of $BaTiO_3$ relative to the other compounds suggests superior shear modulus and stiffness for $BaTiO_3$ compared to $CaTiO_3$ and $SrTiO_3$.

4.5 Optical Properties (Dielectric Function, Absorption)

Firstly, practically, a material's internal microstructure changes are the major causes to optical properties to change. The changes may be implemented through the material's physical stress application or through a chemical doping technique.(Radziuk & Möhwald, 2016). Consequently, the optical features of the materials can be enhanced through a proper doping or with the aid of physical alterations. Thus, it is imperative to study the optical features of perovskite materials so as to determine their potential application in various fields such as photorefractive (M. Hasan & Akther Hossain, 2021a), optoelectronics (Amin et al., 2012; Wong et al., 2017), solar cells (photovoltaic), and others (Ning & Gao, 2019). In order to explore and model the various optical properties of structures and the impact of doping or element substitution on the optical characteristics, the following samples were obtained using a variety of programs, including Castep, WIEN2k, Quantum Espresso, VASP, etc.

Absorption coefficient: is a fundamental metric of photovoltaic performance, as the optical absorption properties of a material govern its ability to harvest incident solar irradiance. When photons traverse a semiconductor, their energy can be transferred to bound electrons, promoting transitions from the valence band to available states in the conduction band. The efficiency with which these Interbrand transitions occur determined by the material's band structure, density of states, and transition matrix elements directly influences the fraction of the solar spectrum that can be converted into mobile charge carriers and thus the overall power-conversion efficiency of a solar cell.

$$\alpha(\mathrm{cm}^{-1}) = \frac{4\pi.\, k(\omega)}{\lambda} \tag{30}$$

Dielectric function: The real and imaginary parts of the dielectric function are key parameters that strongly determine a material's optical behavior. These components are directly linked to properties such as reflectivity $R(\omega)$, the extinction coefficient $k(\omega)$, and the refractive index $n(\omega)$. In the optical calculations for the three structures, both the real and imaginary parts of the dielectric function were thoroughly considered, as expressed by the following equation: the imaginary part $\varepsilon_2(\omega)$ corresponds to the material's conductivity and indicates how much energy is dissipated, while the real part $\varepsilon_1(\omega)$ represents the material's ability to store energy through polarization.

$$\varepsilon(\omega) = \varepsilon_1(\omega) + i\varepsilon_2(\omega) \tag{31}$$

$$\varepsilon_1(\omega) = n^2 - k^2(\omega) \tag{32}$$

$$\varepsilon_2(\omega) = 2nk(\omega) \quad (33)$$

Refractive index: The extinction coefficient k(ω) and refractive index n(ω) are shown versus frequency. They come from the complex refractive index. Which is related to the dielectric function. In practice n(ω) and k(ω) are obtained by taking the complex square root of ε(ω) (or by fitting measured reflectance/transmittance with models such as Drude–Lorentz), so the resulting expressions are frequency dependent and can be algebraically involved eq .35 & 36.

$$N(\omega) = n + ik \quad (34)$$

$$k(\omega) = \frac{1}{2}\left[\sqrt{\varepsilon_1(\omega)^2+\varepsilon_2(\omega)^2} - \varepsilon_1(\omega)^2\right]^{\frac{1}{2}} \quad (35)$$

$$n(\omega) = \frac{1}{\sqrt{2}}\left[\sqrt{\varepsilon_1(\omega)^2+\varepsilon_2(\omega)^2} - \varepsilon_1(\omega)^2\right]^{\frac{1}{2}} \quad (36)$$

to learn more about the optical characteristics. Regarding perovskite doping for various qualities, such as optical, I suggested reading the following reference (Kacimi-Naciri et al., 2024).

4.6 Thermoelectrical Properties of the Materials (Seebeck Coefficient...)

Because of the Seebeck effect, thermoelectric materials can convert a temperature decrease into electrical energy (Hatzikraniotis et al., 2024). They can also create a temperature gradient from electrical energy by using the Peltier effect. Perovskite ABO_3 compounds, such as $BaTiO_3$, $SrTiO_3$, $NaNbO_3$, and $BiTiO_3$ etc..., are becoming more and more well-known due to their significant structural flexibility, their electrically modulable properties, and their stability in the presence of heat. By applying the Boltztrap code (Madsen & Singh, 2006a) with varying temperature ranges, we can simulate the various thermoelectrical properties of any material based on the DFT method and the Boltztrap package (Madsen & Singh, 2006b) implemented in the WIEN2k package, for instance, to observe the impact of applying varying temperatures on the material and their properties (Hamideddine et al., 2022).

Seebeck coefficient S (V/K): The Seebeck phenomenon is based on the movement of charge carriers (also known as electrons or " holes") within a material (Hatzikraniotis et al., 2024). When a conductor's extremity is heated, the charge carriers present in this manner absorb the kinetic energy and move toward the fresher extremity. Tension results from this action's creation of a charge imbalance. The tension created by one unit of temperature difference is measured by the Seebeck coefficient (S). It is commonly expressed in microvolts per kelvin (μV/K) or volts per kelvin (V/K) (Malki & El Farh, 2020).

$$S = -\frac{\Delta V}{\Delta T} \quad (37)$$

ΔV: difference in potential (V)

ΔT: variations in temperature (K)

The unit of measurement is volts per Kelvin (V/K), or more commonly, microvolts per Kelvin (μV/K).

Thermal conductivity (κ): is the ability of a material to transfer heat, that is, its tendency to transfer thermal energy from a hotter region to a colder region when the two are in contact or connected by the material in question. This property is measured with the use of grandeurs physiques and is dependent on the intrinsic nature of the material, its density, its microscopical structure, and occasionally its temperature. She has two contributions: phonons (κ_l) and electrons (κ_e) (Darhi et al., 2023).

$$\kappa = \kappa_e + \kappa_l \tag{38}$$

κ_E: electronic contribution (from the responsible porters).
κ_l: phononic contribution (originating from network vibrations).
The following is the relationship between electron thermal conductivity (κ_e) and electrical conductivity:

$$\kappa_e = L.\sigma.T \tag{39}$$

With L represents the Lorentz constant or the equivalent Lorentz number 2.45×10^{-8} V·K^{-2}.

Electrical conductivity (σ): It measures the ability of a material to conduct electricity, in other words the ease with which electric charges can move through this material under the effect of a potential difference. This property depends on the intrinsic nature of the material, its electronic structure, its temperature and the presence of impurities or defects, which can increase or decrease its conductivity. It depends on the density of charge carriers (n), their charge (e) and their mobility (μ) (Hatzikraniotis et al., 2024).

$$\sigma = n.e.\,\mu \tag{40}$$

n: density of charge carriers (m^{-3}).
e: elementary charge (1.6 10^{-19} C).
μ: mobility of carriers (m^2/V s).

Thermoelectric power (power factor PF): This is a measure of the electrical power generated per unit temperature gradient, expressing how much electrical energy can be produced when a thermal differential is maintained between two points of a given material or system. An intermediate criterion for evaluating performance, relatively related to the Seebeck coefficient (S) and electrical conductivity. (σ/τ) (Abd El Qader et al., 2013).

$$PF = \frac{S^2.\,\sigma}{\tau} \tag{41}$$

5 CASE STUDIES AND APPLICATIONS

In this part we will discuss an example of a perovskite material that was investigated using density functional theory (DFT) methods to predict and analyze the various properties of its crystal structure. The study focuses on how DFT calculations can be employed to determine structural parameters, electronic band structure, density of states, and related physical characteristics. By applying different functions and methods we can predict a lot of properties and get more information about the material and how he acting for more uses in different applications.

5.1 Oxide Perovskites ($KNbO_3$)

Potassium niobate ($KNbO_3$) (You et al., 2021) is a vital one that is an inorganic compound which is particularly famous for the Perovskite type of the ferroelectric crystal structure (Dolhen et al., 2015). Due to its significant nonlinear optical characteristics, it finds a wide range of cutting-edge photonics and optoelectronics applications, inter alia, blue lasers production by frequency doubling of infrared lasers. $KNbO_3$ definitely has structural phase transitions that are temperature-dependent, going from cubic to tetragonal, orthorhombic, and rhombohedral symmetries on the decrease of temperature (Nakamura et al., 2002). The crystal has a very large nonlinear optical coefficient, a response time of the order of one millisecond, and practically zero scattering loss, all of which make it a very efficient nonlinear optical process. The compound's chemical stability, excellent phase matching properties, as well as its high photorefractive stability during the light exposure, are the features that make it indispensable in the micro-lasers and blue light generation devices fields (Saito et al., 2004).

On the physicochemical side, $KNbO_3$ is orthorhombic at room temperature with lattice parameters, and a Curie temperature around 498 K (Qi et al., 2025; Tokay & Yazıcı, 2022). All these characteristics together are making $KNbO_3$ a nonlinear optics material, quantum information processing, and memory storage research material (Zhang et al., 2020). This section is dedicated to starting the calculation of $KNbO_3$ theoretical DFT with the help of the WIEN2k package to figure out how the structural parameters, electronic band structure, density of states, and other physical characteristics can be derived from the calculations. Using various functions and methods, we can foresee numerous properties and obtain more data about the material $KNbO_3$ and how it behaves if used in different applications.

To demonstrate modeling with DFT clearly, the $KNbO_3$ example briefly illustrates the steps researchers take to optimize the crystal structure for different applications. The structure of $KNbO_3$ was created using the WIEN2k package, with the XCrySDen program (Kokalj, 1999) Figure 6 employed for visualization and optimized by employing the GGA-PBE function. The initial lattice parameters were set as follows: a = b = 3.9810 Å, c = 4.010 Å, and angles $\alpha = \beta = \gamma = 90.0000°$. This configuration represents a tetragonal structure belonging to the P4mm (99) space group. The atomic sphere radii (Ri) for the constituent atoms were defined as follows: 2.5 for potassium (K), 1.94 for niobium (Nb), and 1.72 for oxygen. The plane wave cut-off, R_{mt}, which determines the size of the basis sets in the calculations, was set at 7.0. For structural optimization, a Brillouin zone of 1500 was used, while a K-point mesh of 3000 was applied for property computations. The numerical calculations were performed iteratively until the energy convergence threshold of 0.001 Ry was met.

Table 4. Different parameter of the $KNbO_3$ structure after optimization.

Composition	$KNbO_3$
a=b(Å)	4.0814
c(Å)	4.1858
c/a	1.0255
Space group (n°)	P4mm (99)

Composition	$KNbO_3$
Density (g/cm³)	4,637
E-g(eV)	1.653 1.42 (D. Wang et al., 2020)

References: (D. Wang et al., 2020).

After optimization, the table presents essential crystallographic and physical characteristics of the pure compound $KNbO_3$. The parameters a = b = 4.0814 Å and c = 4.1858 Å define the dimensions of the unit cell within a tetragonal structure, as evidenced by the ratio c/a = 1.0255, which indicates a slight deviation from ideal cubic symmetry. The space group is identified as P4mm, a typical configuration for this category of ferroelectric materials. The density is notably high at 4.637 g/cm³, consistent with the mass and compactness of the crystal lattice. Furthermore, the band gap (Eg) is 1.653 eV, signifying a relatively wide energy band gap that affects the electronic and optical properties, including conductivity and light absorption. In conclusion, these parameters affirm that $KNbO_3$ possesses a slightly distorted tetragonal crystal structure, characterized by substantial compactness and a moderate energy gap, rendering it suitable for ferroelectric and piezoelectric applications.

Fig. 8 Present the band gap of the structure after optimization using the GGA-PBE method. And by definition in a material without any electron states, the energy difference between the valence band (V.B.) and the conduction band (C.B.) is known as the band gap. It determines the properties of the material as insulator, semi-conductor or conductor, that is related to the electronical properties of each material. This variation in the value of the band gap influences the different properties related to the absorption of light in the material or the conductivity of the electrical charges. Which is essential for the ferroelectrical, piezoelectrical and optical properties for different application such as the photovoltaic or optoelectronic devises.

Fig. 8 illustrates the electronic band structure of $KNbO_3$, calculated along various symmetry directions within the Brillouin zone. A well-defined band gap is observed between the highest valence band and (V.B) the lowest conduction band (C.B), signifying that $KNbO_3$ functions as a semiconductor. This band gap, estimated to 1.653 eV situated near the Γ and R point, suggests an indirect electronic transition band gap (D. Wang et al., 2020) that essential characteristic for numerous optoelectronic applications. The lower bands, exhibiting a pronounced flatness, correspond to strongly localized states, whereas the dispersion of the upper bands indicates the delocalized nature of the electrons within the crystal structure. This electronic configuration renders $KNbO_3$ particularly suitable for electro-optic and ferroelectric devices, due to its combination of a direct band gap and unique crystallographic properties.

Figure 8. Band Structure of the $KNbO_3$ structure after optimization.

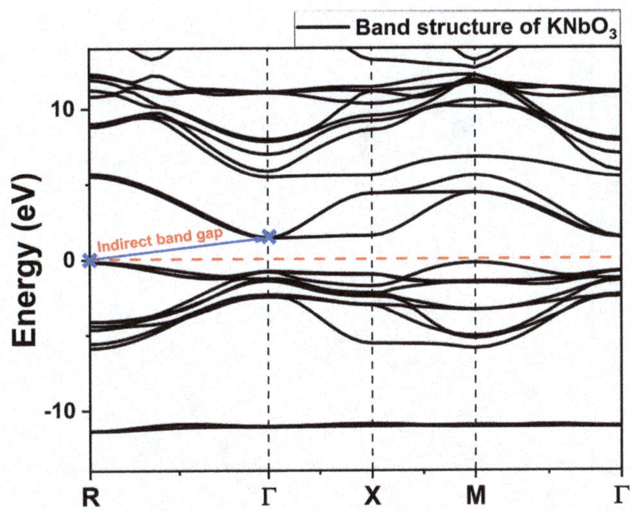

The Total density of state (TDOS) reflects the electronic states that are accessible at each energy level to the entire system, or the material in its entirety. It is a basic characteristic that represents the general distribution of permitted energy levels. Because the existence or lack of states at the Fermi energy (0 eV) indicates the material's conductivity (metallic, semiconducting, or insulating), the TDOS is essential for figuring out the material's characteristics same as the band gap. The main contributors to $KNbO_3$ are oxygen atoms, with Nb and K contributing well but insignificantly. With a bandgap energy of 1.653 eV, the compound's semiconducting nature is demonstrated by the lack of DOS throughout the Fermi energy state (EF=0 eV) (H. Li et al., 2021). The conduction band (CB) of the structure is the consequence of most of the Nb-d and K-p contributions. However, the O-p contributes just a little. The valence band (VB) was often composed of the O-p contribution plus a little amount of Nb-p (H. Li et al., 2021). Due to its substantial contribution to the valence band (VB) and conduction band (CB), niobium has a stronger effect on the electrical structure of $KNbO_3$ than do K and O (Jameel et al., 2023).

Figure 9. TDOS and PDOS of the $KNbO_3$ structure.

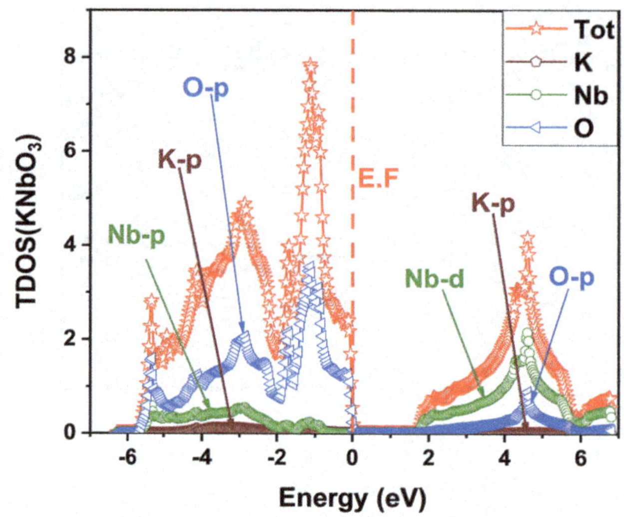

5.2 Effects of Doping and Substitutions (Sr-doped $BaTiO_3$.)

5.2.1 Optimization Case of Sr-Doped $BaTiO_3$

Lead-based PZT ceramics (R. Wang et al., 2019) are highly effective in ferroelectric, piezoelectric, and dielectric functions, but their PbO component has raised environmental concerns, thus the demand for lead-free alternatives has been initiated. Although $BaTiO_3$ (Hu et al., 2022) is less piezoelectric, it is a focus as doping can significantly improve its electrical properties (A. Sharma et al., 2009). Replacing Ba^{2+} with Sr^{2+} in $BaTiO_3$ to create $Ba_{1-X}Sr_XTiO_3$ (BST) is a very effective way (Dai et al., 2020). The addition of Sr improves dielectric and piezoelectric properties, raises the dielectric constant (Sonia et al., 2012), (Emadi et al., 2019), (M. Hasan & Akther Hossain, 2021b), and moves diffraction peaks to higher angles showing better crystallinity and microstructural order(Jamaluddin et al., 2016) (Yang et al., 2017). As the Sr content increases, the Curie temperature (T_c) decreases since Sr replaces Ba, thus shifting the ferroelectric–paraelectric transition (Panomsuwan & Manuspiya, 2018). The ability to change Tc is very important for devices operating at specific temperatures such as microwave phase shifters, actuators, and capacitors. Sr doping also changes the atomic bonding and electronic structure of $BaTiO_3$, which is supported by DFT studies (Datta et al., 2024) (M. Hasan & Akther Hossain, 2021b), thereby providing a detailed atomic-level understanding and guiding experiment (Tihtih, Ibrahim, Basyooni, En-nadir, et al., 2023). Additionally, Sr-doped $BaTiO_3$ exhibits enhanced piezoelectric responses, and co-doping (such as with Ni) (Z. Li et al., 2022) further modifies the lattice parameters and may lead to the appearance of multiferroic behavior. Doping that results in smaller grain size and fewer oxygen vacancies contributes to higher energy storage and lower losses, thus the material's application as a relaxor ferroelectric device can

continued on following page

Table 5. Continued

be broadened (X. Zhou et al., 2023). Using theoretical modeling alongside experiments helps to define in detail the changes in Sr doping in $BaTiO_3$ that affect the structure, electronics, and piezoelectricity. Modeling eliminates a lot of the trial-and-error work by behavior prediction even before synthesis, thus speeding up material development. BST's high dielectric strength, polarization, and dielectric constant make it very suitable for large energy storage (Khandelwal et al., 2019). This research focuses on tetragonal Sr-doped $BaTiO_3$ analyzed by DFT WIEN2k software to forecast the impact of Sr concentration, verified by local experiments, with the resulting applications to be optoelectronics (Miran et al., 2023), sonars (Arul & Nithya, 2020), and energy storage. The mission is to produce lead-free, ecofriendly materials for the electronic and energy needs of the future.

To illustrate the effect of doping we were based on the example of doping $BaTiO_3$ with Sr on different properties electronic structural mechanic and piezoelectric according to kacimi-naciri et al 2025 (Kacimi-Naciri et al., 2025). The electronic structure calculations for $Ba_{1-x}Sr_xTiO_3$ were performed using density functional theory (DFT) within the WIEN2k package (Blaha et al., 2020). Structural optimizations were carried out using different approximations, including LDA, GGA-PBE, and GGA-PBEsol, with GGA-PBE chosen for exchange-correlation potential due to its effectiveness in shape optimization. Atomic sphere radii for Ba, Sr, Ti, and O were set appropriately. Valence electron configurations corresponded to Ba (6s 5p), Sr (5s 4p), Ti (4s 3d), and O (2s 2p). With a convergence energy criterion of 0.0001 Ry. The lattice parameters corresponded to a tetragonal cell with a=b=3.9850Å, c=4.0170Å, and all angles at 90°. To handle the complexity of Sr and Ba ion distribution in $Ba_{1-X}Sr_XTiO_3$, a 2×2×2 supercell (composed of eight primitive $BaTiO_3$ unit cells) was used for electronic structure calculations across doping levels x=0, 0.125 and 0.25. For more details about the methods and materials utilized in this study I recommended to consult the reference (Kacimi-Naciri et al., 2025).

Table 5. Comparison of Band Gap, Density, Elastic Modulus, and Lattice Constants for $BaTiO_3$, $Ba_{0.875}Sr_{0.125}TiO_3$ and $Ba_{0.75}Sr_{0.25}TiO_3$.

	$BaTiO_3$	$Ba_{0.875}Sr_{0.125}TiO_3$	$Ba_{0.75}Sr_{0.25}TiO_3$
Lattice constants. Å			
a=b	**3.910** 3.960 (Moreira & Dias, 2007)	**3.935**	**4.088**
c	**4.017** 4.058 (Boochani et al., 2019)	**4.012**	**4.121**
c/a	**1.027**	**1.0195**	**1.008**
B(GPa)	**163.264** 157.659(Boochani et al., 2019)	**211.068**	**289.830**
Simulated theoretical density (g/cm^3)	**6.0702** 5.90 (M. Hasan & Akther Hossain, 2021b)	**5.7156** 5.77 (M. Hasan & Akther Hossain, 2021b)	**5.6857**
Band gap. (eV)	**2.01** 2.06 (Z. Hasan et al., s. d.)	**1.80** 1.71 (M. Hasan & Akther Hossain, 2021b)	**1.70**

References: (Moreira & Dias, 2007), (Boochani et al., 2019), (M. Hasan & Akther Hossain, 2021b), (Z. Hasan et al., s. d.).

Table 5. summarizes how increasing strontium (Sr) doping in $Ba_{1-x}Sr_xTiO_3$ affects its structural, mechanical, and electrical properties. As Sr concentration rises, lattice constants a and c increase slightly due to changes in bond lengths and angles, leading to a minor expansion of the unit cell despite Sr's smaller ionic size. The tetragonality ratio (c/a) decreases from 1.027 to 1.008, indicating reduced structural distortion and a shift toward a more cubic phase. The bulk modulus (B) increases significantly, showing that Sr doping makes the material stiffer and more resistant to compression, likely due to stronger bonding. The theoretical density decreases as Sr, being lighter than Ba, replaces Ba ions.

5.2.2 Electronic Properties

For $Ba_{0.875}Sr_{0.125}TiO_3$, the partial substitution of Ba by 12.5% Sr slightly alters the electronic structure of $BaTiO_3$ without significantly changing its semiconducting nature. The Ba 6s states contribute to the conduction band, maintaining a direct band gap, while Sr introduces smaller states with less pronounced Sr-O hybridization compared to Ba-O, as evidenced by fewer valence band peaks from Sr-4p versus Ba-5p. Titanium remains the dominant contributor to both valence and conduction bands, preserving the material's overall electronic stability. This modest Sr doping allows minor tunability in conduction and dielectric properties while ensuring the stability of the electronic structure, making $Ba_{0.875}Sr_{0.125}TiO_3$ suitable for applications requiring stable yet slightly adjustable electronic behavior. For the other percentage of doping and the interpretation. Finally, the band gap narrows from 2.01 eV to 1.70 eV with higher Sr content, suggesting improved suitability for optoelectronic applications by requiring less energy for electron excitation, attributed to changes in electronic structure caused by Sr substitution. For more details about the band gap of the tree systems, consult the reference (Kacimi-Naciri et al., 2025).

Figure 10. Example of TDOS of the doped structure of $BaTiO_3$ with 12.5% of Sr.

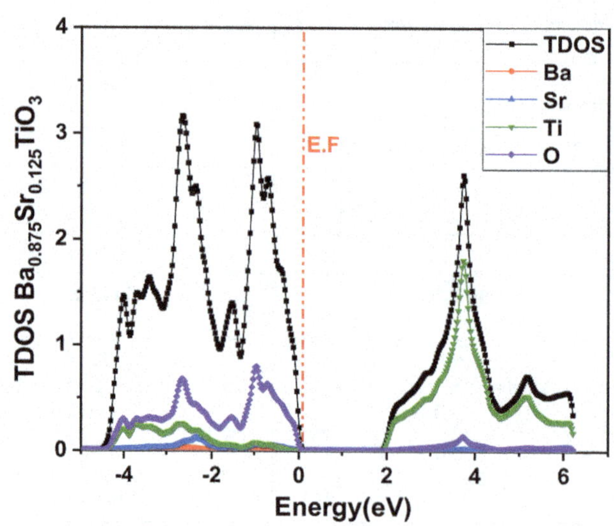

5.2.3 Optical Properties

After optimization of the structure, we can do the different calculation to predict properties. For optical properties the following figure represents one of the most important properties in optic we talk about the absorption.

Figure 11. The absorption of the pure and doped structure of BST.

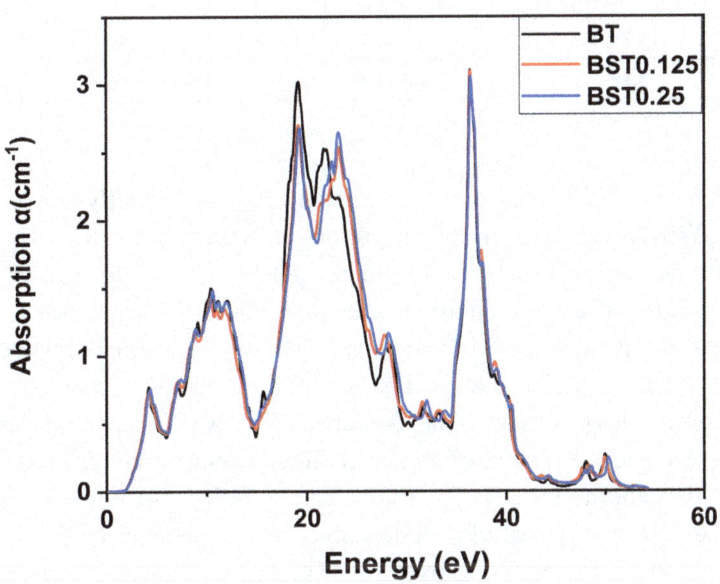

The optical absorption of $Ba_{1-x}Sr_xTiO_3$ (BST) shows that increasing the Sr concentration from x = 0 to x = 0.125 and x = 0.25 slightly enhances the intensity of the absorption peaks, particularly near 20 eV and 40 eV, while the overall profile remains essentially unchanged. This change is likely due to the doping effect of Sr^{2+} ions compared with Ba^{2+} ions, which alters the electronic structure and the Ti–O bonding environment in the crystal lattice. These adjustments suggest that Sr substitution subtly modifies the density of electronic states and the material's response to electromagnetic fields, enabling fine tuning of BST's optical properties. This controlled modification makes Sr-doped BST a promising candidate for applications requiring specific optical characteristics. A similar effect has been reported for Ca doping, an element with a smaller ionic radius than Ba, as noted by Hasan et al (Z. Hasan et al., s. d.).

5.2.4 Mechanical Properties

The effects of Sr doping on the elastic characteristics of $Ba_{1-x}Sr_xTiO_3$ at 0 K are displayed in Table 5. As Sr increases, C_{11}, C_{12}, and C_{13} typically decrease, signifying less stiffness and more lattice flexibility. Because increased C_{33} opposes piezoelectric constant d_{33}, it marginally increases resistance to axial compression along the c-axis and decreases the piezoelectric response (d_{33}). Shear in the (100) plane is described by C_{44}. In general, Sr doping balances flexibility and stability for particular device requirements by reducing isotropic stiffness and improving certain directional characteristics according to Kacimi-Naciri et al (Kacimi-Naciri et al., 2025).

Table 6. $Ba_{1-x}Sr_xTiO_3$ elastic constants (C_{ij}) for various doping compositions (Kacimi-Naciri et al., 2025).

Composition	**C_{ij} (GPa)**			
	C_{11}	**C_{12}**	**C_{13}**	**C_{33}**
$BaTiO_3$	**337.48** 300(K. et al., 2024) 295(Mubarak, 2016)	**130.03** 109(K. et al., 2024) 100(Mubarak, 2016)	**62.67** 90(K. et al., 2024)	**183.44** 149(K. et al., 2024)
$Ba_{0.87}5Sr_{0.125}TiO_3$	**288.59**	**91.26**	**90.72**	**224.85**
$Ba_{0.75}Sr_{0.25}TiO_3$	**238.98**	**87.965**	**87.965**	**238.98**

References: (Mubarak, 2016), (K. et al., 2024).

Strontium (Sr) doping in $Ba_{1-X}Sr_XTiO_3$ perovskite materials significantly impacts their elastic and piezoelectric properties. With increasing Sr concentration, elastic constants C_{11}, C_{12}, and C_{13} generally decrease, indicating softened lattice structures and increased flexibility, typical of related perovskites (Mubarak, 2016). However, C_{33} slightly increases, reflecting enhanced resistance to axial compression along the c-axis. This particular increase in C_{33} inversely affects the piezoelectric constant d_{33}, which decreases with higher C_{33}. The opposite relationship between C_{33} and d_{33} means enhanced mechanical stability in specific directions but a diminished piezoelectric response. This balance brought by Sr doping allows for both reduced isotropic stiffness and improved directional mechanical properties, tailoring the material for applications demanding specific flexibility and stability. Furthermore, Sr doping modifies structural phases, moving from tetragonal to pseudo-cubic, and influences dielectric and piezoelectric behaviors. Moderate Sr doping enhances domain stability and polarization, improving piezoelectric response, whereas excessive doping can disrupt domain alignment and reduce ferroelectric performance. This complex interplay between doping level, lattice structure, and functional properties makes Sr-doped $BaTiO_3$ a versatile material for electronic and electromechanical device applications where controlled mechanical and piezoelectric properties are essential (Quan et al., 2018).

5.3 Comparative Analysis of Theoretical vs. Experimental Results

In order to validate computations, this section compares the theoretical and actual piezoelectric characteristics of $Ba_{1-x}Sr_xTiO_3$. Predicting the effects of Sr doping on $BaTiO_3$'s piezoelectricity will help with composition decisions prior to synthesis. The study connects theory and experiment to evaluate the correctness of the model for optimizing real-world piezoelectric materials by comparing identical compositions. Doping's effect on d_{33} (pC/N) is depicted in Fig. 12 below, confirming the theoretical approach. And for more details consult the reference (Kacimi-Naciri et al., 2025).

Figure 12. Theoretical and experimental values of d_{33} (pC/N) for pure and doped structure.

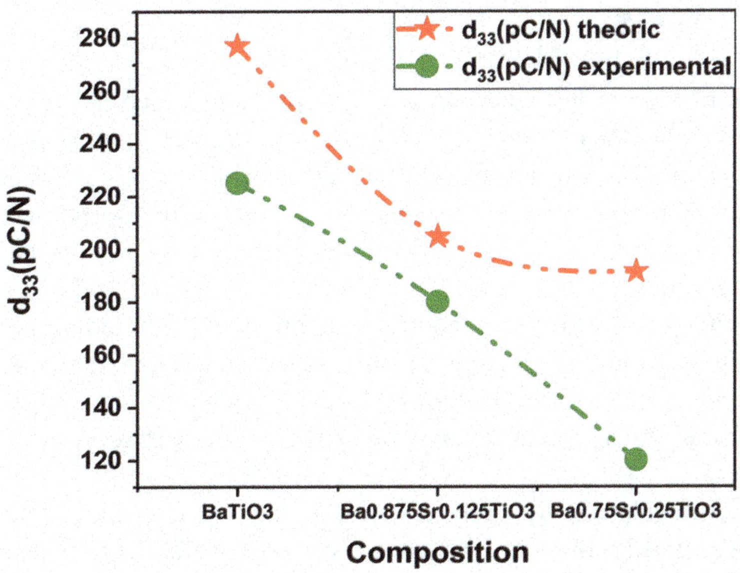

Fig. 12 presents a comparison between the experimental and theoretical values of d_{33} (pC/N) for undoped $BaTiO_3$ and $Ba_{1-x}Sr_xTiO_3$ (x = 0.125, 0.25). Idealized modeling versus experimental constraints is reflected in the fact that theoretical d_{33} values are greater than practical ones. According to both theory and observation, d_{33} decreases as Sr concentration rises, suggesting that Sr-doping reduces the piezoelectric response, most likely by changing the lattice and lessening polarization under stress.

6 APPLICATION AND TECHNOLOGIES OF THE MATERIALS

Density Functional Theory (DFT) provides accurate structural details of the material by forecasting the changes, phase transition, and modification caused by doping. The method also predicts electronic structures with high accuracy that makes it possible to a band-gap be tailored for solar and electronic applications. In addition, DFT can explain the link between very small changes in the structure of the material (atomic-scale displacements) and large changes in the material's properties such as ferroelectricity, piezoelectricity, and optical behavior. Besides that, it is a helpful instrument for experiments as it identifies potential dopants by screening, predicts tolerance to defects, and predicts the stability of the material before it is synthesized.

On top of this, DFT can also calculate the energy barriers and reaction pathways, enabling the identification of transformations that are kinetically accessible and metastable states. Moreover, it assigns the calculation of vibrational spectra, dielectric responses, and carrier masses to the charge, which are direct links to the experimental observables. The application of DFT in high-throughput and machine learning workflows as a result of materials discovery and optimization, the number of costly or time-consuming experimental trials is greatly reduced, and the mechanistic insights that inform targeted synthesis and device design are provided.

7 FUTURE PERSPECTIVES

Theoretical modeling of perovskite materials with Density Functional Theory (DFT) has been identified as a method that is both very effective and multi-faceted to carry out investigations of those materials' properties, which include structural, electronic, mechanical, thermoelectrical, piezoelectrical and optical. By rigorously implementing DFT as a standard way from geometry optimization to advanced property calculations, researchers have obtained profound insights on the basic mechanisms regulating the behavior of oxide perovskites.

Despite that, density functional theory limitations, including band-gap underestimation, problems with the treatment of strongly correlated electrons, and the omission of temperature-dependent effects, indicate the necessity for novel and improved methods. More correct models like hybrid functionals, GW corrections, and ab-initio molecular dynamics could be used for precise descriptions, but they require more computational effort.

Exploring Co-Doping Strategies: Studying the effects of co-doping with other elements can provide valuable insights into synergistically enhancing properties such as thermal stability, piezoelectric response, and durability. For example, combining Sr with elements like yttrium (Y) may lead to improved multi-functional characteristics. This approach is supported by studies examining the roles of A-site (Sr), B-site (Y), and combined A, B-site (Sr, Y) substitution in lead-free $BaTiO_3$ ceramics, focusing on structural, optical, microstructural, mechanical, and thermal features (Tihtih, Ibrahim, Basyooni, Kurovics, et al., 2023). Investigating Sr Doping at the Nanoscale: Examining Sr doping at the nanoscale could uncover new properties and potential applications. Such studies may improve control over dopant distribution and phase stability, especially in tuning the band gap and emission wavelength of strontium-doped barium titanate ($BaTiO_3$, BST) perovskites. This could facilitate the development of high-efficiency visible-light emitters and solar cells (Boubaia et al., 2021).

One can also anticipate that the combination of DFT with high-throughput computational screening and machine learning methods will allow the faster browsing of the perovskite materials with customized properties. Moreover, the coupling of DFT simulations with in-situ experimental characterization will not only improve the predictive accuracy but also extend the application range from theory to practice.

The global workflow of the ML (Schleder et al., 2019) can be summarized in the next steps:

- Collection of data and curation.
- Data preprocessing.
- Data representation and transformation.
- Learning algorithm training.
- Model testing and optimization.
- Application of model.

CONCLUSION

Perovskite materials continue to attract substantial interest due to their exceptional multifunctional properties, which span from ferroelectricity and piezoelectricity to outstanding optical and photovoltaic performance. These unique characteristics make them highly promising for a wide range of technological applications. Density Functional Theory (DFT) has emerged as an indispensable tool for probing these materials at the atomic scale, allowing researchers to predict and analyze their structural stability, intricate electronic structures, and behavior under various external stimuli with remarkable detail and accuracy.

The two cases examined in this chapter provide a comprehensive understanding of the material and the DFT calculation method, as well as how to forecast the various material properties for the $KNbO_3$ structure. The total density of states (TDOS) shows that the band gap, which is 1,653 eV and located close to the Γ and R points, is semi-conducting, mostly because to niobium contributions in the valence and conduction bands. It is proposed that this is an indirect band gap. Although they have a smaller role, the oxygen atoms are also involved. According to these findings, the material exhibits excellent behavior for a variety of optoelectronic device applications. Despite the Sr-doped $BaTiO_3$ structure. The findings demonstrated that strontium (Sr) has an impact on the initial structure, particularly by causing a little rise in network constants and a drop in the tetragonality ratio (c/a), which suggests a transition to a region that is more cubic. Furthermore, when Sr concentration rises, the band gap falls from 2.01 eV to 1.70 eV, indicating increased potential for optoelectronic uses. Additionally, we compare the practical and theoretical results, particularly the piezoelectrical characteristics d_{33} (pC/N) for the clean $BaTiO_3$ and doped $Ba_{1-x}Sr_xTiO_3$ (x = 0.125, 0.25), in order to confirm the theoretical model. Theoretical d_{33} values are higher than real ones, which reflects modeling vs experimental limits. A drop in d_{33} with increasing Sr concentration is consistent with both theory and observation, indicating that Sr-doping lowers the piezoelectric response, most likely via altering the lattice and reducing polarization under stress. The model theory is validated and confirmed by these outcomes.

By employing advanced theoretical modeling, scientists have been able to unravel the microscopic mechanisms responsible for key phenomena such as ferroelectric polarization, the critical role of lattice instabilities in driving phase transitions, and the origin of defect tolerance in halide perovskites. This deeper understanding has significantly advanced the fundamental science of perovskites while simultaneously providing practical insights for experimentalists seeking to optimize material properties and tailor perovskites for specific applications(Hao et al., 2019).

Despite its successes, DFT is not without limitations. Challenges remain in accurately capturing band gaps, describing strongly correlated electron systems, and accounting for temperature-dependent effects all of which are critical factors for a full understanding of material behavior. Nevertheless, continuous improvements in computational methodologies, including hybrid functionals, many-body perturbation approaches, and coupling with molecular dynamics, have steadily enhanced DFT's predictive power. When integrated with experimental results and emerging data-driven approaches such as machine learning, DFT forms a cornerstone of modern materials research, enabling the rational design and accelerated development of next-generation perovskite-based devices with tailored functionalities (Schleder et al., 2019).

Finally the piezoelectric materials like $Ba_{1-x}Sr_xTiO_3$, $KNbO_3$... and lead zirconate titanate (PZT) convert mechanical energy to electrical signals and vice versa (Panda & Sahoo, 2015)., making them vital across many industries. In aerospace they enable vibration damping and structural health monitoring; in naval systems they serve as sonar transducers for underwater communication and navigation. In automotive and industrial settings, they act as precise actuators for fuel injection and flow control.

Medically, the power high-resolution ultrasound imaging and therapeutic devices. They also support energy harvesting from ambient vibrations, and are used in accelerometers, acoustic emission detectors, and micro-positioning devices for optics and robotics. In consumer electronics, piezo ceramics provide haptic feedback in touchscreens and operate in inkjet printers. Although lead-based piezoelectric raise environmental concerns, active research seeks lead-free alternatives with comparable performance. These capabilities make piezoelectric materials key enablers of technological innovation across multiple sectors (Webber et al., 2017).

REFERENCES

Abd El Qader, M., Venkat, R., Kumar, R., Hartmann, T., Ginobbi, P., Newman, N., & Singh, R. (2013). Structural, electrical, and thermoelectric properties of CrSi 2 thin films. *Thin Solid Films*, *545*, 100–105. DOI: 10.1016/j.tsf.2013.07.040

Acosta, M., Novak, N., Rojas, V., Patel, S., Vaish, R., Koruza, J., Rossetti, G. A.Jr, & Rödel, J. (2017). BaTiO3-based piezoelectrics: Fundamentals, current status, and perspectives. *Applied Physics Reviews*, *4*(4), 041305. Advance online publication. DOI: 10.1063/1.4990046

Ahmed, I. (2021). Synthesis, characterization, ferroelectric, and piezoelectric properties of (1 – x)BaTiO 3 – x (BaNi 0.5 Nb 0.5 O 3) perovskite ceramics. *Journal of Materials Science Materials in Electronics*, *32*(8), 10769–10777. DOI: 10.1007/s10854-021-05734-8

Ahmed, S. J., Kivinen, J., Zaporzan, B., Curiel, L., Pichardo, S., & Rubel, O. (2013). BerryPI : A software for studying polarization of crystalline solids with WIEN2k density functional all-electron package. *Computer Physics Communications*, *184*(3), 647–651. DOI: 10.1016/j.cpc.2012.10.028

Ali, M. K. M., Mohsen, A. A., & Allam, N. K. (2026). Lead-free perovskite materials for optoelectronic and solar energy applications. *Solar Energy Materials and Solar Cells*, *295*, 114025. DOI: 10.1016/j.solmat.2025.114025

Amin, B., Khenata, R., Bouhemadou, A., Ahmad, I., & Maqbool, M. (2012). Opto-electronic response of spinels MgAl2O4 and MgGa2O4 through modified Becke-Johnson exchange potential. *Physica B, Condensed Matter*, *407*(13), 2588–2592. DOI: 10.1016/j.physb.2012.03.075

Arul, N. S., & Nithya, V. D. (Eds.). (2020). *Revolution of Perovskite : Synthesis, Properties and Applications*. Springer Singapore., DOI: 10.1007/978-981-15-1267-4

Bechmann, R. (1956). Elastic, Piezoelectric, and Dielectric Constants of Polarized Barium Titanate Ceramics and Some Applications of the Piezoelectric Equations. *The Journal of the Acoustical Society of America*, *28*(3), 347–350. DOI: 10.1121/1.1908324

Berlincourt, D., & Jaffe, H. (1958). Elastic and Piezoelectric Coefficients of Single-Crystal Barium Titanate. *Physical Review*, *111*(1), 143–148. DOI: 10.1103/PhysRev.111.143

Blaha, P., Schwarz, K., Tran, F., Laskowski, R., Madsen, G. K. H., & Marks, L. D. (2020). WIEN2k : An APW+lo program for calculating the properties of solids. *The Journal of Chemical Physics*, *152*(7), 074101. Advance online publication. DOI: 10.1063/1.5143061

Boochani, A., Rasoolian, G., Kafi, F., Aminian, A., & Elahi, S. M. (2019). Electronic, optical, vibrational and EFG properties of tetragonal BaTiO3 under pressure : By first-principles study. *Zhongguo Wuli Xuekan*, *59*, 357–371. DOI: 10.1016/j.cjph.2019.02.023

Boubaia, A., Assali, A., Berrah, S., Bennacer, H., Zerifi, I., & Boukortt, A. (2021). Band gap and emission wavelength tuning of Sr-doped BaTiO3 (BST) perovskites for high-efficiency visible-light emitters and solar cells. *Materials Science in Semiconductor Processing*, *130*, 105837. DOI: 10.1016/j.mssp.2021.105837

Chakraborty, A., Liton, M. N. H., Sarker, M. S. I., Rahman, M. M., & Khan, M. K. R. (2023). A comprehensive DFT evaluation of catalytic and optoelectronic properties of BaTiO3 polymorphs. *Physica B, Condensed Matter*, *648*, 414418. DOI: 10.1016/j.physb.2022.414418

Dai, G., Wang, S., Huang, G., Chen, G., Lu, B., Li, D., Tao, T., Yao, Y., Liang, B., & Lu, S. (2020). Direct and indirect measurement of large electrocaloric effect in barium strontium titanate ceramics. *International Journal of Applied Ceramic Technology*, *17*(3), 1354–1361. DOI: 10.1111/ijac.13384

Darhi, Z., Malki, S., Abbadi, H., El Farh, L., Guesmi, I., & Challioui, A. (2023). Ab-initio calculation of the structural, electronic, mechanical, optical, and thermoelectric properties of orthorhombic ZnAs compound. *Physica B, Condensed Matter*, *654*, 414722. DOI: 10.1016/j.physb.2023.414722

Datta, A. K., & Rahman, Md. S., Mia, N., & Ahmed, S. (2024). Exploration of Structural, Electronic and Optical Properties of Be-doped $BaTiO_3$ Perovskite : First Principles DFT Calculations. 2024 3rd International Conference on Advancement in Electrical and Electronic Engineering (ICAEEE), 1-6. DOI: 10.1109/ICAEEE62219.2024.10561704

Dolhen, M., Mahajan, A., Pinho, R., Costa, M. E., Trolliard, G., & Vilarinho, P. M. (2015). Sodium potassium niobate ($K_{0.5}$ $Na_{0.5}$ NbO_3, KNN) thick films by electrophoretic deposition. *RSC Advances*, *5*(6), 4698–4706. DOI: 10.1039/C4RA11058G

Eglitis, R. I., Piskunov, S., Heifets, E., Kotomin, E. A., & Borstel, G. (2004). Ab initio study of the SrTiO3, BaTiO3 and PbTiO3 (001) surfaces. *Ceramics International*, *30*(7), 1989–1992. DOI: 10.1016/j.ceramint.2003.12.176

Emadi, F., Nemati, A., Hinterstein, M., & Adabifiroozjaei, E. (2019). Microstructural, optical, and electrical characteristics of Ni/C doped BST thin films. *Ceramics International*, *45*(5), 5503–5510. DOI: 10.1016/j.ceramint.2018.12.007

Fan, B., Yang, G., Li, M.-Y., Liu, P., Qiu, S., Shen, M., Liu, H., Zhang, G., Wang, Q., & Jiang, S. (2018). Lead-free Ba(1-x)SrxTiO3 ceramics for room-temperature pyroelectric energy conversion. *Ceramics International*, *44*(7), 8270–8276. DOI: 10.1016/j.ceramint.2018.02.009

Garza, A. J., & Scuseria, G. E. (2016). Predicting Band Gaps with Hybrid Density Functionals. *The Journal of Physical Chemistry Letters*, *7*(20), 4165–4170. DOI: 10.1021/acs.jpclett.6b01807

Glazer, A. M. (1972). The classification of tilted octahedra in perovskites. *Acta Crystallographica. Section B, Structural Crystallography and Crystal Chemistry*, *28*(11), 3384–3392. DOI: 10.1107/S0567740872007976

Glazer, A. M. (2002). **Perovskites modern and ancient**. By Roger H. Mitchell.Thunder Bay, Ontario : AlmazPress, 2002. Price USD 70.00. ISBN0-9689411-0-9. *Acta Crystallographica. Section B, Structural Science*, *58*(6), 1075–1075. DOI: 10.1107/S0108768102020220

Hamideddine, I., Tahiri, N., Bounagui, O. E., & Ez-Zahraouy, H. (2022). Ab initio study of structural and optical properties of the halide perovskite KBX3 compound. *Journal of the Korean Ceramic Society*, *59*(3), 350–358. DOI: 10.1007/s43207-021-00178-6

Hao, J., Li, W., Zhai, J., & Chen, H. (2019). Progress in high-strain perovskite piezoelectric ceramics. *Materials Science and Engineering R Reports*, *135*, 1–57. DOI: 10.1016/j.mser.2018.08.001

Hasan, M., & Akther Hossain, A. K. M. (2021a). Structural, electronic and optical properties of strontium and nickel co-doped BaTiO3 : A DFT based study. *Computational Condensed Matter*, *28*, e00578. DOI: 10.1016/j.cocom.2021.e00578

Hasan, M., & Akther Hossain, A. K. M. (2021b). Structural, electronic and optical properties of strontium and nickel co-doped BaTiO3 : A DFT based study. *Computational Condensed Matter*, *28*, e00578. DOI: 10.1016/j.cocom.2021.e00578

Hasan, Z., Rahman, M. A., Das, D. K., & Rouf, H. K. (s. d.). OPEN Influence of Ca doping in structural, electronic, optical and mechanical properties of B a1−xCaxTiO3 perovskite from first-principles investigation. Scientific Reports.

Hatzikraniotis, E. S., Polymeris, G., & Kyratsi, T. (2024). Thermal Conductivity in Thermoelectric Materials. In R. Vargas-Bernal & R. Palma Guerrero (Éds.), Novel Applications of Piezoelectric and Thermoelectric Materials. IntechOpen. DOI: 10.5772/intechopen.106168

Hu, J., Fan, H., Wu, S., Tang, L., Qin, L., & Luo, W. (2022). Characterization of temperature dependence of dielectric, elastic and piezoelectric properties of BaTiO3 ceramics. *Ceramics International*, *48*(18), 25741–25746. DOI: 10.1016/j.ceramint.2022.05.245

Jamal, M., Bilal, M., Ahmad, I., & Jalali-Asadabadi, S. (2018). IRelast package. *Journal of Alloys and Compounds*, *735*, 569–579. DOI: 10.1016/j.jallcom.2017.10.139

Jamaluddin, A., Suwarni, , Supriyanto, A., & Iriani, Y. (2016). Properties of strontium doped barium titanate powder prepared by solid state reaction. *Journal of Physics: Conference Series*, *776*, 012052. DOI: 10.1088/1742-6596/776/1/012052

Jameel, M. H., Rehman, A., Bin Roslan, M. S., & Agam, M. A. B. (2023). To investigate the structural, electronic, optical and magnetic properties of Sr-doped $KNbO_3$ for perovskite solar cell applications : A DFT study. *Physica Scripta*, *98*(5), 055802. DOI: 10.1088/1402-4896/acc6fb

Janesko, B. G., Henderson, T. M., & Scuseria, G. E. (2009). Screened hybrid density functionals for solid-state chemistry and physics. *Physical Chemistry Chemical Physics*, *11*(3), 443–454. DOI: 10.1039/B812838C

Jayalakshmi, D. S., & Nivetha, B. (2023). First-principles calculations for comparative band structure study of SrTiO3 perovskite on bulk and layered phases for efficient optoelectronic conversion. *Computational & Theoretical Chemistry*, *1220*, 114006. DOI: 10.1016/j.comptc.2022.114006

Johnsson, M., & Lemmens, P. (2007). Crystallography and Chemistry of Perovskites. In H. Kronmüller & S. Parkin (Éds.), Handbook of Magnetism and Advanced Magnetic Materials (1re éd.). Wiley. DOI: 10.1002/9780470022184.hmm411

K., P. M., B., B., T.P., V., T.V., R. M., & M., V. G. B. (2024). Experimental and first-principles investigation of structural, electronic, optical, elastic and ferroelectric properties of BaTiO 3. Physica B: Condensed Matter, 674, 415548. DOI: 10.1016/j.physb.2023.415548

Kacimi-Naciri, H., Rguiti, M., Mabrouk, A., Ben Achour, M. A., Lorgouilloux, Y., Courtois, C., Faska, N., & Bachar, A. (2024). Perovskite Progress : The Role of Sr-Doping in Advancing BaTiO3-Based Technologies. In Amrousse, R., Bachar, A., Mabrouk, A., & Azat, S. (Eds.), (pp. 201–246). Advances in Chemical and Materials Engineering. IGI Global., DOI: 10.4018/979-8-3373-0669-8.ch006

Kacimi-Naciri, H., Rguiti, M., Mabrouk, A., Courtois, C., Ben Achour, M. A., Lorgouilloux, Y., Amrousse, R., Faska, N., & Bachar, A. (2025). DFT-based and experimental study on Sr-doped BaTiO3 : Impacts on piezoelectric and ferroelectric performance. *Ceramics International*, *51*(17), 23801–23813. DOI: 10.1016/j.ceramint.2025.03.069

Khalal, A., Khatib, D., & Jannot, B. (1999). Elastic and piezoelectric properties of BaTiO3 at room temperature. *Physica B, Condensed Matter*, *271*(1), 343–347. DOI: 10.1016/S0921-4526(99)00202-1

Khandelwal, P., Ahlawat, N., Kundu, R. S., Sangwan, K. M., & Bajia, S. (2019). Influence of Sr-doping on structural and ferroelectric properties of BaTiO3 ceramics. 030599. DOI: 10.1063/1.5113438

Koch, W., & Holthausen, M. C. (2001). A Chemist's Guide to Density Functional Theory (1re éd.). Wiley. DOI: 10.1002/3527600043

Kohn, W., & Sham, L. J. (1965). Self-Consistent Equations Including Exchange and Correlation Effects. *Physical Review*, *140*(4A), A1133–A1138. DOI: 10.1103/PhysRev.140.A1133

Kokalj, A. (1999). XCrySDen A new program for displaying crystalline structures and electron densities. *Journal of Molecular Graphics & Modelling*, *17*(3-4), 176–179. DOI: 10.1016/S1093-3263(99)00028-5

Li, H., Wang, L., Xu, L., Li, A., Mao, P., Wu, Q., & Xie, Z. (2021). First-principles study on the structural, elastic, piezoelectric and electronic properties of (BaTiO3, LiTaO3)-modified KNbO3. *Materials Today. Communications*, *26*, 102092. DOI: 10.1016/j.mtcomm.2021.102092

Li, Z., Yu, J., Hao, S., & Janolin, P.-E. (2022). Enhancing properties of lead-free ferroelectric BaTiO3 through doping. *Journal of the European Ceramic Society*, *42*(12), 4693–4701. DOI: 10.1016/j.jeurceramsoc.2022.05.023

Lv, K., Li, Z., Huang, X., Cheng, Z., Wang, Z., & Zhao, H. (2025). $CsPbBr_3$ Perovskite-Based Heterostructures in Photocatalysis : Mechanisms, Stability, and Multifunctional Performance. *Advanced Science (Weinheim, Baden-Wurttemberg, Germany)*, *12*(34), e07747. DOI: 10.1002/advs.202507747

Machado, R., Sepliarsky, M., & Stachiotti, M. G. (2012). First-Principles Calculations of Structural Properties of $NaNbO_3$. *Ferroelectrics*, *427*(1), 98–104. DOI: 10.1080/00150193.2012.674419

Madsen, G. K. H., & Singh, D. J. (2006a). BoltzTraP. A code for calculating band-structure dependent quantities. *Computer Physics Communications*, *175*(1), 67–71. DOI: 10.1016/j.cpc.2006.03.007

Madsen, G. K. H., & Singh, D. J. (2006b). BoltzTraP. A code for calculating band-structure dependent quantities. *Computer Physics Communications*, *175*(1), 67–71. DOI: 10.1016/j.cpc.2006.03.007

Malki, S., & El Farh, L. (2020). First-Principles Investigation on Thermoelectric Properties of $$VSb_{2}$$ Material. *International Journal of Thermophysics*, *41*(5), 58. DOI: 10.1007/s10765-020-02630-x

Miran, H. A. J., Abdullah, Z. N., Altarawneh, M., Rahman, M. M., Al-Bayati, A. T., & Salman, E. M.-T. (2023). First-Principles Analysis of Cr-Doped SrTiO3 Perovskite as Optoelectronic Materials. *Iranian Journal of Materials Science and Engineering*, *20*(1).

Mishra, R., Jain, S., & Prasad, C. D. (2015). A review on piezoelectric material as a source of generating electricity and its possibility to fabricate devices for daily uses of army personnel. International Journal of Systems. *Control and Communications*, *6*(3), 212. DOI: 10.1504/IJSCC.2015.068908

Moreira, R. L., & Dias, A. (2007). Comment on "Prediction of lattice constant in cubic perovskites". *Journal of Physics and Chemistry of Solids*, *68*(8), 1617–1622. DOI: 10.1016/j.jpcs.2007.03.050

Mubarak, A. A. (2016). The first-principle study of the electronic, optical and thermoelectric properties of XTiO3 (X = Ca, Sr and Ba) compounds. *International Journal of Modern Physics B*, *30*(20), 1650141. DOI: 10.1142/S0217979216501411

Nakamura, K., Higuchi, S., & Ohnuma, T. (2018). Density Functional Perturbation Theory to Predict Piezoelectric Properties. In Bakırtaş, İ. (Ed.), *Perturbation Methods with Applications in Science and Engineering*. InTech., DOI: 10.5772/intechopen.76827

Nakamura, K., Tokiwa, T., & Kawamura, Y. (2002). Domain structures in KNbO3 crystals and their piezoelectric properties. *Journal of Applied Physics*, *91*(11), 9272–9276. DOI: 10.1063/1.1476078

Ning, W., & Gao, F. (2019). Structural and Functional Diversity in Lead-Free Halide Perovskite Materials. *Advanced Materials*, *31*(22), 1900326. DOI: 10.1002/adma.201900326

Paier, J., Marsman, M., & Kresse, G. (2007). Why does the B3LYP hybrid functional fail for metals? *The Journal of Chemical Physics*, *127*(2), 024103. DOI: 10.1063/1.2747249

Panda, P. K., & Sahoo, B. (2015). PZT to Lead Free Piezo Ceramics : A Review. *Ferroelectrics*, *474*(1), 128–143. DOI: 10.1080/00150193.2015.997146

Panomsuwan, G., & Manuspiya, H. (2018). Structural and dielectric properties of sol–gel derived $Ba_{1-x}Sr_xTiO_3$ ($0 \leq x \leq 0.5$) ceramics for energy storage applications. *Materials Research Express*, *6*(2), 026310. DOI: 10.1088/2053-1591/aaedf2

Parmar, V. B., Raval, D., Gupta, S. K., Gajjar, P. N., & Vora, A. M. (2023). BaTiO3 perovskite for optoelectronics application : A DFT study. Materials Today: Proceedings, S2214785323005059. DOI: 10.1016/j.matpr.2023.01.410

Perdew, J. P., Burke, K., & Ernzerhof, M. (1996). Generalized Gradient Approximation Made Simple. *Physical Review Letters*, *77*(18), 3865–3868. DOI: 10.1103/PhysRevLett.77.3865

Qi, X., Ren, P., Tong, X., Wang, X., & Zhuo, F. (2025). Enhanced piezoelectric properties of KNN-based ceramics by synergistic modulation of phase constitution, grain size and domain configurations. *Journal of the European Ceramic Society*, *45*(1), 116874. DOI: 10.1016/j.jeurceramsoc.2024.116874

Quan, Y., Ren, W., Niu, G., Wang, L., Zhao, J., Zhang, N., Liu, M., Ye, Z., Liu, L., & Karaki, T. (2018). Large Piezoelectric Strain with Superior Thermal Stability and Excellent Fatigue Resistance of Lead-Free Potassium Sodium Niobate-Based Grain Orientation-Controlled Ceramics. *ACS Applied Materials & Interfaces*, *10*(12), 10220–10226. DOI: 10.1021/acsami.8b01554

Radziuk, D., & Möhwald, H. (2016). Ultrasonically treated liquid interfaces for progress in cleaning and separation processes. *Physical Chemistry Chemical Physics, 18*(1), 21–46. DOI: 10.1039/C5CP05142H

Saito, Y., Takao, H., Tani, T., Nonoyama, T., Takatori, K., Homma, T., Nagaya, T., & Nakamura, M. (2004). Lead-free piezoceramics. *Nature, 432*(7013), 84–87. DOI: 10.1038/nature03028

Sakhya, A. P., Maibam, J., Saha, S., Chanda, S., Dutta, A., Sharma, B. I., Thapa, R. K., & Sinha, T. P. (2015). Electronic structure and elastic properties of ATiO3 (A = Ba, Sr, Ca) perovskites : A first principles study. *APPL PHYS*, •••, 53.

Schleder, G. R., Padilha, A. C. M., Acosta, C. M., Costa, M., & Fazzio, A. (2019). From DFT to machine learning : Recent approaches to materials science–a review. *JPhys Materials, 2*(3), 032001. DOI: 10.1088/2515-7639/ab084b

Schwarz, K. (2003). DFT calculations of solids with LAPW and WIEN2k. *Journal of Solid State Chemistry, 176*(2), 319–328. DOI: 10.1016/S0022-4596(03)00213-5

Sharma, A., Tyagi, V. V., Chen, C. R., & Buddhi, D. (2009). Review on thermal energy storage with phase change materials and applications. *Renewable & Sustainable Energy Reviews, 13*(2), 318–345. DOI: 10.1016/j.rser.2007.10.005

Sharma, S., & Khan, S. (2023). Enhancement of dielectric and ferroelectric properties of (Sr, Ni) Co-doped $BaTiO_3$ ceramics. *Ferroelectrics, 617*(1), 18–24. DOI: 10.1080/00150193.2023.2271131

Shaukat, H., Ali, A., Bibi, S., Mehmood, S., Altabey, W. A., Noori, M., & Kouritem, S. A. (2023). Piezoelectric materials : Advanced applications in electro-chemical processes. *Energy Reports, 9*, 4306–4324. DOI: 10.1016/j.egyr.2023.03.077

Sonia, P., Patel, R. K., Kumar, P., Prakash, C., & Agrawal, D. K. (2012). Low temperature synthesis and dielectric, ferroelectric and piezoelectric study of microwave sintered BaTiO3 ceramics. *Ceramics International, 38*(2), 1585–1589. DOI: 10.1016/j.ceramint.2011.09.045

Tariq, S., Ahmed, A., Saad, S., & Tariq, S. (2015). Structural, electronic and elastic properties of the cubic CaTiO3 under pressure : A DFT study. *AIP Advances, 5*(7), 077111. DOI: 10.1063/1.4926437

Terentjev, A. V., Constantin, L. A., & Pitarke, J. M. (2018). Dispersion-corrected PBEsol exchange-correlation functional. *Physical Review. B, 98*(21), 214108. DOI: 10.1103/PhysRevB.98.214108

The fundamental equations of quantum mechanics. (1925). Proceedings of the Royal Society of London. Series A, Containing Papers of a Mathematical and Physical Character, 109(752), 642-653. DOI: 10.1098/rspa.1925.0150

Thi, M. P., Marchet, P., Levassort, F., & Blantignie, C. (2014). Matériaux piézoélectriques sans plomb pour transducteurs acoustiques.

Tihtih, M., Ibrahim, J. E. F. M., Basyooni, M. A., En-nadir, R., Belaid, W., Abdelfattah, M. M., Hussainova, I., Pszota, G., & Kocserha, I. (2023). Enhanced optical and thermal conductivity properties of barium titanate ceramic via strontium doping for thermo-optical applications. *Optical and Quantum Electronics, 55*(3), 226. DOI: 10.1007/s11082-022-04516-8

Tihtih, M., Ibrahim, J. E. F. M., Basyooni, M. A., Kurovics, E., Belaid, W., Hussainova, I., & Kocserha, I. (2023). Role of A-site (Sr), B-site (Y), and A, B sites (Sr, Y) substitution in lead-free BaTiO3 ceramic compounds : Structural, optical, microstructure, mechanical, and thermal conductivity properties. *Ceramics International*, *49*(2), 1947–1959. DOI: 10.1016/j.ceramint.2022.09.160

Tokay, O., & Yazıcı, M. (2022). A review of potassium sodium niobate and bismuth sodium titanate based lead free piezoceramics. *Materials Today. Communications*, *31*, 103358. DOI: 10.1016/j.mtcomm.2022.103358

Tolba, S. A., Gameel, K. M., Ali, B. A., Almossalami, H. A., & Allam, N. K. (2018). The DFT+U : Approaches, Accuracy, and Applications. In Yang, G. (Ed.), *Density Functional Calculations Recent Progresses of Theory and Application*. InTech., DOI: 10.5772/intechopen.72020

Wang, D., Wang, G., Lu, Z., Al-Jlaihawi, Z., & Feteira, A. (2020). Crystal Structure, Phase Transitions and Photoferroelectric Properties of KNbO3-Based Lead-Free Ferroelectric Ceramics : A Brief Review. *Frontiers in Materials*, *7*, 91. DOI: 10.3389/fmats.2020.00091

Wang, R., Wang, J., Tan, S., Duan, Y., Wang, Z.-K., & Yang, Y. (2019). Opportunities and Challenges of Lead-Free Perovskite Optoelectronic Devices. *Trends in Chemistry*, *1*(4), 368–379. DOI: 10.1016/j.trechm.2019.04.004

Webber, K. G., Vögler, M., Khansur, N. H., Kaeswurm, B., Daniels, J. E., & Schader, F. H. (2017). Review of the mechanical and fracture behavior of perovskite lead-free ferroelectrics for actuator applications. *Smart Materials and Structures*, *26*(6), 063001. DOI: 10.1088/1361-665X/aa590c

Wong, K. M., Irfan, M., Mahmood, A., & Murtaza, G. (2017). First principles study of the structural and optoelectronic properties of the A2InSbO6 (A = Ca, Sr, Ba) compounds. *Optik (Stuttgart)*, *130*, 517–524. DOI: 10.1016/j.ijleo.2016.10.139

Yang, F., Lin, S., Yang, L., Liao, J., Chen, Y., & Wang, C.-Z. (2017). First-principles investigation of metal-doped cubic BaTiO 3. *Materials Research Bulletin*, *96*, 372–378. DOI: 10.1016/j.materresbull.2017.03.023

You, J., Li, G., Zhang, S., Zhang, X., Luo, J., Rao, M., & Peng, Z. (2021). Synthesis, characterization and thermodynamic properties of KNbO3. *Journal of Alloys and Compounds*, *882*, 160641. DOI: 10.1016/j.jallcom.2021.160641

Zhang, N., Zheng, T., & Wu, J. (2020). Lead-Free (K,Na)NbO_3-Based Materials : Preparation Techniques and Piezoelectricity. *ACS Omega*, *5*(7), 3099–3107. DOI: 10.1021/acsomega.9b03658

Zhou, A. (2012). Hohenberg–Kohn theorem for Coulomb type systems and its generalization. *Journal of Mathematical Chemistry*, *50*(10), 2746–2754. DOI: 10.1007/s10910-012-0061-3

Zhou, A. (2019). A mathematical aspect of Hohenberg-Kohn theorem. *Science China. Mathematics*, *62*(1), 63–68. DOI: 10.1007/s11425-018-9337-2

Zhou, X., Xu, D., Yu, T., Duan, X., Zhao, H., & Zhao, W. (2023). Effects of Mn doping and sintering condition on the microstructure, dielectric, and energy storage properties of Ba0.8Sr0.2TiO3 ceramics. *Journal of Materials Science Materials in Electronics*, *34*(20), 1511. DOI: 10.1007/s10854-023-10918-5

Chapter 4

Computational Design of Efficient Lead-Free Double Perovskites for Photovoltaic Applications

Mossab Oublal
https://orcid.org/0009-0007-0198-9294
Laboratory of Condensed Matter Physics and Nanomaterials for Renewable Energy, Faculty of Sciences, Ibn Zohr University, Agadir, Morocco

Essediq Youssef El-Yakoubi
https://orcid.org/0009-0005-9096-3402
Laboratory of Condensed Matter Physics and Nanomaterials for Renewable Energy, Faculty of Sciences, Ibn Zohr University, Agadir, Morocco

ABSTRACT

The global demand for sustainable energy is driving the search for efficient, non-toxic materials for next-generation optoelectronic and photovoltaic devices. This chapter first gives a state-of-the-art overview of perovskite materials, then presents a computational study of lead-free halide double perovskites Cs2NaInX6 (X = Br, I) using density functional theory (DFT). Structural, mechanical and thermodynamic stability are analysed via Goldschmidt's tolerance and octahedral factors, elastic constants and formation energies. Cs2NaInBr6 and Cs2NaInI6 are found to have direct band gaps (TB-mBJ) in the visible and near-IR ranges, respectively, identifying them as promising absorber layers. Their photovoltaic performance is assessed using the spectroscopic limited maximum efficiency model and SCAPS-1D device simulations, which highlight the strong potential of Cs2NaInI6 as an efficient lead-free absorber. The chapter concludes by outlining how machine learning and high-throughput screening could accelerate the discovery and optimisation of related materials for energy applications.

DOI: 10.4018/979-8-3373-6058-4.ch004

1. INTRODUCTION

In 1839, Alexàndre Edmond Becquerel (E. Becquerel, 1839) made the observation of the first photovoltaic effect, thereby demonstrating that electrodes composed of platinum and gold, when immersed in an electrolyte, were capable of generating electricity when exposed to sunlight. The first solar cell was fabricated in 1883 by Charles Fritts, through the addition of a thin layer of gold to selenium. In 1905, Albert Einstein provided a theoretical realisation of the photoelectric effect, based on quantum theory. It was posited that light consists of photons, which can excite electrons into higher energy states, resulting in the generation of an electric current.

Russell Ohl developed the first silicon-based solar cell (SSC) in 1941, though it had only a power conversion efficiency (PCE) of around 1%. Later in 1954, Bell Laboratories have increased the efficiency of SSCs to approximately 6%, thereby marking a significant milestone in the field. Initially, their utilisation is restricted to space missions due to the high production cost; however, the oil crisis in the 1970s led to a decline in solar cell technology prices, facilitating wider adoption of solar cell technology. This spurred research and industrial development aimed at expanding the use of renewable energy and reducing environment pollution.

Photovoltaic technology is generally classified as either inorganic or organic solar cells. Wafer-based designs, comprised of gallium arsenide (GaAs) and crystalline silicon (Si), are representative of first-generation cells. Second-generation cells feature cadmium telluride (CdTe) and copper indium gallium selenide (CIGS), and amorphous silicon, while third-generation cells include organic, perovskite, and quantum dot (QD) solar cells. Second- and third-generation solar cell devices are both derived from thin-film technology. SSCs are the most widely used inorganic type and can exceed 25% efficiency. Gallium arsenide cells currently have the highest efficiency among single-junction devices at nearly 30%, while CdTe and CIGS cells have efficiencies of around 23% and 22%, respectively (*Best Research-Cell Efficiency Chart | Photovoltaic Research | NREL*, n.d.).

Third-generation cells encompass a range of photovoltaic technologies, including dye-sensitised solar cells (DSSCs), perovskite solar cells, QD solar cells and organic photovoltaic cells (OPVs). DSSCs and OPVs typically achieve efficiencies near 12%. Since their introduction in 2009 (Kojima et al., 2009), perovskite solar cells have shown remarkable progress, attaining a record efficiency of 25.2% in just five years. QD solar cells have been shown to possess analogous device structures, and are undergoing parallel development with other emerging technologies (*Best Research-Cell Efficiency Chart | Photovoltaic Research | NREL*, n.d.).

This chapter starts with a delineation of the fundamental physical principles that govern semiconductors and solar cell technology. The subsequent section introduces Density Functional Theory (DFT), which is a sophisticated computational method employed to study solid materials such as perovskites. This section provides a discussion on the background behind DFT, exchange-correlation functionals, including LDA, GGA, and meta-GGA functionals. DFT is utilised to determine the ground-state geometries and material characteristics through calculations of total energy, forces, and electronic structures. Such calculations may include band structure, density of states (DOS), elastic, and optical properties. The chapter provides a comprehensive overview of the role of perovskites in photovoltaic technologies, and an analysis of the challenges faced in this field, with a particular focus on the development of lead-free absorber materials. The study provides insights into accelerating research and fostering innovation in emerging photovoltaic materials.

2. GENERALITIES ON PEROVSKITES

The perovskite structure, named after the mineralogist L. A. Perovksi, is a well-established crystalline framework commonly encountered in oxides such as $CaTiO_3$. Due to their broad range of physical properties, including ferroelectricity ($BaTiO_3$), antiferroelectricity ($PbZrO_3$), ferroelasticity ($SrTiO_3$), antiferromagnetism ($LaTiO_3$), and ferromagnetism ($YTiO_3$), oxide perovskites have long attracted attention. A material that exhibits more than one of these properties simultaneously is classified as multiferroic.

Figure 1. Solar cell efficiency of different technologies adopted from the National Renewable Energy Laboratory (NREL) up to 2025 (Best Research-Cell Efficiency Chart | Photovoltaic Research | NREL, n.d.).

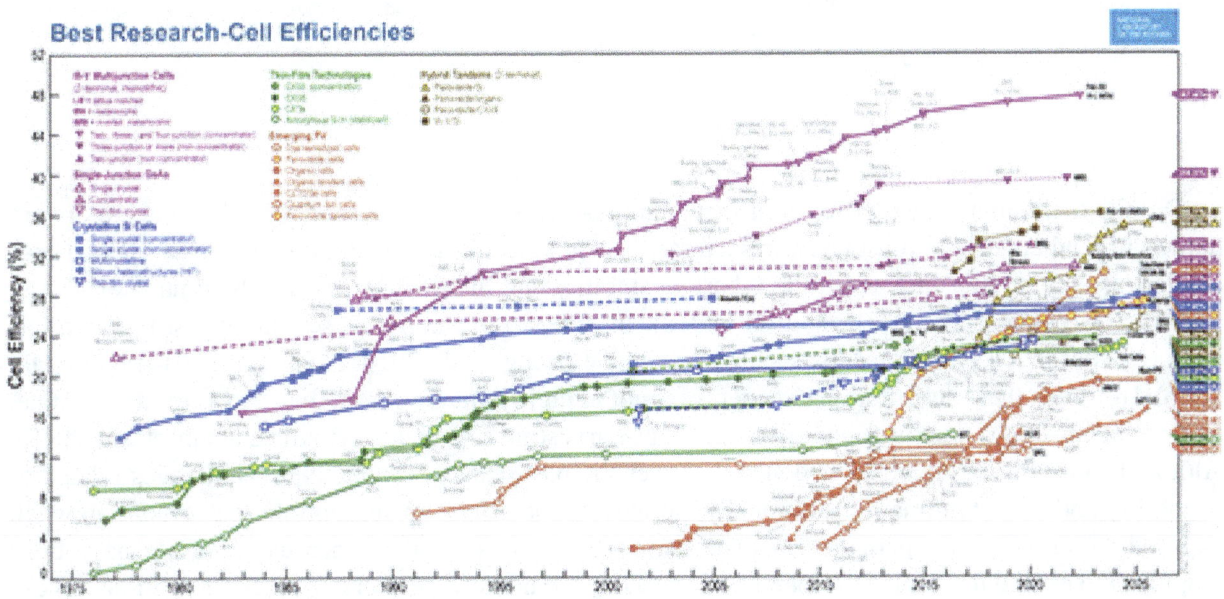

The general chemical formula of perovskites is ABX_3 (see Figure 2 (a)), where A and B are cations, and X is an anion. Structural stability requires that the B-site cation be smaller than the A-site cation. In the ideal cubic perovskite structure, each B cation is surrounded by six anions in an octahedral coordination, while the A cation occupies a 12-fold coordinated cuboctahedral site formed by adjacent octahedra.

Møller first identified the semiconducting behaviour of halide perovskites in 1959 when he observed photoconductivity in caesium lead halides ($CsPbX_3$). In 1978, Weber pioneered hybrid organic–inorganic perovskites by substituting the A-site cation with the organic methylammonium ion ($CH_3NH_3^+$, or MA^+) and successfully synthesising $CH_3NH_3PbX_3$ (X = Cl, Br, or I). Although Weber reported on their optical properties and colour variations depending on the halide composition, it was not until 2009 that Miyasaka and his colleagues used $CH_3NH_3PbI_3$ as a light absorber in DSSCs, achieving an efficiency of about 3.8% (Kojima et al., 2009). This marked the first use of hybrid perovskites in photovoltaic devices. Since their discovery, the efficiency has increased drastically to reach 27% (see Figure 1) (*Best Research-Cell Efficiency Chart | Photovoltaic Research | NREL*, n.d.).

Figure 2. Unit cell of traditional (a) and double (b) perovskites.

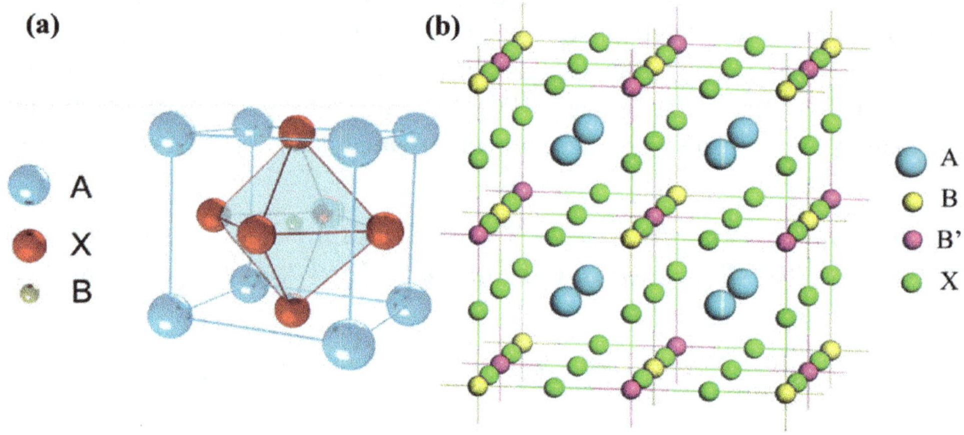

Despite the demonstrated efficacy of lead halide perovskites in solar energy applications, there are numerous challenges that must be addressed for their practical implementation on a large scale. These challenges include moisture sensitivity, thermal stability and the toxicity of lead (Kojima et al., 2009; Shahbazi & Wang, 2016; P. Zhang et al., 2018). In order to overcome these challenges, research has explored alternative IVA elements, such as Sn and Ge (Leng et al., 2016; Noel et al., 2014; Sun et al., 2016). However, these substitutions frequently result in rapid oxidation from the 2^+ to 4^+ oxidation state, compromising both structural integrity and light-harvesting efficiency (Roknuzzaman et al., 2018). Although substantial research was conducted over the past decade, fundamental challenges related to instability and toxicity remain. Additionally, perovskite materials are susceptible to degradation when exposed to humidity, temperature fluctuations, or continuous light. Numerous strategies have been employed to extend the device lifespan from a few days to nearly a year in laboratory conditions. These approaches include compositional adjustments, the incorporation of inorganic charge transport layers, and encapsulation techniques. However, the majority of these devices encounter difficulties in sustaining consistent performance under real-world conditions.

Concerns regarding the toxicity of lead have prompted international initiatives to develop lead-free alternatives, despite only a limited number of these have attained a close PCE to lead-based perovskite solar cells (PSCs). A particularly promising avenue for further research is the design of halide double perovskites (DP), where the Pb^{2+} ions are substituted by trivalent and monovalent metal cations. This substitution results in double perovskites with the general formula $A_2BB'X_6$, as illustrated in Figure 2 (b) (McClure et al., 2016; Slavney et al., 2016). Such materials offer a potential pathway for environmentally sustainable and structurally stable photovoltaic absorbers.

In the structural framework under consideration, X is a halogen anion (F^-, Cl^-, Br^-, or I^-). The A-site cation can be an organic molecule as in $MA_2BiM'Cl_6$ (M'=Cu, K, Ag, Tl) (Deng et al., 2016; Wei et al., 2016), or formamidinium (FA^+), such as $FASnX_3$ (X=Br, I) (Hao et al., 2014; Noel et al., 2014). Additionally, A may be an alkali metal cation, such as rubidium (Rb^+) and caesium (Cs^+). The B-site is occupied by an alkali metal ion (K^+ or Na^+, etc), or a monovalent transition metal ion (Cu^+ or Ag^+, etc). B' corresponds to a trivalent transition metal cation, such as Cr^{3+}, Bi^{3+}, and In^{3+}.

significant progress has been made in the field of perovskite solar cells (PSCs) through various approaches, including surface passivation (Hwang et al., 2015; Jiang et al., 2024), tandem architectures, and band gap engineering via anion and cation substitution (Miah et al., 2024), as well as the application of external pressure (Mursaleen et al., 2025). These strategies have paved the way for high-efficiency, next-generation photovoltaic technologies. Consequently, numerous lead-free halide DPs have been successfully synthesised, demonstrating excellent structural stability and highly promising optical properties (Filip et al., 2016; McClure et al., 2016).

Various compounds, however, have indirect band gaps, which limits their usefulness in photovoltaics because thicker absorber layers are needed to capture enough light. For example, $Cs_2AgBiBr_6$ and $Cs_2AgBiCl_6$ have indirect band gaps of 2.19 eV and 2.77 eV, respectively (McClure et al., 2016). Although these values lie in the visible range, they are still not ideal for efficient solar cell technology. In these materials, the valence and conduction band edges are mainly formed from the electronic states of Ag, Bi and the halide ions. Carefully substituting these elements could therefore help convert the band gap from indirect to direct while also reducing its magnitude. This would improve light absorption and exciton generation, and, ultimately, boost overall solar cell performance.

Noor et al. (Noor et al., 2021) carried out a theoretical investigation of A_2ScInI_6 (A = Cs, Rb) and found that it has direct band gaps, as well as good structural, mechanical, and thermodynamic stability in the cubic phase. Even so, the fact that both In and Sc are present in trivalent states is likely to make experimental realisation difficult. Attention then shifted to Cu-Sb-based DPs of the type A_2CuSbX_6 (A = K, Rb, Cs; X = Cl, Br, I), which were reported to have attractive mechanical, elastic and thermodynamic properties (T. Y. Tang & Tang, 2023). All of these compounds, with the exception of Cs_2CuSbI_6, are both mechanically and thermodynamically stable, and the chloride-based members are more stable than their bromide and iodide variants. Most of this family also behaves in a ductile manner except Cs_2CuSbI_6 making them promising candidates for flexible and wearable devices. However, the indirect band gaps of these materials, which span from 0.30 eV (Cs_2CuSbI_6) to 1.163 eV ($Cs_2CuSbCl_6$), are not ideal for photovoltaic use. Among them, $K_2CuSbCl_6$ stands out, showing higher thermal stability, and better optoelectronic performance, including stronger optical absorption, lower carrier effective masses and longer carrier lifetimes.

A series of studies has highlighted the potential and limitations of several DP families. Tl-based systems, such as Cs_2TlSbX_6 (X= Cl, Br, I) and Cs_2TlYF_6 (Y=Ag, Co), exhibit ductile mechanical behaviour, direct band gaps within the range of 1.17-3.55 eV, and low carrier effective masses. These properties render these materials promising candidates for optoelectronic and photovoltaic applications. Cs_2TlAgF_6, in particular, is more suitable for solar cells, whereas its cobalt analogue exhibits superior thermoelectric performance, with a figure of merit (ZT) value of 1.46 at 300 K in comparison to 0.74 for the silver variant. However, the sever toxicity and environmental impact of thallium significantly limit the practical application of these materials (Alshomrany et al., 2025; T. Tang et al., 2022). Consequently, alternative indium-based DPs have been investigated. For instance, Cs_2InSbF_6 was found to have a direct band gap of 2.3 eV, and has shown both mechanical robustness and dynamical stability (Yaseen et al., 2025). Further research on Cs_2InCoX_6 (X=F, Cl. Br) has demonstrated direct band gaps of 0.865, 1.129, and 3 eV, respectively, in to addition to structural, dynamic and thermodynamic stability (T. Y. Tang et al., 2022). It is noteworthy that the absorption profile of $Cs_2InCoCl_6$ closely matches the AM1.5G solar spectrum (ASTM G173–03, 2012), thereby suggesting its potential for solar energy conversion. Nonetheless, the use of indium in the rarely stable In^+ instead of its trivalent state (In^{3+}) raises concerns regarding experimental feasibility and charge balance, similar to issues previously identified in Cs_2ScInI_6 (Noor et al., 2021).

In parallel, double perovskites based on bromine (Br) and rubidium (Rb) have attracted attention. It has been predicted that $Rb_2AgGaBr_6$ will demonstrate attractive optoelectronic properties and achieve a maximum photovoltaic efficiency of 32.2% within the spectroscopic limited maximum efficiency (Kibbou et al., 2023; Yu & Zunger, 2012). However, the use of silver substantially increases fabrication costs. To address this issue, the recently proposed $Rb_2NaInBr_6$ compound (Ayyaz et al., 2023; Oublal, Dib, Khalal, et al., 2025) replaces Ag with Na. This compound has a direct band gap ranging from 2.88 eV to 3.2 eV, strong visible-ultraviolet absorption, favourable mechanical, and thermodynamic stability. Nevertheless, its relatively wide band gap means that absorption is concentrated mainly in the UV, indicating that further optimisation is required for it to be fully exploited in photovoltaic devices.

A recent theoretical study of $Cs_2NaInCl_6$ reported direct band gaps of 4.16 eV and 4.74 eV using the mBJ and HSE06 functionals, respectively, alongside mechanically brittle behaviour (Shakeel et al., 2024). The synthesis of $Cs_2NaInCl_6$ has been previously accomplished by solution-based methods. High-quality single crystals were obtained through the process of hydrothermal crystallisation, whereby CsCl, $InCl_3$, and NaCl were dissolved in concentrated HCl, and the resultant mixture was subsequently heated in a sealed autoclave at 180°C for a period of 12 hours (Mopoung et al., 2024). In this approach, 1 mmol of CsCl, 0.5 mmol of NaCl, and 0.5 mmol of $InCl_3$ were dissolved in 4 mL of 37% HCl. The solution was then autoclaved at 180 °C for 12 hours, before being gradually cooled to produce transparent $Cs_2NaInCl_6$ crystals. Alternatively, Zhang et al. used a solvothermal method to prepare Mn-dope $Cs_2NaInCl_6$, incorporating $MnCl_2$ as a dopant (L. Zhang et al., 2021). Furthermore, low-temperature chemical reflux (boiling solution at approximately 100-105°C) has been utilised in a comparable manner for analogous double perovskites. Jeevaraj et al. documented the synthesis of $Cs_2NaInCl_6$ and Mn^{2+}:$Cs_2NaInCl_6$ double perovskite microcrystals by means of chemical reflux at 105°C under ambient pressure (Jeevaraj et al., 2024)(Jeevaraj et al., 2024). The compound's comparatively wide band gap results in its primary absorption in the ultraviolet range, thus rendering it unsuitable for utilisation in single-junction solar cells. Nevertheless, such materials could still find applications in multi-junction architectures and UV optoelectronic devices. It is evident that $Cs_2NaInBr_6$ demonstrates superior properties, exhibiting a reduced direct band gap of approximately 2.7 eV and favourable mechanical and optoelectronic characteristics. By targeting atomic substitution, such as replacing Cl with Br and I to adjust the band gaps, visible light absorption can be enhanced, increasing the density of photogenerated carriers and improving photovoltaic performance (Oublal, Dib, & El-yakoubi, 2025; Saeed et al., 2020).

3. GENERALITIES ON PHYSICAL PROPERTIES

3.1. Geometry Optimisation

Geometric optimisation is a pivotal phase in computational chemistry and mate- rials science, as it determines the most stable atomic configuration of a system. It is essential to achieve a state of complete relaxation of all atomic position, thereby minimising the total energy to the lowest possible level consistent with the equilibrium geometry. Within DFT, this objective is realised through an iterative calculation of the forces exerted on individual atoms, accompanied by the adjustment of their positions until the forces attain a negligible magnitude. This final state corresponds to the ground state energy. The forces acting on an atom α at position R_α is defined by the Hellmann-Feynman theorem (Feynman, 1939):

$$F_\alpha = -\nabla_{R_\alpha} E_{tot} \tag{1}$$

This theoretical framework enables the derivation of forces directly from the total energy of the system. The popularity of DFT-based optimisation can be attributed to its ability to efficiently and accurately describe electron density and exchange-correlation effects. The geometries obtained through DFT optimisation form the basis for subsequent investigations, such as electronic structure, mechanical, and optical property calculations. These investigations provide essential insights into material stability, bonding, and reactivity.

3.2. Structural, Mechanical and Thermodynamic Properties

Lead-free halide DPs crystalise in a face cantered cubic structure with Fm$\bar{3}$m symmetry (space group No. 225). In this structure, the atoms occupy the following Wyckoff positions: A at 8c (0.25 0.25 0.25); B at 4b (0.5 0.5 0.5); B at 4a (0 0 0); and X at 24e (0.25 0 0). The structural stability and formability of such compounds can be examined using Goldschmidt's tolerance (τ) and octahedral (μ) factors, defined by the expressions (Goldschmidt, 1926; Li et al., 2004):

$$\tau = \frac{(r_A + r_X)}{\left[\sqrt{2}(r_{M,M'} + r_x)\right]} \tag{2}$$

$$\mu = \frac{r_{M,M'}}{r_X} \tag{3}$$

Here, r_A and r_X represent the ionic radii of A and X ions, respectively. Meanwhile, $r_{B,B'}$ is the average ionic radius of B and B' cations. For a stable DP structure, τ typically falls within the range 0.81-1.11 and μ within the range 0.41-0.90 (Zhao et al., 2017). These geometric parameters are critical indicators of perovskite lattice's potential to form without significant distortion.

The elastic properties of materials are determined by their response to external forces, which renders these materials suitable for specific engineering applications. It is necessary to note that the following properties are of great importance: elasticity, ductility, strength, hardness, and toughness. Elasticity facilitates the restoration of shape following deformation, while ductility plays a role in preventing fractures. Strength and hardness describe the resistance to fracture and wear, and toughness represents the capacity to absorb energy before fracture, i.e., the materials resistance to mechanical failure. These characteristics are important for selecting materials for components that will be subject to mechanical stress.

Young's modulus (Y) and shear modulus (G) are examples of elastic constants that measure a material's tendency to deform. For materials that necessitate shaping or formation, ductility emerges as an indispensable characteristic. Conversely, brittle materials are more prone to sudden and unexpected failure, exhibiting minimal deformation prior to fracture. It is important that materials exhibit adequate toughness and fatigue resistance she subjected to variable or elevated levels of stress. The durability of such materials depends on mitigating crack propagation and failure under repeated stresses. A comprehensive understanding of these properties is essential for designing materials, predicting their behaviour and ensuring safety and reliability of these devices.

The determination of elastic constants for materials with cubic phase can be accomplished through the use of DFT as the foundational approach. The utilisation of computational tools, such as IRELAST in WIEN2k (Jamal et al., 2018), has been instrumental in facilitating this procedure. This method is predicated on the stress-strain relationship, whereby minor deformations are applied to the crystal lattice, resulting in variations in its volume and total energy. The elastic constants are obtained using the second derivative of total energy with respect to strain. The elastic constants are expressed as follows:

$$C_{ij} = \frac{1}{V_0}\left[\frac{\partial^2 E_{tot}}{\partial\varepsilon_i \partial\varepsilon_j}\right] \tag{4}$$

Here, ε_i and ε_j are the strain components. Cubic crystals exhibit three independent elastic constants (C_{11}, C_{12}, and C_{44}), and are arranged in the following matrix:

$$C_{ij} = \begin{pmatrix} C_{11} & C_{12} & C_{12} & 0 & 0 & 0 \\ C_{12} & C_{11} & C_{12} & 0 & 0 & 0 \\ C_{12} & C_{12} & C_{11} & 0 & 0 & 0 \\ 0 & 0 & 0 & C_{44} & 0 & 0 \\ 0 & 0 & 0 & 0 & C_{44} & 0 \\ 0 & 0 & 0 & 0 & 0 & C_{44} \end{pmatrix} \tag{5}$$

Here, C_{11} is indicative of longitudinal stress resistance, C_{12} represents transverse deformation under longitudinal loading, and corresponds to shear deformation resistance. The born-Huang stability criteria (Born et al., 1954) can be used to ascertain the mechanical stability of cubic crystals. The following criterial must be met: $(C_{11} - C_{12}) > 0, C_{11} > 0, (C_{11} + 2C_{12}) > 0, C_{44} > 0$ and $C_{11} > B > C_{12}$, where B is the bulk modulus, which is a measurement of the material's resistance to uniform compression. This quantity can be determined using the Birch-Murnaghan equation of state (Birch, 1947; Murnaghan, 1944) or through the elastic constants:

$$E_{tot}\left(V\right) = E_0 + \frac{9B.V_0}{16}\left\{\left[\left(\frac{V_0}{V}\right)^{2/3} - 1\right]^3 B' + \left[\left(\frac{V_0}{V}\right)^{2/3} - 1\right]^2\left[6 - 4\left(\frac{V_0}{V}\right)^{2/3}\right]\right\} \tag{6}$$

$$B = \frac{C_{11} + 2C_{12}}{3} \tag{7}$$

Where, E_0 and V_0 are the ground-state energy equilibrium volume, respectively, and B' is the pressure derivative of B.

The stiffness of a solid material can be quantified by analysing its shear modulus (G), which is defined as the resistance of the material to shape distortion under the application of stress. G is calculated using the Voigt (G_V) and Reuss (G_R) moduli:

$$G = \frac{G_V + G_R}{2} \tag{8}$$

G_V and G_R are calculated using the expressions:

$$G_V = \frac{C_{11} - C_{12} + 3C_{44}}{5} \quad (9)$$

$$G_R = \frac{5C_{44}(C_{11} - C_{12})}{3(C_{11} - C_{12}) + 4C_{44}} \quad (10)$$

The ductile or brittle nature of a solid can be defined by the Pugh's ratio (*B/G*). A ratio greater than 1.75 is indicative of ductility, defined as the ability of the material to deform plastically before failure. Conversely, a ratio below 1.75 signifies brittleness, characterised by the material's propensity to fracture with minimal plastic deformation.

Another significant mechanical parameter is the Young's modulus (*Y*), which measures the material's rigidity and its capacity to resist elastic deformation when a stress is exerted. The relationship between stress and strain in the elastic region is represented by the following equation, which relates to shear modulus (*G*) and bulk modulus (*B*):

$$Y = \frac{9BG}{3B + G} \quad (11)$$

In the event of the application of a tensile or compressive load to a material, the changes in axial and lateral strain are related to each other through the Poisson ratio (ν). This constant is of paramount importance in quantifying the deformation behaviour of materials and mechanical stability under load. In general, materials are categorised as either ductile ($0.25 < \nu < 0.5$) or brittle ($\nu < 0.25$). The following equation is provided for this purpose:

$$\nu = \frac{3B - 2G}{2(3B + G)} \quad (12)$$

Thermodynamic stability of solid materials can be assessed by the analysis formation energy ΔH_f, determined by the expression:

$$\Delta H_f = E_{tot}(Cs_2NaInX_6) - (2E_{Cs} + E_{Na} + E_{In} + 6E_X) \quad (13)$$

$E_{tot}(Cs_2NaInX_6)$ denotes the total ground-state energy of the compound Cs_2NaInX_6 (X = Br, I). The symbols E_{Cs}, E_{Na}, E_{In} and E_X represent the total energies of the constituent atoms. A positive value of ΔH_f is indicative of a thermodynamically unstable substance, whilst a negative value is suggestive of stability. A negative sign, on the other hand, indicates the energy released during the compound's formation.

The Debye temperature (θ_D) is defined as the characteristic temperature of the vibrational behaviour of atoms in a crystal lattice. It is important to elucidate the manner in which phonons contribute to vibrational entropy, in addition to several pivotal thermodynamic properties, including thermal expansion, heat capacity and thermal conductivity (Qin et al., 2020). From a physical perspective, it marks the temperature at which atoms in the crystal reach their highest normal vibrational mode (Hattabi et al., 2019). Below this threshold, quantum effects largely govern atomic motion, whereas at higher temperatures the behaviour of the lattice is well described by classical physics.

In addition to reflecting the extent of lattice vibrations, the Debye temperature, θ_D, provides insight into the strength of chemical bonding and the mechanical rigidity of materials. An elevated Debye temperature is often associated with stronger covalent bonding, great lattice stiffness, and heightened material hardness (Bao et al., 2018; Kim & Suh, 2017; Yang et al., 2019).

The melting temperature (T_m) is defined as the temperature at which a solid changes into a liquid under standard pressure. This phenomenon is indicative of the strength of the bonding between atoms in a crystal structure, with materials exhibiting stronger interatomic forces having higher melting points. The sound velocities are indicative of the velocity at which elastic waves propagate through a crystal. There are three primary categories of sound velocities: longitudinal (v_l), transverse (v_t), and average (v_a).

$$v_l = \left(\left(B + \tfrac{4}{3}G\right)/\rho\right)^{1/2},\ v_t = (G/\rho)^{1/2},\ v_a = \left[\frac{1}{3}\left(\frac{2}{v_t^3} + \frac{1}{v_l^3}\right)\right]^{-1/3} \tag{14}$$

$$\theta_D = \frac{\hbar}{k_B}\left[\frac{3nN_A\rho}{4\pi M}\right]^{1/3} v_a \tag{15}$$

$$T_m = 533 + 5.91\,C_{11} \tag{16}$$

The extraction of Debye temperature and sound velocities is achieved through the implementation of Anderson's method, as outlined in Equations (14) and (15) (Anderson, 1963). Equation (16) was utilised to ascertain the melting temperature, with the C_{11} elastic constant being the dependent variable (Hussain et al., 2025).

3.3. Electronic Properties

According to the principles of quantum physics, electrons in an isolated atom are capable of occupying only specific, discrete energy levels. Each of these energy levels is associated with a distinct quantum state. It is evident that these levels are exclusive to each element, and are the result of the quantisation of electronic energy.

When a large number of similar atoms bond together to form a solid, their atomic orbitals overlap and interact. This interaction produces discrete energy levels in individual atoms, which subsequently convert into a large variety of closely spaced states. In a crystal with a large number of atoms, these states combine to generate continuous energy bands. The electronic band structure of a crystal is the result of the hybridisation of the atomic energy levels of its constituent atoms.

Within this structure, allowed energy bands are defined as the regions within which electrons are capable of existing, whereas forbidden bands (or band gaps) signify energy ranges devoid of electronic states. In the context of these bands, two are of particular significance: the valence band (VB) and the conduction band (CB). The VB is defined as the highest energy range occupied by electrons, while CB is the next higher (and unoccupied) energy range. The energy difference between these two bands defines the band gap (E_g), a critical parameter distinguishing conductors, semiconductors, and insulators. Semiconductors possess a band gap in the range of 0.1-3 eV, which renders them particularly well-suited for photovoltaic applications due to their capacity for efficient absorption of visible light within this spectrum. The Fermi level, delineated by the separation between the occupied and unoccupied states at 0 K, signifies the extent of electron occupation.

In the case of a photon with an energy equal to or higher than the band gap (E_g), an electron in the VB is excited to the CB, resulting in the creation of a hole in the VB. Consequently, the process of formation of an electron-hole pair (or exciton) is initiated. In the event that the photon energy is lower than E_g, excitation does not occur (see Figure 3 (a), case i). In the event that the energy is greater than E_g, the electron is promoted to a higher energy state within the CB (Figure 3 (a), case ii) and then rapidly relaxes to the conduction band minimum through thermalisation (Figure 3 (a), case iii). This is an ultrafast process that occurs within femtoseconds (fs). The absence of electronic states within the band gap results in a delay in recombination, thereby extending the survival time of the charge carriers from microseconds to milliseconds. This extended duration is sufficient to enable the collection of the charge carriers within a solar cell.

The excitation under discussion may be generated by a number of processes. Firstly, it may be generated by the absorption of a photon. Secondly, it may be generated by phonons (i.e. lattice vibrations). Thirdly, it may be generated by the transfer of energy from another carrier. The generation, recombination, and extraction of photogenerated charge carriers are fundamental to the operation of solar energy systems based on semiconductors.

Figure 3. (a) Energy band diagrams of Insulators, semiconductors and metals, (b) and illustration of direct and indirect band gaps.

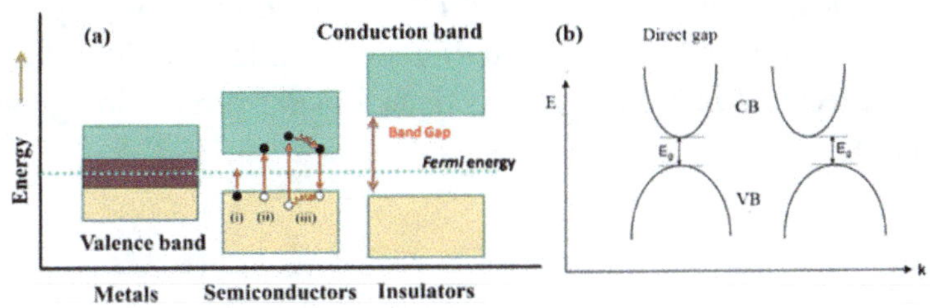

A photon with sufficient energy ($E = h\nu$) can extract an electron from the VB and excite it to the CB, resulting in an electron-hole pair. According to the definitions outlined, an electron in the CB has the capacity to release energy and subsequently recombine with a hole in the VB, thereby generating a photon.

The nature of the electronic transition is determined by the momentum (k-vector) of the electrons. The occurrence of the VB maximum (VBM) and CB minimum (CBM) at the same k-point within the Brillouin zone indicates the material's direct band gap. Optical transitions (absorption and emission) occur without the assistance of phonon, which have been shown to be more efficient.

In contrast, indirect band gap semiconductors exhibit the VBM and CBM at different k-points (see Figure 3 (b)). The conservation of momentum is only possible when a photon is associated with a phonon, i.e. a quantum of lattice vibration. Excitation occurs when an electron is provided with energy by a photon and a phonon prior to reaching the CB. Recombination requires the simultaneous emission of a photon and a phonon in order to maintain momentum conservation.

Indirect band gap materials exhibit a reduced number of optical transitions and a lower absorption coefficient in comparison to direct band gap semiconductors. This property may enable enhanced light absorption and emission, thereby increasing their performance in optoelectronic and photovoltaic applications (Nelson, 2003; Schmidt-Mende et al., 2022).

3.4. Optical Properties

The optical properties of solids are of critical importance in the probing of electronic structures, including energy bands, impurity levels, excitons, localised defect levels, and light-matter interactions. various optical properties for materials, including the dielectric function $\varepsilon(\omega)$, reflectivity $R(\omega)$, absorption $\alpha(\omega)$, and optical conductivity $\sigma(\omega)$, can be studied.

3.4.1. Dielectric Function

The optical properties of a material are typically expressed in terms of the real ($\varepsilon_1(\omega)$) and imaginary ($\varepsilon_2(\omega)$) parts of the complex dielectric function ($\varepsilon(\omega) = \varepsilon_1(\omega) + i\varepsilon_2(\omega)$) (Ehrenreich & Cohen, 1959). The real part describes the phenomenon of photon dispersion. In contrast, the imaginary part is associated with the process of energy absorption within the material. The imaginary part of the dielectric function is derived from the formula:

$$\varepsilon_2\left(\omega\right) = \frac{4\pi^2 e^2}{\omega^2}\sum_{vc}\sum_{k}\left|\int d^3 r \Psi^*_{c,k}(r)(\hat{e}.v)\Psi_{v,k}\left(r\right)\right|^2 \delta\left[E_c(k) - E_v(k) - \hbar\omega\right] \quad (17)$$

Here, ω is the frequency of the incident photon. The wave functions of the valence and conduction band quasiparticles (electrons) of energies $E_v(k)$ and $E_c(k)$, respectively, are represented by $\Psi_{v,k}$ and $\Psi_{c,k}$. $\hat{e}$ is the unit vector which specify the polarisation direction, while v denotes the velocity operator.

The real part of the dielectric function, $\varepsilon_1(\omega)$, is determined from the imaginary part, $\varepsilon_2(\omega)$, using the Kramer-Kronig relation.

$$\varepsilon_1\left(\omega\right) = 1 + \frac{2}{\pi}P\int_0^\infty \frac{\omega'\varepsilon_2(\omega')}{(\omega'^2 - \omega^2)}d\omega' \quad (18)$$

Here, P represents the principal integral value.

3.4.2. Optical Absorption

The light absorption coefficient, denoted by $\alpha(\omega)$, is a material property that quantifies the degree to which a material absorbs electromagnetic radiation. The determination of $\alpha(\omega)$ is achieved through the following relation:

$$\alpha\left(\omega\right) = \frac{\sqrt{2}\omega}{c}\left[\sqrt{\varepsilon_1^2(\omega) + \varepsilon_2^2(\omega)} - \varepsilon_1\left(\omega\right)\right]^{1/2} \quad (19)$$

Where, c refers to the speed of light in a vacuum. The absorption is influenced by a number of factors, including the electronic band structure, impurity levels, and localised defects within the material. It provides insights into the material's electronic transitions and interactions with incident electromagnetic radiation.

3.4.3. Optical Conductivity

Optical conductivity has been demonstrated to be a reliable indicator of photoconductivity, thus offering significant insights into electrical conductivity. The following formula represents the real part of optical conductivity, denoted by $\sigma(\omega)$:

$$\sigma\left(\omega\right) = \frac{\omega\, \varepsilon_2(\omega)}{4\pi} \tag{20}$$

The optical conductivity of a material is defined as its response to an external electromagnetic field, particularly with regard to light absorption and photon-electron interaction. The material's free carrier dynamics are closely linked to the material's ability to conduct under light exposure, and this ability is critical in understanding the photoconductive properties of the material and its applications in opto-electronic devices.

3.5. Photovoltaic Properties

A solar cell is a device that converts part of the incident photon energy into electrical power. The efficiency of a semiconductor absorber is determined by two key parameters: its band gap energy (E_g) and the spectrum of incoming light. The power conversion efficiency (PCE) of photovoltaic devices is contingent upon three primary factors: the absorption coefficient, the charge separation efficiency, and the extent of carrier recombination losses.

The theoretical upper limit of efficiency for a single-junction solar cell is defined by the Shockley-Queisser limit, which establishes the maximum achievable efficiency as a function of the absorber's band gap. Under standard AM1.5G illumination (representing sunlight filtered through 1.5 atmospheric layers at a zenith angle of 48.2°), this limit is approximately 33% for materials with band gaps between 1.0 and 1.5 eV at room temperature (Shockley & Queisser, 1961).

Figure 4. (a) Equivalent circuit of an Ideal solar cell and its and its J-V curve.

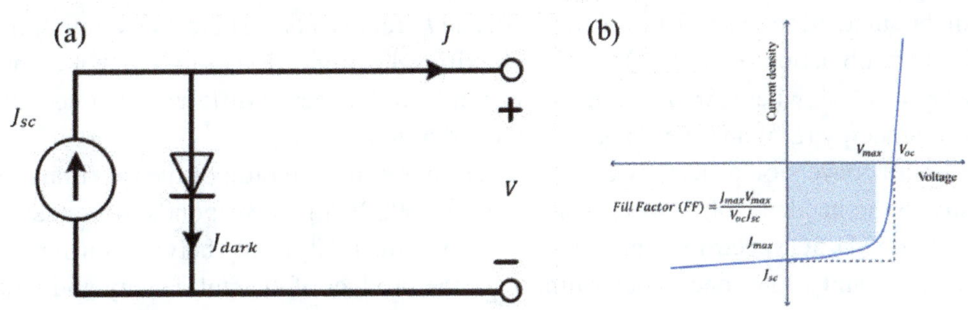

The electrical circuit of an ideal solar cell, as illustrated in Figure 4 (a), consists of a diode and a current source. According to the Shockley model, the total current is the result of the sum of a dark diode current (J_{dark}).

The operation of the device is governed by the diode equation, and a voltage-independent photocurrent (J_{sc}) is generated under illumination. The photocurrent is distributed between the diode's internal resistance and the external load, with the diode providing the voltage required to facilitate the current's flow through the circuit. The current–voltage (J-V) characteristics of the device, both in the dark and under illumination, are shown in Figure 4 (b).

The efficiency of semiconductors can be predicted using theoretical models, such as the Spectroscopic Limited Maximum Efficiency (SLME) model (Yu & Zunger, 2012). This model considers the absorption coefficient (α), the energy band gap, and the absorber thickness (L). Standard irradiation, such as AM1.5G, can be used (ASTM G173–03, 2012). The current density (J) of this ideal solar cell is expressed as follows:

$$J = J_{sc} - J_0\left(e^{\frac{qV}{k_B T_c}} - 1\right) \tag{21}$$

Here, q is the elementary charge, k_B is the Boltzmann constant, J_{sc} and J_0 represent the short-circuit current density and the current recombination rate, respectively. This model only considers radiative recombination processes. However, non-radiative recombination mechanisms, which contribute to additional dark current density and consequently lower the open-circuit voltage, are neglected (Morales-Acevedo, 2023). J_{sc} and J_0 are represented by the expression:

$$J_{sc} = q\int_0^{\infty} a(E) I_{sun}(E) dE \tag{22}$$

$$J_0 = \frac{q\pi}{f_r}\int_0^{\infty} a(E) I_{bb}(E) dE \tag{23}$$

Here, $a(E)$ is the absorptivity of the absorber, and is described as a function of the thin-film thickness: $a(E) = 1 - e^{-f_g \alpha(E) L}$, where f_g represents the geometry that accounts for the solar cell's optical configuration. In this study, $f_g = 1$ is assumed (an idealised configuration with a perfectly reflecting rear surface that eliminates photon losses through the back of the device). In the Shockley-Queisser (SQ) limit, absorptivity is simplified as $a(E) = 1$ for photon energies $E \geq E_g$ and $a(E) = 0$ for energies smaller than the band gap, where E_g denotes the material's fundamental band gap. I_{sun} represents the photon flux density from the standard solar spectrum (AM1.5G), and I_{bb} represents the blackbody emission spectrum. The radiative recombination fraction (f_r), which quantifies the proportion of radiative recombination, is expressed as $f_r = e^{\Delta/k_B T}$, where $\Delta = E^{da} - E_g$ corresponds to the energy difference between the allowed direct transition energy (E^{da}) and the fundamental band gap (E_g).

The maximum conversion efficiency of a solar cell represents the ratio of the maximum extractable power density to the incident solar power density. $P_{in} = 1000$ W/m^2 corresponds to standard solar irradiance under the AM1.5G spectrum. The J–V curve shows the relationship between current and voltage under illumination and allows one to determine P_{max}, the product of current density and voltage at the point of maximum power output.

The fill factor (FF), an essential performance indicator that describes the "squareness" of the J–V curve, is calculated using the expression $FF = P_{max}/\left(V_{oc}J_{sc}\right)$ (Green, 1981).

Open-circuit voltage (V_{oc}), representing the maximum voltage generated by the solar cell when the current density is zero, is obtained from the formula:

$$V_{oc} = \frac{k_B T_c}{q} \ln\left(1 + \frac{J_{sc}}{J_0}\right) \tag{24}$$

The performance of solar cells can be evaluated using simulation tools such as the one-dimensional Solar Capacitance Simulator (SCAPS-1D) (Burgelman et al., 2000) for simulated perovskite solar cells. The p-i-n junction can be modulated by placing the perovskite material between a hole transport layer (HTL), and an electron transport layer (ETL). This architecture is written as Substrate/ETL/ $A_2BB'X_6$/ HTL/back-contact. The AM1.5G solar spectrum first strikes the substrate, initiating photovoltaic process. SCAPS-1D considers several key material and device parameters that govern the solar cell's electrical and optical behaviour. These include the band gap energy, absorber layer thickness, and electron affinity (χ). Additionally, the effective density of states in the valence and conduction bands (N_V and N_C, respectively) are considered, as well as the electron and hole mobilities (μ_n and μ_p, respectively). The thermal velocities of charge carriers (v_{th}^n for electrons and v_{th}^p for holes) are also included. To capture the influence of doping and defect levels, the donor (N_D) and acceptor (N_A) concentrations and the total defect density (N_t) are incorporated. Together, these parameters define the fundamental electronic structure and transport characteristics of the simulated photovoltaic device.

The effective density of states at the conduction (N_C) and valence (N_V) band edges determine the carrier concentration and distribution in a semiconductor. It is expressed as:

$$N_{C,V} = 2g_d\left(\frac{2\pi m_{e,h}^* k_B T}{h^2}\right)^{3/2} \tag{25}$$

In this study, g_d is set to 1, $m_{e,h}^*$, represent the effective mass of electrons or holes, and h is Planck's constant. This expression reflects how the electronic band structure influences the number of available energy states for charge carriers.

The thermal velocity of charge carriers significantly affects their mobility and diffusion, thereby influencing overall device performance. It can be calculated using equation (Sze & Ng, 2006).

$$v_{th}^{n,p} = \sqrt{\frac{3k_B T}{m_{e,h}^*}} \tag{26}$$

The mobility of charge carriers under an applied electric field is further related to the relaxation time (τ) through the expression $\mu_{n,p} = q\tau/m_{e,h}^*$. The aforementioned relationships collectively delineate the manner in which intrinsic material parameters govern charge transport, diffusion, and ultimately, the photovoltaic performance of the absorber material.

Semiconductor devices exhibit two primary mechanisms governing the flow of electrical current: drift and diffusion currents. Drift current is generated by the movement of charge carriers in response to the applied electric field. In this field, both majority and minority carriers are accelerated, with their motion being influenced by the material's doping levels. Conversely, diffusion current occurs when carriers move due to a concentration gradient, diffusing from regions of higher to lower concentration until equilibrium is reached. The SCAPS-1D transport model is an integrated framework that accounts for these mechanisms, employing a combination of drift and diffusion processes to simulate the transport of charge carriers within the context of semiconductor devices. For further details, readers are referred to the relevant references (Burgelman et al., 2000; Oublal, Dib, & El-yakoubi, 2025).

4. DENSITY FUNCTIONAL THEORY

The advent of quantum mechanics in the early 20th century precipitated a paradigm shift within the domain of physics, engendering a unifying theory that encompassed the description of matter at the atomic scale, primarily through the medium of the Schrödinger equation. However, the complexities inherent in electron interactions render many-body Schrödinger equation computations challenging for real-world materials, necessitating approximations. The process of identifying an approximate solution is typically characterised by three distinct stages. Firstly, the Born-Oppenheimer approximation (BOA) is employed to dissociate nuclear and electronic motions, thereby facilitating the solution to the problem. Second stage involves a variety of electronic structure approaches, encompassing Hartree, Hartree-Fock (HF), DFT, and Kohn-Sham (KS) formalism. These approaches allow for efficient approximations of complex electron interactions, thereby reducing computational costs.

The pioneering work of Hohenberg and Kohn in 1964 initiated the development of DFT, subsequently leading to the derivation of the Kohn-Sham equations. These equations provided a realistic and computationally efficient approach for modelling electron interactions. These breakthroughs established density functional theory (DFT) as the foundation of current computational materials science, allowing for precise approximations while significantly reducing the computational requirements. In this context, the general Schrödinger equation for a material containing N electrons is presented:

$$\left[-\sum_{i}\frac{\hbar^2}{2m_e}\nabla_{r_i}^{2}-\sum_{i=1}^{N}\frac{\hbar^2}{2M_i}\nabla_{R_i}^{2}+\frac{1}{4\pi\varepsilon_0}\sum_{i\neq j}\frac{e^2}{2|r_i-r_j|}-\frac{1}{4\pi\varepsilon_0}\sum_{i,j}\frac{Z_ie^2}{|R_i-r_j|}+\frac{1}{4\pi\varepsilon_0}\sum_{i,j}\frac{Z_iZ_je^2}{2|R_i-R_j|}+\right]\Psi$$

$$= E\Psi \tag{27}$$

The masses of the nuclei (at positions R_i) and electrons (at positions r_i) are denoted by M_i and m_e, respectively. The system's total energy comprises the kinetic energy of electrons and nuclei, as well as potential energy terms for electron-electron, nucleus-nucleus, and nucleus-electron interactions. Given the complexity of this many-body problem, which is challenging to address with exactitude, approximations must be employed to facilitate its solvability.

4.1. Born–Oppenheimer Approximation (BOA)

The Born-Oppenheimer approximation (BOA) is a foundational concept in condensed matter physics and materials science, underpinning numerous calculations. The simplification of the fundamental equation (27) is intrinsic to this method, and this is based on the assumption that atomic nuclei, which exhibit a significantly greater mass compared to electrons, move much more slowly. It is therefore posited that the nuclei may be regarded as stationary at their instantaneous positions, with electrons moving in the potential created by these fixed nuclei and the potential from other electrons (Born & Oppenheimer, 1927).

The separation of electronic and nuclear motions is known as the Born-Oppenheimer approximation. It is evident that within the model, the nuclei are regarded as fixed entities. Consequently, all kinetic energy associated with their motion is neglected. Consequently, the nucleus-nucleus potential energy is rendered constant. The total Hamiltonian of the system is now reduced to an electronic Hamiltonian that describes the new dynamics of the electrons. The simplified Hamiltonian is expressed as follows:

$$\left[-\sum_{i}\frac{\hbar^2}{2m_e}\nabla_{r_i}^{\ 2}+\frac{1}{4\pi\varepsilon_0}\sum_{i\neq j}\frac{e^2}{2|r_i-r_j|}-\frac{1}{4\pi\varepsilon_0}\sum_{i,j}\frac{Z_ie^2}{|R_i-r_j|}\right]\psi = E\psi \tag{28}$$

Notwithstanding the utilisation of the Born-Oppenheimer approximation (BOA), the electronic Hamiltonian remains complex. This is due to the fact that the equation under consideration continues to delineate a many-body problem, characterised by electron-electron interactions, the solution of which is intractable in exact terms. It is evident that, due to the inherent complexity of the subject, further approximations must be employed. Such approximations include the Hartree method, Hartree-Fock theory and density functional theory (DFT). The aforementioned approximations permit the problem to be simplified, thereby enabling the Schrödinger equation to be solved more efficiently and making it feasible to study real materials.

4.2. Hartree Approximation

In the year 1928, Hartree established an approximation with the objective of simplifying the complexity of many-electron systems. This method is predicated on the assumption that each electron functions independently within an average or mean field established by the other electrons and the nuclei of the system (Hartree, 1928). Consequently, the N-electron wavefunction $\psi(\mathrm{r}_1,\mathrm{r}_2,...,\mathrm{r}_N)$ is estimated as a product of single-electron wavefunctions, referred to as the Hartree Product (HP). This approximation simplifies the many-body problem by treating the electrons as non-interacting, except through the mean field they generate.

$$\psi(r_1,r_2,\ldots,r_N) = \psi_1(r_1)\cdot\psi_2(r_2)\ldots\psi_N(r_N) = \prod_{i=1}^{N}\psi_i(r_i) \tag{29}$$

Within this framework, each electron is influenced by an effective potential V_{eff} (Kohn, 1999). The Schrödinger equation for a single electron can be written as such:

$$-\frac{\hbar^2}{2m}\nabla^2\psi_i(\vec{r}) + V_{eff}(\vec{r})\psi_i(\vec{r}) = \varepsilon_i\psi_i(\vec{r}) \tag{30}$$

The effective potential, $V_{eff}(\vec{r})$, is hereby defined as the sum of two terms: the external potential, $V_{ext}(\vec{r})$, and the Hartree potential, $V_H(\vec{r})$. The former is the Coulomb attraction between the electron and the nuclei, whilst the latter is the electrostatic repulsion between electrons. Accordingly, the Hartree potential is expressed as follows:

$$V_H(\vec{r}) = -e\int\frac{\rho(\vec{r'})}{|\vec{r}-\vec{r'}|}d\vec{r'} \tag{31}$$

Here, the charge density of all electrons is denoted by $\rho(\vec{r'})$. Each electron is subject to an average electrostatic potential $V_H(\vec{r})$, which is derived from the Coulomb repulsion of all other electrons. This approach facilitates the treatment of electron-electron interactions by considering the effect of all electrons through an average potential.

$$\rho(\vec{r}) = \sum_i \psi_i^*(\vec{r})\psi_i(\vec{r}) \tag{32}$$

4.3. Hartree-Fock Approximation

The Hartree product is inherently ineffective due to its failure to accurately account for the antisymmetry of electron orbitals, as it did not fully account for the indistinguishability of electrons. This deficiency results in an inaccurate representation of electron occupancy in quantum states, thereby neglecting the fundamental rule that electrons cannot occupy the same quantum state (Pauli exclusion principle). The Hartree–Fock (HF) approximation (Fock, 1930) was utilised to address this shortcoming by ensuring the generation of an antisymmetric wavefunction (Slater, 1930). In this method, the wavefunction for n electrons is required to obey both antisymmetry and the Pauli exclusion principle. Consequently, the Slater determinant is utilised to describe the system:

$$\psi_e(\boldsymbol{r}) = \frac{1}{\sqrt{N!}}\begin{pmatrix} \psi_1(\boldsymbol{r}_1) & \cdots & \psi_n(\boldsymbol{r}_1) \\ \vdots & \ddots & \vdots \\ \psi_1(\boldsymbol{r}_n) & \cdots & \psi_n(\boldsymbol{r}_n) \end{pmatrix} \tag{33}$$

The Hartree–Fock approximation is a method employed for the purpose of determining the ground state of a system by applying the variational principle to minimise the total energy of the system. The objective is to ascertain the wavefunction ψ_e that corresponds to the lowest possible energy. This is achieved by solving the electronic Schrödinger equation. The expectation value of the Hamiltonian can be expressed as the sum of the kinetic energy (T), the Coulomb interaction energy (V_{ee}), and the external potential energy (V_{ext}):

$$E_{HF} = min\langle \psi_e T + V_{ee} + V_{ext} \psi_e \rangle \tag{34}$$

The Coulomb interaction contributes two key terms within the Slater determinant: the Hartree contribution ($E_H[\psi_e]$) and the exchange integral ($E_x[\psi_e]$).

$$\langle \psi_e V_{ee} \psi_e \rangle = E_H[\psi_e] + E_x[\psi_e] \tag{35}$$

The Hartree–Fock approximation is a method that can be used to calculate accurate molecular and atomic wavefunctions. However, this method becomes computationally intensive for larger systems due to the large number of basis functions required. Additionally, the model exhibits deficiencies in its ability to predict the behaviour of metallic systems. It underpredicts the infinite electron velocities that are observed near the Fermi energy (E_F) and neglects to consider the full electron correlation effects, with the exception of those that arise from the antisymmetric nature of the wavefunction ψ_e.

4.4. Theoretical Background of DFT

Conventionally, the many-electron Schrödinger equation has been solved using methods such as Hartree and Hartree-Fock (HF). However, HF considers only exchange interactions, ignoring electron correlation. This is a shortcoming that affects its accuracy, especially in solids. Conversely, Density

Functional Theory (DFT), proposed in 1964 by Hohenberg and Kohn, is a more sophisticated approach. DFT is a theoretical approach that replaces the complex many-electron wavefunction with the electron density $\rho(\vec{r})$, thereby reducing the problem's complexity from 3N variables to just 3. It is evident that this simplification renders the problem significantly more computationally feasible. This reduction rendered DFT a potent instrument within the domain of materials science. Kohn and Pople were awarded the 1998 Nobel Prize in Chemistry, in part due to their significant contributions to the field of density functional theory (DFT) and quantum chemical methods.

The DFT is predicated on two fundamental theorems. Firstly, the ground state energy is a functional of the electron density. Secondly, the true density is such that it minimises this functional. In order to facilitate a computational implementation of the DFT, Kohn and Sham (1965) developed a set of single-electron equations, incorporating an effective potential derived from the electron density. DFT represents a development of the Thomas–Fermi model (1927) (Fermi, 1928; Zangwill, 2013), which utilised local density in a uniform electron gas as an approximation for the kinetic energy. Subsequent to this, a local exchange term was incorporated (Dirac, 1930), and this was followed by a series of refinements to the model (e.g., Slater, 1951) (Dirac, 1930).

4.5. Electron Density $\rho(\vec{r})$

Electrons are indistinguishable and inseparable particles, meaning they cannot be observed as distinct entities. However, the probability of finding an electron in a specific volume can be calculated and is described by the electron density $\rho(\vec{r})$ (see Figure 5).

Figure 5. Electron density $\rho(\vec{r})$.

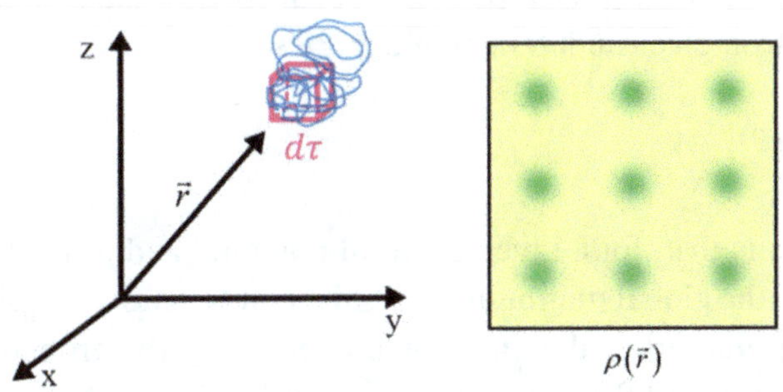

The electron density is indicative of the collective behaviour of electrons, which are modelled as an electron cloud. Figure 5 describes of the spatial domains in which electron occurrence is most probable, emphasising their consistent proximity to the nucleus. It is evident that the electron density is a positive function of only three spatial coordinates (x, y and z). The integration of the electron density over all space gives the total number of electrons (N), assuming the electron density approaches zero as it moves towards spatial infinity.

$$\rho(r \rightarrow \infty) = 0, \int \rho(r)dr = N \tag{36}$$

The electron density $\rho(\vec{r})$ is sufficient to fully determine the various properties of an atomic system, providing a comprehensive description of its behaviour and interactions.

4.5.1. Hohenberg-Kohn theorem of density

The theoretical basis of density functional theory (DFT) was established in two papers by Hohenberg and Kohn in 1964. The theorems previously mentioned are not confined to condensed matter systems, which are comprised of electrons and stationary nuclei, but are applicable to any interacting particle system in an external potential V_{ext}. The Hamiltonian of such a system can be expressed as:

$$H = -\sum_i \frac{\hbar^2}{2m_e}\nabla_{r_i}^2 + \frac{1}{4\pi\varepsilon_0}\sum_{i\neq j}\frac{e^2}{2|r_i - r_j|} - \frac{1}{4\pi\varepsilon_0}\sum_{i,j}\frac{Z_i e^2}{|R_i - r_j|} \tag{37}$$

The Hamiltonian now comprises three fundamental terms. The first term represents the kinetic energy of electrons, the second accounts for electron-electron interaction, and the third term describes the electron-nucleus attraction via the external potential V_{ext}.

The Hohenberg–Kohn theorems form the basis of DFT. The initial theorem posits that a one-to-one correspondence exists between the ground-state electron density $\rho(\vec{r})$ of a multi-electron system and the external potential (V_{ext}). This suggests that the electron density is the primary determining factor in the full Hamiltonian of the system. Therefore, all properties of the system including both ground and excited states can be calculated. The total energy of the system can be expressed as:

$$E[\rho(\vec{r})] = F[\rho(\vec{r})] + \int \rho(\vec{r})V_{\text{ext}}(\vec{r})d^3\vec{r} \tag{38}$$

The Hohenberg–Kohn functional F[ρ(r)] represents the sum of all system-independent terms, including the kinetic energy and electron-electron interaction energies.

$$F[\rho(\vec{r})] = T[\rho(\vec{r})] + V_{ee}[\rho(\vec{r})] \tag{39}$$

The second theorem posits that, for a given external potential and a fixed number of electrons, the ground state corresponds to the global minimum of a universal functional $F[\rho(\vec{r})]$ of the electron density. The electron density that minimises this functional is precisely the ground-state density. Therefore, the precise ground-state energy and the exact ground-state density are fully determined by the energy functional $E[\rho(\vec{r})]$. The minimum energy is expressed as follows:

$$E_0 = E(\rho_0) = \min E[\rho(r)] \tag{40}$$

$$F_{HK}[\rho(\vec{r})] = T[\rho(\vec{r})] + V_{ee}[\rho(\vec{r})] \tag{41}$$

4.5.2. Kohn-Sham Equations

In principle, the knowledge of the exact universal Hohenberg–Kohn functional $F_{HK}[\rho(\vec{r})]$ would enable an exact description of a system's ground-state properties. However, given the unavailability of the exact functional, the Hohenberg-Kohn theorems possess limited practical applicability in isolation. This challenge has been addressed through significant advancements in Density Functional Theory (DFT) in 1965 by Kohn and Sham. The researchers proposed a novel approach, utilising a system of non-interacting particles to ascertain the precise properties of a real, interacting many-body system (Kohn & Sham, 1965).

The Kohn–Sham (KS) formalism is a method in Density Functional Theory (DFT) that introduces a theoretical system of non-interacting electrons in an effective potential. The potential is selected in such a manner that the electron density of the non-interacting system is equivalent to the density of the true interacting system. The objective of the Kohn–Sham equations is to ascertain the wavefunctions (φ_i) that minimise the total energy of the system. This provides an approximation to the ground-state energy and electronic properties of a real system. The Kohn–Sham Hamiltonian is expressed as follows:

$$H = H_{KS} = -\frac{\hbar}{2m}\Delta + V_{KS}(r) \tag{42}$$

Here, $V_{KS}(r)$ is the effective potential, and is defined as:

$$V_{KS} = V_{\text{eff}}(r) = V_{\text{ext}}(r) + V_H[\rho(r)] + V_{xc}[\rho(r)] \tag{43}$$

Where, $V_{\text{ext}}(r)$ represents the external potential due to the nuclei, $V_H[\rho(r)]$ denotes the Hartree potential (Coulomb repulsion), and $V_{xc}[\rho(r)]$ represents the exchange-correlation potential:

$$V_{xc}(r) = \frac{\partial E_{xc}[\rho(r)]}{\partial \rho(r)} \tag{44}$$

In summary, the theoretical framework of Density Functional Theory (DFT) was initially proposed in 1964 by Hohenberg and Kohn. This theoretical framework builds upon earlier works, notably those by Hartree and Fock in the 1920s, which introduced methods for treating electron-electron interactions. However, the approach pioneered by Hartree and Fock entailed the resolution of many-electron equations for the wavefunctions of all electrons, a process that is computationally intensive. DFT, on the other hand, simplifies this problem by focusing on the electron density rather than the many-body wavefunction, thus reducing the complexity of the problem. This approach enables DFT to provide a practical and viable method for calculating ground-state properties of materials with reasonable computational cost. The model requires the atomic numbers and the electron density as inputs, a configuration that ensures both computational efficiency and physical robustness.

The success of DFT can be attributed to its capacity to predict a broad spectrum of material properties with a reasonable degree of accuracy, whilst demanding relatively low computational resources in comparison to methods that explicitly solve for wavefunctions, such as Hartree-Fock or post-Hartree- Fock. The fundamental equations that govern DFT are as follows:

$$H\varphi_i(r) = \left(-\frac{\hbar^2}{2m}\nabla^2 + V_{KS}(r)\right)\varphi_i(r) = \varepsilon_i\varphi_i(r) \tag{45}$$

The mono-electronic functions $\varphi_i(r)$ are employed to calculate the electronic density.

$$\rho(r) = \sum_{i=1}^{N} |\varphi_i(r)|^2 \quad (46)$$

The Kohn–Sham approach constitutes a theoretical framework that facilitates the simplification of the complex many-electron problem into N simpler one-electron problems, thereby enabling the accurate calculation of electronic structure.

4.5.3. Exchange-Correlation Treatment

In the context of Density Functional Theory (DFT), the exchange-correlation functional $E_{xc}[\rho(r)]$ represents a complex and pivotal component, playing a crucial role in the manifestation of exchange and correlation effects. Exchange is underpinned by the Pauli exclusion principle, while correlation is attributed to Coulomb repulsion. Furthermore, it encompasses the disparity in kinetic energy between the authentic interacting electron system and the auxiliary, non-interacting Kohn-Sham system. The absence of true data concerning the nature of $E_{xc}[\rho(r)]$ has resulted in numerous exciting developments in the approximation method. In this series of density functional approximations, the following are included: local density approximation (LDA), generalised gradient approximation (GGA), the modified Becke-Johnson (mBJ) potential, and hybrid functionals (e.g., HSE06). Such models endeavour to account for absent interaction effects, thereby rendering the Kohn-Sham equations solvable. In essence, the exchange-correlation functional is the element that serves to establish a connection between the mathematical simplicity of the Kohn–Sham model and the physical reality of interacting electrons. This approximation has a direct impact on the reliability of the DFT calculations.

4.5.4. Local Density Approximation

The Local Density Approximation (LDA) is the simplest method for approximating the exchange-correlation energy in Density Functional Theory (DFT). The present framework extends the Kohn-Sham framework by providing an algebraic form for E_{xc} that account for the correlations between the motions of different electrons. The fundamental principle underlying LDA is to conceptualise the exchange–correlation potential as a local quantity, determined at each point r r in space and only marginally influenced by the surrounding electron density variations. In this approximation, the exchange–correlation potential at any point r is assumed to be equivalent to that of a homogeneous electron gas with the same local density $\rho(r)$ (Kohn & Sham, 1965). The total exchange–correlation energy ε_c^{LDA} is subsequently derived by integrating the product of the local electron density and the exchange-correlation energy per particle ε_c^{LDA}:

$$E_{XC}^{LDA}[\rho(r)] = \int \rho(r)\, \varepsilon_{xc}^{LDA}[\rho(r)]dr \quad (47)$$

The exchange–correlation energy per particle in a homogeneous gas, $\varepsilon_{xc}^{LDA}[\rho(r)]$, is expressed as:

$$\varepsilon_{xc}^{LDA}[\rho(r)] = \varepsilon_x^{LDA}[\rho(r)] + \varepsilon_c^{LDA}[\rho(r)] \quad (48)$$

The exchange energy per particle is given by:

$$\varepsilon_x^{LDA}[\rho(r)] = -C_x \cdot \int \rho^{4/3}(r)dr \ ; C_x = \frac{3}{4}\left(\frac{3}{\pi}\right)^{1/3} \quad (49)$$

Even though there is no specific analytical form for the correlation component of ε_{xc}, LDA is effective for many systems, primarily due to its simplicity. Nevertheless, this methodology has its limitations (Zunger & Freeman, 1977). These include overestimation of binding and correlation energies, underestimation of ionisation energies and band gaps, and poor treatment of systems with rapid electron density variations. This is due to the fact that electron density is treated locally in the model. The Generalised Gradient Approximation (GGA) was devised to address these issues. GGA generalises LDA by making the exchange-correlation energy dependent on the local electron density and its gradient. This allows GGA to capture more of the inhomogeneity in electron density. It has been demonstrated that this enhancement can improve the precision of essential properties such as lattice constants, bond lengths and cohesive energies, particularly in materials comprising heavy or transition elements. Consequently, GGA has become a much more reliable and popular method in DFT calculations.

4.5.5. Generalised Gradient Approximation (GGA)

The Local Density Approximation (LDA) is effective in calculations involving slowly varying electron densities (e.g. metals and crystals), but is ineffective in describing systems with steep changes in electron density, such as radicals, metals, and optimised molecular structures. The LDA assumes the electron gas to be homogeneous and depends only on the local electron density $\rho(\vec{r})$. In contrast, the GGA improves upon this model by incorporating the gradient of the electron density, $\nabla\rho(\vec{r})$. This additional term enables the exchange–correlation energy to account for variations in electron density, providing a more accurate representation of real systems, particularly when density distributions vary (Perdew et al., 1996). Perdew and his colleagues developed GGA by adding both local and gradient-dependent terms to the exchange-correlated functional. GGA generalises the exchange–correlation energy as follows:

$$E_{xc}^{GGA} = \int f_{xc}^{GGA}[\rho(r), \nabla\rho(r)]d^3 r \tag{50}$$

The inclusion of the density gradient GGA provides a more realistic description of electronic structures, allowing for better predictions of molecular geometries, lattice parameters, reaction energies, ground-state energies, and binding properties than LDA. All of this improved precision comes at a reasonable additional computational expense. Consequently, GGA has become a widely used approach in the context of both solid-state and molecular systems. However, like LDA, GGA has difficulty describing electronic and optical properties precisely, such as band gaps. To increase accuracy, many versions of GGA have been developed, such as Becke-Perdew (BP), Perdew-Wang (PW91) (Perdew et al., 1993), Perdew-Burke-Ernzerhof (PBE) (Perdew et al., 1996) and its variant PBEsol (for solids) (Perdew et al., 2008), Engel-Vosko (EV-GGA) (Engel & Vosko, 1993), and Wu-Cohen (WC) (Wu & Cohen, 2006). PBE and its various forms are the most popular as they provide better estimates for atomic energies, bond strengths and thermochemical properties.

4.5.6. Modified Becke-Johnson Potential

The modified Becke-Johnson (mBJ) potential is a meta-GGA functional developed to give more reliable band gap predictions in semiconductors. The mBJ potential, initially introduced by Becke and Johnson (BJ) in the WIEN2K code, has been shown to significantly enhance the calculated band gaps, thereby bringing them closer to experimental values. This has led to the widespread adoption of mBJ as a highly effective method for studying semiconductor electronic structures (Tran & Blaha, 2009).

The mBJ potential is mathematically expressed as:

$$V_{x,\sigma}^{TB-mBJ}(r) = c\,V_{x,\sigma}^{BR}(r) + \left(3c - 2\right)\frac{1}{\pi}\sqrt{\frac{5}{6}\frac{t_\sigma(r)}{\rho_\sigma(r)}} \tag{51}$$

Here, c is given by equation (52), $V_x^{BR}(r)$ denotes the Becke-Roussel potential and is expressed by formula (53) (Becke & Roussel, 1989), $t_\sigma(r) = (1/2)\sum_{i=1}^{N_\sigma}\nabla\psi_{i,\sigma}^{*}\psi_{i,\sigma}$ is the kinetic-energy density, and $\rho_\sigma(r) = \sum_{i=1}^{N_\sigma}\left|\psi_{i,\sigma}\right|^2$ represents the electron density.

$$c = \alpha + \beta\left(\frac{1}{V_{cell}}\int\frac{\left|\nabla\rho_\sigma(r)\right|}{\rho_\sigma(r)}d^3r\right)^{1/2} \tag{52}$$

$$V_{x,\sigma}^{BR}(r) = -\frac{1}{b_\sigma(r)}\left(1 - e^{-x_\sigma(r)} - \frac{1}{2}x_\sigma\left(r\right)e^{-x_\sigma(r)}\right) \tag{53}$$

Here, $\nabla\rho_\sigma(r)$ denotes the gradient of electron density, $\alpha = -0.012$ and $\beta = 1.023$ bohr$^{1/2}$ were chosen by Tran and Blaha by minimising the mean absolute deviation between calculated and experimental band gaps, and $b_\sigma = \left[x_\sigma^3(r)\,e^{-x_\sigma(r)}/\left(8\pi\rho_\sigma(r)\right)\right]^{1/3}$, where x_σ is related to $t_\sigma\,\rho_\sigma$, $\nabla\rho_\sigma$, and $\nabla^2\rho_\sigma$.

The mBJ functional is recognised for its capacity to predict exchange and correlation energies with a high degree of accuracy, thereby significantly enhancing the precision of band gap calculations in semiconductors. Unlike LDA and GGA, mBJ provides a reliable estimate of band gap values that closely matched the experimental results. Furthermore, mBJ provides these enhancements at a lower computational cost than hybrid functionals, establishing it as a robust and efficient method for accurately predicting the band gaps of semiconductors.

4.5.7. Basis Sets

In the context of Density Functional Theory (DFT), obtaining accurate and efficient electronic wavefunctions is one of the most crucial aspects. The basis set is one of the most significant factors affecting the accuracy and computational expense of calculations, as Kohn–Sham orbitals and associated quantities can be described using a finite basis set. For solid materials, a wide range of techniques are often employed. Plane-wave (PW) basis sets are a regular tool in calculations for periodic systems, particularly in solid-state applications. This method expresses wave-functions as a sum of plane waves in k-space, where the wave vectors correspond to the periodicity of the crystal lattice. Plane-wave (PW) methods offer several advantages, including systematic convergence by increasing the kinetic energy cutoff, respect for the translational symmetry of crystalline solids, and suitability for Fourier-transform-based techniques. While PW basis sets are highly efficient for bulk materials, they are less effective near atomic nuclei. In such cases, the use of pseudopotentials or Projector Augmented-Wave (PAW) methods is necessary to reduce computational costs. These methods have been incorporated into prominent codes such as VASP, CASTEP, and Quantum ESPRESSO.

The Augmented Plane Wave (APW) method, first introduced by Slater, was among the earliest all-electron techniques for solving the Kohn–Sham equations. The model subdivides space into atomic spheres (muffin-tin regions) surrounding nuclei, with an interstitial region separating these regions. Within the confines of the spheres, the wavefunction is represented by radial solutions of the Schrödinger equation, multiplied by spherical harmonics. In the interstitial region, however, plane waves are utilised instead. A major limitation of the APW method is that its radial functions are energy dependent, necessitating an energy-dependent, nonlinear and computationally intensive approach. To overcome this, the LAPW (linearised augmented plane wave) method was proposed. In this approach, the energy-dependent radial functions are replaced with a linear approximation, yielding a linear, numerically stable eigenvalue problem. It has been demonstrated that the LAPW method can be highly accurate for all-electron calculations when the effect of core electrons is significant. This has been implemented in codes such as WIEN2k.

The Full-Potential LAPW (FP-LAPW) technique is a broadening of LAPW, where the spherical approximation of the potential and charge density within the muffin-tin spheres is removed. This is of particular significance for materials characterised by anisotropic electron density or low-symmetry crystal structures, as it ensures highly accurate measurements for complex systems and surfaces. FP-LAPW is widely regarded as a reference method and is also implemented in WIEN2k.

5. Case STUDY: Cs_2NaInX_6 (X=Br, I) LEAD-FREE DOUBLE PEROVSKITES

5.1. Computational Methods

In this study, the electronic and optical properties of halide perovskites were studied using the FP-LAPW method within the DFT framework. The GGA-PBE was used to treat the exchange-correlation potential, while more accurate band gap values were achieved using the TB–mBJ potential. A series of convergence tests were performed to ensure the numerical reliability of the results for both *k*-point sampling and the plane-wave cut-off parameter $R_{MT}K_{max}$. These tests indicated that the total energy, band gap and electric field gradient (EFG) are stable once the calculation parameters have sufficiently converged. Based on this, a 17×17×17 *k*-point grid was chosen for the structural relaxation and electronic structure calculations, and a denser 36×36×36 grid was employed for the density of states (DOS) and optical response calculations. The plane-wave cut-off $R_{MT}K_{max}$ was set to 8 and an energy cut-off of -6.0 Ry was used to distinguish between core and valence electrons. The muffin-tin radii chosen were R_{MT} = 2.45 a.u. (Cs), 2.15 a.u. (Na), 2.15 a.u. (In), 2.40 a.u. (Br) and 2.50 a.u. (I), such that none of the atomic spheres overlapped. Angular momentum was defined out to 10 and the reciprocal lattice vector out to 12 bohr^{-1}. The self-consistent field (SCF) cycles were converged to 0.0001 Ry in total energy and 0.001e in charge density to ensure accurate, self-consistent results.

The Spectroscopic Limited Maximum Efficiency (SLME) model was applied to estimate the theoretical power conversion limits of the absorber materials. Additionally, full device simulations of FTO/TiO_2/Cs_2NaInX_6 (X = Br, I)/Cu_2O/Au perovskite solar cells (PSCs) were performed using SCAPS-1D. Lead-free double perovskites were modeled on a fluorine-doped tin oxide (FTO) substrate. The model incorporated *n*-type TiO_2 as the electron transport layer (ETL) and *p*-type Cu_2O as the hole transport layer (HTL). A gold (Au) back contact was also included. Device performance was analysed as a function of absorber thickness, which typically ranges from 0.4 to 1.4 μm to achieve high efficiency at a moderate cost. For consistency, the SCAPS-1D simulations maintained identical layer thicknesses: 0.023 μm for the ETL, 0.24 μm for the HTL, and 0.5 μm for the FTO layer.

5.2. Structural, Elastic and Thermodynamic Properties

Lead-free halide double perovskites (DPs) crystallise in a face-centred cubic structure with $Fm\bar{3}m$ symmetry. Their atomic arrangement gives rise to strongly distinct structural features. Geometry optimisation was performed to obtain ground-state structural parameters, including ground-state energy, lattice constants, volume, bulk modulus, and its pressure derivative. Substituting bromine (Br) with iodine (I) increased the lattice parameter from 11.175 Å for $Cs_2NaInBr_6$ to 12.139 Å for Cs_2NaInI_6 due to iodine's larger ionic radius. This expansion leads to a reduction of the density and solidity. The bulk modulus decreased from 21.296 GPa in $Cs_2NaInBr_6$ to 15.42 GPa in Cs_2NaInI_6. Also, as the electronegativity of bromine (2.96 Pauling units) differs from that of iodine (2.66 Pauling units), bonding tendencies result. Iodine forms weaker bonds with indium and sodium than bromine does. The structural stability of these materials satisfies the definitions of the Goldschmidt's tolerance and octahedral factors. Furthermore, as shown in Table 1, both materials have negative formation energies, indicating that they are thermodynamically stable. A negative formation energy indicates that a compound is energetically favourable and can form spontaneously without external energy input. This property is very important in DFT studies, as it confirms the structural stability of the materials prior to examining their mechanical, electronic and optical properties for solar cell applications.

Table 1. Lattice parameter (a), ground state energy (E_0), volume at ground state (V_0), bulk modulus (B) and its pressure derivative (B'), Goldschmidt's tolerance factor (τ), and the octahedral factor (μ), and formation energy (ΔH_f) of Cs_2NaInX_6 (X=Br, I).

Compounds	a=b=c (Å)	E_0 (Ry)	V_0 (Å^3)	B (GPa)	B' (GPa)	τ	μ	ΔH_f (eV/atom)
$Cs_2NaInBr_6$	11.193	-74533.4018	1402.29	21.296	5.052	0.946	0.46	-1.794
Ref. (Saeed et al., 2020)	10.97	-	-	20.64	-	-	-	-
Cs_2NaInI_6	12.139	-128682.2086	1788.74	15.42	5.21	0.927	0.413	-1.377

The elastic constants of Cs_2NaInX_6 (X = Br, I) are summarised in Table 1. The Born–Huang criteria of mechanical stability for the cubic compounds and the corresponding (C_{ij}) elastic constants are obtained. The calculated bulk moduli of $Cs_2NaInBr_6$ and Cs_2NaInI_6 are 21.51 and 15.9 GPa, respectively, suggesting that the bromide compound is more resistive to compression than the iodide analogue. These bulk moduli values match well with those obtained using the energy-volume curve. The shear modulus (*G*) of $Cs_2NaInBr_6$ (11.74 GPa) is also higher than that of Cs_2NaInI_6 (8.21 GPa), suggesting that the bromide-based compound is stiffer. The computed Young's moduli (*Y*) also match this trend, with $Cs_2NaInBr_6$ yielding 29.81 GPa, relatively higher than the 20.92 GPa of Cs_2NaInI_6 (see Table 2).

Table 2. Elastic constants and elastic moduli for Cs_2NaInX_6 (X=Br, I).

DPs	C_{11} (GPa)	C_{12} (GPa)	C_{44} (GPa)	B (GPa)	Y (GPa)	G (GPa)	ν	B/G
$Cs_2NaInBr_6$	39.39	12.57	10.75	21.51	29.81	11.74	0.2690	1.83
Cs_2NaInI_6	28.2	9.0	7.4	15.4	20.92	8.21	0.2736	1.875

The computed elastic constants indicate that the Br-based compound is stiffer, harder, and more resistant to compressive forces. Furthermore, the ductility of both materials is verified by the *B/G* ratio and Poisson's ratio (ν), as the *B/G* values for both materials are higher than 1.75, and ν for both materials lie between 0.25 and 0.5.

Table 3. The longitudinal (v_l), transverse (v_t) and average (v_a), Debye temperature (θ_D) and melting temperature (T_m) of Cs_2NaInX_6 (X=Br, I).

Compounds	v_l (m/s)	v_t (m/s)	v_a (m/s)	θ_D (K)	T_m (K)
$Cs_2NaInBr_6$	2981.280	1675.851	1864.726	169.619	785.794
Cs_2NaInI_6	2468.255	1378.006	1534.161	128.672	719.662

Table 3 summarises the key physical parameters of both compounds, including the sound velocities, Debye temperatures (θ_D) and melting temperatures (T_m). The longitudinal (v_l), transverse (v_t), and average (v_m) sound velocities of $Cs_2NaInBr_6$ (2981.280 m/s, 1675.851 m/s and 1864.726 m/s, respectively) are all higher than those of Cs_2NaInI_6 (2468.255 m/s, 1378.006 m/s and 1534.161 m/s, respectively). This points to a stronger interatomic bonding and stiffer lattice in $Cs_2NaInBr_6$. The same tendency is reflected in the Debye temperatures, which are 169.619 K for $Cs_2NaInBr_6$ and 128.672 K for Cs_2NaInI_6. The calculated melting temperatures are 785.794 K and 719.662 K for $Cs_2NaInBr_6$ and Cs_2NaInI_6, respectively. These results show that the bromide analogue possesses higher thermal stability than the iodide.

5.3. Band Structure and Density of States

The electronic band structures of Cs_2NaInX_6 (X=Br, I) are depicted in Figure 6 as calculated using GGA-PBE and TB-mBJ exchange-correlation potentials. The calculated band gaps show that all of the compounds have direct band gaps at the Γ point. The energy band gaps are 1.226 eV for $Cs_2NaInBr_6$ and 0.226 eV for Cs_2NaInI_6 under GGA-PBE calculations. Using the TB-mBJ potential, the band gap increases to 2.712 eV and 1.268 eV for $Cs_2NaInBr_6$ and Cs_2NaInI_6, respectively. A similar band gap value of 2.68 eV was found for $Cs_2NaInBr_6$ using HSE06 (Dong et al., 2023). Both band gaps fall within the visible and near-infrared regions, indicating their potential for use in optoelectronic and solar cell applications.

Figure 6. Electronic band structures of $\mathbf{Cs_2}$*NaIn*X_6 *(X=Br, I) using (a) GGA-PBE and (b) TB-mBJ.*

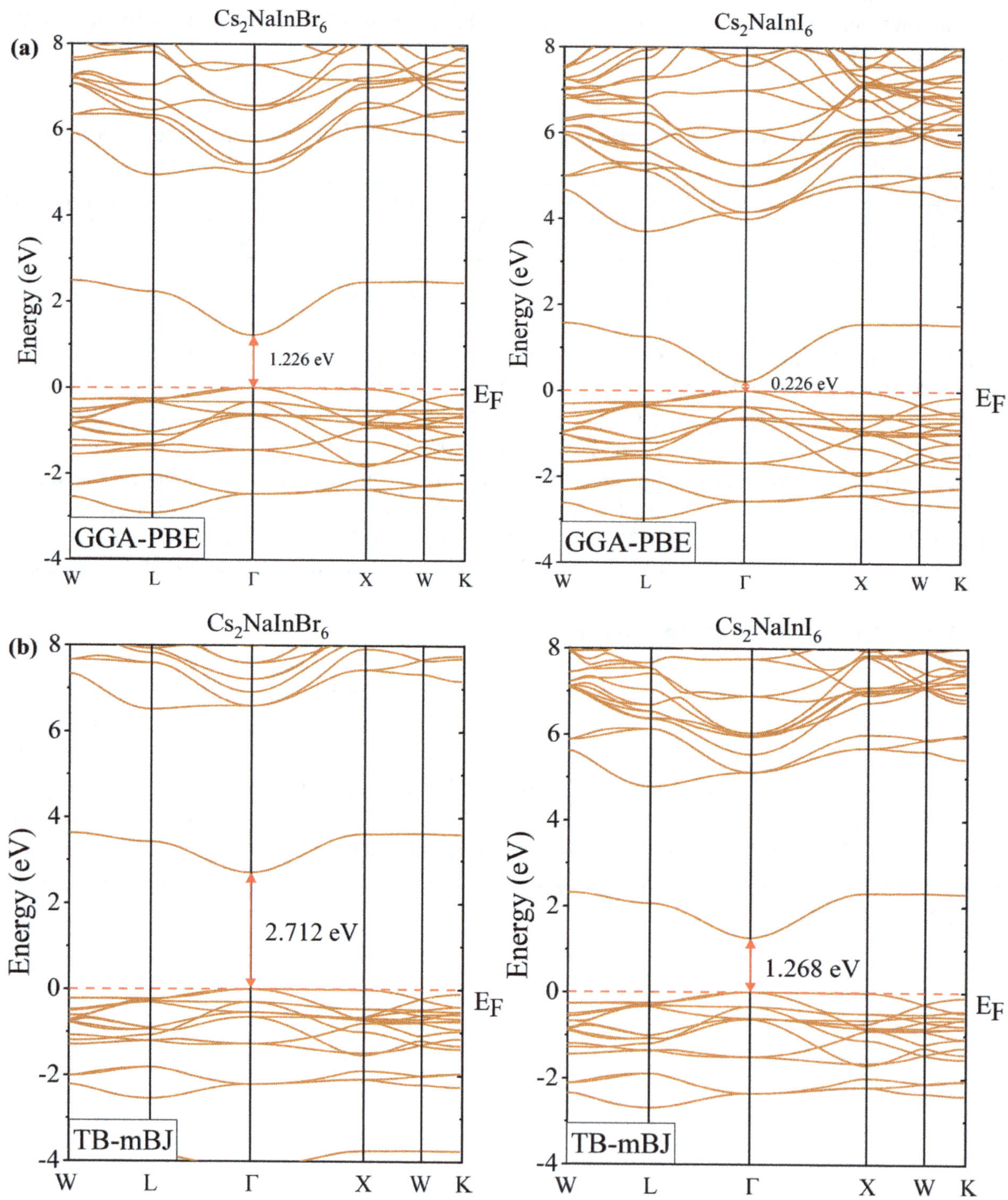

The total DOS (TDOS) and partial DOS (PDOS) provide detailed insights into the contributions of atomic orbitals and the electronic behaviour of the material. As the band gaps obtained from the TB-mBJ method are in close agreement with experimental energy band gap measurements, TB-mBJ was used to compute the TDOS and PDOS. In Figure 7 (a) and (b), TDOS and PDOS of Cs_2NaInX_6 (X = Br, I) are depicted over the energy range -4 to 8 eV. In this context, the vertical dashed line marking the Fermi level E_F. The calculation reveals that the VBM is primarily composed of Br-*p* states (consistent with previous reports (Oublal, Dib, Khalal, et al., 2025)), with smaller contributions from Cs(*s-p-d*), Na(*s-p*) and In(*s-p*) orbitals for $Cs_2NaInBr_6$. The In-*p* and In-*s* states are mainly located at lower energies (around -2 eV and -4 eV). A similar pattern is observed for Cs_2NaInI_6, where the VBM is again dominated by I-*p* states, but with lower intensity compared to the bromide (3.54 states/eV for Br-*p* versus 2.67 states/eV for I-*p* near 0.4 eV). Weaker contributions from Cs(*s-p-d*), In(*s-p*), and Na(*s-p*) are present in the valence band. In both compounds, the CBM is predominantly made up of unoccupied In-*s* and halogen-*p* states, with only slight contributions from sodium-3*s* states. Substituting Br with I shifts the In(*s-p*) and Na(*s-p*) densities closer to E_F, which can be attributed to weaker hybridisation between the X-*p* orbitals and the In-*s* state. This decreased hybridisation and the higher energy of the I-*p* orbitals, increases the valence-band edge and results in a narrower band gap.

Additionally, the carrier transport properties were examined by calculating the effective masses of electrons and holes from the curvature of the band edges (*54*). These effective masses provide crucial insight into carrier mobility within the material, a factor that plays a crucial role in charge extraction efficiency and solar cell performance. Improving solar energy conversion efficiency in DPs requires optimising effective carrier transport.

$$\frac{1}{m^*} = \frac{1}{\hbar^2}\frac{\partial^2 E}{\partial k^2} \tag{54}$$

The effective electron masses for $Cs_2NaInBr_6$ and Cs_2NaInI_6, were found to be $0.2m_0$ and $0.15m_0$ along (Γ-L) direction and $0.184m_0$ and $0.136m_0$ along (Γ-X) direction, respectively. The effective hole masses were $0.752m_0$ and $0.55m_0$ along (Γ-L) direction and $0.34m_0$ and $0.25m_0$ along (Γ-X) direction. The average effective masses of the electrons and holes were found to be $0.192m_0$ and $0.546m_0$ for $Cs_2NaInBr_6$ and $0.143m_0$ for electrons and $0.40m_0$ for holes in Cs_2NaInI_6, respectively. These results suggest favorable carrier transport properties.

Figure 7. TDOS and PDOS of (a) $Cs_2NaInBr_6$ and (b) Cs_2NaInI_6 using TB-mBJ.

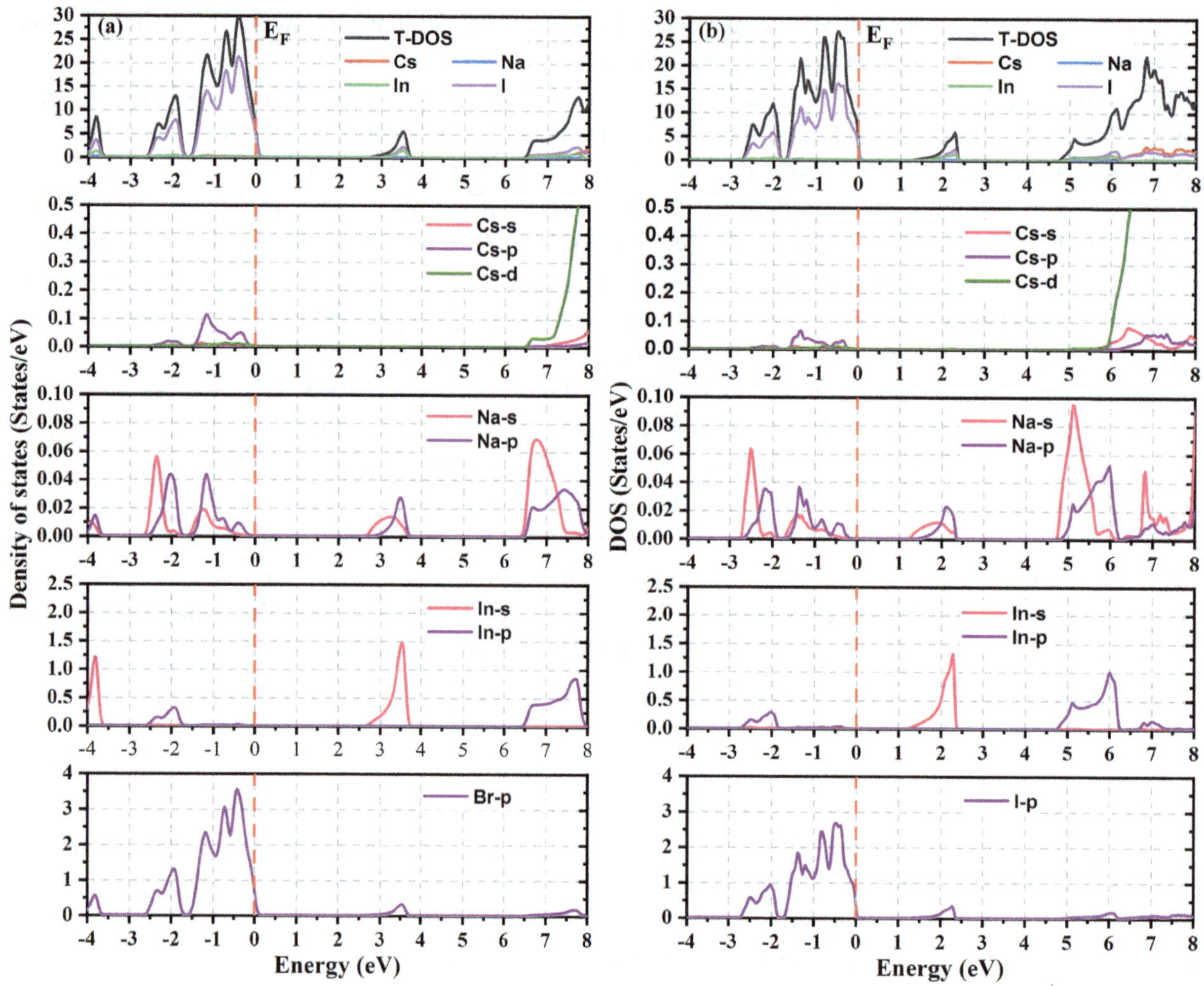

5.4. Optical Properties

Figure 8 (a) shows the real part of the dielectric function, $\varepsilon_1(\omega)$, for Cs_2NaInX_6 (X = Br, I) in the photon energy range 0-12 eV. The static dielectric constant $\varepsilon_1(0)$ increases from 2.62 for the bromide to 3.53 for the iodide. As the photon energy rises, $\varepsilon_1(\omega)$ initially increases and reaches resonance maxima of 4.19 at 5.10 eV for $Cs_2NaInBr_6$ and 4.93 at 3.85 eV for Cs_2NaInI_6. These peaks correspond to maximum dispersion and strongest polarisation of the crystal. At higher energies, $\varepsilon_1(\omega)$ decreases and eventually becomes negative at about 11.17 eV (Br) and 8.95 eV (I), signalling a transition to metallic-like behaviour in which incident photons are fully absorbed.

Figure 8 (b) presents the imaginary part, $\varepsilon_2(\omega)$. For $Cs_2NaInBr_6$, $\varepsilon_2(\omega)$ begins to rise from zero at around 2.75 eV and reaches a peak of 0.90 at 3.75 eV. Cs_2NaInI_6, it starts to increase at about 1.3 eV and attains a maximum of 1.21 at 2.40 eV. These features are associated with optical transitions from occupied X-*p* states to unoccupied In-5s states. Overall, the imaginary part of the dielectric function confirms that both compounds exhibit strong absorption across the visible and ultraviolet regions of the spectrum.

Figure 8. Real (a) and (b) imaginary parts of the dielectric function, absorption coefficient (c) and reflectivity (d) of $\mathbf{Cs_2NaInX_6}$ *(X = Br, I).*

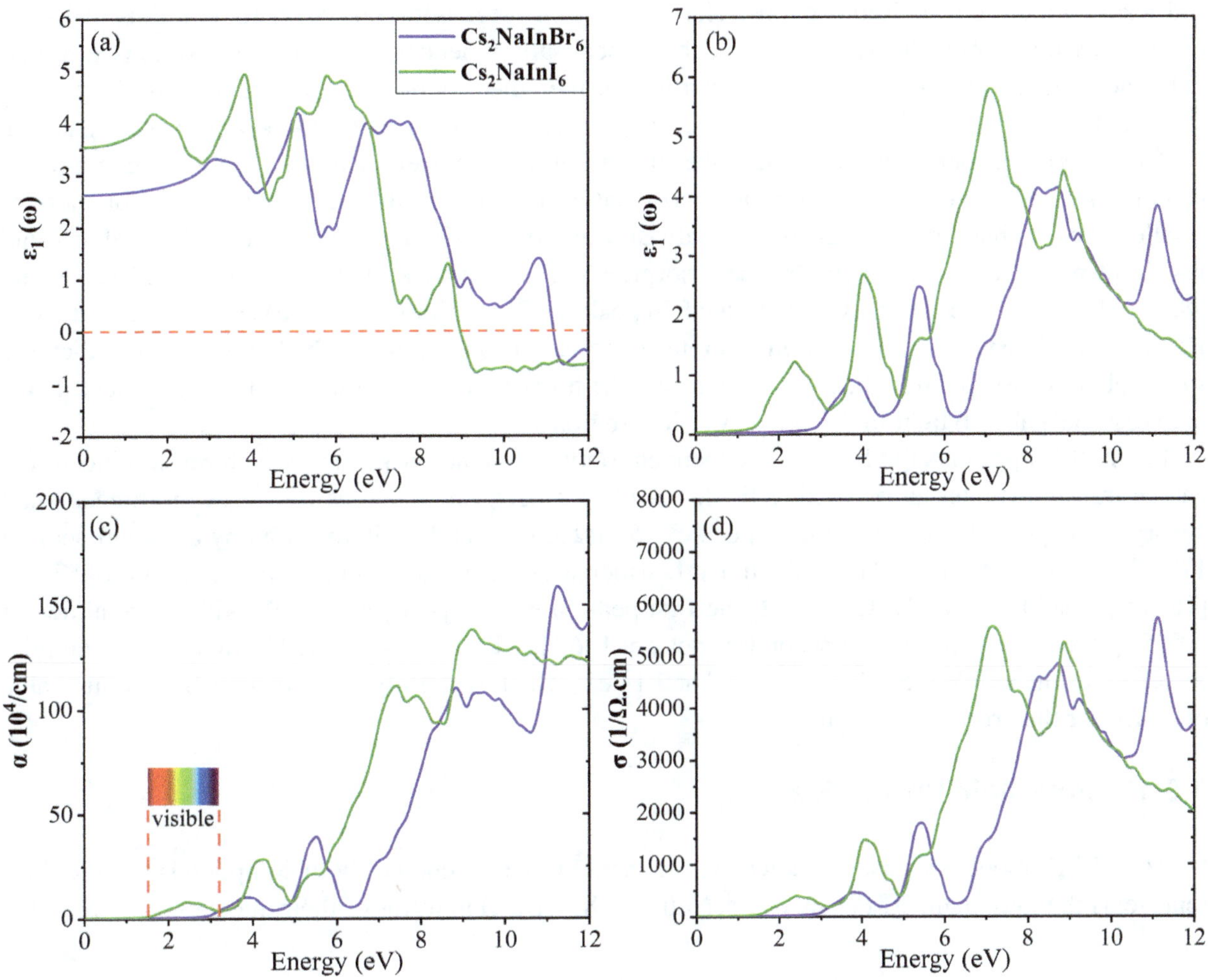

Additionally, the exciton binding energies (E_b) were determined using the Wannier-Mott model, which treats the exciton in analogy to a hydrogen atom. In this picture, the effective masses of the electron (m_e^*) and hole (m_h^*) play roles similar to the electron and nuclear masses in hydrogen. The exciton binding energy is given by (Ashcroft & Mermin, 1976; Kinani et al., 2022):

$$E_b = \frac{m_e e^2}{2(4\pi\varepsilon_0)^2\hbar^2}\frac{\mu_r^*}{m_e\varepsilon^2} \approx 13.56\frac{\mu_r^*}{m_e\varepsilon^2}\left(\text{eV}\right) \tag{55}$$

Here, μ_r^* denotes the reduced effective mass ($\mu_r^* = (m_e^*.m_h^*)/(m_e^* + m_h^*)$), and ε_0 is the vacuum dielectric constant. The dielectric constant of a material plays a key role in determining how effectively it screens external electric fields via internal charge polarisation. This screening can be either static or dynamic, depending on the frequency of the applied field. For Wannier-Mott excitons, screening is chiefly dictated by the static dielectric constant (Bechstedt et al., 2005; Kinani et al., 2022). The exciton binding energies decreases from 280 meV to 114 meV when bromine is replaced by iodine. This reduction is linked to the stronger dielectric screening in the iodide, which weakens the Coulomb interaction and leads to more delocalised excitons. The lower electronegativity and larger ionic radius of iodine compared with bromine leads to reduced carrier localisation; hence, the iodide perovskite is expected to be more advantageous for charge separation. For the other double perovskites, such as Cs_2AgYX_6 (Y=Bi, Sb; X=Cl, Br) exhibit exciton binding energies ranging from 170 to 434 meV (Biega et al., 2021).

The absorption coefficient, $\alpha(\omega)$, describes the rate at which a material absorbs light of a given wavelength. Figure 8 (c) shows the absorption coefficient for both materials. $\alpha(\omega)$ remains zero for energies smaller than the band gap energy. $\alpha(Cs_2NaInBr_6)$ and $\alpha(Cs_2NaInI_6)$ start to rise above 2.9 eV and 1.4 eV, respectively. For $Cs_2NaInBr_6$, the absorption coefficient peaks at 9.93x 10^4 cm^{-1} at 3.82 eV and 38.9x 10^4 cm^{-1} at 5.5 eV. For Cs_2NaInI_6, the first peak is 7.75x 10^4 cm^{-1} at 2.43 eV, and the second 28.2x 10^4 cm^{-1} at 4.25 eV. These results demonstrate an increase in absorption with photon energy, indicating strong photon absorption in the 2-6 eV range and the formation of electron-hole pairs (excitons). Thus, both materials absorb light in the UV and visible regions.

Figure 8 (d) presents the frequency-dependent optical conductivity, $\sigma(\omega)$, which reflects electronic charge transport under an oscillating field. For $Cs_2NaInBr_6$, $\sigma(\omega)$ begins to rise at the fundamental absorption edge and reaches an initial peak of 457 $[\Omega.cm]^{-1}$ at 3.8 eV, followed by a second peak of 1784.3 $[\Omega.cm]^{-1}$ at 5.4 eV. The conductivity then increases further, attaining a maximum value of 5708.4 $[\Omega.cm]^{-1}$ at 11.11 eV. For Cs_2NaInI_6 the first peak in $\sigma(\omega)$ appears at 2.40 eV with a magnitude of 392.54 $[\Omega.cm]^{-1}$, followed by a secondary peak of 1463 $[\Omega.cm]^{-1}$ at 4.07 eV. The highest conductivity recorded for the iodide is 5545.3 $[\Omega.cm]^{-1}$ at 7.14 eV, which is slightly lower than the maximum value obtained for the bromide analogue.

5.5. Photovoltaic Properties

The SLME model was used to calculate the maximum efficiency of both compounds. Figure 9 (a) represents the maximum efficiency of Cs_2NaInX_6 (X = Br, I) in terms of their thickness.

Figure 9. (a) Maximum efficiency as a function of thickness, and (b) SQ-limit and SLME of $\mathbf{Cs_2NaIn}$ $\mathbf{X_6}$ *(X = Br, I) for 0.8 µm at 298.15 K.*

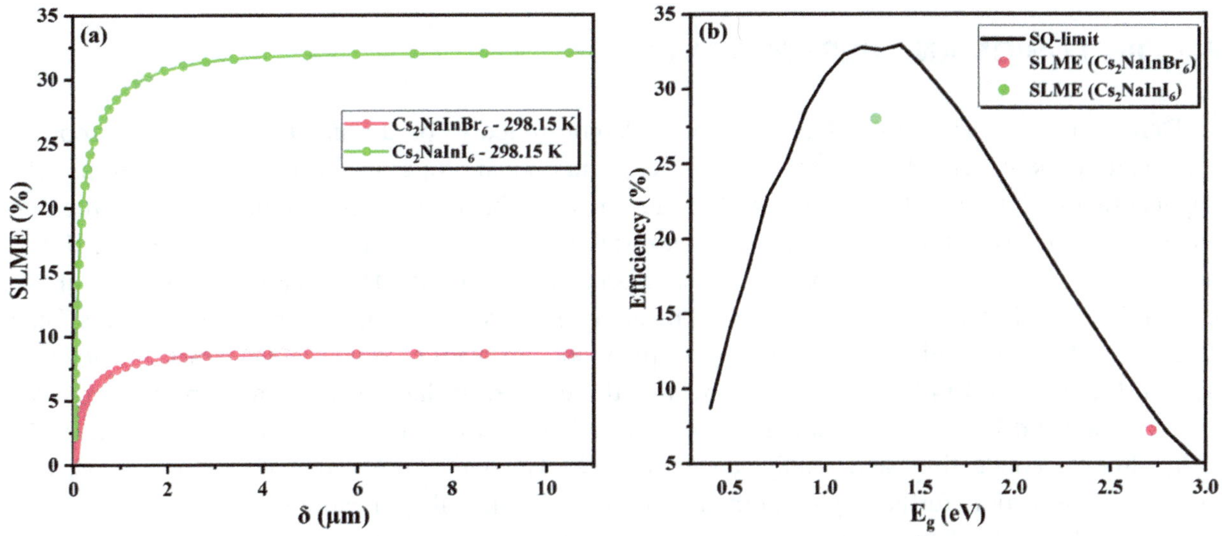

A maximum PCE of 31.95% was obtained for Cs_2NaInI_6 at 10.48 µm and a cell temperature of 298.15 K. This efficiency exceeds the 8.55% recorded for $Cs_2NaInBr_6$ under the same temperature conditions for 5.96 µm. This difference in efficiency is essentially due to the disparity in energy band gaps. Cs_2NaInI_6 exhibits an optimal gap of 1.268 eV (see Figure 9 (b)), placing it within the range where the Shockley–Queisser (SQ) limit is maximised. Both materials display conversion efficiencies slightly below the SQ limit corresponding to their respective band gaps ($Cs_2NaInBr_6$: 8.55% < SQ: 8.57%; Cs_2NaInI_6: 31.95% < SQ: 32.6%), as expected, since the SQ model assumes ideal photon absorption and complete exciton collection. $Cs_2NaInBr_6$'s larger band gap (2.712 eV) accounts for its lower efficiency. The $Cs_2NaInBr_6$ absorber reaches maximum efficiency for 5.96 µm, whereas Cs_2NaInI_6 requires 10.48 µm. At T = 298.15 K, Cs_2NaInI_6 achieves 31.90% efficiency at 5.96 µm and 30.18% at a smaller 1.59 µm thick-film. This indicates that thinner layers can yield high performance while significantly lowering fabrication costs.

Table 4. Solar cell parameters of FTO/Ti O_2*/*Cs_2NaInX_6 *(X=Br, I)/*Cu_2O*/Au perovskite solar cells at 298.15 K.*

Absorber	J_{sc} (A/m²)	V_{oc} (V))	J_m (A/m²)	V_m (V)	P_m (W/m²)	*FF*	η(%)
$Cs_2NaInBr_6$	51.331	1.466	51.331	1.115	54.82	0.728	5.48
Cs_2NaInI_6	348.325	0.902	334.30	0.799	267.10	0.850	26.71

SCAPS-1D simulations were used to confirm the results of the previous SLME analysis. The simulation results are summarised in Table 4. Although the bromide-based perovskite solar cell (PSC) has a higher open-circuit voltage, its lower short-circuit current density reduces its overall efficiency to 5.48%, compared to 26.71% for the iodide-based PSC. This difference in performance is primarily due to the iodide-based compound's more suitable band gap. Similar behaviour is observed in power density:

the Br-based PSC reaches a maximum of 54.82 W/m^2, while the I-based PSC achieves a significantly higher value of 267.10 W/m^2.

6. CONCLUSION AND PERSPECTIVES

Density Functional Theory (DFT) has been demonstrated to be a crucial and adaptable tool for investigating the structural, electronic, mechanical, and optical properties of perovskite materials. The application of DFT has led to a substantial advancement in the researchers' comprehension of the fundamental mechanisms that govern the behaviour, stability, and performance of these materials. Halide double perovskites (Cs_2NaInX_6; X = Br, I) have been shown through DFT analysis to be structurally, mechanically and thermodynamically stable, meaning they are amenable to experimental study. Their calculated elastic moduli suggest that these compounds are ductile, which is necessary for practical use. Furthermore, TB-mBJ calculations show that these materials have direct band gaps of 2.712 eV ($Cs_2NaInBr_6$) and 1.268 eV (Cs_2NaInI_6). The analysis reveals that the valence band mainly consists of Br/I, while the conduction band mainly consists of In/X. The optical properties of these materials, such as UV-Vis absorption, further highlight their viability for solar cell applications.

Among these, Cs_2NaInI_6 is particularly noteworthy, with a predicted solar power conversion efficiency of 26.71% for a 0.8 μm film at 298.15 K. This performance far exceeds that of $Cs_2NaInBr_6$, which exhibits a modest efficiency of 5.48%. This marked difference in performance underscores the necessity of adapting material properties to ensure optimal functionality in photovoltaic devices. Nevertheless, wide band gap materials of this kind remain attractive for use in multi-junction perovskite solar cells and light-emitting diodes (LEDs). These double perovskites can also, in principle, be synthesised using common experimental methods, including hydrothermal, solvothermal, and chemical reflux approaches (Chen et al., 2021; Gray et al., 2020; Jeevaraj et al., 2024; Mahraj et al., 2025; L. Zhang et al., 2021). In the future, perovskite research will likely be driven by the integration of density functional theory (DFT) with emerging techniques such as machine learning (ML) and high-throughput screening. This combination will facilitate the expeditious identification and customisation of novel materials, thereby accelerating the development of perovskite-based technologies for solar energy conversion and other optoelectronic applications. This approach holds great promise in the field of materials science, as it has the potential to yield new avenues for the development of more efficient and scalable solar cell technologies.

REFERENCES

Alshomrany, A. S., Bousbih, R., Sajid, M., Jamil, M., Munir, J., Bani-Fwaz, M. Z., Jabir, M. S., Majdi, H., Assem, E. E., Shaban, M., & Nazar, M. (2025). DFT-Based Tailoring of the Thermoelectric and Photovoltaic Response of the Halide Double Perovskite Cs2TlYF6 (Y = Ag, Co). *Journal of Inorganic and Organometallic Polymers and Materials*, *35*(3), 1591–1602. DOI: 10.1007/s10904-024-03187-0

Anderson, O. L. (1963). A simplified method for calculating the debye temperature from elastic constants. *Journal of Physics and Chemistry of Solids*, *24*(7), 909–917. DOI: 10.1016/0022-3697(63)90067-2

Ashcroft, N., & Mermin, N. D. (1976). Electron Dynamics in Solids. *Solid State Physics*, *116*(117). https://www.scirp.org/reference/referencespapers?referenceid=1476522

ASTM G173–03. (2012). Standard Tables for Reference Solar Spectral Irradiances: Direct Normal and Hemispherical on 37° Tilted Surface. In G173–03. https://www.astm.org/g0173-23.html

Ayyaz, A., Murtaza, G., Shafiq, M., Qasim Shah, M., Sfina, N., & Ali, S. (2023). Exploring structural, thermodynamic, elastic, electro-optic, and thermoelectric characteristics of double perovskites Rb2X-InBr6 (X = Na, K) for photovoltaic applications: A DFT approach. *Solar Energy*, *265*, 112131. DOI: 10.1016/j.solener.2023.112131

Bao, W., Liu, D., & Duan, Y. (2018). A first-principles prediction of anisotropic elasticity and thermal properties of potential superhard WB3. *Ceramics International*, *44*(12), 14053–14062. DOI: 10.1016/j.ceramint.2018.05.002

Bechstedt, F., Seino, K., Hahn, P. H., & Schmidt, W. G. (2005). Quasiparticle bands and optical spectra of highly ionic crystals: AlN and NaCl. *Physical Review B: Condensed Matter and Materials Physics*, *72*(24), 245114. DOI: 10.1103/PhysRevB.72.245114

Becke, A. D., & Roussel, M. R. (1989). Exchange holes in inhomogeneous systems: A coordinate-space model. *Physical Review A: General Physics*, *39*(8), 3761–3767. DOI: 10.1103/PhysRevA.39.3761

Becquerel, E. (1839). On electron effects under the influence of solar radiation. *Comptes Rendus Hebdomadaires des Séances de l'Académie des Sciences*, *9*(9), 561–567.

Best Research-Cell Efficiency Chart | Photovoltaic Research | NREL. (n.d.). Retrieved November 3, 2025, from https://www.nrel.gov/pv/cell-efficiency

Biega, R. I., Filip, M. R., Leppert, L., & Neaton, J. B. (2021). Chemically Localized Resonant Excitons in Silver-Pnictogen Halide Double Perovskites. *The Journal of Physical Chemistry Letters*, *12*(8), 2057–2063. DOI: 10.1021/acs.jpclett.0c03579

Birch, F. (1947). Finite elastic strain of cubic crystals. *Physical Review*, *71*(11), 809–824. DOI: 10.1103/PhysRev.71.809

Born, M., Huang, K., & Lax, M. (1954). Dynamical Theory of Crystal Lattices. Clarendon Press, Oxford, 141. DOI: 10.1119/1.1934059

Born, M., & Oppenheimer, R. (1927). Zur Quantentheorie der Molekeln. *Annalen der Physik*, *389*(20), 457–484. DOI: 10.1002/andp.19273892002

Burgelman, M., Nollet, P., & Degrave, S. (2000). Modelling polycrystalline semiconductor solar cells. *Thin Solid Films*, *361–362*, 527–532. DOI: 10.1016/S0040-6090(99)00825-1

Chen, L., Yang, W., Fu, H., Liu, W., Shao, G., Tang, B., & Zheng, J. (2021). Mn2+-doped Cs2NaInCl6 double perovskites and their photoluminescence properties. *Journal of Materials Science*, *56*(13), 8048–8059. DOI: 10.1007/s10853-021-05822-4

Deng, Z., Wei, F., Sun, S., Kieslich, G., Cheetham, A. K., & Bristowe, P. D. (2016). Exploring the properties of lead-free hybrid double perovskites using a combined computational-experimental approach. *Journal of Materials Chemistry. A, Materials for Energy and Sustainability*, *4*(31), 12025–12029. DOI: 10.1039/C6TA05817E

Dirac, P. A. M. (1930). Note on Exchange Phenomena in the Thomas Atom. *Mathematical Proceedings of the Cambridge Philosophical Society*, *26*(3), 376–385. DOI: 10.1017/S0305004100016108

Dong, C., Guan, X., Wang, Z., Zhao, H., Kuai, Y., & Gao, S. (2023). The effects of cation and halide anion on the stability, electronic and optical properties of double perovskite Cs2NaMX6 (M = In, Tl, Sb, bi; X = Cl, Br, I). 220, 112058. DOI: 10.1016/j.commatsci.2023.112058

Ehrenreich, H., & Cohen, M. H. (1959). Self-consistent field approach to the many-electron problem. *Physical Review*, *115*(4), 786–790. DOI: 10.1103/PhysRev.115.786

Engel, E., & Vosko, S. H. (1993). Exact exchange-only potentials and the virial relation as microscopic criteria for generalized gradient approximations. *Physical Review B: Condensed Matter*, *47*(20), 13164–13174. DOI: 10.1103/PhysRevB.47.13164

Fermi, E. (1928). Eine statistische Methode zur Bestimmung einiger Eigenschaften des Atoms und ihre Anwendung auf die Theorie des periodischen Systems der Elemente. *The European Physical Journal A*, *48*(1–2), 73–79. DOI: 10.1007/BF01351576

Feynman, R. P. (1939). Forces in Molecules. *Physical Review*, *56*(4), 340–343. DOI: 10.1103/PhysRev.56.340

Filip, M. R., Hillman, S., Haghighirad, A. A., Snaith, H. J., & Giustino, F. (2016). Band Gaps of the Lead-Free Halide Double Perovskites Cs2BiAgCl6 and Cs2BiAgBr6 from Theory and Experiment. *The Journal of Physical Chemistry Letters*, *7*(13), 2579–2585. DOI: 10.1021/acs.jpclett.6b01041

Fock, V. (1930). Näherungsmethode zur Lösung des quantenmechanischen Mehrkörperproblems. *The European Physical Journal A*, *61*(1–2), 126–148. DOI: 10.1007/BF01340294

Goldschmidt, V. M. (1926). The Laws of Crystallochemistry. *Naturwissenschaften*, *14*(21), 477–485. DOI: 10.1007/BF01507527

Gray, M. B., Hariyani, S., Strom, T. A., Majher, J. D., Brgoch, J., & Woodward, P. M. (2020). High-efficiency blue photoluminescence in the Cs2NaInCl6:Sb3+ double perovskite phosphor. *Journal of Materials Chemistry. C, Materials for Optical and Electronic Devices*, *8*(20), 6797–6803. DOI: 10.1039/D0TC01037E

Green, M. A. (1981). Solar cell fill factors: General graph and empirical expressions. *Solid-State Electronics*, *24*(8), 788–789. DOI: 10.1016/0038-1101(81)90062-9

Hao, F., Stoumpos, C. C., Cao, D. H., Chang, R. P. H., & Kanatzidis, M. G. (2014). Lead-free solid-state organic–inorganic halide perovskite solar cells. Nature Photonics 2014 8:6, 8(6), 489–494. DOI: 10.1038/nphoton.2014.82

Hartree, D. R. (1928). The Wave Mechanics of an Atom with a Non-Coulomb Central Field. Part I. Theory and Methods. *Mathematical Proceedings of the Cambridge Philosophical Society*, *24*(1), 89–110. DOI: 10.1017/S0305004100011919

Hattabi, I., Abdiche, A., Naqib, S. H., & Khenata, R. (2019). First-principles calculations of elastic and thermodynamic properties under hydrostatic pressure of cubic InNxP1-x ternary alloys. *Zhongguo Wuli Xuekan*, *59*, 449–464. DOI: 10.1016/j.cjph.2019.04.011

Hussain, S., Aftab, S., Muneer, S., Rehman, J. U., Tahir, M. B., Tirth, V., & Algahtani, A. (2025). Computational predictions of physical properties of perovskite CsGeR3 (R = F, Cl, Br, and I) materials for optoelectronic applications. *Inorganic Chemistry Communications*, *173*, 113811. DOI: 10.1016/j.inoche.2024.113811

Hwang, I., Jeong, I., Lee, J., Ko, M. J., & Yong, K. (2015). Enhancing Stability of Perovskite Solar Cells to Moisture by the Facile Hydrophobic Passivation. *ACS Applied Materials & Interfaces*, *7*(31), 17330–17336. DOI: 10.1021/acsami.5b04490

Jamal, M., Bilal, M., Ahmad, I., & Jalali-Asadabadi, S. (2018). IRelast package. *Journal of Alloys and Compounds*, *735*, 569–579. DOI: 10.1016/j.jallcom.2017.10.139

Jeevaraj, M., Sivaganesh, D., Saravanakumar, S., Bahadur, S. A., Sudhahar, S., & Kumar, M. K. (2024). Extrinsic electronic states to tune the luminescence and bonding nature of Cs2NaInCl6 double perovskite. Materials Chemistry and Physics, 311(June 2023), 128569. DOI: 10.1016/j.matchemphys.2023.128569

Jiang, X., Qin, S., Meng, L., He, G., Zhang, J., Wang, Y., Zhu, Y., Zou, T., Gong, Y., Chen, Z., Sun, G., Liu, M., Li, X., Lang, F., & Li, Y. (2024). Isomeric diammonium passivation for perovskite–organic tandem solar cells. *Nature*, *635*(8040), 860–866. DOI: 10.1038/s41586-024-08160-y

Kibbou, M., Haman, Z., Essaoudi, I., & Ainane, A. (2023). Designing new halide double perovskite materials Rb2AgGaX6 (X: Br, Cl) with direct band gaps and high power conversion efficiency. *Journal of Solid State Chemistry*, *317*, 123698. DOI: 10.1016/j.jssc.2022.123698

Kim, J., & Suh, Y. J. (2017). Temperature- and pressure-dependent elastic properties, thermal expansion ratios, and minimum thermal conductivities of ZrC, ZrN, and Zr(C0.5N0.5). *Ceramics International*, *43*(15), 12968–12974. DOI: 10.1016/j.ceramint.2017.06.195

Kinani, M. A., Lekdadri, A., Mir, Y., & Zazoui, M. (2022). First-principles study of double perovskites Cs2Sn(Br1−xIx)6 for the design of high-efficiency thin-film photovoltaics. Computational Condensed Matter, 30(September 2021), e00634. DOI: 10.1016/j.cocom.2021.e00634

Kohn, W. (1999). Nobel Lecture: Electronic structure of matter wave functions and density functionals. *Reviews of Modern Physics*, *71*(5), 1253–1266. DOI: 10.1103/RevModPhys.71.1253

Kohn, W., & Sham, L. J. (1965). Self-Consistent Equations Including Exchange and Correlation Effects. *Physical Review*, *140*(4A), A1133–A1138. DOI: 10.1103/PhysRev.140.A1133

Kojima, A., Teshima, K., Shirai, Y., & Miyasaka, T. (2009). Organometal halide perovskites as visible-light sensitizers for photovoltaic cells. *Journal of the American Chemical Society*, *131*(17), 6050–6051. DOI: 10.1021/ja809598r

Leng, M., Chen, Z., Yang, Y., Li, Z., Zeng, K., Li, K., Niu, G., He, Y., Zhou, Q., & Tang, J. (2016). Lead-Free, Blue Emitting Bismuth Halide Perovskite Quantum Dots. *Angewandte Chemie International Edition*, *55*(48), 15012–15016. DOI: 10.1002/anie.201608160

Li, C., Soh, K. C. K., & Wu, P. (2004). Formability of ABO3 perovskites. *Journal of Alloys and Compounds*, *372*(1–2), 40–48. DOI: 10.1016/j.jallcom.2003.10.017

Mahraj, I., Oublal, M., & Ptok, A. (2025). Theoretical investigation of the photovoltaic properties of MgSnN2 for multi-junction solar cells. *Physica B, Condensed Matter*, *713*, 417261. DOI: 10.1016/j.physb.2025.417261

McClure, E. T., Ball, M. R., Windl, W., & Woodward, P. M. (2016). Cs2AgBiX6 (X = Br, Cl): New Visible Light Absorbing, Lead-Free Halide Perovskite Semiconductors. *Chemistry of Materials*, *28*(5), 1348–1354. DOI: 10.1021/acs.chemmater.5b04231

Miah, M. H., Khandaker, M. U., Rahman, M. B., Nur-E-Alam, M., & Islam, M. A. (2024). Band gap tuning of perovskite solar cells for enhancing the efficiency and stability: Issues and prospects. *RSC Advances*, *14*(23), 15876–15906. DOI: 10.1039/D4RA01640H

Mopoung, K., Dávid, A., Liu, X., Fahlman, M., Buyanova, I. A., Chen, W. M., & Puttisong, Y. (2024). Spin Centers in Vanadium-Doped Cs2NaInCl6 Halide Double Perovskites. *ACS Materials Letters*, *6*(2), 566–571. DOI: 10.1021/acsmaterialslett.3c01590

Morales-Acevedo, A. (2023). Fundamentals of solar cell physics revisited: Common pitfalls when reporting calculated and measured photocurrent density, open-circuit voltage, and efficiency of solar cells. *Solar Energy*, *262*, 111774. DOI: 10.1016/j.solener.2023.05.051

Murnaghan, F. D. (1944). The Compressibility of Media under Extreme Pressures. *Proceedings of the National Academy of Sciences of the United States of America*, *30*(9), 244–247. DOI: 10.1073/pnas.30.9.244

Mursaleen, I., Ain, Q., Kumar, A., Kanabar, B., Rajiv, A., Thatoi, D. N., Ojha, M. K., Karamti, H., Munir, J., & Oza, A. D. (2025). Exploring the effect of pressure on Rb2CsGaCl6 perovskite halide: A pathway for enhanced optoelectronic performance. *Inorganic Chemistry Communications*, *179*, 114723. DOI: 10.1016/j.inoche.2025.114723

Nelson, J. (2003). The Physics of Solar Cells. The Physics of Solar Cells. DOI: 10.1142/p276

Noel, N. K., Stranks, S. D., Abate, A., Wehrenfennig, C., Guarnera, S., Haghighirad, A. A., Sadhanala, A., Eperon, G. E., Pathak, S. K., Johnston, M. B., Petrozza, A., Herz, L. M., & Snaith, H. J. (2014). Lead-free organic-inorganic tin halide perovskites for photovoltaic applications. *Energy & Environmental Science*, *7*(9), 3061–3068. DOI: 10.1039/C4EE01076K

Noor, N. A., Iqbal, M. W., Zelai, T., Mahmood, A., Shaikh, H. M., Ramay, S. M., & Al-Masry, W. (2021). Analysis of direct band gap A2ScInI6 (A=Rb, Cs) double perovskite halides using DFT approach for renewable energy devices. *Journal of Materials Research and Technology*, *13*, 2491–2500. DOI: 10.1016/j.jmrt.2021.05.080

Oublal, M., Dib, M., & El-yakoubi, E. Y. (2025). First-principles study of optoelectronic properties of new lead-free double perovskites Cs2NaInX6 (X=Br, I) for next generation thin-film photovoltaics. *Micro and Nanostructures*, *209*, 208435. DOI: 10.1016/j.micrna.2025.208435

Oublal, M., Dib, M., Khalal, A., Chibane, F., Es-Salhi, E.-S., Nafidi, A., & El-Yakoubi, E. Y. (2025). Prediction of optoelectronic properties and efficiency of halide double perovskites Rb2NaInX6 (X=Br, I) for solar cell applications: A density functional theory approach. *Materials Science in Semiconductor Processing*, *200*, 109917. DOI: 10.1016/j.mssp.2025.109917

Perdew, J. P., Burke, K., & Ernzerhof, M. (1996). Generalized gradient approximation made simple. *Physical Review Letters*, *77*(18), 3865–3868. DOI: 10.1103/PhysRevLett.77.3865

Perdew, J. P., Chevary, J. A., Vosko, S. H., Jackson, K. A., Pederson, M. R., Singh, D. J., & Fiolhais, C. (1993). Erratum: Atoms, molecules, solids, and surfaces: Applications of the generalized gradient approximation for exchange and correlation. *Physical Review B: Condensed Matter*, *48*(7), 4978. DOI: 10.1103/PhysRevB.48.4978.2

Perdew, J. P., Ruzsinszky, A., Csonka, I., Vydrov, O. A., Scuseria, G. E., Constantin, L. A., Zhou, X., & Burke, K. (2008). Restoring the Density-Gradient Expansion for Exchange in Solids and Surfaces. *Physical Review Letters*, *100*(13), 136406. DOI: 10.1103/PhysRevLett.100.136406

Qin, M., Gild, J., Hu, C., Wang, H., Hoque, M. S., Braun, J. L., Harrington, T. J., Hopkins, P. E., Vecchio, K. S., & Luo, J. (2020). Dual-Phase High-Entropy Ultra-High Temperature Ceramics. *Journal of the European Ceramic Society*, *40*(15), 5037–5050. DOI: 10.1016/j.jeurceramsoc.2020.05.040

Roknuzzaman, M., Ostrikov, K., Chandula Wasalathilake, K., Yan, C., Wang, H., & Tesfamichael, T. (2018). Insight into lead-free organic-inorganic hybrid perovskites for photovoltaics and optoelectronics: A first-principles study. *Organic Electronics*, *59*, 99–106. DOI: 10.1016/j.orgel.2018.04.051

Saeed, Y., Amin, B., Khalil, H., Rehman, F., Ali, H., Khan, M. I., Mahmood, A., & Shafiq, M. (2020). Cs2NaGaBr6: A new lead-free and direct band gap halide double perovskite. *RSC Advances*, *10*(30), 17444–17451. DOI: 10.1039/D0RA01764G

Schmidt-Mende, L., Fakharuddin, A., & Kraner, S. (2022). Organic and Hybrid Solar Cells: An Introduction. Organic and Hybrid Solar Cells: An Introduction, 2nd Edition, 1–348. https://doi.org/DOI: 10.1515/9783110736939/XML

Shahbazi, M., & Wang, H. (2016). Progress in research on the stability of organometal perovskite solar cells. *Solar Energy*, *123*, 74–87. DOI: 10.1016/j.solener.2015.11.008

Shakeel, S., Song, P., Shah, S. H., Zada, Z., Huang, T., Laref, A., Hakimi, N., & Faizan, M. (2024). DFT analysis of elastic and optoelectronic properties of Cs2NaXCl6 (X=In, La, Sc, Y) double perovskite compounds. *Materials Chemistry and Physics*, *324*, 129683. DOI: 10.1016/j.matchemphys.2024.129683

Shockley, W., & Queisser, H. J. (1961). Detailed balance limit of efficiency of p-n junction solar cells. *Journal of Applied Physics*, *32*(3), 510–519. DOI: 10.1063/1.1736034

Slater, J. C. (1930). Atomic Shielding Constants. *Physical Review*, *36*(1), 57–64. DOI: 10.1103/PhysRev.36.57

Slavney, A. H., Hu, T., Lindenberg, A. M., & Karunadasa, H. I. (2016). A Bismuth-Halide Double Perovskite with Long Carrier Recombination Lifetime for Photovoltaic Applications. *Journal of the American Chemical Society*, *138*(7), 2138–2141. DOI: 10.1021/jacs.5b13294

Sun, P.-P., Li, Q.-S., Yang, L.-N., & Li, Z.-S. (2016). Theoretical insights into a potential lead-free hybrid perovskite: Substituting Pb2+ with Ge2+. *Nanoscale*, *8*(3), 1503–1512. DOI: 10.1039/C5NR05337D

Sze, S. M., & Ng, K. K. (2006). Physics of Semiconductor Devices. In *Physics of Semiconductor Devices*. Wiley., DOI: 10.1002/0470068329

Tang, T., Hu, D., Zhao, X., Li, L., & Tang, Y. (2022). First-principles study on the mechanical, electronic and optical properties of double halide perovskite Cs2TlSbX6 (X = Cl, Br, I). *Physica Scripta*, *97*(12), 125821. DOI: 10.1088/1402-4896/ac9ff7

Tang, T. Y., & Tang, Y. L. (2023). Physical and optoelectronic properties of double halide perovskites A2CuSbX6 (A = Cs, Rb, K; X = Cl, Br, I) based on first principles calculations. *Chemical Physics*, *570*, 111897. DOI: 10.1016/j.chemphys.2023.111897

Tang, T. Y., Zhao, X. H., Hu, D. Y., Liang, Q. Q., Wei, X. N., & Tang, Y. L. (2022). An ab-initio investigation of novel double halide perovskite Cs2InCoX6(X=F, Cl, Br) materials with direct band structure and broadband light absorption. *Materials Science in Semiconductor Processing*, *152*, 107047. DOI: 10.1016/j.mssp.2022.107047

Tran, F., & Blaha, P. (2009). Accurate band gaps of semiconductors and insulators with a semilocal exchange-correlation potential. *Physical Review Letters*, *102*(22), 5–8. DOI: 10.1103/PhysRevLett.102.226401

Wei, F., Deng, Z., Sun, S., Xie, F., Kieslich, G., Evans, D. M., Carpenter, M. A., Bristowe, P. D., & Cheetham, A. K. (2016). The synthesis, structure and electronic properties of a lead-free hybrid inorganic-organic double perovskite (MA)2KBiCl6 (MA = methylammonium). *Materials Horizons*, *3*(4), 328–332. DOI: 10.1039/C6MH00053C

Wu, Z., & Cohen, R. E. (2006). More accurate generalized gradient approximation for solids. *Physical Review B: Condensed Matter and Materials Physics*, *73*(23), 235116. DOI: 10.1103/PhysRevB.73.235116

Yang, Y., Ma, L., Gan, G. Y., Wang, W., & Tang, B. Y. (2019). Investigation of thermodynamic properties of high entropy (TaNbHfTiZr)C and (TaNbHfTiZr)N. *Journal of Alloys and Compounds*, *788*, 1076–1083. DOI: 10.1016/j.jallcom.2019.02.254

Yaseen, H., Ain, Q., Murtaza, H., Munir, J., Ahmed, A. A. A., Qaid, S. M. H., & Aldwayyan, A. S. (2025). The Influence of Multiple Potentials on the Bandgap Improvement and the Physical Properties of Cs2InSbF6 Fluoroperovskite: A DFT Study. *Journal of Inorganic and Organometallic Polymers and Materials*, *35*(12), 1–14. DOI: 10.1007/s10904-025-03903-4

Yu, L., & Zunger, A. (2012). Identification of Potential Photovoltaic Absorbers Based on First-Principles Spectroscopic Screening of Materials. *Physical Review Letters, 108*(6), 068701. DOI: 10.1103/PhysRevLett.108.068701

Zangwill, A. (2013). Hartree and Thomas: the forefathers of density functional theory. Archive for History of Exact Sciences 2013 67:3, 67(3), 331–348. DOI: 10.1007/s00407-013-0114-4

Zhang, L., Xie, X., Li, D., Yuan, Y., Xue, X., Li, Q., Xu, J., Wang, H., Hu, F., & Zhang, X. (2021). Investigation on lead-free Mn-doped Cs2NaInCl6 double perovskite phosphors and their optical properties. *Optical Materials, 122*, 111802. DOI: 10.1016/j.optmat.2021.111802

Zhang, P., Yang, J., & Wei, S.-H. (2018). Manipulation of cation combinations and configurations of halide double perovskites for solar cell absorbers. *Journal of Materials Chemistry. A, Materials for Energy and Sustainability*, *6*(4), 1809–1815. DOI: 10.1039/C7TA09713A

Zhao, X.-G., Fu, Y., Yang, D., Xu, Q., Zhang, L., Yang, J.-H., Yu, L., Wei, S.-H., & Zhang, L. (2017). Design of Lead-Free Inorganic Halide Perovskites for Solar Cells via Cation-Transmutation. *Journal of the American Chemical Society, 139*(7), 2630–2638. DOI: 10.1021/jacs.6b09645

Zunger, A., & Freeman, A. J. (1977). Ground-state electronic properties of diamond in the local-density formalism. *Physical Review. B, Solid State, 15*(10), 5049–5065. DOI: 10.1103/PhysRevB.15.5049

Chapter 5

Computational Engineering of Direct-Bandgap Double Perovskites for Solar Cells

El-Yakoubi Essediq Youssef
https://orcid.org/0009-0005-9096-3402
Faculty of Applied Physics, Ait Melloul, Université Ibn Zohr, Morocco

ABSTRACT

The pursuit of sustainable, lead-free perovskites has fueled the growing interest in halide double perovskites. This study utilizes first-principles calculations to explore the structural, elastic, electronic, and optical properties of Rb2NaInX6 (X=Br, I) and to evaluate their photovoltaic potential. The structural stability and mechanical strength of Rb2NaInBr6 have been substantiated through the application of Goldschmidt's tolerance factors, elastic constants, and bulk moduli. These findings collectively indicate that Rb2NaInBr6 exhibits superior mechanical performance. Electronic analysis indicates direct band gaps of 2.879 eV and 1.35 eV (TB-mBJ) for $Rb_2NaInBr_6$ and Rb_2NaInI_6, respectively, thereby enabling robust absorption across the ultraviolet (UV) to visible (Vis) spectrum. The simulated solar cell architectures (FTO/TiO2/Rb2NaInX6/Cu2O/Au) demonstrate efficiencies of 4.81% for the bromide and 26.6% for the iodide compound, with an absorber thickness of 0.8 μm. These findings identify Rb2NaInI6 as a promising eco-friendly material for high-efficiency, lead-free solar cells.

1. INTRODUCTION

Energy is the key for modern life and the functioning of society, and its accessibility as well as the effectiveness of its conversion and constitute an important challenge to sustainable development (Green et al., 2014). Renewable energy sources have become increasingly important in this context because they provide means to meet our growing energy needs while reducing their harmful ecological impacts. Photovoltaic solar energy is the greatest prominent renewable energy generators. It is widely available, plentiful, and cost-free. However, for us to fully utilize sunlight's immense potential, solar cell technology must continue to innovate in favor to produce systems that are both highly efficient and commercially feasible (Khan et al., 2025; Papon, 2021).

DOI: 10.4018/979-8-3373-6058-4.ch005

Figure 1. Shares of global renewable electricity generation and outlook by technology, 2014-2030 (Updated 4 Dec 2025).

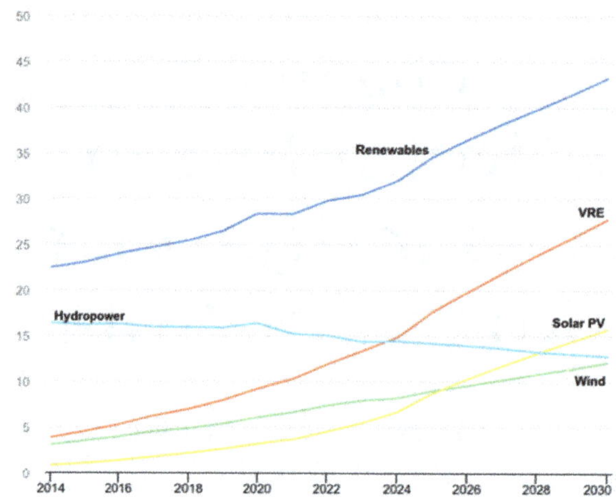

From Becquerel's discovery of the photovoltaic effect in 1839 (BECQUEREL, 1839) to modern thin-film semiconductor systems, innovations in light-harvesting materials have affected the evolution of solar cell technology. Due to their stability and high efficiencies, silicon-based devices have dominated commercial deployment (Green, 2001). However, due to their high manufacturing costs and strict processing requirements, interest in alternative semiconductors that allow for low-temperature, solution-based fabrication has persisted. 2009 saw the disruptive development of metal-halide perovskite solar cells (Kojima et al., 2009), which exhibited the promise of solution-processed semiconductors and achieved quick efficiency gains. However, the seek for stable, lead-free perovskite substitutes has been motivated by worries about instability, environmental deterioration, and Pb^{2} toxicity (Z.-J. Zhao et al., 2020).

Double perovskites $A_2BB'X_6$ have evolved a potential option within these substitutes for other-generation solar cell technology. These materials keep the perovskite structural pattern but show significantly improved thermal and chemical stability by substituting combinations of monovalent B = Ag ,Cu ,N $a^{+}$and trivalent B′= Bi^{3} ,Sb^{3} ,In^{3} cations for divalent Pb^{2+} (Kale et al., 2022; Usman & Yan, 2020). Prototypical compounds like $Cs_2AgBiBr_6$ exhibit remarkable resistance to heat, humidity, and extended illumination, which directly remedies the main drawbacks of hybrid lead halide perovskites (Du et al., 2017; P. Zhang et al., 2018). a study by (Ghosh et al., 2022), their operational endurance, low toxicity, and inherent flaw tolerance make them ideal for ecologically friendly optoelectronic applications. along with these benefits, the device performance of double perovskites is generally limited by their indirect or wide bandgaps, moderate absorption coefficients, and low carrier mobilities. Their optoelectronic behavior has been significantly augmented by recent advancements such dimensional reduction, strain engineering, halide substitution, and B-site alloying, which have allowed more competitive device topologies (Ghosh et al., 2022; H. Li et al., 2021; Mao et al., 2025; Shao et al., 2025; Volonakis et al., 2017). The field has made major strides because to the widespread use of ab-initio computational tools, particularly density functional theory (DFT) and several methods for body perturbation (GW, Bethe-Salpeter equation). These strategies led to an extensive knowledge of the optical, electrical, and structural characteristic of double perovskites. Computational studies have identified bandgap character, predicted carrier effective masses, noticed defect formation energies, and confirmed the critical role of spin–orbit coupling in heavy element systems such as Bi–Sb-based perovskites (Fu et al., 2008; Ünlü et al., 2022).

As well by employing high-throughput DFT screening and machine learning-assisted computational methods, new $A_2BB'X_6$ compositions with optimized bandgaps, optimized carrier transport, and heightened thermodynamic stability have been found quickly (Bartel et al., 2020; Choe et al., 2025; Hung et al., 2025; S. Sun et al., 2019). Ab-initio methodologies are vital for correlating atomic-scale processes to solar cells' overall performance, which directs experimental synthesis and device advancement. They are ideal choices for high-stability single-junction devices, multiple structures, radiation-resistant space photovoltaics, and next-generation optoelectronic systems because of their compositional tunability, inherent reliability, and compatibility with scalable processing techniques (Priya & Stonier, 2025). A thorough understanding of these materials that is validated by both experimental qualification and ab-initio modeling is needed in order to successfully employ them in future solar cell technologies.

2. OVERVIEWS OF THE CHAPTER

This chapter lays the groundwork for comprehensive research by exploring the basic principles, theoretical frameworks, and computational methods that direct the study of double perovskite materials. It starts by examining the structural and chemical flexibility of perovskites in order to establish the foundation to appreciate their many physical properties. The chapter then addresses (DFT) as an essential computational framework, focusing on important factors in identifying appropriate exchange-correlation functionals LDA, meta-GGA, GGA, and hybrid formulations. It also illustrates how DFT employs calculations of atomic forces, total energies, and electronic-structure descriptors (similar as band structures, densities of states, charge densities, optical responses, electronic behavior, elastic parameters, and photovoltaic indicators) to accurately determine epitaxial ground-state geometries, energetics, and intrinsic material properties. likewise, the need of selecting reliable software packages and validating against experimental results is highlighted.

The chapter also provides illustrative case studies, covering the double perovskites Rb_2NaInX_6 ($X = Br, I$), in an attempt to clarify how compositional changes affect material behavior. Aside from that, the chapter exhibits the synergy between device-level performance analysis and first-principles modeling by combining DFT with SCAPS-1D device simulations as an effective multiscale approach for assessing these compounds' practical viability in solar devices. The chapter ends by evaluating the benefits and drawbacks of DFT and highlighting novel strategies including high-throughput computational screening and machine learning-assisted materials discovery.

These elements collectively form a solid basis for detailed evaluations, with DFT providing a balance of computing efficiency and accuracy for predicting ground states and material properties. However, it may struggle with complex or strongly correlated perovskites. The document also explores new research directions, including enhanced computational screening and machine learning-assisted material layout, laying the groundwork for further investigations.

3. BASIC OF THE PEROVSKITE

3.1. Discovery

One of the broadest and most structurally diverse groups of crystalline oxides can be illustrated by the perovskite family compound. The mineral $CaTiO_3$ (Figure 2), whose distinctive framework identifies the classic perovskite structure, is the source of its name. The geologist Gustav Rose recognized this mineral in 1839 and named it after the prominent Russian mineralogist Count Lev Aleksevich Von Perovski (1792–1856) (Bhalla et al., 2000). This revelation established the groundwork for a key structural motif in contemporary materials science.

Figure 2. Perovskite oxide of calcium titanate ($\mathbf{CaTiO_3}$*).*

In their most basic form, which can be expressed by the ABX_3 formula (Figure 3 (a)), a bigger cation occupies the corners of an anion octahedron while a small metal ion is situated in the center. This structural dance is crucial for stability.

3.2. The Rise of Halide Double Perovskites

When Møller discovered in 1959 that halide perovskites, especially cesium lead salts $(CsPbX_3)$, could conduct light-induced charge, the perovskite tale took an unexpected turn (Trihalide S B et al., 1959). Twenty years later, Weber added an organic ion to the A-site to generate hybrid perovskites like $MAPbX_3$ X = Cl, Br, or I) colorful crystals (Weber, 2018). Until 2009, when Miyasaka applied $MAPbI_3$ as a light harvester and ignited the current perovskite-solar-cell revolution (Kojima et al., 2009), the full potential of these materials.

However, there were drawbacks to the evolution of lead-based halide perovskites, including the toxicity of lead itself, thermal fragility, and moisture sensitivity (Kojima et al., 2009; Shahbazi & Wang, 2016; P. Zhang et al., 2018). Germanium (Ge) and tin (Sn) were promising substitutes (Leng et al., 2016; Noel et al., 2014; P.-P. Sun et al., 2016), but their propensity to oxidize compromised stability and effectiveness (Roknuzzaman et al., 2018). Despite confined laboratory settings, most devices still perform poorly despite layered structures, compositional tuning, and protective encapsulation.

Double perovskites are a new approach motivated by this ongoing instability. Researchers synthesized lead-free halide double perovskites compounds the $A_2BB'X_6$ by updating Pb^2 with paired monovalent and trivalent metals (McClure et al., 2016; Slavney et al., 2017). This architectural redesign opened the way for stable, eco-friendly perovskite photovoltaics.

3.3. Structure

The lead-free halide double perovskites $A_2BB'X_6$ crystallize in a face centered cubic structure (Fm $\bar{3}$m) where the A cation sits in the 8c (0.25, 0.25, 0.25) site, the B cation instead 4b (0.5, 0.5, 0.5), the B′ cation instead 4a (0, 0, 0), and halides anion X in 24e (0.25, 0, 0) positions (Figure 3 (b)).

Figure 3. Unit cell of (a) single $\mathbf{ABX_3}$ *and (b) double* $\mathbf{A_2BB'X_6}$ *perovskites.*

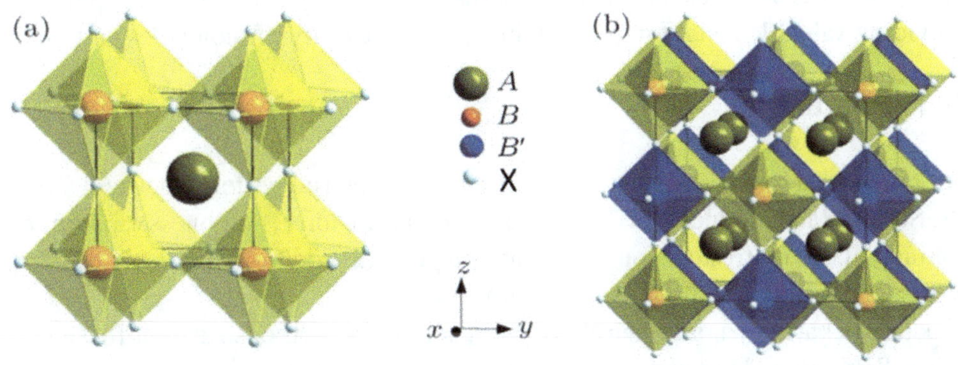

3.4. Stability Criteria (of Double Perovskites)

The favourability of DPs is indicated by the Goldschmidt's tolerance factor (τ) (Goldschmidt, 1926) and the octahedral factor (μ) (C. Li et al., 2004), which are given by equations (1) and (2), respectively. τ should lie between 0.81 and 1.11, while μ should range from 0.41 and 0.9 (X.-G. Zhao et al., 2017).

$$\tau = \frac{r_A + r_X}{\sqrt{2}(\overline{r_B} + r_x)} \tag{1}$$

$$\mu = \overline{r_B}/r_X \tag{2}$$

The ionic radii of the X-site anion and the A-site cation are named by r_X and r_A, correspondingly, the average ionic radius of the cations B and B′ is shown by the expression ($\overline{r_B}$), which is determined as $\overline{r_B} = (r_B + r_{B'})/2$.

4. FUNDAMENTAL MATERIAL CHARACTERISTICS

4.1. Optimised Crystal Structure

In first-principles calculations, obtaining a stable atomic configuration requires relaxing the structure to its lowest-energy state. Within (DFT), this is accomplished through an iterative procedure in which Mechanisms acting in every atom are evaluated and the atomic positions are progressively updated. The relaxation process continues until all forces are reduced below predefined convergence criteria, indicating that the system has arrived in balance. Using the Hellmann–Feynman theorem (Feynman, 1939), by analyzing the negative gradient of the total energy, the force on atom i at point R_i can be found:

$$\mathbf{F}_i = -\frac{d\mathbf{E}_{tot}}{d\mathbf{R}_i} = -\left(\frac{\partial \mathbf{E}_{tot}}{\partial \mathbf{x}_i}, \frac{\partial \mathbf{E}_{tot}}{\partial \mathbf{y}_i}, \frac{\partial \mathbf{E}_{tot}}{\partial \mathbf{z}_i}\right) \tag{3}$$

This method is commonly used within DFT because of it investigates the balance between computational cost and accuracy when treating electron density and exchange-correlation. The optimised geometries form the basis for subsequent investigations into electronic structures and other material properties, providing valuable insight into the stability and reactivity of molecules and materials.

4.2. Elastic Properties

In engineering, mechanical characteristics are crucial for material performance and are impacted by external loads. Strength and hardness (resistance to wear and fracture), toughness (absorption of energy before to breaking), elasticity (restoration of original shape during deformation), and ductility (capacity for deformation without fracture) are important parameters.

These characteristics are essential when choosing materials for parts or structures that could encounter different stresses. Strength and hardness, which suggest resistance to wear and fracture, and elastic constants like Young's modulus and shear modulus, which characterize deformation, are important factors to consider when selecting materials for stressed components. Also, as brittle materials collapse quickly causing little deformation, ductility is crucial for materials employed in forming operations.

For materials used in dynamic or worst-cast load scenarios, where performance is employed by the capacity to slow crack formation and prevent failure from repeated loads, toughness and low resistance as significant.

In engineering applications, determining properties is essential for material design, performance estimation, and guaranteeing safety and dependability. The elastic constants of materials with cubic crystal structures, such as ($Fm\bar{3}m$), can be found using first principles calculations.

Software such as WIEN2k's IRELAST simplifies this process easier (Jamal et al., 2018). The stress-strain relationship is identified elastic constants when one or more deformations are applied to the crystal lattice. The system's volume and total energy change as a result of these deformations. From the energy-strain relationship, the elastic constants are obtained. The basic postulates underlying this procedure is predicated during the evaluation of second derivative the total energy concerning strain. This principle is reflective of symmetric the crystal structure under study.

In an isotropic, linear elastic material, the elastic stiffness constants (C_{ij}), are defined through Hooke's law (Mavko et al., 2020; Timoshenko & Goodier, n.d.), which relates stress and strain:

$$\sigma_i = \sum_j C_{ij}\, \varepsilon_j \tag{4}$$

They can be also obtained from the strain-energy density:

$$\boldsymbol{C}_{ij} = \frac{1}{\boldsymbol{V}_0}\left\{\frac{\partial}{\partial \varepsilon_i}\left(\frac{\partial \boldsymbol{E}_{tot}}{\partial \varepsilon_j}\right)\right\} \tag{5}$$

Where σ_i represents the stress component, while ε_i and ε_j represent the strain components. For cubic crystal, C_{11}, C_{12}, and C_{44}, are the free elastic constants shown by the stiffness matrix by:

$$\begin{bmatrix} \boldsymbol{C}_{11} & \boldsymbol{C}_{12} & \boldsymbol{C}_{12} & 0 & 0 & 0 \\ \boldsymbol{C}_{12} & \boldsymbol{C}_{11} & \boldsymbol{C}_{12} & 0 & 0 & 0 \\ \boldsymbol{C}_{12} & \boldsymbol{C}_{12} & \boldsymbol{C}_{11} & 0 & 0 & 0 \\ 0 & 0 & 0 & \boldsymbol{C}_{44} & 0 & 0 \\ 0 & 0 & 0 & 0 & \boldsymbol{C}_{44} & 0 \\ 0 & 0 & 0 & 0 & 0 & \boldsymbol{C}_{44} \end{bmatrix} \tag{6}$$

The material's resistance to longitudinal stress, transverse deformation in reaction to longitudinal stress, and resistance to shear stress are expressed by the elastic constants C_{11}, C_{12}, and C_{44}, subsequently. Mechanical stability of cubic systems can be verified through Born and Huang criteria (Born et al., 1954): $(C_{11} - C_{12}) > 0, C_{11} > 0, (C_{11} + 2C_{12}) > 0, C_{44} > 0$ and $C_{11} > B > C_{12}$. Here, the bulk modulus, B, represents the resistance of a material to compressive stress and can be calculated by fitting of the total energy-volume curve to the Birch-Murnaghan equation of state (7) (Birch, 1947; Murnaghan, 1944), or through elastic constants (8).

$$\boldsymbol{E}_{tot}\left(V\right) = \boldsymbol{E}_0 + \frac{9\boldsymbol{B}.\boldsymbol{V}_0}{16}\left\{\left[\left(\frac{\boldsymbol{V}_0}{\boldsymbol{V}}\right)^{2/3} - 1\right]^3 \boldsymbol{B}' + \left[\left(\frac{\boldsymbol{V}_0}{\boldsymbol{V}}\right)^{2/3} - 1\right]^2\left[6 - 4\left(\frac{\boldsymbol{V}_0}{\boldsymbol{V}}\right)^{2/3}\right]\right\} \tag{7}$$

$$\boldsymbol{B} = \frac{\boldsymbol{C}_{11} + 2\boldsymbol{C}_{12}}{3} \tag{8}$$

In this case, B' is the pressure derivative of B, while E_0 , V_0 are the ground-state energy and volume.

The stiffness of a material can be analysed through the shear modulus ($G = (G_V + G_R)/2$), where G_V and G_R are the Voigt and Reuss shear moduli, and is expressed by the equations:

$$\boldsymbol{G}_V = \frac{\boldsymbol{C}_{11} - \boldsymbol{C}_{12} + 3\boldsymbol{C}_{44}}{5} \tag{9}$$

$$\boldsymbol{G}_R = \frac{5\boldsymbol{C}_{44}\left(\boldsymbol{C}_{11} - \boldsymbol{C}_{12}\right)}{3\left(\boldsymbol{C}_{11} - \boldsymbol{C}_{12}\right) + 4\boldsymbol{C}_{44}} \tag{10}$$

The ratio (B/G) gives insights on the brittle/ductile nature of a solid material. Values greater than 1.75 signifies a ductile nature. Conversely, a ratio below 1.75 indicates the material's brittle nature.

Additionally, one important mechanical metric that measures a material's stiffness its capability to resist deformation in applied stress is its Young's modulus (*Y*). It is described as the elastic area of deformation's stress to strain ratio and has an associated relationship with the bulk modulus (*B*) and shear modulus (*G*):

$$Y = \frac{9BG}{3B + G} \tag{11}$$

Poisson's ratio (ν) describes the relationship between the axial and lateral deformation of a material when subjected to mechanical stress. This parameter provides essential information about a material's deformation characteristics, helping to distinguish between ductile ($0.5 > \nu > 0.25$) and brittle ($0.25 > \nu$) behaviour and to predict its mechanical stability under stress. It is expressed as:

$$\nu = \frac{3B - 2G}{2(3B + G)} \tag{12}$$

Equation (13) can be employed to get the anisotropic factor (*A*), serves as a key indicator of the of a material's nature anisotropic (Kube, 2016; Ledbetter & Migliori, 2006; Ranganathan & Ostoja-Starzewski, 2008). When (*A*) equals unity, the solid displays perfectly isotropic behaviour, with uniform properties in all directions. The presence and degree of anisotropy within the material are indicated by any deviation from the perfect value.

$$A = \frac{2C_{44}}{C_{11} - C_{12}} \tag{13}$$

4.3. Electronic Properties of Semiconductors

4.3.1. Energy Bands

Quantum mechanics shows that electrons in an isolated atom can occupy only a limited number of well-defined states, each corresponding to a discrete energy level. These levels arise from the quantisation of electronic energy and are characteristic of each type of atom.

When a number of identical atoms are brought together to form a solid, their atomic orbitals begin to interact. This interaction profoundly alters the energy structure: each discrete energy level of an isolated atom splits into a large number of closely spaced states. In a crystal containing an enormous number of atoms, these states become so numerous and so close in energy that they form continuous energy bands. Hence, the electronic bands in a crystal result from the hybridisation of the energy levels of atoms that make up the crystal lattice.

The allowed bands represent the ranges of energy that electrons may occupy, while the forbidden bands (or band gaps) correspond to energy intervals that no electron can occupy. Among the allowed bands, two are particularly important; The valence band (VB) and the conduction band (CB). The energy dislocation among these bands defines the band gap (Eg), a fundamental property that distinguishes conductors, semiconductors, and insulators (Figure 4).

Figure 4. Schematic band diagram of metals, semiconductors and Insulators.

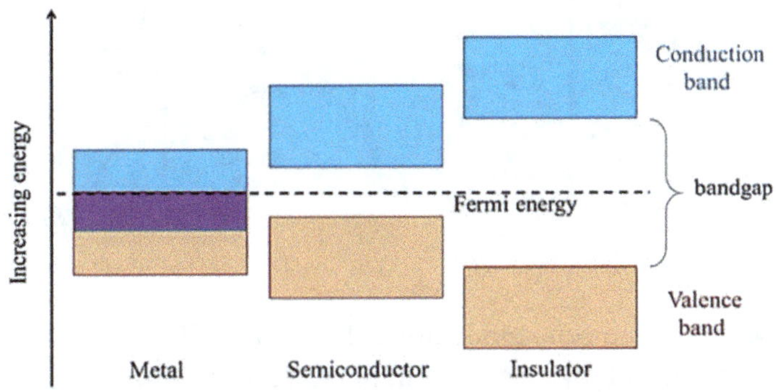

According to (Morab et al., 2023), semiconductors generally have band gaps between 0.1 to 3 eV, which makes them appropriate for photovoltaic applications because this range permits effective absorption of visible light. The difference between the energy levels that are typically inhabited and those that are unoccupied at absolute zero is displayed by the Fermi level.

While photons with energy above the band gap ($h\nu < Eg$) excite electrons from the valence band (VB) to high conduction band (CB) states, generating electron–hole pairs, or excitons, sub-bandgap photons ($h\nu > Eg$) aren't absorbed and no electron excitation takes place. Through ultrafast processes, these excited electrons quickly thermalize to the conduction band minimum (CBM) (f_s). The lack of states in the band gap delays electron hole recombination, permitting for longer carrier lifetimes (μs - ms) that are advantageous for solar extraction. Energy required for excitation can be supplied by various sources, including photon absorption, phonon interactions (lattice vibrations), or energy transfer from other charge carriers. This ability to sustain and extract photogenerated carriers underlies the way of semiconductor-based solar cells.

4.3.2. Type of Gaps

An electron in the CB can release energy and recombine with a hole in the valence band by realizing a photon. The nature of this transition depends on the momentum (k-vector) of the electrons involved. If the maximum of the valence band (MVB) and the minimum of the conduction band (MCB) occur at the same k-vector, the material has a direct band gap. In this case, optical transitions (absorption and emission of photons) can produce without assistance (Hui & O'sullivan, 2023).

Figure 5. Illustration of direct (a) and indirect (b) band gaps.

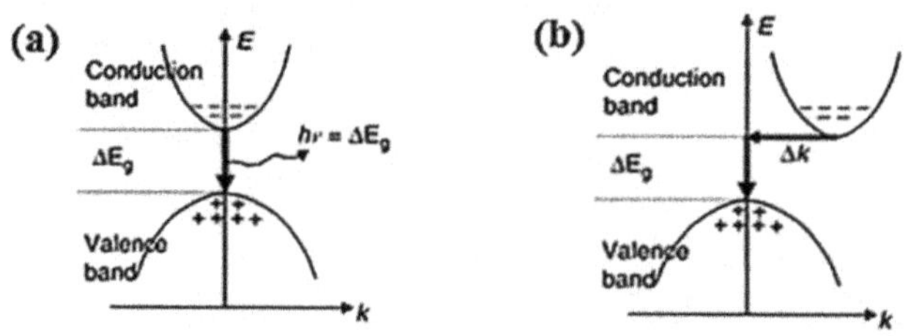

In contrast, an indirect band gap semiconductor exhibits a MVB and a MCB at distinct k-points in the Brillouin zone (Figure 5. (a) and (b)). As a result, photon absorption or emission alone is insufficient to conserve momentum. These transitions require the simultaneous involvement of a phonon to satisfy momentum conservation. For instance, an electron can absorb a photon and a phonon at once to transition to the CB, or emit both a photon and a phonon when recombining. Thus, direct band gap materials are more efficient for optoelectronic applications, while indirect band gap semiconductors exhibit weaker optical transitions due to phonon involvement (Fan et al., 2026; Schmidt-Mende et al., 2022).

4.4. Optical Characteristics

The optical parameters of solids are a valuable object for investigating the electronic structure of energy bands, excitons, impurity levels, localized defects, and the interplay of light with matter. In this paper, we emphasize the calculate various optical parameters for each material under examination, including the dielectric function $\varepsilon(\omega)$, reflectivity $R(\omega)$, and absorption $\alpha(\omega)$.

4.4.1. Dielectric Function

Typically, the optical properties of a material are described using the real $\varepsilon_1(\omega)$ and imaginary $\varepsilon_2(\omega)$ parts of the complex dielectric function, expressed as (Ehrenreich & Cohen, 1959): $\varepsilon(\omega) = \varepsilon_1(\omega) + i\varepsilon_2(\omega)$, where $\varepsilon_1(\omega)$ corresponds to photon dispersion and $\varepsilon_2(\omega)$ is associated with the material's energy absorption. The imaginary part of the dielectric function is determined from the elements of the momentum matrix within orbitals that are occupied and vacant.

$$\varepsilon_2\left(\omega\right) = \frac{4\pi^2 e^2}{\omega^2}\sum_{vc}\sum_{k}\left|\int d^3 r \Psi^*_{c,k}(r)\left(\hat{e}\,.\,v\right)\Psi_{v,k}\left(r\right)\right|^2 \delta\left[E_c(k) - E_v(k) - \hbar\omega\right] \tag{14}$$

Here, ω is frequency of incident photon, $\Psi_{v,k}$ and $\Psi_{c,k}$ symbolizes the wave functions of valence and conduction band quasiparticle (electron) of energies $E_v(k)$ and $E_c(k)$, respectively. The unit vector $\hat{e}$ specifies the polarisation direction, while P represents principal integral value and v denotes velocity operator.

The imaginary part of the dielectric function is determined from the real part using Kramer-Kronig relation:

$$\varepsilon_1\left(\omega\right) = 1 + \frac{2}{\pi}P\int_0^{\infty}\frac{\omega'\varepsilon_2(\omega')}{(\omega'^2 - \omega^2)}d\omega' \tag{15}$$

4.4.2. Reflectivity

The optical reflectivity, $R(\omega)$, represents the quotient between the energy of the reflected wave from a surface when the first the incident wave striking the second.

$$R\left(\omega\right) = \left|\frac{\varepsilon_1(\omega) + i\varepsilon_2(\omega) - 1}{\varepsilon_1(\omega) + i\varepsilon_2(\omega) + 1}\right|^2 \tag{16}$$

Changes in reflectivity can also provide valuable insights into the transition points within the crystal structure. This includes information about the nature of electronic transitions, such as those between the VB and CB, which can be observed as changes in the optical answer of the material at specific frequencies. Such variations are often linked to the presence of excitons, band gaps, and the effects of impurities or defects within the crystal lattice.

4.4.3. Optical Absorption

The light absorption factor $\alpha(\omega)$ quantifies how strongly a material absorbs incident electromagnetic radiation. It is determined by relation:

$$\alpha\left(\omega\right) = \frac{\sqrt{2}\,\omega}{c}\left[\sqrt{\varepsilon_1^2(\omega) + \varepsilon_2^2(\omega)} - \varepsilon_1\left(\omega\right)\right]^{1/2} \tag{17}$$

Here, c denotes the momentum of light in vacuum. $\alpha(\omega)$ depends on a lot of variables, containing the electronic band structure, impurity levels, and localized defects within the material. It provides essential insight into the material's electronic transitions and its interaction with incident electromagnetic radiation.

4.5. Solar Cell Generalities

A portion of the incoming photon energy transforms into electrical power by a solar cell. A semiconductor absorber's bandgap energy (Eg) and the incident light spectrum are two crucial factors that affect its efficiency. The absorption factor of the material, the efficiency of charge dislocates, and recombination losses are important variables controlling the power conversion efficiency (PCE) of solar cell devices.

The Shockley- Queisser (SQ) extent, which establishes a highest efficiency possible for a specific band gap energy of the absorber material, guarantees the physical limit efficiency of a single-junction solar cell. This limit peaks at about 33% under conventional AM 1.5 G illumination, which resembles sunlight filtered by 1.5 atmospheric thicknesses, or a 48.2° zenith angle (Shockley & Queisser, 1961).

A solar cell's absorption of sunlight creates excitons, or electron–hole pairs, which need to be collected and separated before they may recombine. An internal electric field denoted as the intrinsic field is what causes this charge separation. A diode and a current source are part of the same as circuit of an ideal solar cell, as illustrated in Figure 6 (a). The Shockley model states that the total current is made up of a voltage-independent photocurrent (J_{sc}) created under illumination and a dark diode current (J_{dark}), which is clarified by the diode equation. The external load and the diode's variable resistance split the generated photocurrent, with the diode providing the voltage required to push the current through the circuit. Figure 6 (b). depict the resulting current-voltage (J-V) specification. The p-n homojunction diode, which is frequently constructed in silicon (*Si*), serves as a foundational model for illustrating the operation of solar cells. With all of its valence electrons, each atom in pure silicon builds four covalent bonds with nearby (*Si*) atoms. A phosphorus (*P*) atom, which has five valence electrons, establishes four bonds with nearby Si atoms when it is added to the Si lattice, leaving one free electron. Phosphorus becomes an n-type dopant when this additional electron enters the conduction band. The n-type silicon, which has a large number of mobile electrons, is the resultant material.

Figure 6. Equivalent Circuit scheme of an ideal Solar Cell (a), current–voltage (J–V) characteristic (b).

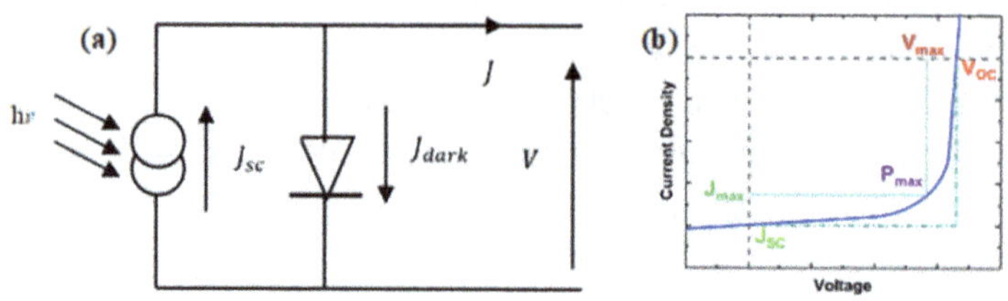

Conversely, when a trivalent atom such as boron (*B*) is incorporated, it forms only three bonds, creating an electron deficiency or "hole." These holes can move freely within the valence band as nearby electrons shift to fill the vacancy. Boron thus plays as a p-type dopant, and the doped compound is referred to as p-type silicon.

In intrinsic (pure) semiconductors, the Fermi level (E_F) is approach a middle of the band gap, equidistant from the VB and CB. Since there are no available electronic states within the gap and thermal excitation is minimal, electrical conductivity is low. In n-type doping, donor atoms such as phosphorus introduce additional electron states proximity to the conduction band, allowing electrons to be thermally motivated into the conduction band where they contribute to current flow. Conversely, p-type doping introduces acceptor states near the valence band, enabling electrons from the valence band to be thermally thrilled into these states, leaving behind mobile holes that carry current.

Perovskite solar cells, which are discussed in detail later, commonly utilize a p-i-n or n-i-p heterojunction structure. These devices combine wide-bandgap emitter layers and a light absorbing material with an appropriate band gap. To enable effective charge separation and collection, strongly doped n-type (such as TiO) and p-type (such as spiro- OMeTAD) transport layers are inserted between an intrinsic (i) layer, which is in charge of light absorption and exciton dissociation.

The efficiency of semiconductors can be predicted by theoretical models such as the spectroscopic limited maximum efficiency model (SLME) (Yu & Zunger, 2012). This model takes into account the absorption coefficient (α), energy band gap, and absorber thickness (L). the standard irradiation such as AM1.5G can be used (Marzo et al., 2016). The current density, J, **of this perfect solar cell is expressed as:**

$$J = J_{sc} - J_0\left(e^{\frac{qV}{k_B T_c}} - 1\right) \tag{18}$$

Here, J_{sc} and J_0 represent the short-circuit current density and the current recombination rate, accordingly, q is charge elementary, and k_B is the Boltzmann constant. This model considers only radiative recombination processes, while non-radiative recombination mechanisms which contribute additional dark current density and consequently lower the open-circuit voltage are neglected (Morales-Acevedo, 2023). J_{sc} and J_0 are symbolized by equations:

$$J_{sc} = q\int_0^{\infty} a(E)I_{sun}(E)dE \quad (19)$$

$$J_0 = \frac{q\pi}{f_r}\int_0^{\infty} a(E)I_{bb}(E)dE \quad (20)$$

Here, the absorptivity of the absorber material, $a(E)$, is described in relation of the nanoparticle thickness $a(E) = 1 - e^{-f_g a(E)L}$, where f_g represent the geometrical that clarification for the optical configuration of the solar cell. Thus research, $f_g = 1$ **is modelled, congruent with idealized configuration with a perfectly reflecting rear surface that eliminates photon losses through the back of the device. In the Shockley-Queisser (SQ) limit, the absorptivity is simplified as** $a(E) = 1$ for photon energies for $E \geq E_g$, and $a(E) = 0$ for photon energies for $E < E_g$, where E_g denotes the stoichiometric band gap of the material. The photon flux density from the standard solar spectrum (AM 1.5 G) is denoted by I_{sun}, while I_{bb} represents the black-body emission spectrum. The radiative recombination fraction, f_r which calculates the proportion of synthesis events that are radiative, is expressed as: $f_r = e^{\Delta/k_BT_c}$, where $\Delta = E^{da} - E_g$, correlates to the energy different aspects of the allowed direct transition energy (E^{da}) and the basic band gap (E_g). Maximum conversion efficiency ($\eta = P_{max}/P_{in}$) of a solar cell is defined as the fraction of the higher extractable power density (P_{max}) to the incident solar power density (P_{in}). Here, $P_{in} = 1000\ W/m^2$ corresponds to the standard solar irradiance inferior of AM1.5G spectrum. The current density-voltage (J-V) curve provides the relationship between current and voltage under illumination, allowing the determination of P_{max}, which corresponds to the results of current density and voltage at the point of highest power results.

The fill factor (FF), an essential performance indicator describing the "squareness" of the J-V curve, is calculated using the expression $FF = P_{max}/(V_{oc}J_{sc})$ (Green, 1981).

The open-circuit voltage, V_{oc}, representing the extremum voltage generated by the solar cell when the current density is zero, and is obtained from the formula:

$$V_{oc} = \frac{k_BT_c}{q}ln\left(1 + \frac{J_{sc}}{J_0}\right) \quad (21)$$

Solar cell performance can be evaluated through simulated perovskite solar cells using simulation tools such as one-dimensional Solar Capacitance Simulator (SCAPS-1D) (Burgelman et al., 2000). The p-i-n junction can be modulated by placing the perovskite material between a Hole Transport Layer (HTL) and an Electron Transport Layer (ETL).

"Substrate/ETL/ $A_2MM'X_6$/HTL/back-contact" is how the architecture is expressed. Several important material and device characteristics that control the electrical and optical behavior of the solar cell are taken into account in SCAPS-1D's simulation.

The band gap energy, absorber layer thickness, and electron affinity (χ) are some of these. The mobilities of electrons and holes (μ_n, μ_p, consecutively), as well as the effective density of states in the valence and conduction bands (N_C, N_V, consecutively), are also taken into account. Indeed, are the thermal velocities of charge carriers (ν_{th}^n for electrons and ν_{th}^p for holes). The donor (N_D) and acceptor (N_A) concentrations, as well as the total defect density (N_t), are included so as to reflect the influence of doping and defect levels. The simulated photovoltaic device's basic electronic structure and transport properties are defined by these factors taken together. The concentration and distribution of carriers in a semiconductor are mostly dictated by the effective density of states at the valence (N_C) and conduction (N_V) band edges. It is defined by:

$$N_{C,V} = 2g_d\left(\frac{2\pi m^*_{e,h}k_BT}{h^2}\right)^{3/2} \tag{22}$$

where T is the temperature, h is Planck's constant, $m^*_{e,h}$ is the effective mass of electrons or holes, and g_d is the degeneracy factor (in this case, set to 1). This formula depicts how the number of usable energy levels for charge carriers is impacted by the electronic band structure.

The following formula can be used for identifying the thermal velocity of charge carriers, which has an enormous effect on their mobility and diffusion as well as the overall performance of the device (Sze & Ng, 2006):

$$v^{n,p}_{th} = \sqrt{\frac{3k_BT}{m^*_{e,h}}} \tag{23}$$

The relaxation period (τ) and the mobility of charge carriers in the presence of an electric field are further associated through an equation $\mu_{n,p} = q\tau/m^*_{e,h}$. Together, these correlations clarify how intrinsic material properties control diffusion, charge transport, and ultimately the absorber material's photovoltaic performance.

The two primary mechanisms that control electrical current flow in semiconductor devices are drift current and diffusion current. Charge carriers' mobility (holes, electrons) as a result of a practice electric field causes drift current. Both majority, minority carriers are accelerated by this field, and the doping levels of the material affect their velocity. Conversely, diffusion current happens when carriers diffuse from areas of higher to lower concentration until equilibrium is established as a result of a concentration gradient. In order to mimic charge carrier transport within semiconductor devices, the SCAPS-1D transport model takes into consideration both of these phenomena by integrating drift and diffusion processes.

5. FUNDAMENTALS OF DENSITY FUNCTIONAL THEORY (DFT)

The development of quantum model in the 1930s made it possible to predict the basic physical characteristics of solids from first principles, which has long been a major goal in condensed matter physics. This chapter covers the basic formulas and approximations of (DFT), the computational process adopted in this chapter. It elucidates how DFT is an effective framework for analyzing a material's optical, electrical, and vibrational characteristics.

5.1. From Quantum Mechanics to DFT

The early 20th century mentioned the scientific revelation of quantum mechanics, represents a major advancement in physics by providing a unified framework for describing matter at the atomic scale through the Schrödinger approach. However, solving the many-body Schrödinger equation for real materials remained computationally intractable, hence, simplifications are necessary. The process of finding an approximate solution typically proceeds in three main stages. First, nuclear and electronic motion are separated via the Born-Oppenheimer approximation (BOA). It distinguishes between electrical and nuclear motion. The second comprises several electronic structural techniques, including (DFT), Hartree, and Hartree-Fock (HF), and the Kohn- Sham (KS) formalism, enabling accurate approximations of complex electron interactions while significantly reducing computational cost.

Until the groundbreaking work of Hohenberg and Kohn in 1964, followed by the Kohn-Sham equations. These developments laid a framework for DFT, enabling accurate approximations of complex electron interactions while significantly reducing computational cost. In this context, the general Schrödinger equation ($H\psi = E\psi$) for a solid containing N electron is introduced:

$$\left[-\sum_{i}\frac{\hbar^2}{2m_e}\nabla_{r_i}^{2}-\sum_{i=1}^{N}\frac{\hbar^2}{2M_i}\nabla_{R_i}^{2}+\frac{1}{4\pi\varepsilon_0}\sum_{i\neq j}\frac{e^2}{2|r_i-r_j|}-\frac{1}{4\pi\varepsilon_0}\sum_{i,j}\frac{Z_ie^2}{|R_i-r_j|}+\frac{1}{4\pi\varepsilon_0}\sum_{i,j}\frac{Z_iZ_je^2}{2|R_i-R_j|}+\right]\psi = E\psi \quad (24)$$

m_e and M_i represent the masses of the nuclei (at positions R_i), and electrons (at positions r_i), respectively. The total energy involves the kinetic energy of both electrons and nuclei, plus potential energy terms statistical for nucleus-nucleus, electron- nucleus and electron-electron interactions. Because an exact solution to this many-body problem is intractable, simplifications are necessary.

5.2. Born–Oppenheimer Approximation

Many computations in condensed matter physics and materials research originated on the Born-Oppenheimer approximation (BOA). It makes the main equation (24) simpler by assuming that atomic nuclei move much more slowly than electrons due to their greater mass. Therefore, the nuclei can be treated as fixed at their instantaneous positions, while electrons move in the external potential created by these stationary nuclei and in the potential due to other electrons (Born & Oppenheimer, 1927).

The Born-Oppenheimer approximation refers to this division of nuclear and electronic movements. Since the nuclei are thought to be stationary, their kinetic energy can be disregarded, and the nucleus–nucleus potential energy becomes a fixed. With approximation, the total Hamiltonian reduces to an electronic Hamiltonian, is now written as:

$$\left[-\sum_{i}\frac{\hbar^2}{2m_e}\nabla_{r_i}^{2}+\frac{1}{4\pi\varepsilon_0}\sum_{i\neq j}\frac{e^2}{2|r_i-r_j|}-\frac{1}{4\pi\varepsilon_0}\sum_{i,j}\frac{Z_ie^2}{|R_i-r_j|}\right]\psi = E\psi \quad (25)$$

However, even after applying the BOA, the electronic Hamiltonian remains complex, as the resulting equation is still a many-body problem with unknown electron–electron interaction terms. Therefore, additional approximations (such as Hartree, Hartree–Fock, or DFT) are necessary for the Schrödinger equation to be solved.

5.3. Hartree Approximation

In 1928, Hartree introduced another approximation to address the problem of multi-electron systems. This approximation assumes that each electron moves independently in an average essential.

The main drawback of the Hartree product was that the electron orbitals lacked antisymmetric properties, meaning they did not properly account for the indistinguishability of electrons. This shortcoming was resolved through the Hartree–Fock (HF) approximation (Fock, 1930).

In this method, the general expression of the Slater determinant for a system of n electrons is used to construct an antisymmetric wavefunction (Slater, 1930).

$$\psi_e(r) = \frac{1}{\sqrt{N!}}\begin{pmatrix} \psi_1(r_1) & \cdots & \psi_n(r_1) \\ \vdots & \ddots & \vdots \\ \psi_1(r_n) & \cdots & \psi_n(r_n) \end{pmatrix} \tag{26}$$

Using the variational principle, this approximation seeks to ascertain a system's ground state. Obtaining the wavefunction that corresponds to the lowest possible energy, or the numerical approximate of the electronic Schrödinger equation, is the first stage. The kinetic energy (T), the Coulomb interaction energy (V_{EE}), and the external potential energy (V_{EX}) can all be added to determine the estimate value of the Hamiltonian:

$$E_{HF} = min\langle \psi_e T + V_{ee} + V_{ext} \psi_e \rangle \tag{27}$$

The Coulomb interaction gives rise to two terms in the Slater determinant: The Hartree contribution ($E_H[\psi_e]$) and exchange integral ($E_x[\psi_e]$).

$$\langle \psi_e V_{ee} \psi_e \rangle = E_H[\psi_e] + E_x[\psi_e] \tag{28}$$

Using the Hartree–Fock approximation, molecular and atomic wavefunctions can be calculated with high numerical accuracy. However, this method becomes computationally expensive for enormous systems due to the large number of basic functions important. It also fails to describe metallic systems accurately, predicting infinite electron velocities near the Fermi energy (E_F). Moreover, it ignores electron correlation effects, except those arising from the antisymmetric nature of the wavefunction ψ_e.

5.4. Formalism of DFT

The sum-electron Schrödinger model has been solved using conventional techniques like Hartree and Hartree–Fock (HF), but HF includes only exchange interactions and neglects electron correlation, limiting its accuracy for solids. In 1964, Hohenberg and Kohn implemented (DFT), offers a more practical alternative by succeeding the many-electron wavefunction with the electron density $\rho(\vec{r})$, reducing the problem's complexity from 3N variables to just 3. This simplification made DFT widely applicable in materials science. For their contributions to DFT and quantum chemical techniques, Kohn and Pople were granted the 1998 Nobel Prize in Chemistry. DFT rests on two theorems: The proper density minimizes the ground-state energy, which is a function of the electron density. To implement DFT computationally, Kohn and Sham (1965) proposed single-electron equations with an effective potential derived from the density. The roots of DFT lie in the Thomas–Fermi model (1927) (Fermi, 1928; Zangwill, 2013), which approximated kinetic energy using local density in a uniform electron gas. Dirac later added a local exchange term (1930), and Slater (1951) refined the model further (Dirac, 1930).

5.4.1. Electron Density $\rho(\vec{r})$

Electrons cannot be seen as distinct entities because they are indistinguishable and inseparable particles. However, the electron density ($\rho(\vec{r})$) describes the likelihood of detecting an electron within a particular volume element.

As a result, electrons are seen as a cloud, and the electron density shows the areas in space where electrons are most likely to be found typically concentrated around the nucleus. The three terms dimensions (x, y, and z) are the sole factors that affect the electron density, which is a positive function. When integrated over all space, it approaches zero at infinity and equals N, the aggregate number of electrons.

$$\rho(r \rightarrow \infty) = 0 , \int \rho(r) dr = N \tag{29}$$

This quantity is sufficient to fully determine the various properties of an atomic system.

5.4.2. Hohenberg-Kohn Theorems

Density Functional Theory (DFT) can be understood through the Hohenberg–Kohn theorems, which were formulated in 1964 (Hohenberg & Kohn, 1964). These theorems establish the theoretical foundation of DFT. The theory is compatible to any system of combining particles in an external potential $V_{ext}(\vec{r})$, not just condensed matter systems with electrons and fixed nuclei.

$$H = -\sum_i \frac{\hbar^2}{2m_e} \nabla_{r_i}^2 + \frac{1}{4\pi\varepsilon_0} \sum_{i \neq j} \frac{e^2}{2|r_i - r_j|} - \frac{1}{4\pi\varepsilon_0} \sum_{i,j} \frac{Z_i e^2}{|R_i - r_j|} \tag{30}$$

The Hamiltonian of such a system consists of three main terms: Three components are noted by the expression: the kinetic energy of the system is indicated by the first term, the electron-electron interaction potential by the second, and the attractive interaction between electrons and the nuclear background (V_{ext}) by the last term.

DFT is inspired on the two Hohenberg-Kohn theorems, conforming to this theorem, a multi-electron system's ground-state charge density $\rho(\vec{r})$ and its exterior potential V_{ext} are identical.

- **First theorem:** The exterior potential $V_{ext}(r)$ is solely established by the ground-state electron density $\rho(\vec{r})$. As a result, every system's characteristic, including its ground and excited states, are exclusively specified by its density. The total energy functional of the system is thus particularly defined by the ground-state density and may be noted as:

$$E[\rho(\vec{r})] = F[\rho(\vec{r})] + \int \rho(\vec{r}) V_{\text{ext}}(\vec{r}) d^3\vec{r} \tag{31}$$

$F[\rho(\vec{r})]$ indicate Hohenberg-Kohn functional, containing all system-independent terms (kinetic and electron–electron interaction energies).

$$F[\rho(\vec{r})] = T[\rho(\vec{r})] + V_{ee}[\rho(\vec{r})] \quad (32)$$

- **Second theorem:** It says that the ground state of the system is the global lowest of a universal functional $F[\rho(\vec{r})]$ of the electron density for a quantized electron occupancy and the existing external potential. The ground-state density is precisely the density reduce this functional. Hence, the energy functional $E[\rho(\vec{r})]$ accomplish both the exact ground-state energy and the exact ground-state density. The energy's lowest value can be characterized by:

$$E_0 = E(\rho_0) = \min E[\rho(r)] \quad (33)$$

$$F_{HK}[\rho(\vec{r})] = T[\rho(\vec{r})] + V_{ee}[\rho(\vec{r})] \quad (34)$$

5.4.3. Kohn-Sham Theorem

In principle, if the exact universal Hohenberg–Kohn functional $F_{HK}[\rho(\vec{r})]$ were known, the corresponding equation would give a precise explanation of the ground-state characteristics of a system. However, since the exact functional is unknown and inaccessible, the Hohenberg-Kohn theorems have limited practical use on their own.

To overcome this limitation, the (DFT) was further developed in 1965 by Kohn and Sham, who introduced a method using non-interacting reference particles to provoke the exact characteristic of a real interacting many-body system (Kohn & Sham, 1965).

The Kohn–Sham (KS) formula proposes a hypothetical system of non-interacting electrons (where electron–electron interaction is set to zero) that recreates the same ground-state electron density as the real interacting system. The main goal of the KS equations is to find the electronic wavefunctions that minimize the several energies of the system. Kohn-Sham Hamiltonian can be defined as:

$$H = H_{KS} = -\frac{\hbar}{2m}\Delta + V_{KS}(r) \quad (35)$$

Where $V_{KS}(r)$, the effective potential, is identified by:

$$V_{KS} = V_{\text{eff}}(r) = V_{\text{ext}}(r) + V_H[\rho(r)] + V_{xc}[\rho(r)] \quad (36)$$

Here, $V_{\text{ext}}(r)$ is external potential due to the nuclei, $V_H[\rho(r)]$ is the Hartree potential representing the average Coulomb repulsion, and $V_{xc}[\rho(r)]$ represents exchange-correlation potential, realized by:

$$V_{xc}(r) = \frac{\partial E_{xc}[\rho(r)]}{\partial \rho(r)} \quad (37)$$

Finalizing, Hartree and Fock's innovative operate from the late 1920s, who developed techniques to simplify electron-electron interactions, is the basis for (DFT), which was invented in 1964. When combined, these advancements have generated a very effective and trustworthy first-principles framework for discovering out a material's ground-state characteristics. DFT requires only the atomic count of the establish atoms as input, making it both computationally practical and physically rigorous. Its fundamental governing equations are:

$$H\varphi_i(r) = \left(-\frac{\hbar^2}{2m}\nabla^2 + V_{KS}(r)\right)\varphi_i(r) = \varepsilon_i\varphi_i(r) \tag{38}$$

The N mono-electronic functions $\varphi_i(r)$ make it possible to obtain the electronic density $\rho(r)$.

$$\rho(r) = \sum_{i=1}^{N} |\varphi_i(r)|^2 \tag{39}$$

The Kohn–Sham approach modified the difficult many-electron problem with N simpler one-electron problems, making accurate electronic structure calculations computationally feasible.

5.4.4. Treatment of Exchange Correlation

The exchange- correlation functional $E_{xc}[\rho(r)]$ is open key in (DFT) because it takes into account the kinetic energy difference between the interacting electron system and the non-interacting Kohn–Sham system, as well as electron exchange resulting from the Pauli exclusion principle and electron correlation generated from Coulomb repulsion.

$E_{xc}[\rho(r)]$ must be approximated since its precise shape is unknown. For this aim, a number of models have been created, including the changed Becke-Johnson (mBJ) potential, the Local Density Approximation (LDA), the Generalized Gradient Approximation (GGA), and hybrid functionals like HSE06. These techniques seek to discover a way the Kohn- Sham equations and contributes a precise description of the missing interaction effects.

Overall, the exchange–correlation functional is essential for connecting the mathematical simplicity of the Kohn–Sham model with the physical reality of interacting electrons, and the quality of its approach determines the accuracy of DFT simulations.

5.4.5. A Local Density Approach (LDA)

The fundamental technique for characterizing exchange-correlation energy in (DFT) that incorporates the Kohn–Sham framework by contribute E_{xc} for associations among various electrons movements is the (LDA). Depending on the local electron density $\rho(r)$, it treats the exchange-correlation potential as a local quantity at each location in space. By integrating the product of local density and exchange-correlation energy per particle (electron) ε_{xc}^{LDA}, the total exchange-correlation energy E_{XC}^{LDA} is determined. LDA is useful for estimating contributions in the Kohn-Sham equations despite being a rough approximation; it is prized for striking a compromise between computing efficiency and precision.

$$E_{XC}^{LDA}[\rho(r)] = \int \rho(r)\, \varepsilon_{xc}^{LDA}[\rho(r)]dr \tag{40}$$

The exchange–correlation energy per particle in a homogenous gas $\varepsilon_{xc}^{LDA}[\rho(r)]$ is named by:

$$\varepsilon_{xc}^{LDA}[\rho(r)] = \varepsilon_{x}^{LDA}[\rho(r)] + \varepsilon_{c}^{LDA}[\rho(r)] \quad (41)$$

The exchange energy per particle is described with:

$$\varepsilon_{x}^{LDA}[\rho(r)] = -C_x \cdot \int \rho^{4/3}(r)dr \; ; C_x = \frac{3}{4}\left(\frac{3}{\pi}\right)^{1/3} \quad (42)$$

There is no specific scientific formula for the correlation part.

Despite being incredibly successful for many systems, the (LDA) has obvious drawback (Zunger & Freeman, 1977). It frequently struggles with systems where the electron density fluctuates quickly, overestimates binding and correlation energies, and underestimates ionization energies and band gaps. Its exclusive local treatment of electron density is the source of these flaws. This led to the formulation of the (GGA). By integrating both the local electron density and its spatial gradient to the exchange-correlation energy, GGA improves LDA and leaves density inhomogeneity more precisely depicted. This breakthrough improves the accuracy of cohesive energies, bond lengths, and lattice constants in DFT calculations, particularly for materials incorporating transition or heavy elements.

5.4.6. GGA Stands for Generalized Gradient Approximation

The (GGA) originated as a solution to the problems of the (LDA), which is unable to accurately estimate the exchange-correlation energy in systems where the electron density varies quickly, such as molecules or inhomogeneous materials. GGA adds a correction by considering the gradient of the electron density $\nabla\rho(r)$, whereas LDA only considers the local electron density $\rho(r)$ and assumes a homogenous electron gas.

This additional term improves LDA's visualization of real systems by permitting the exchange–correlation energy to adjust for variation in electron density (Kohn & Sham, 1965). The (GGA), designed by Perdew and colleagues, integrates gradient-dependent and local aspects to improve the exchange-correlation functional. A general equation for the GGA exchange-correlation energy can be found:

$$E_{xc}^{GGA} = \int f_{xc}^{GGA}[\rho(r), \nabla\rho(r)]d^3r \quad (43)$$

By introducing the density gradient $\rho(r)$, GGA extends LDA by giving a more accurate description of electronic structures. This results in greater predictability of molecule geometries, reaction energy, and attachment characteristics while keeping high computational efficiency. Because GGA can rectify LDA's incorrect estimate of lattice constants, it performs better than LDA in predicting physical characteristic notably cohesive energies, bond lengths, and lattice factors, making it a preferred option for solid-state and molecular systems.

Band gaps and other optical and electrical characteristics are difficult for both GGA and LDA to precisely define. Becke-Perdew (BP), Perdew-Wang (PW91) (Perdew et al., 1993), Perdew-Burke-Ernzerhof (PBE) (Perdew et al., 1996), its variant PBEsol (for solides) (Terentjev et al., 2018), Engel-Vosko (EV-GGA) (Engel & Vosko, 1993), and Wu-Cohen (WC) (Wu & Cohen, 2006), have all been developed in an effort to increase accuracy. With better forecasts of atomic energies, bond strengths, and thermochemical characteristics, PBE and its modified variants are the most popular among them.

5.4.7. Hybrid Functionals

Hybrid functionals enhance Density Functional Theory (DFT) by integrating specific non-local exchange energy from (HF) theory into the standard DFT exchange-correlation approach. This method analyzes exchange interactions by applying Hartree-Fock analysis with either (LDA) or (GGA). To estimate the total exchange- correlation energy, it employs a hybrid functional which includes HF and DFT exchange and correlation components. The creation of hybrid functionals that integrate non-local exchange energy from Hartree–Fock (HF) theory into the DFT exchange–correlation framework has advanced (DFT). While DFT, which frequently uses (GGA) or (LDA), successfully characterizes correlation effects, this technique uses HF to capture exchange interactions. A weighted combination of contributions from both HF and DFT is the computation of the complete exchange-correlation energy in hybrid functionals. Heyd et al.,(Heyd et al., 2003) invented a famous examined hybrid functional for solids known as the Heyd-Scuseria-Ernzerhof (HSE) functional. It comes from the (GGA) functional of the Perdew- Burke- Ernzerhof (PBE):

$$E_{xc} = \alpha E_x^{SR,HF} + (1 - \alpha)E_x^{SR,PBE} + E_x^{LR,PBE} + E_c^{PBE} \quad (47)$$

Short-range (SR) and long-range (LR) components constitute the exchange energy in the HSE functional. Only a small portion of the Hartree-Fock exchange is contained in the short-range portion; the long-range component is still essentially semi-local. The mixing parameter (α), typically set to 0.25, accomplish the proportion of exact exchange included in the short-range part. Another important factor is the screening parameter (λ), it regulates how far apart short and long-range interactions through the error function. In the commonly used HSE06 version (Krukau et al., 2006), λ is set to 0.11 bohr^{-1}. This formulation provides a balance between computational efficiency and improved accuracy for solid-state systems.

5.4.8. Modified Becke-Johnson Potential (mBJ)

The adjusted Becke-Johnson (mBJ) potential is a type of meta-GGA functional designed to provide highly accurate predictions of band gaps in semiconductors. Originally introduced by Becke and Johnson (BJ) in the WIEN2K code, the mBJ potential improves calculated band gaps, bringing them much closer to experimental values. This makes mBJ a widely recognized and effective approach for studying semiconductor electronic structures (Tran & Blaha, 2009). Mathematically, the mBJ potential takes the from:

$$V_{x,\sigma}^{TB-mBJ}(r) = c\,V_{x,\sigma}^{BR}(r) + \left(3c - 2\right)\frac{1}{\pi}\sqrt{\frac{5}{6}\frac{t_\sigma(r)}{\rho_\sigma(r)}} \quad (44)$$

Where, c is named by equation (49), $V_x^{BR}(r)$ formulated the Becke-Roussel (BR) potential and is simulated by equation (50) (Becke & Roussel, 1989), $t_\sigma(r) = (1/2)\sum_{i=1}^{N_\sigma} \nabla\psi_{i,\sigma}^{*}\psi_{i,\sigma}$ denotes the kinetic-energy density, and $\rho_\sigma(r) = \sum_{i=1}^{N_\sigma} |\psi_{i,\sigma}|^2$ is the electron density.

$$c = \alpha + \beta\left(\frac{1}{V_{cell}}\int\frac{|\nabla\rho_\sigma(r)|}{\rho_\sigma(r)}d^3r\right)^{1/2} \quad (45)$$

$$V^{BR}_{x,\sigma}(r) = -\frac{1}{b_\sigma(r)}\left(1 - e^{-x_\sigma(r)} - \frac{1}{2}x_\sigma\left(r\right)e^{-x_\sigma(r)}\right) \quad (46)$$

With $\alpha = -0.012$ and $\beta = 1.023\ bohr^{1/2}$, refer by Tran and Blaha through diminishing the average measurement error between theorical and data band gaps, $\nabla\rho_\sigma(r)$ depict the gradient of electron density, and $b_\sigma = [x_\sigma^3(r)\, e^{-x_\sigma(r)}/(8\pi\rho_\sigma(r))]^{1/3}$, where x_σ is related to $t_\sigma\, \rho_\sigma$, $\nabla\rho_\sigma$, and $\nabla^2\rho_\sigma$.

The main feature of the changed Becke-Johnson (mBJ) functional is its ability to precisely predict exchange and correlation energies, which significantly improves the simulate of band gaps in semiconductors. Suggested by Tran and Blaha, mBJ addresses the limitations of LDA and GGA, which tend to underestimate and overestimate lattice parameters, respectively. By averaging lattice parameters from LDA and GGA and using them in a self-consistent field (SCF) determination with the mBJ potential, the functional achieves band gap values in excellent agreement with experiment. Moreover, mBJ provides these improvements with lower computational cost compared to hybrid functionals, making it an efficient and reliable choice for semiconductor band gap predictions.

6. BASIS SETS AND PSEUDOPOTENTIALS

(DFT) relies on an efficient and accurate representation of the electron wavefunctions. In practice, the Kohn–Sham orbitals and other relevant quantities require discretization, manageable basis set. The choice of basis set strongly influences both the computational cost and the precision of the estimations. Several widely used approaches in solid-state and materials simulations are outlined below.

6.1. Plane-Wave (PW)

Plane-wave basis collections consist of the most popular representations in periodic systems, particularly for solid-state calculations. In this manner, the wavefunctions are expanded as a total of plane waves with wavevectors that are compatible with the periodic boundary conditions of the crystal lattice. Plane-wave methods are particularly attractive because:

- They allow systematic convergence by simply increasing the kinetic energy cutoff, ensuring a controllable approximation.
- They naturally respect the translational symmetry of crystalline solids.
- They are well-suited for Fourier-transform-based techniques, which facilitate efficient evaluation of integrals in reciprocal space.

PW basis sets are widely implemented in codes such as VASP, CASTEP, and Quantum ESPRESSO, which often combine them with pseudopotentials or the Projector Augmented-Wave (PAW) approach to reduce the computational cost attached to core electrons. One limitation of PW basis sets is that they are inefficient for describing rapidly varying wavefunctions near atomic nuclei, which is why pseudopotentials or augmented schemes are typically required.

6.2. Augmented Plane Wave (APW)

The Augmented Plane Wave (APW) approach, previously mentioned by Slater, is one of the earliest all-electron techniques for solving the Kohn–Sham equations in periodic solids. In APW, space is subdivided into two regions: atomic (muffin-tin) spheres centred per nucleus and the interstitial region between them.

- Contained to the spheres: radial solutions of the Schrödinger equation multiplied by spherical harmonics, accurately capturing the speedy variations of wavefunction approach the nucleus.
- In the interstitial region: plane waves, suitable for the smooth behaviour of the wavefunction in the space between atoms.

A limitation of the original APW method is that the radial functions are energy-dependent, rendering the secular equation non-linear and computationally demanding, as it requires iterative solutions for each eigenvalue. To overcome this, Andersen developed the Linearized Augmented Plane Wave (LAPW) simulation.

6.3. Augmented Linearized Plane Wave (LAPW)

Method was developed to overcome this limitation. LAPW replaces the energy-dependent radial functions with a linear approximation around a reference energy, making the eigenvalue problem linear and numerically stable. LAPW retains the high accuracy of all-electron calculations and is suitable for systems where core electrons play an important role. The LAPW method is used in codes like WIEN2k. In this way, space is subdivided into non-overlapping atomic spheres (muffin-tin regions) and the outside region between atoms. The wavefunction is explain as:

- Inside the spheres: Radial functions enhanced by spherical harmonics, accurately capturing the rapid variation near nuclei.
- Outside region: Plane waves, suitable for smoother part of the wave function.

This dual representation allows LAPW to achieve extremely high accuracy for properties sensitive to the core and valence electrons, such as electric field gradients, and precise electronic band structures. This method is computationally more expensive than plane-wave pseudopotential methods, especially for large systems and assumes spherical symmetry for the potential inside muffin-tin spheres (addressed by FP-LAPW).

6.4. Full-Potential LAPW (FP-LAPW)

The Ful-potential LAPW is an evolution of LAPW that removes the spherical approximation of the potential and charge density into the muffin-tin spheres. This is particularly important for materials where the electron density is highly anisotropic, or where low-symmetry crystal structures are present. This allows the potential and charge density to vary freely, providing very high accuracy for low-symmetry systems, surfaces, and complex materials. FP-LAPW is considered a benchmark method and is widely implemented in WIEN2k.

6.5. The Wien2k Calculation Code

We implemented the FP-LAPW approach in this work, which is implemented in the Wien2k code. Several distinct programs coupled by C make up the Wien2k code (Blaha et al., 2001). Shell Script:

NN: This application assists in estimating the atomic radius of a sphere by offering the distances between nearest neighbors.

LSTART: A software that creates atomic densities and decides how various orbitals, as in core states with or absence local orbitals, are handled in the band structure calculation.

SYMMETRY: It identifies the point group of specific atomic sites, establishes the local rotation matrices, creates the LM expansion for the lattice harmonics, and performs the invariance transformations of the space group. Thus, until the convergence criteria are satisfied, a self-consistent cycle is initiated and continued. The steps in this cycle are as follows:

- LAPW0: Establishes the potential density.
- LAPW1: Determines the eigenvectors, eigenvalues, and valence bands.
- LAPW2: Computes the valence densities of the eigenvectors.
- LCORE: Identifies the densities and core states.

7. COMPUTATIONAL INVESTIGATION OF Rb_2NaInX_6 $(X = Br, I)$

In this section of the chapter, two examples of lead-free double perovskites Rb_2NaInX_6 $(X = Br, I)$ will be studied applying DFT and SCAPS-1D models. The consequences of structural, elastic, electronic, optical, and photovoltaic characteristics will be presented and discussed.

7.1. Computational Details

The Full- Potential Linearized Augmented Plane wave (FP-LAPW) approach adhere assessment structure of (DFT) was used in the research presented to interpret the electrical and optical proprieties of perovskite materials. Along with the (GGA) utilizing the Perdew-Burke-Ernzerhof (PBE) functional for the exchange-correlation potential, Tran and Blanha's changed Becke-Johnson (TB-mBJ) potential was further employed to increase the precision of band gap estimates. This led to the usage of a denser mesh of 36 × 36 × 36 (40,000 k-points) for the optical characteristic and a 17 × 17 × 17 k-points mesh (5000k-point) for structural optimization and electronic calculations. To determine core versus valence character, the plane-wave cut off $R_{MT}K_{max}$ was adjusted to 8 the energy cut-off was stabilized at -6.0Ry. To avoid atomic sphere overlap, muffin-tin radii (R_{MT}) of 2.45, 2.15, 2.15, 2.40, and 2.50 a.u. were used for Rb, Na, In, Br, and I, accordingly. The extremum angular momentum ℓ_{max} was assumed to be 10 and the highest reciprocal lattice vector G_{max} to be 12 $bohr^{-1}$. To guarantee the accuracy and consistency of the computational results, the convolution requirements for the self-consistent field (SCF) computations were set at 0.0001 Ry for energy and 0.001 e for charge density.

Figure 7. The typical architecture of the ***FTO/TiO_2/Rb$_2$NaInX_6 (X = Br, I)/Cu$_2$O/Au*** *PSC devices.*

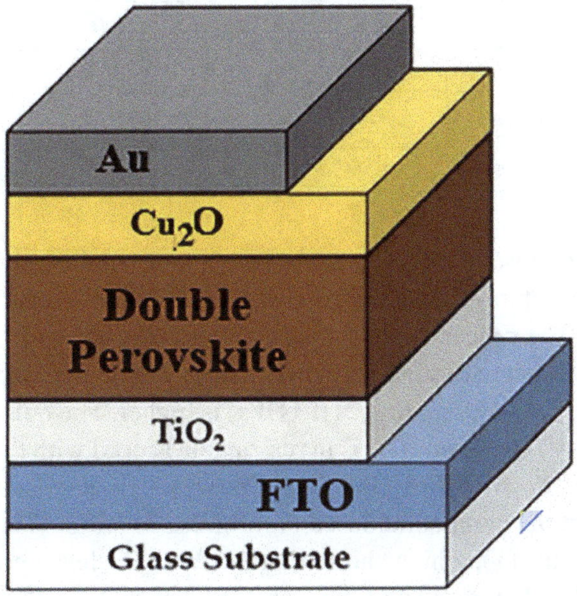

$FTO/TiO_2/Rb_2NaInX_6\ (X = Br, I)/Cu_2O/Au$ perovskite solar cells (PSCs) full-device simulations were performed utilizing SCAPS-1D (Figure **7**). A fluorine-doped tin oxide (FTO) substrate is employed to generate the lead-free double perovskites, with p-type Cu_2O acting as the hole transport layer (HTL) and n-type Ti_2O operating as the electron transport layer (ETL). The back contact is made of gold (Au). The absorber thickness affects how well solar cell devices work, with typical values ranging from 0.4 to 1.4 μm for excellent efficiency at a favourable cost. For consistency, the ETL, HTL, and FTO layers were fixed at 0.023 μm, 0.24 μm, and 0.5 μm, correspondingly, in the SCAPS-1D study, guaranteeing good efficiency at a fair price. Additional information exists in citation (Oublal et al., 2025).

7.2. Structural and Elastic Characteristics

The lead-free halide double perovskites, with a face-centred cubic structure ($Fm\bar{3}m$ symmetry), show important structural features based on their atomic positions. A geometry optimisation was completed to verify the ground-state structural factors, containing lattice parameter, ground-state energy, volume, bulk modulus, and other factors like Goldschmidt's tolerance factor (Table 1). It was found that replacing bromine (Br) with iodine (I) increases the lattice parameter from 11.107 Å to 12.059 Å due to the bigger size ionic radius of iodine. This results in larger inter-atomic distances, reducing rigidity and density, which is signified in the bulk modulus decreasing from 21.72 GPa to 15.54 GPa for $Rb_2NaInBr_6$ and Rb_2NaInI_6, respectively. The difference in electronegativity between bromine and iodine also affects the bonding, with iodine forming weaker bonds with indium and sodium compared to bromine.

Table 1. Goldschmidt's tolerance factor (τ), bulk modulus (B), pressure derivative (B'), lattice parameter (a), ground state energy (E_0), bulk at ground state (V_0), and octahedral factor (μ) of Rb_2NaInX_6 (X=Br, I)

Compounds	a=b=c (Å)	E_0 (Ry)	V_0 ($Å^3$)	B (GPa)	B' (GPa)	τ	μ
$Rb_2NaInBr_6$	11.107	-55298.1568	1370.22	21.77	5.0136	0.906	0.464
Ref.(Ayyaz et al., 2023)	11.18	-55297.9	–	23.0	7.17	0.41	0.89
Rb_2NaInI_6	12.06	-109446.9552	1754.05	15.76	5.129	0.891	0.413

The norms of both the Goldschmidt's tolerance factor and octahedral factor is satisfied for both materials, indicating their structural stability. The various elastic constants of Rb_2NaInX_6 ($X = Br, I$) are summarized in Table 2. Their mechanical stability in the cubic structure is demonstrated by the obtained C_{ij} elastic constant, which satisfies Born-Huang criteria. Compared to the iodide-based compound (15.9 GPa), the bulk modulus of $Rb_2NaInBr_6$ (21.13 GPa) is higher. The bulk modulus qualities realized from the energy-volume (E–V) curve analysis is in reasonable accord with the values established. When comparing the shear moduli, $Rb_2NaInBr_6$ (10.65 GPa) proves a greater shear modulus than Rb_2NaInI_6 (7.955 GPa), suggesting that the compound based on bromide is stiffer. This pattern is also evident in the computed Young's modulus (*Y*), which shows that $Rb_2NaInBr_6$ has a greater value of (24.31 GPa) than Rb_2NaInI_6, which has a value of (20.45 GPa). This implies that the *Br*-based compound has better hardness, stiffness, and compressive force resistance. Also, both compounds' B/G ratios are surpassing 1.75, revealing that both materials are ductile.

Table 2. Elastic constants for Rb_2NaInX_6 (X=Br, I).

Compounds	C_{11} (GPa)	C_{12} (GPa)	C_{44} (GPa)	B (GPa)	G (GPa)	Y (GPa)	B/G
$Rb_2NaInBr_6$	38.8	12.3	9.2	21.13	10.65	24.31	1.98
Rb_2NaInI_6	29.3	9.2	6.8	15.9	7.955	20.45	1.99

For both materials, the calculated values full below unity, revealing the presence of elastic anisotropy despite their cubic symmetry. To further assess the directional, rely on the physical parameters revealed Young's modulus (*Y*), linear compressibility (β), shear modulus (*G*) and Poisson's ratio (ν) tow- and three-dimensional graphical representations are shown in Figure 8.

Figure 8. (Y), (β), (G) and (v) Young's modulus, linear compressibility, shear modulus and Poisson's ratio are illustrated in (2D) and (3D) for (a) $Rb_2NaInBr_6$ *and (b)* Rb_2NaInI_6, *correspondingly.*

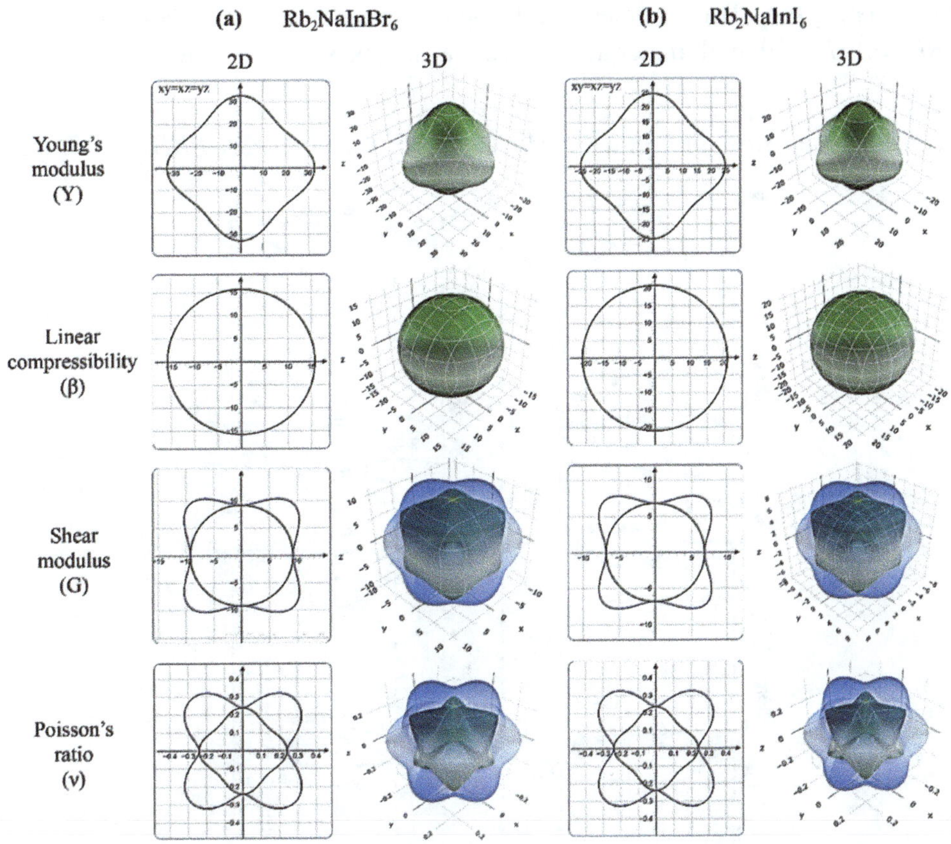

These plots were generated using the ELATE code (Gaillac et al., 2016). In such visualizations, perfectly circular (2D) and spherical (3D) contours correspond to isotropic behavior, whereas distortions from these ideal shapes reflect anisotropic characteristics. Notably, similar trends are observed across the (*xy*), (*xz*) and (*yz*) crystallographic planes, confirming the consistency of the elastic response.

Analysis of Young's modulus (*Y*) reveals pronounced anisotropy across all three crystallographic planes for both compositions. The highest stiffness is observed along the principal axes $x\ (-x)$, $y\ (-y)$ and z $(-z)$ whereas the lowest values occur along the diagonal orientations. When a uniaxial load is applied, the resulting axial contraction or expansion of a solid is quantified by its linear compressibility (β). For materials intended for solar cell applications, a low β is particularly desirable, as it enhances resistance to pressure-induced degradation (Ayyaz et al., 2023). The two- and three-dimensional visualizations indicate that linear compressibility remains essentially isotropic in both materials.

7.3. Electronic Characteristics

Figure 9 illustrates the electronic band structures of Rb_2NaInX_6 (X=Br, I) applying the GGA-PBE and TB-mBJ exchange-correlation potentials. The calculated band gaps, reveal that each elements have direct semiconductor band gaps at the Γ point, with values from 1.3 eV for $Rb_2NaInBr_6$ to 0.248 eV for Rb_2NaInI_6 under GGA-PBE calculations. The band gap of $Rb_2NaInBr_6$ is slightly smaller than the previously reported 1.6 eV (Noor et al., 2021), which could be due to differences in computational parameters. Upon coating the TB-mBJ potential, the band gap increases to 2.879 eV for $Rb_2NaInBr_6$ and 1.35 eV for Rb_2NaInI_6, with both materials behaving as p-type semiconductors.

Figure 9. Electronic band structures of $\boldsymbol{Rb_2NaInX_6}$ *(X=Br, I) using GGA-PBE and TB-mBJ.*

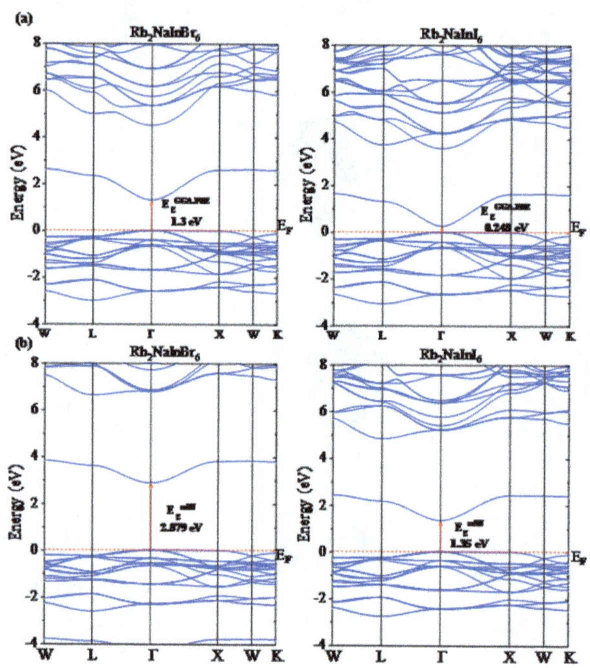

The replacement of *Br* with *I* leads narrow in the band gap due to greater ionic radius of iodine. The band gaps of $Rb_2NaInBr_6$ and Rb_2NaInI_6 descend in the visible and near-infrared regions, suggesting potential applications in optoelectronic and photovoltaic devices.

Figure 10. TDOS and PDOS of $\boldsymbol{Rb_2NaInX_6}$ *(X=Br, I) using TB-mBJ.*

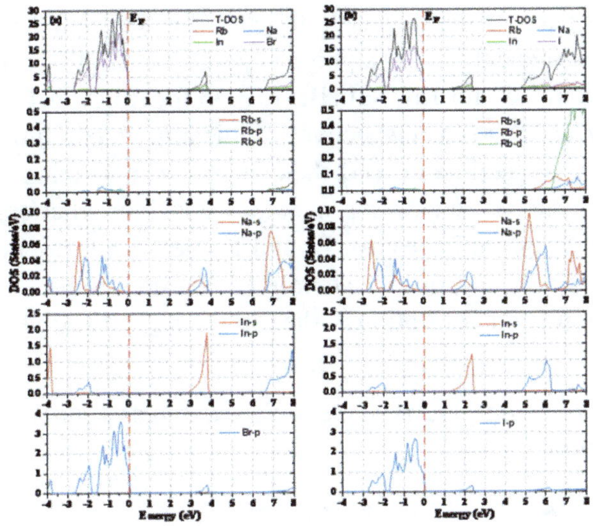

The total and partial density of states (TDOS and PDOS) provide additional details regarding how atomic orbitals affect the electrical structure of the material. The TB-mBJ approach, which produces band gaps comparable to those seen in tests, was used to compute these densities. In Figures 10. (a) and (b), which reflect the TDOS and PDOS for both compounds, the Fermi level is presented by orthogonal dashed lines.

For $Rb_2NaInBr_6$, with lowest contributions from Na(s-p), Rb(p-d-s), and In(s-p) states, Br-p states make up the majority of the valence band highest (VBM). The VBM in Rb_2NaInI_6 is dominated by I-p states, but at a slightly lower intensity than Br-p. With an insignificant contribution from Na-s states, In-s and X-p states dominate the conduction band minimum (CBM). The band gap diminishes due to less hybridization between the X-p and In-s orbitals when Br is substituted with I, shifting the In(s-p) and Na(s-p) densities closer to the Fermi level. This change lowers the band gap and rises the valence band edge in conjunction with rising in p-orbital energy.

$$\frac{1}{\mathbf{m}^*} = \frac{1}{\hbar^2}\frac{\partial^2 \mathbf{E}}{\partial \mathbf{k}^2} \tag{47}$$

The effective carrier masses were determined by analysing the curvature of the band edges utilizing model (51). For $Rb_2NaInBr_6$ and Rb_2NaInI_6, the effective electron masses were found to be $0.2\,m_0$ and $0.15\,m_0$ along Γ-L orientation, and $0.186\,m_0$, $0.135\,m_0$ along Γ-X orientation, respectively. The effective hole masses are $0.75\,m_0$ and $0.54\,m_0$ along the Γ-L orientation, and $0.34\,m_0$ and $0.25\,m_0$ along Γ-X orientation for $Rb_2NaInBr_6$ and Rb_2NaInI_6, respectively. The average effective masses of electrons and holes were $0.192m_0$ ($0.142\,m_0$) and $0.545m_0$ ($0.394\,m_0$) for $Rb_2NaInBr_6$ (Rb_2NaInI_6), respectively, this suggests valuable carrier transport characteristics. But, the heavy hole masses in Rb_2NaInX_6 (X=Br, I) are less significant in this study due to the flat band along the Γ-X orientation. The lack of dislocating in this orientation results in higher effective masses, which lower carrier movement and hinder electrical conductivity.

7.4. Optical Properties

The real and imaginary parts of the dielectric function for Rb_2NaInX_6 (X=Br, I) are highlighted in Figure 11. (a) and (b), subsequently, over a photon energy range of 0 to 12 eV. The static dielectric constant $\varepsilon_1(0)$, increases from 2.53 for $Rb_2NaInBr_6$ to 3.45 for Rb_2NaInI_6, and both materials exhibit resonance peaks at 4.14 eV and 4.96 eV, respectively, indicating maximum light dispersion and polarization at these frequencies. The real part $\varepsilon_1(\omega)$ then decreases, becoming negative at higher energies (11.33 eV for Br and 9.15 eV for I), suggesting metallic behaviour where photons are fully absorbed. The imaginary part of $\varepsilon_2(\omega)$, rises at 2.9 eV for Br and 1.36 eV for I, reaching peak values at 3.98 eV (0.90) and 2.5 eV (1.19), respectively, due to electron excitations from the X-p states to In-s orbitals. This rise indicates excellent photon absorption in the ultraviolet and visible spectra.

*Figure 11. (a) Real and (b) imaginary parts of the dielectric function, (c) optical reflectivity R(**ω**), and (d) absorption coefficient **α(ω)**.*

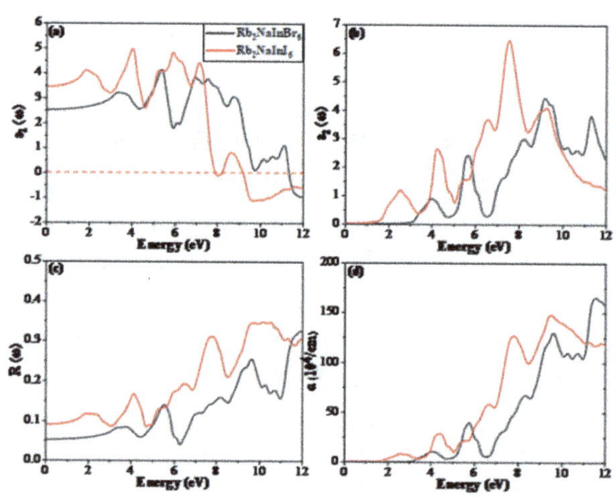

The reflectivity, $R(\omega)$, illustrated in Figure 11. (c) shows how much light is summarized by the material's surface, with static values R(0) of 0.052 for $Rb_2NaInBr_6$ and 0.09 for Rb_2NaInI_6. Reflectivity increases with photon energy. The absorption coefficient $\alpha(\omega)$ (Figure 11. (d)) starts at zero below the energy gap, then rises above 3 eV for Br and 1.6 eV for I, obtaining peaks of 10.66×10^4 cm^{-1} at 4 eV and 40.2×10^4 cm^{-1} at 5.75 eV for $Rb_2NaInBr_6$, while Rb_2NaInI_6 reaches peaks of 8×10^4 cm^{-1} at 2.57 eV and 28.6×10^4 cm^{-1} at 4.4 eV. The absorption consistently increases with photon energy, suggesting dominant absorption in the UV and visible regions, and efficient electron-hole pair generation (excitons) in this range.

7.5. Photovoltaic Properties

The SCAPS-1D simulation results are presented in Table 3. Although the bromide-based PSC exhibits a better open-circuit voltage, its lowest-circuit current density leads to reduced overall efficiency (4.81%) compared to the iodide-based PSC (26.6%). This performance difference is primarily linked the optimal band gap of the iodide-based compound. A similar trend is observed in the power density as the Br-PSC recorded a maximum power density of 48.17 W/m^2 compared to I-PSC (266.03 W/m^2).

Table 3. Solar cell parameters of FTO/Ti O_2/Rb_2NaInX_6 (X=Br, I)/Cu_2O/Au perovskite solar cells at 298.15 K.

Absorber	J_{sc} (A/m^2)	J_m (A/m^2)	V_{oc} (V))	V_m (V)	P_m (W/m^2)	FF	η(%)
$Rb_2NaInBr_6$	44.902	42.86	1.480	1.124	48.17	0.724	4.81
Rb_2NaInI_6	325.405	313.02	0.954	0.849	266.03	0.856	26.60

8. CORRELATION WITH EXPERIMENTAL AND PREDICTIVE STUDIES FOR Rb_2NaInX_6 (X = Br, I)

As of now, there are no published experimental studies directly validating the predicted optoelectronic properties—such as band gaps and absorption spectra—of Rb_2NaInX_6 (X = Br, I). The literature to date is largely theoretical, with first-principles DFT calculations reporting band gaps and solar-cell performance metrics (e.g., SLME and SCAPS-1D efficiencies) that indicate promising optoelectronic behavior, particularly for the iodide analogue (Cordeiro et al., 2025; Mahedi et al., 2025). These studies predict visible-light absorption and relatively high theoretical power-conversion efficiencies for Rb_2NaInI_6; however, experimental synthesis and optical or electronic characterization of either $Rb_2NaInBr_6$ or Rb_2NaInI_6 have not yet been reported in peer-reviewed literature. Consequently, no UV–Vis absorption, photoluminescence, photoelectrical, or device-level measurements are currently available to confirm the predicted optical gaps, absorption onset, or charge-transport properties.

Despite the lack of direct experimental validation, the theoretical trends predicted for Rb_2NaInX_6 are consistent with experimental observations in related lead-free rubidium-based double perovskites, where iodide substitution typically results in narrower band gaps and enhanced visible-light absorption compared with bromide counterparts (Apon et al., 2025). Previous theory–experiment studies on analogous double perovskite systems further demonstrate that device engineering, including the selection of appropriate charge-transport layers and interface alignment, can strongly influence photovoltaic performance (Siekmann et al., 2023). Together, these considerations highlight both the potential of Rb_2NaInX_6—particularly the iodide variant—for optoelectronic applications and the need for future experimental efforts to bridge the gap between theoretical predictions and experimental validation.

9. LIMITATIONS OF DFT

Density functional theory (DFT) remains a foundational tool for predicting structural and electronic properties of lead-free halide double perovskites, but it has well-recognized limitations that can affect quantitative accuracy. Standard semilocal exchange–correlation functionals such as PBE typically underestimate band gaps, particularly for systems with indirect or weak optical transitions (Butler et al., 2015; Perdew et al., 1996), and fail to capture fine features of the band structure without hybrid or beyond-

DFT methods (e.g., optimally-tuned hybrid or range-separated approaches) that improve agreement with experiment (Sagredo et al., 2024; Z. Zhang et al., 2025). The inclusion of spin–orbit coupling (SOC) is important for heavy elements (e.g., Bi, Sb) but its interplay with approximate functionals still leads to uncertainties in band dispersion, effective masses, and band-edge alignment (Perdew et al., 1996). Furthermore, conventional DFT often assumes ideal periodic order and neglects B/B′ cation disorder, local symmetry breaking, mixed oxidation states, and lattice anharmonicity, all of which can significantly influence electronic states and carrier dynamics in these materials presenting a structural and chemical complexity (Slavney et al., 2017). Soft lattice dynamics and strong electron–phonon coupling typical of halide perovskites, along with finite-temperature effects, are also not fully captured in standard zero-temperature DFT calculations, further limiting predictive accuracy (Sendner et al., 2016). While defect energetics and transition levels can be sensitive to functional choice, supercell size, and charge correction schemes, leading to uncertainties that may contribute to discrepancies between theoretical predictions and experimental observations in lead-free systems (Freysoldt et al., 2014). These challenges have motivated the use of advanced functionals and post-DFT techniques to obtain more reliable predictions of bandgaps and optical properties in lead-free double perovskites.

10. PERSPECTIVE OUTLOOK

For reviewing the structural, electrical, mechanical, and optical parameters of perovskite materials, (DFT) has emerged as a crucial and adaptable instrument. By using DFT, scientists have gained a deeper understanding of the underlying mechanisms that control these molecules' stability and behavior. The approach does have certain drawbacks, though. It frequently ignores temperature-dependent effects, underestimates band gaps, and has trouble with systems that have tightly coupled electrons. More sophisticated methods, including as hybrid functionals, GW approximations, and ab-initio molecular dynamics, have been developed to overcome these obstacles. Although these sophisticated techniques provide more accuracy, their effectiveness requires more processing power.

Future double perovskite research will be based on (DFT), machine learning (ML), and high-throughput screening methods. This will make it easier to quickly identify materials with certain features. Recent advances in machine learning–density functional theory (ML–DFT) frameworks have significantly transformed halide perovskite research by enabling faster, more accurate, and more scalable materials discovery. ML models trained on DFT-generated datasets can efficiently predict key electronic, structural, and thermodynamic properties, allowing rapid screening of large compositional spaces that would be computationally prohibitive using DFT alone. Beyond surrogate modeling, emerging ML–DFT approaches now incorporate active learning, uncertainty quantification, and closed-loop optimization, where ML guides the selection of new candidate materials for DFT evaluation, iteratively improving model performance and discovery efficiency. These integrated frameworks are particularly impactful for halide perovskites, where complex structure–property relationships, compositional flexibility, and defect chemistry demand data-efficient and adaptive computational strategies to accelerate photovoltaic material design.

Looking ahead, further advances in ML–DFT integration, combined with autonomous discovery workflows and experimental validation, are expected to play a key role in guiding the rational design of high-performance, stable halide double perovskites for next-generation solar cells.

11. CONCLUSION

The halide double perovskites Rb_2NaInX_6 (X = Br, I) exhibit structural, mechanical, and thermodynamic stability according to DFT analysis, which qualifies them for experimental investigation. Elastic moduli reveal anisotropy and ductility. TB-mBJ was employed to establish direct band gaps of 2.879 eV for $Rb_2NaInBr_6$ and 1.35 eV for Rb_2NaInI_6. Conduction bands are dominated by In/X and valence bands by Br/I. UV-Vis absorption can be examined by optical analysis, indicating potential for solar cells. Compared to $Rb_2NaInBr_6$ (4.81%), Rb_2NaInI_6 with a film width of 0.8 μm at 298.15 K was estimated by SCAPS-1D to have a power conversion efficiency of 26.60%. In the same settings, Rb_2NaInI_6 exhibited a higher power generation density of 266.03 W/m^2 compared to 48.17 W/m^2 for $Rb_2NaInBr_6$.

REFERENCES

Apon, I. A., Kriaa, K., Rahman, M. A., Hossain, M. A., Maatki, C., Assadi, A. A., & Elboughdiri, N. (2025). Rb 2 BX 6 double perovskites: unlocking 22% efficiency through structural, electronic, mechanical, and optical insights. DOI: 10.1039/D5RA03981A

Ayyaz, A., Murtaza, G., Shafiq, M., Qasim Shah, M., Sfina, N., & Ali, S. (2023). Exploring structural, thermodynamic, elastic, electro-optic, and thermoelectric characteristics of double perovskites Rb2X-InBr6 (X = Na, K) for photovoltaic applications: A DFT approach. *Solar Energy*, *265*, 112131. DOI: 10.1016/j.solener.2023.112131

Bartel, C. J., Clary, J. M., Sutton, C., Vigil-Fowler, D., Goldsmith, B. R., Holder, A. M., & Musgrave, C. B. (2020). Inorganic Halide Double Perovskites with Optoelectronic Properties Modulated by Sublattice Mixing. *Journal of the American Chemical Society*, *142*(11), 5135–5145. DOI: 10.1021/jacs.9b12440

Becke, A. D., & Roussel, M. R. (1989). Exchange holes in inhomogeneous systems: A coordinate-space model. *Physical Review A: General Physics*, *39*(8), 3761–3767. DOI: 10.1103/PhysRevA.39.3761

BECQUEREL. E. (1839). On Electric Effects under the Influence of Solar Radiation. Compt. Rend., 9, 561. https://cir.nii.ac.jp/crid/1570009750806226944.bib?lang=ja

Bhalla, A. S., Guo, R., & Roy, R. (2000). The perovskite structure - A review of its role in ceramic science and technology. *Materials Research Innovations*, *4*(1), 3–26. DOI: 10.1007/s100190000062

Birch, F. (1947). Finite elastic strain of cubic crystals. *Physical Review*, *71*(11), 809–824. DOI: 10.1103/PhysRev.71.809

Blaha, P., Schwarz, K., Madsen, G. K. H., Kvasnicka, D., & Luitz, J. (2001). wien2k. An Augmented Plane Wave+ Local Orbitals Program for Calculating Crystal Properties, 60(1), 155–169.

Born, M., Huang, K., & Lax, M. (1954). Dynamical Theory of Crystal Lattices. Clarendon Press, Oxford, 141. DOI: 10.1119/1.1934059

Born, M., & Oppenheimer, R. (1927). Zur Quantentheorie der Molekeln. *Annalen der Physik*, *389*(20), 457–484. DOI: 10.1002/andp.19273892002

Burgelman, M., Nollet, P., & Degrave, S. (2000). Modelling polycrystalline semiconductor solar cells. *Thin Solid Films*, *361–362*, 527–532. DOI: 10.1016/S0040-6090(99)00825-1

Butler, K. T., Frost, J. M., & Walsh, A. (2015). Band alignment of the hybrid halide perovskites CH 3 NH 3 PbCl 3, CH 3 NH 3 PbBr 3 and CH 3 NH 3 PbI 3. *Materials Horizons*, *2*(2), 228–231. DOI: 10.1039/C4MH00174E

Choe, H., Yoon, H., Kim, I., & Kim, S. Y. (2025). Machine Learning Unlocks New Directions in Halide Perovskite Research. *ChemElectroChem*. Advance online publication. DOI: 10.1002/celc.202500282

Cordeiro, D. S., Wang, J., Wang, Y., Liu, X., & Wang, X. (2025). Academic Editor: M. Natália Prediction and Screening of Lead-Free Double Perovskite Photovoltaic Materials Based on Machine Learning. DOI: 10.3390/molecules30112378

Dirac, P. A. M. (1930). Note on Exchange Phenomena in the Thomas Atom. *Mathematical Proceedings of the Cambridge Philosophical Society*, *26*(3), 376–385. DOI: 10.1017/S0305004100016108

Du, K., Meng, W., Wang, X., Yan, Y., & Mitzi, D. B. (2017). Bandgap Engineering of Lead-Free Double Perovskite Cs 2 AgBiBr 6 through Trivalent Metal Alloying. *Angewandte Chemie*, *129*(28), 8270–8274. DOI: 10.1002/ange.201703970

Ehrenreich, H., & Cohen, M. H. (1959). Self-Consistent Field Approach to the Many-Electron Problem. *Physical Review*, *115*(4), 786–790. DOI: 10.1103/PhysRev.115.786

Engel, E., & Vosko, S. H. (1993). Exact exchange-only potentials and the virial relation as microscopic criteria for generalized gradient approximations. *Physical Review B: Condensed Matter*, *47*(20), 13164–13174. DOI: 10.1103/PhysRevB.47.13164

Fan, B., Jiang, Q., Dai, X., Xu, X., Wu, Y., & Peng, Q. (2026). Coordinating energy-level alignment and morphology in perovskite–organic heterostructures for efficient and stable solar cells. *Materials Science and Engineering R Reports*, *168*, 101165. DOI: 10.1016/j.mser.2025.101165

Fermi, E. (1928). Eine statistische Methode zur Bestimmung einiger Eigenschaften des Atoms und ihre Anwendung auf die Theorie des periodischen Systems der Elemente. Zeitschrift F r Physik, 48(1–2), 73–79. DOI: 10.1007/BF01351576

Feynman, R. P. (1939). Forces in Molecules. *Physical Review*, *56*(4), 340–343. DOI: 10.1103/PhysRev.56.340

Fock, V. (1930). N herungsmethode zur L sung des quantenmechanischen Mehrk rperproblems. Zeitschrift F r Physik, 61(1–2), 126–148. DOI: 10.1007/BF01340294

Freysoldt, C., Grabowski, B., Hickel, T., Neugebauer, J., Kresse, G., Janotti, A., & Van De Walle, C. G. (2014). First-principles calculations for point defects in solids. *Reviews of Modern Physics*, *86*(1), 253–305. DOI: 10.1103/RevModPhys.86.253

Fu, H., Li, D., Peng, F., Gao, T., & Cheng, X. (2008). Ab initio calculations of elastic constants and thermodynamic properties of NiAl under high pressures. *Computational Materials Science*, *44*(2), 774–778. DOI: 10.1016/j.commatsci.2008.05.026

Gaillac, R., Pullumbi, P., & Coudert, F.-X. (2016). ELATE: An open-source online application for analysis and visualization of elastic tensors. *Journal of Physics Condensed Matter*, *28*(27), 275201. DOI: 10.1088/0953-8984/28/27/275201

Ghosh, S., Shankar, H., & Kar, P. (2022). Recent developments of lead-free halide double perovskites: A new superstar in the optoelectronic field. *Materials Advances*, *3*(9), 3742–3765. DOI: 10.1039/D2MA00071G

Goldschmidt, V. M. (1926). The Laws of Crystallochemistry. *Naturwissenschaften*, *14*(21), 477–485. DOI: 10.1007/BF01507527

Green, M. A. (1981). Solar cell fill factors: General graph and empirical expressions. *Solid-State Electronics*, *24*(8), 788–789. DOI: 10.1016/0038-1101(81)90062-9

Green, M. A. (2001). Third generation photovoltaics: Ultra-high conversion efficiency at low cost. *Progress in Photovoltaics: Research and Applications*, *9*(2), 123–135. DOI: 10.1002/pip.360

Green, M. A., Ho-Baillie, A., & Snaith, H. J. (2014). The emergence of perovskite solar cells. *Nature Photonics*, *8*(7), 506–514. DOI: 10.1038/nphoton.2014.134

Heyd, J., Scuseria, G. E., & Ernzerhof, M. (2003). Hybrid functionals based on a screened Coulomb potential. *The Journal of Chemical Physics*, *118*(18), 8207–8215. DOI: 10.1063/1.1564060

Hohenberg, P., & Kohn, W. (1964). Inhomogeneous Electron Gas. *Physical Review*, *136*(3B), B864–B871. DOI: 10.1103/PhysRev.136.B864

Hui, R., & O'sullivan, M. (2023). Fundamentals of optical devices. In *Fiber-Optic Measurement Techniques* (pp. 1–135). Elsevier., DOI: 10.1016/B978-0-323-90957-0.00002-3

Hung, C., Chih, C., Huang, K., Xue, Y., Chu, H., Tseng, C., Li, C., Chen, J., Chen, B., Yang, S., Cheng, Y., & Chou, P. (2025). Perovskite-Coupled NIR Organic Hybrid Solar Cells Achieving an 84.2% Fill Factor and a 25.2% Efficiency: A Comprehensive Mechanistic Exploration. *Angewandte Chemie International Edition*, *64*(21), e202501375. Advance online publication. DOI: 10.1002/anie.202501375

Jamal, M., Bilal, M., Ahmad, I., & Jalali-Asadabadi, S. (2018). IRelast package. *Journal of Alloys and Compounds*, *735*, 569–579. DOI: 10.1016/j.jallcom.2017.10.139

Kale, A. J., Chaurasiya, R., & Dixit, A. (2022). Lead-Free Cs 2 BB′X 6 (B: Ag/Au/Cu, B′: Bi/Sb/Tl, and X: Br/Cl/I) Double Perovskites and Their Potential in Energy Conversion Applications. *ACS Applied Energy Materials*, *5*(9), 10427–10445. DOI: 10.1021/acsaem.2c00672

Khan, S. Y., Rauf, S., Liu, S., Chen, W., Shen, Y., & Kumar, M. (2025). Revolutionizing the solar photovoltaic efficiency: A comprehensive review on the cutting-edge thermal management methods for advanced and conventional solar photovoltaics. *Energy & Environmental Science*, *18*(3), 1130–1175. DOI: 10.1039/D4EE03525A

Kohn, W., & Sham, L. J. (1965). Self-Consistent Equations Including Exchange and Correlation Effects. *Physical Review*, *140*(4A), A1133–A1138. DOI: 10.1103/PhysRev.140.A1133

Kojima, A., Teshima, K., Shirai, Y., & Miyasaka, T. (2009). Organometal Halide Perovskites as Visible-Light Sensitizers for Photovoltaic Cells. *Journal of the American Chemical Society*, *131*(17), 6050–6051. DOI: 10.1021/ja809598r

Krukau, A. V., Vydrov, O. A., Izmaylov, A. F., & Scuseria, G. E. (2006). Influence of the exchange screening parameter on the performance of screened hybrid functionals. *The Journal of Chemical Physics*, *125*(22), 224106. Advance online publication. DOI: 10.1063/1.2404663

Kube, C. M. (2016). Elastic anisotropy of crystals. *AIP Advances*, *6*(9), 095209. DOI: 10.1063/1.4962996

Ledbetter, H., & Migliori, A. (2006). A general elastic-anisotropy measure. *Journal of Applied Physics*, *100*(6), 063516. DOI: 10.1063/1.2338835

Leng, M., Chen, Z., Yang, Y., Li, Z., Zeng, K., Li, K., Niu, G., He, Y., Zhou, Q., & Tang, J. (2016). Lead-Free, Blue Emitting Bismuth Halide Perovskite Quantum Dots. *Angewandte Chemie International Edition, 55*(48), 15012–15016. DOI: 10.1002/anie.201608160

Li, C., Soh, K. C. K., & Wu, P. (2004). Formability of ABO3 perovskites. *Journal of Alloys and Compounds, 372*(1–2), 40–48. DOI: 10.1016/j.jallcom.2003.10.017

Li, H., Li, F., Shen, Z., Han, S.-T., Chen, J., Dong, C., Chen, C., Zhou, Y., & Wang, M. (2021). Photoferroelectric perovskite solar cells: Principles, advances and insights. *Nano Today, 37*, 101062. DOI: 10.1016/j.nantod.2020.101062

Mahedi, A. H., Uddin, S., Yasin, M., Khan, H., Tarekuzzaman, M., Alsalmi, O., Rasheduzzaman, M., Mostafa, S. M. G., Hossen, M. M., & Hasan, M. Z. (2025). Multiscale analysis of Rb 2 NaInI 6 : from electronic structure to device performance for next-generation perovskite solar cells. DOI: 10.1039/D5RA04739K

Mao, P., Bi, W., Lv, J., Zhang, Z., Wang, B., & Zhong, Y. (2025). Decoding Buried Interfaces in Perovskite Solar Cells: Core Issues, Strategic Engineering, and Prospects for High-Efficiency Stable Devices. *Advanced Science (Weinheim, Baden-Wurttemberg, Germany), 12*(40), e12523. Advance online publication. DOI: 10.1002/advs.202512523

Marzo, A., Ferrada, P., Beiza, F., Alonso-Montesinos, J., Ballestrín, J., & Román, R. (2016). Comparison of atacama desert solar spectra vs. ASTM G173-03 reference spectra for solar energy application. International Energy Society. Conference Proceedings.

Mavko, G., Mukerji, T., & Dvorkin, J. (2020). Elasticity and Hooke's Law. The Rock Physics Handbook, 37–120. DOI: 10.1017/9781108333016.003

McClure, E. T., Ball, M. R., Windl, W., & Woodward, P. M. (2016). Cs2AgBiX6 (X = Br, Cl): New Visible Light Absorbing, Lead-Free Halide Perovskite Semiconductors. *Chemistry of Materials, 28*(5), 1348–1354. DOI: 10.1021/acs.chemmater.5b04231

Morab, S., Sundaram, M. M., & Pivrikas, A. (2023). Review on Charge Carrier Transport in Inorganic and Organic Semiconductors. *Coatings, 13*(9), 1657. DOI: 10.3390/coatings13091657

Morales-Acevedo, A. (2023). Fundamentals of solar cell physics revisited: Common pitfalls when reporting calculated and measured photocurrent density, open-circuit voltage, and efficiency of solar cells. *Solar Energy, 262*, 111774. DOI: 10.1016/j.solener.2023.05.051

Murnaghan, F. D. (1944). The Compressibility of Media under Extreme Pressures. *Proceedings of the National Academy of Sciences of the United States of America, 30*(9), 244–247. DOI: 10.1073/pnas.30.9.244

Noel, N. K., Stranks, S. D., Abate, A., Wehrenfennig, C., Guarnera, S., Haghighirad, A. A., Sadhanala, A., Eperon, G. E., Pathak, S. K., Johnston, M. B., Petrozza, A., Herz, L. M., & Snaith, H. J. (2014). Lead-free organic-inorganic tin halide perovskites for photovoltaic applications. *Energy & Environmental Science, 7*(9), 3061–3068. DOI: 10.1039/C4EE01076K

Noor, N. A., Iqbal, M. W., Zelai, T., Mahmood, A., Shaikh, H. M., Ramay, S. M., & Al-Masry, W. (2021). Analysis of direct band gap A2ScInI6 (A=Rb, Cs) double perovskite halides using DFT approach for renewable energy devices. *Journal of Materials Research and Technology*, *13*, 2491–2500. DOI: 10.1016/j.jmrt.2021.05.080

Oublal, M., Dib, M., Khalal, A., Chibane, F., Es-Salhi, E.-S., Nafidi, A., & El-Yakoubi, E. Y. (2025). Prediction of optoelectronic properties and efficiency of halide double perovskites Rb_2NaInX_6 (X=Br, I) for solar cell applications: A density functional theory approach. *Materials Science in Semiconductor Processing*, *200*, 109917. DOI: 10.1016/j.mssp.2025.109917

Papon, P. (2021). AIE (Agence internationale de l'énergie), Net Zero by 2050: A Roadmap for the Global Energy Sector, Paris: AIE / OCDE, mai 2021, 224 p. Futuribles, N° 444(5), VI–VI. DOI: 10.3917/futur.444.0113f

Perdew, J. P., Burke, K., & Ernzerhof, M. (1996). Generalized Gradient Approximation Made Simple. *Physical Review Letters*, *77*(18), 3865–3868. DOI: 10.1103/PhysRevLett.77.3865

Perdew, J. P., Chevary, J. A., Vosko, S. H., Jackson, K. A., Pederson, M. R., Singh, D. J., & Fiolhais, C. (1993). Erratum: Atoms, molecules, solids, and surfaces: Applications of the generalized gradient approximation for exchange and correlation. *Physical Review B: Condensed Matter*, *48*(7), 4978–4978. DOI: 10.1103/PhysRevB.48.4978.2

Priya, P., & Stonier, A. A. (2025). Emerging innovations in solar photovoltaic (PV) technologies: The perovskite solar cells and more. *Energy Reports*, *14*, 216–242. DOI: 10.1016/j.egyr.2025.06.003

Ranganathan, S. I., & Ostoja-Starzewski, M. (2008). Universal elastic anisotropy index. *Physical Review Letters*, *101*(5), 055504. DOI: 10.1103/PhysRevLett.101.055504

Roknuzzaman, M., Ostrikov, K., Chandula Wasalathilake, K., Yan, C., Wang, H., & Tesfamichael, T. (2018). Insight into lead-free organic-inorganic hybrid perovskites for photovoltaics and optoelectronics: A first-principles study. *Organic Electronics*, *59*, 99–106. DOI: 10.1016/j.orgel.2018.04.051

Sagredo, F., Gant, S. E., Ohad, G., Haber, J. B., Filip, M. R., Kronik, L., & Neaton, J. B. (2024). Electronic structure and optical properties of halide double perovskites from a Wannier-localized optimally-tuned screened range-separated hybrid functional. *Physical Review Materials*, *8*(10), 105401. DOI: 10.1103/PhysRevMaterials.8.105401

Schmidt-Mende, L., Kraner, S., & Fakharuddin, A. (2022). *Organic and hybrid solar cells*. Walter de Gruyter GmbH & Co KG. DOI: 10.1515/9783110736939

Sendner, M., Pabitra Nayak, ab K., Egger, D. A., Beck, S., Christian Mü ller, ab, Bernd Epding, abe, Wolfgang Kowalsky, ae, Leeor Kronik, abe, Snaith, H. J., & Pucci abf, A. (2016). Optical phonons in methylammonium lead halide perovskites and implications for charge transport †. 3, 613. DOI: 10.1039/C6MH00275G

Shahbazi, M., & Wang, H. (2016). Progress in research on the stability of organometal perovskite solar cells. *Solar Energy*, *123*, 74–87. DOI: 10.1016/j.solener.2015.11.008

Shao, B., Song, X., Zhu, H., Bioud, Y. A., Wu, W., Abulikemu, M., Saiari, H., Aqeel, S., Gereige, I., Mohammed, O. F., & Bakr, O. M. (2025). Single-crystal perovskites for photovoltaic and high-energy detection applications. *Chemical Society Reviews*, *54*(21), 9939–9977. DOI: 10.1039/D5CS00625B

Shockley, W., & Queisser, H. J. (1961). Detailed Balance Limit of Efficiency of p-n Junction Solar Cells. *Journal of Applied Physics*, *32*(3), 510–519. DOI: 10.1063/1.1736034

Siekmann, J., Kulkarni, A., Akel, S., Klingebiel, B., Saliba, M., Rau, U., & Kirchartz, T. (2023). Characterizing the Influence of Charge Extraction Layers on the Performance of Triple-Cation Perovskite Solar Cells. *Advanced Energy Materials*, *13*(32), 2300448. Advance online publication. DOI: 10.1002/aenm.202300448

Slater, J. C. (1930). Atomic Shielding Constants. *Physical Review*, *36*(1), 57–64. DOI: 10.1103/PhysRev.36.57

Slavney, A. H., Smaha, R. W., Smith, I. C., Jaffe, A., Umeyama, D., & Karunadasa, H. I. (2017). Chemical Approaches to Addressing the Instability and Toxicity of Lead–Halide Perovskite Absorbers. *Inorganic Chemistry*, *56*(1), 46–55. DOI: 10.1021/acs.inorgchem.6b01336

Sun, P.-P., Li, Q.-S., Yang, L.-N., & Li, Z.-S. (2016). Theoretical insights into a potential lead-free hybrid perovskite: Substituting Pb2+ with Ge2+. *Nanoscale*, *8*(3), 1503–1512. DOI: 10.1039/C5NR05337D

Sun, S., Hartono, N. T. P., Ren, Z. D., Oviedo, F., Buscemi, A. M., Layurova, M., Chen, D. X., Ogunfunmi, T., Thapa, J., Ramasamy, S., Settens, C., DeCost, B. L., Kusne, A. G., Liu, Z., Tian, S. I. P., Peters, I. M., Correa-Baena, J.-P., & Buonassisi, T. (2019). Accelerated Development of Perovskite-Inspired Materials via High-Throughput Synthesis and Machine-Learning Diagnosis. *Joule*, *3*(6), 1437–1451. DOI: 10.1016/j.joule.2019.05.014

Sze, S. M., & Ng, K. K. (2006). LEDs and lasers. *Physics of Semiconductor Devices*, *3*, 601–657.

Terentjev, A. V., Constantin, L. A., & Pitarke, J. M. (2018). Dispersion-corrected PBEsol exchange-correlation functional. *Physical Review. B*, *98*(21), 214108. DOI: 10.1103/PhysRevB.98.214108

Timoshenko, S. P., & Goodier, J. N. (n.d.). Theory of Elasticity THIRD EDITION PROFESSOR EMERITUS OF. *Gongcheng Lixue*.

Tran, F., & Blaha, P. (2009). Accurate Band Gaps of Semiconductors and Insulators with a Semilocal Exchange-Correlation Potential. *Physical Review Letters*, *102*(22), 226401. DOI: 10.1103/PhysRevLett.102.226401

Trihalide S B, P., Christian, Y., & Mølle, K. N. (1959). Matematisk-fysiske Meddelelser udgivet af Det Kongelige Danske Videnskabernes Selskab THE STRUCTUR E OF PEROVSKITE-LIKE CAESIUM Introductio n. Mat. Fys. Medd. Dan. Vid. Selsk, 32(2).

Ünlü, F., Deo, M., Mathur, S., Kirchartz, T., & Kulkarni, A. (2022). Bismuth-based halide perovskite and perovskite-inspired light absorbing materials for photovoltaics. *Journal of Physics. D, Applied Physics*, *55*(11), 113002. DOI: 10.1088/1361-6463/ac3033

Usman, M., & Yan, Q. (2020). Recent Advancements in Crystalline Pb-Free Halide Double Perovskites. *Crystals*, *10*(2), 62. DOI: 10.3390/cryst10020062

Volonakis, G., Haghighirad, A. A., Milot, R. L., Sio, W. H., Filip, M. R., Wenger, B., Johnston, M. B., Herz, L. M., Snaith, H. J., & Giustino, F. (2017). Cs 2 InAgCl 6 : A New Lead-Free Halide Double Perovskite with Direct Band Gap. *The Journal of Physical Chemistry Letters*, *8*(4), 772–778. DOI: 10.1021/acs.jpclett.6b02682

Weber, O. J. (2018). Structural Chemistry of Hybrid Halide Perovskites for Thin Film Photovoltaics.

Wu, Z., & Cohen, R. E. (2006). More accurate generalized gradient approximation for solids. *Physical Review B: Condensed Matter and Materials Physics*, *73*(23), 235116. DOI: 10.1103/PhysRevB.73.235116

Yu, L., & Zunger, A. (2012). Identification of Potential Photovoltaic Absorbers Based on First-Principles Spectroscopic Screening of Materials. *Physical Review Letters*, *108*(6), 068701. DOI: 10.1103/PhysRevLett.108.068701

Zangwill, A. (2013). Hartree and Thomas: The forefathers of density functional theory. *Archive for History of Exact Sciences*, *67*(3), 331–348. DOI: 10.1007/s00407-013-0114-4

Zhang, P., Yang, J., & Wei, S.-H. (2018). Manipulation of cation combinations and configurations of halide double perovskites for solar cell absorbers. *Journal of Materials Chemistry. A, Materials for Energy and Sustainability*, *6*(4), 1809–1815. DOI: 10.1039/C7TA09713A

Zhang, Z., Fung, V., & Hu, G. (2025). Pushing the boundary of the stability and band gap Pareto front by going towards high-entropy perovskites † Lead-free Cs 2 BX 6 (B = Zr 4+, Sn 4+, Te 4+, Hf 4+, Re 4+, Os 4+, Ir 4+, and Pt 4+ and X = Cl. 17, 11376. DOI: 10.1039/D4NR05013D

Zhao, X.-G., Fu, Y., Yang, D., Xu, Q., Zhang, L., Yang, J.-H., Yu, L., Wei, S.-H., & Zhang, L. (2017). Design of Lead-Free Inorganic Halide Perovskites for Solar Cells via Cation-Transmutation. *Journal of the American Chemical Society*, *139*(7), 2630–2638. DOI: 10.1021/jacs.6b09645

Zhao, Z.-J., Liu, S., Zha, S., Cheng, D., Studt, F., Henkelman, G., & Gong, J. (2020). Publisher Correction: Theory-guided design of catalytic materials using scaling relationships and reactivity descriptors. *Nature Reviews. Materials*, *7*(8), 671–671. DOI: 10.1038/s41578-020-0192-2

Zunger, A., & Freeman, A. J. (1977). Ground-state electronic properties of diamond in the local-density formalism. *Physical Review. B, Solid State*, *15*(10), 5049–5065. DOI: 10.1103/PhysRevB.15.5049

Chapter 6
Theoretical and Computational Advances in Solvent Effect on NMR Spectroscopy: A Comprehensive Review

Nadia Faska
https://orcid.org/0000-0001-6203-6904
Faculty of Applied Sciences of Ait-Melloul, Ibn Zohr University, Agadir, Morocco

Zaina Izghri
https://orcid.org/0000-0003-3205-556X
Faculty of Sciences Semlalia-Marrakech, Cadi Ayyad university, Morocco

Soukayna Maitouf
https://orcid.org/0009-0000-4619-7863
LGP, Faculty of Sciences Agadir. Eco-Process, Optimization, and Decision Support, Jules Verne University of Picardie, France

Said El Hizazi
Higher School of Technology of Agadir, Ibn Zohr University, Morocco

ABSTRACT

Accurate computational prediction of NMR chemical shifts requires proper treatment of solvent effects, which influence nuclear shielding through electrostatic polarization, hydrogen bonding, and dispersion interactions. This review examines computational methodologies from continuum solvation models (PCM, COSMO) to explicit approaches incorporating ab initio molecular dynamics and QM/MM schemes. We evaluate method performance and provide practical guidelines for methodology selection based on solute-solvent interaction types and system size. Emerging machine learning approaches are discussed. While significant progress has been achieved, quantitative accuracy requires careful integration of quantum mechanical theory, statistical mechanics, and validation against experimental data.

DOI: 10.4018/979-8-3373-6058-4.ch006

I. INTRODUCTION

Nuclear magnetic resonance (NMR) spectroscopy has emerged as an indispensable analytical tool in contemporary chemistry, providing unparalleled insights into molecular structure, dynamics, and interactions across diverse chemical systems (Lodewyk et al., 2011; Willoughby et al., 2014). The advent of computational quantum chemistry has revolutionized the field, enabling the accurate prediction and interpretation of NMR parameters through first-principles calculations. Among the myriad factors influencing NMR parameters, solvent effects constitute a particularly complex yet critical component, profoundly modulating the electronic environment of nuclei and, consequently, their magnetic shielding properties (Aslam et al., 2017; Patrahau et al., 2024)

The accurate treatment of solvation has been established as a crucial factor in the quantum chemical prediction of NMR chemical shifts. Solvent molecules alter the electronic environment of nuclei through multiple mechanisms: dielectric polarization, hydrogen bonding, van der Waals interactions, and short-range specific interactions. These perturbations can strongly influence nuclear shielding, particularly for labile nuclei such as ^{1}H, ^{15}N, and ^{31}P (Dračínský et al., 2019; Faska et al., 2004). The magnitude of solvent-induced chemical shift variations can range from fractions of a part per million to several tens of ppm, depending on the nucleus, molecular structure, and the nature of solute-solvent interactions (Cohen et al., 2012).

Over the past two decades, substantial progress has been achieved in developing and refining theoretical frameworks for modeling solvent effects on NMR parameters. These advances encompass a spectrum of approaches, from computationally efficient implicit continuum models to sophisticated explicit solvation schemes incorporating molecular dynamics simulations (Tomasi et al., 2005; Marenich et al., 2009). The Polarizable Continuum Model (PCM) and its variants, including the Integral Equation Formalism PCM (IEF-PCM) and the Conductor-like Screening Model (COSMO), have become standard tools for accounting for bulk electrostatic effects (Tomasi et al., 2005; Klamt & Schüürmann, 1993). However, such continuum approaches often prove insufficient for capturing directional interactions and dynamic solvent fluctuations near polar functional groups (Cancès et al., 1997; Faska et al., 2007).

To address these limitations, hybrid quantum mechanics/molecular mechanics (QM/MM) methodologies and explicit solvation models have been increasingly adopted (Cui & Karplus, 2000; He et al., 2009; Swails et al., 2015). More recently, the integration of ab initio molecular dynamics (AIMD) with NMR calculations has emerged as a powerful approach for incorporating thermal averaging and conformational sampling (Dračínský et al., 2013; Autschbach et al., 2010; Charpentier et al., 2013). Furthermore, the application of machine learning techniques to NMR chemical shift prediction has opened new avenues for rapid and accurate predictions, particularly when combined with quantum mechanical descriptors (Guan et al., 2021; Yang et al., 2021).

This comprehensive review examines the theoretical foundations, computational methodologies, and practical applications of solvent effect modeling in NMR spectroscopy. We critically evaluate the performance of various solvation models, discuss the integration of density functional theory (DFT) with molecular dynamics simulations, and explore emerging approaches including machine learning-enhanced predictions. Through analysis of representative case studies spanning small organic molecules, transition metal complexes, and biomolecular systems, we provide practical guidelines for methodology selection and highlight current challenges and future directions in this rapidly evolving field.

II. THEORETICAL FRAMEWORK OF SOLVENT EFFECTS

1. Fundamental Mechanisms

The influence of solvation on NMR chemical shifts arises from multiple interconnected mechanisms operating at different length scales with distinct physical origins. Understanding these mechanisms is essential for selecting appropriate computational approaches and interpreting solvent-dependent chemical shift variations.

The primary mechanism involves the solvent modulating the electron density distribution of the solute through long-range electrostatic interactions (Casabianca, 2020; Siskos et al., 2017). This bulk electrostatic effect depends on the solvent dielectric constant (ε) and can be effectively captured by continuum solvation models. The polarization of the solute electron density by the solvent reaction field modifies nuclear magnetic shielding tensors, typically producing systematic shifts that correlate with solvent polarity parameters. For polar solvents with high dielectric constants ($\varepsilon > 20$), this mechanism can contribute 1-3 ppm to ^{1}H chemical shifts and 5-15 ppm to ^{13}C shifts in molecules containing carbonyl or other polarizable functional groups (Tomasi et al., 2005).

Specific short-range interactions, particularly hydrogen bonding, often dominate solvent effects on polar functional groups (Mari et al., 2019). Hydrogen bond formation induces substantial electron density redistribution at both donor and acceptor sites, leading to pronounced deshielding effects on participating nuclei. The magnitude depends on both the strength and geometry of the hydrogen bond. Experimental studies demonstrate that hydroxyl proton chemical shifts can vary by more than 5 ppm between non-hydrogen-bonding solvents like chloroform and strongly hydrogen-bonding solvents like dimethyl sulfoxide (DMSO) (Sarotti & Pellegrinet, 2009). These directional interactions cannot be adequately described by isotropic continuum models and necessitate explicit consideration of solute-solvent hydrogen bonding geometries (Faska et al., 2011).

Van der Waals dispersion interactions and exchange-repulsion effects constitute additional contributions to solvent-induced chemical shifts (Kayal et al., 2024). These short-range quantum mechanical effects arise from electron density overlap between solute and solvent molecules and typically exhibit weaker magnitude compared to electrostatic and hydrogen bonding effects. Nevertheless, for aromatic systems and halogen-containing compounds, dispersion interactions can produce measurable chemical shift perturbations of 0.2-0.8 ppm for ^{1}H and 2-5 ppm for ^{13}C nuclei. Accurate treatment requires either high-level ab initio methods or density functionals incorporating empirical dispersion corrections such as Grimme's D3 scheme (Grimme et al., 2010).

Dynamic averaging represents a fourth fundamental aspect of solvent effects, particularly relevant for systems exhibiting conformational flexibility or rapid intermolecular exchange (Casabianca, 2020). NMR observables represent time-averaged properties over picosecond to nanosecond timescales. Consequently, thermal fluctuations, solvent reorganization, and conformational interconversion must be properly sampled to obtain accurate chemical shift predictions. Ab initio molecular dynamics simulations have demonstrated that static structure calculations can deviate significantly from experimental values due to neglect of thermal averaging. For flexible molecules, the difference between static and dynamically averaged chemical shifts can exceed 2 ppm for ^{1}H and 10 ppm for ^{13}C (De Gortari et al., 2010).

Temperature influences NMR chemical shifts primarily through changes in conformational state populations and modifications to hydrogen bonding equilibria (Trainor et al., 2019). As temperature increases, populations of higher-energy conformers increase according to Boltzmann statistics, and hydrogen bond strength typically decreases. These temperature-dependent phenomena produce observable chemical shift variations that can be quantitatively reproduced by computational methods incorporating proper conformational sampling and thermal averaging. Dračínský et al. (2013) demonstrated that including vibrational motion through ab initio molecular dynamics significantly improves the accuracy of chemical shift predictions, particularly for nuclei participating in hydrogen bonds or located in flexible molecular regions.

2. Mathematical Foundation

The theoretical description of solvent effects on NMR parameters requires a quantum mechanical framework accounting for both intramolecular electronic structure and intermolecular solute-solvent interactions. The observable NMR parameters chemical shifts (δ), spin-spin coupling constants (J), and relaxation times are fundamentally determined by the electronic environment surrounding magnetic nuclei, which is profoundly altered by solvation (Das & Merz, 2025).

The nuclear magnetic shielding constant σ, which determines the observed chemical shift, relates fundamentally to the electron density distribution and the electronic system's response to an applied magnetic field. The diamagnetic shielding component (σ_d) arises from the local diamagnetic response of molecular electrons:

$$\sigma_d = \frac{e^2}{6m_e c^2} \sum_i \langle \psi r_i^2 \psi \rangle \tag{1}$$

where e is the elementary charge, m_e is the electron mass, c is the speed of light, and the sum runs over all electrons. The paramagnetic shielding component (σ_p) involves transitions between occupied and virtual orbitals:

$$\sigma_p = -\frac{2\mu_0 e^2}{3\pi} \sum_i \sum_{n \neq i} \frac{\langle iL_z n \rangle \langle nL_z i \rangle}{\Delta E_{ni}} \tag{2}$$

where ΔE_{ni} represents orbital energy differences, and L_z is the orbital angular momentum operator. The total shielding constant is expressed as:

$$\sigma_{total} = \sigma_d + \sigma_p + \sigma_{spin-orbit} \tag{3}$$

In the presence of a solvent, the nuclear shielding constant becomes dependent on the electrostatic potential generated by that solvent. Within the polarizable continuum model framework:

$$\sigma_{solv} = \sigma_{vac} + \Delta\sigma_{solv} \tag{4}$$

where σ_{vac} is the gas-phase shielding constant and $\Delta\sigma_{solv}$ represents solvent-induced perturbation. Solvent correction typically involves both the reaction field generated by the bulk dielectric and specific solute-solvent interactions (Merrill & Tantillo, 2019).

For hydrogen-bonded systems, the shielding constant can be decomposed as:

$$\sigma = \sigma_0 - \alpha_{HB} . n_{HB} - \beta \Delta_{solv} \quad (5)$$

where σ_0 is the baseline shielding, α_{HB} represents the hydrogen bonding effect, n_{HB} is the number of hydrogen bonds, and $\beta \Delta_{solv}$ accounts for additional solvation effects.

The isotropic chemical shift (δ) relates to the absolute shielding constant (σ) through:

$$\delta = \sigma_{ref} - \sigma_{sample} \quad (6)$$

where σ_{ref} and σ_{sample} are the shielding constants of the reference compound and sample, respectively. This relative reporting partially cancels systematic errors in quantum chemical calculations, particularly when reference and sample are calculated using identical methodology (Shenderovich, 2023).

III. COMPUTATIONAL SOLVATION MODELS

1. Implicit Continuum Models

Implicit continuum solvation models represent the solvent as a homogeneous polarizable medium characterized by its dielectric constant ε, avoiding explicit treatment of individual solvent molecules. This approximation offers substantial computational efficiency while capturing dominant bulk electrostatic effects of solvation (Provorse et al., 2016).

The Polarizable Continuum Model (PCM) framework constructs a molecular cavity using either a solvent-accessible surface (SAS) or solvent-excluded surface (SES) and computes apparent surface charges that reproduce the solvent reaction field (Figure 1). The molecular cavity is typically defined by interlocking spheres centered on atomic nuclei, with radii scaled from van der Waals radii by a factor of 1,1 – 1,2. The cavity surface is discretized into small tesserae (surface elements), and apparent surface charges on each tessera are computed to satisfy electrostatic boundary conditions at the dielectric interface (Mennucci, 2012).

Figure 1. The solvent-accessible surface (SAS) and solvent-excluded surface (SES).

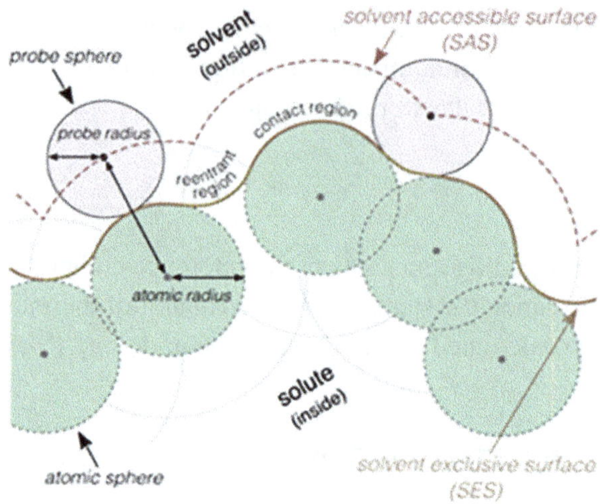

The Integral Equation Formalism variant (IEF-PCM) has become the most widely implemented version, employing exact dielectric boundary conditions that ensure numerical stability and accurate treatment of both neutral and ionic species. Recent benchmark studies demonstrate that IEF-PCM provides reliable predictions for ^{1}H and ^{13}C chemical shifts in polar solvents when combined with appropriate DFT functionals (Klamt et al., 2015). Cohen et al. (2023) evaluated IEF-PCM performance across diverse drug-like molecules in various solvents, reporting root-mean-square deviations (RMSD) typically in the range of $0{,}15 - 0{,}30\ ppm$ for ^{1}H and $1{,}5 - 3{,}0\ ppm$ for ^{13}C chemical shifts. The accuracy proves sufficient for most applications involving conformational analysis and structure elucidation, though limitations emerge for systems with strong specific solute-solvent interactions.

The Conductor-like Screening Model (COSMO) simplifies the PCM approach by initially treating the solvent as an ideal conductor (infinite dielectric constant) and subsequently scaling the resulting surface charges by a factor $f(\varepsilon) = (\varepsilon - 1)/(\varepsilon + x)$ to account for the finite dielectric constant ε. The parameter x typically equals 0.5 for neutral molecules and 0.0 for ions (Klamt et al., 2015). This conductor boundary condition simplifies the mathematical formulation and reduces computational cost while maintaining comparable accuracy to PCM for most applications. Comprehensive comparisons demonstrate that with correct parametrization, no significant differences emerge between COSMO and IEF-PCM predictions for large datasets of neutral and ionic solutes (Liu. F et al., 2018; Silva et al., 2016).

The Solvation Model based on Density (SMD) represents a universal solvation approach that augments continuum electrostatics with empirical terms accounting for cavitation, dispersion, and solvent structure effects (Mirzaei et al., 2019). SMD employs optimized atomic radii specific to particular elements and functional groups, yielding improved accuracy for systems with strong hydrogen bonding interactions. The model includes separate parametrizations for aqueous and organic solvents, with parameters derived from extensive fitting to experimental solvation free energies. Systematic evaluations demonstrate that SMD exhibits superior performance compared to IEF-PCM, especially for functional groups involved in hydrogen bonding, reducing RMSD by approximately $15 - 25\%$ for hydroxyl, amino, and carboxylic acid groups (Stahn et al., 2023).

Despite their computational efficiency, continuum models exhibit inherent limitations. They cannot adequately describe directional interactions such as specific hydrogen bonds, which require explicit consideration of individual solvent molecules and their geometric orientations (Faska et al., 2011). Continuum models fail to capture local solvent structure and dynamics, including formation of solvent shells and cooperative hydrogen bonding networks. They may provide inaccurate results for systems where solute-solvent charge transfer or quantum mechanical exchange-repulsion effects are significant (Dooley & Vyas, 2025; Zhu et al., 2013). The choice of cavity radii and surface discretization can introduce systematic errors, particularly for unusual molecular shapes or extended conjugated systems.

2. Explicit Solvation Models

Explicit solvation models treat individual solvent molecules quantum mechanically or classically, enabling precise representation of specific solute-solvent interactions including hydrogen bonding, π-stacking, and other directional effects. The microsolvation or cluster model approach incorporates a small number of explicit solvent molecules in the first solvation shell, proving remarkably effective for capturing hydrogen bonding effects while maintaining computational tractability.

Recent investigations demonstrate that strategic placement of just 5-15 explicit solvent molecules can substantially improve NMR chemical shift predictions for polar functional groups (Boereboom et al., 2018; Hernandes et al., 2023). Da Silva et al. (2025) examined azithromycin in chloroform, water, and DMSO using an increasing number of explicit solvent molecules (n = 5, 15, 25, and 50) combined with the PCM model for bulk solvation. Their results revealed that 5 explicit solvent molecules strategically positioned at hydrogen bonding sites, supplemented by PCM for bulk solvation, achieved chemical shift predictions approaching experimental accuracy. The ^{1}H chemical shift RMSD decreased from 0.52 ppm (PCM only) to 0.18 ppm (5 explicit + PCM), and ^{13}C RMSD improved from 3.8 ppm to 1.9 ppm. Further increasing explicit molecules to 15 or 25 provided only marginal additional improvements, suggesting that the first solvation shell captures the majority of specific interaction effects (Da Silva et al. 2025).

The optimal number and positioning of explicit solvent molecules require careful consideration. For molecules with multiple hydrogen bonding sites, systematic protocols involve identifying all potential hydrogen bond donor and acceptor positions through molecular electrostatic potential analysis or natural bond orbital calculations. Solvent molecules are then placed at these sites with geometries optimized to maximize hydrogen bonding interaction energies. Alternative approaches employ molecular dynamics simulations to generate realistic solvent configurations, followed by extraction of representative snapshots for quantum mechanical NMR calculations.

In a representative case study of hydrogen bonding effects, Mari et al. (2019) investigated solvent-dependent structures of anthraquinone natural products including chrysophanol, emodin, and physcion using combined explicit-implicit solvation. Temperature-dependent ^{1}H NMR experiments in DMSO, acetone, and chloroform revealed dramatically different chemical shifts for hydroxyl groups depending on their involvement in intramolecular versus intermolecular hydrogen bonding. The computational protocol employed geometry optimization with explicit DMSO molecules positioned at intermolecular hydrogen bonding sites, followed by NMR calculations with C-PCM representing bulk solvent. The B3LYP/6-311+G(2d,p) level provided excellent agreement with experiment, validating the combined explicit-implicit approach (Mari et al., 2019). Chemical shifts of OH groups participating in strong intramolecular hydrogen bonds exhibited minimal solvent dependence ($|\Delta\delta| < 0.3$ ppm), remaining relatively constant across different solvents (Charisiadis et al., 2014). In contrast, OH groups engaged in intermolecular hydrogen bonding showed large variations exceeding 5 ppm between DMSO and chloroform. This dichotomy arises because intramolecular hydrogen bonds remain intact regardless of solvent, while intermolecular interactions exhibit strong solvent-specific character (Abraham et al., 2006).

The explicit solvent approach, however, encounters several methodological challenges requiring attention for reliable predictions. Identifying optimal solvent molecule positions requires sophisticated sampling techniques or chemical intuition regarding likely hydrogen bonding geometries. The number of explicit solvent molecules necessary for convergence varies substantially depending on the system and property of interest. Computational cost escalates rapidly as additional solvent molecules are incorporated into the quantum mechanical region. These considerations motivate development of hybrid QM/MM methodologies combining quantum mechanical treatment of the solute and key solvent molecules with classical force field descriptions of the extended solvation environment.

3. Hybrid QM/MM Methodologies

Hybrid quantum mechanics/molecular mechanics (QM/MM) methodologies partition the system into distinct regions treated at different levels of theory, enabling accurate modeling of localized quantum mechanical phenomena within extended molecular environments. For NMR chemical shift calculations, the solute molecule and critical proximal solvent molecules are treated quantum mechanically (the QM region), while the remainder is described using classical molecular mechanics force fields (the MM region). This partitioning reduces computational cost by several orders of magnitude compared to fully quantum mechanical treatment while preserving accuracy for the chemically relevant region (Boereboom et al., 2018; Cui et al., 2000).

The theoretical framework underlying QM/MM methodology is based on the fundamental energy expression that takes the form:

$$E_{QM/MM} = E_{QM} + E_{MM} + E_{coupling} \tag{7}$$

where the coupling term accounts for electrostatic, van der Waals, and sometimes covalent interactions between QM and MM parts. For NMR calculations, the QM-MM electrostatic interaction proves particularly important, as MM point charges polarize the QM electron density and consequently affect nuclear shielding tensors (Boulanger & Harvey, 2018). Modern implementations employ electrostatic embedding, wherein MM atomic charges are included in the QM Hamiltonian, ensuring that the QM wavefunction properly responds to the electrostatic field generated by the MM environment (Lafiosca et al., 2023).

Among the various QM/MM implementations, the Automated Fragmentation QM/MM (AF-QM/MM) approach represents a particularly powerful method for biomolecular NMR calculations. The system is automatically divided into overlapping fragments, each small enough (typically ~200 atoms) for efficient DFT calculations (Zhu et al., 2013). The fragmentation scheme ensures that each chemically distinct environment is captured within at least one fragment. NMR shielding tensors are computed for each fragment and subsequently combined to yield complete system properties through a self-consistent procedure eliminating double-counting errors (He et al., 2009).

AF-QM/MM incorporates solvent effects through Poisson-Boltzmann continuum models or explicit treatment of water molecules in the first solvation shell. Benchmarking studies on proteins and nucleic acids demonstrated that AF-QM/MM predictions provide valuable insights for systems containing non-standard residues, ligands, or metal centers outside empirical training sets. The method has been successfully applied to predict ^{1}H, ^{13}C, ^{15}N, and ^{17}O chemical shifts in proteins with typical RMSD values of 0.4-0.6 ppm for ^{1}H, 2-4 ppm for ^{13}C, and 8-15 ppm for ^{15}N (He et al., 2009). The QM/MM framework has proven particularly valuable for systematic studies of ^{31}P NMR chemical shifts in nucleotides and phosphorylated biomolecules. Szántó et al. (2024) developed a comprehensive protocol combining MM molecular dynamics sampling with QM/MM structure optimizations and NMR calculations to investigate ATP hydrolysis in the p97 AAA+ ATPase protein. The study revealed that ^{31}P chemical shifts exhibit extreme sensitivity to local geometry, with bond length variations as small as 0.10 Å inducing shifts exceeding 10 ppm. Classical MD simulations, even when initiated from high-resolution X-ray crystal structures, sampled configurations producing ^{31}P chemical shifts deviating by 15-30 ppm from experimental values. QM/MM geometry optimizations of MD snapshots significantly improved agreement with experiment, reducing RMSD from 12.4 ppm to 4.8 ppm (Szántó et al., 2024).

Advanced QM/MM implementations can incorporate polarizable force fields that allow MM atoms to respond to changes in the QM electron density. Such polarizable QM/MM schemes provide more consistent treatment of mutual polarization between QM and MM regions, potentially improving accuracy for systems where the MM environment exhibits significant polarizability. However, the computational overhead and increased complexity of parameter

IV. DENSITY FUNCTIONAL THEORY METHODOLOGY

1. Functional Selection and Performance

The selection of an appropriate exchange-correlation functional represents a critical decision in DFT-based NMR chemical shift calculations, as functional choice profoundly influences both accuracy and computational efficiency. Performance varies across functional classes from generalized gradient approximations (GGA) through hybrid and meta-hybrid to double-hybrid functionals, with optimal selection depending on the molecular system, target nuclei, and available computational resources (De Oliveira et al., 2021; Ramos et al., 2024).

Among the hybrid functionals, B3LYP has historically dominated NMR calculations due to its favorable balance of accuracy and computational cost. Introduced in the mid-1990s, B3LYP combines Becke's three-parameter exchange functional with the Lee-Yang-Parr correlation functional, incorporating 20% exact Hartree-Fock exchange (Becke, 1993). Systematic investigations demonstrate that B3LYP combined with moderately sized basis sets achieves chemical accuracy (mean absolute error <0.5 ppm) for ^{1}H chemical shifts in organic molecules. The functional's success arises from fortuitous error cancellation between the approximate exchange-correlation treatment and basis set incompleteness (Paier et al., 2007).

Wiitala et al. (2007) evaluated various DFT protocols for distinguishing diastereomeric methylcyclohexanols through ^{13}C NMR chemical shifts, confirming that B3LYP/6-311+G(2d,p) calculations provide reliable relative chemical shift predictions with RMSD values of 1.8-2.4 ppm. While absolute chemical shift predictions may exhibit systematic errors of 5-10 ppm, relative shifts between conformers or isomers typically achieve much higher accuracy (±1-2 ppm), enabling confident stereochemical assignments (Wiitala et al., 2007). However, B3LYP exhibits known deficiencies for systems involving substantial dispersion interactions, charge-transfer excitations, or transition metal complexes with significant spin-orbit coupling effects. The functional systematically underestimates hydrogen bonding strengths, overestimates conjugation effects in extended π-systems, and provides inconsistent performance for transition metal compounds (Castro et al., 2020).

The Minnesota family of meta-hybrid functionals, particularly M06-2X, has emerged as a competitive alternative for NMR applications. M06-2X incorporates 54% Hartree-Fock exchange and was specifically parametrized for main-group thermochemistry and noncovalent interactions through extensive fitting to experimental and high-level ab initio benchmark data. The functional's high exact-exchange admixture improves the description of hydrogen bonding, dispersion interactions, and barrier heights while reducing self-interaction errors (Zhao & Truhlar, 2007; Wang et al., 2020). Figueirinhas and Dong (2014) demonstrated that M06-2X calculations accurately reproduce ^{1}H and ^{13}C chemical shifts in toluene solution when combined with IEF-PCM solvation, achieving RMSD values of 0.22 ppm and 2.1 ppm respectively for a test set of 15 organic molecules. The functional's improved treatment of medium-range correlation effects yields superior performance for conjugated systems and molecules with extensive π-electron delocalization (Figueirinhas & Dong, 2014).

Long-range corrected functionals such as ωB97X-D and CAM-B3LYP partition the exchange interaction into short-range and long-range components, with different fractions of exact exchange applied at different interelectronic separations. The ωB97X-D functional additionally includes Grimme's empirical D2 dispersion correction, ensuring accurate treatment of non-covalent interactions (Iron, 2017). Hernandes et al. (2023) employed ωB97X-D for azithromycin NMR calculations in multiple solvents, demonstrating excellent agreement with experimental data when combined with explicit-implicit solvation models. The functional achieved ^{1}H RMSD of 0.16 ppm and ^{13}C RMSD of 1.7 ppm, representing improvement over B3LYP (0.24/2.4 ppm) and M06-2X (0.19/2.0 ppm) for this challenging macrolide system.

For transition metal systems, specialized considerations arise due to relativistic effects, spin-orbit coupling, and potential multireference character. The PBE0 functional, combining the PBE exchange-correlation functional with 25% exact exchange, has emerged as a reliable choice for transition metal NMR calculations. Vícha et al. (2015) systematically benchmarked DFT functionals for NMR chemical shifts in heavy metal complexes, demonstrating that PBE0 coupled with relativistic treatments provides superior accuracy. Crucially, they found optimal results for nuclei like ^{195}Pt using 40-50% exact exchangesignificantly higher than PBE0's standard 25%to compensate for missing exchange-correlation response in spin-orbit formalisms.

2. Basis Set Considerations

The choice of basis set critically affects the accuracy of gauge-independent atomic orbital (GIAO) NMR calculations. NMR properties depend sensitively on the description of electronic density in regions far from the nucleus, requiring basis sets with diffuse functions to adequately describe long-range magnetic perturbation. NMR calculations typically employ higher-quality basis sets than those used for geometry optimization (De Souza et al., 2018).

The Pople-style polarized triple-zeta basis sets, particularly 6-311+G(2d,p), have become de facto standards for routine organic molecule NMR calculations. This basis set incorporates diffuse functions essential for describing tail regions of electron density distributions, along with multiple sets of polarization functions (two sets of d functions on heavy atoms, one set of p functions on hydrogen) that enable accurate representation of magnetic shielding tensors. The split-valence triple-zeta formalism provides flexibility in the valence region while maintaining computational efficiency through minimal basis functions for core electrons (Pierens, 2014; Cohen et al., 2023). Pierens et al. (2017) systematically derived

^{1}H and ^{13}C NMR scaling factors for commonly used solvents with various DFT methods and basis sets, demonstrating that 6-311+G(2d,p) achieves convergence for typical organic compounds. Comparison with larger basis sets including 6-311++G(3df,3pd) and aug-cc-pVTZ showed that chemical shift differences typically remain below 0.1 ppm for ^{1}H and 1.0 ppm for ^{13}C, indicating that 6-311+G(2d,p) provides an excellent compromise between accuracy and computational efficiency for molecules containing up to 50-100 heavy atoms.

The Jensen polarization-consistent (pc) basis set family was specifically designed for property calculations including NMR parameters. The pcS-n series (n = 0, 1, 2, 3, 4) provides systematic convergence toward the complete basis set limit through balanced augmentation of polarization and diffuse functions. Unlike Pople basis sets optimized primarily for energy calculations, the pcS basis sets target property convergence, achieving faster basis set convergence for magnetic shielding tensors (Kjær & Sauer, 2011). Benassi (2016) introduced the molecular motif-specific scaling (MOSS-DFT) protocol using the B97-2 functional with pcS-1, pcS-2, and pcS-3 basis sets for aqueous solution NMR predictions. The pcS-3 basis set achieved root-mean-square deviations of 1.93 ppm and 0.154 ppm for ^{13}C and ^{1}H chemical shifts respectively in a dataset of 176 metabolite molecules. The pcS-2 basis set provides an excellent balance for routine applications, achieving approximately 95% of the pcS-3 accuracy at roughly one-third the computational cost.

Basis set superposition error (BSSE) represents a potential concern for calculations involving explicit solvent molecules or intermolecular complexes. BSSE arises when one fragment's basis functions improve the description of another fragment's electron density, leading to artificial stabilization of the complex (Goldey et al., 2013). For NMR calculations, BSSE can affect both optimized geometries and computed shielding tensors. Counterpoise corrections can be applied to estimate and mitigate BSSE effects, though their importance diminishes rapidly with increasing basis set size and becomes negligible for triple-zeta quality basis sets with diffuse functions (Laun & Bredow, 2021).

V. MOLECULAR DYNAMICS INTEGRATION

1. Classical Molecular Dynamics Approaches

Classical molecular dynamics simulations provide configurational sampling at nanosecond to microsecond timescales, enabling exploration of conformational space and solvent reorganization dynamics relevant to NMR observables. When combined with subsequent quantum mechanical NMR calculations on trajectory snapshots, classical MD offers a computationally tractable approach for incorporating thermal averaging effects (Hollingsworth & Dror, 2018; De Vivo et al., 2016). The standard computational protocol for combining molecular dynamics with NMR calculations involves generating an equilibrated MD trajectory using a validated force field (AMBER, CHARMM, or OPLS), extracting representative snapshots at regular intervals (typically 10-50 ps), and computing NMR shielding tensors quantum mechanically for each snapshot (Koes & Vries, 2017). The time-averaged chemical shift is obtained as a Boltzmann-weighted or arithmetic average over the ensemble. For systems exhibiting rapid conformational equilibration on the MD timescale, arithmetic averaging suffices, while systems with well-separated conformational states require proper Boltzmann weighting based on relative free energies (Mocci & Laaksonen, 2015).

Lodewyk et al. (2011) demonstrated the utility of conformational averaging for strychnine and other rigid natural products, showing that including multiple low-energy conformers substantially improves agreement between calculated and experimental ^{13}C chemical shifts. Their DP4 probability analysis, which combines calculated chemical shifts with experimental data to assign stereochemistry, achieves success rates exceeding 95% when proper conformational averaging is employed.

Classical MD-based protocols, however, suffer from inherent limitations arising from force field inaccuracies. Force fields employ fixed atomic charges and approximate bonded and non-bonded interaction potentials, which may inadequately describe polarization, charge transfer, and subtle geometric preferences that critically influence NMR parameters. Hernandes et al. (2023) demonstrated that for azithromycin, MM-optimized structures can differ substantially from QM-optimized geometries, particularly for hydrogen bonding geometries and flexible torsional angles involving heteroatoms. These structural discrepancies propagate into NMR calculations, potentially introducing systematic errors of 1-3 ppm for ^{1}H and 5-15 ppm for ^{13}C chemical shifts. Performing QM/MM geometry optimizations on MM-generated snapshots significantly improved chemical shift predictions, reducing RMSD by approximately 30-40%.

Szántó et al. (2024b) encountered similar challenges investigating ^{31}P NMR chemical shifts in ATP-protein complexes. Their MM-MD simulations, employing AMBER force fields with extensively validated phosphate parameters, sampled configurations producing ^{31}P chemical shifts deviating by 15-30 ppm from experimental values. Detailed analysis revealed that force field limitations in describing P-O bond lengths (errors of 0.05-0.10 Å) and O-P-O angles (errors of 2-5 degrees) accounted for the majority of the discrepancy. Given that bond length variations as small as 0.10 Å induce ^{31}P chemical shift changes exceeding 10 ppm, classical MD provides insufficient structural accuracy for quantitative predictions.

The choice of force field significantly impacts MD trajectory quality and subsequent NMR predictions (Mu et al., 2021). Modern force fields like AMBER ff14SB, CHARMM36, and OPLS-AA/M have been extensively parametrized against quantum mechanical and experimental data, providing improved accuracy for proteins, nucleic acids, and drug-like organic molecules (Abdullah et al., 2025). Specialized force fields exist for particular systems: lipid bilayers (Slipids, CHARMM36-lipid), carbohydrates (GLYCAM06), and ionic liquids (CL&P, OPLS-IL) (Kirschner et al., 2007; Goloviznina et al., 2021). Selecting an appropriate, well-validated force field represents a critical methodological decision.

2. Ab Initio Molecular Dynamics

Ab initio molecular dynamics circumvents force field limitations by propagating nuclear trajectories on potential energy surfaces computed directly from electronic structure calculations (Marx & Hutter, 2009). The two principal AIMD variants Born-Oppenheimer molecular dynamics (BOMD) and Car-Parrinello molecular dynamics (CPMD) differ in their treatment of electronic structure convergence during propagation, but both provide quantum mechanically consistent descriptions of atomic forces and molecular geometries throughout the simulation (Tuckerman, 2002).

In Born-Oppenheimer molecular dynamics (BOMD), the electronic structure is fully converged to the Born-Oppenheimer surface at each MD step before computing forces and advancing nuclear positions. This approach ensures that forces are computed on the true electronic ground state, maintaining rigorous adherence to the Born-Oppenheimer approximation. Modern implementations employ efficient SCF convergence algorithms and wavefunction extrapolation from previous steps to minimize computational overhead. BOMD has become the preferred AIMD variant for most applications due to its robustness and straightforward interpretation (Niklasson, 2008).

An alternative approach, Car-Parrinello molecular dynamics (CPMD), introduces fictitious electronic degrees of freedom with an associated fictitious mass, allowing simultaneous propagation of electronic and nuclear coordinates (Car & Parrinello, 1985). The electronic and nuclear equations of motion are coupled such that the electrons remain near their instantaneous ground state if the fictitious mass is chosen appropriately. CPMD offers computational advantages through reduced SCF iterations but requires careful selection of the fictitious electron mass and smaller timesteps to maintain adiabaticity (Blöchl, 1994).

Dračínský et al. (2019) pioneered the systematic combination of AIMD with NMR calculations, demonstrating that inclusion of fast, small-amplitude motions through AIMD significantly improves accuracy for common nuclei including ^{1}H, ^{13}C, ^{14}N, ^{17}O, and ^{35}Cl. Their benchmark study on glycine in the solid state revealed that BOMD-averaged chemical shifts reproduce experimental values substantially better than static structure calculations, with improvements of 0.3-0.8 ppm for ^{1}H and 2-8 ppm for ^{13}C (Dračínský et al., 2019). The investigation emphasized that neglecting dynamics can lead to very inaccurate chemical shift predictions, especially for temperature-dependent shifts dominated by low-frequency motions.

The AIMD approach provides mechanistic insights into solvent effects inaccessible through static calculations or continuum models. Truflandier et al. (2010) identified a solvent-assisted inverse hydration phenomenon in square-planar Pt(II) complexes, wherein water molecules preferentially accumulate near axial positions (perpendicular to the molecular plane) despite the absence of direct coordination. This distinctive solvation pattern arises from favorable electrostatic interactions with the positive electrostatic potential above and below the Pt(II) plane, inducing systematic ^{195}Pt chemical shift perturbations of 50-100 ppm that cannot be captured by implicit solvation models alone (Truflandier & Autschbach, 2010).

Kaplanskiy et al. (2025) investigated ^{1}H NMR chemical shifts in strong hydrogen bond complexes, demonstrating that nuclear quantum effects and proton delocalization represent dominant factors determining chemical shifts in short, symmetric hydrogen bonds. Their multiscale approach combined AIMD for conformational sampling with path integral molecular dynamics (PIMD) for nuclear delocalization effects. The study revealed that conventional static structure calculations systematically underestimate deshielding in strong hydrogen bonds by 5-10 ppm because static calculations place the bridging proton at the potential energy minimum, whereas nuclear quantum effects result in significant zero-point delocalization (Kaplanskiy et al., 2025).

Despite its advantages, AIMD remains computationally demanding, typically limiting simulations to systems containing fewer than 100-200 atoms and trajectories spanning 20-50 ps with DFT-level electronic structure (Marx & Hutter, 2009). For a 100-atom system, a 25 ps AIMD trajectory requires approximately 25,000-50,000 DFT calculations, consuming 500-2000 CPU hours depending on the functional and basis set. Accelerated AIMD methods applying bias potentials to enhance sampling of rare events show promise for extending AIMD-NMR applications to slower conformational transitions and more complex solvation environments (Ceriotti et al., 2016).

VI. APPLICATIONS AND CASE STUDIES

1. Small Organic Molecules and Natural Products

The prediction of NMR chemical shifts for small organic molecules and natural products represents one of the most successful applications of computational chemistry, enabling confident stereochemical assignments and structure determination for complex natural products (Lodewyk et al., 2011). The combination of DFT methods with continuum solvation models has revolutionized natural product structure elucidation, reducing the need for lengthy chemical derivatizations and providing rapid computational validation of proposed structures (Grimblat & Sarotti, 2016).

Sarotti and Pellegrinet introduced the DP4 probability method, which combines calculated ^{1}H and ^{13}C chemical shifts with experimental data to determine the probability that each candidate structure represents the true molecular constitution. The method employs Bayes' theorem to compute posterior probabilities based on the agreement between calculated and experimental shifts, properly accounting for both expected calculation errors and experimental uncertainty. Applied to over 100 challenging natural product structure determination problems, DP4 achieves success rates exceeding 95% in distinguishing between diastereomers and constitutional isomers (Smith & Goodman, 2010). The DP4 approach has been extended to DP4+, which incorporates both scaled and unscaled chemical shifts and demonstrates improved performance for flexible molecules and systems with extensive hydrogen bonding (Grimblat & Sarotti, 2016). The methodology employs empirical scaling factors derived from large training sets to correct for systematic DFT errors. For a benchmark set of 72 natural products with known stereochemistry, DP4+ correctly assigned 100% of cases when only two candidate structures were considered, and 96% when four or more candidates were evaluated (Marcarino et al., 2021).

Table 1 summarizes benchmark accuracies across multiple computational methods and functional choices, demonstrating the progression of methodological improvements over the past decade.

Table 1. Benchmark Accuracies for NMR Chemical Shift Prediction Methods in Organic Molecules

Method	Nucleus	RMSD (ppm)	n	Uncertainty (95% CI)	Reference
B3LYP/6-311+G(2d,p)/PCM	^{1}H	0.15	80	±0.03	Lodewyk et al., 2012
B3LYP/6-311+G(2d,p)/PCM	^{13}C	1.8	80	±0.4	Lodewyk et al., 2012
mPW1PW91/6- 311+G(2d,p)/PCM	^{1}H	0.12	50	±0.02	Willoughby et al., 2014
mPW1PW91/6-311+G(2d,p)/PCM	^{13}C	1.5	50	±0.3	Willoughby et al., 2014
ωB97X-D/def2-SVP/SMD	^{13}C	1.93	143	±0.21	Iron & Janes, 2023
WP04/6-311++G(2d,p)/PCM	^{1}H	0.11	114	±0.02	Iron & Janes, 2023
CASCADE (ML)	^{1}H	0.21	13000	±0.01	Guan et al., 2021
CASCADE (ML)	^{13}C	2.1	13000	±0.1	Guan et al., 2021

As shown in Table 1, machine learning methods (CASCADE) achieve comparable accuracy to traditional DFT approaches while reducing computational time by several orders of magnitude.

Willoughby et al. (2014) demonstrated the power of ^{1}H NMR shift calculations for stereochemical determination in complex natural products, showing that computed ^{1}H chemical shifts often provide more discriminating power than ^{13}C shifts for distinguishing diastereomers. Their comprehensive study evaluated 52 natural products including alkaloids, terpenes, and polyketides, revealing that ^{1}H shift predictions with modern DFT methods achieve mean absolute deviations of 0.08-0.15 ppm when appropriate solvation models and conformational averaging are employed. The study emphasized the critical importance of including all low-energy conformers in the Boltzmann-weighted average, recommending conformational searches at the molecular mechanics level followed by DFT reoptimization of all conformers within 2-3 kcal/mol of the global minimum (Willoughby et al., 2014).

Smith and Goodman (2010) pioneered the application of DFT chemical shift calculations to structural revision cases, successfully identifying errors in over 30 published natural product structures. Their systematic approach involves computing chemical shifts for all plausible structural alternatives, comparing with experimental data using statistical metrics (RMSD, correlation coefficients, DP4 probabilities), and identifying the structure with best overall agreement (Smith & Goodman, 2010). The case of hexacyclinol illustrates both the power and potential pitfalls of NMR-based structure determination originally assigned a hexacyclic structure, it was later revised to a pentacyclic structure after computational chemical shift predictions and total synthesis conclusively demonstrated errors in the original assignment (Rychnovsky, 2006).

2. Hydrogen Bonding and Solvation Effects

Hydrogen bonding represents one of the most important and challenging aspects of solvent effects on NMR chemical shifts. The formation of hydrogen bonds induces substantial electron density redistribution and deshielding of participating nuclei, with magnitudes ranging from 1-2 ppm for weak C-H···O contacts to more than 10 ppm for strong O-H···O and N-H···O hydrogen bonds.

Sarotti et al. (2012) conducted a comprehensive investigation of solvent-dependent structures of anthraquinone natural products including chrysophanol, emodin, and physcion. Temperature-dependent ^{1}H NMR experiments in DMSO, acetone, and chloroform revealed dramatically different chemical shifts for hydroxyl groups depending on their involvement in intramolecular versus intermolecular hydrogen bonding. OH groups participating in strong intramolecular hydrogen bonds exhibited minimal solvent dependence ($|\Delta\delta| < 0{,}3$ ppm), while OH groups engaged in intermolecular hydrogen bonding showed large variations exceeding 5 ppm between DMSO (δ ~9-10 ppm) and chloroform (δ ~4-5 ppm) (Sarotti & Pellegrinet, 2012).

The computational protocol successfully reproduced experimental trends using B3LYP/6-311+G(2d,p) calculations with C-PCM implicit solvation combined with discrete DMSO molecules for intermolecular hydrogen bonds. For chrysophanol, calculated ^{1}H chemical shifts for the intramolecular hydrogen-bonded 1-OH group remained at $12{,}1 - 12{,}4\ ppm$ across all solvents, agreeing with experimental values of $12{,}0 - 12{,}2\ ppm$. The free 8-OH group showed calculated shifts of 4.8 ppm ($CHCl_3$) to $9{,}6\ ppm$ (DMSO), matching experimental values of 4,9 and $9{,}8\ ppm$ respectively (Lomas, 2015).

In studies of nucleobase solvation, Guan et al. investigated aqueous solvation effects on nucleobases and nucleotides, demonstrating that explicit water molecules are essential for accurate ^{1}H and ^{15}N chemical shift predictions. Their benchmark calculations on adenine, guanine, cytosine, thymine, and uracil employed microsolvation models with 5-15 water molecules positioned at hydrogen bonding sites, supplemented by PCM for bulk solvation effects. This hybrid explicit-implicit approach achieved ^{1}H

RMSD of 0.22 ppm and ^{15}N RMSD of 4.8 ppm, representing substantial improvement over PCM-only calculations (0.58/12,3 *ppm*). The study revealed that amino group ^1H chemical shifts exhibit extreme sensitivity to hydrogen bonding, varying by 3 – 5 *ppm* depending on whether one or both hydrogen atoms participate in hydrogen bonds to water (Exner et al., 2012; Victora et al., 2014).

Kaplanskiy et al. (2025) examined strong hydrogen bonds in protonated dimers and biological complexes, revealing that nuclear quantum effects become dominant when hydrogen bonds are short and symmetric. For the [pyridine-H-pyridine]$^+$ cation, the bridging proton exhibits extreme deshielding (δ ~18-20 ppm) arising from delocalization across the N···N distance. Classical calculations systematically underpredict this deshielding by 5-8 ppm. Including nuclear quantum effects through path integral molecular dynamics correctly captures proton delocalization, yielding chemical shifts in excellent agreement with gas-phase and cryogenic matrix isolation experiments (Kaplanskiy et al., 2025).

3. Ionic Systems and Counterion Effects

Ionic species present unique challenges for NMR chemical shift predictions due to strong long-range electrostatic interactions with the solvent and counterions, potential ion pairing, and the high sensitivity of chemical shifts to charge distribution. Continuum solvation models generally perform better for ions than neutral molecules due to the dominant role of electrostatic solvation, yet explicit treatment of the first solvation shell often proves necessary for quantitative accuracy (Collins, 2012; Chen & Izgorodina, 2017).

Some work investigated aqueous solvation effects on ^{17}O chemical shifts in various oxyanions including carbonate, sulfate, phosphate, and nitrate (Wang et al., 2018; Pruitt et al., 2015; Steiner et al., 2021). Their systematic study employed microsolvation models with 4-12 water molecules combined with PCM, revealing that 6-8 water molecules in the first shell provide converged chemical shift predictions. For carbonate anion (CO_3^{2-}), the calculated ^{17}O chemical shift varied from 201 ppm (gas phase) to 168 ppm (6 H_2O + PCM), in excellent agreement with the experimental value of 170 ppm in aqueous solution. The large 33 ppm solvation shift arises from hydrogen bonding between water and carbonate oxygen atoms, which decreases the C-O bond order and modulates the electronic environment at oxygen (Pascual-Borràs et al., 2014).

For lithium coordination environments, Sethio et al. studied lithium ion solvation and ^7Li NMR chemical shifts in various organic solvents, finding that Li^+ chemical shifts vary by more than 3 ppm depending on the donor strength of coordinating solvent molecules. In weakly coordinating solvents like acetonitrile, Li^+ coordinates 4 solvent molecules in a tetrahedral arrangement (δ ~0 ppm), while in DMSO or other strong donors, octahedral coordination with 6 solvent molecules predominates (δ ~2.5 ppm) (Sethio et al., 2025). Computational predictions require accurate treatment of the first coordination shell, with at least 4-6 explicit solvent molecules essential for reproducing experimental trends.

Counterion effects represent another important consideration for chemical shift predictions in ionic solutions. Deshielding or shielding induced by nearby counterions can reach 0.5-2 ppm for ^1H and ^{13}C nuclei within 5-8 Å of the counterion (Wang et al., 2018). For systems where ion pairing is significant, explicit inclusion of counterions in the QM region or QM/MM treatment becomes necessary. Continuum models that assume complete charge screening by the dielectric medium fail to capture these short-range counterion effects (Chen & Izgorodina, 2017).

4. Transition Metal Complexes

Transition metal complexes present formidable challenges for NMR chemical shift prediction due to the combination of relativistic effects, spin-orbit coupling, potential multireference character, and strong sensitivity to ligand coordination geometry. Nevertheless, modern relativistic DFT methods combined with appropriate solvation treatments have achieved remarkable success for a wide range of metal centers and nuclei.

Autschbach and Zheng (2008) performed four-component relativistic DFT calculations on square-planar *Pt(II)* complexes in aqueous solution, combining AIMD sampling with spin-orbit ZORA calculations. Their results demonstrated that relativistic effects contribute approximately 7000-8000 ppm to ^{195}Pt shielding constants, with spin-orbit coupling accounting for roughly 25-30% of this contribution. The study revealed that explicit solvation profoundly affects ^{195}Pt chemical shifts through direct coordination of water molecules in trans positions and second-shell solvation effects arising from water molecules positioned above and below the square-planar coordination plane (Autschbach & Zheng, 2008).

Castro et al. (2020) extended these methodologies to iridium polyhydride complexes, investigating the classical hydride chemical shifts observed for $[IrH_5(PPh_3)_2]$. The calculated ^{1}H chemical shifts for the five hydride ligands ranged from −10 to −15 ppm after AIMD averaging, in excellent agreement with experimental values. Analysis of individual trajectory snapshots revealed rapid hydride exchange processes occurring on the picosecond timescale, with instantaneous chemical shifts varying from +5 to −25 ppm depending on the instantaneous coordination geometry and trans influence from phosphine ligands (Castro et al., 2020).

Gahlawat et al. (2024) investigated ^{19}F NMR in nickel-fluoride complexes, combining two-component ZORA calculations with explicit solvation and AIMD. Their study revealed that ^{19}F chemical shifts in metal-fluoride bonds exhibit exceptional sensitivity to hydrogen bonding between coordinated fluoride and solvent molecules. In benzene solution, explicit benzene molecules positioned near the Ni-F bond through C-H···F hydrogen bonding interactions induced ^{19}F shift changes of 12-15 ppm compared to PCM-only calculations. AIMD simulations showed that C-H···F hydrogen bonds form and break on picosecond timescales, with trajectory-averaged ^{19}F chemical shifts achieving RMSD of 3.2 ppm relative to experimental values (Gahlawat et al., 2024).

Table 2 consolidates benchmark results for transition metal NMR predictions, highlighting the superior performance of AIMD-coupled relativistic methods across diverse metal centers and nuclei

Table 2. Benchmark Accuracies for Transition Metal NMR Predictions

Complex Type	Nucleus	Method	RMSD (ppm)	Range (ppm)	Rel. Error (%)	n	Reference
Pt(II) square-planar	^{195}Pt	AIMD + SO- ZORA	127	6000	2.1	8	Komorovský et al., 2015
Ni-F complexes	^{19}F	AIMD + ZORA	3.2	45	7.1	6	Baishya et al., 2024
Ir polyhydrides	^{1}H (hydride)	AIMD + SO- DFT	0.8	15	5.3	5	Castro et al., 2020
Cisplatin derivatives	^{195}Pt	AIMD + ZORA	52	1200	4.3	4	Castro et al., 2011

The consistent ~2-7% relative errors shown in Table 2 for heavy nuclei represent remarkable accuracy given the 1000-6000 ppm chemical shift ranges characteristic of these systems, demonstrating that modern computational approaches achieve quantitative reliability for transition metal NMR.

VII. MACHINE LEARNING AND EMERGING METHODS

1. Machine Learning for Chemical Shift Prediction

Machine learning (ML) approaches have emerged as powerful alternatives or complements to traditional quantum mechanical NMR prediction methods, offering dramatically improved computational efficiency while achieving comparable or superior accuracy for molecules within their training domains. The past five years have witnessed explosive growth in ML-based NMR prediction, driven by advances in graph neural networks, attention mechanisms, and the availability of large experimental NMR databases (St John et al., 2019; Han et al., 2024; Z. Yang et al., 2021b).

Guan et al. (2021) developed CASCADE, a message-passing neural network trained on approximately 13,000 organic molecules with experimental ^{1}H and ^{13}C chemical shifts. The model learns relationships between molecular graph topology, atomic features, and NMR chemical shifts without explicit quantum mechanical calculations (Guan et al., 2021). CASCADE achieves ^{1}H RMSD of 0.21 *ppm* and ^{13}C RMSD of 2,1 *ppm* on held-out test molecules performance comparable to DFT methods (B3LYP/6-311+G(2d,p) with PCM) while reducing prediction time from hours to milliseconds per molecule (Gao et al., 2020).

The CASCADE architecture employs graph convolutions to propagate information between connected atoms, capturing both local chemical environment and longer-range through-bond effects. Attention mechanisms weight the importance of different neighboring atoms, enabling the model to learn chemical principles such as electronegativity effects, aromaticity influences, and through-space coupling. The model implicitly learns solvent effects by training on experimental data measured in specific solvents, though it cannot generalize to solvent conditions absent from the training set (Guan et al., 2021).

Xin et al. (2017) introduced a hybrid approach combining machine learning with quantum mechanical descriptors, achieving superior performance to purely ML or QM methods alone. Their model employs DFT-calculated atomic charges, molecular orbital energies, and local electron density features as inputs to a neural network trained to predict residual errors in DFT chemical shift calculations. This correction network learns to compensate for systematic DFT deficiencies, reducing ^{1}H RMSD from 0.18 ppm (DFT alone) to 0.11 ppm (hybrid ML-DFT) and ^{13}C RMSD from 2,1 to 1,3 *ppm* (Xin et al., 2017).

The hybrid approach offers several advantages: it generalizes better to molecular scaffolds absent from training data by retaining the physics-based DFT foundation, it requires smaller training sets since it learns corrections rather than absolute shifts, and it provides interpretability through analysis of learned correction patterns. The method identifies that DFT systematically underpredicts hydroxyl ^{1}H shifts by 1,5 – 2,5 *ppm* and overpredicts conjugated aromatic ^{13}C shifts by 2 – 4 *ppm* corrections that the ML model learns to apply automatically (Guan et al., 2021).

Transfer learning enables extension of ML-NMR models to new nuclei or chemical spaces with minimal retraining. Paruzzo et al. (2018b) demonstrated that a model pretrained on ^{1}H and ^{13}C shifts can be fine-tuned for ^{15}N predictions using just 500 experimental ^{15}N shifts, achieving performance comparable to models trained on 5000+ examples from scratch. The pretrained model captures general principles of chemical shift dependencies on electronic structure, which transfer across different nuclei despite their distinct magnetic properties (Paruzzo et al., 2018).

Despite impressive performance on diverse organic molecules, ML models face challenges for systems exhibiting strong solvation effects, conformational flexibility, or heavy-element-induced chemical shifts. The models generally assume a single dominant conformation and cannot explicitly account for Boltzmann averaging over multiple conformers unless explicitly trained on conformer-averaged data. Future ML-NMR developments will likely incorporate 3D molecular structure directly through graph neural networks operating on conformer ensembles, enable multi-task learning across different nuclei and solvents simultaneously, and integrate uncertainty quantification to flag predictions requiring validation.

2. Polarizable Force Fields and Advanced Sampling

Polarizable force fields represent an important methodological advance for molecular dynamics simulations, incorporating explicit treatment of induced dipoles and higher-order multipole moments that respond to the instantaneous electric field environment. These enhanced force fields provide more realistic descriptions of electrostatic interactions in heterogeneous environments, potentially improving the quality of MD-generated structures for subsequent NMR calculations.

The AMOEBA (Atomic Multipole Optimized Energetics for Biomolecular Applications) force field employs permanent atomic multipoles through quadrupole order combined with mutual induced dipoles. This sophisticated electrostatic model captures polarization, anisotropic charge distributions, and many-body effects that fixed-charge force fields cannot describe (Ponder et al., 2010). For proteins and nucleic acids, AMOEBA demonstrates improved agreement with quantum mechanical interaction energies and experimental solvation free energies compared to fixed-charge models. Preliminary applications show that simulations with AMOEBA produce protein backbone φ,ψ angle distributions in better agreement with NMR J-coupling data, suggesting improved sampling of conformational space (Shi et al., 2013).

The Drude oscillator model represents an alternative polarization scheme employed in the CHARMM Drude force field. Polarizability is implemented through auxiliary charged particles (Drude oscillators) attached to polarizable atoms via harmonic springs (Baker et al., 2010). The Drude particles respond to the local electric field, inducing dipole moments that capture polarization effects. This classical mechanical implementation offers computational efficiency advantages over the self-consistent AMOEBA approach while retaining most of the physical accuracy (Lemkul et al., 2016).

Linear-scaling DFT methods based on localized orbitals or density matrix approaches offer potential solutions to computational bottlenecks, reducing scaling to O(N) for sufficiently large systems (Kussmann & Ochsenfeld, 2007). The ONETEP program implements linear-scaling DFT with GIAO-based NMR calculations, enabling quantum mechanical predictions for systems containing 1000+ atoms. Applications to proteins in explicit water solvent demonstrated feasibility of full-QM calculations for systems containing 300-500 atoms, though computational cost remains substantial.

Fragment-based quantum mechanical methods decompose large systems into overlapping fragments, compute properties for each fragment quantum mechanically, and recombine the results to obtain system-level properties. The fragment molecular orbital (FMO) approach and related methods enable QM treatment of biomolecules with thousands of atoms by exploiting locality principles (Jose & Raghavachari, 2017). However, NMR chemical shifts require careful treatment of long-range magnetic field effects, and standard fragment methods may not capture all contributions correctly unless special boundary conditions are implemented.

VIII. CURRENT LIMITATIONS AND FUTURE PERSPECTIVES

1. Present Challenges

Despite remarkable progress over the past two decades, computational prediction of NMR chemical shifts in solution faces several fundamental and practical limitations that constrain accuracy and applicability.

Electron correlation treatment in DFT remains approximate, with all practical functionals employing simplifications to the exact exchange-correlation energy. Systematic errors arise from self-interaction, delocalization failures, and inadequate treatment of dispersion and static correlation (Gräfenstein & Cremer, 2009). For challenging systems including diradicals, strongly correlated transition metal complexes, and charge-transfer states, DFT predictions may exhibit errors exceeding 5 ppm (^{1}H) or 30 ppm (^{13}C). Higher-level wave function methods (CCSD(T), CASPT2) provide more rigorous electron correlation but remain computationally prohibitive except for small molecules (<30 atoms) (Polo et al., 2002).

Conformational sampling presents both conceptual and computational challenges, particularly for flexible molecules with 5+ rotatable bonds. The number of possible conformers scales exponentially with rotatable bond count, rapidly exceeding the capacity of systematic grid searches. Enhanced sampling methods (metadynamics, replica exchange MD, umbrella sampling) accelerate exploration but introduce methodological parameters requiring careful validation. Determining which conformers contribute significantly to the observed NMR spectrum distinguishing thermal equilibrium from kinetically trapped populations remains non-trivial for systems with high barriers (Bussi et al., 2006).

Solvation modeling continues to rely on approximations limiting accuracy for complex solvent environments. Continuum models fundamentally cannot describe specific hydrogen bonds, π-π interactions, or structured solvation shells. Explicit solvation requires educated guesses regarding the number and placement of solvent molecules, with few systematic approaches for optimizing these choices (Marenich et al., 2009). Ionic liquids, deep eutectic solvents, and mixed solvent systems present additional challenges due to complex composition-dependent equilibria that current models handle poorly. Nuclear quantum effects (zero-point motion, tunneling) are neglected in standard AIMD simulations and static structure calculations, introducing errors for light nuclei (particularly ^{1}H and ^{2}H) in shallow potential energy wells. Strong hydrogen bonds with delocalized protons represent extreme cases where classical nuclear treatment fails qualitatively (Dračínský & Hodgkinson, 2013). Path integral molecular dynamics (PIMD) incorporates nuclear quantum effects but increases computational cost by factors of 10-50, limiting routine application (Dračínský et al., 2015).

Computational cost remains a practical limitation for large molecules, extensive conformational searches, and high-level methods. Routine DFT calculations scale as $O(N^3)$ with system size, restricting application to hundreds of atoms. AIMD simulations require thousands of DFT evaluations, limiting trajectory lengths to 20-50 ps for systems >100 atoms. Biomolecular applications demand either empirical methods (sacrificing first-principles rigor) or fragmentation schemes (introducing approximations in treating long-range effects) (Ceriotti et al., 2016).

2. Future Directions

Several promising research directions aim to overcome current limitations and extend the accuracy, efficiency, and applicability of computational NMR methods.

Machine learning-accelerated quantum chemistry represents a transformative direction with potential to reduce computational cost by orders of magnitude while maintaining quantum mechanical rigor (Unke et al., 2021). Neural network force fields trained on DFT data enable AIMD-quality sampling at classical MD cost (Smith et al., 2017; Kovács et al., 2025), extending accessible timescales from picoseconds to microseconds. Hybrid ML-QM approaches combining physics-based calculations with data-driven corrections promise unprecedented accuracy by learning to compensate for systematic DFT errors (Li et al., 2024). Concurrently, developing density functionals specifically optimized for magnetic shielding tensors rather than energetic properties could substantially improve accuracy, particularly for paramagnetic shielding contributions and spin-orbit coupling effects in heavy-element systems. Multiscale methods seamlessly integrating quantum mechanics, classical mechanics, and continuum electrostatics (Senn & Thiel, 2009) will provide hierarchical treatment of solvation across different length scales, enabling accurate predictions in complex heterogeneous environments including membrane-bound proteins and enzymatic active sites. Enhanced sampling techniques such as metadynamics and replica exchange will extend accessible conformational space, while machine learning models trained to identify slow collective variables may guide sampling strategies, potentially reducing simulation time by one to two orders of magnitude.

These computational advances will reach full potential through genuine experimental-computational data fusion, where Bayesian approaches incorporating computational uncertainties and experimental errors yield consensus structures maximally consistent with all available information. Real-time computational guidance could dynamically optimize NMR measurement parameters, creating adaptive experimental designs that maximize information content. Underpinning these developments, comprehensive open-source databases (Steinbeck et al., 2003) containing experimental chemical shifts, computed structures, and computational parameters will accelerate method development through machine learning model training, systematic benchmarking, and meta-analysis of errors. Community-driven standardization and quality control will democratize access to reference data and facilitate reproducible workflows across the field.

IX. CONCLUSION

The computational prediction of NMR chemical shifts in solution has matured into a robust and versatile tool for structural chemistry, achieving quantitative accuracy across diverse molecular systems when appropriate methodologies are employed. This review has demonstrated that successful predictions require careful consideration of multiple interdependent factors: functional and basis set selection, solvation model choice, conformational sampling strategy, and treatment of specialized effects including relativistic contributions and nuclear quantum dynamics.

For routine applications involving organic molecules, the combination of hybrid DFT functionals (B3LYP, M06-2X, ωB97X-D) with triple-zeta basis sets and continuum solvation models provides reliable predictions within chemical accuracy. However, systems exhibiting strong hydrogen bonding, transition metal coordination, or heavy-element effects demand more sophisticated treatments incorporating explicit solvation, relativistic methods, or ab initio molecular dynamics. The field increasingly recognizes that static structure calculations represent an oversimplification, with dynamic averaging and conformational sampling emerging as essential components for quantitative accuracy.

Machine learning approaches now complement traditional quantum mechanical methods, offering dramatic computational speedups while achieving comparable accuracy within their training domains. The integration of ML with physics-based calculations represents a particularly promising direction, combining data efficiency with interpretability and transferability.

Looking forward, the convergence of improved density functionals, linear-scaling quantum chemistry, enhanced sampling algorithms, and machine learning acceleration promises to extend computational NMR predictions to larger, more complex systems while maintaining or improving accuracy. The ultimate goal seamless integration of experimental and computational NMR data within unified statistical frameworks remains within reach, potentially transforming structure determination from an interpretive art into a quantitative science with rigorous uncertainty quantification and automated validation protocols.

REFERENCES

Abdullah, A. S., Wang, Y., Menger, M. F., Sami, S., & Head-Gordon, T. (2025). Improved treatment of 1–4 interactions in force fields for molecular dynamics simulations. *Journal of Chemical Theory and Computation, 21*(22), 11669–11678. DOI: 10.1021/acs.jctc.5c01458 PMID: 41197020

Abraham, R. J., Byrne, J. J., Griffiths, L., & Perez, M. (2006). 1H chemical shifts in NMR: Part 23, the effect of dimethyl sulphoxide versus chloroform solvent on 1H chemical shifts. *Magnetic Resonance in Chemistry, 44*(5), 491–509. DOI: 10.1002/mrc.1747 PMID: 16395732

Aslam, N., Pfender, M., Neumann, P., Reuter, R., Zappe, A., De Oliveira, F. F., Denisenko, A., Sumiya, H., Onoda, S., Isoya, J., & Wrachtrup, J. (2017). Nanoscale nuclear magnetic resonance with chemical resolution. *Science, 357*(6346), 67–71. DOI: 10.1126/science.aam8697 PMID: 28572453

Autschbach, J., & Zheng, S. (2008). Analyzing Pt chemical shifts calculated from relativistic density functional theory using localized orbitals: The role of Pt 5d lone pairs. *Magnetic Resonance in Chemistry, 46*(S1), S45–S55. DOI: 10.1002/mrc.2289 PMID: 18855339

Autschbach, J., Zheng, S., & Schurko, R. W. (2010). Analysis of electric field gradient tensors at quadrupolar nuclei in common structural motifs. *Concepts in Magnetic Resonance. Part A, Bridging Education and Research., 36A*(2), 84–126. DOI: 10.1002/cmr.a.20155

Baker, C. M., Anisimov, V. M., & MacKerell, A. D.Jr. (2010). Development of CHARMM Polarizable Force Field for nucleic acid bases based on the classical Drude Oscillator model. *The Journal of Physical Chemistry B, 115*(3), 580–596. DOI: 10.1021/jp1092338 PMID: 21166469

Becke, A. D. (1993). Density-functional thermochemistry. III. The role of exact exchange. *The Journal of Chemical Physics, 98*(7), 5648–5652. DOI: 10.1063/1.464913

Benassi, E. (2016). Benchmarking of density functionals for a soft but accurate prediction and assignment of 1H and 13C NMR chemical shifts in organic and biological molecules. *Journal of Computational Chemistry, 38*(2), 87–92. DOI: 10.1002/jcc.24521 PMID: 27796077

Boereboom, J. M., Fleurat-Lessard, P., & Bulo, R. E. (2018). Explicit solvation matters: Performance of QM/MM solvation models in nucleophilic addition. *Journal of Chemical Theory and Computation, 14*(4), 1841–1852. DOI: 10.1021/acs.jctc.7b01206 PMID: 29438621

Boulanger, E., & Harvey, J. N. (2018). QM/MM methods for free energies and photochemistry. *Current Opinion in Structural Biology, 49*, 72–76. DOI: 10.1016/j.sbi.2018.01.003 PMID: 29414514

Cancès, E., Mennucci, B., & Tomasi, J. (1997). A new integral equation formalism for the polarizable continuum model: Theoretical background and applications to isotropic and anisotropic dielectrics. *The Journal of Chemical Physics, 107*(8), 3032–3041. DOI: 10.1063/1.474659

Car, R., & Parrinello, M. (1985). Unified Approach for Molecular Dynamics and Density-Functional Theory. *Physical Review Letters, 55*(22), 2471–2474. DOI: 10.1103/PhysRevLett.55.2471 PMID: 10032153

Casabianca, L. B. (2020). Calculating nuclear magnetic resonance chemical shifts in solvated systems. *Magnetic Resonance in Chemistry, 58*(7), 611–624. DOI: 10.1002/mrc.4994 PMID: 31916612

Castro, A. C., Balcells, D., Repisky, M., Helgaker, T., & Cascella, M. (2020). First-Principles Calculation of 1H NMR chemical shifts of complex metal polyhydrides: The essential inclusion of relativity and dynamics. *Inorganic Chemistry*, *59*(23), 17509–17518. DOI: 10.1021/acs.inorgchem.0c02753 PMID: 33226791

Ceriotti, M., Fang, W., Kusalik, P. G., McKenzie, R. H., Michaelides, A., Morales, M. A., & Markland, T. E. (2016). Nuclear quantum effects in water and aqueous Systems: Experiment, theory, and current challenges. *Chemical Reviews*, *116*(13), 7529–7550. DOI: 10.1021/acs.chemrev.5b00674 PMID: 27049513

Charisiadis, P., Kontogianni, V., Tsiafoulis, C., Tzakos, A., Siskos, M., & Gerothanassis, I. (2014). 1H-NMR as a structural and analytical tool of intra- and intermolecular hydrogen bonds of Phenol-Containing natural products and model compounds. *Molecules (Basel, Switzerland)*, *19*(9), 13643–13682. DOI: 10.3390/molecules190913643 PMID: 25185070

Charpentier, T., Menziani, M. C., & Pedone, A. (2013). Computational simulations of solid state NMR spectra: A new era in structure determination of oxide glasses. *RSC Advances*, *3*(27), 10550. DOI: 10.1039/c3ra40627j

Chen, S., & Izgorodina, E. I. (2017). Prediction of 1H NMR chemical shifts for clusters of imidazolium-based ionic liquids. *Physical Chemistry Chemical Physics*, *19*(26), 17411–17425. DOI: 10.1039/C7CP02951A PMID: 28650016

Cohen, A. J., Mori-Sánchez, P., & Yang, W. (2012). Challenges for density Functional Theory. *Chemical Reviews*, *112*(1), 289–320. DOI: 10.1021/cr200107z PMID: 22191548

Cohen, R. D., Wood, J. S., Lam, Y., Buevich, A. V., Sherer, E. C., Reibarkh, M., Williamson, R. T., & Martin, G. E. (2023). DELTA50: A highly accurate database of experimental 1H and 13C NMR chemical shifts applied to DFT benchmarking. *Molecules (Basel, Switzerland)*, *28*(6), 449. DOI: 10.3390/molecules28062449 PMID: 36985422

Collins, K. D. (2012). Why continuum electrostatics theories cannot explain biological structure, polyelectrolytes or ionic strength effects in ion–protein interactions. *Biophysical Chemistry*, *167*, 43–59. DOI: 10.1016/j.bpc.2012.04.002 PMID: 22608112

Cui, Q., & Karplus, M. (2000). Molecular properties from combined QM/MM methods. 2. Chemical shifts in large molecules. *The Journal of Physical Chemistry B*, *104*(15), 3721–3743. DOI: 10.1021/jp994154g

Da Silva, H. C., Hernandes, I. S., & De Almeida, W. B. (2025). Modeling solvent effects in quantum chemical calculation of relative energies and NMR chemical shifts for azithromycin. *The Journal of Physical Chemistry A*, *129*(9), 2200–2216. DOI: 10.1021/acs.jpca.4c08015 PMID: 39985808

Das, S., & Merz, K. M.Jr. (2025). Exploring the frontiers of computational NMR: Methods, applications, and challenges. *Chemical Reviews*, *125*(19), 9256–9295. DOI: 10.1021/acs.chemrev.5c00259 PMID: 40923454

De Gortari, I., Portella, G., Salvatella, X., Bajaj, V. S., Van Der Wel, P. C. A., Yates, J. R., Segall, M. D., Pickard, C. J., Payne, M. C., & Vendruscolo, M. (2010). Time averaging of NMR chemical shifts in the MLF peptide in the solid state. *Journal of the American Chemical Society*, *132*(17), 5993–6000. DOI: 10.1021/ja9062629 PMID: 20387894

De Oliveira, M. T., Alves, J. M. A., Braga, A. C., Wilson, D. J. D., & Barboza, C. A. (2021). Do Double-Hybrid Exchange–Correlation functionals provide accurate chemical shifts? A benchmark assessment for proton NMR. *Journal of Chemical Theory and Computation*, *17*(11), 6876–6885. DOI: 10.1021/acs.jctc.1c00604 PMID: 34637284

De Souza, L. A., Da Silva, H. C., & De Almeida, W. B. (2018). Structural Determination of Antioxidant and Anticancer Flavonoid Rutin in Solution through DFT Calculations of 1H NMR Chemical Shifts. *ChemistryOpen*, *7*(11), 902–913. DOI: 10.1002/open.201800209 PMID: 30460171

De Vivo, M., Masetti, M., Bottegoni, G., & Cavalli, A. (2016). Role of molecular dynamics and related methods in drug discovery. *Journal of Medicinal Chemistry*, *59*(9), 4035–4061. DOI: 10.1021/acs.jmedchem.5b01684 PMID: 26807648

Dračínský, M., Čechová, L., Hodgkinson, P., Procházková, E., & Janeba, Z. (2015). Resonance-assisted stabilisation of hydrogen bonds probed by NMR spectroscopy and path integral molecular dynamics. *Chemical Communications*, *51*(73), 13986–13989. DOI: 10.1039/C5CC05199A PMID: 26245988

Dračínský, M., & Hodgkinson, P. (2013). A molecular dynamics study of the effects of fast molecular motions on solid-state NMR parameters. *CrystEngComm*, *15*(43), 8705–8712. DOI: 10.1039/c3ce40612a

Dračínský, M., Möller, H. M., & Exner, T. E. (2013). Conformational sampling by AB initio molecular dynamics simulations improves NMR chemical shift predictions. *Journal of Chemical Theory and Computation*, *9*(8), 3806–3815. DOI: 10.1021/ct400282h PMID: 26584127

Dračínský, M., Unzueta, P., & Beran, G. J. O. (2019). Improving the accuracy of solid-state nuclear magnetic resonance chemical shift prediction with a simple molecular correction. *Physical Chemistry Chemical Physics*, *21*(27), 14992–15000. DOI: 10.1039/C9CP01666J PMID: 31237586

Exner, T. E., Frank, A., Onila, I., & Möller, H. M. (2012). Toward the quantum chemical calculation of NMR chemical shifts of proteins. 3. Conformational sampling and explicit solvents model. *Journal of Chemical Theory and Computation*, *8*(11), 4818–4827. DOI: 10.1021/ct300701m PMID: 26605634

Faska, N., Auhmani, A., & Esseffar, M. (2004). Theoretical solvent effects on 13C NMR shielding of some aromatic hydrocarbons. Linear regression analysis. *Journal of Molecular Structure THEOCHEM*, *709*(1–3), 171–181. DOI: 10.1016/j.theochem.2004.05.038

Faska, N., Auhmani, A., & Esseffar, M. (2007). Solvent effects on 13C and 1H NMR shielding of cyclic ketones: An experimental and theoretical study. *Journal of Molecular Structure THEOCHEM*, *811*(1–3), 203–213. DOI: 10.1016/j.theochem.2006.12.057

Faska, N., Auhmani, A., Esseffar, M., & Abboud, J. L. (2011). Computational (solute–solvent cluster + PCM) study of medium effects on the experimental 13C and 1H NMR chemical shifts of lactones and lactams. *Journal of Physical Organic Chemistry*, *24*(12), 1209–1221. DOI: 10.1002/poc.1849

Figueirinhas, J. L., & Dong, R. Y. (2014). NMR of bent-core nematogens: A mini-review. *Magnetic Resonance in Chemistry*, *52*(10), 614–624. DOI: 10.1002/mrc.4097 PMID: 24975632

Gahlawat, S., Hopmann, K. H., & Castro, A. C. (2024). Advancing 19F NMR Prediction of Metal-Fluoride Complexes in Solution: Insights from Ab Initio Molecular Dynamics. *The Journal of Physical Chemistry A, 128*(49), 10498–10506. DOI: 10.1021/acs.jpca.4c05408 PMID: 39626023

Goldey, M., Dutoi, A., & Head-Gordon, M. (2013). Attenuated second-order Møller–Plesset perturbation theory: Performance within the aug-cc-pVTZ basis. *Physical Chemistry Chemical Physics, 15*(38), 15869. DOI: 10.1039/c3cp51826d PMID: 23942866

Goloviznina, K., Gong, Z., & Padua, A. H. (2021). The CL&Pol polarizable force field for the simulation of ionic liquids and eutectic solvents. *Wiley Interdisciplinary Reviews. Computational Molecular Science, 12*(3), e1572. Advance online publication. DOI: 10.1002/wcms.1572

Gräfenstein, J., & Cremer, D. (2009). The self-interaction error and the description of non-dynamic electron correlation in density functional theory. *Theoretical Chemistry Accounts, 123*(3–4), 171–182. DOI: 10.1007/s00214-009-0545-9

Grimblat, N., & Sarotti, A. M. (2016). Computational Chemistry to the Rescue: Modern Toolboxes for the assignment of complex molecules by GIAO NMR Calculations. *Chemistry (Weinheim an der Bergstrasse, Germany), 22*(35), 12246–12261. DOI: 10.1002/chem.201601150 PMID: 27405775

Grimme, S., Antony, J., Ehrlich, S., & Krieg, H. (2010). A consistent and accurate ab initio parametrization of density functional dispersion correction (DFT-D) for the 94 elements H-Pu. *The Journal of Chemical Physics, 132*(15), 154104. DOI: 10.1063/1.3382344 PMID: 20423165

Guan, Y., Shree Sowndarya, S. V., Gallegos, L. C., St. John, P. C., & Paton, R. S. (2021). Real-time prediction of ^{1}H and ^{13}C chemical shifts with DFT accuracy using a 3D graph neural network. *Chemical Science (Cambridge), 12*(36), 12012–12026. DOI: 10.1039/D1SC03343C PMID: 34667567

He, X., Wang, B., & Merz, K. M.Jr. (2009). Protein NMR chemical shift calculations based on the automated fragmentation QM/MM approach. *The Journal of Physical Chemistry B, 113*(31), 10380–10388. DOI: 10.1021/jp901992p PMID: 19575540

Hernandes, I. S., Da Silva, H. C., Santos, H. F. D., Ávila, E. P., De Almeida, M. V., & De Almeida, W. B. (2023). Quantum chemical investigation of predominant conformation of the antibiotic azithromycin in water and DMSO solutions: Thermodynamic and NMR analysis. *Royal Society Open Science, 10*(10), 230409. DOI: 10.1098/rsos.230409 PMID: 37830015

Hollingsworth, S. A., & Dror, R. O. (2018). Molecular dynamics simulation for all. *Neuron, 99*(6), 1129–1143. DOI: 10.1016/j.neuron.2018.08.011 PMID: 30236283

Iron, M. A. (2017). Evaluation of the Factors Impacting the Accuracy of 13C NMR Chemical Shift Predictions using Density Functional Theory: The Advantage of Long-Range Corrected Functionals. *Journal of Chemical Theory and Computation, 13*(11), 5798–5819. DOI: 10.1021/acs.jctc.7b00772 PMID: 29016125

Jose, K. V. J., & Raghavachari, K. (2017). Fragment-Based approach for the evaluation of NMR chemical shifts for large biomolecules incorporating the effects of the solvent environment. *Journal of Chemical Theory and Computation, 13*(3), 1147–1158. DOI: 10.1021/acs.jctc.6b00922 PMID: 28194972

Kaplanskiy, M. V., Shitov, D. A., Tolstoy, P. M., & Tupikina, E. Y. (2025). Deconstructing 1H NMR chemical shifts in strong hydrogen bonds: A computational investigation of solvation, dynamics, and nuclear delocalization effects. *Journal of Chemical Information and Modeling*, *65*(10), 5019–5034. DOI: 10.1021/acs.jcim.5c00566 PMID: 40350596

Kayal, R., Baldinelli, L., Harden, I., Neese, F., & Bistoni, G. (2024). Understanding and quantifying the impact of solute–solvent van der Waals interactions on the selectivity of asymmetric catalytic transformations. *Chemical Science (Cambridge)*, *16*(6), 2700–2709. DOI: 10.1039/D4SC04329D PMID: 39802687

Kirschner, K. N., Yongye, A. B., Tschampel, S. M., González-Outeiriño, J., Daniels, C. R., Foley, B. L., & Woods, R. J. (2007). GLYCAM06: A generalizable biomolecular force field. Carbohydrates. *Journal of Computational Chemistry*, *29*(4), 622–655. DOI: 10.1002/jcc.20820 PMID: 17849372

Kjær, H., & Sauer, S. P. A. (2011). Pople Style Basis Sets for the Calculation of NMR Spin–Spin Coupling Constants: The 6-31G-J and 6-311G-J Basis Sets. *Journal of Chemical Theory and Computation*, *7*(12), 4070–4076. DOI: 10.1021/ct200546q PMID: 26598352

Klamt, A., Moya, C., & Palomar, J. (2015). A Comprehensive Comparison of the IEFPCM and SS(V)PE Continuum Solvation Methods with the COSMO Approach. *Journal of Chemical Theory and Computation*, *11*(9), 4220–4225. DOI: 10.1021/acs.jctc.5b00601 PMID: 26575917

Klamt, A., & Schüürmann, G. (1993). COSMO: A new approach to dielectric screening in solvents with explicit expressions for the screening energy and its gradient. *Journal of the Chemical Society, Perkin Transactions 2: Physical Organic Chemistry*, (5), 799–805. DOI: 10.1039/P29930000799

Koes, D. R., & Vries, J. K. (2017). Evaluating amber force fields using computed NMR chemical shifts. *Proteins*, *85*(10), 1944–1956. DOI: 10.1002/prot.25350 PMID: 28688107

Kovács, D. P., Moore, J. H., Browning, N. J., Batatia, I., Horton, J. T., Pu, Y., Kapil, V., Witt, W. C., Magdău, I., Cole, D. J., & Csányi, G. (2025). MACE-OFF: Short-Range Transferable Machine Learning Force Fields for Organic Molecules. *Journal of the American Chemical Society*, *147*(21), 17598–17611. DOI: 10.1021/jacs.4c07099 PMID: 40387214

Kussmann, J., & Ochsenfeld, C. (2007). Linear-scaling method for calculating nuclear magnetic resonance chemical shifts using gauge-including atomic orbitals within Hartree-Fock and density-functional theory. *The Journal of Chemical Physics*, *127*(5), 054103. DOI: 10.1063/1.2749509 PMID: 17688330

Lafiosca, P., Rossi, F., Egidi, F., Giovannini, T., & Cappelli, C. (2023). Multiscale Frozen Density Embedding/Molecular Mechanics approach for simulating magnetic response properties of solvated systems. *Journal of Chemical Theory and Computation*, *20*(1), 266–279. DOI: 10.1021/acs.jctc.3c00850 PMID: 38109486

Laun, J., & Bredow, T. (2021). BSSE-corrected consistent Gaussian basis sets of triple-zeta valence with polarization quality of the sixth period for solid-state calculations. *Journal of Computational Chemistry*, *42*(15), 1064–1072. DOI: 10.1002/jcc.26521 PMID: 33792062

Lemkul, J. A., Huang, J., Roux, B., & MacKerell, A. D.Jr. (2016). An empirical polarizable force field based on the classical Drude oscillator model: Development history and recent applications. *Chemical Reviews*, *116*(9), 4983–5013. DOI: 10.1021/acs.chemrev.5b00505 PMID: 26815602

Li, J., Liang, J., Wang, Z., Ptaszek, A. L., Liu, X., Ganoe, B., Head-Gordon, M., & Head-Gordon, T. (2024). Highly Accurate Prediction of NMR Chemical Shifts from Low-Level Quantum Mechanics Calculations Using Machine Learning. *Journal of Chemical Theory and Computation*, *20*(5), 2152–2166. DOI: 10.1021/acs.jctc.3c01256 PMID: 38331423

Liu, F., Sanchez, D. M., Kulik, H. J., & Martínez, T. J. (2018). Exploiting graphical processing units to enable quantum chemistry calculation of large solvated molecules with conductor-like polarizable continuum models. *International Journal of Quantum Chemistry*, *119*(1), e25760. Advance online publication. DOI: 10.1002/qua.25760

Lodewyk, M. W., Siebert, M. R., & Tantillo, D. J. (2011). Computational prediction of ^{1}H and ^{13}C chemical shifts: A useful tool for natural product, mechanistic, and synthetic organic chemistry. *Chemical Reviews*, *112*(3), 1839–1862. DOI: 10.1021/cr200106v PMID: 22091891

Lomas, J. S. (2015). 1H NMR spectra of alcohols in hydrogen bonding solvents: DFT/GIAO calculations of chemical shifts. *Magnetic Resonance in Chemistry*, *54*(1), 28–38. DOI: 10.1002/mrc.4312 PMID: 26256675

Marcarino, M. O., Cicetti, S., Zanardi, M. M., & Sarotti, A. M. (2021). A critical review on the use of DP4+ in the structural elucidation of natural products: The good, the bad and the ugly. A practical guide. *Natural Product Reports*, *39*(1), 58–76. DOI: 10.1039/D1NP00030F PMID: 34212963

Marenich, A. V., Cramer, C. J., & Truhlar, D. G. (2009). Universal solvation model based on solute electron density and on a continuum model of the solvent defined by the bulk dielectric constant and atomic surface tensions. *The Journal of Physical Chemistry B*, *113*(18), 6378–6396. DOI: 10.1021/jp810292n PMID: 19366259

Mari, S. H., Varras, P. C., Atia-Tul-Wahab, N., Choudhary, I. M., Siskos, M. G., & Gerothanassis, I. P. (2019). Solvent-Dependent structures of natural products based on the combined use of DFT calculations and 1H-NMR chemical shifts. *Molecules (Basel, Switzerland)*, *24*(12), 2290. DOI: 10.3390/molecules24122290 PMID: 31226776

Marx, D., & Hutter, J. (2009). AB Initio Molecular Dynamics. In Cambridge University Press eBooks. DOI: 10.1017/CBO9780511609633

Mennucci, B. (2012). Polarizable continuum model. *Wiley Interdisciplinary Reviews. Computational Molecular Science*, *2*(3), 386–404. DOI: 10.1002/wcms.1086

Merrill, A. T., & Tantillo, D. J. (2019). Solvent optimization and conformational flexibility effects on 1H and 13C NMR scaling factors. *Magnetic Resonance in Chemistry*, *58*(6), 576–583. DOI: 10.1002/mrc.4986 PMID: 31883397

Mirzaei, S., Ivanov, M. V., & Timerghazin, Q. K. (2019). Improving performance of the SMD solvation model: BoNdi Radii improve predicted aqueous solvation free energies of ions and PKA values of thiols. *The Journal of Physical Chemistry A*, *123*(44), 9498–9504. DOI: 10.1021/acs.jpca.9b02340 PMID: 31318553

Mocci, F., & Laaksonen, A. (2015). Combining MD simulations and NMR spectroscopy for molecular insight and methodological synergy: the integrated MD-NMR method. In Nuclear magnetic resonance (pp. 592–616). DOI: 10.1039/9781782622758-00592

Mu, J., Liu, H., Zhang, J., Luo, R., & Chen, H. (2021). Recent force field strategies for intrinsically disordered proteins. *Journal of Chemical Information and Modeling*, *61*(3), 1037–1047. DOI: 10.1021/acs.jcim.0c01175 PMID: 33591749

Niklasson, A. M. (2008). Extended born-oppenheimer molecular dynamics. *Physical Review Letters*, *100*(12), 123004. DOI: 10.1103/PhysRevLett.100.123004 PMID: 18517861

Paruzzo, F. M., Hofstetter, A., Musil, F., De, S., Ceriotti, M., & Emsley, L. (2018). Chemical shifts in molecular solids by machine learning. *Nature Communications*, *9*(1), 4501. DOI: 10.1038/s41467-018-06972-x PMID: 30374021

Pascual-Borràs, M., López, X., Rodríguez-Fortea, A., Errington, R. J., & Poblet, J. M. (2014). 17O NMR chemical shifts in oxometalates: From the simplest monometallic species to mixed-metal polyoxometalates. *Chemical Science (Cambridge)*, *5*(5), 2031. DOI: 10.1039/c4sc00083h

Patrahau, B., Piejko, M., Mayer, R. J., Antheaume, C., Sangchai, T., Ragazzon, G., Jayachandran, A., Devaux, E., Genet, C., Moran, J., & Ebbesen, T. W. (2024). Direct observation of polaritonic chemistry by nuclear magnetic resonance spectroscopy. *Angewandte Chemie International Edition*, *63*(23), e202401368. DOI: 10.1002/anie.202401368 PMID: 38584127

Pierens, G. K. (2014). 1H and 13C NMR scaling factors for the calculation of chemical shifts in commonly used solvents using density functional theory. *Journal of Computational Chemistry*, *35*(18), 1388–1394. DOI: 10.1002/jcc.23638 PMID: 24854878

Pierens, G. K., Venkatachalam, T. K., & Reutens, D. C. (2017). NMR and DFT investigations of structure of colchicine in various solvents including density functional theory calculations. *Scientific Reports*, *7*(1), 5605. DOI: 10.1038/s41598-017-06005-5 PMID: 28717218

Polo, V., Kraka, E., & Cremer, D. (2002). Electron correlation and the self-interaction error of density functional theory. *Molecular Physics*, *100*(11), 1771–1790. DOI: 10.1080/00268970110111788

Ponder, J. W., Wu, C., Ren, P., Pande, V. S., Chodera, J. D., Schnieders, M. J., Haque, I., Mobley, D. L., Lambrecht, D. S., DiStasio, R. A.Jr, Head-Gordon, M., Clark, G. N. I., Johnson, M. E., & Head-Gordon, T. (2010). Current status of the AMOEBA polarizable force field. *The Journal of Physical Chemistry B*, *114*(8), 2549–2564. DOI: 10.1021/jp910674d PMID: 20136072

Provorse, M. R., Peev, T., Xiong, C., & Isborn, C. M. (2016). Convergence of excitation energies in mixed quantum and classical solvent: Comparison of continuum and point charge models. *The Journal of Physical Chemistry B*, *120*(47), 12148–12159. DOI: 10.1021/acs.jpcb.6b09176 PMID: 27797196

Pruitt, S. R., Brorsen, K. R., & Gordon, M. S. (2015). Ab initio investigation of the aqueous solvation of the nitrate ion. *Physical Chemistry Chemical Physics*, *17*(40), 27027–27034. DOI: 10.1039/C5CP04445F PMID: 26412597

R., Peev, T., Xiong, C., & Isborn, C. M. (2016). Convergence of excitation energies in mixed quantum and classical solvent: Comparison of continuum and point charge models. The Journal of Physical Chemistry B, 120(47), 12148–12159. DOI: 10.1021/acs.jpcb.6b09176

Ramos, S. A., Mueller, L. J., & Beran, G. J. O. (2024). The interplay of density functional selection and crystal structure for accurate NMR chemical shift predictions. *Faraday Discussions*, *255*(0), 119–142. DOI: 10.1039/D4FD00072B PMID: 39258864

Rychnovsky, S. D. (2006). Predicting NMR spectra by computational methods: Structure revision of hexacyclinol. *Organic Letters*, *8*(13), 2895–2898. DOI: 10.1021/ol0611346 PMID: 16774284

Sarotti, A. M., & Pellegrinet, S. C. (2009). A multi-standard approach for GIAO ^{13}C NMR calculations. *The Journal of Organic Chemistry*, *74*(19), 7254–7260. DOI: 10.1021/jo901234h PMID: 19725561

Sarotti, A. M., & Pellegrinet, S. C. (2012). Application of the multi-standard methodology for calculating 1H NMR chemical shifts. *The Journal of Organic Chemistry*, *77*(14), 6059–6065. DOI: 10.1021/jo3008447 PMID: 22713105

Senn, H. M., & Thiel, W. (2009). QM/MM methods for biomolecular systems. *Angewandte Chemie International Edition*, *48*(7), 1198–1229. DOI: 10.1002/anie.200802019 PMID: 19173328

Sethio, D., Bopp, P., & Hermansson, K. (2025). ^{7}Li NMR chemical shifts in Battery-Relevant electrolytes: Ligand-Induced shifts cancel in both experiment and calculation. *ChemSusChem*, *18*(21), e202500798. DOI: 10.1002/cssc.202500798 PMID: 40971230

Shenderovich, I. G. (2023). Experimentally established 15N NMR absolute shielding scale for theoretical calculations. *The Journal of Physical Chemistry A*, *127*(26), 5547–5555. DOI: 10.1021/acs.jpca.3c02350 PMID: 37344362

Shi, Y., Xia, Z., Zhang, J., Best, R., Wu, C., Ponder, J. W., & Ren, P. (2013). Polarizable Atomic Multipole-Based AMOEBA force field for proteins. *Journal of Chemical Theory and Computation*, *9*(9), 4046–4063. DOI: 10.1021/ct4003702 PMID: 24163642

Silva, N. M., Deglmann, P., & Pliego, J. R.Jr. (2016). CMIRS Solvation Model for Methanol: Parametrization, Testing, and Comparison with SMD, SM8, and COSMO-RS. *The Journal of Physical Chemistry B*, *120*(49), 12660–12668. DOI: 10.1021/acs.jpcb.6b10249 PMID: 27973833

Siskos, M., Choudhary, M., & Gerothanassis, I. (2017). Hydrogen Atomic Positions of O–H···O Hydrogen Bonds in Solution and in the Solid State: The Synergy of Quantum Chemical Calculations with 1H-NMR Chemical Shifts and X-ray Diffraction Methods. *Molecules (Basel, Switzerland)*, *22*(3), 415. DOI: 10.3390/molecules22030415 PMID: 28272366

Smith, J. S., Isayev, O., & Roitberg, A. E. (2017). ANI-1: An extensible neural network potential with DFT accuracy at force field computational cost. *Chemical Science (Cambridge)*, *8*(4), 3192–3203. DOI: 10.1039/C6SC05720A PMID: 28507695

Smith, S. G., & Goodman, J. M. (2010). Assigning stereochemistry to single diastereoisomers by GIAO NMR calculation: The DP4 probability. *Journal of the American Chemical Society*, *132*(37), 12946–12959. DOI: 10.1021/ja105035r PMID: 20795713

St John, P. C., Phillips, C., Kemper, T. W., Wilson, A. N., Guan, Y., Crowley, M. F., Nimlos, M. R., & Larsen, R. E. (2019). Message-passing neural networks for high-throughput polymer screening. *The Journal of Chemical Physics*, *150*(23), 234111. DOI: 10.1063/1.5099132 PMID: 31228909

Stahn, M., Ehlert, S., & Grimme, S. (2023). Extended Conductor-like Polarizable Continuum Solvation Model (CPCM-X) for semiempirical methods. *The Journal of Physical Chemistry A*, *127*(33), 7036–7043. DOI: 10.1021/acs.jpca.3c04382 PMID: 37567769

Steinbeck, C., Krause, S., & Kuhn, S. (2003). NMRShiftDBConstructing a Free Chemical Information System with Open-Source Components. *Journal of Chemical Information and Computer Sciences*, *43*(6), 1733–1739. DOI: 10.1021/ci0341363 PMID: 14632418

Steiner, M., Holzknecht, T., Schauperl, M., & Podewitz, M. (2021). Quantum chemical microsolvation by automated water placement. *Molecules (Basel, Switzerland)*, *26*(6), 1793. DOI: 10.3390/molecules26061793 PMID: 33806731

Swails, J., Zhu, T., He, X., & Case, D. A. (2015). AFNMR: Automated fragmentation quantum mechanical calculation of NMR chemical shifts for biomolecules. *Journal of Biomolecular NMR*, *63*(2), 125–139. DOI: 10.1007/s10858-015-9970-3 PMID: 26232926

Szántó, J. K., Dietschreit, J. C. B., Shein, M., Schütz, A. K., & Ochsenfeld, C. (2024). Systematic QM/MM study for predicting 31P NMR chemical shifts of adenosine nucleotides in solution and stages of ATP hydrolysis in a protein environment. *Journal of Chemical Theory and Computation*, *20*(6), 2433–2444. DOI: 10.1021/acs.jctc.3c01280 PMID: 38497488

Tomasi, J., Mennucci, B., & Cammi, R. (2005). Quantum mechanical continuum solvation models. *Chemical Reviews*, *105*(8), 2999–3094. DOI: 10.1021/cr9904009 PMID: 16092826

Trainor, K., Palumbo, J. A., MacKenzie, D. W. S., & Meiering, E. M. (2019). Temperature dependence of NMR chemical shifts: Tracking and statistical analysis. *Protein Science : a Publication of the Protein Society*, *29*(1), 306–314. DOI: 10.1002/pro.3785 PMID: 31730280

Truflandier, L. A., & Autschbach, J. (2010). Probing the Solvent Shell with 195Pt Chemical Shifts: Density Functional Theory Molecular Dynamics Study of PtII and PtIV Anionic Complexes in Aqueous Solution. *Journal of the American Chemical Society*, *132*(10), 3472–3483. DOI: 10.1021/ja9096863 PMID: 20166712

Tuckerman, M. E. (2002). Ab initiomolecular dynamics: Basic concepts, current trends and novel applications. *Journal of Physics Condensed Matter*, *14*(50), R1297–R1355. DOI: 10.1088/0953-8984/14/50/202

Unke, O. T., Chmiela, S., Sauceda, H. E., Gastegger, M., Poltavsky, I., Schutt, K. T., Tkatchenko, A., & Muller, K. R. (2021). Machine learning force fields. *Chemical Reviews*, *121*(16), 10142–10186. DOI: 10.1021/acs.chemrev.0c01111 PMID: 33705118

Vícha, J., Novotný, J., Straka, M., Repisky, M., Ruud, K., Komorovsky, S., & Marek, R. (2015). Structure, solvent, and relativistic effects on the NMR chemical shifts in square-planar transition-metal complexes: Assessment of DFT approaches. *Physical Chemistry Chemical Physics*, *17*(38), 24944–24955. DOI: 10.1039/C5CP04214C PMID: 26344822

Victora, A., Möller, H. M., & Exner, T. E. (2014). Accurate ab initio prediction of NMR chemical shifts of nucleic acids and nucleic acids/protein complexes. *Nucleic Acids Research*, *42*(22), e173. DOI: 10.1093/nar/gku1006 PMID: 25404135

Wang, H., Vlcek, L., Neuefeind, J. C., Page, K., Irle, S., Simonson, J. M., & Stack, A. G. (2018). Decoding oxyanion aqueous solvation structure: A potassium nitrate example at saturation. *The Journal of Physical Chemistry B*, *122*(30), 7584–7589. DOI: 10.1021/acs.jpcb.8b05895 PMID: 29991255

Wang, J., Liu, Y., Li, W., & Gao, G. (2018). Prediction of 1H NMR chemical shifts for ionic liquids: Strategy and application of a relative reference standard. *RSC Advances*, *8*(50), 28604–28612. DOI: 10.1039/C8RA04822C PMID: 35548389

Wang, Y., Verma, P., Zhang, L., Li, Y., Liu, Z., Truhlar, D. G., & He, X. (2020). M06-SX screened-exchange density functional for chemistry and solid-state physics. *Proceedings of the National Academy of Sciences of the United States of America*, *117*(5), 2294–2301. DOI: 10.1073/pnas.1913699117 PMID: 31953258

Wiitala, K. W., Al-Rashid, Z. F., Dvornikovs, V., Hoye, T. R., & Cramer, C. J. (2007). Evaluation of various DFT protocols for computing 1H and 13C chemical shifts to distinguish stereoisomers: Diastereomeric 2-, 3-, and 4-methylcyclohexanols as a test set. *Journal of Physical Organic Chemistry*, *20*(5), 345–354. DOI: 10.1002/poc.1151

Willoughby, P. H., Jansma, M. J., & Hoye, T. R. (2014). A guide to small-molecule structure assignment through computation of (1H and 13C) NMR chemical shifts. *Nature Protocols*, *9*(3), 643–660. DOI: 10.1038/nprot.2014.042 PMID: 24556787

Xin, D., Sader, C. A., Chaudhary, O., Jones, P., Wagner, K., Tautermann, C. S., Yang, Z., Busacca, C. A., Saraceno, R. A., Fandrick, K. R., Gonnella, N. C., Horspool, K., Hansen, G., & Senanayake, C. H. (2017). Development of a 13C NMR chemical shift prediction procedure using B3LYP/CC-PVDZ and empirically derived systematic error correction terms: A computational small molecule structure elucidation method. *The Journal of Organic Chemistry*, *82*(10), 5135–5145. DOI: 10.1021/acs.joc.7b00321 PMID: 28398046

Yang, Z., Chakraborty, M., & White, A. D. (2021). Predicting chemical shifts with graph neural networks. *Chemical Science (Cambridge)*, *12*(32), 10802–10809. DOI: 10.1039/D1SC01895G PMID: 34476061

Zhao, Y., & Truhlar, D. G. (2007). The M06 suite of density functionals for main group thermochemistry, thermochemical kinetics, noncovalent interactions, excited states, and transition elements: Two new functionals and systematic testing of four M06-class functionals and 12 other functionals. *Theoretical Chemistry Accounts*, *120*(1–3), 215–241. DOI: 10.1007/s00214-007-0310-x

Zhu, T., Zhang, J. Z. H., & He, X. (2013). Automated Fragmentation QM/MM Calculation of Amide Proton Chemical Shifts in Proteins with Explicit Solvent Model. *Journal of Chemical Theory and Computation*, *9*(4), 2104–2114. DOI: 10.1021/ct300999w PMID: 26583557

Chapter 7
Molecular Docking and Virtual Screening:
Advances, Applications, and Integration With AI in Modern Drug Discovery

Bancha Yingngam
https://orcid.org/0000-0001-7215-9123
Faculty of Pharmaceutical Sciences, Ubon Ratchathani University, Thailand

Charuwan Thanawiroon
https://orcid.org/0009-0004-9076-3219
Faculty of Pharmaceutical Sciences, Ubon Ratchathani University, Thailand

Pichit Sudta
Faculty of Science and Technology, Phetchaburi Rajabhat University, Thailand

ABSTRACT

Molecular docking and virtual screening are computational methods for drug discovery that can help predict the interaction between ligands and targets at high speed. This chapter describes recent developments with a particular focus on artificial intelligence (AI) integration, experimental validation, and applications: cancers, neurodegenerative diseases, and cardiovascular diseases. The presented cases demonstrate that hybrid workflows, consensus scoring, and AI-driven library design can generate targets with potential clinical relevance while simultaneously lowering time-related costs and driving molecular docking toward scalable containerized cloud discovery. Finally, the chapter discusses viewpoints on future hurdles and scalability in addition to the democratization of docking with cloud-based and automated platforms.

1. INTRODUCTION

Molecular docking-based in silico methods of structure-based screening have provided essential tools for drug development in the contemporary era. They make it possible for researchers to quickly screen how drug candidate molecules interact with biological targets and to help identify potential compounds

DOI: 10.4018/979-8-3373-6058-4.ch007

to be tested in the laboratory (da Rocha et al., 2025; Paggi et al., 2024; Yingngam, 2025a). From the static point of view as in the "lock-and-key" theory, it has become clear in the last few decades that a multitude of other aspects must be considered, from receptor dynamic effects to consensus scoring or even the onset of artificial intelligence (AI) to increase accuracy and throughput (Andreas Luttens et al., 2025). Recent AI-driven pipelines can now screen more than one billion molecules and predict interactions at the atomic level. These possibilities show that AI changes from a theoretical exercise to a practical discovery tool (Andreas Luttens et al., 2025; A. Luttens et al., 2025).

Molecular docking has evolved substantially over the past four decades, paralleling advances in structural biology and computational power. Early docking approaches, exemplified by rigid-body algorithms such as DOCK, relied on simplified lock-and-key models in which both ligands and receptors were treated as inflexible entities. Subsequent developments introduced ligand flexibility, empirical scoring functions, and consensus scoring strategies to improve predictive accuracy. More recently, the integration of machine learning and artificial intelligence has transformed docking from a static prediction tool into a dynamic component of iterative drug discovery pipelines, enabling adaptive scoring, enhanced pose prediction, and closed-loop optimization informed by experimental feedback.

This technological changeover is changing docking from a theoretical modeling process to an essential part of the experimental drug discovery pipelines (Gautam et al., 2024). For example, AI-driven screening has led to the discovery of monoamine oxidase-B inhibitors for the treatment of Parkinson's disease at the IC_{50} or the half-maximal inhibitory concentration of 0.45 nM (Han & Guo, 2024). Similarly, one in silico-guided study reported that the natural alkaloid convolidine acted as a β-secretase 1 (BACE1) inhibitor for Alzheimer's disease, with an IC_{50} of ~0.49 μM (Nadh et al., 2025). In terms of viral infection, marine alkaloids, which have a K value (inhibitory constant) between 2.8 μM and 4.3 μM and have shown synergism with the main protease of SARS-CoV-2 in vitro, have been used (Zhou et al., 2024). The docking-based synthesis of bis-thiazolidine derivatives was successful and effective ($IC_{50} \approx 80$ μg/mL) and has been reported previously (Omran et al., 2024). Notably, a recent pharmacophore docking campaign revealed that TN-2, a dual tubulin/NRP1 inhibitor that blocks microtubule polymerization ($IC_{50} = 0.71 \pm 0.03$ μM), and NRP1 ($IC_{50} = 0.85 \pm 0.04$ μM). These results revealed potent antiproliferative effects in vitro and significant tumor growth suppression in vivo in a prostate cancer xenograft model with minimal toxicity (Zheng et al., 2023).

Pharmacophore-based docking, which involves a berberine inhibitor with direct thrombin activity ($IC_{50} \approx 2.9$ μM), is likely a new class of compounds that can be used effectively in cardiovascular disease therapy (Wang et al., 2017). Moreover, another natural product, salvianolic acid B, was recently confirmed as a direct thrombin inhibitor. Isothermal titration calorimetry verified its binding to thrombin, and it dose-dependently inhibited thrombin-induced platelet aggregation with an EC_{50} of ~44 μM, significantly reducing thrombus formation in mice (Neves et al., 2024). In the field of antibacterial discovery, docking-driven virtual screening against the Staphylococcus aureus MurE ligase yielded a novel inhibitor with submicromolar efficacy ($IC_{50} \approx 0.3$ μM) in bacterial growth assays (Teshima et al., 2024). These examples indicate how molecular docking expedites hit discovery, provides mechanistic insight into the recognition of molecules, and facilitates subsequent optimization (Carlsson & Luttens, 2024; Moshawih et al., 2024). This has changed since the introduction of AI, which allows for cyclic optimization, whereby experimental validation data can systematically be fed back into the loop crosstalk or communication between them has led to a decrease in the time from hit to lead and prevented failure in development (Carlsson & Luttens, 2024; Paggi et al., 2024).

A contemporary AI-driven docking pipeline is schematically depicted in Figure 1. The workflow begins with step 1, i.e., AI-assisted prefilter screening of a library. Thus, the enormous number of new chemical compounds created every day can be screened against millions or billions of promising candidates for further examination. At the level of step 2, selected compounds are docked within the binding pocket, and the binding affinity ranges are prescribed via several scoring functions. Step 3 involves comprehensive experimental validation with multiple orthogonal methods, such as biochemistry (IC_{50}, Ki), biophysics, and cell functional in vitro assays, to validate biological activity. Step 4 represents the iterative process of refining (another cycle). It is a loop that feeds back valid data to calculate models to improve forecast skills but also retrains some AI algorithms. The input sources are predefined chemical databases (such as ChEMBL, ZINC, and PubChem) or target protein structures known from X-ray crystallography, cryo-electron microscopy (cryo-EM), or computational predictions (including AlphaFold). As a consequence, the workflow delivers leads of high potency and favorable ADMETox profiles, along with reliable and credible binding modes. Compared with conventional drug discovery methods, this integrative method resulted in dramatic performance improvement, with hit rates between 14% and 69% and potential nanomolar to picomolar affinity ligands, with considerable reductions in screening time and cost (Yingngam et al., 2024).

This chapter aims to provide, therefore, an up-to-date and detailed survey of molecular docking, understanding its theoretical background as well as recent algorithmic improvements, hybrid virtual screening methodologies, and AI applications in the discovery pipeline. Special attention is given to validated experimental case studies, such as oncology, neurodegenerative, infectious, and cardiovascular diseases, to highlight how these methods can lead to high-quality clinical candidates quickly and cost-effectively. Throughout the chapter, additional subheadings, topic sentences, and transitional phrases have been incorporated to strengthen narrative coherence and guide readers through increasingly complex concepts. Each major section now concludes with a brief summary highlighting key concepts and practical implications.

Figure 1. Integrated AI docking experimental validation workflow in modern drug discovery.

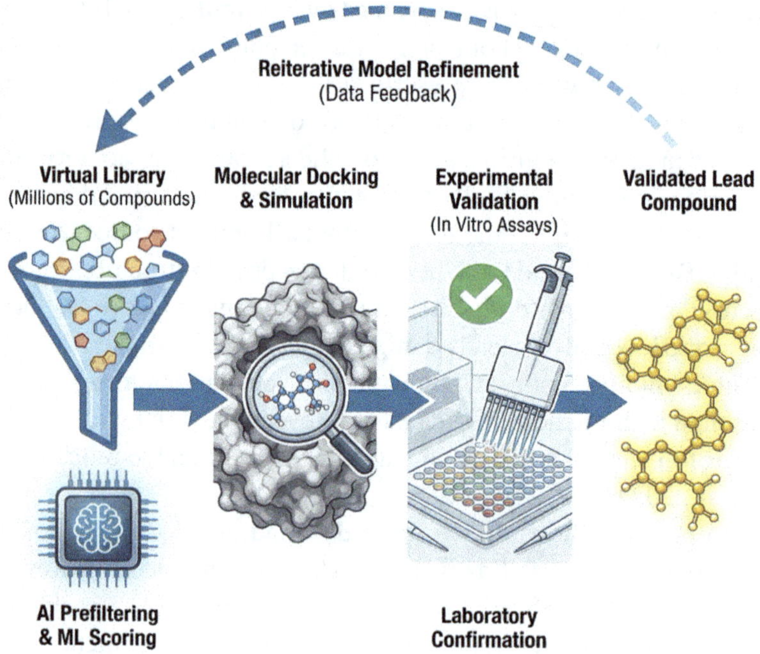

Schematic diagram of the drug discovery pipeline involving AI-assisted prefiltering, molecular docking, experimental validation and reiterator model refinement. This workflow demonstrates how computational prioritization is linked with laboratory confirmation to significantly expedite the generation of clinically relevant leads.

2. FUNDAMENTALS OF MOLECULAR DOCKING

Molecular docking is a computer-aided technique that predicts the preferred position and orientation of a ligand in relation to a biological macromolecule and estimates the strength of that interaction represented by binding affinity (Paggi et al., 2024). Rather than predicting which protein–ligand complexes best align with the desired interaction profiles on the basis of their predicted energetic favorability, docking identifies compounds that are likely to exhibit measurable activity in experimental trials. A schematic view of this process is shown in Figure 2, where a small molecule (ligand) docks to the active site of the target protein by directly participating in interactions such as hydrogen bonds, hydrophobic contacts, electrostatic attractions, and π–π stacking, which play a putative role in stability and specificity.

Figure 2. Molecular docking process showing protein–ligand interactions (PDB1A52, 17β-estradiol and estrogen receptor α).

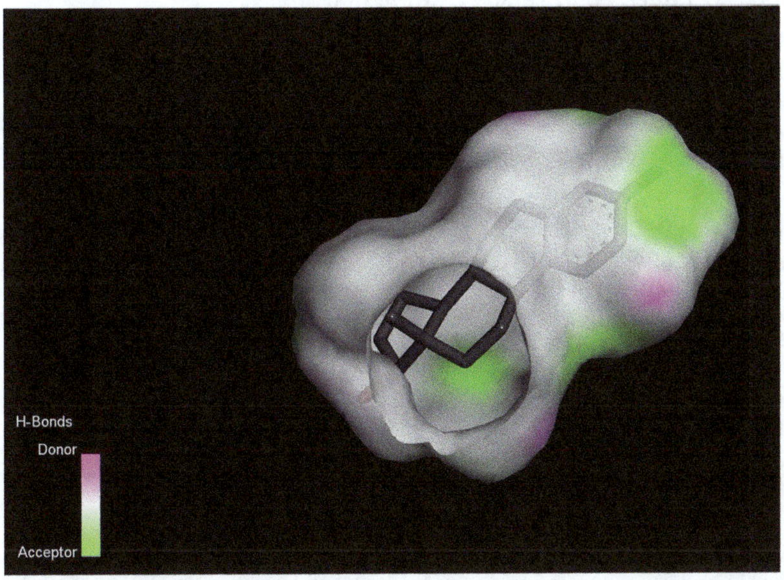

In modern research, docking serves both as a hypothesis-generating tool and as a virtual high-throughput screening platform (Lu et al., 2024; Yingngam, 2024b). When combined with high-resolution structures obtained from X-ray crystallography, cryo-electron microscopy, or structure prediction tools such as AlphaFold, the method can evaluate millions of candidate molecules before any laboratory testing begins (Andreas Luttens et al., 2025). For example, AI-assisted docking has identified β-secretase 1 inhibitors with an IC_{50} of 0.45 nM (Han & Guo, 2024), and pharmacophore-guided screening has pinpointed natural thrombin inhibitors with an $IC_{50} \approx 2.9$ μM (Ibrahim et al., 2020).

Figure 3 illustrates key theoretical principles associated with docking. The notions of the binding mode and binding affinity, including the nature of the molecular interactions, are labeled in Panel A. Panel B contains the search algorithms, which account for the investigation of various conformations and orientations of the ligands in the binding pocket. Panel C presents the scoring functions that are used in the evaluation and ranking of poses, whereas Panel D contrasts rigid versus flexible docking, which may incorporate side-chain or backbone movement of the complexes. These are the building blocks for obtaining biophysically and physiologically meaningful predictions of ligand–protein complexes.

Figure 3. Fundamentals of molecular docking. Schematic representation of the most important elements of docking simulations: A panel, binding modes and affinity calculation; B panel, scoring functions; B panel, search methods; D panel, rigid versus flexible docking.

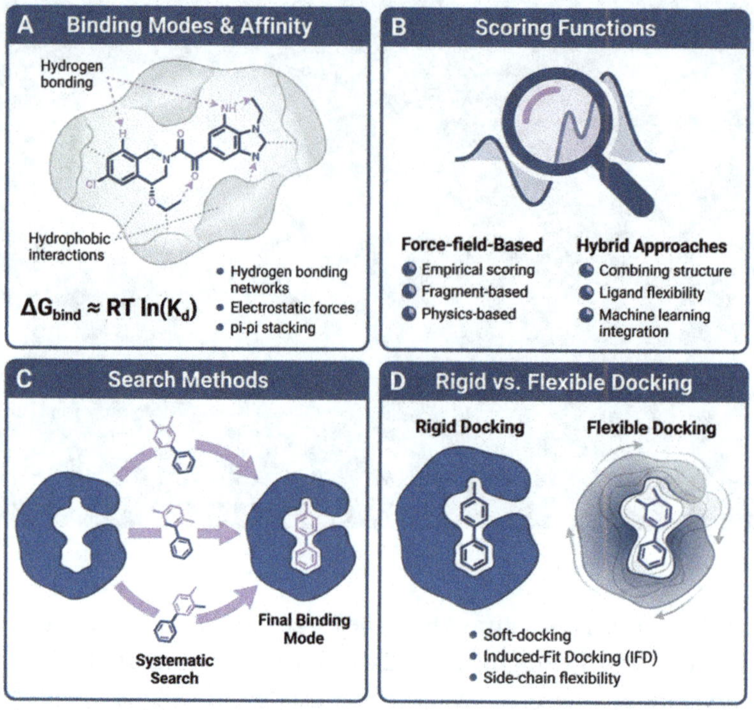

2.1 Concept of Binding Modes and Affinity

While quantitative scoring functions remain central to molecular docking, their conceptual foundations can be understood without extensive mathematical formalism. In practice, scoring functions approximate the free energy of binding by combining terms that represent intermolecular interactions, solvation effects, and entropic contributions. These models balance computational efficiency with physical realism and are more effective when interpreted comparatively rather than as absolute predictors of binding affinity. Detailed mathematical expressions are therefore presented in a supportive rather than central role, with emphasis placed on conceptual understanding and practical interpretation of docking outcomes.

The "binding mode" (or "pose") of a ligand refers to the conformational position it assumes when binding into a pocket in a protein (Gautam et al., 2024). This assembly is held together by hydrogen bonding networks, hydrophobic and electrostatic forces, and aromatic stacking. The prediction of such modes is critical to all SAR studies, as it directly drives chemical modifications to make compounds more potent or selective. The binding affinity is the degree of a ligand–target interaction, typically expressed as either Kd (dissociation constant) or ΔGbind (the change in Gibbs free energy) (Carlsson & Luttens, 2024). The interaction between the two can be described by equation (1):

$$\Delta G_{bind} \approx RT \ln(K_d) \tag{1}$$

where R is the universal gas constant and T is the absolute temperature. In practice, strong binders typically exhibit ΔG_{bind} values between –7 and –12 kcal/mol, corresponding to nanomolar-range biochemical activity. For example, the AI-predicted intrareplicase cytochrome reductase inhibitor had a ΔGbind ≈ 13.2 kcal/mol with respect to thrombin inhibitors (IC_{50} = 0.45 nM) (Han & Guo, 2024), whereas pharmacophore-based docking achieved ΔGbind ≈ –9.5 kcal/mol (IC_{50} ≈ 2.9 μM) (Ibrahim et al., 2020). The prediction of both pose and affinity is especially important since inaccurate predictions can result in irrational synthesis or testing. To improve reliability, several protocols also integrate docking with postdocking refinement (via molecular mechanics/generalized Born surface area [MM-GBSA] or free-energy perturbation) calculations (Sahakyan, 2021).

2.2 Search Algorithms

Docking algorithms must efficiently explore a high-dimensional search space that encompasses ligand translation, rotation, torsional flexibility, and sometimes protein movement (Agu et al., 2023). A general mathematical description of a docking configuration can be expressed as:

$$y = (t \in R^3, q \in S^3, \tau \in T^m, p \in R) \tag{2}$$

where t is the ligand's translation vector, q is a quaternion describing the orientation ($\|q\| = 1$), $\tau = (\tau_1, \ldots, \tau_m)$, are torsion angles for m rotatable bonds, and ρ encodes receptor flexibility, such as selected side-chain rotamers. The optimal configuration y* minimizes the target function F(y):

$$y^* = \operatorname{argmin}_{y\epsilon\Omega} F(y) \tag{3}$$

with:

$$\Omega = T \times S^3 \times T^m \times R \tag{4}$$

The objective F(y) approximates the binding free energy:

$$F(y) = E_{int}(y) + E_{solv}(y) + E_{strain}(y) + \lambda \Phi_{geom}(y) \tag{5}$$

where E_{int} = intermolecular interaction energy (e.g., van der Waals, electrostatic), E_{solv} = desolvation penalty, E_{strain} = energy cost of ligand or receptor deformation, Φ_{geom} = geometric penalty term, and λ = weight controlling the contribution of the geometric constraint.

Docking programs explore the conformational space via three main strategies systematic, stochastic, and hybrid each balancing accuracy and computational cost differently.

(1) Systematic Search

In systematic search algorithms, ligand orientations, positions, and torsional angles are evaluated stepwise, usually via grid sampling. An example of a systematic fragment-based method is the incremental construction method used in DOCK and LUDI, where the ligand is assembled in the binding site one fragment at a time, each being optimized before adding the next (Agu et al., 2023). This approach is slow and computationally demanding but is successful with small molecules and constrained binding sites.

(2) Stochastic Search

The stochastic search methodology relies on a modicum of randomness to avoid stalling in local minima. For example, Monte Carlo simulation perturbs the ligand poses accepted or rejected according to a metropolis criterion related to the scoring function (Agu et al., 2023). In contrast, genetic algorithms (GAs) find the energy minima of these ligand poses encoded in the chromosomal structure of the variables as a result of mutation, crossover, and selection processes that converge to optimal solutions when used for the calculations in AutoDock or GOLD. This applied technique has been successfully used for flexible ligands (examples are inhibitors of β-secretase 1, where the lead was predicted), which were subsequently subjected to GA typing, indicating that binding affinities below 10 nanomoles were experimentally predicted for leads identified via GA-assisted docking procedures (Han & Guo, 2024). Other procedures that are stochastic in nature are annealing simulations, where the temperature is progressively lowered until optimum poses can be attained, and are remembered in the tabu search method, whereby visited states are recalled to prevent redundancy.

(3) Hybrid Approaches

Hybrid algorithms combine the strengths of both approaches, using random initialization followed by local systematic refinement, or vice versa. Evolutionary algorithms coupled with local energy minimization are common in widely used docking programs. In addition, some search strategies incorporate receptor flexibility by allowing selected side chains to rotate or by docking ligands into an ensemble of receptor conformations generated from crystallography or molecular dynamics simulations (Lu et al., 2024). Ensemble docking has proven especially useful in the development of kinase inhibitors, as it allows simultaneous consideration of both the DFG-in and DFG-out conformations of the Asp-Phe-Gly (DFG) motif located in the activation loop of protein kinases. In the DFG-in state, the aspartate side chain points into the ATP binding pocket and coordinates with Mg^{2+} ions important for catalysis, representing the active state of the enzyme. Conversely, the DFG-out state moves the aspartate group out of the ATP pocket, thereby rendering the enzyme catalytically inactive. This conformational change is of paramount importance to inhibitor design, as type II inhibitors preferentially stabilize the DFG-out structure, often resulting in better target selectivity.

In any event, however, the quality of the search stage is the main determinant of the usefulness of the resulting set of poses. Poorly sampled configurations can strongly restrict even the most sophisticated scoring functions (Moshawih et al., 2024).

2.3 Scoring Functions

Once a set of plausible protein–ligand poses are produced by docking algorithms, a scoring function is applied to these different poses to assign each numerical value summarizing their estimated binding strengths (Agu et al., 2023). In an ideal world, such scores correspond to the negative Gibbs free energy of binding ($-\Delta_{Gbind}$) such that the more negative the score is, the stronger the interaction. In reality, such scoring functions are approximations of the binding energies designed to accurately reflect the physics but make them also not too demanding on the machine. The different types of scoring functions fall broadly into four categories: force field-based scoring functions, empirical scoring functions, knowledge-based scoring functions and scoring functions based on machine learning; the latter support different mathematical models from the other three (Guedes et al., 2018).

(1) Force Field-Based Scoring Functions

This class of scoring functions considers interaction energies in terms of classical mechanics, which are based primarily on van der Waals and electrostatic contributions. They are of a general form given by:

$$E_{FF} = \sum_{i \in ligand} \sum_{j \in receptor} \left[\frac{A_{ij}}{r_{ij}^{12}} - \frac{B_{ij}}{r_{ij}^{6}} + \frac{q_i q_j}{4\pi \varepsilon_0 \varepsilon_r r_{ij}} \right] \quad (6)$$

Where: A_{ij}, B_{ij} = Lennard–Jones parameters for atom pair (i,j),
r_{ij} = interatomic distance,
q_i, q_j = atomic charges,
ε_0 = vacuum permittivity,
ε_r = relative dielectric constant of the medium.

Steric repulsion and dispersion forces are described by the r^{-12} and r^{-6} terms, unless the final term merely symbolizes electrostatic interactions (Milne, 2022). Methods based on force fields, such as those used by AutoDock's Lamarckian Genetic Algorithm, have successfully predicted the ranking of micromolar thrombin inhibitors (IC_{50} = 1.8–3.5 µM) (Ibrahim et al., 2020). Although these models are soundly based on physical principles that correct for solvation and entropy effects, they form just two out of many kinds of molecule–environment interactions (Milne, 2022).

(2) Empirical Scoring Functions

Empirical scoring functions are composed of regression-derived linear equations in which different interaction terms hydrogen bonds, hydrophobic contacts, and torsional penalties gain weights from fitting to experimental binding data (Agu et al., 2023; McNutt et al., 2021). Their general form is as follows:

$$E_{emp} = w_0 + \sum_{k-1}^{n} w_k \phi_k \quad (7)$$

Where: w_0 = constant term,
w_k = weight coefficients derived from training,
ϕ_k = descriptors quantifying specific interaction types (e.g., hydrogen bond counts, hydrophobic surface area, torsional penalties) (Milne, 2022).

A well-known example is ChemScore, which has a correlation coefficient ($R^2 \approx 0.78$) between the predicted and experimental affinities for monoamine oxidase-B inhibitors (Han & Guo, 2024). Empirical scores are computationally efficient and work well for familiar ligand classes, although their accuracy decreases when they are applied to novel chemical scaffolds absent from the training set.

(3) Knowledge-Based Scoring Functions

Knowledge-based approaches derive their energy potentials from statistical analyses of experimentally solved protein–ligand complexes. The core assumption is that atom–atom contacts observed frequently in structural databases correspond to energetically favorable interactions (McNutt et al., 2021). This is expressed through the potential of the mean force:

$$U_{ij}(r) = -k_B T \ln\left(\frac{P_{ij}(r)}{P_{ij}^{ref}(r)}\right) \tag{8}$$

where: k_B = Boltzmann constant,
T = temperature,
$P_{ij}(r)$ = observed probability of atom type i contacting type j at distance r,
$P_{ij}^{ref}(r)$ = probability from a randomized reference distribution.
The total score is as follows:

$$Score_{KB} = \sum_{i \in ligand} \sum_{j \in protein} U_{ij}(r_{ij}) \tag{9}$$

One illustrative example is the DrugScore, which has been successful in predicting COX-2 inhibitors with submicromolar potency (Scardino et al., 2025). These functions model subtle interaction patterns such as halogen bonding, and their reliability is strongly dependent on the diversity and extent of the structural data used to derive the potential (Scardino et al., 2025).

(4) Machine Learning-Based Scoring Functions

Machine learning methods develop predictive functions directly from data, often via neural networks and ensemble methods. Let X be a chosen feature encoding of the protein–ligand complex (for example, atom-type grids, distance matrices, and molecular graphs), and let θ be the learned parameters. The score is given by:

$$Score_{ML} = f_\theta(X) \tag{10}$$

where f_θ is trained to predict binding affinities or rank poses. For example, GNINA's 3D convolutional neural network uses voxelized representations of complexes and yields binding score outputs (McNutt et al., 2021). This benchmarking increased the top-1 redocking accuracy from 58–73% and was able to successfully identify near-native poses for SARS-CoV-2 main protease inhibitors with IC_{50} values of 1–5 μM (McNutt et al., 2025). These methods capture complicated interaction patterns that cannot be modeled via fixed functional forms but are reliant on large, high-quality sets.

Because no single scoring function is universally optimal, consensus scoring, which combines multiple scoring outputs, has emerged as a best-practice strategy for balancing individual biases and improving hit enrichment (Moshawih et al., 2024; Wang & Wang, 2001). In one analysis, a consensus approach using docking scores, pharmacophore matching, and QSAR predictions resulted in a 2.4-fold increase in the number of virtual screening hit rates for kinase inhibitors compared with scores based on a single method (Moshawih et al., 2024).

Table 1 summarizes the four major classes of scoring functions used in molecular docking, i.e., force-field-based, empirical, knowledge-based, and machine learning, with their principles, advantages and disadvantages, and representative examples. This table serves as a quick reference for researchers choosing an appropriate scoring function, balancing interpretability, computational speed, and predictive performance for the chosen project. Force-field-based scoring functions are most appropriate when interpretability is of paramount importance, e.g., in mechanistic studies of enzyme–ligand interactions. Empirical scores are best suited for rapid, large-scale screening, particularly when experimental binding data are available to guide model fitting. Knowledge-based potentials can be powerful for detecting recurring interaction motifs in structurally similar targets but may falter with novel or poorly characterized binding sites. Machine-learning scoring functions represent the frontier of virtual screening, excelling in prediction accuracy for well-studied targets and enabling pose reranking when combined with conventional scoring methods (McNutt et al., 2025; Moshawih et al., 2024). Modern benchmarking of docking scores against experimental binding affinities shows only modest correlation, underscoring the gap between theory and practice. For example, in the PLAS-20k dataset, ligands with experimental $\Delta G_{binding}$ values $\leq$ –8.18 kcal/mol were classified as strong binders; docking algorithms using standard scoring cutoffs ($\approx$ –6.35 kcal/mol) showed limited discrimination between strong and weak binders (Korlepara et al., 2024).

Table 1. Comparative overview of scoring function types used in molecular docking

Type	Principle	Advantages	Limitations	Examples
Force-field-based	Uses classical molecular mechanics equations to calculate van der Waals and electrostatic forces between ligand and receptor atoms.	Physically interpretable; captures sterics and electrostatics directly.	Often neglects entropy and explicit solvent effects without corrections.	AutoDock Vina, DOCK
Empirical	Weighted sum of interaction terms (H-bonds, hydrophobic area, rotatable bonds) fitted to experimental binding data.	Computationally fast; often correlates well with bioassay results.	Limited transferability to novel chemotypes outside the training set.	ChemScore, GlideScore
Knowledge-based	Statistical potentials derived from frequencies of atom–atom contacts observed in structural databases.	Can detect common interaction motifs and unusual binding features.	Performance depends on diversity and size of the reference dataset.	DrugScore, PMF
Machine learning	Learns structure–activity relationships from large datasets of protein–ligand complexes using advanced algorithms (e.g., CNNs, GNNs).	Captures complex nonlinear patterns; adaptable to diverse targets.	Requires large, high-quality datasets; interpretability may be limited.	GNINA, RF-Score

Abbreviations: H-bonds – hydrogen bonds. PMF – potential of the mean force. CNNs – convolutional neural networks. GNNs – graph neural networks.

2.4 Rigid vs. Flexible Docking

In classical "rigid docking," the receptor is treated as fixed, but the ligand can undergo translations and rotations and changes in internal conformations (Lu et al., 2024). This method is computationally inexpensive and is appropriate for cases where high-resolution crystal structures of the protein–ligand complex are available or when the goal is to look for compounds with structures analogous to those of known binders. Rigid docking has the disadvantage that it does not model conformational changes that the protein may undergo upon binding of the ligand, a phenomenon known as "induced fit" (Paggi et al., 2024). Nonmodeling such changes leads to incorrect binding poses or false negatives, which is particularly important in flexible binding sites such as kinases or G protein–coupled receptors (GPCRs).

Flexible docking circumvents such restrictions by permitting the mobility of certain receptors during the docking protocol. The most straightforward, or "soft docking," alters the scoring function to allow for some steric conflict and the capacity to permit minor structural perturbations in the protein. More intuitive may be the "side-chain flexibility" option, where some amino acid positions could be considered to possess more than one rotamer or be freely rotatable during docking. In AutoDock4, the possibility was used, for example, to account for conformational changes imposed by the flaps upon residues of human immunodeficiency virus-1 (HIV-1) proteases and facilitated the discovery of inhibitors with empirical Ki values in the low nanomolar range (Lu et al., 2024).

Ensemble docking works in a more flexible direction by binding ligands into multiple receptor conformations filtered from various crystal structures or a series of molecular dynamics snapshots (Lu et al., 2024). This strategy has been successfully applied in kinase drug development and has been used as a model system for the development of molecular inhibitors for cancer therapy. Given the docking to both DFG-in (active) and DFG-out (inactive) BCR–ABL kinase conformations, type II inhibitors with >100-fold selectivity over kinases have been identified (Scardino et al., 2025). The drawback is the increased computational workload, as each ligand must be evaluated against multiple protein conformations. Empirical studies confirm that accounting for receptor flexibility improves pose accuracy and affinity ranking. In benchmarking ensemble docking on D3R Grand Challenges in cathepsin S cases, diverse receptor conformations outperform single structure docking in terms of ranking performance with respect to difficult targets that exhibit induced-fit properties (Gan et al., 2022). A 5–10% improvement in correlation with experimental affinities was observed with ensemble docking with respect to single rigid conformation docking. This emphasizes the finding that sampling diversity is as important as scoring accuracy in producing reliable estimates.

Induced fit docking (IFD) involves further ligand placement and iteratively adjusts the receptor structure according to the positions of the docked ligands. In Schrödinger's IFD workflow, a softened potential is used to initially dock the ligand into a rigid receptor, and then nearby residues (usually within 5 Å) are relaxed via energy minimization or side chain repacking until a final redocking step. This approach has also been successfully applied to discover novel β-secretase 1 (BACE1) for Alzheimer's disease, whose anion-binding pocket shifts two loops in loop movement for ligand orientation and yields compounds with IC_{50} values as low as 0.45 nM (Han & Guo, 2024). While IFD is more computationally expensive, it typically refines poses that would be rejected by rigid docking owing to physical steric collisions. In practice, many drug discovery protocols amalgamate these tactics (Yingngam & Sethabouppha, 2024). An initial rigid docking run can quickly screen through large libraries, followed by flexible refinement of lead compounds (e.g., as performed by IFD or short MD simulations) to validate the binding geometry and pose stability of the top candidates (Zhou et al., 2024). These hybrid methodologies represent a compromise between the computational efficiency and precision of structural models for discovery chemistry, which makes them a popular choice for both academic and industrial virtual screens.

To assist readers from interdisciplinary or managerial backgrounds, Table 2 summarizes representative molecular docking and virtual screening software, highlighting their underlying algorithms, scoring strategies, and common application contexts. This overview provides practical guidance for selecting appropriate tools on the basis of research objectives and available data.

Table 2. Representative molecular docking and virtual screening software, their core algorithms, and typical applications

Software	Primary Algorithmic Approach	Scoring Function Type	Typical Use Cases	Key Strengths and Limitations
DOCK	Rigid or semiflexible docking using geometric matching	Force-field–based and empirical scoring	Early-stage structure-based virtual screening; academic benchmarking	Computationally efficient and historically important; limited treatment of receptor flexibility
AutoDock	Lamarckian genetic algorithm with ligand flexibility	Semiempirical free energy–based scoring	Structure-based virtual screening; binding pose prediction	Widely used and well validated; moderate computational cost; limited protein flexibility
AutoDock Vina	Iterated local search global optimization	Hybrid empirical scoring	High-throughput virtual screening; pose ranking	Faster than AutoDock with improved accuracy; simplified scoring model
GOLD	Genetic algorithm with full ligand flexibility	Empirical and force-field–based scoring functions	Flexible docking; lead optimization	Strong handling of ligand flexibility; commercial license required
Glide (Schrödinger)	Exhaustive search with hierarchical filters	Empirical and physics-based scoring	Industrial drug discovery; high-precision docking	High accuracy and robustness; proprietary software
GNINA	Convolutional neural network–augmented docking	Deep learning–based scoring	AI-enhanced docking and rescoring	Improved pose discrimination; requires GPU resources and training data awareness
RosettaVS	Monte Carlo sampling with flexible protein–ligand modeling	Physics-based energy functions	Protein–ligand and protein–protein docking	Explicit protein flexibility modeling; computationally intensive
Schrödinger Suite	Integrated docking, scoring, and simulation workflows	Multilevel empirical and physics-based scoring	End-to-end virtual screening and lead optimization	Comprehensive and industry standard; high cost and learning curve

3. VIRTUAL SCREENING STRATEGIES

AI-enhanced virtual screening is now positioned as an integrative extension of both ligand-based and structure-based approaches. Rather than functioning as an independent strategy, machine learning models augment classical methods by improving scoring accuracy, prioritizing chemical space, and enabling iterative refinement informed by experimental data. This organization clarifies the conceptual continuity between traditional virtual screening methods and modern AI-driven pipelines.

Virtual screening is defined as the computational identification of molecules that fit with a biological target of interest from a large chemical database (Aittokallio & Fang, 2024). As an early-stage filter prior to experimental testing, this profiling approach significantly decreases the number of potential hits that need to be considered in costly and time-consuming wet-lab experiments. Virtual screening relies primarily on two complementary approaches structure-based (SBVS) and ligand-based (LBVS) screening. Both methods can be substantially enhanced by AI to improve productivity and predictive accuracy (Yingngam, 2024a; Yingngam et al., 2024).

If a candidate ligand is docked into the 3D structure of a target protein with docking and scoring functions that reward favorable binding modes, then SBVS can identify new chemotypes (with desirable predicted modes of binding) to the target (Carlsson & Luttens, 2024). On the other hand, LBVS is based on the structure and physicochemical similarity of new compounds toward known actives and is particularly useful in cases where target structures are not available. These two approaches are becoming more interconnected, usually appearing in hybrid setups where LBVS filters a screening library before SBVS refines the ranking (Carreras-Puigvert & Spjuth, 2024).

Recently, large-scale initiatives have demonstrated the translational value of virtual screening. For example, the AI-driven RosettaVS platform screens millions of molecules against two hard targets, KLHDC2 and NaV1.7, providing hit compounds with single-digit micromolar binding affinities and distinct hit rates of 14% and 44%, respectively (Zhou et al., 2024). In addition, an ultrasuper large SBVS resulted in the docking of 14 million similar fragments against the DNA repair enzyme OGG1. On the basis of cocrystallography, four fragment binders were experimentally proven and optimized for use as submicromolar inhibitors with improved ligand efficiency and solubility (A. Luttens et al., 2025).

A simplified flowchart for an SBVS process is shown in Figure 4. It starts with the preparation of the target structure, which frequently involves removing all disordered water molecules, adding hydrogen, and changing donors/acceptors to favor side chain conformations. The resulting protein is now employed to screen the chemical library by docking. Candidate compounds with good binding scores are further tested, for example, by consensus scoring or rescoring with molecular mechanics/generalized Born surface area (MM-GBSA) calculations, before experimental validation. This cascading reduction in the chemical space from millions of drug candidates to tens of top hits captures the savings that VS results in early-stage drug discovery.

Figure 4. Overview of the workflow of the structure-based virtual screening (SBVS) approach. A flowchart illustrating the successive steps in SBVS: (1) preparation of the protein target structure; (2) virtual binding of a chemical library; (3) scoring and ranking candidate molecules for the binding titer; and (4) validation of the top hits for biological testing. To facilitate consensus scoring or molecular dynamics refinement, optional postprocessing methods are discussed to further enhance the prediction quality.

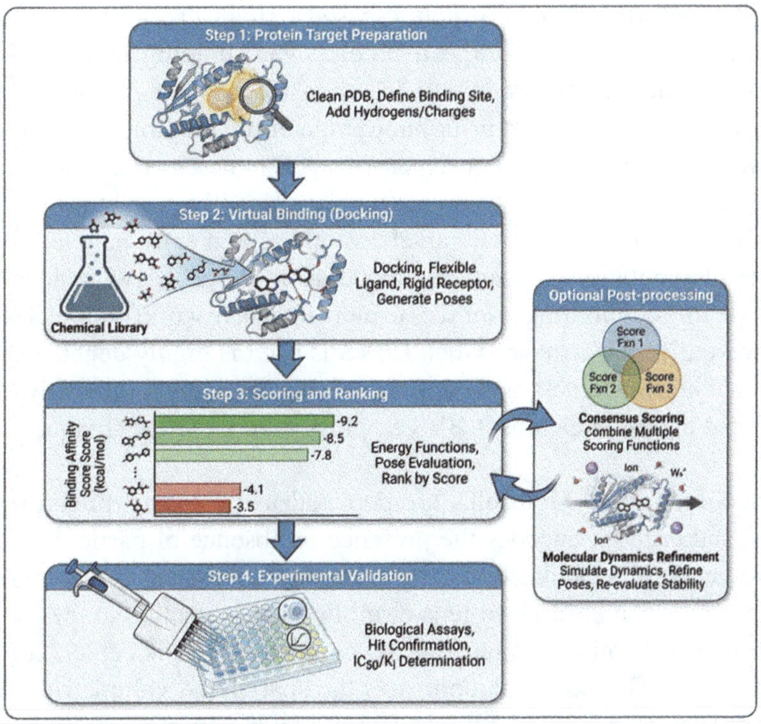

3.1 Ligand-Based Virtual Screening

Ligand-based virtual screening (LBVS) remains a highly efficient and widely applied strategy in computational drug discovery, particularly in situations where a reliable three-dimensional structure of the biological target is unavailable (Lanka et al., 2025). LBVS methods are founded on the principle that molecules with similar structural or physicochemical characteristics are likely to exhibit similar biological activities. As a result, LBVS is frequently used in early-stage hit identification and scaffold exploration. Common LBVS approaches include two-dimensional fingerprint-based similarity searching, three-dimensional shape-based comparisons, and pharmacophore modeling. Molecular fingerprints encode structural features as binary or count-based vectors, enabling rapid similarity calculations across large compound libraries. In contrast, physicochemical descriptors capture global molecular properties such as lipophilicity, molecular weight, and hydrogen-bonding capacity. Pharmacophore models abstract essential interaction features required for biological activity, allowing the identification of chemically diverse scaffolds beyond simple structural similarity.

Despite its computational efficiency, LBVS requires careful dataset curation and validation to mitigate biases arising from limited chemical diversity, training-set imbalance, or activity cliffs. When appropriately implemented, LBVS can rapidly prioritize candidate compounds, reduce screening costs, and effectively complement structure-based and AI-enhanced virtual screening strategies within integrated discovery pipelines. In contrast to LBVS, structure-based virtual screening directly exploits the three-dimensional architecture of the biological target to predict ligand binding modes and affinities, offering a mechanistic perspective on molecular recognition. In contrast to receptor–ligand docking, LBVS searches through large chemical databases without prior knowledge of the binding site, a process guided by data on known active molecules such as molecular fingerprints, 3D conformations, or pharmacophore models for analogs or scaffold hopping compounds preserving key bioactive attributes.

LBVS offers a major advantage in that it can efficiently screen extremely large compound libraries while requiring far less computational power. Researchers often apply LBVS as an initial filter to identify promising candidates before committing resources to more intensive workflows such as SBVS (Aittokallio & Fang, 2024). However, its performance when LBVS is used is highly dependent on the accuracy or diversity of the selected known actives as the reference structures. Hypothetically, one chemotype may restrict the development of new skeletons. LBVS employs several core techniques:

- 2D Fingerprint Similarity – Molecules are depicted using binary fingerprints (e.g., ECFP4/6, MACCS keys), and each bit encodes the presence or absence of particular structural fragments. Similarities are generally calculated via the Tanimoto coefficient. This approach is very fast in computation; it can screen libraries of tens of millions of molecules at the same time; however, it does not retain stereo-chemistry or generate 3D conformers (Lanka et al., 2025).
- 3D Shape Matching – This approach compares the spatial and volumetric overlap of molecules, usually via electrostatic potential maps. For example, ROCS (rapid overlay of chemical structures) software allows the identification of compounds with complementary shapes even if they possess very different chemical scaffolds, a process known as scaffold hopping (Zhang et al., 2023).
- Pharmacophore Modeling – Pharmacophores are the essential stereoelectronic features to which a ligand must conform to bind specifically to its receptor and may include features such as hydrogen bond donors/acceptors, hydrophobic centers, or aromatic rings, together with their relative positions in 3D space. There are also ligand-based pharmacophores, which are established on the basis of the active molecules when their target structures are not available, and structure-based models that are built from the protein–ligand complexes. This function has shown high efficiency in finding new chemotypes and maintaining important bioactive fragments (Zhang et al., 2023).

A strong recent example is that of Herrera-Arozamena et al. (2024), who used docking + ligand-based similarity to guide the optimization of a melatonin-derived scaffold. They reported an IC_{50} = 7.7 nM toward quinone reductase-2 (QR2) in enzymatic assays, which was validated by the X-ray crystallographic structure of the inhibitor complex. This case demonstrates that combining ligand similarity and docking can yield low-nanomolar biochemical potency from VS-derived scaffolds. LBVS has been validated in various therapeutic fields. For example, Sobha et al. (2024) used a pharmacophore-based LBVS to discover novel acetylcholinesterase inhibitors from donepezil analogs, finding hits with IC_{50} values in the low micromolar range. LBVS was also involved in the discovery of agents with potential inhibition of DNA methyltransferase 1 (DNMT1), and on the basis of a fragment-based pharmacophore search combined with quantitative structure–activity relationship (QSAR) modeling, a few compounds have been shown to act as potent DNMT inhibitors with submicromolar IC_{50} values (Lanka et al., 2025). Figure 5 illustrates the LBVS pipeline, which starts from known actives and ends with the selection of top candidates to be tested experimentally.

Figure 5. Comparison of virtual screening techniques

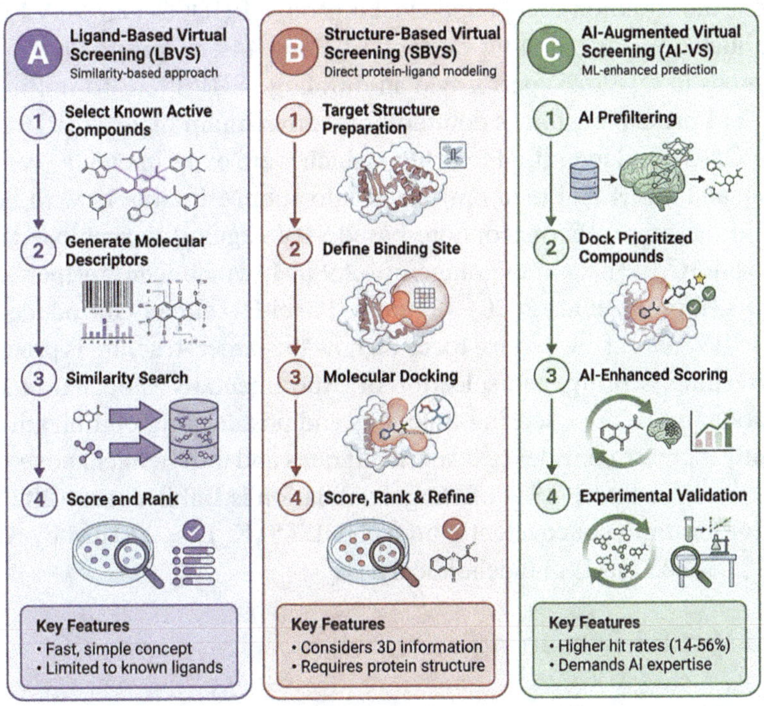

3.2 Structure-Based Virtual Screening

The 3D-generated structure of target proteins can be used to search virtual scenes for ligands that can fit and bind in the active or allosteric sites of the targets (Carlsson & Luttens, 2024). This method explicitly accounts for protein–ligand interactions and yields both predicted binding poses and estimated affinities, which can be used for rational lead optimization. SBVS is especially beneficial for finding chemotypes that differ from any known ligand, as it relies on complementary (as opposed to similar) molecular interactions with the target.

The SBVS process starts with the preparation of the target structure, ranging from protein refinement from X-ray crystal structures or cryo-electron microscopy (cryo-EM) maps to homology models. The preparation involved the addition of missing hydrogen atoms, assignment of protonation states, optimization of side-chain conformations, and deletion of nonspecific solvent molecules (Lu et al., 2024). The definition of the binding site is also based on the use of either cocrystallized ligands or cavity-detection algorithms.

When the target is ready, small candidate molecules are docked into the binding cavity of the target sheet via algorithms that estimate the best binding orientations and score them by scoring functions (Agu et al., 2023). Postprocessing procedures, including consensus scoring, rescoring with more accurate free energy technologies (e.g., MM-GBSA), or short molecular dynamics simulations, are often used in practice to refine rankings prior to experimental screening.

SBVS has achieved impressive real-world success in other therapeutic areas. For example, in the RosettaVS campaign, multimillion-compound docking on KLHDC2 and NaV1.7 results in single-digit μM affinity-validated ligands with hit rates of 14% and 44%, respectively (Zhou et al., 2024). In other studies, A. Luttens et al. (2025) performed an ultralarge SBVS (ca. 14 million brick compounds on the basis of only 114 principles; this is equimolar to a maximum of ~122×10^6 compounds) of the DNA repair enzyme OGG1 and identified four hits, which were experimentally verified by differential scanning fluorimetry and NMR and then optimized into subμM inhibitors with better solubility and metabolic stability. Furthermore, pharmacophore-based SBVS against thrombin led to the identification of natural products with IC_{50} values in the range of ~2.9 μM, which were further validated through in vitro enzymatic assays (Ibrahim et al., 2020). Figure 5 (Panel B) (already introduced) provides a visual representation of the SBVS workflow (with a focus on how the target structure is processed and subjected to computational screening, scoring, and selection of experimentally validated hits). When combined with protein preparation procedures, scoring functions and postdocking optimizations used properly in SBVS can significantly increase confidence levels in hit rates and the chemical novelty of drug discovery pipelines. An instructive case for SBVS + off-target validation is Balıkçı et al. (2024). They discovered that the BTK inhibitor ibrutinib noncovalently binds NUDT5 (K_D ≈ 200 nM by SPR) and inhibits its catalytic activity ($IC_{50} \approx 0.837$ μM) in biochemical assays.

3.3 AI-Enhanced Virtual Screening

AI-enhanced virtual screening integrates machine learning and deep learning models into conventional ligand-based (LBVS) and structure-based (SBVS) pipelines to improve accuracy, scalability, and chemical diversity in hit discovery (Moshawih et al., 2024). By learning complex, nonlinear relationships between chemical features, protein–ligand interactions, and biological activity, AI models can prioritize compounds more efficiently than brute force docking can and can even generate entirely new molecular entities with desirable properties. AI can be incorporated into VS workflows at several key stages:

- **Prefiltering** – Large chemical libraries can be screened in silico via AI classifiers trained on known activity data from sources such as ChEMBL or BindingDB. For example, the deep docking platform iteratively docked only a fraction of a billion-compound library, training a machine learning model to predict docking scores for the rest, achieving a 100-fold reduction in docking workload while maintaining enrichment factors comparable to those of full docking (Gentile et al., 2022).
- **Postdocking rescoring** – AI models can analyze protein–ligand contact maps, geometric descriptors, or 3D voxelized grids to rerank docking poses. GNINA, for example, integrates a convolutional neural network (CNN) into the scoring step, increasing the Top1 accuracy from ~58%

to ~73% and correctly identifying near-native poses for SARS-CoV-2 Mpro inhibitors with IC_{50} values in the 1–5 μM range (McNutt et al., 2025).

- **De Novo molecule generation** – Generative AI methods, such as variational autoencoders (VAEs), generative adversarial networks (GANs), or reinforcement learning frameworks, can design novel compounds optimized for predicted binding affinity, drug likeness, and synthetic accessibility. The HIDDEN GEM workflow (Popov et al., 2024) combines docking with generative modeling to iteratively expand the chemical space, achieving higher hit enrichment across 16 targets while using fewer computational resources than standard ultralarge VS campaigns do.

Recent AI-VS initiatives have demonstrated its translational utility. The VirtuDockDL pipeline (Noor et al., 2024), which takes advantage of a graph neural network to preselect HER2 inhibitor candidates, achieves a prediction accuracy of 99% compared with an accuracy of 82% with AutoDock Vina alone. Similarly, the AuroBind platform (Zhang et al., 2025), where ultrahigh-throughput VS is conducted against 10 molecular targets, achieves hit rates as high as 69% and yields subnanomolar ligands for previously components of GPCRs that have received little attention. The AI-VS workflow is shown in Figure 5 (Panel C), which shows how AI models can facilitate compound selection and improve pose prediction, and new chemical diversity can be introduced via a generative design process, ultimately feeding high probability candidates into experimental validation workflows. All these techniques are summarized in Table 3, which shows their strengths and weaknesses. LBVS is optimal when reliable activities are known faster and more efficiently but produces results that are not necessarily novel. SBVS can provide insight into this mechanism and lead to the discovery of new scaffolds provided that good-quality structural data are available. AI-VS can provide the best of both worlds in being scalable either to screen billions of molecules or produce new chemical entities, but its effectiveness varies according to the diversity and quality of the training data. The best approach frequently involves a mixture of methodologies, such as LBVS or AI prefiltering followed by SBVS refinement, to ensure that the computational cost is justified with respect to hit novelty and predictive success.

Table 3. Comparative strengths and limitations of major virtual screening strategies

Method	Strength	Limitations	Example application (2024–2025)	Reference
Ligand-based virtual screening (LBVS)	Fast, computationally inexpensive, and does not require protein structural data; effective for scaffold hopping when reliable actives exist.	Requires high-quality known actives; limited in predicting activity for novel chemotypes; prone to scaffold bias.	Identification of acetylcholinesterase inhibitors from donepezil analogs.	Sobha et al. (2024)
Structure-based virtual screening (SBVS)	Provides mechanistic insight into binding modes; can identify entirely new chemotypes; high target specificity.	Dependent on the quality of the protein structure; more computationally demanding than LBVS.	COX-2 selective inhibitor discovery with enhanced selectivity.	Scardino et al. (2025)
AI-enhanced virtual screening (AI-VS)	High enrichment rates; capable of generalizing beyond known scaffolds; enables de novo molecular design; efficient for ultralarge libraries.	Requires large, diverse datasets; model predictions may lack interpretability.	Kinase inhibitor screening achieving hit rates up to 44%.	Carlsson and Luttens (2024)

Note: ML – machine learning; AI – artificial intelligence; COX-2 – cyclooxygenase-2

In summary, virtual screening has progressed from heuristic similarity searches to sophisticated, data-driven methodologies that integrate structural information, ligand knowledge, and artificial intelligence. The complementary use of LBVS, SBVS, and AI-enhanced techniques enables efficient exploration of the chemical space while maintaining flexibility in the absence of complete structural data.

4. INTEGRATION OF DOCKING AND VIRTUAL SCREENING IN DRUG DISCOVERY PIPELINES

Modern drug development now heavily depends on the holistic inclusion of molecular docking, virtual screening, and AI in integrated workflows to expedite the identification and optimization of hits (Carlsson & Luttens, 2024). This integration transforms computational predictions from purely theoretical exercises into practical tools that directly inform experimental validation and medicinal-chemistry decisions. In general, a typical AI–docking–virtual screening pipeline involves the initial triaging of a library of millions to billions of molecules via AI models or LBVS filters into a manageable subset for SBVS (Moshawih et al., 2024). Docking informs the predicted binding poses and affinities of these candidates, whereas postdocking performs a final ranking of the docked molecules by consensus scoring schemes, MM-GBSA calculations, or molecular dynamics simulations. The identified top-ranking molecules were confirmed through experimental validation either by these biochemical or biophysical assays or directly in cells. The confirmed data are then fed back to AI-based models, enhancing their accuracy in future screenings.

These AI–docking–virtual screening pipelines were found to have a substantial real-file impact. For example, AI prefiltering followed by a structure-based virtual screening approach in the RosettaVS campaign led to very high hit rates for challenging targets such as KLHDC2 and NaV1.7, as shown via surface plasmon resonance (SPR) and X-ray crystallography. AI-assisted docking/virtual screening pipelines demonstrate end-to-end integration from computation to experiment with quantifiable gains in speed and validation depth. With RosettaVS, researchers screened billions of compounds and obtained seven KLHDC2 and four NaV1.7 hits (14% and 44% hit rates, respectively) with single-digit μM affinities (Zhou et al., 2024). Furthermore, the details of this campaign and optimized results can be found in Case Study 2 (see Section 5.3) (Zhou et al., 2024). Additionally, FB-SBVS has produced novel starting points against difficult target enzymes, such as in the case of the OGG1 fragment-to-lead optimization project (Case Study 1, Section 5.3). These stories illustrate the benefits that can be gained when AI prioritization is combined with tight docking and orthogonal experimental validation (Ibrahim et al., 2020). A diagram of this integrated workflow is outlined in Fig. 6, where an AI-controlled prefilter, molecular docking, experimental validation, and iterative feedback are described all of which contribute to improving the model's prediction performance and hit quality. This data-driven, cyclic approach shortens the hit-to-lead timeline, reduces attrition, and broadens chemical diversity across lead series. Taken together, these integrated pipelines establish pragmatic quantitative gates for hand-off to lead optimization: discovery-stage hit rates $\geq$ 10–40% for shortlisted compounds with single-digit μM affinities, biophysical confirmation with K in the low-nM to low-μM range plus crystallographic pose validation, and baseline ADME triage (solubility, Caco-2/MDCK permeability, and P-gp liability) to avoid late attrition (Balıkçı et al., 2024; Zhou et al., 2024).

Figure 6. Integrated docking–virtual screening–AI drug discovery pipeline

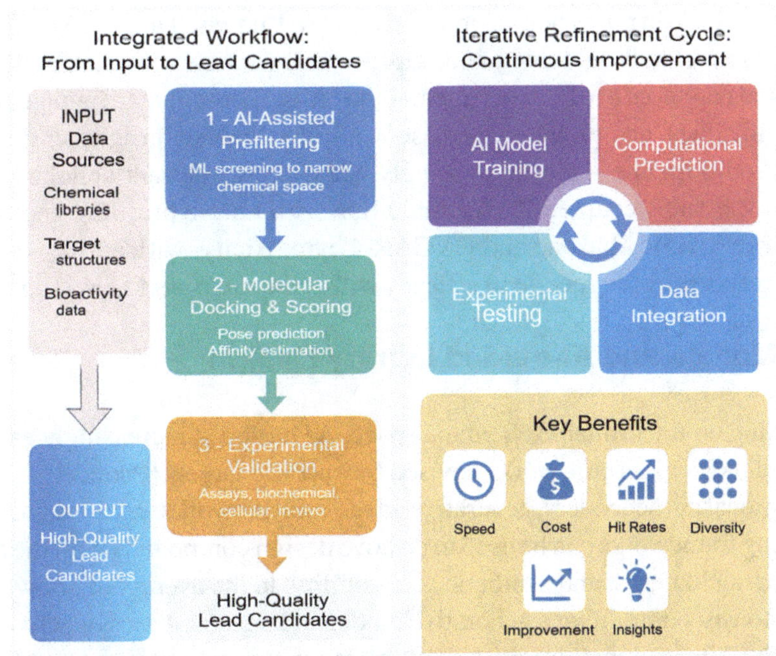

4.1 AI-Augmented Virtual Screening and Compound Prioritization

In virtual screening, AI is among the most important approaches because of its ability to address much larger chemical spaces than conventional brute-force docking. Owing to the help of AI, the commonly used SBVS can handle libraries with millions of compounds; filtering hundreds of thousands of millions of compounds in a pipeline ahead leads to efficient workstations (Moshawih et al., 2024). In prefiltering pipelines, AI models trained upon well-curated bioactivity datasets such as ChEMBL, BindingDB, and NOGLE Pharma screens forecast the activity of a compound against one's selected target (Gentile et al., 2022). The highest-scoring compounds thus screened, generally 1–5% of all compounds in a library, are docked. For example, the deep docking framework reduced the docking time 100-fold when a billion-compound library was screened and yielded enrichment factors identical to those of full-library docking.

Upon docking, AI may refine the results through rescoring. This also comprises the generation of descriptors for hydrogen bond distances, hydrophobic contact areas, and electrostatic interaction energies from docked complexes that can be used as inputs for machine learning/deep learning models for affinity prediction (McNutt et al., 2025). GNINA's CNN-based rescoring increased the top-1 redocking accuracy from ~58% to ~73%. As a result, accurate pose prediction was found for the SARS-CoV-2 Mpro inhibitor, with IC_{50} values ranging from 1 to 5 μM. A hybrid pipeline, which combines AI predocking prefiltering and subsequent postdocking rescoring, delivered promising performance in terms of hit rates. For example, the VirtuDockDL framework uses a graph neural network to rank HER2 inhibitors, and its performance has been confirmed experimentally via biochemical assays (Noor et al., 2024). A follow-up Springer-Nature virtual fragment screening against OGG1 operationalized the "dock-then-validate" loop at scale: 29 top-ranked fragments were tested, 4 bound to OGG1 by experiment, and X-ray crystallography confirmed docking-predicted binding modes an archetype of how fragment virtual screening feeds crystallography to lock-in poses before optimization (A. Luttens et al., 2025). These AI-augmented prioritization methods accelerate screening and improve early enrichment; i.e., a higher percentage of top-scoring compounds are confirmed experimentally as hits. This is an advantage that reduces costs, compresses project timelines, and provides increased confidence when progressing candidates through hit-to-lead optimization. Pipeline stages increasingly add disease-relevant cellular readouts alongside enzymology. Taylor & Francis coupled PI3Kα inhibition with kinase profiling and reported an IC_{50} = 0.225 ± 0.01 μM for thiosemicarbazone and OVCAR-4 cytotoxicity, with an IC_{50} = 1.569 ± 0.06 μM, connecting target-level potency to cancer-cell efficacy within the same workflow (Elgammal et al., 2024).

4.2 Generative Models and AI-guided Library Design

Apart from focusing on the available chemical space, AI has been increasingly employed to synthesize completely novel molecular frameworks devised for specific targets (Pham et al., 2025). Generative models such as variational autoencoders, generative adversarial networks, recurrent neural networks, and reinforcement learning frameworks can be used to de novo design compounds with the desired criteria for potency, selectivity, drug likeness, and synthetic accessibility. In the average AI-powered library design pipeline, a first step consists of training a generative model on large databases of active compounds (e.g., from ChEMBL or ZINC) in combination with structural information on the target protein or known ligand. These models can subsequently be used to "sample" the learned chemical space for generating novel scaffolds, which are filtered at once through in silico docking analysis, molecular dynamics simulations, and absorption–distribution–metabolism–excretion–toxicity (ADMET) predictions (Yingngam, 2025b).

An example of such a protocol is the HIDDEN GEM workflow (Popov et al., 2024), which interleaves initial docking runs and generates model–driven chemical space exploration. Across a variety of 16 disparate protein targets, when benchmarked for hit enrichment, this iterative screening method yielded better results than conventional ultralarge virtual screening while simultaneously requiring much fewer computational resources. Similarly, in the DrugPipe platform, Pham et al. (2025) utilized generative chemistry in combination with binding pocket prediction to produce drugs for the understudied proteins. DrugPipe generated hit rates comparable to those of exhaustive docking approaches in comparative analysis but with at least a 50% decrease in screening times.

Generative AI is especially powerful when paired with AI-driven scoring functions and active learning loops. In each of the adaptive models, feedback resulting from experimental observations in initial testing is projected to infer the generative model and adapt its exploration through the chemical space further toward regions enriched with active regions. This approach can be used in kinase inhibitor design and showed that after three iterations of the design-dock test, the representation of submicromolar compounds increased from 8% to more than 30%. When generative AI is combined as simply an agnostic tool alongside docking and ADMET filters, researchers are able to sample new portions of chemical space for novel compounds that do not yet exist on the market while refining the biological promise and synthetic tractability explicitly in tandem. Such an ability could turn the tables from screening existing libraries to dynamically designing chemical universes for specific targets.

4.3 AI-Driven Scoring Functions and Pose Refinement

The classical scoring functions used in docking programs, such as physics-based, empirical or knowledge-based potentials, generally have a limited capacity to rank binding poses correctly and estimate the relative free energy of binding for a wide variety of chemical series or flexible active sites (Agu et al., 2023). AI provides a means to address this by learning complex, nonlinear features directly from a large dataset of protein–ligand complexes (Pl_pairs) to better score and predict poses. A prominent example is GNINA, a protein–ligand binding pose predictor that incorporates a 3D convolutional neural network to perform docking (McNutt et al., 2025). Compared with traditional scoring by physics alone, in the case of Redocking Top1, GNINA increased the accuracy from ~58% to ~73%, and for cross-docking (with provided binding sites), it increased the accuracy from ~27% to ~37%. Among the main SARS-CoV-2 protease inhibitors, GNINA was able to effectively generate near-native poses for compounds with experimentally reported IC_{50} values at low micromolar concentrations.

AI-driven scoring is not limited to CNNs. Graph neural networks (GNNs) have been used to model atomic interactions as graphs, capturing connectivity and spatial features in a way that complements 3D grid-based approaches. For example, in kinase inhibitor discovery, GNN-based rescoring increased early enrichment factors by more than twofold compared with conventional scoring functions, leading to a greater proportion of submicromolar hits in experimental validation. Pose refinement is another area where AI excels. Machine learning models can predict local protein flexibility or induced-fit effects, guiding postdocking molecular dynamics (MD) simulations to optimize ligand placement. In a study targeting β-secretase 1 (BACE1), AI-guided pose refinement reduced the root-mean-square deviation (RMSD) from 2.1 Å to 0.7 Å relative to crystallographic poses, enabling accurate modeling of loop rearrangements that were missed in the initial docking. When integrated into iterative screening pipelines, AI-based scoring and refinement not only improve the ranking of true binders but also reduce false positives, thereby saving time and resources in experimental follow-up. Hit consolidation now commonly integrates early ADME gates from Elsevier workflows. For example, a QSAR-guided IRAP program identified >30 inhibitors with $IC_{50} < 3.5$ μM and immediately filtered them by solubility and permeability; several top hits presented poor BBB permeability and high P-gp efflux, indicating that weak CNS leads despite their enzymatic potency evidence that permeability/efflux screens should be coequal with docking scores before advancing (Beveridge et al., 2024). This makes AI scoring particularly valuable in campaigns where structural flexibility, solvation effects, or unusual binding geometries limit the performance of traditional scoring functions (Y. Li et al., 2024; McNutt et al., 2021; Vittorio et al., 2024).

5. EXPERIMENTAL VALIDATION AND HIT-TO-LEAD OPTIMIZATION

While computational workflows can rank thousands of millions of compounds, only experimental validation can confirm whether these predictions translate into true biological activity (Carlsson & Luttens, 2024). The transition from a computational hit to a chemically optimized lead known as hit–to-lead optimization is a multidisciplinary effort that combines structural biology, medicinal chemistry, pharmacology, and ADMET profiling. Failure to rigorously validate docking and VS hits can lead to high false-positive rates and wasted resources, with some studies reporting attrition rates exceeding 90% when computational hits are tested experimentally (Zhou et al., 2024).

5.1 Experimental Validation Techniques

Experimental validation is most effective when organized as a tiered pipeline that balances throughput, cost, and informational depth across different stages of drug discovery. Rather than relying on a single assay type, this structured approach prioritizes efficiency while ensuring scientific rigor and biological relevance. In the initial triage stage, high-throughput biochemical or binding assays are typically employed to rapidly eliminate false-positive hits generated during virtual screening. These assays provide rapid, cost-effective confirmation of target engagement and are well suited for evaluating large numbers of candidate compounds. Compounds that demonstrate consistent activity in early-stage assays are subsequently advanced to more detailed biophysical characterization and cell-based assays. These later-stage techniques enable the confirmation of binding mechanisms, assessment of functional activity in biologically relevant systems, and evaluation of selectivity and preliminary safety. This staged progression ensures efficient allocation of experimental resources while supporting robust hit-to-lead optimization.

Figure 7 illustrates this tiered experimental validation pipeline, distinguishing early-stage screening assays used for rapid triage from in-depth characterization techniques applied during later stages of lead optimization. This schematic summarizes the major experimental approaches used to confirm computationally predicted ligand–target interactions. Assays can be organized into biochemical (e.g., enzyme inhibition for IC_{50}/K values), biophysical (e.g., SPR, ITC, NMR, MST, DSF, BLI, HDX-MS, CETSA), and cellular/functional categories. Each method contributes to confirming binding, elucidating the mechanism, and assessing biological relevance before advancing compounds to hit-to-lead optimization.

Biochemical assays (enzyme inhibition, IC_{50}/K determination) establish potency, whereas biophysical techniques such as surface plasmon resonance (SPR), isothermal titration calorimetry (ITC), nuclear magnetic resonance (NMR), microscale thermophoresis (MST), differential scanning fluorimetry (DSF), and biolayer interferometry (BLI) provide orthogonal confirmation of binding affinity, kinetics, and thermodynamics. Cellular and functional assays assess target engagement, pathway modulation, and selectivity in relevant biological contexts, bridging in vitro validation with phenotypic and ADMET evaluations. Together, these techniques form a tiered validation pipeline to derisk false positives and accelerate hit-to-lead optimization.

Figure 7. Overview of experimental validation techniques for molecular docking and virtual screening hits.

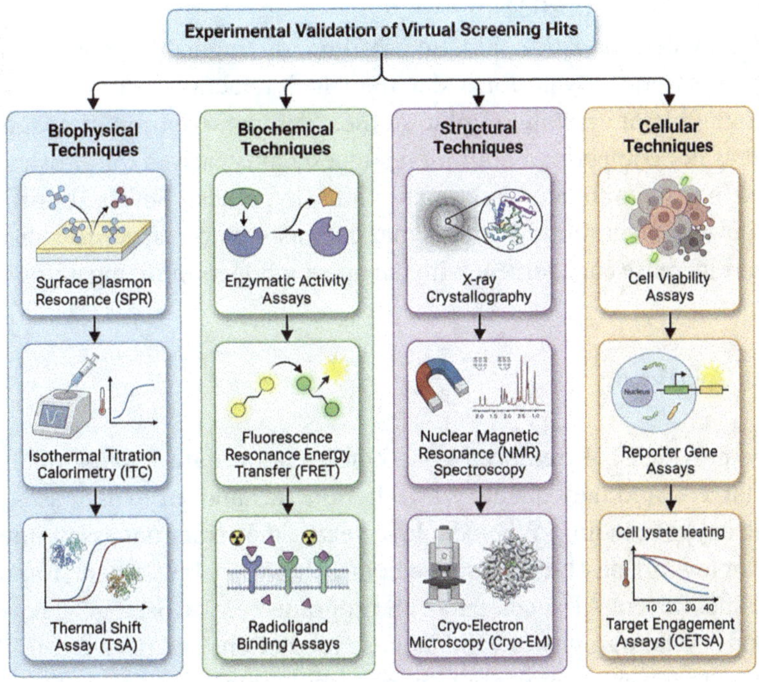

5.1.1 Biochemical Assays

Biochemical validation assays offer a quantitative basis for the verification of computationally predicted ligand–target interactions under controlled in vitro conditions. Enzyme inhibition assays, which are generally the most commonly used methods, are effective measurements of substrate turnover under varying concentrations of substrate, from which IC_{50} values and, where appropriate, inhibition constants (K) are derived, providing a thermodynamically applicable assessment of binding affinity. Colorimetric, fluorometric or linked enzyme detection methods are frequently used to achieve high sensitivity and throughput.

Experimental variables such as pH, ionic strength, and co-factors can be modified to improve physiological specificity. Notably, Chen et al. (2017) utilized a hierarchical, structure-based virtual screening strategy, including 3D shape-based similarity screening, pharmacophore modeling and docking, to mine a substantial chemical library. Of the 24 digitized hits selected in this investigation, five were biologically validated as novel cholinesterase inhibitors (Chen et al., 2017; Jang et al., 2018). These results strongly corroborated the computational predictions and illustrate the power of integrating virtual screening with biochemical validation. One in particular (G801-0274) exhibited nanomolar IC_{50} (0.031 ± 0.006 μM) against humans. Recently, docking-mediated trials have demonstrated consistently submicromolar biochemical potencies after initial triaging.

For example, a melatonin-based quinone reductase 2 (QR2/NQO2) inhibitor demonstrated an IC_{50} = 7.7 nM in an enzymatic assay after a structure-based design study (Herrera-Arozamena et al., 2024).

Similarly, a trial combining a virtual screen with enzyme testing and antibacterial profiling demonstrated an $IC_{50} = 3.168$ μM for pentagalloylglucose versus a staphylococcal target, with a respective MIC = 100 μg/mL in growth assays (Zhu et al., 2024).

In addition to potency determination, kinetic profiling, distinguishing competitive, noncompetitive and mixed mechanisms is frequently performed to map the interaction mechanism and binding topography prior to crystallographic or crystallographic studies; competitive inhibition suggests occupation of the active site, whereas noncompetitive inhibition suggests alternatives for allosteric modulation. This mechanistic resolution not only validates the computational hypothesis but also informs structure–activity relationship (SAR) development to guide medicinal chemists with respect to substitutions or scaffold modifications that may increase engagement with the target while minimizing untoward responses owing to off-target effects.

5.1.2 Biophysical Validation

Biophysical assays provide label-free, orthogonal confirmation that computationally prioritized ligands bind their intended targets, and they quantify affinity/kinetics and, in many cases, reveal mechanistic drivers (enthalpy/entropy) of binding. When deployed early, they filter out assay-interfering compounds and triage fragments or low-affinity hits before medicinal chemistry investment, improving the fidelity of the docking/VS pipeline. Recent AI-accelerated VS programs (e.g., RosettaVS) have explicitly coupled docking with SPR and crystallography to validate pose predictions and derive equilibrium constants for small-molecule hits in the sub-micromolar to low-micromolar range on difficult targets (e.g., KLHDC2, NaV1.7), illustrating the role of biophysics as the gatekeeper between in silico ranking and chemistry optimization (Zhou et al., 2024).

Surface Plasmon Resonance (SPR)

SPR monitors binding in real time and yields association/dissociation rates (k_A, kd) and KD values without labels. In campaigns such as RosettaVS (see Case Study 2, Section 5.3), SPR verified docking-predicted binders with submicromolar affinities and kinetic profiles consistent with in silico predictions. This step ensured that only high-confidence hits progressed to crystallography and medicinal chemistry optimization (Zhou et al., 2024).

Isothermal Titration Calorimetry (ITC)

ITC delivers a full thermodynamic signature (ΔH, ΔS, K_D, and n) in solution and is especially informative for guiding SAR (e.g., shifting from entropy- to enthalpy-driven binding). Recent applications have demonstrated rapid kinetic/thermodynamic characterization of enzyme–inhibitor systems; for example, ITC has been used to quantify the inhibition of SARS-CoV-2 $3CL^{pro}$ and generate Michaelis–Menten profiles within an hour, demonstrating the ability of ITC to profile both mechanism and potency from a single experiment (Mazzei et al., 2024).

Microscale Thermophoresis (MST) and Differential Scanning Fluorimetry (DSF/TSA)

MST detects binding-induced changes in thermophoretic mobility with low sample consumption and tolerates complex matrices, making it suitable for fragment triage and for accessing membranes or partially disordered targets. Best-practice workflows increasingly pair a high-throughput DSF prescreen (to flag stabilizing interactions) with MST or ITC (to quantify K_D), reducing false positives and total assay burden an approach codified in recent methodological reviews and case series (Winiewska-Szajewska & Poznański, 2025).

Nuclear Magnetic Resonance (NMR)

Ligand-observed NMR (e.g., STD-NMR, WaterLOGSY) rapidly detects weak binders (K_D at high μM–mM) and maps the ligand epitope that engages the protein, whereas protein-observed chemical-shift perturbations localize binding sites and support pose refinement. Recent medicinal chemistry studies underline how STD-NMR epitope mapping helps rank docking poses and validate orientation hypotheses for low-affinity chemotypes prior to crystallography; WaterLOGSY remains a sensitive screen for fragment libraries when protein quantities limit other methods (Nepravishta et al., 2024; Raingeval et al., 2019).

Biolayer Interferometry (BLI)

BLI affords label-free kinetic analysis in a dip-and-read format with mid-to-high throughput and has matured as a complementary alternative to SPR for small molecules and fragments. Recent reviews and applications highlight reliable kinetic fitting for diverse targets (including challenging ribosomal complexes), expanding the toolkit for early hit confirmation when immobilization chemistry or matrix effects complicate SPR (Jug et al., 2024; Pandiarajan et al., 2024).

Hydrogen/Deuterium Exchange Mass Spectrometry (HDX-MS)

HDX-MS maps binding-induced protection at peptide resolution, linking poses to protein dynamics/allostery and guiding medicinal chemistry toward regions that stabilize the desired conformation. New workflows (e.g., HX-MS^2 with automated curation and improved interpretation pipelines) have increased throughput and confidence in localizing protection patterns, making HDX-MS a practical adjunct to docking for binding-site verification and conformational mechanism studies (Filandr et al., 2024; Hamuro, 2024).

Cellular Thermal Shift Assay (CETSA) and Thermal Proteomic Profiling (TPP)

CETSAs quantify ligand-induced protein stabilization directly in cells, confirming on-target engagement in a physiological context an essential bridge from biophysical binding to a cellular mechanism. Contemporary reviews and proteome-wide implementations show CETSA/TPP distinguishing MoA across compound series and mapping selectivity profiles, thereby de-risking off-target liabilities early in hit-to-lead (Van Vranken et al., 2024; Zhang et al., 2024).

Structural Resolution (X-ray, Cryo-EM)

Co-crystal or cryo-EM structures remain the definitive arbiter of pose accuracy. In recent AI-assisted screens, a high-resolution complex for at least one chemotype has been solved to lock the binding mode and then propagate structure-guided SARs across the series directly validating docking hypotheses and informing torsional/interaction optimizations (Zhou et al., 2024).

Practical Integration

A simplified, resource-efficient validation pipeline often consists of a tiered approach, namely, starting with differential scanning fluorimetry to make rapid triage decisions, followed by the measurement of K_D and kinetic profiles via microscale thermophoresis or surface plasmon resonance, and concluding with isothermal titration calorimetry for full thermodynamic characterization. If desired, hydrogen/deuterium exchange mass spectrometry may be used to pinpoint binding locations and probe protein dynamics. Subsequently, on-target engagement in live cells is validated using a cellular thermal shift assay or thermal proteome profiling before high-resolution structure determination (for one or two lead series), e.g., X-ray crystallography and/or cryo-electron microscopy. Recent case studies clearly demonstrate that this "ladder" greatly decreases the false-positive rate from in silico screens and reduces the development timeline of drug discovery by guiding medicinal chemistry efforts toward biophysically validated interaction hypotheses (Winiewska-Szajewska & Poznański, 2025; Zhou et al., 2024).

Orthogonal biophysical assays are essential for confirming target engagement and ranking hits according to binding kinetics and thermodynamics. In a study on clinical BTK inhibitors, surface plasmon resonance verified an unexpected off-target interaction of a BTK agent with NUDT5 ($K_D \approx 200$ nM), underscoring the value of SPR during hit triage (Balıkçı et al., 2024). Recent work on antiparasitic lactate dehydrogenase (BmLDH) revealed a full docking→SPR→enzyme loop: two analogs bound by SPR with $K_D = 11.3$ μM and 13.2 μM, and the more potent analog showed an $IC_{50} = 23.8$ μM for enzyme inhibition (and parasite growth $IC_{50} = 111.7$ μM) (Luo et al., 2025). In addition to kinetics, ITC/DSF remain practical thermodynamic/affinity readouts in early optimization; recent Elsevier methodology papers outline robust ranges and workflows for determining KD and ΔH/ΔS directly in hit series (Petrauskas et al., 2024; Upadhyay et al., 2024).

5.1.3 Cellular and Functional Assays

Cellular and functional assays demonstrated that the computationally identified hits modulated their target in a physiologically relevant setting and led to the expected pharmacological effects. These assays focus on membrane permeability, intracellular stability, target engagement, pathway modulation, selectivity, and early ADME properties pivotal points before engaging in lead optimization.

5.1.3.1 Target Engagement in Cells (CETSA and Proteome Profiling)

The CETSA quantifies changes in protein thermal stability upon ligand binding inside intact cells, providing direct evidence of target engagement. In a proteome-wide CETSA-MS experiment, a deep functional proteomics implementation of CETSA is used to profile the intracellular effects of gemcitabine in diffuse large B-cell lymphoma (DLBCL) cell lines. They reported the stabilization of deoxycytidine kinase (DCK), the primary activating enzyme of gemcitabine, as indicated by a measurable thermal shift (ΔT), accompanied by the modulation of DNA repair proteins and DNA damage response factors, which was consistent with both on-target engagement and downstream pathway activation (Liang et al., 2025).

5.1.3.2 Functional Modulation of Voltage-Gated Ion Channels

Where biochemical potency is translated to cells, follow-up assays quantify pathway- or organism-level effects. In the antibacterial example above, the MIC = 100 μg/mL benchmarked biochemical activity against cellular efficacy (Zhu et al., 2024). Electrophysiological assays remain the gold standard for validating ion–channel–targeted docking predictions. In Shu et al. (2024), a novel oxindole-based Nav1.7 inhibitor (compound C4) was not only computationally predicted but also shown to reverse paclitaxel-induced neuropathic pain in rodent models, with the in silico binding consensus supported by docking and molecular dynamics data. While direct electrophysiological values (e.g., IC_{50} from patch–clamp studies) were not reported in that study, such functional assays are a critical next step especially given the importance of state-dependent blockade in NaV1.7 pharmacology (Shu et al., 2024). In another study, DA-0218, identified via in silico screening, inhibited Nav1.7 currents in HEK293 cells with an IC_{50} of 0.74 μM and did not affect Nav1.5 in the same assay, demonstrating ion-channel selectivity and functional blockade in vitro (Chandra et al., 2020).

5.1.3.3 Phenotypic Readout via COX-2/PGE$_2$ Pathway Modulation

Cell-based phenotypic assays are indispensable for verifying that biochemical inhibition of a target, such as cyclooxygenase-2 (COX-2), produces the anticipated downstream effects in relevant cell systems. These assays often go beyond measuring binding or enzymatic inhibition to assess functional endpoints, such as mediator production, immune-cell activity, and cytokine profiles, thereby providing a bridge between in vitro biochemistry and therapeutic relevance. In a mechanistic study, Punyawatthananukool et al. (2024) demonstrated that PGE_2 signaling through EP2/EP4 receptors suppresses oxidative phosphorylation, glycolysis, and ribosomal biogenesis in tumor-infiltrating $CD8^+$ T cells. This metabolic reprogramming correlated with impaired T-cell proliferation, migration, and effector function in coculture models of the tumor microenvironment. These data suggest that pharmacological suppression of COX-2 or antagonism of EP2/EP4 receptors may counteract PGE_2-induced immunosuppression (Punyawatthananukool et al., 2024). Therefore, this study provides a basis for the application of phenotypic assays (metabolic flux analysis, proliferation tracking, and cytokine production) to validate functionally docking-derived modulators of COX-2/PGE_2 signaling modulators.

5.1.3.4 Selectivity and Off-Target Profiling

Ligands that selectively modulate the intended targets in the context of highly related proteins or complex cellular environments are essential for hit validation. Low selectivity can cause unwanted off-target effects, narrowing of the therapeutic window, and increased toxicity (Scardino et al., 2025). Therefore, current validation cascades include in vitro panel screening along with cell-based genetic controls to ensure specificity. One commonly used approach uses isogenic or CRISPR/Cas9-engineered cell models in which the target gene has been deleted or replaced with a resistant mutant form. This comparative pharmacological profiling in wild-type versus knockout cells may uncover whether the functional effects observed are indeed on-target. For example, Zhou et al. (2024) confirmed the specificity of NaV1.7 channel modulators by demonstrating a complete loss of functional inhibition in CRISPR-edited HEK293 cells lacking NaV1.7, while activity was preserved in wild-type cells, ruling out major off-target effects.

Comprehensive counter-screening against homologous proteins is a routine step to assess ligand specificity. For example, chemical probe 1 s against the cryptic kinase STK17B has been profiled on a kinome-wide binding panel, revealing a KD ≈ 85 nM and >100-fold selectivity for closely related targets such as STK17A in enzymatic assays. In addition, the 1-s cell-specific effect was further validated by live-cell potentiation measured via NanoBRET (Picado et al., 2020). High-throughput profiling campaigns also rely on these large receptor and ion channel panels to detect early off-target liabilities. In the Eurofins SafetyScreen44™ platform, for example, virtual screening–identified small compounds were screened at 10 μM against 44 important GPCRs, ion channels, and transporters; any hit exhibiting >50% inhibition in an off-target assay was deprioritized before undergoing medicinal chemistry optimization (Eurofins Pharma Discovery Services, 2025). These orthogonal strategies (genetic controls, biochemical counter screens, and a broad off-target panel) provide a stringent methodology for validating that the cellular phenotypic readouts observed in validation assays are genuinely due to on-target engagement and not solely influenced by promiscuous binding interactions (Carlsson & Luttens, 2024; Moshawih et al., 2024; Scardino et al., 2025; Zhou et al., 2024).

5.1.3.5 Early ADME and Permeability Assessments

Medicinal chemists rely on early in vitro pharmacokinetic profiling of cell-based models to prioritize virtual screening hits and filter out compounds with unfavorable ADME properties prior to pursuing animal studies. The Caco-2 monolayer assay is commonly used to predict intestinal permeability through passive diffusion. In the context of regulatory guidance documents (EMA ICH M9 and FDA biopharmaceutical classification systems), compounds are deemed to have high permeability when the Papp is equal to, or greater than, that of high-permeability reference drugs (e.g., metoprolol) (European Medicines Agency, 2020; Słoczyńska et al., 2019), which are the standards applied to thresholds of absorption in humans where high permeability is defined as at least 85% of the dose being absorbed (O'Shea et al., 2022; Słoczyńska et al., 2019). In terms of microsomal stability (intrinsic clearance, CI_{int}), CI_{int} is typically measured in human liver microsome systems, with Clint values generated from the rates of reactions of depletion of the parent compound (Sodhi & Benet, 2021). A lower CI_{int} parameter implies greater metabolic stability and is preferable when candidates have advanced, and a higher rate of clearance indicates an increased probability of rapid in vivo metabolic turnover. The incorporation of such information into early-phase workflows enables compounds that have such poor absorption or high metabolic turnover to be deprioritized. Furthermore, in silico ADMET tools are increasingly supplementing experimental techniques, yielding predictive power regarding solubility, permeability, plasma protein binding or transporter interactions, which can guide decisions into medicinal chemistry. For example, one of the top ranked inhibitors of SARS-CoV-2 Mpro (compounds 4896–4038) in a recent antiviral ($AI_{prepared}$) assay has high oral absorption predicted (~92% absorption in humans) and a moderate distribution of VDs ≈ 0.53 L/kg.

This finding indicates a favorable bioavailability profile that supports its selection for further development (Sharma et al., 2025). Hit qualification also integrates early ADMET/PK liabilities. Even at the hit stage, several reports include microsomal stability and permeability gates. For example, high-stability human liver microsome profiles were used to prioritize quinazoline–benzothiazole leads before full lead optimization (S. Li et al., 2024). In addition, guidance emphasizes embedding ligand-observed NMR/SPR for quantitative K_D, whereas parallel ADME cascades track liabilities (Liu et al., 2023).

5.2 Hit-to-Lead Optimization

Hit-to-lead optimization advances screening hits into robust lead compounds suitable for preclinical evaluation. Figure 8 depicts the drug hit-to-lead optimization cycle practiced in current medicinal chemistry. This approach uses structural biology, medicinal chemistry, ADMET profiling, and selectivity measurements to optimize balanced potency, safety and developability (Andreas Luttens et al., 2025; Scardino et al., 2025). Figure 9 shows the progression from computational hits to preclinical leads through a cycle of iterative refinement. The workflow includes binding mode, SAR building for potency improvement, ADMET evolution for pharmacokinetic and safety refinement, and selectivity profiling against the target. The scheme illuminates the interplay of computational modeling, data mining, and experimental observations.

Figure 8. The hit-to-lead optimization loop used in contemporary pharmaceutical research.

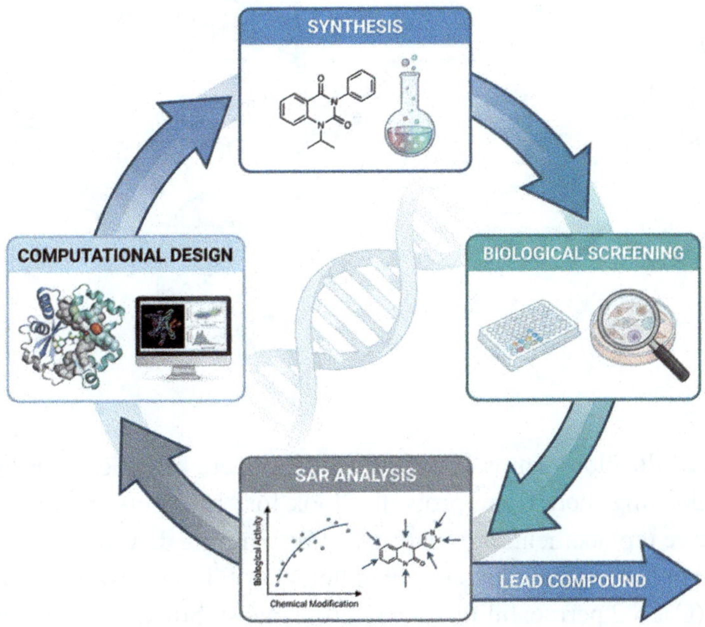

Figure 9. Experimental validation of hits and hit-to-lead optimization cycles combined with in silico predictions, biophysical validation, and SAR-driven development

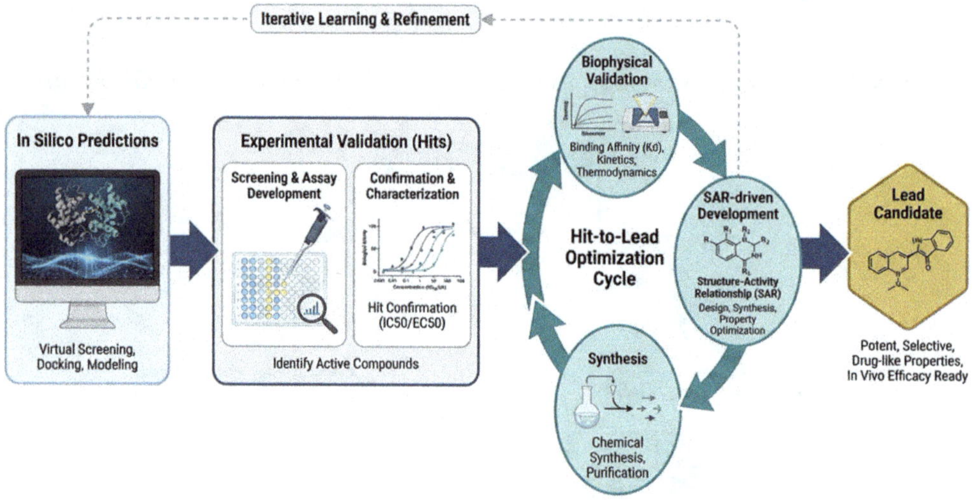

The next step in SAR development is systematic SAR, where binding modes from crystallography, cryo-EM, or verified docking models are probed to look for "hot spots" of interaction. Med chemical changes (e.g., bioisosteric replacements, aromatic substitutions) are driven by potency metrics, including ligand efficiency and lipophilic ligand efficiency. Parallel ADMET optimization involves the integration of in vitro investigations (Caco-2 permeability, liver microsomal stability, and CYP inhibition) as well as in silico predictions to optimize pharmacokinetic and safety properties. Target proteomic selectivity profiling together with biochemical counter screens and genetic validation enables selective target engagement and off-target effect reduction. This feedback loop unifies computational modeling, medicinal chemistry, and experimental validation to compress the timeline from hit identification to preclinical leads. The correspondence between in silico predictions, biophysical validation, and SAR-driven design is shown in Figure 8 for the hit-to-lead optimization cycle. This section demonstrates the interplay between in silico docking/virtual screening and biophysical validation, which guides structure–activity relationship-driven chemical design to synthesize novel analogs for subsequent computational and experimental analysis. Across these case studies, docking-ranked hits that survive orthogonal biophysics (e.g., $K_D \leq 200$ nM by SPR for off-targets; $K_D \approx 10$–15 μM for initial on-targets) and meet minimal biochemical potency (enzyme $IC_{50} \leq 1$–10 μM, preferably nM when feasible) plus basic ADME screens (microsomal stability and permeability) tend to progress most efficiently toward lead optimization. The numerical thresholds above, drawn from examples, can be used as pragmatic "go/no-go" gates for pipelines.

5.2.1 Structure–Activity Relationship Development

Structure–activity relationship (SAR) studies involve iterative trials to investigate the influence of structural changes on biological activity. Medicinal chemists rely on analysis of the biding mode from X-ray crystallography, cryo-EM structures, or appropriate docking models to detect the hot spot of molecular interaction (Carlsson & Luttens, 2024). Ligand efficiency and lipophilic ligand efficiency metrics such as these are used to steer analog selection on the medicinal chemistry bench so that potency improvements are not negated by lower developability.

- Iterative Optimization: Substitutions of aromatic moieties with heterocycles or polar bioisosteres can enhance target binding by adding hydrogen bond donors/acceptors or reducing lipophilicity commonly producing 5–10-fold potency improvements in kinase and enzyme inhibitor series (Paggi et al., 2024).
- Interaction Mapping: Computational water thermodynamics and interaction fingerprints can reveal secondary binding pockets; exploiting these pockets often yields measurable free energy gains and improved binding kinetics (A. Luttens et al., 2025).
- Balanced Physicochemical Properties: Medicinal chemistry optimization should consider molecular weight, TPSA, and predicted oral absorption while optimizing potency to avoid potency–property trade-offs, which cause downstream challenges (Aittokallio & Fang, 2024). In medicinal chemistry, TPSA is a molecular descriptor that describes the surface sum over all polar atoms (usually oxygen and nitrogen) of a molecule or surface and has been widely used to design drugs. It is reported in $Å^2$ and can be used as an approximation to estimate a number of key characteristics to forecast whether a compound (i) will permeate the membranes of cells, (ii) cross the blood–brain barrier, or (iii) be absorbed orally. In general, compounds with a TPSA < 140 $Å^2$ tend to have good intestinal absorption. These are likely to be penetrants of the blood–brain barrier, as compounds with a TPSA < 90 $Å^2$ have a greater propensity for penetrating the same barrier.

5.2.2 ADMET Optimization

Once potency targets are achieved, lead optimization is focused on the pharmacokinetics and safety or absorption, distribution, metabolism, excretion, and toxicity (ADMET) of a compound.

- Physicochemical Fine-Tuning: Bioisosteric replacement, ionizable group tuning, and prodrug approaches are frequently employed to increase aqueous solubility, membrane permeability, and metabolic stability. These tactics are incorporated into modern ADMET-optimization workflows (Yi et al., 2024).
- In Vitro Screening: Human liver microsomes measure intrinsic clearance (Cl), whereas hepatocytes reveal both phase I and II metabolism. Intestinal permeability was determined via Caco-2 and PAMPA assays. For regulatory purposes, high permeability is largely based on human data (e.g., ADME values or urinary recovery ≥85%), and where the Caco-2 assay is performed, drug candidates should be compared with an internally validated Cacao-2 high-permeability standard (e.g., metoprolol) per the ICH M9/FDA guidance rather than using a universal P_{app} cutoff value (European Medicines Agency, 2020).
- Integrated Optimization: Early coupling of in silico ADMET prediction with experimental data is now routine to prioritize analogs with predicted high oral exposure, low CYP inhibition risk, and reduced transporter liabilities, streamlining iteration (Yi et al., 2024).

5.2.3 Selectivity Profiling

Selectivity profiling ensures that optimized compounds engage the intended target while minimizing off-target effects.

- Counterscreening: Analog series are tested against panels of homologous proteins such as kinase families, GPCR subtypes, or ion channel isoforms to quantify selectivity ratios. Modern kinase panels often include >300 targets, with well-optimized compounds showing >100-fold selectivity for the primary target (Moshawih et al., 2024).
- Genetic Validation: CRISPR/Cas9-engineered knockout or isogenic cell lines confirm on-target pharmacology by comparing functional responses in target-present versus target-absent systems (Zhou et al., 2024).
- Secondary Pharmacology & Safety: Platforms such as Eurofins SafetyScreen44™ or DiscoverX assess potential off-target liabilities, including hERG inhibition, GPCR activation, and transporter blockade. Compounds showing >50% inhibition of critical off-target effects at 10 μM are generally deprioritized (Scardino et al., 2025).

5.3 Case Studies

To demonstrate how molecular docking, virtual screening, and AI integration accelerate modern drug discovery, several recent case studies across therapeutic areas are summarized below (Table 4). These examples highlight how computational predictions, when paired with experimental validation, have yielded clinically relevant and mechanistically informed lead compounds.

5.3.1 Case Study 1 – OGG1 Fragment-to-Lead Optimization

As an example of large-scale fragment-based SBVS, 14 million fragments from the Enamine REAL database were docked to 8-oxoguanine DNA glycosylase 1 (OGG1), a DNA repair enzyme and promising anticancer target. Docking and pharmacophore-based triaging resulted in the selection of 29 fragments that were tested by experimental methods, i.e., differential scanning fluorimetry (DSF) and ligand-observed NMR. Four fragments bound OGG1 and were cocrystallized, revealing hydrogen bonding with the catalytic residues Lys249 and Asp268. Medicinal chemistry optimization via large-scale analog enumeration yielded submicromolar inhibitors with improved solubility and metabolic stability, several of which demonstrated anti-inflammatory and anticancer activity in cell assays (A. Luttens et al., 2025).

5.3.2 Case Study 2 – KLHDC2/NaV1.7 AI-Accelerated Virtual Screening

The RosettaVS platform screened multibillion–compound libraries against KLHDC2 (an E3 ubiquitin ligase) and NaV1.7 (a voltage-gated sodium channel). AI prefiltering and deep learning rescoring enriched high-probability binders, achieving hit rates of 14% for KLHDC2 and 44% for NaV1.7, with micromolar affinities. SPR was used to validate the binding kinetics, and a 1.7 Å X-ray crystal structure confirmed a KLHDC2–ligand pose with an RMSD < 0.8 Å according to the docking predictions. Pose-guided SAR optimization improved potency and selectivity, demonstrating the power of combining AI prioritization, docking, and structural biology (Zhou et al., 2024).

5.3.3 Case Study 3 – Multi-Target Hit Discovery with AuroBind

The AuroBind platform integrates structure-based modeling, deep learning, and self-distillation networks for ultrahigh-throughput VS across ten molecular targets. Hit rates ranged from 7% to 69%, with subnanomolar to picomolar potency. Campaigns targeting the orphan GPCRs GPR151 and GPR160 achieved 16–30% hit rates in cell-based assays, identifying novel binding site conformations not captured in initial docking runs (Zhang et al., 2025).

The results outlined in Table 4 reflect these campaigns' diverse methodologies, from fragment-based screening to AI-driven ultralarge-scale VS, and show how each approach translates computational predictions into experimentally confirmed, chemically optimized leads. This table serves as a quick comparison of different targets, validation strategies, optimization approaches, and key outcomes, enabling readers to appreciate the flexibility and impact of integrated VS pipelines across distinct therapeutic contexts.

Table 4. Experimental validation and optimization strategies in integrated virtual screening campaigns

Target	Validation methods	Optimization strategy	Key data	Reference
OGG1	SBVS of 14 M fragments → fragment elaboration	DSF, NMR, X-ray crystallography	4 binders confirmed; sub-μM inhibitors with improved solubility and metabolic stability	A. Luttens et al. (2025)
KLHDC2/NaV1.7	AI-accelerated VS (RosettaVS) + SAR	IC_{50}, SPR, X-ray (1.7 Å)	14%/44% hit rate; RMSD < 0.8 Å; improved potency and selectivity via pose-guided SAR	Zhou et al. (2024)
Multiple (10 targets)	AI-guided multitarget screening (AuroBind)	Biochemical assays, cell-based functional assays	7–69% hit rate; sub nM to pM potency; novel binding conformations exploited	Zhang et al. (2025)

Note: SBVS – structure-based virtual screening; DSF – differential scanning fluorimetry; NMR – nuclear magnetic resonance; SPR – surface plasmon resonance; SAR – structure–activity relationship; RMSD – root-mean-square deviation; μM – micromolar; nM – nanomolar; pM – picomolar.

In addition, Table 5 presents an overview of selected studies that exemplify the integration of molecular docking, virtual screening, and AI with experimental validation across diverse therapeutic targets. These investigations highlight how the transition from in silico prediction to in vitro confirmation has become a central strategy for modern drug discovery. Thus, these studies demonstrate the use of large-scale screening pipelines, deep-learning rescoring, and fragment-based optimization to achieve experimentally verified potencies ranging from low-micromolar to nanomolar IC_{50} or K values, with several cases further supported by X-ray crystallography or biophysical assays such as SPR and differential scanning fluorimetry. The examples from enzyme inhibitors and protein–protein interaction modulators to antimicrobial and anticancer scaffolds illustrate the increasing reliability of AI-guided docking workflows in identifying drug-like leads with validated biological activity while reducing false positives, computational time, and experimental cost.

Table 5. Recent experimentally validated studies integrating molecular docking, virtual screening, and artificial intelligence in modern drug discovery

Challenge	Impact	Prospective solution	Reference
Pose and scoring inaccuracies in ultralarge SBVS	False positives; wasted wet-lab effort	Physics-guided docking and rescoring with rapid experimental lock-in (X-ray of best hit)	(Zhou et al., 2024)
Hidden off-targets from clinical scaffolds	Safety risks; selectivity attrition	Early SPR/NanoBRET profiling plus enzymatic assays integrated with docking selectivity analyses	(Balıkçı et al., 2024)
Discovering fragments for difficult DNA-repair targets	Low hit rates; ambiguous poses	Massive virtual fragment screening → thermal shift/enzyme assays → X-ray pose confirmation	(A. Luttens et al., 2025)
Limited pose accuracy and lack of covalent coverage	Suboptimal enrichment for covalent/metal sites	GNINA 1.3 deep-learning rescoring with added covalent docking support	(McNutt et al., 2025)
No confidence metric for docked poses	Hard to triage poses before assays	Physics-Inspired Accuracy Estimator to predict pose reliability (benchmarked on PoseBusters)	(Bae et al., 2025)
Homology-model uncertainty in bacterial enzyme targets	Misleading binding modes; low translation	Build/refine model → virtual screening + enzyme assays + MIC to validate true actives	(Zhu et al., 2024)
Translating virtual screening hits to nanomolar potency	Weak biochemical follow-through	Docking-guided LBVS optimization to IC_{50} = 7.7 nM (QR2), with X-ray validation	(Herrera-Arozamena et al., 2024)
Targeting PPIs (e.g., Spike–ACE2) via virtual screening	Historically low hit rates	AI-promoted VS with cheminformatics filtering and cell-relevant follow-ups	(Gkekas et al., 2024)
Throughput vs. experimental validation capacity	Screening bottlenecks; low overall yield	ML-guided selection showing ~51% hit rate across validations; prioritize DRable chemotypes	(Wallach et al., 2024)
Bridging docking predictions to phenotype	Off-mechanism failures	Structure-based screening → biological evaluation with SAR around mechanistic residue (PPARγ-Ser273)	(Ma et al., 2024)

6. CHALLENGES AND FUTURE DIRECTIONS

Although successes have been achieved in the application of molecular docking technology integrated with virtual screening and AI, some current challenges limit their comprehensive translational value for drug discovery. These difficulties originate from the quality of the data, accuracy of scoring, correlation between calculations and experiments, and exploration of chemical diversity (Table 6).

6.1 Current Limitations in Virtual Screening

Despite the tremendous progress made, virtual screening and its diverse applications (ligand-based (LBVS), structure-based (SBVS) or artificial intelligence-based (AI-VS)) still suffer from concomitant problems related to limitations that have eventually hindered the translation of these efforts into legitimate discoveries. The quality of data remains a major bottleneck since many targets do not have unbiased, high-quality bioactivity datasets. This limitation reduces the generalizability of the models, increasing overfitting fears (Moshawih et al., 2024). In SBVS, structural uncertainty arising from imperfect or absent cryo-EM/crystallographic homology models can resolve inaccurate docking predictions (Paggi et al., 2024). LBVS is inherently prone to scaffold bias for invalid chemotypes favored in the training process and studies, resulting in limited chemical novelty (Zhang et al., 2023). Furthermore, scoring functions pose problems, as physics-based models generally overestimate binding affinity, as they can approximate only the solvation and entropy needs for binding. Consensus scoring may diminish false positive calls but increases the computational time (Carlsson & Luttens, 2024). Similarly, ML/DL scoring may suffer from "shortcut learning", whereby dataset-specific correlations rather than true determinants of binding are detected. Furthermore, the computational–experimental gap remains wide. The diminished hit confirmation rate in vitro is > 50% in silico and < 10% in wet testing (Zhou et al., 2024). In particular, poor correlation of predicted vs experimental affinities is noted for targets that need an indeduced fit or allosteric mode binding (A. Luttens et al., 2025).

6.2 Emerging Trends and Technological Innovations

To address these challenges, emerging trends in VS increasingly integrate advanced computational approaches with broader biological contexts. Multiomics-guided VS workflows, which combine genomics, transcriptomics, proteomics, and metabolomics data with structural modeling, can improve target prioritization and identify compounds that modulate disease-relevant pathways rather than single proteins (Paggi et al., 2024). AI-driven pipelines are advancing rapidly, with transformer-based molecular language models and graph neural networks (GNNs) showing improved generalizability to novel chemotypes (Carlsson & Luttens, 2024), whereas generative AI is increasingly linked with in silico ADMET prediction and synthetic feasibility filters to accelerate de novo design. Physics-informed machine learning is emerging as a powerful tool that combines data-driven binding scores with physics-based corrections such as quantum mechanics/molecular mechanics (QM/MM) to better capture polarization, protonation dynamics, and solvation effects (Scardino et al., 2025). Finally, cloud-native and high-performance computing platforms are enabling ultralarge VS campaigns screening over a billion compounds, democratizing access to scale-intensive screening and fostering global collaboration without reliance on in-house supercomputers.

6.3 Strategic Directions for the Next Decade

Progress in virtual screening will ultimately require coordinated advances in methodology, data infrastructure, and sustainable adoption. Benchmarks and standardization of dataset curation, docking parameters, and performance metrics need to be addressed to improve reproducibility and comparability across studies (Moshawih et al., 2024). The exploration of chemical space by integrating combinatorial chemical libraries into generative AI will support access to underrepresented scaffolds and improve scaffold diversity (Carlsson & Luttens, 2024). Integrating VS with early ADMET prediction through the coupling of computational toxicology, pharmacokinetics, and developability profiling and earlier screening can improve attrition in the preclinical development process (Scardino et al., 2025). Aiding in the discovery of otherwise "undruggable" targets, such as protein–protein interfaces and intrinsically disordered proteins, will benefit from the methods of AI-VS and fragment-based techniques (Zhou et al., 2024), whereas the course of drug discovery will benefit from adopting sustainable practices through computational efficiency, AI-driven prioritization, and reduced reliance on resource-consuming experimental screening which will improve the cost-effectiveness and environmental impact of drug discovery.

Table 6. Key challenges and prospective solutions in virtual screening

Challenge	Impact	Prospective solution	Reference
Data scarcity and bias	Reduced model generalization	Expand datasets via public–private partnerships; apply active learning to iteratively improve models	Moshawih et al. (2024)
Scoring inaccuracies	High false-positive rates	Develop hybrid scoring functions combining physics-based and ML approaches	Scardino et al. (2025)
Poor correlation with experimental activity	Resource waste in downstream validation	Implement early orthogonal validation assays (biophysical, biochemical) before medicinal chemistry investment	Zhou et al. (2024)
Limited scaffold novelty	Narrow exploration of chemical space	Employ generative AI for de novo design targeting underrepresented scaffolds	Carlsson and Luttens (2024)

Note: ML – machine learning; AI – artificial intelligence.

7. CONCLUSIONS

Over the past two decades, molecular docking and virtual screening have become integral components of modern drug discovery pipelines. Their increasing connection to AI, molecular dynamics, and high-throughput experimental validation has shifted the focus from empirical science toward data-driven modeling. Today, AI enables docking and virtual screening to search for more than one billion compounds, predict binding modes with high accuracy, and prioritize molecules that balance physicochemical and pharmacological properties. Those searches have already generated actionable leads in more than a dozen disease areas encompassing antivirals and neurodegenerative diseases, as well as oncology and antimicrobial resistance proving that computational approaches can deliver clinically relevant insights faster than ever. The predictive reliability is still hampered by the quality of the scoring functions, the protein flexibility treatment applied, and the quality of the training sets. Moreover, appropriate ethical reflection and interpretability of treatment with AI-based methods would also need to be clarified prior to full regulatory acceptance of these applications. However, this wave toward standardized, transparent (also hybrid AI–physics) frameworks is rapidly restoring the trust in models and in their ability to further science.

Future directions, expectations for the next generation of 'computational drug discovery' include innovations such as (i) closed-loop interconnectivity between in silico prediction and in vitro experiments providing adaptive learning cycles; (ii) explainable AI architectures revealing molecular logic behind predictions for rational chemical design; (iii) quantum-accelerated simulations capable of sampling complex conformational landscapes arbitrarily well to facilitate the search for novel targets with a dedicated ligand or drug; and (iv) multiscale integration with omics data capable of tying molecular binding events to cellular or systemic responses. The integration of AI, computational chemistry, and experimental science will not replace human creativity; instead, it will amplify it allowing researchers to make faster, better-informed, and more sustainable decisions. While this convergence takes place, we expect that molecular docking and in silico screening (with the assistance of AI) will continue to offer the foundation for innovative, cheaper, smart drug discovery for several more years.

REFERENCES

Agu, P. C., Afiukwa, C. A., Orji, O. U., Ezeh, E. M., Ofoke, I. H., Ogbu, C. O., Ugwuja, E. I., & Aja, P. M. (2023). Molecular docking as a tool for the discovery of molecular targets of nutraceuticals in diseases management. *Scientific Reports*, *13*(1), 13398. DOI: 10.1038/s41598-023-40160-2 PMID: 37592012

Aittokallio, T., & Fang, E. F. (2024). Editorial overview–artificial intelligence methodologies in structural biology: Bridging the gap to medical applications. *Current Opinion in Structural Biology*, *87*, 102862. DOI: 10.1016/j.sbi.2024.102862 PMID: 38870638

Bae, B.-H., Choi, J., Seok, C., & Park, H. (2025). Physics-inspired accuracy estimator for model-docked ligand complexes. *Journal of Chemical Theory and Computation*, *21*(4), 2140–2152. DOI: 10.1021/acs.jctc.4c01373 PMID: 39928910

Balıkçı, E., Marques, A.-S. M. C., Bauer, L. G., Seupel, R., Bennett, J., Raux, B., Buchan, K., Simelis, K., Singh, U., Rogers, C., Ward, J., Cheng, C., Szommer, T., Schützenhofer, K., Elkins, J. M., Sloman, D. L., Ahel, I., Fedorov, O., Brennan, P. E., & Huber, K. V. M. (2024). Unexpected noncovalent off-target activity of clinical BTK inhibitors leads to discovery of a dual NUDT5/14 antagonist. *Journal of Medicinal Chemistry*, *67*(9), 7245–7259. DOI: 10.1021/acs.jmedchem.4c00072 PMID: 38635563

Beveridge, J., Söderström, M., Prieto-Díaz, R., Gutierrez-de-Teran, H., Odell, L. R., Hallberg, M., Larhed, M., & Gising, J. (2024). Benzylhydroxamic acids as inhibitors of insulin regulated aminopeptidase (IRAP). *European Journal of Medicinal Chemistry Reports*, *12*, 100215. DOI: 10.1016/j.ejmcr.2024.100215

Carlsson, J., & Luttens, A. (2024). Structure-based virtual screening of vast chemical space as a starting point for drug discovery. *Current Opinion in Structural Biology*, *87*, 102829. DOI: 10.1016/j.sbi.2024.102829 PMID: 38848655

Carreras-Puigvert, J., & Spjuth, O. (2024). Artificial intelligence for high content imaging in drug discovery. *Current Opinion in Structural Biology*, *87*, 102842. DOI: 10.1016/j.sbi.2024.102842 PMID: 38797109

Chandra, S., Wang, Z., Tao, X., Chen, O., Luo, X., Ji, R.-R., & Bortsov, A. V. (2020). Computer-aided discovery of a new Nav1.7 inhibitor for treatment of pain and itch. *Anesthesiology*, *133*(3), 611–627. DOI: 10.1097/ALN.0000000000003427 PMID: 32788559

Chen, Y., Lin, H., Yang, H., Tan, R., Bian, Y., Fu, T., Li, W., Wu, L., Pei, Y., & Sun, H. (2017). Discovery of new acetylcholinesterase and butyrylcholinesterase inhibitors through structure-based virtual screening. *RSC Advances*, *7*(6), 3429–3438. DOI: 10.1039/C6RA25887E

da Rocha, M. N., de Sousa, D. S., da Silva Mendes, F. R., dos Santos, H. S., Marinho, G. S., Marinho, M. M., & Marinho, E. S. (2025). Ligand and structure-based virtual screening approaches in drug discovery: Minireview. *Molecular Diversity*, *29*(3), 2799–2809. DOI: 10.1007/s11030-024-10979-6 PMID: 39223358

Elgammal, W. E., Shaban, S. S., Eliwa, E. M., Halawa, A. H., Abd El-Gilil, S. M., Hassan, R. A., Abdou, A. M., Elhagali, G. A. M., & Reheim, M. A. M. A. (2024). Thiazolation of phenylthiosemicarbazone to access new thiazoles: Anticancer activity and molecular docking. *Future Medicinal Chemistry*, *16*(12), 1219–1237. DOI: 10.1080/17568919.2024.2342668 PMID: 38989988

Eurofins Pharma Discovery Services. (2025). SafetyScreen44™ Panel. Retrieved 11 August 2025 from https://cdnmedia.eurofins.com/corporate-eurofins/media/safetyscreen44_epdsfl420june16.pdf?utm_source=chatgpt.com

European Medicines Agency. (2020). ICH M9 guideline on biopharmaceutics classification system-based biowaivers - questions and answers. Retrieved 11 August 2025 from https://www.ema.europa.eu/en/documents/other/ich-m9-biopharmaceutics-classification-system-based-biowaivers-step-5-questions-and-answers_en.pdf?utm_source=chatgpt.com

Filandr, F., Sarpe, V., Raval, S., Crowder, D. A., Khan, M. F., Douglas, P., Coales, S., Viner, R., Syed, A., Tainer, J. A., Lees-Miller, S. P., & Schriemer, D. C. (2024). Automating data analysis for hydrogen/deuterium exchange mass spectrometry using data-independent acquisition methodology. *Nature Communications*, *15*(1), 2200. DOI: 10.1038/s41467-024-46610-3 PMID: 38467655

Gan, J. L., Kumar, D., Chen, C., Taylor, B. C., Jagger, B. R., Amaro, R. E., & Lee, C. T. (2022). Benchmarking ensemble docking methods in D3R Grand Challenge 4. *Journal of Computer-Aided Molecular Design*, *36*(2), 87–99. DOI: 10.1007/s10822-021-00433-2 PMID: 35199221

Gautam, S., Pathak, S., & Dubey, S. H. (2024). The role of molecular docking in modern drug discovery and development: A comprehensive review. *Journal of Drug Discovery and Health Sciences*, *1*(03), 129–137. DOI: 10.21590/jddhs.01.03.02

Gentile, F., Yaacoub, J. C., Gleave, J., Fernandez, M., Ton, A.-T., Ban, F., Stern, A., & Cherkasov, A. (2022). Artificial intelligence–enabled virtual screening of ultralarge chemical libraries with deep docking. *Nature Protocols*, *17*(3), 672–697. DOI: 10.1038/s41596-021-00659-2 PMID: 35121854

Gkekas, I., Katsamakas, S., Mylonas, S., Fotopoulou, T., Magoulas, G. E., Tenchiu, A. C., Dimitriou, M., Axenopoulos, A., Rossopoulou, N., Kostova, S., Wanker, E. E., Katsila, T., Papahatjis, D., Gorgoulis, V. G., Koufaki, M., Karakasiliotis, I., Calogeropoulou, T., Daras, P., & Petrakis, S. (2024). AI promoted virtual screening, structure-based hit optimization, and synthesis of novel COVID-19 S-RBD domain inhibitors. *Journal of Chemical Information and Modeling*, *64*(22), 8562–8585. DOI: 10.1021/acs.jcim.4c01110 PMID: 39535926

Guedes, I. A., Pereira, F. S. S., & Dardenne, L. E. (2018). Empirical scoring functions for structure-based virtual screening: Applications, critical aspects, and challenges. *Frontiers in Pharmacology*, *9*, 1089. DOI: 10.3389/fphar.2018.01089 PMID: 30319422

Hamuro, Y. (2024). Interpretation of hydrogen/deuterium exchange mass spectrometry. *Journal of the American Society for Mass Spectrometry*, *35*(5), 819–828. DOI: 10.1021/jasms.4c00044 PMID: 38639434

Han, X., & Guo, K. (2024). Pharmacophore-based virtual screening of commercial databases against β-secretase 1 for drug development against Alzheimer's disease. *Frontiers in Chemistry*, *12*, 1412349. DOI: 10.3389/fchem.2024.1412349 PMID: 39045333

Herrera-Arozamena, C., Estrada-Valencia, M., García-Díez, G., Pérez, C., León, R., Infantes, L., Morales-García, J. A., Pérez-Castillo, A., del Sastre, E., López, M. G., & Rodríguez-Franco, M. I. (2024). Discovery of a potent melatonin-based inhibitor of quinone reductase-2 with neuroprotective and neurogenic properties. *European Journal of Medicinal Chemistry*, *277*, 116763. DOI: 10.1016/j.ejmech.2024.116763 PMID: 39146834

Ibrahim, R. S., Mahrous, R. S. R., Fathy, H. M., Omar, A. A., & Abu El-Khair, R. M. (2020). Anticoagulant activity screening of an in-house database of natural compounds for discovering novel selective factor Xa inhibitors; a combined in silico and in vitro approach. *Medicinal Chemistry Research*, *29*(4), 707–726. DOI: 10.1007/s00044-020-02516-5

Jang, C., Yadav, D. K., Subedi, L., Venkatesan, R., Venkanna, A., Afzal, S., Lee, E., Yoo, J., Ji, E., Kim, S. Y., & Kim, M. (2018). Identification of novel acetylcholinesterase inhibitors designed by pharmacophore-based virtual screening, molecular docking and bioassay. *Scientific Reports*, *8*(1), 14921. DOI: 10.1038/s41598-018-33354-6 PMID: 30297729

Jug, A., Bratkovič, T., & Ilaš, J. (2024). Biolayer interferometry and its applications in drug discovery and development. *Trends in Analytical Chemistry*, *176*, 117741. DOI: 10.1016/j.trac.2024.117741

Korlepara, D. B., C. S, V., Srivastava, R., Pal, P. K., Raza, S. H., Kumar, V., . . . Priyakumar, U. D. (. (2024). PLAS-20k: Extended dataset of protein–ligand affinities from MD simulations for machine learning applications. *Scientific Data*, *11*(1), 180. DOI: 10.1038/s41597-023-02872-y PMID: 38336857

Lanka, G., Banerjee, S., Adhikari, N., & Ghosh, B. (2025). Fragment-based discovery of new potential DNMT1 inhibitors integrating multiple pharmacophore modeling, 3D-QSAR, virtual screening, molecular docking, ADME, and molecular dynamics simulation approaches. *Molecular Diversity*, *29*(1), 117–137. DOI: 10.1007/s11030-024-10837-5 PMID: 38637479

Li, S., Liu, S., Yuan, D., Liu, R., Hu, L., & Zhu, X. (2024). Discovery of quinazoline-benzothiazole derivatives as novel potent protease-activated receptor 4 antagonists with improved pharmacokinetics and low bleeding liability. *European Journal of Medicinal Chemistry*, *280*, 116980. DOI: 10.1016/j.ejmech.2024.116980 PMID: 39442337

Li, Y., Lin, H., Yang, H., Yuan, Y., Zou, R., Zhou, G., Zhang, L., & Zheng, H. (2024). Synergistic application of molecular docking and machine learning for improved binding pose. *National Science Open*, *3*(2). Advance online publication. DOI: 10.1360/nso/20230058

Liang, Y. Y., Khalid, K., Le, H. V., Teo, H. M. V., Raitelaitis, M., Gerault, M.-A., Lee, J. J. H., Lyu, J., Chan, A., Jeyasekharan, A. D., Tam, W. L., Nordlund, P., & Prabhu, N. (2025). MS CETSA deep functional proteomics uncovers DNA repair programs leading to gemcitabine resistance. *Nature Communications*, *16*(1), 4234. DOI: 10.1038/s41467-025-59505-8 PMID: 40335468

Liu, M., Mirza, A., McAndrew, P. C., Thapaliya, A., Pierrat, O. A., Stubbs, M., Hahner, T., Chessum, N. E. A., Innocenti, P., Caldwell, J., Cheeseman, M. D., Bellenie, B. R., van Montfort, R. L. M., Newton, G. K., Burke, R., Collins, I., & Hoelder, S. (2023). Determination of ligand-binding affinity (Kd) using transverse relaxation rate (R_2) in the ligand-observed ^{1}H NMR experiment and applications to fragment-based drug discovery. *Journal of Medicinal Chemistry*, *66*(15), 10617–10627. DOI: 10.1021/acs.jmedchem.3c00758 PMID: 37467168

Lu, W., Zhang, J., Huang, W., Zhang, Z., Jia, X., Wang, Z., Shi, L., Li, C., Wolynes, P. G., & Zheng, S. (2024). DynamicBind: Predicting ligand-specific protein–ligand complex structure with a deep equivariant generative model. *Nature Communications*, *15*(1), 1071. DOI: 10.1038/s41467-024-45461-2 PMID: 38316797

Luo, W., Yu, L., Lu, S., Yu, Y., Bai, Y., Wang, S., Li, D., Han, Z., Zheng, Y., Li, F., Zhao, J., & He, L. (2025). Design and screening of novel molecular compounds targeting lactate dehydrogenase of Babesia microti. *Parasites & Vectors*, *18*(1), 69. DOI: 10.1186/s13071-024-06623-9 PMID: 39988659

Luttens, A., Cabeza de Vaca, I., Sparring, L., Brea, J., Martínez, A. L., Kahlous, N. A., Radchenko, D. S., Moroz, Y. S., Loza, M. I., Norinder, U., & Carlsson, J. (2025). Rapid traversal of vast chemical space using machine learning-guided docking screens. *Nature Computational Science*, *5*(4), 301–312. DOI: 10.1038/s43588-025-00777-x PMID: 40082701

Luttens, A., Vo, D. D., Scaletti, E. R., Wiita, E., Almlöf, I., Wallner, O., Davies, J., Košenina, S., Meng, L., Long, M., Mortusewicz, O., Masuyer, G., Ballante, F., Michel, M., Homan, E., Scobie, M., Kalderén, C., Warpman Berglund, U., Tarnovskiy, A. V., & Carlsson, J. (2025). Virtual fragment screening for DNA repair inhibitors in vast chemical space. *Nature Communications*, *16*(1), 1741. DOI: 10.1038/s41467-025-56893-9 PMID: 39966348

Ma, L., Tang, J., Chen, F., Liu, Q., Huang, J., Liu, X., Zhou, Z., & Yi, W. (2024). Structure-based screening, optimization and biological evaluation of novel chrysin-based derivatives as selective PPARγ modulators for the treatment of T2DM and hepatic steatosis. *European Journal of Medicinal Chemistry*, *276*, 116728. DOI: 10.1016/j.ejmech.2024.116728 PMID: 39089002

Mazzei, L., Ranieri, S., Silvestri, D., Greene-Cramer, R., Cioffi, C., Montelione, G. T., & Ciurli, S. (2024). An isothermal calorimetry assay for determining steady state kinetic and Ensitrelvir inhibition parameters for SARS-CoV-2 3CL-protease. *Scientific Reports*, *14*(1), 32175. DOI: 10.1038/s41598-024-81990-y PMID: 39741150

McNutt, A. T., Francoeur, P., Aggarwal, R., Masuda, T., Meli, R., Ragoza, M., Sunseri, J., & Koes, D. R. (2021). GNINA 1.0: Molecular docking with deep learning. *Journal of Cheminformatics*, *13*(1), 43. DOI: 10.1186/s13321-021-00522-2 PMID: 34108002

McNutt, A. T., Li, Y., Meli, R., Aggarwal, R., & Koes, D. R. (2025). GNINA 1.3: The next increment in molecular docking with deep learning. *Journal of Cheminformatics*, *17*(1), 28. DOI: 10.1186/s13321-025-00973-x PMID: 40025560

Milne, B. F. (2022). Scoring functions, ScotChem protein–ligand docking course documentation. Retrieved 12 August 2025 from https://arts.st-andrews.ac.uk/scotchem/pldc/scoring_functions.html#:~:text=%5C%5BE_%7Bbind%7D%20%3D%20%5Csum_%7Bi%3D1%7D,6%7D%20%2B%20%5Cfrac%7Bq_iq_j%7D%7B%5Cepsilon%20r_%7Bij%7D%7D%20%5Cright

Moshawih, S., Bu, Z. H., Goh, H. P., Kifli, N., Lee, L. H., Goh, K. W., & Ming, L. C. (2024). Consensus holistic virtual screening for drug discovery: A novel machine learning model approach. *Journal of Cheminformatics*, *16*(1), 62. DOI: 10.1186/s13321-024-00855-8 PMID: 38807196

Nadh, A. G., Kunhikrishnan, M. J., Ravi, V., Ramakrishnan, K., Rehman, N., Adithya, K. S. B., Revikumar, A., Sudhakaran, P. R., & Raju, R. (2025). Convolidine as potent BACE1 inhibitor for Alzheimer's disease; in-silico coupled with in vitro assessment. *Journal of Computer-Aided Molecular Design, 39*(1), 13. DOI: 10.1007/s10822-025-00592-6 PMID: 40208466

Nepravishta, R., Ramírez-Cárdenas, J., Rocha, G., Walpole, S., Hicks, T., Monaco, S., Muñoz-García, J. C., & Angulo, J. (2024). Fast quantitative validation of 3D models of low-affinity protein–ligand complexes by STD NMR spectroscopy. *Journal of Medicinal Chemistry, 67*(12), 10025–10034. DOI: 10.1021/acs.jmedchem.4c00204 PMID: 38848103

Neves, M. A. D., Ni, T. T., Mackeigan, D. T., Shoara, A. A., Lei, X., Slavkovic, S., Yu, S.-Y., Stratton, T. W., Gallant, R. C., Zhang, D., Xu, X. R., Fernandes, C., Zhu, G., Hu, X., Chazot, N., Donaldson, L. W., Johnson, P. E., Connelly, K., Rand, M., & Ni, H. (2024). Salvianolic acid B inhibits thrombosis and directly blocks the thrombin catalytic site. *Research and Practice in Thrombosis and Haemostasis, 8*(4), 102443. DOI: 10.1016/j.rpth.2024.102443 PMID: 38993621

Noor, F., Junaid, M., Almalki, A. H., Almaghrabi, M., Ghazanfar, S., & Tahir ul Qamar, M. (2024). Deep learning pipeline for accelerating virtual screening in drug discovery. *Scientific Reports, 14*(1), 28321. DOI: 10.1038/s41598-024-79799-w PMID: 39550439

O'Shea, J. P., Augustijns, P., Brandl, M., Brayden, D. J., Brouwers, J., Griffin, B. T., Holm, R., Jacobsen, A.-C., Lennernäs, H., Vinarov, Z., & O'Driscoll, C. M. (2022). Best practices in current models mimicking drug permeability in the gastrointestinal tract - An UNGAP review. *European Journal of Pharmaceutical Sciences, 170*, 106098. DOI: 10.1016/j.ejps.2021.106098 PMID: 34954051

Omran, H. A., Majed, A. A., Hussein, K., Abid, D. S., Abdel-Maksoud, M. A., Elwahsh, A., Aufy, M., & Kotob, M. H. (2024). Anticancer activity, DFT and molecular docking study of new bisthiazolidine amide. *Results in Chemistry, 12*, 101835. DOI: 10.1016/j.rechem.2024.101835

Paggi, J. M., Pandit, A., & Dror, R. O. (2024). The art and science of molecular docking. *Annual Review of Biochemistry, 93*(1), 389–410. DOI: 10.1146/annurev-biochem-030222-120000 PMID: 38594926

Pandiarajan, I., Walunj, S. B., Banerjee, N., Rout, J., Srivastava, S., Patankar, S., & Kaledhonkar, S. (2024). Application of biolayer interferometry for the analysis of ribosome-protein interactions. *Frontiers in Molecular Biosciences, 11*, 1398964. DOI: 10.3389/fmolb.2024.1398964 PMID: 39148630

Petrauskas, V., Kazlauskas, E., Gedgaudas, M., Baranauskienė, L., Zubrienė, A., & Matulis, D. (2024). Thermal shift assay for protein–ligand dissociation constant determination. *Trends in Analytical Chemistry, 170*, 117417. DOI: 10.1016/j.trac.2023.117417

Pham, P., Nguyen, V. T. D., Cho, K. H., & Hy, T.-S. (2025). DrugPipe: Generative artificial intelligence-assisted virtual screening pipeline for generalizable and efficient drug repurposing. *Biology Methods and Protocols, 10*(1), bpaf038. Advance online publication. DOI: 10.1093/biomethods/bpaf038 PMID: 40486179

Picado, A., Chaikuad, A., Wells, C. I., Shrestha, S., Zuercher, W. J., Pickett, J. E., Kwarcinski, F. E., Sinha, P., de Silva, C. S., Zutshi, R., Liu, S., Kannan, N., Knapp, S., Drewry, D. H., & Willson, T. M. (2020). A chemical probe for dark kinase STK17B derives its potency and high selectivity through a unique P-loop conformation. *Journal of Medicinal Chemistry*, *63*(23), 14626–14646. DOI: 10.1021/acs.jmedchem.0c01174 PMID: 33215924

Popov, K. I., Wellnitz, J., Maxfield, T., & Tropsha, A. (2024). Hit discovery using docking ENriched by GEnerative modeling (HIDDEN GEM): A novel computational workflow for accelerated virtual screening of ultra-large chemical libraries. *Molecular Informatics*, *43*(1), e202300207. DOI: 10.1002/minf.202300207 PMID: 37802967

Punyawatthananukool, S., Matsuura, R., Wongchang, T., Katsurada, N., Tsuruyama, T., Tajima, M., Enomoto, Y., Kitamura, T., Kawashima, M., Toi, M., Yamanoi, K., Hamanishi, J., Hisamori, S., Obama, K., Charoensawan, V., Thumkeo, D., & Narumiya, S. (2024). Prostaglandin E2-EP2/EP4 signaling induces immunosuppression in human cancer by impairing bioenergetics and ribosome biogenesis in immune cells. *Nature Communications*, *15*(1), 9464. DOI: 10.1038/s41467-024-53706-3 PMID: 39487111

Raingeval, C., Cala, O., Brion, B., Le Borgne, M., Hubbard, R. E., & Krimm, I. (2019). [1]D NMR WaterLOGSY as an efficient method for fragment-based lead discovery. *Journal of Enzyme Inhibition and Medicinal Chemistry*, *34*(1), 1218–1225. DOI: 10.1080/14756366.2019.1636235 PMID: 31286785

Sahakyan, H. (2021). Improving virtual screening results with MM/GBSA and MM/PBSA rescoring. *Journal of Computer-Aided Molecular Design*, *35*(6), 731–736. DOI: 10.1007/s10822-021-00389-3 PMID: 33983518

Scardino, V., Galarce, M. J., Mignone, M. E., & Cavasotto, C. N. (2025). Enhancing the reliability of integrated consensus strategies to boost docking-based screening campaigns using publicly available docking programs. *Molecular Informatics*, *44*(5-6), e2445. DOI: 10.1002/minf.2445 PMID: 40537017

Sharma, G., Kumar, N., Sharma, C. S., Alqahtani, T., Tiruneh, Y. K., Sultana, S., Rolim Silva, G. V., de Lima Menezes, G., Zaki, M. E. A., & Nobre Oliveira, J. I. (2025). Identification of promising SARS-CoV-2 main protease inhibitor through molecular docking, dynamics simulation, and ADMET analysis. *Scientific Reports*, *15*(1), 2830. DOI: 10.1038/s41598-025-86016-9 PMID: 39843610

Shu, J., Wang, Y., Guo, W., Liu, T., Cai, S., Shi, T., & Hu, W. (2024). Carbenoid-involved reactions integrated with scaffold-based screening generates a Nav1.7 inhibitor. *Communications Chemistry*, *7*(1), 135. DOI: 10.1038/s42004-024-01213-3 PMID: 38866907

Słoczyńska, K., Gunia-Krzyżak, A., Koczurkiewicz, P., Wójcik-Pszczoła, K., Żelaszczyk, D., Popiół, J., & Pękala, E. (2019). Metabolic stability and its role in the discovery of new chemical entities. *Acta Pharmaceutica (Zagreb, Croatia)*, *69*(3), 345–361. DOI: 10.2478/acph-2019-0024 PMID: 31259741

Sobha, A., Ganapathy, A., Mohan, S., Madhusoodanan, N., Babysulochana, A. D., Alaganandan, K., & Somappa, S. B. (2024). Novel small molecule-based acetylcholinesterase (AChE) inhibitors: From biological perspective to recent developments. *European Journal of Medicinal Chemistry Reports*, *12*, 100237. DOI: 10.1016/j.ejmcr.2024.100237

Sodhi, J. K., & Benet, L. Z. (2021). Successful and unsuccessful prediction of human hepatic clearance for lead optimization. *Journal of Medicinal Chemistry*, *64*(7), 3546–3559. DOI: 10.1021/acs.jmedchem.0c01930 PMID: 33765384

Teshima, M., Monobe, K., Okubo, S., & Aoki, S. (2024). Discovery of antibacterial compounds with potential multipharmacology against Staphylococcus mur ligase family members by in silico structure-based drug screening. *Molecules (Basel, Switzerland)*, *29*(16), 3792. DOI: 10.3390/molecules29163792 PMID: 39202871

Upadhyay, V., Lucas, A., Patrick, C., & Mallela, K. M. G. (2024). Isothermal titration calorimetry and surface plasmon resonance methods to probe protein–protein interactions. *Methods (San Diego, Calif.)*, *225*, 52–61. DOI: 10.1016/j.ymeth.2024.03.007 PMID: 38492901

Van Vranken, J. G., Li, J., Mintseris, J., Wei, T.-Y., Sniezek, C. M., Gadzuk-Shea, M., Gygi, S. P., & Schweppe, D. K. (2024). Large-scale characterization of drug mechanism of action using proteome-wide thermal shift assays. *eLife*, *13*, RP95595. Advance online publication. DOI: 10.7554/eLife.95595 PMID: 39526730

Vittorio, S., Lunghini, F., Morerio, P., Gadioli, D., Orlandini, S., Silva, P., Jan Martinovic, , Pedretti, A., Bonanni, D., Del Bue, A., Palermo, G., Vistoli, G., & Beccari, A. R. (2024). Addressing docking pose selection with structure-based deep learning: Recent advances, challenges and opportunities. *Computational and Structural Biotechnology Journal*, *23*, 2141–2151. DOI: 10.1016/j.csbj.2024.05.024 PMID: 38827235

Wallach, I., Bernard, D., Nguyen, K., Ho, G., Morrison, A., Stecula, A., Rosnik, A., O'Sullivan, A. M., Davtyan, A., Samudio, B., Thomas, B., Worley, B., Butler, B., Laggner, C., Thayer, D., Moharreri, E., Friedland, G., Truong, H., van den Bedem, H., & Heifets, A. (2024). AI is a viable alternative to high throughput screening: A 318-target study. *Scientific Reports*, *14*(1), 7526. DOI: 10.1038/s41598-024-54655-z PMID: 38565852

Wang, R., & Wang, S. (2001). How does consensus scoring work for virtual library screening? an idealized computer experiment. *Journal of Chemical Information and Computer Sciences*, *41*(5), 1422–1426. DOI: 10.1021/ci010025x PMID: 11604043

Wang, X., Zhang, Y., Yang, Y., Wu, X., Fan, H., & Qiao, Y. (2017). Identification of berberine as a direct thrombin inhibitor from traditional Chinese medicine through structural, functional and binding studies. *Scientific Reports*, *7*(1), 44040. DOI: 10.1038/srep44040 PMID: 28276481

Winiewska-Szajewska, M., & Poznański, J. (2025). Differential scanning fluorimetry followed by microscale thermophoresis and/or isothermal titration calorimetry as an efficient tool for ligand screening. *Biophysical Reviews*, *17*(1), 199–223. DOI: 10.1007/s12551-025-01280-3 PMID: 40060009

Yi, J.-C., Yang, Z.-Y., Zhao, W.-T., Yang, Z.-J., Zhang, X.-C., Wu, C.-K., Lu, A.-P., & Cao, D.-S. (2024). ChemMORT: An automatic ADMET optimization platform using deep learning and multiobjective particle swarm optimization. *Briefings in Bioinformatics*, *25*(2), bbae008. Advance online publication. DOI: 10.1093/bib/bbae008 PMID: 38385872

Yingngam, B. (2024a). AI Models for Biopharmaceutical Property Prediction. In Khanna, A., Barachi, M. E., Jain, S., Kumar, M., & Nayyar, A. (Eds.), *Artificial Intelligence and Machine Learning in Drug Design and Development* (pp. 407–449). Wiley., DOI: 10.1002/9781394234196.ch13

Yingngam, B. (2024b). Introduction to Alzheimer's Disease, Biomarkers, and the AI Revolution. In Lilhore, U. K., Anand, A., Kumar, A., Yadav, S. P., & Vyas, N. (Eds.), *AI-Driven Alzheimer's Disease Detection and Prediction* (pp. 1–23). IGI Global Scientific Publishing., DOI: 10.4018/979-8-3693-3605-2.ch001

Yingngam, B. (2025a). AI-Driven Drug Discovery and Repurposing for Neurodegenerative Disorders. In Machine Learning for Neurodegenerative Disorders: Advancements and Applications (pp. 80-124). CRC Press. DOI: 10.1201/9781032661025-4

Yingngam, B. (2025b). AI in Predictive Toxicology. In Shaik, A. (Ed.), *AI-Powered Advances in Pharmacology* (pp. 79–134). IGI Global Scientific Publishing., DOI: 10.4018/979-8-3693-3212-2.ch005

Yingngam, B., Navabhatra, A., & Sillapapibool, P. (2024). AI-Driven Decision-Making Applications in Pharmaceutical Sciences. In Nguyen, T. V. T., & Vo, N. T. M. (Eds.), *Using Traditional Design Methods to Enhance AI-Driven Decision Making* (pp. 1–63). IGI Global Scientific Publishing., DOI: 10.4018/979-8-3693-0639-0.ch001

Yingngam, B., & Sethabouppha, B. (2024). Advanced AI Applications for Drug Discovery. In Shah, I. A., & Sial, Q. (Eds.), *Advances in Computational Intelligence for the Healthcare Industry 4.0* (pp. 42–86). IGI Global Scientific Publishing., DOI: 10.4018/979-8-3693-2333-5.ch003

Zhang, L., Wang, Y., Zheng, C., Zhou, Z., & Chen, Z. (2024). Cellular thermal shift assay: An approach to identify and assess protein target engagement. *Expert Review of Proteomics*, *21*(9-10), 387–400. DOI: 10.1080/14789450.2024.2406785 PMID: 39317941

Zhang, W.-X., Huang, J., Tian, X.-Y., Liu, Y.-H., Jia, M.-Q., Wang, W., Jin, C.-Y., Song, J., & Zhang, S.-Y. (2023). A review of progress in o-aminobenzamide-based HDAC inhibitors with dual targeting capabilities for cancer therapy. *European Journal of Medicinal Chemistry*, *259*, 115673. DOI: 10.1016/j.ejmech.2023.115673 PMID: 37487305

Zhang, Z., Rao, J., Zhong, J., Bai, W., Wang, D., Ning, S., . . . Hua, W. (2025). Fitness aligned structural modeling enables scalable virtual screening with AuroBind. arXiv preprint arXiv:2508.02137. https://doi.org//arXiv.2508.02137DOI: 10.48550

Zheng, L., Zou, Y., Xie, T., Wu, X., Tan, Y., Mei, S., Geng, Y., Chen, S., Xu, S., & Niu, M.-M. (2023). Discovery of a dual tubulin and neuropilin-1 (NRP1) inhibitor with potent in vivo anti-tumor activity via pharmacophore-based docking screening, structure optimization, and biological evaluation. *Journal of Medicinal Chemistry*, *66*(23), 16187–16200. DOI: 10.1021/acs.jmedchem.3c01572 PMID: 38093696

Zhou, G., Rusnac, D.-V., Park, H., Canzani, D., Nguyen, H. M., Stewart, L., Bush, M. F., Nguyen, P. T., Wulff, H., Yarov-Yarovoy, V., Zheng, N., & DiMaio, F. (2024). An artificial intelligence accelerated virtual screening platform for drug discovery. *Nature Communications*, *15*(1), 7761. DOI: 10.1038/s41467-024-52061-7 PMID: 39237523

Zhu, M., Qu, J., & Deng, Q. (2024). Identification of potential inhibitors against Staphylococcus aureus shikimate dehydrogenase through virtual screening and susceptibility test. *Journal of Enzyme Inhibition and Medicinal Chemistry*, *39*(1), 2301768. DOI: 10.1080/14756366.2024.2301768 PMID: 38234148

KEY TERMS AND DEFINITION

AI-Enhanced Virtual Screening: Integrates machine learning, deep learning, or generative AI techniques with classical docking or similarity-based approaches to accelerate hit discovery.

Consensus Scoring: Combines multiple scoring functions to reduce false positives and improve hit identification reliability.

Ensemble Docking: Involves docking ligands into multiple conformations of a target protein to capture structural flexibility.

Hit Compounds: Are molecules identified through screening that show measurable biological activity.

Induced-Fit Docking: Accounts for conformational changes in the protein binding site upon ligand binding, improving the prediction accuracy for flexible targets.

Lead Compounds: Are optimized hits with improved potency, selectivity, and drug-like properties that are suitable for further development.

Ligand-Based Virtual Screening (LBVS): Relies on known active compounds to identify new candidates on the basis of molecular similarity, physicochemical descriptors, or pharmacophore features, particularly when a reliable target structure is unavailable.

Machine Learning–Based Scoring Functions: Use statistical or artificial intelligence models trained on experimental data to improve binding affinity predictions.

Molecular Docking: Refers to a computational method used to predict the preferred orientation and binding affinity of a small molecule (ligand) when it interacts with a biological macromolecule, typically a protein or nucleic acid.

Structure-Based Virtual Screening (SBVS): Uses the three-dimensional structure of a biological target to predict ligand binding modes and affinities through molecular docking and scoring functions.

The Scoring Function: Is a mathematical or empirical model used to estimate the binding free energy between a ligand and a target protein during docking simulations.

Virtual Screening (VS): Is a computational approach for evaluating large libraries of chemical compounds to identify candidates with a high likelihood of biological activity against a specific target.

APPENDIX

Further Reading

For foundational concepts in molecular docking and virtual screening:

Kitchen, D. B., Decornez, H., Furr, J. R., & Bajorath, J. (2004). Docking and scoring in virtual screening for drug discovery: Methods and applications. Nature Reviews Drug Discovery, 3(11), 935–949. DOI: 10.1038/nrd1549.
Morris, G. M., & Lim-Wilby, M. (2008). Molecular docking. In A. Kukol (Ed.), Molecular modeling of proteins (Methods in Molecular Biology, Vol. 443, pp. 365–382). Humana Press. https://doi.org/10.1007/978-1-59745-177-2_19.

For advances in structure- and ligand-based virtual screening:

Lionta, E., Spyrou, G., Vassilatis, D. K., & Cournia, Z. (2014). Structure-based virtual screening for drug discovery: Principles, applications, and recent advances. Current Topics in Medicinal Chemistry, 14(16), 1923–1938. DOI: 10.2174/1568026614666140929124445.
Cherkasov, A., Muratov, E. N., Fourches, D., Varnek, A., Baskin, I. I., Cronin, M., Dearden, J., Gramatica, P., Martin, Y. C., Todeschini, R., Consonni, V., Kuz'min, V. E., Cramer, R., & Tropsha, A. (2014). QSAR modeling: Where have you been? Where are you going to? Journal of Medicinal Chemistry, 57(12), 4977–5010. DOI: 10.1021/jm4004285.

For AI and machine learning in drug discovery:

Zhavoronkov, A., Ivanenkov, Y. A., Aliper, A., Veselov, M. S., Aladinskiy, V. A., Aladinskaya, A. V., … Lee, L. H. (2019). Deep learning enables rapid identification of potent DDR1 kinase inhibitors. Nature Biotechnology, 37(9), 1038–1040. DOI: 10.1038/s41587-019-0224-x.
Walters, W. P., & Murcko, M. A. (2020). Assessing the impact of generative AI on medicinal chemistry. Nature Biotechnology, 38(2), 143–145. DOI: 10.1038/s41587-020-0418-2.

For the experimental validation and integration pipelines:

Copeland, R. A. (2016). Drug–target residence time model: A 10-year retrospective study. Nature Reviews Drug Discovery, 15(2), 87–95. DOI: 10.1038/nrd.2015.18.
Holdgate, G. A., & Ward, W. H. J. (2005). Measurements of binding thermodynamics in drug discovery. Drug Discovery Today, 10(22), 1543–1550. DOI: 10.1016/S1359-6446(05)03610-X.

Abbreviations

3D – Three-dimensional
AChE – Acetylcholinesterase
ADMET – Absorption, distribution, metabolism, excretion, and toxicity

AI – Artificial intelligence
BLI – Biolayer interferometry
BuChE – Butyrylcholinesterase
CETSA – Cellular thermal shift assay
Clint – Intrinsic clearance
CNN – Convolutional neural network
COX-2 – Cyclooxygenase-2
CRISPR – Clustered regularly interspaced short palindromic repeats
DNA – Deoxyribonucleic acid
DSF – Differential scanning fluorimetry
EP2/EP4 – Prostaglandin E_2 receptors 2 and 4
FDA – U.S. Food and Drug Administration
GAN – Generative adversarial network
GNN – Graph neural network
GPCR – G protein-coupled receptor
HDX-MS – Hydrogen/deuterium exchange mass spectrometry
HLM – Human liver microsomes
IC_{50} – Half-maximal inhibitory concentration
IFD – Induced-fit docking
ITC – Isothermal titration calorimetry
K – Inhibition constant
KD – Dissociation constant
KLHDC2 – Kelch domain-containing protein 2
LBVS – Ligand-based virtual screening
LLE – Lipophilic ligand efficiency
MACCS Keys – Molecular Access System keys
MD – Molecular dynamics
ML – Machine learning
MM-GBSA – Molecular mechanics/generalized surface area
MST – Microscale thermophoresis
NMR – Nuclear magnetic resonance
OGG1 – 8-Oxoguanine DNA glycosylase 1
Papp – Apparent permeability coefficient
QSAR – Quantitative structure–activity relationship
SAR – Structure–activity relationship
SBVS – Structure-based virtual screening
SPR – Surface plasmon resonance
STD-NMR – Saturation transfer difference NMR
TPP – Thermal proteome profiling
TPSA – Topological polar surface area
VS – Virtual screening
ΔGbind – Gibbs free energy of binding

Chapter 8

Multiscale Simulation of Reactive Polymeric Systems With RMD and SPH Transport Modeling

Amirhosein Yazdanbakhsh
https://orcid.org/0000-0002-1263-5105
Department of Polymer Engineering and Color Technology, AmirKabir University of Technology, Iran

Sajede Narjili
https://orcid.org/0009-0007-7363-4281
Department of Polymer Engineering and Color Technology, AmirKabir University of Technology, Iran

Vahid Haddadi-Asl
Department of Polymer Engineering and Color Technology, AmirKabir University of Technology, Iran

Mehrnoosh Airam
https://orcid.org/0009-0003-6084-829X
Department of Polymer Engineering and Color Technology, AmirKabir University of Technology, Iran

Elahe Abdollahi
Department of Polymer Engineering and Color Technology, AmirKabir University of Technology, Iran

Hadi Rostamipoor
Department of Polymer Engineering and Color Technology, AmirKabir University of Technology, Iran

ABSTRACT

Advancing polymeric materials requires understanding how molecular reactions couple with macroscopic transport during processing and service. This chapter reviews multiscale modeling of reactive polymer systems, focusing on coupling Reactive Molecular Dynamics (RMD) with Smoothed Particle Hydrodynamics (SPH). RMD resolves bond breaking, formation, and reaction kinetics, while SPH represents large-scale flow, heat transfer, and rheology. Their integration enables a unified framework for simulating polymerization, cross-linking, degradation, and flow-driven reactions under realistic conditions. Theoretical foundations and coupling strategies including domain decomposition, bridging-scale methods, adaptive resolution, and concurrent simulations are discussed. Applications such as reactive

DOI: 10.4018/979-8-3373-6058-4.ch008

extrusion, phase separation, gelation, and pyrolysis highlight predictive capability. Key challenges remain in temporal coupling, reactivity transfer, force-field transferability, and reliable validation. Future directions include AI-enhanced coupling and quantum-informed reactive potentials.

1. INTRODUCTION TO MULTISCALE MODELING IN POLYMER SYSTEMS

To develop advanced polymeric materials with desired properties, it is essential to comprehend the complex interactions between chemistry at the molecular level and macroscopic behavior. Polymer science today represents an inherent challenge because the properties that matter to engineers (mechanical strength, thermal stability, processing, and longevity) ultimately derive from emergent phenomena at distinct length and time scales (Fish, 2006). At the nanoscale, individual chemical bonds break and form during polymerization, cross-linking, or degradation. At the mesoscale, polymer chains become entangled, oriented, and crystallized. And at the macroscale, and in bulk form, flow, thermal conductivity, and stresses that distribute determine processing and ultimate product performance (Peter & Kremer, 2009a). Due to the multiscale nature of polymer behavior, traditional simulations, employing a single scale, cannot accurately capture this behavior; hence, a critical gap exists between molecular understanding and engineering success (Miller & Tadmor, 2009).

During the last 20 years, we have made tremendous leaps in computational polymer science enabled by an explosion in the amount of computing power available and the emergence of modern simulation approaches (Kremer & Müller-Plathe, 2002). Quantum mechanical calculations can accurately describe small molecular systems, but they are simply too costly for polymer-scale problems. Classical molecular dynamics (CMD) allows for calculations on millions of atoms over nanoseconds with chain dynamics and local structure but is unable to reach the microsecond-to-second timescales of polymer processing (Frenkel & Smit, 1996). Continuum approaches like Finite Element Method (FEM) or computational fluid dynamics (CFD) are better suited to modeling macroscopic behavior but cannot provide the chemical specificity needed for reactive transformations or molecular mechanisms. This is the essence of our challenge that no one approach is suitable for the full range of length or time scales necessary and has led to a concerted effort to develop multiscale modeling methods that fit different simulation approaches in a coherent structure that allows for the exchange of information and techniques (Wagner & Liu, 2003).

Within reactive polymeric systems, coupling chemical reactions with flow and transport introduces some of the more complicated challenges. In reactive extrusion, resin transfer molding, or as in the case of thermoplastics while injection molding, the resin goes through chemical transformations at the same time as the fluid is flowing in the mold, heating of the resin is occurring, and diffusion of species in the material is taking place. The local reaction kinetics depend on temperature, pressure, and the local molecular environment, while the flow field in the mold itself is altered by viscosity changes resulting from molecular weight increase, crosslinking, or degradation (Odian, n.d.). This means that the close coupling of chemistry and transport results in feedback loops, which are often unsuitable for a single-scale model representation.

Reactive Molecular Dynamics (RMD) has emerged as a significant form of CMD to simulate reactions at the atomic scale, with explicit consideration of bond formation and breaking. In contrast to CMD methods, which utilize fixed-topology force fields, RMD can dynamically change bond connectivity through flexible potentials (Senftle, Hong, Islam, Kylasa, Zheng, Shin, Junkermeier, Engel-Herbert, Janik, Aktulga, et al., 2016) (most commonly ReaxFF). ReaxFF emphasizes the total system energy as a summation of bond, angle, torsion, and non-bonded (van der Waals and Coulombic) contributions that all monotonically depend on a bond-order parameter that varies continuously between zero (no bond) and integer values (single, double, or triple bonds) (Nayir et al., 2019). This construction allows the ReaxFF framework to directly capture not only the reaction pathways, but also the transition states and distributions of products without any a priori information regarding the reaction mechanism. RMD has been successfully implemented to simulate process such as polymer pyrolysis, oxidative degradation, mechanochemical scission of bonds, and during curing reactions (S. P. Patil et al., 2016; W. Zhang et al., 2022).

Conversely, Smoothed Particle Hydrodynamics (SPH) is a meshless, Lagrangian method that is particularly useful for simulating large scale fluid flows, especially flows with free surfaces, large deformation or complex geometry (Monaghan, 2005). In SPH, the continuum domain is divided into a set of individual moving particles that define the local flow field by carrying physical properties such as mass, density, velocity, and energy. Field variables and their spatial derivatives are computed via weighted summation over neighboring particles with kernel functions (Morris et al., 1997). Without a fixed mesh, SPH is naturally adaptive to moving boundaries or topological changes, which is advantageous in polymer processing scenarios, such as mold filling, fiber suspension flows, and material jetting (Hu & Adams, 2006). SPH is capable of efficiently simulating the macroscopic transport of momentum, energy, and species, and can also incorporate complex rheological models to simulate non-Newtonian polymer melts (Yasuda & Yamamoto, 2008). However, standard SPH does not contain any representation of molecular structure or chemical reactivity, treating the material as a continuum with phenomenological constitutive laws and reaction kinetics given from external sources (Fatehi & Manzari, 2011).

The primary rationale for coupling RMD and SPH is to develop a computational framework that maintains the chemical fidelity of atomistic simulation in regions where reactions occur, and utilizes the efficiency of continuum methods in the bulk domain where molecular detail is less important. This spatial decomposition strategy (often referred to as domain decomposition or concurrent coupling) involves splitting the simulation domain into an atomistic region governed by RMD, a continuum region governed by SPH, and an overlap or handshake region where information is exchanged between scales (Werder et al., 2005a). The atomistic region usually arises in reactive sites, crack tips, or interfaces where bond-breaking processes dictate behavior. The continuum region contains the remaining domain, where transport and flow are accurately described by macroscopic conservation laws. The handshake region acts as a buffer zone where atomistic information (local reaction rates, heat released, species concentrations) is coarse-grained and passed to SPH as source terms, while the continuum fields (temperature, pressure, velocity) are downscaled to provide boundary conditions or constraints for the atomistic simulation (Praprotnik et al., 2005).

There are some significant technical hurdles that arise in developing this coupling which has been the focus of substantial investigation for the last ten years. The first question is one of time synchronization: RMD requires femtosecond time steps to capture the atomic vibrations, while SPH can run at appropriate time steps for continuum transport in picoseconds and nanoseconds (Wagner et al., 2004). If the two methods are run at the same time, then the SPH program will necessarily be forced to run at the RMD time step, and this defeats the purpose of SPH which has a clear computational advantage. The alternative is time integration where RMD integrates on multiple timescales, meaning that many RMD time steps are run at each simulation step of SPH, and time-averaged information about the atoms is generated before passing it to the continuum program (J. Liu et al., 2007). The second question is one of spatial scale: individual atoms in RMD will also have to be mapped to continuum fields in SPH. In other words, the coupling requires some restriction of the operators that coarse-grain atomistic data (e.g., what local density, temperature, or reaction rate is by averaging atomic positions and velocities), and lifting operators that reconstruct configurations of atoms that are consistent with imposed continuum fields (Fritsch et al., 2012). Finally, the third significant hurdle is to ensure that fundamental quantities (mass, momentum, and energy) are conserved across the interface of the two models. If the coupling algorithm is not constructed with care, then artificial sources or sinks could exist in the handshake (or overlapping) region, which would not conserve whatever quantity the handshake region controls, and this could destabilize the numerical solutions (Fish & Chen, 2004)

In addition to these number issues, the coupling has to confront a more subtle challenge: the transfer of chemical information. RMD simulates, at an atomic level, molecular collisions, bond rearrangements, and energy transfers that comprise reaction events. From this microscopic information, one needs to extract macroscopic reaction rates, heat release coefficient, and species production terms that are appropriate values to be inserted into SPH continuum reaction-transport equations (Potestio et al., 2013). This extraction process, often referred to as homogenization, or coarse-graining, requires, and must average over sufficient numbers of molecules, and sufficient time to produce smooth representative fields. However, reactive events tend to be rare, as well as local, particularly in polymer systems, and thus statistics may not converge (A. Agarwal et al., 2014). Furthermore, the rates extracted should not only depend on temperature and concentration (as they would, along with their thermodynamic equations in traditional kinetic models), but also local stress, strain rate, and molecular conformation, among other cumbersome dependencies (Mukherji et al., 2014).

Notwithstanding these challenges, the possible advantages of RMD-SPH coupling in the area of reactive polymer systems are considerable and have resulted in increasing research activity in recent years. A bibliometric analysis of publications indexed in Web of Science indicates that multiscale modeling approaches in polymer science have drastically increased over the past two decades(Senftle, Hong, Islam, Kylasa, Zheng, Shin, Junkermeier, Engel-Herbert, Janik, Aktulga, et al., 2016). Between 2005 and 2024, the number of published works combining molecular dynamics and continuum methods increased by about 340%, with a notable jump in publications after 2015 that coincided with improvements in computing and increased development of open-source coupling frameworks. Specifically, there has been a growth in the number of research papers published on reactive force field methods from fewer than 50 papers per year in 2005 to over 420 papers in 2024, equating to an average growth rate of 18% per year (Violeau & Rogers, 2016). Crude comparisons show that there has been a change from around 30 publications of meshless algorithms like SPH in polymer processing during 2005 to almost 180 in 2024. These two areas of research (coupling atomistic reactive simulations and continuum transport models) are relatively young but are growing quickly, increasing the count of published papers from perhaps 1 to 5 papers in 2010 to over 45 papers in 2024 (Monaghan, 2011).

This growth path reflects both the refinement of specific simulation techniques or modeling approaches and the growing acknowledgement that the complicated phenomena taking place during polymer processing cannot be captured sufficiently using single-scale approaches. In recent years, the ability to develop efficient coupling algorithms, better coarse-graining techniques, and enhanced computational capabilities have opened the door to exploring problems that were previously thought to be intractable in reactive polymer processing. With the emergence of machine learning methods to accelerate atomistic simulations or extract reaction parameters for polymers, a new area for multiscale modeling has opened up, in fact, there are likely many opportunities for multiscale modeling to capture some of the statistical convergence issues collected to rare-event chemistry (Unke et al., 2021). With continued expansion of computational ability as well as new methodologies, coupling RMD and SPH is an exciting direction to connect molecular chemistry to engineering scale polymer processing, with the potential of reshaping how we develop fundamental understanding and industry practices for design and manufacture of modern polymeric materials (Schütt et al., 2018).

Figure 1. illustrates the growth trends in multiscale polymer modeling publications over the past two decades, highlighting the accelerating adoption of coupled simulation approaches.

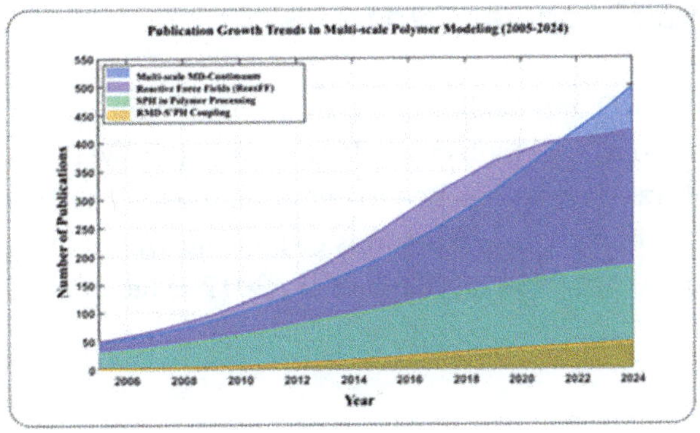

2. THEORY AND METHODOLOGY OF RMD AND SPH

The conceptual origins of Reactive Molecular Dynamics and Smoothed Particle Hydrodynamics come from two rather different scientific traditions. RMD was originally introduced with the development of ReaxFF by Van Duin, Goddard, and coworkers in 2001 to carry out simulations at the interface between quantum chemistry and classical molecular dynamics. ReaxFF developed a bond-order-dependent reactive potential that had the capability to reproduce bond breaking, bond formation, and charge redistribution in large condensed-phase systems. For the first time, it provided the enabling capability to simulate chemically reactive processes at the atomistic scale without resorting to fully quantum mechanical calculations. Since its inception, this ReaxFF framework has been extended to a wide range of chemistries and interfaces and represents today the standard methodology for investigating chemical reaction networks, material degradation, and advanced material fabrication, involving polymers and heterogeneous composites (van Duin et al., 2001; Senftle et al., 2016). Its capability to simulate thousands of atoms with an accuracy comparable to semiempirical quantum methods but at a fraction of the computational cost makes it the cornerstone of reactive molecular simulations.

In contrast, SPH was originally developed in astrophysics for the simulation of unsteady, non-axisymmetric, self-gravitating flows based on a mesh-free Lagrangian point-mass formulation. The method has evolved over the decades beyond its astrophysical origins to become a versatile computational tool for complex fluids, free-surface flows, multiphase systems, and recently non-Newtonian and polymeric materials. Its mesh-free nature and capability of dealing with large deformations make it very suitable for applications both in the physics of soft matters and polymer processing. Methodological advances within SPH include stabilization techniques, boundary treatments, and multiphysics coupling methods.

SPH has also been applied to fluid-structure interaction problems, viscoelastic flows, and free-surface dynamics (Lucy, 1977; Gingold & Monaghan, 1977; Monaghan, 1992; Ye et al., 2019; Zhang et al., 2022).

Taken together, these historical developments form the intellectual underpinning for contemporary RMD-SPH multiscale coupling, wherein chemically accurate atomistic simulations may be seamlessly coupled with large-scale continuum transport. The resultant hybrid paradigm now permits the study of complex material phenomena that involve many orders of magnitude in both space and time and link molecular-level reactivity with macroscopic mechanical and transport response.

2.1. Theory of Reactive Molecular Dynamics

As discussed earlier, the RMD approach is a computational simulation method used to analyze the time-dependent behavior of atoms and molecules. It is particularly well suited for investigating chemical reactions and structural transformations occurring at the nanoscale. The method is fundamentally based on the rigorous solution of Newton's equations of motion, where the force acting on each atom is determined as follows(Aktulga et al., 2012a; Farah et al., 2012; Mao et al., 2023):

$$F_i = -\nabla_i U(r(t), BO(t), q(t)) \tag{1}$$

$$a_i = \frac{F_i}{m_i} \tag{2}$$

where U, F_i, a_i, and m_i denote the potential energy of the system, the acting force, the acceleration, and the mass of atom i, respectively. In RMD, unlike CMD where the potential energy depends solely on the atomic coordinates within the system the potential energy is a dynamic function of the instantaneous atomic positions r(t), bond orders BO(t), and partial charges q(t). This dynamic dependence allows the system topology to evolve during the simulation, enabling the automatic formation and breaking of chemical bonds in a physically realistic manner (Mao et al., 2023).

In RMD simulations, the potential energy corresponding to interatomic or intermolecular interactions is calculated through reactive force fields, which are sets of mathematical relationships describing the total potential energy of the system as a function of the instantaneous atomic configuration. As atoms move, bonds may form or dissociate accordingly. Various reactive force fields have been developed for this purpose, including the Embedded Atom Model (EAM)(Daw & Baskes, 1984), Modified Embedded Atom Model (MEAM)(Baskes, 1987), Tersoff (Tersoff, 1989), Stillinger-Weber (SW)(Stillinger & Weber, 1985), Reactive Empirical Bond Order (REBO)(Tersoff, 1988), Adaptive Intermolecular REBO (AIREBO)(Violi, 2004), Charge Optimized Many-Body (COMB)(Yu et al., 2007), Electron Force Field (eFF)(Su & Goddard III, 2007), and ReaxFF(Van Duin et al., 2001). Table 1 summarizes these reactive force fields and their main applications. Among them, ReaxFF serves as a general-purpose framework capable of describing a wide range of chemical elements and reactions. Therefore, this section focuses on the theoretical formulation and parameterization of the ReaxFF method.

Table 1. Applications of various force fields in RMD(Mao et al., 2023)

Force-field	Abbreviation	Application
Embedded Atom Model	EAM	Metallic systems: bulk properties, grain boundaries, surfaces, alloys, and mechanical properties
Modified Embedded Atom Model	MEAM	Metallic systems with angular dependence; improved description of thermodynamic and structural properties
Charge Optimized Many-Body	COMB	Systems with variable charge transfer (e.g., Si/SiO_2, metals, oxides, and organic molecules); interface and oxidation studies
Tersoff Potential	–	Covalent systems such as Si, Si–C, and Si–Ge; elastic and thermodynamic properties of semiconductors
Stillinger–Weber Potential	SW	Melting and liquid behavior of Si and related systems (e.g., Si–F, Si–O); nanoparticle studies
Reactive Empirical Bond Order	REBO	Carbon-based systems (graphite, diamond, hydrocarbons);
Adaptive Intermolecular REBO	AIREBO	Liquids, polymers, and thin-film carbon systems
Electron Force Field	eFF	Systems involving electrons and excited states; studies of plasma etching, shock compression, and charged reactions
Reactive Force Field	ReaxFF	Universal reactive model for covalent, metallic, and ionic systems across the periodic table; simulates bond formation and breaking

2.1.1. ReaxFF

The total energy of the system in ReaxFF is defined as the sum of various physical and chemical terms, each representing a type of interaction or energy correction within the system: bond energy, over coordination penalty, undercoordination reward, lone-pair effect, valence angle energy, dihedral torsion energy, and nonbonded interactions including van der Waals and Coulombic terms, along with specific terms for special conditions. The simultaneous dependence of these terms on bond order and atomic charge allows ReaxFF to naturally reconstruct the reaction pathway from reactants to products, through the formation and disappearance of new bonds and charge transfers. The total energy relation of this system is expressed as follows(Aktulga et al., 2012b; Mao et al., 2023; Senftle, Hong, Islam, Kylasa, Zheng, Shin, Junkermeier, Engel-Herbert, Janik, & Aktulga, 2016a; Van Duin et al., 2001):

$$U = U_{bond} + U_{lp} + U_{over} + U_{under} + U_{angle} + U_{torsion} + U_{vdw} + U_{coulombic} + U_{specific} \quad (3)$$

The bond energy term, as the most important part of the equation, is expressed based on a modified Morse potential that depends on the bond order (Bo). In this formulation, the distance between two atoms and the type of bond (σ, π, or double-π) determine the equilibrium parameters, while the bond order decays exponentially with empirical coefficients. This formulation ensures that as the bond is stretched to larger distances, its energy continuously approaches zero, allowing smooth bond dissociation without any energy discontinuity (Equations 4 and 5 in Table 2)(Aktulga et al., 2012b; Mao et al., 2023).

The lone pair energy term in reactive force field models such as ReaxFF represents the effects of unpaired or nonbonding electron pairs on atomic behavior. Unlike CMD, which typically does not explicitly include such a term, this component plays a crucial role in reactive simulations since the presence of lone pairs can influence molecular geometry, polarization, and even the reactivity of active sites. In ReaxFF, the calculation of lone pair energy is generally based on the local geometry and the bond order of surrounding atoms. The algorithm compares the expected number of lone pairs for each atom (according to its bonding capacity and electronic state) with its actual configuration, applying an energy correction when excess or deficiency of electron pairs occurs. This approach ensures that even during complex reactions, in which bonds break and reform, the effects of nonbonding electrons are consistently and continuously accounted for (Equation 6 in Table 2)(Aktulga et al., 2012b; Mao et al., 2023).

The over coordination penalty term is introduced to prevent the formation of more bonds than the stable capacity of an atom. This term adds a significant positive energy contribution to the system whenever the total bond order of an atom exceeds its equilibrium value, thereby driving structures back toward physically valid states. The opposite effect is represented by the undercoordination term, which adds negative energy to provide extra stability and maintain the reactivity of undercoordinated centers (Equations 7-10 in Table 2)(Aktulga et al., 2012b).

The valence angle energy term in the ReaxFF reactive force field plays an essential role in preserving molecular geometry and resisting structural deformation due to changes in bond angles. This energy term specifically arises when three atoms, i-j-k, are connected by bonds i-j and j-k, and the angle $\theta_{(\ )}$ deviates from its equilibrium value. In ReaxFF, this term is formulated based on a modified Morse potential to ensure that angular variations are treated continuously without abrupt discontinuities during bond breaking or formation (Equations 11-14 in Table 2)(Aktulga et al., 2012b, 2012a; Mao et al., 2023).

The torsion energy term in the ReaxFF model represents the resistance of bonds against the relative rotation of two parts of a molecule around an axis bond. This term typically involves four atoms i-j-k-l, where the torsion angle $\varphi_{(\ )}$ is defined between the i-j-k and j-k-l planes. In ReaxFF, torsional energy is defined considering its simultaneous dependence on the dynamic bond order (BO), adjacent valence angles ($\theta_{(\ )}$ and $\theta_{(\ )}$), and the over- or undercoordination state of the central atoms (Equation 15 in Table 2) (Aktulga et al., 2012b, 2012a; Mao et al., 2023).

In the ReaxFF model, the van der Waals energy term is not expressed using the simple classical form such as the Lennard–Jones potential, but rather through a modified Morse-type potential as a function of interatomic distance. This formulation enables continuous representation of reactive behavior and structural changes during bond formation and dissociation. To moderate the extremely strong repulsion at short distances, a shielding function (f_5) is introduced to adjust the van der Waals effect and prevent the emergence of nonphysical forces. In addition, a Taper function ensures that the van der Waals energy smoothly approaches zero when the distance between two atoms exceeds the cutoff radius for nonbonded interactions ($r_{(}non_{)}$) (Equations 16-18 in Table 2)(Aktulga et al., 2012a; Mao et al., 2023).

In the ReaxFF model, the Coulombic energy term, similar to the van der Waals interactions, is computed for all atomic pairs both bonded and nonbonded. This treatment ensures that no numerical discontinuities or abrupt changes appear in the potential energy surface during bond formation or breaking. A fundamental feature of the Coulombic formulation in ReaxFF is its independence from bond order; in other words, regardless of the bonding state between two atoms, the interaction is evaluated solely based on their partial charges and interatomic distance (Equation 19 in Table 2)(Mao et al., 2023).

Table 2. Equations of potential energy components in the ReaxFF method(Aktulga et al., 2012b; Mao et al., 2023)

Term Name	Formula	Parameters Definition
Bond Energy (Eq. 4,5)	$U_{bond} = D_e^{\sigma} BO_{ij}^{\sigma} \exp\left[p_{be1}\left(1 - BO_{ij}^{\sigma\, p_{be2}}\right)\right] - D_e^{\pi} BO_{ij}^{\pi} - D_e^{\pi\pi} BO_{ij}^{\pi\pi}$ $BO_{ij} = \exp\left[p_{bo1} \cdot \left(\frac{r_{ij}}{r_o^{\sigma}}\right)^{p_{bo2}}\right] + \exp\left[p_{bo3} \cdot \left(\frac{r_{ij}}{r_o^{\pi}}\right)^{p_{bo4}}\right] + \exp\left[p_{bo5} \cdot \left(\frac{r_{ij}}{r_o^{\pi\pi}}\right)^{p_{bo6}}\right]$	$r_o^{\sigma}, r_o^{\pi}, r_o^{\pi\pi}$ representequilibrium bond lengths. p_{bo1} to p_{bo6} terms are parameters to model the bonds. $D_e^{\sigma}, D_e^{\pi}, D_e^{\pi\pi}$ are the well depth parameters and p_{be1} is the bond parameter for σ bond.
Lone pair Energy (Eq. 6)	$U_{lp} = \frac{p_{lp2} . \Delta_i^{lp}}{1 + \exp\{-75 . \Delta_i^{lp}\}}$	$\Delta_i^{lp} = n_{opt}^{lp} - n_i^{lp}$ corresponds to the number of unpaired electrons.
Over & Under-coordination Energy (Eqs. 7-10)	$U_{over} = \Delta_i^{lpcorr} . \frac{\sum_{j \in nbrs(i)} p_{ovun1} . D_e^{\sigma} . Bo_{ij}}{(\Delta_i^{lpcorr} + Val_i)(1 + \exp\{p_{ovun2} . \Delta_i^{lpcorr}\})}$ $U_{under} = -p_{ovun5} . f_6(i, p_{ovun7}, p_{ovun8}) . \frac{1 - \exp\{p_{ovun6} . \Delta_i^{lpcorr}\}}{1 + \exp\{p_{ovun2} . \Delta_i^{lpcorr}\}}$ $\Delta_i^{lpcorr} = \Delta_i - \Delta_i^{lp} . f_6(i, p_{ovun3}, p_{ovun4})$ $f_6(i, p_1, p_2) = \left(1 + p_1 . \exp\left\{p_2 . \left[\sum_{j \in nbrs(i)} (\Delta_j - \Delta_j^{lp}) . (BO_{ij}^{\pi} + BO_{ij}^{\pi\pi})\right]\right\}\right)^{-1}$	
Valence Angle Energy (Eqs. 11-14)	$U_{angle} = f_1(Bo_{ij}) f1 (Bo_{jk}) f_2(\Delta_j)\{p_{angle1} - p_{angle1} \exp[-p_{angle2} (\theta_o - \theta_{ijk})^2]$ $\Delta_j = -Val_j + \sum_{j=1}^{neighbours(i)} \sum BO_{ij}$ $f_1(Bo_{ij}) = 1 - \exp(-p_{angle3} Bo_{ij}^{angle4})$ $f_2(\Delta_j) = p_{angle5} - (p_{angle5} - 1) \frac{2 + \exp(p_{angle6} \Delta_j)}{1 + \exp(\exp(p_{angle6} \Delta_j) + \exp(-p_{angle7} \Delta_j))}$	θ_o is the equilibrium angle for θ_{ijk}. Δ_j is The overcoordination for the difference between the total bond order around the atom j and the number of its bonding electrons Valj. p_{angle1} to p_{angle7} are fitting parameters to the angle
Torsion Angle Energy (Eq. 15)	$U_{torsion} = f_3(BO_{ij}, BO_{jk}, BO_{kl}) \sin\theta_{ijk} \sin\theta_{jkl} \left[\frac{1}{2} V1 (1 + \cos\varphi_{ijkl}) + \frac{1}{2} V2 \exp\{ p_{tor1} (BO_{jk}^{\pi} - 1 + f11(\Delta_j \Delta_k))^2 \} . (1 - \cos 2\varphi_{ijkl}) + \frac{1}{2} V3 (1 + \cos 3\phi_{ijkl})\right.$	f_3 and f_4 are similar to f1 and f_2. V1, V2 and V3 are torsional rotational barriers
van der Waal's Interaction (Eqs. 16-18)	$U_{vdw} = Tap\, D_{ij} \left\{ \exp\left[\alpha_{ij} \left(\frac{1 - f_{5(r_{ij})}}{r_{vdw}}\right)\right] - 2\exp\left[\frac{1}{2} \alpha_{ij} \left(\frac{1 - f_{5(r_{ij})}}{r_{vdw}}\right)\right] \right\}$ $Tap(r_{ij}) = Tap_7 . r_{ij}^7 + Tap_6 . r_{ij}^6 + Tap_5 . r_{ij}^5 + Tap_4 . r_{ij}^4 + Tap_3 . r_{ij}^3 + Tap_2 . r_{ij}^2 + Tap_1 . r_{ij}^1 + Tap_0$ $f_5(r_{ij}) = \left[r_{ij}^{pvdw1} + \gamma_{ij}^{-pvdw1}\right]^{\frac{1}{pvdw1}}$	Tap is the Taper correction, Dij is the well path of atoms i, j. f_5 is shielded interaction, γij and p_{vdW1} are the van der Waals radius and exponent parameter.
Coulomb Interaction (Eq. 19)	$U_{coulomb} = Tap(r_{ij}) . C . \frac{q_i q_j}{\left[r_{ij}^3 + \left(\frac{1}{\gamma_{ij}}\right)^3\right]^{\frac{1}{3}}}$	C is the Coulomb constant, $q_i$and $q_j$are the partial charges of atoms iand j, and $r_{ij}$is the distance between the atoms. Tap is the Taper correction, γ_{ij} is the van der Waals radius

The partial charge of each atom (q) is one of the key dynamic variables updated at every time step to reflect changes in the chemical environment and atomic polarization. These charges are determined based on electrostatic equilibrium and chemical potential equalization principles. Two main approaches are commonly used for this calculation(Mao et al., 2023; Mortier et al., 1986; Rappe & Goddard III, 1991):

i. Charge Equilibration (QEq) Method
ii. Electronegativity Equalization Method (EEM)

The QEq method proceeds by minimizing the total electrostatic energy of the system, which consists of the sum of the self-energy contribution of each atom and the Coulombic interactions between all atomic pairs. For atom i, the energy is expanded as a second-order function of its partial charge q around the reference state (q = 0):

$$E_i(q_i) = E_i^0 + q_i\chi_i^0 + \frac{1}{2}J_{ii}^0 q_i^2 \tag{20}$$

where χ^0 is the intrinsic electronegativity, and J^0 as the self-Coulomb term is derived from the ionization potential (IP) and the electron affinity (EA) as follows:

$$\chi_i^0 = \frac{IP_i + EA_i}{2} \tag{21}$$

$$J_{ii}^0 = IP_i - EA_i \tag{22}$$

The combination of these terms gives the general form of the QEq energy as follows:

$$E_{QEq}(q_1,\ldots,q_N) = \sum_{i=1}^{N}\left[E_i^0 + \chi_i^0\, q_i + \frac{1}{2}J_{ii}^0\, q_i^2\right] + \sum_{i<j} J_{ij}\, q_i q_j \tag{23}$$

The minimization of this energy must be carried out such that the sum of the partial charges of all atoms exactly equals the net charge of the system:

$$\sum_{i=1}^{N} q_i = Q_{net} \tag{24}$$

This condition ensures that the resulting charge distribution is not only stable and consistent with the instantaneous structure of the system but also that the total charge of the simulation remains conserved without any deviation from its real physical value. The solution of this system is performed at each time step based on the instantaneous geometry, allowing the charges to remain synchronized with bonding changes and variations in the electrostatic environment(Rappe & Goddard III, 1991).

2.1.2. Parameterization and Optimization of the ReaxFF

As discussed earlier, in ReaxFF each energy term is computed using a specific equation, and these equations contain parameters that must be adjusted (optimized) in order to reproduce the correct behavior for the chemical system of interest. Determining accurate parameter values is fundamentally important because these parameters control the overall shape of the potential energy surface and therefore govern the reactive behavior of the system. A typical ReaxFF force field defines on the order of one hundred parameters for each element; some of these parameters represent well-defined physical properties (such as van der Waals parameters, atomic radii, or electronegativity-related quantities), but many are empirical in nature and can only be obtained by fitting to reference data. Consequently, the reliability of ReaxFF depends entirely on the quality of the parameterization procedure and the data on which it is based. Since intra- and interatomic interactions are directly controlled by these parameters, they must be optimized such that the force field reproduces the reference values with acceptable accuracy. These reference data form a collection known as the FF training set(Senftle, Hong, Islam, Kylasa, Zheng, Shin, Junkermeier, Engel-Herbert, Janik, & Aktulga, 2016b; Sengul et al., 2021).

The data contained in the training set typically cover a broad range of atomic and molecular properties, including bond lengths and bond energies, valence and torsional angles, charge distributions, bond-stretching curves, formation energies of species, and critical points along reaction pathways. Beyond simple bond-breaking and bond-forming reactions, more complex processes (such as vacancy dynamics and ion diffusion) are included to ensure that the force field remains reliable under different conditions and environments. The majority of the data in this training set are obtained from high-level quantum mechanical (QM) calculations or experimental measurements, and ReaxFF is parameterized to reproduce these reference values. However, due to its inherent structural characteristics, ReaxFF faces a fundamental challenge known as "transferability." Unlike classical force fields, ReaxFF does not employ atom types, meaning that each element must be described by a single set of parameters across all chemical environments, phases, oxidation states, and molecular contexts. Consequently, if the training set covers only a limited subset of the relevant chemical space (for example, a few simple gas-phase reactions or a small collection of molecular clusters) the resulting parameter set will lose accuracy when applied to different conditions, such as phase transitions, changes in chemical domain, polar environments, or species with different oxidation states. In other words, a parameter set that performs well in one chemical domain may "break down" when transferred to another domain or more complex conditions, simply because the initial training data did not sufficiently capture the full range of chemical behavior. This limitation is intrinsic and independent of the specific optimization algorithm used, highlighting the necessity of global optimization strategies and more comprehensive training sets(Senftle, Hong, Islam, Kylasa, Zheng, Shin, Junkermeier, Engel-Herbert, Janik, & Aktulga, 2016b)

Despite variations in algorithmic strategies and search schemes, all ReaxFF parameterization methods share a common principle: they minimize an error function. This error function (Eq.26) is usually defined as the weighted sum of squared differences between the property calculated by the force field and the corresponding reference value. In other words, the quality of a parameter set is assessed through its ability to reduce this error function, and the goal of all parameterization approaches is to find a set of parameters that minimizes the deviation between ReaxFF predictions and the reference data.(Iype et al., 2013)

$$E_{rr} = \sum_{i}^{N}\left(\frac{Vi^{ReaxFF} - Vi^{ref}}{\sigma_i}\right)^2 \tag{25}$$

In this expression, Vi^{ref} is the reference value (from Qm or experiment), Vi^{ReaxFF} is the model prediction, and σi is a weighting factor that specifies the relative importance of each data type. Typical values are $\sigma \approx 0.05$ Å for bond lengths, 3° for valence angles, 0.01 for atomic charges, and 0.1–1.0 kcal/mol for energies.

Initially, ReaxFF parameters were optimized exclusively using the Single-Parameter Parabolic Search method. This approach assumes that the variation of the error with respect to a given parameter can be approximated by a parabola. To achieve this, the error is evaluated at three distinct values of the parameter, and a parabolic curve is fitted to identify the minimum. However, this method faces several fundamental issues: (i) strong coupling among parameters causes the optimal value of one parameter to shift when others are modified; thus, the method often requires numerous time-consuming cycles to reach a reasonable point. (ii) The assumption of parabolic behavior is generally invalid because the ReaxFF search space is highly complex, rugged, and filled with local minima. As a result, SOPPE is inherently a local method and relies heavily on user expertise and the ability to simplify the search region near a good initial guess, which makes it difficult to use and inefficient for large-scale parameterizations. For these reasons, the single-parameter approach becomes ineffective(Hubin et al., 2016; Iype et al., 2013).

To overcome these limitations, Iype and co-workers introduced a global optimization method known as Monte Carlo Simulated Annealing (MCSA). In this approach, all parameters are optimized simultaneously. At each iteration, a subset of parameters is randomly selected and perturbed slightly; the error is then recomputed over the entire training set. Acceptance or rejection of this move is determined by a Metropolis-type probability expression(Iype et al., 2013):

$$P_{acc} = \min\left[1, \exp\left(-\frac{\Delta E_{rr}}{k_B T}\right)\right] \tag{26}$$

where ΔE_{rr} is the change in error between the new and the last accepted state, k_B is the Boltzmann constant, and T is the temperature. This algorithm is currently implemented in the ReaxFF module of the commercial SCM software package.

Alongside MCSA, another major global optimization approach for ReaxFF parameterization is the Genetic Algorithm (GA). Instead of a single search trajectory, GA operates on a population of candidate parameter sets, each representing a complete ReaxFF parameter vector. The quality of each individual is evaluated using the same error function described earlier. Population evolution proceeds through three main operators: (1) selection, in which individuals with lower error values have a higher probability of contributing to the next generation; (2) crossover, in which two parent vectors exchange segments of their parameters to produce offspring with mixed characteristics; and (3) mutation, whereby a small fraction of parameters is randomly modified to maintain population diversity and prevent trapping in local minima. This population-based nature of GA enables simultaneous exploration of multiple regions of parameter space and allows the algorithm to better handle the strong inter-parameter correlations inherent in ReaxFF(Larsson & Hartke, 2013; Senftle, Hong, Islam, Kylasa, Zheng, Shin, Junkermeier, Engel-Herbert, Janik, & Aktulga, 2016b).

An important extension of this approach is the multi-objective GA, first introduced by Jaramillo-Botero and co-workers. Instead of minimizing a single scalar error, several error components (such as errors in energies, geometries, and partial charges) are optimized simultaneously in a Pareto-based framework. This enables a balanced representation of different property types and prevents overfitting toward any single class of data. Moreover, Weingarten and colleagues applied multi-objective evolutionary strategies to ReaxFF parameterization for explosive materials and reported that this approach achieves better compromise solutions across large and heterogeneous training sets(Larsson & Hartke, 2013; Sengul et al., 2021).

Despite its advantages, GA still faces several technical challenges: (1) premature convergence in early generations, which reduces population diversity and restricts exploration; (2) strong sensitivity to the choice of mutation and crossover rates, which can significantly affect search quality; and (3) high computational cost, since evaluating each individual requires a full ReaxFF calculation over the entire training set. This cost escalates rapidly for large training sets containing reaction pathways, energy curves, charge distributions, and extensive geometrical data(Larsson & Hartke, 2013).

To address these challenges, recent approaches have combined GA with machine learning models. In these hybrid strategies, a surrogate model (typically a neural network or a nonparametric regression model) is trained on data generated by ReaxFF and then used during the GA search as a low-cost estimator of the error. Only the most promising candidates (elite individuals) are evaluated using the true ReaxFF model. This significantly reduces computational cost while maintaining the quality of the search(Larsson & Hartke, 2013).

Overall, both the classical single-objective GA and its multi-objective variants provide powerful global search capabilities, benefiting from recombination mechanisms and natural parallelism. Machine-learning-based extensions further enable efficient use of large training sets and more accurate parameterization, paving the way toward modern, transferable, and broadly applicable reactive force fields.

2.2. Theory of Smoothed Particle Hydrodynamics

As mentioned in the introduction, the SPH method is a mesh-free numerical approach formulated in the Lagrangian framework, developed for simulating physical phenomena involving large deformations, moving boundaries, and multiphase flows(Lind et al., 2020; M. B. Liu & Liu, 2010; Rajapriyadharshini, 2022). In this approach, the continuous medium is represented by a set of discrete particles, each carrying certain physical quantities such as mass, density, pressure, and energy. These particles act as sampling points of the continuous field, and different physical fields are reconstructed based on their corresponding particle values(M. B. Liu & Liu, 2010).

The mesh-free nature of SPH differentiates it from classical grid-based methods such as the FEM or the (FVM), as it does not require the definition of inter-nodal connectivity or a structured computational mesh. This feature makes SPH particularly well suited for analyzing problems involving moving boundaries, material fracture, and topological changes(Bagheri et al., 2024; Rajapriyadharshini, 2022). The SPH formulation involves two fundamental steps to transform the governing (PDEs) into a mesh-free numerical form. These are known as the kernel approximation and the particle approximation (Rajapriyadharshini, 2022).

In the first step, each physical variable f(or its derivatives) defined over a continuous domain Ω at a point xis expressed in an integral form. Along this process, the value of the function f(x) at a point is approximated by a weighted average of the same function values within its neighboring region, using a smoothing kernel function $W(x - x', h)$, where his the smoothing length (Lind et al., 2020; Rajapriyadharshini, 2022; C. Zhang et al., 2022). Comprehensive studies on the properties and various forms of the smoothing (kernel) functions have been reported by Liu et al(M. B. Liu & Liu, 2010).

$$\int_{\Omega} f(x) \approx f(x')W(x - x', h)dx' \tag{27}$$

In this formulation, the parameter h denotes the smoothing length, which determines the size of the support domain of the kernel function, while Wrepresents the smoothing function, also known as the kernel function. The smoothing function is generally chosen to be a continuous, isotropic, and symmetric function with compact support, ensuring that only particles located within a finite neighborhood are included in the weighted averaging process. As a result of this approximation, all differential operators required in the governing equations such as the gradient, divergence, and Laplacian can be expressed without the use of a fixed mesh, relying solely on the smoothing of the surrounding region of each particle. This principle provides the mathematical foundation for transforming the continuous form of the equations into a discrete form within the SPH framework(Lind et al., 2020; C. Zhang et al., 2022). In the second step, the continuous integral representation of the computational domain is numerically discretized. This process is performed by replacing the integral with a summation over a set of neighboring particles. In this manner, the domain is described by a collection of particles, each of which represents a volume equal to the ratio of its mass to its density ($V_j = m_j/\rho_j$)[18].

$$\langle f(x)\rangle = \sum_{j}^{N} f\left(x_j\right) W\left(x - x_j, h\right) V_j \tag{29}$$

$$f(x_i) = \sum_{j}^{N} \frac{m_j}{\rho_j} f_j W_{ij} \tag{29}$$

In these expressions, the subscripts i and j denote the target particle and its neighboring particles, respectively. N represents the number of particles within the support domain of the kernel function, and W_{ij} refers to the smoothing kernel between the target particle i and its neighbor j. By following the same procedure, the particle approximation of the derivative can be expressed as follows(M. B. Liu & Liu, 2010)

$$< \nabla . f\left(x_j\right) > = -\sum_{j=1}^{N} \frac{m_j}{\rho_j} f\left(x_j\right) . \nabla W\left(x - x_j, h\right) \tag{30}$$

As a result of the two steps described earlier, the PDEs are reformulated in a particle-based, mesh independent form and thus transformed into a set of Ordinary Differential Equations (ODEs) expressed in terms of the positions and properties of the particles. By numerically integrating these equations, the spatial and temporal trajectories of the particles and consequently, the variations of physical quantities such as density, pressure, and velocity throughout the computational domain can be determined (Bagheri et al., 2024; C. Zhang et al., 2022).

2.2.1. Boundary Conditions

In the SPH method, imposing boundary conditions is considered one of the most challenging aspects of the numerical formulation. This difficulty originates from the fact that the original SPH scheme was developed for astrophysical simulations, where distinct physical boundaries are absent and particles are freely distributed in open space. In engineering applications particularly in fluid-flow analyses near solid walls a portion of the kernel's support domain is truncated near the boundary, leading to loss of accuracy and numerical inconsistency.

A widely used solution is the Dynamic Boundary Particle (DBP) approach. In this method, the boundary surface is covered with two or more layers of virtual particles to ensure complete kernel support for fluid particles. These boundary particles participate in the SPH equations similarly to fluid particles; however, their positions are fixed, and their velocities are set equal to that of the solid surface (or zero for a static wall). This method is straightforward and stable but cannot reproduce normal pressure gradients and shear stresses on the wall with sufficient accuracy. Nevertheless, it yields acceptable results in systems dominated by inertial forces. Subsequently, Adami and his co-workers introduced a modification to this method, in which physical quantities such as pressure in the boundary particles are adjusted so as to reflect the influence of the surrounding fluid flow. This adjustment led to a notable improvement in both the stability and accuracy of the method. Another approach, known as the Mirror Particle Method, is inspired by vortex methods. In this technique, for each particle located near a wall, a corresponding mirror particle is placed outside the domain at a position symmetric with respect to the boundary surface. For no slip conditions, the velocity components of the two particles are equal in magnitude and opposite in direction, and the normal pressure gradient is set to zero; for slip conditions, only the tangential velocity components are taken to be equal. Although mathematically accurate, this method is computationally expensive and difficult to implement for complex geometries particularly in three dimensions and is therefore generally limited to simple configurations.

In recent years, semi-analytical correction methods have been developed to address the loss of kernel support near boundaries. These methods compute precise correction coefficients associated with the truncated support and achieve very high accuracy; however, due to their strong dependence on the kernel type and the complexity of the analytical formulations, they are typically restricted to two-dimensional problems. Even so, three-dimensional and generalized variants for both compressible and incompressible flows have also been reported (Lind et al., 2020). Other boundary treatment techniques and their refinements are discussed in detail in (Lind et al., 2020).

2.3. Multiscale Coupling Approaches for Atomistic-to-Continuum Modeling

In many materials problems, the physical behavior cannot be represented within a single scale, because phenomena such as fracture, plasticity, reactivity, and structure-dependent deformation simultaneously depend on atomistic mechanisms and on stress and deformation fields at much larger scales. For this reason, several multiscale coupling strategies have been developed to connect different length and time scales in a manner that preserves atomistic fidelity in critical regions while remaining computationally tractable for the whole system. One such approach is Coarse-Grained Molecular Dynamics (CGMD), in which groups of atoms are represented as larger effective particles and the interactions between these coarse units are defined such that the thermodynamic and mechanical response of the underlying material is retained as closely as possible. CGMD enables the simulation of very large systems, including wave propagation, large deformations, and mechanical response of nanoscale structures over long time spans, while critical regions can still be described by full MD. It is therefore a suitable multiscale method that enhances efficiency without sacrificing essential atomistic physics(Rudd & Broughton, 1998).

Another widely used strategy for combining high-fidelity physics with computational efficiency is the Quantum Mechanics/Molecular Mechanics (QM/MM) hybrid framework. In this approach, the region requiring electronic-level accuracy (such as reactive sites, breaking bonds, active defects, or the highly stressed region near a crack tip) is treated using quantum mechanical methods like DFT, while the surrounding environment, which constitutes the majority of the material volume, is modeled using classical MD or non-reactive force fields. This method is especially important in reactive or mechanochemical systems, because it allows for accurate determination of activation energies, charge transfer, and bond-breaking processes without incurring the prohibitive computational cost of fully quantum simulations. The key advantage of QM/MM is that quantum accuracy is applied only where necessary, enabling simulations of much larger systems (Vreven & Morokuma, 2006).

For modeling crystalline plasticity and dislocation behavior, an advanced and well-established method known as Coupled Atomistic-Discrete Dislocation Dynamics (CADD) has been developed. In CADD, the near-core region of dislocations (including dislocation nucleation, pinning interactions, jog formation, and other short-range processes) is resolved using full molecular dynamics (MD), while collective motion of dislocations, long-range stress interactions, network formation, and strain hardening are described by discrete dislocation dynamics (DDD). The coupling between these two domains is achieved through an interface region in which displacements and stresses are exchanged bidirectionally to ensure full mechanical consistency. This method is particularly suited for studying the onset of plasticity at crack tips, size-dependent deformation in micropillars, dislocation nucleation under high stress, and the fundamental mechanisms of strain hardening. Since MD alone cannot handle large volumes and DDD cannot capture core-level physics, CADD provides a rigorous and physically consistent bridge between these two scales and is now recognized as a standard multiscale tool in plasticity research(Shilkrot et al., 2002).

3. COUPLING STRATEGIES AND COMPUTATIONAL IMPLEMENTATION

Using both chemical precision of RMD and configurability and low-complexity (geometrical flexibility) of SPH, the capacity for Multiscale Simulation of Polymers has emerged. RMD provides an atomic-level description of Bond Breaking, Bond Forming and complex reaction pathways due to the ability to utilize Reactive Force Fields (like ReaxFF). However, the time and size restrictions on this technique limit the size and scope of what can realistically be studied (Van Duin et al., 2001; Senftle et al., 2016). On the other hand, as SPH Models treat the materials they are modeling more as continuums and utilize the unique geometric flexibility and large domain capabilities of this modelling Approach to model very complex free-surface flow behaviour, large deformation of bodies and multi-phase flow behaviour while inherently preserving mass, momentum and energy (Violeau & Rogers, 2016). The unique benefit of Using the SPH modelling Approach is that because it does not consider the explicit molecular nature of the materials; it does not currently account for capturing the molecular events of Chemical Reaction processes or molecular level structural changes.

The coupling strategies between RMD and SPH utilize the complementary features of each method to create a cohesive system wherein RMD handles the numerous reactive events and material changes taking place at critical locations while SPH models the materials' hydrodynamics, thermal energy transfer and overall deformations, etc. Depending on what type of modeling problems are being studied, the coupling of these two methodologies can take on varied forms (i.e., loose coupling, wherein relative information exchange occurs sporadically, to tightly-coupled schemes, wherein specified regions are repeatedly examined together and where the exchange of all relevant information occurs continually and reciprocally). It is essential that the transfer of stresses, heat fluxes, reaction rates and densities between the two methodologies preserves the integrity of global conservation laws, as well as providing numerical stability throughout the entire simulation. The goal of this chapter is to discuss the above-noted coupling methodologies, demonstrate how they may be used within reactive polymer systems, and ultimately identify their respective benefits/limitations, together with providing useful criteria for selecting a particular coupling methodology.

3.1 Conceptual Classification of RMD–SPH Coupling Strategies

There are three ways in which RMD-SPH coupling strategies can be categorized based on how the atomistic and continuum scales interact conceptually (Broughton et al., n.d.; Ghoniem et al., 2003). Static decomposition methods provide an example of the first of these three types. Static decomposition methods use a fixed decomposition of the computational domain into three subregions: an atomistic region, a continuum region and typically a fixed handshake region. Each subregion is performed using a different simulation method (RMD or SPH) and the assignment of each subregion remains unchanged during the simulation. The static decomposition approach is particularly suited for applications where it is known in advance that the region of the domain requiring atomistic resolution exists.

Adaptive Resolution Methods create regions where local resolution and sometimes even the type of active computational model can vary over time and space. Local criteria such as density gradients, stress/strain fields, strain rate, temperature and reaction progress can change the shape and size of these different resolution regions. The Adaptive Resolution Scheme (AdResS) has been developed to allow a particle to move between a fully atomistic representation and a Coarse-Grained representation while maintaining thermodynamic consistency. When combined with the SPH/Continuum method, adaptive resolution methods allow for the accurate tracking of moving cracks, evolving reaction fronts, or changing polymer damage zones (Praprotnik et al., 2005; Praprotnik et al., 2008).

Concurrent bidirectional methods belong to the third category of multi-scale approaches since both RMD and SPH are employed in overlap areas to provide two-way communication. With these models, additional coupling terms, weights (or blended energy functions) are used so that both atomistic and continuum components exist in the overlap region, thereby providing one set of representation for the material state. The Bridging Scale Method (Wagner & Liu, 2003), Arlequin-type models (Dhia & Rateau, 2005a; Prudhomme et al., 2008), hierarchical multiscale models (Peter & Kremer, 2009b) and concurrent multiple-scale methods (O & Thompson, 1995; Borg et al., 2015) are each example of concurrent bidirectional methods. Generally speaking, they produce more accurate results in coupling regions because they are able to combine storage amounts of information found within two different models; however, they require more computation time than other approaches (Belytschko & Xiao, 2003).

In the following sections, these three categories are discussed in more detail, with emphasis on their application to reactive polymer systems and on how they may be combined with RMD and SPH.

3.2 Static Decomposition Methods

Historically, Static decomposition mechanisms are usually the earliest developed hybrid atomistic-continuum models and continue to be extensively employed due to their simplicity of conceptualization, along with their relatively low commitments of technical effort associated with the implementations (Weinan & Engquist, 2003; Rudd & Broughton, 1998). Static decompositions describe a partitioned domain, consisting of an atomistically resolved region (RMD) and a continuum resolved region which incorporates the capabilities of SPH to model the outflow of energy in conjunction with subsequent thermal or mechanical loading and deformations of the material. Both of these regions connect at an interface, forming a bridge between the two areas and assuring the proper global energy conservation is maintained at both regions (Hadjiconstantinou, 1999).

The many advantages associated with static decomposition are mainly a result of having fixed boundaries for each of the regions being processed, which allows for a pre-planned approach to load balancing and parallelisation of processing, as well as re-use with minimal alteration of pre-existing RMD and SPH codes. The RMD region may be kept to a relatively small area where molecular resolution is essential. For example, it could be used in the modelling of a reactive interface, a catalyst bed or a small area of damage. In each case, all other parts of the system are accurately captured using the SPH method. A major disadvantage is that if the region of interest moves or fluctuates, the static decomposition method is restricted to the original defined RMD region. Therefore, although it is suitable for many cases, the static decomposition approach is more useful for solving problems with stationary critical regions (Fogarty et al., 2010; Gartner & Jayaraman, 2019).

The use of domain decomposition is especially relevant for reactive polymers that contain a smaller region of reactivity surrounded by larger volumes of material that are mostly fluid (non-solid) or have viscosity and elasticity (i.e., viscoelastic). Fixed catalyst regions, stationary hot spots, and immobile reactive interfaces serve as some examples of reactive polymers in which domain decomposition methods may be utilized (Baeurle et al., 2006; Carvalho et al., 2022). Conversely, adaptive or concurrent approaches may be preferable when a particular reactive zone has the potential to migrate considerable distances over time.

3.3 Adaptive Methods

Adaptive techniques enhance the multiscale coupling capability by enabling dynamic changes in both the resolution and governing models to suit the continually changing states of the system. Rather than providing fixed partitions between regions, these adaptive methods create regions of interest that can expand or contract as needed, and move around during operation. This is critical in polymer systems because critical events (such as crack initiation or extension; localized shear; or chemical reaction initiation) can occur without prior knowledge of the location of those events (Praprotnik et al., 2008; Brini et al., 2013).

3.3.1 Adaptive Resolution Scheme (AdResS)

The Adaptive Resolution Scheme (AdResS), developed by Praprotnik and others (Praprotnik et al., 2005), is one of the oldest and widest in application among adaptive multiscale methodologies. At its core, this technique allows for the dynamic adjustment of the number of degrees of freedom in a simulation based on where in space the particles are and what has occurred prior to being at that location. In the AdResS framework, the domain is divided into three regions, one of which is entirely atomistic, the second of which is coarse grain (CG) or continuum-like, and an intermediate region where the particles transition between atomistic and CG representations (Praprotnik et al., 2008).

In the atomistic region of the domain, all interactions between atoms are described in full detail through the ReaxFF or other similar reactive force fields when simulating reactive polymers (Van Duin et al., 2001; Senftle et al., 2016). Meanwhile, the coarse-grained region replaces groups of atoms with effective beads that act through CG and continuum-like potentials, thus greatly reducing computational expense (Potestio et al., 2013; Deng & Shi, 2016). Finally, the transition region has a smooth weighting function that is continuous (in a thermodynamically consistent manner) and seamlessly transitions between the atom and CG representation, thereby reducing unphysical artifacts (Poblete et al., 2010; Petsev et al., 2017).

To extend AdResS to the coupling of RMD with SPH is conceptually simple; the outer region of the discretization of the continuum is represented as an SPH particle, which represents the continuum (the bulk) readout of the polymer, while the transition region acts as a buffer zone where the atomistic degrees of freedom are gradually transferred to the continuum representation (Delgado-Buscalioni et al., 2008; Delle Site et al., 2020). The main problem is how to establish a robust criterion that can be used to trigger the transition from one resolution to another, at the points where it is required, and to only those points. The existing criteria that can trigger a change in resolution have been suggested include the presence of sharp density gradients, which indicates the presence of either an interface or crack tip (Agarwal et al., 2014); large stress or strain gradients, which occur during bond breaking or structure changes (Fritsch et al., 2012); large strain rates, which indicate a nonlinear rheological response (Mukherji et al., 2018); and the presence of localised high temperatures or energy, which signals a potential reaction hot spot or a phase change (Junghans et al., 2007). The selection of thresholds for each of these criteria should be carefully considered. A threshold that is overly conservative may result in large portions of the system being taken into the expensive atomistic regime; conversely, an excessively aggressive threshold will result in the system losing the ability to resolve important local phenomena (Potestio et al., 2013).

AdResS-type methodology has proven to work very well for polymer applications, particularly in regard to dynamic crack growth studies. In this case, the RMD automatically moves with the crack-tip without having the user pre-define the RMD's future trajectory. Furthermore, this allows the user to make more accurate crack propagation path and overall failure behaviour predictions (Kremer & Müller-Plathe, 2002; Praprotnik et al., 2008; Müller-Plathe, n.d.). This characteristic is a clear example of one of the major advantages of using adaptive techniques when studying regions that are localized and mobile at the same time.

3.4 Concurrent Bidirectional Methods

The class of RMD-SPH strategies which most tightly couple RMD and SPH, respectively, are the concurrent bidirectional methods. They differ from both the static and adaptive methods, which tend to treat RMD and SPH independently and exchange boundary conditions; with concurrent methods, both are created such that they exist together in a region of overlap, with the two providing mutual constraints on each other, at each time increment. Thus, one, combined representation of all fields (i.e. displacements, stresses etc.) is generated from atomistic as well as continuum elements.

This method relies on an overlap domain where the RMD and SPH definitions reside next to each other. Also, the operations that allow for the mapping and reconstructing quantities between both representations need to be defined also (Park et al., 2005; Prudhomme et al., 2008). The atomistic representations of virial stress, microscopic velocity, and temperature will be coarse-grained and mapped onto the SPH particles. These continuum-level definitions will provide the constraints (velocity or traction), which will be reconstructed and imposed on the atomistic representation (Dhia & Rateau, 2005a; Delle Site et al., 2020). Energy or momentum correction terms need to be applied to prevent double counting and ensure conservation laws remain valid (Belytschko & Xiao, 2003; Gholami et al., 2022). As the coupling operations are performed at each time step in a time-dependent coupling approach, there is a computational cost associated with these methods compared to static or adaptive formulations; however, these approaches tend to be far more accurate and accurate when considering situations in which there is a strong two-way coupling between the mechanical and chemical systems (Borg et al., 2015; Sun et al., 2012).

3.4.1 Bridging Scale Method (BSM)

Introduced by Wagner and Liu (Wagner & Liu, 2003), the Bridging Scale Method, or BSM, represents an approach for concurrently bidirectionally coupling models of different scales. In BSM, displacement fields are decomposed into two distinct components: the continuum-part being smooth, while the atomistic-part comprises of unfiltered fluctuations. Weighting functions are then defined for both parts, such that only the overlapping regions between the two scales are responsible for the effective continuum displacement at any point within that overlap region. Effective continuum displacements throughout a given model can be determined as a linear combination of both types of displacements and are used in determining the equations of motion that govern the behaviour of a system at both scales, based upon the coupling forces that are associated with this blended continuum-atomistic displacement field (Park et al., 2005).

BSM includes a critical component in the form of smooth weighting functions which determine how much influence RMD has on any location in the solution domain relative to how much influence SPH will have (Wagner et al., 2004). In an area where RMD is the only method of analysis, the atomistic weight will be 1 and the continuum weight will be 0; in an area where SPH is the only method of analysis, the weights will be reversed. The overlap area contains both weights having a value between 0 and 1; as well as the sum of both weights equal to 1. This design ensures a gradual transition between methods and prevents the introduction of abrupt changes in the method of analyzing an area. There are several benefits to using BSM in comparison to simply breaking the solution domain into two parts and solving each separately: the smooth transition eliminates the potential for spurious reflections and numerical artefacts at the interface of the two methods (Belytschko & Xiao, 2003); both methods are subject to conservation of mass, momentum and energy simultaneously at the discrete level of resolution; and by varying the shape and size of the overlap area, optimal performance can be achieved for different solving problems (Fish & Chen, 2004).

For investigations into polymer behaviour, particularly the behaviour of molten polymers as they flow past solid surfaces, BSM has been used as a tool (along with other applications). In these studies, the RMD region is located near the solid surface where the effect of surface interactions (such as chain adsorption/surface friction) and wall slip are of particular importance. In contrast, the SPH method describes the bulk behaviour of the flow. A recent study by Yasuda & Yamamoto (2010) has demonstrated that coupling of the two methods can provide an accurate simulation of the shear rate dependent viscosity and wall slip behaviours of polyethylene melts.

3.4.2 Hierarchical Multiscale Method

Hierarchical multiscale techniques extend the bridging concept by adding more scaling levels between continuum and fully detailed, atomistic systems (Peter & Kremer, 2008). Rather than moving from atomistic systems to continuum systems without any intermediary representation, hierarchical multiscale techniques provide an adequate representation of a system through multiple stages (Karimi-Varzaneh et al., 2011). The first stage is an all-atom RMD level. In the next stage, one or more coarse-grained stages combine groups of atoms into a single 'bead' (Karimi-Varzaneh et al., 2012). Finally, the continuum representation, formed from constitutive equations that were developed from lower-resolution stages (Padding & Briels, 2011; Brini et al., 2011), is reached through the use of SPH.

There are two types of information flow in the hierarchy: upward and downward. Upward information flows from RMD simulation to CG Model. The RMD is used to obtain the CG potential for a certain molecular system, and then input into either a continuum or coarse-grained model. Downward information can be defined through the SPH simulations of temperature, pressure and local strain rate; these can be used as boundary conditions for new RMD simulations in microdomains that are representative of the original structure and chemistry (and therefore their macroscopic response). The hierarchical view is particularly beneficial when developing constitutive models or reaction rate laws which are valid on a broad range of conditions because it enables a systematic evaluation of how a material's molecular characteristics (or chemistry) creates macroscopic characteristics through the use of predictive models (Baeurle et al., 2006; Carvalho et al., 2022; Deringer et al., 2019).

3.4.3 Concurrent Multiscale Method (CMM)

The Concurrent Multiscale Method (CMM) is another way to achieve a complete coupled multiscale simulation, as atomistic and continuum domains are both solved and provide real-time communications during each time increment (O & Thompson, 1995). With CMM, the main difference between it and basic domain decomposition is that in CMM, the interactive nature is an ongoing bi-directional relationship, in which the RMD and SPH subdomains communicate efficiently through the inclusion of new source and constraint terms (Kharazmi et al., n.d.; J. Liu et al., 2007).

CMM uses coupling forces in its RMD motion equations to model how continuum fields affect RMD systems (e.g., using continuum field variables, such as pressure gradients or prescribed velocity profiles). While the SPH formulations of mass, momentum and energy include source terms to account for atomistic processes that produce the localised behaviour of the system (e.g., the effects of local stress fluctuations or heat release due to chemical reactions), these source terms are derived from averaging the atomistic quantities over appropriate control volumes (J. Liu et al., 2007). The primary benefit of this method is that it provides much greater accuracy in the coupling region than do methods without this coupling because the two approaches stay in step with each other throughout very rapid dynamics and intense temporal variability (Borg et al., 2015). However, there is also a significant overhead associated with the repeated interpolations, projections, and constraint applications (Sun et al., 2012).

3.5 Comparative Assessment of Coupling Strategies

There is not a universal optimal coupling technique between RMD and SPH, thus selecting the best technique for a given polymer system involves evaluating multiple factors (Fish, 2006; Ghoniem et al., 2003). Four factors that play a large role in determining the appropriate technique are accuracy, computational efficiency, flexibility and ease of use.

From the accuracy perspective, concurrent bidirectional techniques, such as BSM and CMM, tend to have the highest accuracy due to the fact they allow both the atomistic and continuum fields to be consistent at every time step (Wagner & Liu, 2003; O & Thompson, 1995). A static decomposition may work well as long as the interface between the RMD and SPH is in a region where gradients are modest and relatively smooth, but will degrade in performance when the interface is subjected to strong gradients or very localized events. Adaptive methods operate between this spectrum by allowing for the atomistic region to travel with the moving hot spots, allowing for a high degree of local accuracy without requiring the entire domain to be modeled with an RMD (Praprotnik et al., 2008; Potestio et al., 2013).

Lastly, which method is easiest to implement and most useful is a question taken into consideration practically when implementing a static domain decomposition. For example, for many existing RMD/SPH codes, only simple data exchange routines need to be created around old code, which doesn't affect the solving process internally (Hadjiconstantinou, 1999; Miller & Tadmor, 2009). In contrast, to develop a concurrent framework using adaptive methods means substantial development time due to the requirement for new data structures, overlapping regions, and projection/reconstruction operators, as well as implementation and validation of a multi-physics solving environment. In the end, for many researchers as well as industrial organizations, an acceptable method exists as a balance of a method that provides sufficient flexibility and accuracy to capture required physical behaviour but at the same time being manageable in both the amount of coding and computational resources consumed. Further, a compact, multidimensional summary of the comparisons of static decomposition, adaptive methods, and bi-directional concurrent strategies based upon their relative position along the axes of accuracy, efficiency, flexibility, and ease of implementation can be found in Figure 2 of this chapter.

Figure 2. Multi-dimensional comparison of coupling strategies across four criteria (Accuracy, Efficiency, Flexibility, Implementation); higher values indicate better performance.

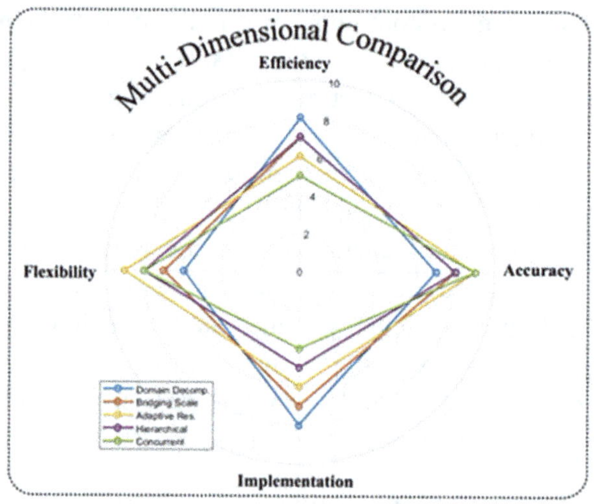

3.6. Software Frameworks and Implementation

Many RMD simulations rely on general-purpose atomistic engines such as LAMMPS at present because of their built-in support for reactive force fields, such as ReaxFF. For large-scale ReaxFF simulation purposes, LAMMPS is now the standard engine and GROMACS was originally created for simulation of biomolecular systems, but has been adopted in hybrid simulation processes. GROMACS has the benefit of having optimised kernels and multiple levels of parallelism that make it worthwhile to use in this way, especially when embedding a reactive core within a larger non-reactive or coarse-grained environment. The mesh free SPH solvers; DualSPHysics and PySPH represent a very mature set of SPH software. DualSPHysics caters to large Free-Surface and coastal flowing Problems and relies on GPU computational acceleration to do this whilst PySPH aims at providing rapid prototyping using Python for complex rheological and various multi-physics formulations (Plimpton, 1995; Abraham et al., 2015; Crespo et al., 2015; Ramachandran et al., 2019).

The linking of RMD engines like LAMMPS or GROMACS with SPH solvers like DualSPHysics, GPUSPH, or PySPH presents numerous challenges regarding the proper implementation of functionalities, including synchronizing the time-stepping between two codes, transferring data back and forth from the different discretizations and memory layouts that they may use, and maintaining the same level of scalability over distributed heterogeneous hardware (CPU/GPU/COP) resources across both codes.

Current methods tend to be loosely coupled co-simulations that only transfer data between different software using file- or socket-based data transfers. A few custom created frameworks exist that have combined particle and continuum kernels into a single piece of software similar to existing tools which couple CFD and particle dynamics such as SediFoam and a unified library SPH library, SPHinXsys, that as well includes the capability to perform multi-physics and multi-resolution simulations, and could be further developed to add RMD-SPH coupling as their initial class feature (Sun & Xiao, 2016; Zhang et al., 2021).

4. CASE STUDIES AND APPLICATIONS

Reactive polymeric systems inherently display multiscale behavior, with atomistic chemical reaction being intricately intertwined with mesoscopic transport phenomena such as diffusion, convection, and rheological flow. It is important to accurately capture this multiscale coupling in the modeling so that advanced polymer processes (such as polymerization, crosslinking, curing, and degradation) are represented in which macroscopic transport and chemical reactivity are closely interconnected. Classical continuum-based approaches such as the FEM and fictitious-domain methods have served to capture macroscopic flow behavior and non-Brownian hydrodynamics well. These models are weak in their inherent deterministic nature in capturing stochastic thermal fluctuations and nanoscale microscopic reaction kinetics, which rule diffusion-controlled processes (Bian et al., 2012).

To avoid such constraints, the combination of RMD with SPH enables a robust multiscale simulation methodology for reactive polymeric systems. In this hierarchical multiscale hybrid approach, RMD dictates atomistic events (bond breaking, bond forming, and reaction energetics), whereas SPH manages mesoscale transport, flow-induced stresses, and rheological evolution. This synergistic combination enables the concurrent prediction of chemical kinetics and macroscopic flow behavior under realistic processing conditions, thereby bridging molecular-scale mechanisms and continuum level dynamics.

Overall, the RMD-SPH coupling method provides a physically meaningful, thermodynamically consistent, and computationally effective means to analyze reactive polymeric systems. Through the correlation of transport on the process scale with molecular reactivity, it enables mechanistic insight of depth into how chemistry and flow work together to determine polymer processing and material performance (Bian et al., 2012).

Figure 3. Schematic illustration of the integrated RMD-SPH framework for polymeric systems. (A) RMD simulation stages: (a) initial polymer chain configuration in the condensed phase with atoms color-coded by element; (b) dynamic bond formation and cleavage during reactive events; (c) evolution of intermediates and products under thermal or mechanical stimuli; and (d) potential energy surface mapping of local minima and transition states.

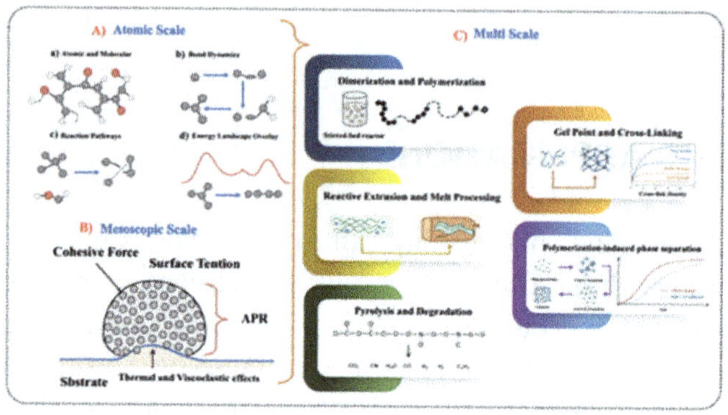

(B) SPH methodology applied to a polymeric droplet interacting with a substrate or fluid phase, where Lagrangian particles represent the continuum system, incorporating cohesive and surface tension forces via the Korteweg stress tensor. APR enables detailed resolution of interfacial and capillary-driven dynamics. (C) Representative phenomena and case studies that can be investigated using the coupled RMD-SPH approach.

4.6. RMD for Polymeric Materials

These last decades have seen remarkable advances in computational algorithms and hardware that have extended first-principles-based atomistic simulation to increasingly advanced engineering systems. Among the many approaches available, RMD is now a revolutionary technique by which scientists can probe chemical and physical processes at the atomistic scale with heretofore impossible temporal and spatial resolution. By the direct simulation of bond-breaking and bond-forming transitions, RMD provides molecular-level insight into reaction mechanisms, energy transformations, and dynamic structural reorganization not obtainable through either experiments or classical continuum-based simulations. Its versatility has seen extensive use in various applications from combustion, catalysis, and energy conversion, to nanoparticle synthesis and more recently, polymer and composite materials (Mao et al., 2023).

Molecular dynamics (MD) simulations have long been a requirement in order to comprehend the mechanical, thermal, and structural material properties of polymers and their composites on a molecular level. MD simulations yield quantitative data about properties that are dictated by chain architecture, functionality, and non-covalent interactions. Traditional (non-reactive) MD, however, will primarily concentrate on mean system responses (e.g., bulk stress-strain behavior) and is largely accused of neglecting localized deformation events. This shortcoming is particularly critical in heterogeneous polymeric systems with fillers or nanofillers, where the chains in the interfacial regions experience non-homogeneous strain fields that lead to inelasticity at the localized points, chain scission, or damage, which have a direct impact on macroscopic mechanical and thermal behaviors (Kawagoe & Okabe, 2023).

RMD generalizes the classical MD concept by providing for direct simulation of chemistry in large atomic and molecular systems with full accounting of intermolecular interactions in condensed phases. Thus, RMD bridges the long-standing gap between quantum chemical computations, which provide high accuracy but poor scalability, and classical molecular simulations, which simulate large-scale dynamics but no chemical reactivity. This combined capability allows RMD to probe thermally activated reactions, cross-linking, oxidation, or degradation pathways unavailable to conventional MD (Nyden et al., 2004).

Even though grounded in Newtonian mechanics, predictive accuracy of RMD simulations is strongly reliant on the quality of reactive force fields, initial and boundary conditions, and temporal resolution with small integration time steps. Though computationally more costly than CMD, recent developments in parallel computing and graphics processing unit acceleration now enable simulations of hundreds of thousands of atoms within timescales of nanoseconds. RMD has successfully been employed to study nanoparticle sintering, droplet evaporation, nanoscale fluid flow, and coupled physicochemical phenomena in energy storage and conversion devices. Safe usage, however, requires well-designed simulation, adequate statistical sampling, and validation against experiment or quantum chemical benchmarks to ensure thermodynamic and kinetic fidelity (Mao et al., 2023).

In polymeric systems, RMD has particularly proven useful in the determination of the mechanism of thermal decomposition and degradation of vinyl polymers including polyethylene (PE), polypropylene (PP), and poly(isobutylene) (PIB). These researches have shown that differences in thermal stability and product distributions among polymers are controlled by differences in activation energies for initiation and propagation reaction parameters themselves controlled by pendant substituents and backbone structure. By establishing a direct connection between molecular-scale reaction events and macroscopic thermal behavior, RMD provides a mechanistic foundation for the rational design and optimization of polymeric materials with tailored thermal, mechanical, and chemical properties (Nyden et al., 2004).

4.7. The SPH for Polymeric Materials

The SPH method has emerged as a versatile and effective mesh-free computation algorithm for the simulation of complex fluid and multiphase phenomena, particularly in systems dominated by deforming interfaces and large deformations. Originally developed in the area of astrophysics to model self-gravitating and rarefied flows, SPH is presently a general numerical framework capable of resolving strongly nonlinear processes, free-surface physics, and multiphase interactions without the geometric constraints or numerical diffusion inherent in fixed-grid methods. Its particle formulation Lagrangian allows for the natural description of deforming domains and moving interfaces, which makes it extremely well-suited to study droplet dynamics, polymer processing, and fluid structure interaction (FSI) problems.

One of the most significant advantages of SPH for multiphase and polymeric systems is that it can incorporate interfacial tension effects explicitly in the form of cohesive force terms derived from thermodynamically consistent equations of state (such as the van der Waals or Peng-Robinson models) in a way that no explicit interface-tracking algorithms need to be employed. Apart from this, advanced SPH formulations typically incorporate the Korteweg stress tensor, which is better at handling capillary stresses and surface tension, and allows for stable simulation of diffuse interfaces during evaporation, spreading, or coalescence of droplets in systems (Charles & Daivis, 2011; Pütz & Nielaba, 2015). These have been used in various engineering issues like droplet impact and spreading on high-temperature substrates, fuel droplet evaporation, and multi-component phase-change systems(Cui et al., 2019; Z. Liu et al., 2022; Sohag et al., 2022; Subedi et al., 2022; D. Wang et al., 2020; X. Xiao et al., 2023; Yang et al., 2019; Yang & Kong, 2020).

The particle nature of SPH also enables a smooth handling of deformable and moving boundaries, a significant requirement of polymeric and multiphase systems. For instance, in substrate-droplet interactions, SPH accurately captures recoiling, spreading, coalescence, and breakup without reinitializing interface functions. In higher-order systems, such as polymer melts, non-Newtonian fluids, and viscoelastic media, SPH was successfully utilized to model rheologically complex flows, such as Oldroyd-B, Maxwell, Herschel-Bulkley, and Bingham-type models (Ellero et al., 2002; Fang et al., 2006). They are highly relevant for polymer processing, additive manufacturing, and suspension research and soft matter systems.

In melt processing of polymers, SPH provides unique benefits to model intricate flow patterns, interface deformation, and surface instabilities during extrusion, injection molding, coating, and film formation. SPH has also been applied effectively in droplet-based polymerization, microdroplet generation, and emulsion systems to study the influence of viscosity, viscoelasticity, and interfacial tension on droplet coalescence, breakup, and spreading. Furthermore, SPH has increasingly been used to design and optimize polymeric thin films, sprays, and composite coatings, where surface morphology, wetting, and mass transport need to be predicted accurately. SPH enables particle-level examination of flow and interface dynamics while enabling direct correlation between processing parameters, polymer microstructure, and macroscopic material properties.

Inherent conservation of energy, mass, and momentum in SPH (used in combination with numerical stability at large deformations and Galilean invariance) renders it a powerful means for multiphysics process simulation, from thermal conduction to phase transition and chemistry. New methodology developments, such as adaptive particle refinement (APR) and splitting/merging schemes, have advanced the accuracy and efficiency of SPH in multiscale simulations. Moreover, hybrid SPH-mesh models have also been suggested, where SPH captures regions of high interface deformation or multiphase coupling successfully and grid-based solvers manage bulk flow areas, thereby reducing computational cost without loss of accuracy (Alexiadis et al., 2018; Amicarelli et al., 2020; Bilotta et al., 2016; Cercos-Pita, 2015; Crespo et al., 2015; Dong et al., 2022; Fuchs et al., 2021a; Kalateh & Koosheh, 2020; Li et al., 2022; Ramachandran et al., 2021)].

Beyond its standard engineering applications, SPH has also found applications in microscale and biological systems. In microfluidic and droplet-based systems, SPH mimics droplet deformation, coalescence, break-up, and bubble dynamics without explicit interface reconstruction. In biological fluid dynamics, SPH has been employed to simulate blood flow, for instance, red blood cell (RBC) deformation, aggregation, and migration in capillaries, giving quantitative information regarding cell-cell interactions and disease states such as malaria (Ye et al., 2019).

Computational efficiency of SPH has been greatly enhanced through open-source, GPU-based implementations that enable simulations involving billions to millions of particles. These advances now enable realistic simulation of polymer-fluid interaction, droplet evaporation, sloshing, jet atomization, and phase-change processes at industrially relevant conditions (Amicarelli et al., 2020; Bilotta et al., 2016; Crespo et al., 2015; Ramachandran et al., 2021). SPH has thus evolved into a physically consistent, stable, and multiscale computing methodology, poised to investigate transient heat and mass transfer, multiphase flow, and interfacial dynamics in polymeric and soft matter systems.

4.8. Multiscale Modeling of Polymeric Systems

Polymeric materials inherently possess multiscale nature, where response at one time or length scale considerably influences others. One of the most popular instances is viscoelasticity and viscoplasticity of polymers, wherein the induced stress results in time-delayed deformation due to intramolecular relaxation of the chain and intermolecular reorganization interactions primarily in entangled polymers. The process of crystallization is another instance of this multiscale coupling from local monomer-scale ordering up to mesoscale lamellar growth and ultimately macroscopic spherulitic structure. Melt memory is a concept that demonstrates how mesoscale structural features, including entanglement density and chain orientation, govern local packing and dictate recrystallization kinetics.

Mechanical response at the macroscopic level is governed in nanoscale fillers in polymer nanocomposites by nanoscale filler-induced refinement of strain distribution and creation of chain extension or orientation at interfaces, which means that atomic-scale interactions must be related to mesoscale morphology and continuum-level properties. Similarly, biological proteins such as spider silk and collagen are hierarchical structures in which molecular conformations, fibrillar assemblies, and macroscopic fibers mutually enable mechanical performance. Multiphase and heterophase system (block copolymers, blends, foams, and filled matrices) also possess this inherent coupling, where the mechanical, rheological, and transport properties are a function of the complex interaction of interfacial resistances, phase morphology, and filler dispersion on various scales [45].

To address these multiscale challenges, hybrid computational models combining SPH-RMD have gained significant interest. Such a dual framework bridges the atomistic and continuum regimes so as to simulate reactive multiphase flow, droplet impact and evaporation, shockwave propagation, and nanoscale physicochemical interactions in complex geometries phenomena inaccessible to conventional MD by virtue of system size or timescale limitation. SPH-RMD methodology has been used successfully in a wide range of representative systems including:

- Combustion and vaporization of fuel droplets, where it captures synchronized hydrodynamic deformation and surface chemical reaction.
- Aggregation and sintering of nanoparticles in flows, shedding light on interactions among diffusion, reactivity, and hydrodynamic forces.
- Shock-induced reactions in energetic materials, allowing direct visualization of reaction initiation and propagation at extreme thermodynamic conditions.

These estimates demonstrate the predictive power of SPH-RMD for systems where fluid dynamics and chemical reactivity are closely intertwined. Although computationally intensive, SPH-RMD permits a new multiscale perspective that harmonizes molecular-level mechanisms with macroscopic transport and rheology (Mao et al., 2023).

Advanced SPH formulations have introduced SPH-RMD to coupled FSI problems, enabling accurate modeling of collision, impact load, and constrained motion in strongly fluid-rigid coupled systems. Some key applications are:

- Sloshing dynamics for partially filled tanks with baffles, predicting pressure forces and moments under transient loading.
- Debris transport in cases of flooding or sedimentation, simulating particle-fluid-structure interaction.

- Droplet impact on moving or deformable surfaces, determining fragmentation, spreading, and recoil behaviors.
- Wave loading on marine structures with realistic force distribution and wave-structure interaction predictions.

Other methodological enhancements have paired RMD with SPH formulations adjusted to solve chemical reactions as well as fluid-structure coupling limitations, allowing the simulation of reactive multiphase processes such as fuel droplet combustion, explosive boiling, chemical vapor deposition (CVD), and polymerization in confined space. In particular, total Lagrangian SPH approaches combined with atomistic simulations enable the direct extraction of pseudo-continuum deformation fields (stretch, local strain, and shear) from atomic trajectories. The enhanced SPH approach gets rid of boundary and interface inaccuracy, enabling accurate determination of local deformation in heterogeneous systems. These experiments determine that atomic-scale strain localization is strongly correlated with inhomogeneous stiffness and the onset of cohesive failure, results that are in superb congruence with experiments (Kawagoe & Okabe, 2023).

On the whole, SPH-RMD presents a multiscale modeling platform with a general-purpose and physically well-behaved character, effectively bridging reactive atomistic responses and continuum-scale hydrodynamics. It enables predictive simulation of polymerization, constrained-geometry curing, droplet evaporation and combustion, shock-ignited reaction, and nanoparticle agglomeration in flowing media. By creating pseudo-continuum deformation fields as well as molecular reaction data, SPH-RMD is a useful tool for the design of polymer composites, energetic materials, and functional nanomaterials, complementing experimental and theoretical efforts (Kawagoe & Okabe, 2023). A few representative case studies illustrate the versatility of SPH and SPH-RMD models:

- Yang et al. (Yang & Kong, 2017, 2020) developed an SPH model consisting of vapor species continuity and a splitting/merging method to preserve mass, momentum, and energy conservation in droplet evaporation. Simulations nicely captured spreading, breakup, splash, partial rebound, and full rebound regimes for varying Weber numbers and surface temperatures.
- Sigalotti et al. (Hermann et al., 2015) applied a non-local thermodynamic SPH model based on van der Waals mean-field theory to simulate spinodal decomposition, homogeneous nucleation, and explosive boiling in superheated states with high correspondence to experimental evidence.
- SPH-RMD simulations of reactive droplet evaporation have deciphered coupled hydrodynamics and chemical kinetics from nanometers (reaction zones) to millimeters (droplet motion) scale, delivering accurate insight into multiscale evaporation, combustion, and interfacial transport phenomena.

In summary, the SPH and SPH-RMD methods provide a sound, mesh-free, and thermodynamically consistent methodology for investigating multiphase flows, phase transformations, and reactive transport in complex polymeric and soft matter materials. Their innovative ability to combine continuum hydrodynamics with atomistic reactivity places them at the forefront of the next generation of heat and mass transfer, multiphase process, and fluid-structure interaction computational tools. Ongoing innovations (like curvature estimation, surface tension simulation, parallel scalability, and reactive transport coupling) keep SPH-RMD a cornerstone of multiscale and multiphysics modeling in the years to come, having revolutionary capabilities within the energy, aerospace, chemical, and polymer engineering disciplines.

Some common RMD-SPH applications in polymeric materials are introduced and discussed below in the following sections, highlighting its strengths in polymer processing, film formation, and interfacial behavior.

4.8.1. Dimerization and Polymerization

Multiscale research of the recent past has indicated the power of coupling RMD with continuum-level modeling to bridge atomistic reaction processes and macroscopic transport behavior of reactive polymer systems. Yazdanbakhsh et al. (Haddadi-Asl et al., 2025) employed coupled RMD-SPH to investigate gas-phase dimerization of propylene (C_3H_6) into dipropylene (C_6H_{12}). At the atomistic level, RMD simulations provided pictorial accounts of bond formation mechanisms, acceleration due to catalysts, and attendant energetic transformations. Simultaneously, SPH simulation gave a mesoscopic image of temperature, pressure, and velocity profiles within a stirred-bed reactor. Their simulations indicated that the use of a Ziegler-Natta catalyst reduced the dimerization time from 15.2 ps to 10.2 ps and changed the endothermic character of the reaction to exothermic with an overall final reaction energy of –97.91 kcal. mol^{-1}. On the reactor scale, SPH simulations presented thermodynamic equilibrium, a flat temperature profile (~348 K), and low particle mobility ($v < 0.02$ $m.s^{-1}$), which suggested stable polymerization conditions with no interparticle stress.

Karim et al. (Farah et al., 2010) in a related work designed a reactive coarse-grained molecular dynamics (CGMD) simulation to simulate the irreversible styrene polymerization to polystyrene. Their work emphasized the critical role played by kinetic and structural parameters, such as the capture radius (R_i, R_p) and reaction delay time (τ_r), in controlling the growth of polymers, chain propagation, and polydispersity. The simulations were found to successfully replicate characteristic equilibrium quantities like density, end-to-end chain distance, and radius of gyration and provide mechanistic insights into the manner in which reactivity parameter tuning captures unrelaxed and fully relaxed regimes of polymerization.

Collectively, these studies reveal the usefulness of multiscale models combining RMD with SPH or CGMD in describing the interplay between reaction kinematics and transport phenomena and the couplings between monomers and catalysts. In bridging molecular-scale reactions with macroscopic structural development, such schemes offer predictive ability most useful for optimization, design, and scale-up of industrial polymerization processes.

4.8.2. Gel Point and Cross-Linking Dynamics

Knowledge of the gelation process and how it affects the formation of thermo-mechanical properties is critical in optimizing cross-linked polymeric systems such as epoxy resins. These materials, used extensively in aerospace and structural composites, must have their cross-linking reactions closely regulated to achieve higher order mechanical integrity and dimensional stability. To address such complex behaviors, an experimentally verified RMD-SPH formulation has been developed in order to accurately model the curing process and offer predictions for property progression as a function of cure progress.

In this multiscale approach, reactive force field calculations were integrated with rheometric and mechanical testing to quantitatively analyze how molecular network formation governs critical material properties (including mass density, volumetric shrinkage, elastic modulus, yield strength, and Poisson's ratio) during curing. Computational predictions were in excellent agreement with experimental data and literature reference values, establishing the capability of reactive force field based RMD to accurately model epoxy property development at a range of cross-link densities.

One of the fundamental concepts of this study was the accurate measurement of the gel point, the onset of a coherent molecular network that permeates the entire simulation space. Using MD-based cluster evolution following, the largest cluster size, second-largest cluster weight, and reduced molecular weight (RMW) were monitored as functions of cross-linking level.

The cluster connectivity development at the molecular level exhibited a distinct structural development: at a cross-link density of 0.2, there is the formation of small aggregates from individual polymer chains; at 0.4, there is some network connectivity; and at 0.7, the largest cluster percolates throughout the periodic simulation box, meaning an uninterrupted, load-carrying covalent network has formed. This microstructural transformation is inseparably connected with an abrupt increase in tensile strength and stiffness, highlighting the gel point as a bifurcation between macroscopic mechanical performance and chemical cross-linking kinetics(S. U. Patil et al., 2021).

Overall, this case study indicates that SPH-RMD simulations, if compared against experimental observation, can provide a reliable multiscale model for thermosetting polymer structure-property evolution prediction. With the inclusion of chemical conversion, gelation response, and mechanical behavior, this hybrid computational approach provides a predictive method for process optimization and material design in epoxy-based composites and cross-linked polymeric materials in general.

4.8.3. Reactive Extrusion and Melt Processing

The reactive extrusion and melt processing of polymer-matrix nanocomposites (PNCs) are highly coupled systems where chemical reactions, flow dynamics, and interfacial interactions are simultaneously occurring at different spatial and temporal scales. In order to decouple such strongly coupled phenomena, multiscale molecular modeling has emerged as an inevitable instrument linking macroscopic flow behavior with molecular-level chemistry.

At the molecular level, RMD simulations provide an atomistically detailed image of bond breaking and formation, reaction rates, and nanofiller-matrix interfacial interactions. Such simulations capture transient chemical processes and local structure development with great fidelity and enable basic insights into cross-linking processes, grafting reactions, and interfacial compatibilization. Nevertheless, because they are chemically correct, atomistic RMD models are computationally limited to nanometer regions and nanosecond timescales and tend to model polymer chains as short oligomers rather than genuinely entangled networks.

These constraints can be addressed through CG and mesoscale techniques that extend the window of simulation to longer time and lengths, enabling studies of chain entanglement, nanoparticle dispersion, viscoelasticity, and flow-induced morphology. These simulations are critically dependent on good parameterization and calibration with atomistic simulations or experiments to be predictive.

The need to couple intensive molecular chemistry with macroscale transport has led to integrated multiscale methods, in which RMD is combined with continuum-scale flow simulations such as SPH or CFD. In these hybrid models, the RMD addresses the advancement of chemical reaction and cross-linking kinetics at the nanoscale level, while the SPH or CFD subdomains address the advancement of temperature, pressure, and shear rate fields during the process of the extruder. This hierarchical coupling allows for simultaneous monitoring of reaction-induced rheological advancement, advancement of viscoelastic behavior, as well as transport processes during reactive extrusion.

Critical processing parameters like shear rate, temperature profile, and nanofiller-matrix interfacial energy are the critical parameters that dictate the resulting product structure and property. Shear rate affects molecular orientation and reaction kinetics, temperature regulates reaction conversion and viscosity build-up, while interfacial energy dictates nanoparticle dispersion and stress transfer efficiency. Coupled SPH-RMD calculations have demonstrated that reaction-induced flow can significantly affect nanocomposite microstructure and viscoelastic behavior through exhibiting a pronounced intercoupling between mechanical response and reaction kinetics.

Multiscale SPH-RMD modeling of reactive extrusion, in general, offers a solid and unifying platform for clarifying and rationalizing simultaneous reaction-flow events in polymer nanocomposites. By bridging atomistic-level chemistry with continuum-scale rheology and transport, these hybrid simulations reduce the experimental trial-and-error load and facilitate predictive understanding of process optimization, melt homogeneity, and material design. Ultimately, this approach opens up the door to rational engineering of high-performance polymer nanocomposites with mechanically, thermally, and rheologically tunable properties tailored to the needs of next-generation industrial and biomedical applications (Vogiatzis & Theodorou, 2018).

4.8.4. Polymerization-Induced Phase Separation

Polymerization-induced phase separation (PIPS) is a key process in organic and inorganic polymeric material synthesis and structuring, whereby polymerization and phase segregation are concurrently executed to create highly complex hierarchical morphologies. Multiscale modeling over the last decade has brought the tool of direct computation of these reaction-assembly processes coupled together for molecular-scale chemistry and mesoscale structure.

A fine example of this progress was provided by Carvalho et al. (Carvalho et al., 2022), who have built a reactive coarse-grained molecular model to simulate water-based silica polymerization under real experimental conditions such as biosilicification and templated porous silica synthesis. The model applies a continuous potential to simulate the reversible cleavage and assembly of Si–O–Si bonds, hence accurately replicating the tetrahedral geometry of bonding inherent in silica. By incorporating this reactivity scheme into the MARTINI coarse-grained force field, the methodology is amenable to large systems with millions of atoms and simulation times that extend up to the microsecond regime.

By means of large-scale simulations, the authors explored key structural and kinetic parameters governing silica condensation, including the emergence of Q^n species (Q^0 to Q^4), cluster growth development, and condensation level in the course of gelation. Results well recapitulated experimentally reported three-step polymerization process: (i) rapid silicic acid oligomerization, (ii) fractal cluster aggregation and growth, and (iii) eventual gelation of the network via coalescence of clusters. As schematically visualized in their simulations at the isoelectric point (pH $\approx$ 2.5), isolated individual Q^0 monomers rapidly dimerize and cyclize to yield ring-like intermediates, eventually forming branched clusters and a silica percolated network.

To assess the transferability of the model, the authors also simulated high-pH micelle-templated silica formation, where negatively charged silicate species electrostatically engage with cationic surfactants cetyltrimethylammonium (CTA^+) groups. The simulation replicated the concurrent reaction self-assembly coupling, and it explained how small surfactant aggregates rapidly coalesce into micelles that further stabilize through silica adsorption onto their surfaces an evolution that is in accord with experimental descriptions of mesostructured silica formation. This hybrid model framework thus provides a strong paradigm for investigating coupled polymerization and phase separation processes in reactive inorganic-organic systems (Carvalho et al., 2022).

Adding to this work, a reactive coarse-grained polymer model was developed in an attempt to study how polydispersity affects polymerization kinetics and ensuing macroscopic properties of polymeric materials. Unlike conventional non-reactive, monodisperse models, this technique allows for in silico simulation of stepwise polymerization with linear, branched, and cross-linked topology. The emergent chain length distribution for linear polymers corresponds with the Flory-Schulz distribution and thus replicates realistic polydisperse molecular assemblies.

The research systematically investigated how the chain length distribution influences kinetic and mechanical properties. Self-diffusivity analysis revealed that polydisperse polymer melts will exhibit lower diffusivity than monodisperse systems when normalized using number-average chain length but higher diffusivity when normalized using weight-average chain length. Moreover, mechanical testing in uniaxial tension revealed considerable sensitivity of Young's modulus, strain hardening character, and UTS to molecular polydispersity and strain rate. Notably, polydisperse polymer glasses exhibited a reduction in Young's modulus but enhanced UTS in comparison to monodisperse analogs under low strain-rate deformation (Deng & Shi, 2016).

Generally, these reactive coarse-grained simulations highlight the growing capability of particle-based simulation platforms (such as SPH-RMD) to mimic polymerization-induced phase separation, chain-growth behavior, and structure property relationships on a series of multiple length and time scales. Through a combination of chemical reactivity with mesoscale transport and mechanical response, SPH-RMD-based reactive coarse-grained models provide a robust platform for investigating how molecular architecture, reaction kinetics, and phase organization together determine the performance of polymeric and hybrid materials under realistic synthesis and processing conditions.

4.8.5. Pyrolysis and Degradation

RMD simulations are now an efficient computational tool for the understanding of the complex chemical mechanisms underlying polymeric material pyrolysis and thermal degradation. In one such typical study, Lu Xu et al. (Lu et al., 2015) employed RMD to investigate Kapton-type polyimide decomposition mechanisms at the atomistic level. A multi-chain reactive polyimide model was constructed to simulate temperature-dependent degradation behavior, enabling analysis in detail of reaction intermediates and volatile products. Simulations showed that CO_2 and CN were dominant gas species, primarily formed by cleavage of C–N bonds in the imide rings. Additionally, transient –C(=O)–O–C– linkages were identified as dominant intermediates for CO_2 evolution and stepwise de-carbonization of aromatic structures culminating in CN formation. From the extraction of kinetic parameters (an apparent activation energy of ~187 $kJ.mol^{-1}$ and corresponding pre-exponential factors) the study demonstrated excellent agreement between simulated predictions and trends observed experimentally. Such findings confirmed that RMD can accurately capture bond dissociation, reaction kinetics, and product evolution, providing atomistic-level insight into the thermal degradation pathways of high-performance polymers.

Aside from molecular-scale research, multiscale coupling techniques have also been adopted to improve the predictive capability of RMD to macroscale phenomena. Gui-Rong Liu et al. (G. R. Liu et al., 2017) successfully integrated SPH with RMD to create a reactive SPH-RMD model with the capability to link microscopic reaction kinetics and continuum-scale material properties in energetic polymers such as HMX. In the hybrid approach, Equations of State (EoS) of condensed-phase reactants and gaseous products were determined from NVT RMD simulations and used subsequently in SPH models. Coupling by this technique enabled realistic simulation of pressure evolution, detonation front propagation, and shock-wave behavior under extreme thermodynamic conditions. The SPH-RMD simulations were reported to agree very well with experimental detonation data and traditional models of ignition-growth, reproducing main parameters such as peak pressure, arrival time, and spatial pressure profiles.

Together, the results confirm the versatility of RMD and SPH-RMD methods for the simulation of reactive polymeric systems at different scales. Whereas RMD provides atomistically resolved data on bond-breaking chemistry and gas-phase growth in pyrolysis, the coupled SPH-RMD method places molecular reactivity onto mechanistic interface with macroscopic transport or mechanical response. The coupled strategy allows for a full overview of combustion, polymer degradation, and detonation phenomena and offers a computationally efficient and mechanistically informative alternative to hazardous or costly experiments.

In addition to pyrolysis and thermal degradation, RMD is singularly well-suited to investigate reactive-intermediate phenomena such as radicals, ions, and electronically excited entities. Applications include, for instance, polymer oxidation, interactions of biomolecules with plasmas, and chemistry induced by radiation. For instance, RMD has been employed to uncover high-performance polymers' pyrolysis mechanisms, such as temperature-dependent decomposition pathways and volatile products such as CO_2, CN, and acetylene. Synergistic with this in plasma medicine, RMD simulations have provided molecular-level insights into the interaction of reactive oxygen and nitrogen species (ROS/RNS) with biomolecular targets (peptidoglycan, phospholipid membranes, and DNA) exhibiting oxidative stress mechanisms, chemical modification, and membrane permeability alteration (Bogaerts et al., 2014).

Besides organic systems, RMD has also been applied to study solid-liquid interfacial reactivity, e.g., with studies of the silica-water interface. These simulations provided atomistic details of interfacial hydrolysis and hydroxylation reactions, observing the dynamic formation of silanol (Si–OH) groups, dangling-bond removal, and water molecule diffusion through hydrogen hopping mechanisms analogous to the Grotthuss proton-transfer process. Interestingly, RMD was able to simulate an effective diffusion coefficient of about 1.68×10^{-6} cm^2/s showed an interfacial potential difference of nearly 10.5 V, attributed to water dipoles' polarization upon the silica surface. With RMD, bonding environments' evolution, angular distributions, and reaction kinetics could be quantified-information that remain largely untapped for classical or ab initio molecular dynamics approaches(Fogarty et al., 2010)

To bridge the inherent spatial and temporal scales of atomistic RMD computations, multiscale coupling strategies have been developed, blending RMD with mesoscale techniques such as SPH. The hybrid SPH-RMD model is capable of dealing with microscopic reactive events and macroscopic transport phenomena simultaneously, thus relating chemical reactivity to mechanical, thermal, and fluid dynamic behavior. In polymeric systems, this bonding is particularly beneficial to probe complex environments such as reactive flows, polymer combustion, plasma-polymer interactions, and degradation caused by stress, where bond rearrangements walk hand in hand with large-scale deformation or heat transfer (Bogaerts et al., 2014; Fogarty et al., 2010).

Through this multiscale-capable integrated perspective, SPH-RMD simulations offer a predictive and physically reconcilable system to describe the interactions between chemistry, mechanics, and thermodynamics in reactive polymeric conditions. Great challenges notwithstanding with computational expense, transferability of parameters, and force field calibration, RMD (with or without the integration of SPH) remains an indispensable tool to unravel and forecast the behavior of reactive polymers and to make novel materials with improved durability, stability, and functionality under extreme or reactive conditions.

4.8.6. Limitations and Uncertainties in Multiscale Modeling

Multiscale simulations of reactive polymer systems, ranging over RMD, CGMD, to coupled SPH-RMD approaches, have clearly illustrated the possibility to connect atomistic reaction mechanisms with mesoscale transport phenomena and macroscopic material behavior. Indeed, these applications have given detailed mechanistic insight into bond formation, gelation, polymerization-induced phase separation, reactive extrusion, and thermal degradation. However, each has intrinsic limitations: atomistic RMD is restricted to nanometer-length scales and nanosecond timescales, which precludes the capture of long-chain entanglement, slow kinetic events, or heterogeneous catalytic environments. Coarse-grained and mesoscale models increase both spatial and temporal reach but at the sacrifice of atomistic detail, which could mask the subtleties of certain reaction intermediates or interfacial effects. Predictive accuracy is also critically dependent upon parameterization, force field transferability, and validation against experiment, introducing uncertainties that need to be carefully weighed.

In spite of these limitations, the concerted use of atomistic and continuum-level methods provides a solid platform for mechanistic understanding and process optimization. By clearly addressing computational constraints, parameter sensitivities, and experimental simplifications, these works present a believable basis for predictive modeling. Future research needs to be directed at stochastic heterogeneities, time and length scales that extend via adaptive or multiresolution approaches, and inclusion of uncertainty quantification to derive confidence in simulation-driven material design. Taken together, this evaluative perspective indicates that while there are a number of challenges remaining, multiscale modeling remains irreplaceable in the rational engineering of polymeric systems with specific chemical, mechanical, and transport properties.

4.8.7. Technical Challenges in RMD-SPH Coupling and Conventional Solutions

Despite the development of various methods, the stable and accurate implementation of coupling between RMD and SPH faces three major technical challenges, which are reviewed below along with their conventional solutions.

The first one is time incoherence and quantity conservation. The primary problem is the inherent time difference between the two models. In RMD simulations, the time step is on the order of femtoseconds (fs), while SPH typically works with larger time steps (picoseconds to nanoseconds). This difference makes it either computationally expensive or numerically unstable if we directly synchronize the two models(Tang et al., 2023). The suggested solutions are: 1) Sub-cycling or Multi-rate Integrators: Perform multiple RMD steps per SPH step, then average filter the data for transfer to SPH. 2) Conservative Flux Exchange: When exchanging quantities such as mass, energy, or momentum, algorithms are used that ensure that the total sum of these quantities remains constant in the overlap region (e.g. in the Arlequin or flux-correction framework(Sockwell et al., 2023).

Another technical challenge is referred to Ghost Forces and Boundary Conditions. At the interface between the two models, the difference in representation (atomic discrete representation and mesoscale continuous representation) may lead to artificial forces or unrealistic reflections (Chamoin et al., 2010). To overcome, researchers recommend the following steps: 1) Using a smooth overlap region with a gradual weight function that makes the changes continuous. 2) Applying low-pass filters to remove unwanted fluctuations in the vector data transfer. 3) Using an energy-based formulation (Arlequin) to correct the energy and force in the connection region, so that ghost forces are minimized(Shapeev, 2011; Tao et al., 2021).

The third issue which may result to some technical challenges is Transfer of chemical reactions and polymerization rates. In reactive polymer systems, the third challenge concerns how to accurately transfer chemical information. RMD calculates the actual reaction rates and molecular paths, while SPH must include this information as source terms in the mass and energy transfer equations. Errors in transferring this data can severely distort the temperature or composition field(Coller et al., 2016). The distortions may be solved through: 1) Rate Extraction: Extract local reaction rates from RMD simulations and convert them into pseudo-Arrhenius relations or data tables for use in SPH. 2) On-the-fly Surrogates: Use surrogate models such as neural networks or Gaussian processes (GP) to predict reaction rates or heat released based on temperature, shear rate, and local composition. This approach allows for large-scale simulations without having to directly run RMD at each point(Ferrari et al., 2024; Jiang & Webb, 2024).

Table 3. summary of Technical Challenges in RMD-SPH Coupling and Solutions.

Challenge	Problem	Conventional solutions
Time incoherence & quantity conservation	RMD uses femtosecond steps while SPH uses much larger steps (ps-ns). Direct synchronization is computationally expensive or unstable; conservation of mass/energy/momentum across scales is nontrivial.	Sub-cycling/multi-rate integrators (many RMD steps per SPH step) and conservative flux-exchange or Arlequin/flux-correction schemes to enforce conservation.
Ghost forces & boundary/ interface conditions	Discrete (atomic)-continuous (mesoscale) representation mismatch at the interface generates artificial forces and wave reflections (ghost forces), degrading accuracy.	Smooth overlap with gradual weight functions, energy-based Arlequin or blended energy formulations, and explicit ghost-force correction schemes. Low-pass filtering of transferred vector fields also advised.
Transfer of chemical reactions & polymerization rates	RMD produces local, state-dependent reaction pathways and rates that must be provided to SPH as source terms; direct transfer errors can distort temperature/composition fields.	**Rate extraction:** tabulate local reaction rates, pseudo-Arrhenius fits from RMD; **On-the-fly surrogates:** train neural networks or Gaussian-process surrogates to predict rates/heat release from local state (T, shear, composition).

4.8.8. Emerging Role of Machine Learning in RMD-SPH Coupling

Machine learning (ML) has transformed the way RMD and Particle-based continuum methods (such as SPH) are implemented for Multi-Scale Simulations. The major issues discussed throughout this document regarding RMD include; heavy computational expenditure associated with this type of simulation, there is a limited amount of time accessible to do so, and there is a challenge associated with appropriate identification of parameter values to use for simulating RMD's reactive Forces; all these challenges can be viewed from the perspective of data and model reduction. By using machine learning, we can develop a set of compact, data-driven models of the microscopic behavior of molecules and incorporate those into SPH simulations to allow for "fast" querying as opposed to running a large number of high cost RMD calculations to complete the continuum loop process (Deringer et al., 2019; Unke et al., 2021).

ML models based on atomistic information can be used to create complementary methods to speed up RMD. These ML atoms have the potential to reproduce energies and forces using data generated from ab initio (first-principles) calculations with similar accuracy as quantum mechanical calculations, but at a significantly lower computational cost. When using ML potentials in RMD simulations, they can replace conventional interatomic potentials in different subsystems, or they can support fine-tuning of ReaxFF by providing a more trustworthy, higher dimensional mapping between local chemical conditions and the reactivity of potential energy surfaces. Therefore, ML interatomic potentials provide a solution to the persistent challenge in reactive simulations of multifaceted polymeric systems and multiphase mixtures of maintaining an appropriate balance of accuracy, transferability and efficiency (Behler & Parrinello, 2007; Mishin, 2021).

The second primary role of ML is as a substitute for using RMD in SPH. Instead of repeatedly contacting RMD for each time step and spatial location, we can establish a database of RMD trajectories that represent all the possible density, temperature, deformation rate, conversion, and composition combinations that could occur. The ML models are trained to connect these state variables with the values produced by the RMD (stress, effective viscosity, reaction rate, or evolving moduli) at that particular SPH state. The SPH will therefore obtain this information from the ML model during a coupled simulation, without the need to continuously contact RMD. Therefore, this approach effectively resolves the mismatch in time scales and the issue of exploring high-dimensional parameter spaces topologically in a way that parallels other hybrid micro–macro-surrogate methods (Hoffer et al., 2022; Khristenko et al., 2022). While these ML components can enhance the RMD-SPH process by substituting computationally expensive RMD simulations with a surrogate model, they have their own challenges concerning validation and robustness. A surrogate model is only as good as the RMD data that it is built upon, meaning that predictions made by a surrogate model can degrade very quickly when the SPH trajectories fall outside of the state space in which the training data are located. For this reason, it will be important for further uncertainty quantification and active learning techniques to be employed by using new RMD simulation triggers in cases where the surrogate is making an extrapolative prediction and by comparing predictions made by the surrogate to actual experimental results and to high-fidelity simulations in an ongoing systematic manner. The issues of reliability and managing error in a workflow of ML-enhanced RMD-SPH will be revisited in detail in the upcoming sections regarding technical challenges and validation (Unke et al., 2021).

5. VALIDATION AGAINST EXPERIMENTAL OR INDUSTRIAL DATA

In the rapidly-evolving field of materials science, validation is the bridge that connects theory and reality. If modeling is the art of prediction, validation is the science of proof, the test of how well predictions match the real world and verify the actual phenomena. In multiscale polymer modeling, this test occurs on three scales simultaneously: atomic, mesoscopic, and macroscale. Each scale possesses a distinct scientific language, but the common goal is to ensure that the model not only produces similar quantitatively results, but also exhibits equivalent trend (M. Agarwal et al., 2024; Gartner III & Jayaraman, 2019).

In multiscale polymer reaction modeling, the main aim of validation goes beyond fitting a curve; it must demonstrate that the set of models (atomic, mesoscopic, macroscopic) has the capability to predict and explain the time- and space-dependent physical behavior within its specific application domain. In systems where chemical reaction and mass transport are interrelated (e.g., polymerization or exothermic decomposition in melt flow), the simultaneous coupling of an atomic reaction model (RMD/ReaxFF) with a mesoscopic continuum particle model (SPH) allows for real-time feedbacks between the reaction and the temperature, velocity and composition fields; but this is the very feature that makes validation more complex and at the same time more demanding (E. R. Smith & Theodorakis, 2024a; S. Wang et al., 2023). From a validation perspective, any coupling effort must demonstrate the ability to meet three main requirements: 1) conservation of fundamental quantities (mass, momentum, energy) over the entire range and especially in the coupled overlap region, 2) exponential reproduction of experimental observations at all scales (atomic, mesoscopic, macroscopic), and 3) transparent reporting of uncertainty and sensitivity that allows industrial users to judge the validity of the model (Capone et al., 2024; Delle Site et al., 2020; Gholami et al., 2022).

Simultaneous coupling requires metrics that go beyond signal matching to ensure the integrity of the information exchange between the two models. Three categories of essential criteria are:

1). **Global conservation error:** The cumulative difference in mass, momentum, and energy over the entire domain, and especially in the overlap region, should not exceed a specified limit; this value can be reported as the ratio of the cumulative error to the reference value and used as an acceptance threshold in patch tests. Conservation is also the basis of many valid coupling methods (Sadjina et al., 2017).
2). **Source consistency (Flux consistency):** The mass and heat rates extracted from RMD and fed into SPH as source terms should be consistent in mean and dispersion with the RMD output. For example, if RMD reports a production rate of species A, the integral of this rate over the overlap volume should be approximately equal to the increase in the species stock in SPH (up to the numerical error of sampling). Obvious inconsistencies indicate errors in the lifting or restricting operators or temporal instability (Petsev et al., 2017; Qi et al., 2023).
3). **Emulator accuracy:** When statistical emulators (Gaussian Process, Polynomial Chaos Expansion (PCE), neural nets) are used to approximate rates or effective functions to reduce computational cost, the accuracy of these emulators should be reported with cross-validation metrics (RMSE, Negative Log-Likelihood (NLL), calibration plots) and the uncertainty extracted from the emulator should be published in the final analysis. Without this reporting, any claim of model-data fit can be misleading. Moreover, reporting local error heatmaps and parametric sensitivity analysis (Sobol indices or Morris screening) is essential to track down the source of discrepancies such as material, boundary conditions, or model structure errors (Manfredi & Trinchero, 2024; Yan et al., 2018).

5.1. Purposeful Experimental Design for Coupling Validation

For validation of the RMD/SPH coupling to be meaningful, experiments must be designed to reveal sensitive points in the physics of the coupling; this way, the experimental results can support statistical decisions about the quality of the coupling (Kumagai et al., 2022; Pannico et al., 2024; Possenti et al., 2021).

- **In situ microscopic chemical experiments:** Techniques such as in-line ATR-FTIR, time-resolved DSC, and TGA-MS or GC-MS can record the timing of chemical transformations and decomposition products with sufficient resolution. These data are critical for validating the local rates and reaction energies extracted from RMD. When planning, the temporal alignment between the simulation and the spectral data must be resolved using time-rate corrections (Arrhenius fit, time-temperature superposition) or increasing the experimental resolution(Kumagai et al., 2022; Pannico et al., 2024; Possenti et al., 2021).

- **Mesoscopic experiments:** DSR and rotational rheometers (steady and oscillatory) are necessary to measure η ($\dot{\gamma}$,T) and $G'(\omega), G''(\omega)$. In addition, micro-PIV or confocal imaging can record the local velocity field and morphology to validate SPH outputs about flow and morphology. The imaging data should be analyzed with quantitative metrics (distribution metrics, EMD) to determine qualitative and quantitative overlap with the simulation(Lu et al., 2018; Sheidaei et al., 2024).
- **Processing Industrial Testing:** For industrial-scale validation, in-mold sensors (pressure sensors, cavity sensors), thermal imaging (IR thermography), and phased-array ultrasonic detectors are used to track the melt-front and temperature profile. The test design should create a reactive and controllable region in which RMD can be performed and the online effects of the reaction on the SPH flow field can be observed(Ageyeva et al., 2019; Kariminejad et al., 2021; Techawinyutham et al., 2024).

Figure 4. Schematic illustration of experimental design for coupling validation in three various scales.

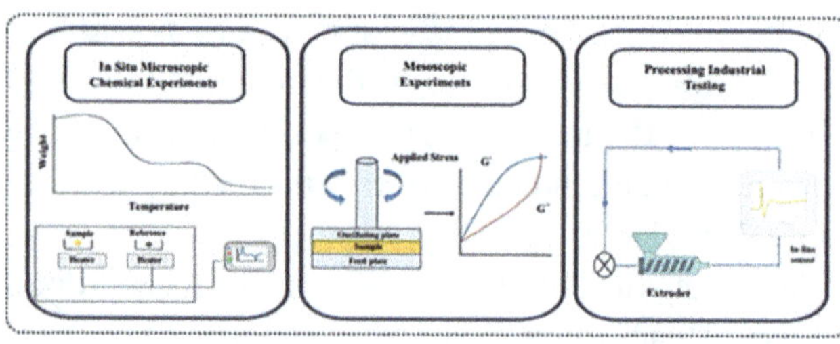

A complete validation should include a set of sensitivity tests designed for coupling to determine which parameters or numerical choices have the greatest impact on the results. Some conventional sensitivity tests include: 1) Handshake width and weight function profile: A parametric sweep over the width of the overlap region and the types of weight functions (linear, soft-tangent, Boolean) shows whether ghost forces or numerical instabilities emerge in different choices (Dhia & Rateau, 2005a). 2) Sub-cycling ratio (RMD steps per SPH step): Changing this ratio has a direct impact on the timeliness accuracy and computational cost; numerical investigation of instability and cumulative error in this parameter is essential (Karimi & Nakshatrala, 2015). 3) Surrogate model discrepancy and its impact on key performance indicators (KPIs): By performing Monte-Carlo sampling from the emulator fluctuation distribution, we can see how much emulator uncertainty affects process KPIs (Hoffer et al., 2022). The conclusions from these tests should not only represent that the model fits the experiment, but also clarify the source of the error such as numerical, structural, or parametric(Wei et al., 2024).

5.2. Conventional Coupling Schemes in Multiscale Simulations

In modeling polymer reactions at different scales, choosing the appropriate pattern for coupling between the molecular domain (RMD) and the mesoscale domain (SPH) is of crucial importance. In practice, there are three main patterns for this coupling, each of which has different advantages, limitations, and areas of application.

1). **Coupling with Domain Decomposition**: In this approach, the simulation domain is divided into two main parts. The first is the atomic region (RMD), which is usually a small region in the center or reaction zone, where chemical reactions, bond formation and breaking, or structural transformations occur. The second is the continuum region (SPH), which covers the bulk of the system and solves the flow, heat transfer, and diffusion equations on a larger scale. At the interface between these two parts, a region of overlap is defined, called the handshake region or overlap zone. In this region, transport phenomena (such as density, energy, and momentum) is occurred and the physical coherence of the system is maintained (S. P. Xiao & Belytschko, 2004). This method is the basis of many concurrent coupling schemes and is widely used in studies of reactive polymers, such as resin injection or reactive extrusion (Kawagoe & Okabe, 2023).

2). **Energy-driven methods (Arlequin/Energy-based Blending):** The Arlequin method was originally proposed by H. Ben Dhia and G. Rateau. It was later extended by Prudhomme, Oden and colleagues for particle-continuum (atomistic-continuum) coupling application. In this method, the coupling between the two models is based on an energy or work functional in the overlap region. In effect, the total energy of the system is written as a combination of the two models, with each model contributing a portion of the total energy based on a local weighting function. This formulation ensures force balance and energy conservation over the entire domain. The Arlequin method has shown particularly effective application in mechanical and thermal problems of polymers, where stress and strain continuity are important(Dhia & Rateau, 2005b; Prudhomme et al., 2008).

3). **Flux/Handshake Coupling:** In this model, data exchange between scales is done explicitly through lifting and restricting operations. Restricting operation is defined as converting atomic (microscale) data into effective mesoscale quantities, such as density, stress, reaction rate, and heat generated, while lifting operation is outlined as reconstructing atomic configurations consistent with SPH continuous fields, based on local density and composition. This model is the basis of many modern coupling models in polymer reaction systems and allows the use of local RMD data to feed the mesoscale equations(E. R. Smith & Theodorakis, 2024b; Weinan & Engquist, 2003).

5.3. Technical Challenges in RMD-SPH Coupling and Conventional Solutions

Despite the development of various methods, the stable and accurate implementation of coupling between RMD and SPH faces three major technical challenges, which are reviewed below along with their conventional solutions.

The first one is time incoherence and quantity conservation. The primary problem is the inherent time difference between the two models. In RMD simulations, the time step is on the order of femtoseconds (fs), while SPH typically works with larger time steps (picoseconds to nanoseconds). This difference makes it either computationally expensive or numerically unstable if we directly synchronize the two models(Tang et al., 2023). The suggested solutions are: 1) Sub-cycling or Multi-rate Integrators: Perform multiple RMD steps per SPH step, then average filter the data for transfer to SPH. 2) Conservative Flux Exchange: When exchanging quantities such as mass, energy, or momentum, algorithms are used that ensure that the total sum of these quantities remains constant in the overlap region (e.g. in the Arlequin or flux-correction framework(Sockwell et al., 2023).

Another technical challenge is referred to Ghost Forces and Boundary Conditions. At the interface between the two models, the difference in representation (atomic discrete representation and mesoscale continuous representation) may lead to artificial forces or unrealistic reflections(Chamoin et al., 2010). To overcome, researchers recommend the following steps: 1) Using a smooth overlap region with a gradual weight function that makes the changes continuous. 2) Applying low-pass filters to remove unwanted fluctuations in the vector data transfer. 3) Using an energy-based formulation (Arlequin) to correct the energy and force in the connection region, so that ghost forces are minimized(Shapeev, 2011; Tao et al., 2021).

The third issue which may result to some technical challenges is Transfer of chemical reactions and polymerization rates. In reactive polymer systems, the third challenge concerns how to accurately transfer chemical information. RMD calculates the actual reaction rates and molecular paths, while SPH must include this information as source terms in the mass and energy transfer equations. Errors in transferring this data can severely distort the temperature or composition field(Coller et al., 2016). The distortions may be solved through: 1) Rate Extraction: Extract local reaction rates from RMD simulations and convert them into pseudo-Arrhenius relations or data tables for use in SPH. 2) On-the-fly Surrogates: Use surrogate models such as neural networks or Gaussian processes (GP) to predict reaction rates or heat released based on temperature, shear rate, and local composition. This approach allows for large-scale simulations without having to directly run RMD at each point(Ferrari et al., 2024; Jiang & Webb, 2024).

Table 4. summary of Technical Challenges in RMD-SPH Coupling and Solutions.

Challenge	Problem	Conventional solutions
Time incoherence & quantity conservation	RMD uses femtosecond steps while SPH uses much larger steps (ps-ns). Direct synchronization is computationally expensive or unstable; conservation of mass/energy/momentum across scales is nontrivial.	Sub-cycling/multi-rate integrators (many RMD steps per SPH step) and conservative flux-exchange or Arlequin/flux-correction schemes to enforce conservation.
Ghost forces & boundary/ interface conditions	Discrete (atomic)-continuous (mesoscale) representation mismatch at the interface generates artificial forces and wave reflections (ghost forces), degrading accuracy.	Smooth overlap with gradual weight functions, energy-based Arlequin or blended energy formulations, and explicit ghost-force correction schemes. Low-pass filtering of transferred vector fields also advised.
Transfer of chemical reactions & polymerization rates	RMD produces local, state-dependent reaction pathways and rates that must be provided to SPH as source terms; direct transfer errors can distort temperature/composition fields.	**Rate extraction:** tabulate local reaction rates, pseudo-Arrhenius fits from RMD; **On-the-fly surrogates:** train neural networks or Gaussian-process surrogates to predict rates/heat release from local state (T, shear, composition).

5.4. Practical Steps for Implementing RMD-SPH Coupling

In this section, a step-by-step framework for the practical implementation of a coupled model between RMD and SPH in reactive polymer systems is presented.

1). First, the reaction region must be defined. This region typically corresponds to an area with a high reactant concentration or a high reaction rate. This region is covered by the RMD model, and the rest of the system is modeled with SPH (Haddadi-Asl et al., 2025).

2). The next step involves design of the handshake region. The width of this region is usually chosen to be in the range of a few nanometers to a few tens of nanometers. The weight function should also be chosen to maintain continuity in the data transfer; smooth functions such as the hyperbolic tangent or smoothed Boolean functions are usually suitable choices (Fuchs et al., 2021b).

3). Afterward, the exchanged quantities must be determined. The quantities exchanged between the two models are: Density, momentum, and velocity field (to preserve flow and fluid dynamics, Mass fraction of species or chemical composition, Energy and heat production rate, Reaction rates or effective reaction coefficients(Haddadi-Asl et al., 2025).

4). Lifting and restricting operators are imperative to be outlined in the succeeding action. Restricting from RMD to SPH computes volume averages from atomic data including density, temperature, velocity, stress, and mass or heat production rates. Lifting from SPH to RMD reconstructs atomic configurations consistent with SPH continuum conditions through methods such as Monte Carlo or constrained packing algorithms(Gear et al., 2003).

5). In order to numerical stability maintenance, it is necessary to specify the data update rate between the two models. In this step, the sub-cycling ratio is determined (e.g., each SPH step is equivalent to 50-100 RMD steps) and the data convergence intervals are determined.

6). During the simulation, the total sum of mass, energy, and momentum over the entire domain and overlap region should be continuously monitored so that cumulative errors can be detected and corrected.

7). Final and crucial step is referred to chemical calibration. The rates extracted from RMD must be corrected with experimental data or quantum calculations (QM/DFT) to increase the realism of the model. These data are then used in the SPH model in tabulated form or through a learning emulator (Jiang & Webb, 2024; Tang et al., 2023).

RMD-SPH coupling models, as the backbone of multiscale simulations of reactive polymers, allow for the connection between molecular reactions and macroscopic fluid behavior. Choosing the appropriate method, paying attention to time inhomogeneity, controlling boundary forces, and being careful in the transfer of chemical reactions are essential prerequisites for achieving stable and reliable results. By following the above steps, it is possible to build models that are not only numerically stable, but also physically reflect the complex reality of polymer systems. Table 3 represents summary of practical steps for implementing RMD-SPH coupling.

Table 5. Summary of Practical Steps for Implementing RMD-SPH Coupling.

Step	Key points	Practical actions and implementation tips	Typical values and examples
Region definition	Select the reactive zone (high reactant concentration or reaction rate). RMD covers this. SPH for the remainder.	Identify reactive hotspots (concentration, rate maps), mark RMD domain and SPH domain, define overlap boundaries.	Localized region around reaction sites (nm scale).
Handshake region design	Overlap region ensures smooth data transfer; use smooth weight functions to avoid discontinuities.	Choose overlap width, pick continuous weight function (e.g., hyperbolic tangent or smoothed Boolean), test/profile continuity of transferred fields.	Width: a few nm to a few tens of nm; weight = tanh or smoothed step.
Exchanged quantities	Exchange conserved and chemical fields: density, momentum, velocity, species mass fractions, energy/heat production, reaction rates, effective coefficients.	Implement consistent interpolation, averaging procedures, ensure unit consistency, apply boundary-force control to avoid artifacts.	Ensure mass, momentum, energy are conserved across exchange.
Lifting & restricting operators	Restrict (RMD→SPH): volume-average atomistic data. Lift (SPH→RMD): rebuild atomistic configurations matching continuum constraints.	Restrict: compute local averages for density, temperature, velocity, stress, mass, heat production. Lift: use Monte Carlo, constrained packing, or structure-reconstruction algorithms; equilibrate locally after lift.	Restrict = volume averaging; Lift methods = Monte Carlo, constrained packing.
Scheduling & sub-cycling	Synchronize time stepping between models to preserve stability; choose sub-cycle ratio and data-update frequency.	Select sub-cycling ratio, define synchronization, update intervals, run stability/convergence tests, adaptively adjust if needed.	Example: 1 SPH step = 50-100 RMD steps (typical starting point).
Conservation monitoring	Continuously monitor global and overlap-region totals of mass, momentum, and energy to detect cumulative errors.	Implement diagnostics, checkpoints, log conserved quantities, apply correction schemes if drift detected (rescaling, redistribution).	Monitoring frequency: regular checkpoints (e.g., every synchronization step).
Chemical calibration	Calibrate RMD-derived rates with experimental data or QM/DFT; supply calibrated kinetics to SPH as tables or via an emulator.	Compare RMD rates to experiments or QM calculations, correct/scale rates, implement tabulated rates or train a learning emulator for use in SPH.	Use QM/DFT or experimental corrections; feed SPH with tabulated or emulator-based kinetics.

5.5. Validation of Coupling RMD-SPH for Polymers

In recent years, valuable efforts have been made to bridge the molecular and mesoscale scales in polymer simulations. However, most of these efforts have not yet been exactly simultaneous coupling of RMD-SPH, but have instead relied on sequential multiscale models or the use of molecular models to extract parameters. The following is a summary of the work that has been done.

Voth and co-workers, in the series of their studies, deeply probed proton exchange membranes (PEMs) based on Perfluoro-sulfonic Acid (PFSA) ionomers such as Nafion as the main system. Their main aim was to develop a multiscale framework for simulating proton transport at the mesoscopic scale and analyzing the effect of morphology and hydration level on proton conductivity. To overcome the time and space limitations of the simulations, the researchers used a combined approach: 1). Coarse Grained structural model, i.e. using the Wescott model in which each Nafion monomer is converted into three particles in which every 10.5 water molecules into one particle. This model allows structures with dimensions of 40 nanometers (much larger than the atomic scale) to be studied. 2). The transport algorithm (SPH-Nernst-Planck), through which the mass and charge transport continuity equations can be solved using the SPH method on this coarse-grained substrate. Professor Voth's 2012 research was a milestone in proving the effectiveness of this method. The main focus was on how adding physical details to the model brought the results closer to experimental reality. In early models that ignored local electrostatic interactions and diffusion coefficient changes, proton conductivity grew nonlinearly and with a very steep slope (20-fold increase) with increasing hydration. This is while experimental data confirm the linearity of the relationship between conductivity and hydration. The researchers succeeded in correcting the nonlinear behavior of the model by applying a local electric potential (caused by sulfonate groups) and a position-dependent diffusion coefficient (PDDC). In the modified model, protons are trapped near the walls (double layer) due to electrostatic attraction, and the conductivity increases only about 2-fold. This result showed a very superb qualitative and trend agreement with the experimental data. Furthermore, spectroscopic studies suggest that the proton concentration in the center of the Nafion pores differs from that near the walls (2-3 times difference). The SPH model showed that the proton concentration increases from about 1.5 molar at the center of the cavity to about 3 to 6 molars near the walls. This concentration gradient is in close agreement with the range predicted in experimental experiments and confirms the mechanism of double layer formation at the mesoscale. Using the developed tool, this research group's 2015 study addressed one of the most controversial experimental topics, namely the change in the geometric shape of water pores with increasing humidity. One of the most important parts of this research was the use of simulation to interpret the laboratory data. Initial simulation results were slightly lower than the experimental numbers (due to ignoring water flow). But when the results were normalized by an "empirical correction factor" (the ratio of the actual permeability to the atomic model permeability), a novel pattern emerged. At low hydration (λ=5), the simulation results for the cluster morphology matched the experimental data very closely. While at high hydration (λ=10,15), the experimental results moved away from the cluster model and closer to the simulation results for the lamellar and cylinder models. This comparative analysis suggests that Nafion undergoes a phase transition with increasing water absorption, from discrete, clustered pores (at low humidity) to continuous, lamellar channels (at high humidity). This finding provides valuable insight into the interpretation of small-angle X-ray scattering (SAXS) data. The simulation showed that the cluster morphology has much lower conductivity than the lamellar models due to the presence of "bottlenecks" and tortuosity. This computational observation explains why in real membranes (which are a combination of these morphologies), conductivity is always lower than the theoretical ideal (in comparison to pure lamellar model). Combining these two studies shows that mesoscale simulation is not just a numerical prediction tool, but can also reveal hidden mechanisms in experimental data. Professor Voth's 2012 research proved that modeling goes wrong without considering electrostatic and diffusion inhomogeneities (which are present in experiments), while his 2015 study strengthened the hypothesis of a change in the internal structure of the membrane due to swelling by quantitatively comparing different geometric models with experimental data. This body of work suggests that to design more efficient membranes, the polymer structure should be engineered such that, at operational hydrations, the morphology is biased towards "lamellar / cylindrical" states to minimize cluster-confining effects. Concentration profiles for lamellar structures, upshots of conductivity vs. water content and Proton conductivity for various morphologies at different hydration levels are all illustrated at Fig. 5. (Jorn & Voth, 2012; S. Liu et al., 2015).

Systems include atomic models of gold crystals, epoxy resin cured between silica walls, and resins containing nanofillers, have been studied to analyze local strain fields with atomic precision. The aim of the research is to develop and validate a multiscale method for calculating local strains with high accuracy by combining MD simulation with the corrected SPH method in the form of a total Lagrangian formulation. To overcome the challenges in analyzing discrete atomic data, the researchers used the cubic B-spline kernel function, which is depicted in Fig. 6(a)). One of the key parameters in this method is the smoothing parameter (h), which was examined in a sensitivity analysis on values of 5, 8, and 10 A^0. The results showed that choosing a value of h greater than or equal to 10 A^0 is essential to eliminate local noise and achieve an accurate match to the overall strain of the simulated cell. Also, to fix the kernel truncation error at system boundaries (such as common seasons), the correction matrix B was used, which ensures the accuracy of calculations near walls and free surfaces. Initial validation of this method was performed on a gold (Au) crystal containing 4000 atoms in a 10×10×10 supercell with a lattice constant of 4.07 Å and under a strain rate of 5×10^8 per second. The findings indicated that at cell strains of 2.53 and 10.5%, the average strain calculated by the SPH method closely matches the applied strain. Although significant noise was observed at 5 Å smoothing, increasing this parameter to 10 Å resulted in a completely uniform strain distribution. For epoxy resin enclosed between silica walls, quantitative differences between the two failure mechanisms of adhesive and cohesive were investigated by varying the interaction coefficient (k). In the case of adhesive failure (k=0.5), after the resin separates from the wall at a cell strain of about 9%, the stress suddenly drops and the internal strain of the resin approaches zero due to elastic recovery. In contrast, in the case of cohesive failure (k = 1.0), even after macroscopic failure at strains above 10%, the resin still retains a residual strain of about 10%, indicating significant plastic and inelastic deformation in the polymer structure. Analysis of the hardness heterogeneity by dividing the resin into 12 pieces showed that the areas adjacent to the wall had a Young's modulus of 4 to 4.5 GPa due to higher density, while the central areas had fluctuations and the fracture initiation points exactly coincided with the weaker areas with a Young's modulus of less than 2 GPa. The effect of gold nanoparticles with diameters of 20 and 30 Å in an epoxy matrix was investigated. Calculations showed that an interphase layer with a constant thickness of 6.3 Å is formed around it, independent of the particle size. Comparison of mechanical properties showed that the presence of filler significantly increased local hardness; so that the hardness of the polar interphase region for 20 and 30 Å particles was calculated to be 3.5 and 2.7 GPa, respectively, which represents an increase of about 22 to 25 percent in comparison to the bulk resin. It was also found that at a total strain of 16.12%, the critical strain concentration occurs exactly at the nanoparticle poles, indicating the onset of the cavitation phenomenon at these points. Colorful snapshots of uniaxial tension in epoxy systems without and with Au filler are presented in Fig. 6(b)). Generally, researchers represented that the local Young's modulus in polymer composites can vary over nanometer ranges from 1.8 to 4.5 GPa, and that the interfacial regions play a key role with an increase in hardness of about 25%. This method has a high ability to quantitatively detect fracture initiation points and analyze the mechanical behavior of heterogeneous materials at the atomic scale. (Kawagoe & Okabe, 2023).

Figure 5. (a) Concentration profiles for lamellar structures with uniform charge distributions along the pore walls, (b) Conductivity of the DPD equilibrated structures from volume averaging across segments of the structure both without electrostatics (red) and including electrostatics with eq 9 (blue), and (c) Upper: Proton conductivity for different morphologies at different hydration levels: lamellar (blue square), cylinder (red circle), and cluster (green cross). Experiment values are from ref 71 (black triangle). Lower: Qualitative scaling of proton conductivities obtained by multiplying by a correction factor.

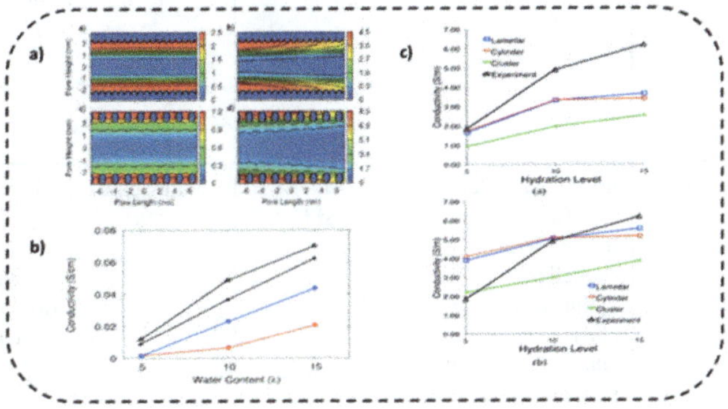

Figure 6. (a) Schematic illustration of the perfect/imperfect kernels in the MD simulations, (b) Snapshots of uniaxial tension in epoxy systems without and with Au filler at $\lambda x = 1.15$, with the systems colored according to the maximum principal stretch $\lambda 1$ calculated using corrected SPH. The left-hand panels show the xz cross-sectional images at the centers of the fillers, and the right-hand panels show only the damaged SPH particles with $\lambda 1 > 1.1\lambda x$.

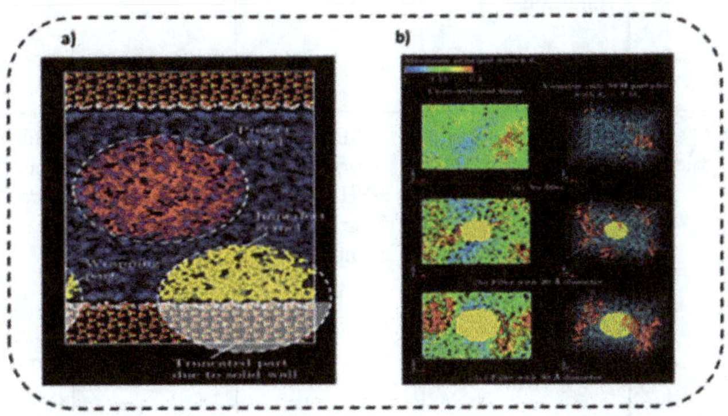

Table 6. summary of conducted researchers on coupling of RMD/SPH approaches.

Authors	System Studied	Research Objective	Simulation Method	Scale	Key Input Parameters	Key Findings
Voth et al.	PEMs (Nafion)	Simulate proton transport and study effect of morphology and hydration	RMD/MS-EVB → SPH	Multiscale	Structure of aqueous proton channels, diffusion coefficient, charge density	Proton conductivity strongly correlates with channel connectivity; lamellar and cylindrical structures show higher conductivity than cluster structures
Voth et al., morphology extension	PEMs (Nafion)	Investigate effect of different morphologies (lamellar, cylindrical, cluster) and water content on proton conductivity	RMD/MS-EVB → SPH	Multiscale	Morphology type, hydration level, site-dependent diffusion coefficients	Lamellar and cylindrical networks with continuous connectivity exhibit higher conductivity; site-dependent diffusion coefficients improve accuracy of transport predictions
Propylene polymerization study	Gas-phase propylene polymerization with $MgCl_2$-supported Ziegler-Natta catalysts	Link chain initiation and growth reactions with mesoscale particle morphology under industrial conditions	RMD → SPH	Multiscale	Reaction rates, surface kinetic parameters, chain growth rates	Mass and heat transfer effects, active site distribution, and catalyst presence determine particle morphology; RMD + SPH enables realistic industrial-scale predictions
Atomic strain evaluation study	Epoxy resin / nanofillers, gold crystals	Evaluate atomic-resolution strain fields	MD → Corrected SPH (Total Lagrangian formulation)	Multiscale / Atomistic → continuum	Atomic coordinates and forces from MD	Accurately maps local strain fields; reveals strain concentration and stiffness variations around nanofillers or phase interfaces; consistent with experiments and continuum mechanics

5.6. Strengths and Gaps in Coupling Validation for Polymers

- **Strengths:**

The ability to combine molecular data with macroscopic flow simulations (e.g. Ren 2022) has shown that the internal structure of chains and their orientation can well predict macroscopic effects under different flow conditions (Khristenko et al., 2022). Development of models to complement features such as entanglements, chain orientation, local stress and the influence of mold geometry on the composite flow, i.e. comparing the molecular change trends at different points in the mold. Statistical and rheological approaches have been used to test temperature and viscosity; they have helped to fine-tune parameters such as the viscosity model or resistance values for non-Newtonian flow in SPH.

- **Knowledge Gaps and Technical Challenges**

Many studies do not have simultaneous (two-way) feedback between MD and SPH; RMD or MD of the molecular component always plays a central role, whereas SPH usually takes data from the molecule and does not provide molecular feedback. This means that there are still few models that can simultaneously couple while preserving molecular and macroscopic information, and that their validation is complete. Local molecular or morphological experimental data are rarely recorded in sufficient detail to be compared with MD output, especially data such as local entanglement or chain orientation at arbitrary points in the mold. Studies that examine the effect of local viscosity on the chemical transformation function and show how molecular changes lead to changes in front-time, pressure, or mold defects are still limited. Consideration of criteria such as mass or energy conservation at junctions or precise time synchronization in models with complex coupling patterns has been less well studied.

Conduct laboratory cases recording local molecular data (e.g. entanglement, orientation, chemical transformation) simultaneously with macroscopic data (velocity field, front-time, temperature). Implement bidirectional coupling between MD / RMD and SPH, including molecular feedback to SPH and the effect of SPH on molecular conditions. Focus on rigorous validation metrics: RMSE/NRMSE, geometric metrics (Haus Dorff), structural metrics (EMD, domain size distribution, orientation) as well as quantitative numerical conservation (mass / energy). Collect local experimental data for reaction and morphology (e.g. with in-situ FTIR, PIV, structural imaging). Use emulator methods to reduce costs but with validation of their uncertainty. Fully publish data, codes and error metrics for Reproducibility.

6. CHALLENGES, OPEN QUESTIONS, AND FUTURE DIRECTIONS

While the unprecedented level of progress achieved by combining RMD and SPH is exemplary, many scientific and technical challenges remain open that perplex the application of predictive multiscale modeling in reactive polymeric systems to its potential. One of the most fundamental of these problems is due to the temporal and spatial disparities between the atomistic and continuum regimes. RMD requires femtosecond time steps to properly resolve bond vibrations and reaction dynamics, but SPH possesses much larger time increments typical of mesoscale transport processes. Such a disparity in the time scale makes synchronization difficult and will most often produce either numerical instability or intolerably high computational overhead. Though multi-time-step and sub-cycling schemes have been suggested to reduce this limitation, these schemes are not yet equipped fully to offer strict mass, momentum, and energy conservation across the coupling interface.

The other difficulty lies in the reliable exchange of information between the two scales. It is not an easy task to transfer atomic-scale information (reaction rates, heat release, and molecular structural change) into something that can be utilized at the continuum level. The current coarse-graining and homogenization procedures are likely to rely on empirical averaging schemes that need not reflect the randomness of reactive events in polymers. Consequently, the critical fluctuations and local phenomena might be lost in data transfer, leading to spurious gradients in thermodynamic fields. Gathering useful, statistically convergent reaction kinetics from RMD remains most difficult, as reactive events in polymers are comparatively infrequent and spatially localized. Emerging machine learning techniques for extrapolating effective reaction parameters or constitutive relations from atomistic data hold significant promise, although generalizability and interpretability of such techniques remain to be proven.

Computationally, the cost of coupled simulations remains a significant constraint. Even with parallel computing environments and GPU acceleration, bidirectionally coupled large-scale RMD-SPH simulations are extremely computationally intensive. Computational intensity restricts system size, simulation run time, and the ability to obtain ensemble averaging for statistically meaningful outcomes. Adaptive resolution methods, in which dynamically the scale of detail is adjusted as a function of local physical conditions, promise a solution but introduce new challenges in thermodynamic consistency and erasure of artificial forces at the advancing interface. Furthermore, transferability of reactive force fields such as ReaxFF continues to be a fundamental issue. Since force field parameters are often fitted for a specific chemistry or temperature range, one cannot expect them to predict well for other polymerization systems or degradation conditions. Development of self-consistent, data-driven parameterization protocols with uncertainty quantification could have a major impact on the quality of future RMD-SPH applications.

Besides methodological and computational problems, validation remains arguably the largest and open problem for multiscale polymer simulation. It is typically difficult to relate simulation results to experiment because the observables extracted from RMD (scission occurrences of chains, molecular alignment, or local entanglement) are qualitatively distinct from the macroscopic variables measured from experiments like viscosity, flow rate, or heat generation. Quantitative correlations between these inhomogeneous length scales require advanced in-situ characterization tools (e.g., time-resolved spectroscopy, high-speed imaging, and synchrotron-based scattering) to monitor structural and thermodynamic evolution both in space and in time. No reproducible protocols and benchmark datasets exist, and the process thus carries the added burden of verification. In order to gain industrial relevance, follow-up research must be focused on building open, cross-referenced databases of reactive polymer systems with experimental and simulated data comparably reported under given conditions.

A multitude of open questions lies at the core of RMD-SPH coupling. How can two-way feedback between the atomistic and the continuum level be created in real-time without destabilizing the simulation? What is the best strategy for adaptive domain decomposition when reactive domains are dynamically resizing and reshaping throughout polymerization, curing, or degradation? Can machine learning surrogates such as neural operators or Gaussian process models simulate chemical kinetics in non-equilibrium states where the traditional rate laws do not hold? And lastly, how can rigorous thermodynamic and mechanical consistency, particularly conservation of mass, momentum, and energy, be enforced across dynamically coupled domains?

Future research roadmaps will have to address these concerns using physics-informed machine learning (PIML) models and adaptive coupling methods that can learn local physical conditions on-the-fly and adaptively vary spatial resolution. Modular, open-source simulation platforms coupling molecular, mesoscopic, and continuum scales tightly would facilitate reproducibility and allow inter-disciplinary collaborative progress. Additionally, hybrid validation approaches that coupling RMD-SPH simulation with concurrent experimental observation offer a seamless bridge between physical measurement and numerical simulation. Quantum-corrected reactive potentials and coarse-grained neural force fields remain to be investigated further to further enhance accuracy and scalability for chemically complex polymer systems.

Last but not least, the synergy of high-level algorithms, computational resources, and model-data integration will transform RMD-SPH coupling into a forecast design tool for developing digital twins of reactive polymer processing. Such designs will not only demystify the complex dynamic interaction of chemical kinetics, heat and mass transfer, and rheological development but also provide us with a rational method for designing future polymers with targeted performance under practical production conditions. To transcend such challenges, further synergy between chemistry, physics, materials science, and computer engineering (this multi-disciplinary effort) will define the coming decade in multiscale polymer simulation.

REFERENCES

Agarwal, A., Wang, H., Schütte, C., & Site, L. D. (2014). Chemical potential of liquids and mixtures via adaptive resolution simulation. *The Journal of Chemical Physics*, *141*(3), 034102. Advance online publication. DOI: 10.1063/1.4886807

Agarwal, M., Pasupathy, P., Wu, X., Recchia, S. S., & Pelegri, A. A. (2024). Multiscale computational and artificial intelligence models of linear and nonlinear composites: A review. *Small Science*, *4*(5), 2300185. DOI: 10.1002/smsc.202300185

Ageyeva, T., Horváth, S., & Kovács, J. G. (2019). In-mold sensors for injection molding: On the way to industry 4.0. *Sensors (Basel)*, *19*(16), 3551. DOI: 10.3390/s19163551

Aktulga, H. M., Pandit, S. A., Van Duin, A. C. T., & Grama, A. Y. (2012a). Reactive molecular dynamics: Numerical methods and algorithmic techniques. *SIAM Journal on Scientific Computing*, *34*(1), C1–C23. DOI: 10.1137/100808599

Aktulga, H. M., Pandit, S. A., Van Duin, A. C. T., & Grama, A. Y. (2012b). Reactive molecular dynamics: Numerical methods and algorithmic techniques. *SIAM Journal on Scientific Computing*, *34*(1), C1–C23. DOI: 10.1137/100808599

Alexiadis, A., Ghraybeh, S., & Qiao, G. (2018). Natural convection and solidification of phase-change materials in circular pipes: A SPH approach. *Computational Materials Science*, *150*, 475–483. DOI: 10.1016/j.commatsci.2018.04.037

Amicarelli, A., Manenti, S., Albano, R., Agate, G., Paggi, M., Longoni, L., Mirauda, D., Ziane, L., Viccione, G., Todeschini, S., Sole, A., Baldini, L. M., Brambilla, D., Papini, M., Khellaf, M. C., Tagliafierro, B., Sarno, L., & Pirovano, G. (2020). SPHERA v. 9.0. 0: A Computational Fluid Dynamics research code, based on the Smoothed Particle Hydrodynamics mesh-less method. *Computer Physics Communications*, *250*, 107157. DOI: 10.1016/j.cpc.2020.107157

Baeurle, S. A., Usami, T., & Gusev, A. A. (2006). A new multiscale modeling approach for the prediction of mechanical properties of polymer-based nanomaterials. *Polymer*, *47*(26), 8604–8617. DOI: 10.1016/j.polymer.2006.10.017

Bagheri, M., Mohammadi, M., & Riazi, M. (2024). A review of smoothed particle hydrodynamics. *Computational Particle Mechanics*, *11*(3), 1163–1219. DOI: 10.1007/s40571-023-00679-7

Baskes, M. I. (1987). Application of the embedded-atom method to covalent materials: A semiempirical potential for silicon. *Physical Review Letters*, *59*(23), 2666–2669. DOI: 10.1103/PhysRevLett.59.2666

Behler, J., & Parrinello, M. (2007). Generalized neural-network representation of high-dimensional potential-energy surfaces. *Physical Review Letters*, *98*(14), 146401. DOI: 10.1103/PhysRevLett.98.146401

Belytschko, T., & Xiao, S. P. (2003). Coupling Methods for Continuum Model with Molecular Model. *International Journal for Multiscale Computational Engineering*, *1*(1), 12. DOI: 10.1615/IntJMultCompEng.v1.i1.100

Ben Dhia, H., & Rateau, G. (2005a). The Arlequin method as a flexible engineering design tool. *International Journal for Numerical Methods in Engineering*, *62*(11), 1442–1462. DOI: 10.1002/nme.1229

Ben Dhia, H., & Rateau, G. (2005b). The Arlequin method as a flexible engineering design tool. *International Journal for Numerical Methods in Engineering*, *62*(11), 1442–1462. DOI: 10.1002/nme.1229

Bian, X., Litvinov, S., Qian, R., Ellero, M., & Adams, N. A. (2012). Multiscale modeling of particle in suspension with smoothed dissipative particle dynamics. *Physics of Fluids*, *24*(1), 012002. DOI: 10.1063/1.3676244

Bilotta, G., Hérault, A., Cappello, A., Ganci, G., & Del Negro, C. (2016). GPUSPH: a Smoothed Particle Hydrodynamics model for the thermal and rheological evolution of lava flows.

Bogaerts, A., Yusupov, M., Van der Paal, J., Verlackt, C. C. W., & Neyts, E. C. (2014). Reactive molecular dynamics simulations for a better insight in plasma medicine. *Plasma Processes and Polymers*, *11*(12), 1156–1168. DOI: 10.1002/ppap.201400084

Borg, M. K., Lockerby, D. A., & Reese, J. M. (2015). A hybrid molecular-continuum method for unsteady compressible multiscale flows. *Journal of Fluid Mechanics*, *768*, 388–414. DOI: 10.1017/jfm.2015.83

Brini, E., Algaer, E. A., Ganguly, P., Li, C., Rodríguez-Ropero, F., & Van Der Vegt, N. F. A. (2013). Systematic coarse-graining methods for soft matter simulations-a review. In Soft Matter (Vol. 9, Issue 7, pp. 2108–2119). Royal Society of Chemistry. DOI: 10.1039/C2SM27201F

Broughton, J. Q., Abraham, F. F., Bernstein, N., & Kaxiras, E. (n.d.). Concurrent coupling of length scales: Methodology and application.

Capone, M., Romanelli, M., Castaldo, D., Parolin, G., Bello, A., Gil, G., & Vanzan, M. (2024). A vision for the future of multiscale modeling. *ACS Physical Chemistry Au*, *4*(3), 202–225. DOI: 10.1021/acsphyschemau.3c00080

Carvalho, A. P., Santos, S. M., Pérez-Sánchez, G., Gouveia, J. D., Gomes, J. R. B., & Jorge, M. (2022). Sticky-MARTINI as a reactive coarse-grained model for molecular dynamics simulations of silica polymerization. *npj Computational Materials*, *8*(1), 49. DOI: 10.1038/s41524-022-00722-w

Cercos-Pita, J. L. (2015). AQUAgpusph, a new free 3D SPH solver accelerated with OpenCL. *Computer Physics Communications*, *192*, 295–312. DOI: 10.1016/j.cpc.2015.01.026

Chamoin, L., Prudhomme, S., Ben Dhia, H., & Oden, T. (2010). Ghost forces and spurious effects in atomic-to-continuum coupling methods by the Arlequin approach. *International Journal for Numerical Methods in Engineering*, *83*(8-9), 1081–1113. DOI: 10.1002/nme.2879

Charles, A. N., & Daivis, P. (2011). A three dimensional smooth particle hydrodynamics model of the nanoscale condensation of water. *19th International Congress on Modelling and Simulation*, 516–522.

Coller, D. H., Vicente, B. C., & Scott, S. L. (2016). Rapid extraction of quantitative kinetic information from variable-temperature reaction profiles. *Chemical Engineering Journal*, *303*, 182–193. DOI: 10.1016/j.cej.2016.05.136

Crespo, A. J. C., Domínguez, J. M., Rogers, B. D., Gómez-Gesteira, M., Longshaw, S., Canelas, R., Vacondio, R., Barreiro, A., & García-Feal, O. (2015). DualSPHysics: Open-source parallel CFD solver based on Smoothed Particle Hydrodynamics (SPH). *Computer Physics Communications*, *187*, 204–216. DOI: 10.1016/j.cpc.2014.10.004

Cui, X., Bakkar, A., & Habashi, W. G. (2019). A multiphase SPH framework for supercooled large droplets dynamics. *International Journal of Numerical Methods for Heat & Fluid Flow*, *29*(7), 2434–2449. DOI: 10.1108/HFF-10-2018-0547

Daw, M. S., & Baskes, M. I. (1984). Embedded-atom method: Derivation and application to impurities, surfaces, and other defects in metals. *Physical Review B: Condensed Matter*, *29*(12), 6443–6453. DOI: 10.1103/PhysRevB.29.6443

Delgado-Buscalioni, R., Kremer, K., & Praprotnik, M. (2008). Concurrent triple-scale simulation of molecular liquids. *The Journal of Chemical Physics*, *128*(11), 114110. Advance online publication. DOI: 10.1063/1.2890729

Delle Site, L., Praprotnik, M., Bell, J. B., & Klein, R. (2020). Particle–continuum coupling and its scaling regimes: Theory and applications. *Advanced Theory and Simulations*, *3*(5), 1900232. DOI: 10.1002/adts.201900232

Deng, B., & Shi, Y. (2016). A reactive coarse-grained model for polydisperse polymers. *Polymer*, *98*, 88–99. DOI: 10.1016/j.polymer.2016.06.018

Deringer, V. L., Caro, M. A., & Csányi, G. (2019). Machine learning interatomic potentials as emerging tools for materials science. *Advanced Materials*, *31*(46), 1902765. DOI: 10.1002/adma.201902765

Dong, X., Hao, G., & Yu, R. (2022). Two-dimensional smoothed particle hydrodynamics (SPH) simulation of multiphase melting flows and associated interface behavior. *Engineering Applications of Computational Fluid Mechanics*, *16*(1), 588–629. DOI: 10.1080/19942060.2022.2026820

Ellero, M., Kröger, M., & Hess, S. (2002). Viscoelastic flows studied by smoothed particle dynamics. *Journal of Non-Newtonian Fluid Mechanics*, *105*(1), 35–51. DOI: 10.1016/S0377-0257(02)00059-9

Fang, J., Owens, R. G., Tacher, L., & Parriaux, A. (2006). A numerical study of the SPH method for simulating transient viscoelastic free surface flows. *Journal of Non-Newtonian Fluid Mechanics*, *139*(1–2), 68–84. DOI: 10.1016/j.jnnfm.2006.07.004

Farah, K., Karimi-Varzaneh, H. A., Müller-Plathe, F., & Böhm, M. C. (2010). Reactive molecular dynamics with material-specific coarse-grained potentials: Growth of polystyrene chains from styrene monomers. *The Journal of Physical Chemistry B*, *114*(43), 13656–13666. DOI: 10.1021/jp107001e

Farah, K., Müller-Plathe, F., & Böhm, M. C. (2012). Classical reactive molecular dynamics implementations: State of the art. *ChemPhysChem*, *13*(5), 1127–1151. DOI: 10.1002/cphc.201100681

Fatehi, R., & Manzari, M. T. (2011). Error estimation in smoothed particle hydrodynamics and a new scheme for second derivatives. *Computers & Mathematics with Applications (Oxford, England)*, *61*(2), 482–498. DOI: 10.1016/j.camwa.2010.11.028

Ferrari, B. S., Manica, M., Giro, R., Laino, T., & Steiner, M. B. (2024). Predicting polymerization reactions via transfer learning using chemical language models. *npj Computational Materials, 10*(1), 119. DOI: 10.1038/s41524-024-01304-8

Fish, J. (2006). Bridging the scales in nano engineering and science. *Journal of Nanoparticle Research, 8*(5), 577–594. DOI: 10.1007/s11051-006-9090-9

Fish, J., & Chen, W. (2004). Discrete-to-continuum bridging based on multigrid principles. *Computer Methods in Applied Mechanics and Engineering, 193*(17–20), 1693–1711. DOI: 10.1016/j.cma.2003.12.022

Fogarty, J. C., Aktulga, H. M., Grama, A. Y., Van Duin, A. C. T., & Pandit, S. A. (2010). A reactive molecular dynamics simulation of the silica-water interface. *The Journal of Chemical Physics, 132*(17), 174704. DOI: 10.1063/1.3407433

Frenkel, D., & Smit, B. (1996). Understanding molecular simulation: From algorithms to applications. In *Understanding molecular simulation: From algorithms to applications*. Academic Press, Incorporated., DOI: 10.1063/1.881812

Fritsch, S., Poblete, S., Junghans, C., Ciccotti, G., Delle Site, L., & Kremer, K. (2012). Adaptive resolution molecular dynamics simulation through coupling to an internal particle reservoir. *Physical Review Letters, 108*(17), 170602. Advance online publication. DOI: 10.1103/PhysRevLett.108.170602

Fuchs, S. L., Meier, C., Wall, W. A., & Cyron, C. J. (2021a). An SPH framework for fluid–solid and contact interaction problems including thermo-mechanical coupling and reversible phase transitions. *Advanced Modeling and Simulation in Engineering Sciences, 8*(1), 15. DOI: 10.1186/s40323-021-00200-w

Fuchs, S. L., Meier, C., Wall, W. A., & Cyron, C. J. (2021b). An SPH framework for fluid–solid and contact interaction problems including thermo-mechanical coupling and reversible phase transitions. *Advanced Modeling and Simulation in Engineering Sciences, 8*(1), 15. DOI: 10.1186/s40323-021-00200-w

Gartner, T. E.III, & Jayaraman, A. (2019). Modeling and simulations of polymers: A roadmap. *Macromolecules, 52*(3), 755–786. DOI: 10.1021/acs.macromol.8b01836

Gear, C. W., Li, J., & Kevrekidis, I. G. (2003). The gap-tooth method in particle simulations. *Physics Letters. [Part A], 316*(3–4), 190–195. DOI: 10.1016/j.physleta.2003.07.004

Gholami, A., Klein, R., & Delle Site, L. (2022). Simulation of a particle domain in a continuum, fluctuating hydrodynamics reservoir. *Physical Review Letters, 129*(23), 230603. DOI: 10.1103/PhysRevLett.129.230603

Ghoniem, N. M., Busso, E. P., Kioussis, N., & Huang, H. (2003). Multiscale modelling of nanomechanics and micromechanics: An overview. *Philosophical Magazine, 83*(31–34), 3475–3528. DOI: 10.1080/14786430310001607388

Gingold, R. A., & Monaghan, J. J. (1977). Smoothed particle hydrodynamics: Theory and application to non-spherical stars. *Monthly Notices of the Royal Astronomical Society, 181*(3), 375–389. DOI: 10.1093/mnras/181.3.375

Haddadi-Asl, V., Abdollahi, E., Rostamipoor, H., Shahrasari, A., & Narjili, S. (2025). Multiscale simulation of gas-phase propylene polymerization initiated by $MgCl_2$-supported Ziegler-Natta catalysts: A reactive molecular dynamics and smoothed particle hydrodynamics approach. *Journal of Polymer Research, 32*(7), 1–21.

Hadjiconstantinou, N. G. (1999). Hybrid Atomistic-Continuum Formulations and the Moving Contact-Line Problem. In Journal of Computational Physics (Vol. 154). http://www.idealibrary.comon

Hermann, J., Lorusso, A., Perrone, A., Strafella, F., Dutouquet, C., & Torralba, B. (2015). Simulation of emission spectra from nonuniform reactive laser-induced plasmas. *Physical Review E: Statistical, Nonlinear, and Soft Matter Physics, 92*(5), 053103. DOI: 10.1103/PhysRevE.92.053103

Hoffer, J. G., Geiger, B. C., & Kern, R. (2022). Gaussian process surrogates for modeling uncertainties in a use case of forging superalloys. *Applied Sciences (Basel, Switzerland), 12*(3), 1089. DOI: 10.3390/app12031089

Hu, X. Y., & Adams, N. A. (2006). A multi-phase SPH method for macroscopic and mesoscopic flows. *Journal of Computational Physics, 213*(2), 844–861. DOI: 10.1016/j.jcp.2005.09.001

Hubin, P. O., Jacquemin, D., Leherte, L., & Vercauteren, D. P. (2016). Parameterization of the ReaxFF reactive force field for a proline-catalyzed aldol reaction. *Journal of Computational Chemistry, 37*(29), 2564–2572. DOI: 10.1002/jcc.24481

Iype, E., Hütter, M., Jansen, A. P. J., Nedea, S. V., & Rindt, C. C. M. (2013). Parameterization of a reactive force field using a Monte Carlo algorithm. *Journal of Computational Chemistry, 34*(13), 1143–1154. DOI: 10.1002/jcc.23246

Jiang, S., & Webb, M. (2024). Physics-Guided Neural Networks for Transferable Prediction of Polymer Properties.

Jorn, R., & Voth, G. A. (2012). Mesoscale simulation of proton transport in proton exchange membranes. *The Journal of Physical Chemistry. C, Nanomaterials and Interfaces, 116*(19), 10476–10489. DOI: 10.1021/jp300040w

Junghans, C., Praprotnik, M., & Kremer, K. (2007). Transport properties controlled by a thermostat: An extended dissipative particle dynamics thermostat. *Soft Matter, 4*(1), 156–161. DOI: 10.1039/B713568H

Kalateh, F., & Koosheh, A. (2020). Simulation of cavitating fluid–Structure interaction using SPH–FE method. *Mathematics and Computers in Simulation, 173*, 51–70. DOI: 10.1016/j.matcom.2020.01.019

Karimi, S., & Nakshatrala, K. B. (2015). A monolithic multi-time-step computational framework for first-order transient systems with disparate scales. *Computer Methods in Applied Mechanics and Engineering, 283*, 419–453. DOI: 10.1016/j.cma.2014.10.003

Karimi-Varzaneh, H. A., Van Der Vegt, N. F. A., Müller-Plathe, F., & Carbone, P. (2012). How good are coarse-grained polymer models? A comparison for atactic polystyrene. In ChemPhysChem (Vol. 13, Issue 15, pp. 3428–3439). Wiley-VCH Verlag. DOI: 10.1002/cphc.201200111

Kariminejad, M., Tormey, D., Huq, S., Morrison, J., & McAfee, M. (2021). Ultrasound sensors for process monitoring in injection moulding. *Sensors (Basel), 21*(15), 5193. DOI: 10.3390/s21155193

Kawagoe, Y., & Okabe, T. (2023). Evaluations of atomic-resolution strain fields using molecular dynamics simulations combined with corrected smoothed particle hydrodynamics. *Computational Materials Science*, *228*, 112333. DOI: 10.1016/j.commatsci.2023.112333

Kharazmi, A., Student, M., & Kamali, R. (n.d.). A Continuum: Molecular Dynamics Hybrid Method for Micro- and Nano-Fluid Flow. http://www.asme.or

Khristenko, U., Constantinescu, A., Le Tallec, P., & Wohlmuth, B. (2022). Statistically equivalent surrogate material models: Impact of random imperfections on the elasto-plastic response. *Computer Methods in Applied Mechanics and Engineering*, *402*, 115278. DOI: 10.1016/j.cma.2022.115278

Kremer, K., & Müller-Plathe, F. (2002). Multiscale simulation in polymer science. *Molecular Simulation*, *28*(8–9), 729–750. DOI: 10.1080/0892702021000002458

Kumagai, S., Sato, M., Ma, C., Nakai, Y., Kameda, T., Saito, Y., Watanabe, A., Watanabe, C., Teramae, N., & Yoshioka, T. (2022). A comprehensive study into the thermo-oxidative degradation of sulfur-based engineering plastics. *Journal of Analytical and Applied Pyrolysis*, *168*, 105754. DOI: 10.1016/j.jaap.2022.105754

Larsson, H. R., & Hartke, B. (2013). Fitting reactive force fields using genetic algorithms. Comput. Method Mater. Sci, 13(1).

Li, L., Yang, X., Sohag, M. M. A., Wang, X., & Liu, Q. (2022). SPH-ASR study of drop impact on a heated surface with consideration of inclined angle and evaporation. *Engineering Analysis with Boundary Elements*, *141*, 235–249. DOI: 10.1016/j.enganabound.2022.05.016

Lind, S. J., Rogers, B. D., & Stansby, P. K. (2020). Review of smoothed particle hydrodynamics: Towards converged Lagrangian flow modelling. *Proceedings - Royal Society. Mathematical, Physical and Engineering Sciences*, *476*(2241), 20190801. DOI: 10.1098/rspa.2019.0801

Liu, G. R., Wang, G. Y., Peng, Q., & De, S. (2017). A micro-macro coupling approach of MD-SPH method for reactive energetic materials. *AIP Conference Proceedings*, *1793*(1), 050006. DOI: 10.1063/1.4971540

Liu, J., Chen, S., Nie, X., & Robbins, M. O. (2007). A continuum-atomistic simulation of heat transfer in micro- and nano-flows. *Journal of Computational Physics*, *227*(1), 279–291. DOI: 10.1016/j.jcp.2007.07.014

Liu, M. B., & Liu, G. (2010). Smoothed particle hydrodynamics (SPH): An overview and recent developments. *Archives of Computational Methods in Engineering*, *17*(1), 25–76. DOI: 10.1007/s11831-010-9040-7

Liu, S., Savage, J., & Voth, G. A. (2015). Mesoscale study of proton transport in proton exchange membranes: Role of morphology. *The Journal of Physical Chemistry. C, Nanomaterials and Interfaces*, *119*(4), 1753–1762. DOI: 10.1021/jp511830f

Liu, Z., Li, S., Pan, X., & Fang, H. (2022). Mechanism study on spreading dynamics of nanofluids droplet coupled with thermal evaporation. *International Journal of Heat and Mass Transfer*, *183*, 122172. DOI: 10.1016/j.ijheatmasstransfer.2021.122172

Lu, X., Wang, X., Li, Q., Huang, X., Han, S., & Wang, G. (2015). A ReaxFF-based molecular dynamics study of the pyrolysis mechanism of polyimide. *Polymer Degradation & Stability*, *114*, 72–80. DOI: 10.1016/j.polymdegradstab.2015.02.004

Lu, X., Zhao, Y., & Dennis, D. J. C. (2018). Flow measurements in microporous media using micro-particle image velocimetry. *Physical Review Fluids*, *3*(10), 104202. DOI: 10.1103/PhysRevFluids.3.104202

Lucy, L. B. (1977). A numerical approach to the testing of the fission hypothesis. *The Astronomical Journal*, *82*(12), 1013–1024. DOI: 10.1086/112164

Manfredi, P., & Trinchero, R. (2024). Nonparametric formulation of polynomial chaos expansion based on least-square support-vector machines. *Engineering Applications of Artificial Intelligence*, *133*, 108182. DOI: 10.1016/j.engappai.2024.108182

Mao, Q., Feng, M., Jiang, X. Z., Ren, Y., Luo, K. H., & van Duin, A. C. T. (2023). Classical and reactive molecular dynamics: Principles and applications in combustion and energy systems. *Progress in Energy and Combustion Science*, *97*, 101084. DOI: 10.1016/j.pecs.2023.101084

Miller, R. E., & Tadmor, E. B. (2009). A unified framework and performance benchmark of fourteen multiscale atomistic/continuum coupling methods. *Modelling and Simulation in Materials Science and Engineering*, *17*(5), 053001. Advance online publication. DOI: 10.1088/0965-0393/17/5/053001

Mishin, Y. (2021). Machine-learning interatomic potentials for materials science. *Acta Materialia*, *214*, 116980. DOI: 10.1016/j.actamat.2021.116980

Müller-Plathe, F. (n.d.). Coarse-Graining in Polymer Simulation: From the Atomistic to the Mesoscopic Scale and Back.

Monaghan, J. J. (1992). Smoothed particle hydrodynamics. *Annual Review of Astronomy and Astrophysics*, *30*(1), 543–574. DOI: 10.1146/annurev.aa.30.090192.002551

Monaghan, J. J. (2005). Smoothed particle hydrodynamics. In Reports on Progress in Physics (Vol. 68, Issue 8, pp. 1703–1759). DOI: 10.1088/0034-4885/68/8/R01

Monaghan, J. J. (2011). Smoothed particle hydrodynamics and its diverse applications. *Annual Review of Fluid Mechanics*, *44*(1), 323–346. DOI: 10.1146/annurev-fluid-120710-101220

Morris, J. P., Fox, P. J., & Zhu, Y. (1997). Modeling Low Reynolds Number Incompressible Flows Using SPH. In JOURNAL OF COMPUTATIONAL PHYSICS (Vol. 136). DOI: 10.1006/jcph.1997.5776

Mortier, W. J., Ghosh, S. K., & Shankar, S. (1986). Electronegativity-equalization method for the calculation of atomic charges in molecules. *Journal of the American Chemical Society*, *108*(15), 4315–4320. DOI: 10.1021/ja00275a013

Mukherji, D., Marques, C. M., & Kremer, K. (2014). Polymer collapse in miscible good solvents is a generic phenomenon driven by preferential adsorption. *Nature Communications*, *5*(1), 4882. Advance online publication. DOI: 10.1038/ncomms5882

Mukherji, D., Marques, C. M., & Kremer, K. (2018). Collapse in two good solvents, swelling in two poor solvents: Defying the laws of polymer solubility? *Journal of Physics Condensed Matter*, *30*(2), 024002. Advance online publication. DOI: 10.1088/1361-648X/aa9c77

Nayir, N., Van Duin, A. C. T., & Erkoc, S. (2019). Development of the ReaxFF Reactive Force Field for Inherent Point Defects in the Si/Silica System. *The Journal of Physical Chemistry A*, *123*(19), 4303–4313. DOI: 10.1021/acs.jpca.9b01481

Nyden, M. R., Stoliarov, S. I., Westmoreland, P. R., Guo, Z. X., & Jee, C. (2004). Applications of reactive molecular dynamics to the study of the thermal decomposition of polymers and nanoscale structures. *Materials Science and Engineering A*, *365*(1–2), 114–121. DOI: 10.1016/j.msea.2003.09.060

O, S. T., & Thompson, P. A. (1995). Molecular dynamics-continuum hybrid computations: A tool for studying complex fluid flows (Vol. 52, Issue 6).

Odian, G. (n.d.). *PRINCIPLES OF POLYMERIZATION* (4th ed.). DOI: 10.1002/047147875X

Padding, J. T., & Briels, W. J. (2011). Systematic coarse-graining of the dynamics of entangled polymer melts: The road from chemistry to rheology. In Journal of Physics Condensed Matter (Vol. 23, Issue 23). DOI: 10.1088/0953-8984/23/23/233101

Pannico, M., Mensitieri, G., & Musto, P. (2024). In-situ FTIR spectroscopy of epoxy resin degradation: Kinetics and mechanisms. *Frontiers in Chemistry*, *12*, 1476965. DOI: 10.3389/fchem.2024.1476965

Park, H. S., Karpov, E. G., Wing, K. L., & Klein, P. A. (2005). The bridging scale for two-dimensional atomistic/continuum coupling. *Philosophical Magazine*, *85*(1), 79–113. DOI: 10.1080/14786430412331300163

Patil, S. P., Heider, Y., Hernandez Padilla, C. A., Cruz-Chú, E. R., & Markert, B. (2016). A comparative molecular dynamics-phase-field modeling approach to brittle fracture. *Computer Methods in Applied Mechanics and Engineering*, *312*, 117–129. DOI: 10.1016/j.cma.2016.04.005

Patil, S. U., Shah, S. P., Olaya, M., Deshpande, P. P., Maiaru, M., & Odegard, G. M. (2021). Reactive molecular dynamics simulation of epoxy for the full cross-linking process. *ACS Applied Polymer Materials*, *3*(11), 5788–5797. DOI: 10.1021/acsapm.1c01024

Peter, C., & Kremer, K. (2009a). Multiscale simulation of soft matter systems. *Faraday Discussions*, *144*, 9–24. DOI: 10.1039/B919800H

Peter, C., & Kremer, K. (2009b). Multiscale simulation of soft matter systems - From the atomistic to the coarse-grained level and back. In Soft Matter (Vol. 5, Issue 22, pp. 4357–4366). DOI: 10.1039/b912027k

Petsev, N. D., Leal, L. G., & Shell, M. S. (2017). Coupling discrete and continuum concentration particle models for multiscale and hybrid molecular-continuum simulations. *The Journal of Chemical Physics*, *147*(23), 234112. DOI: 10.1063/1.5001703

Poblete, S., Praprotnik, M., Kremer, K., & Delle Site, L. (2010). Coupling different levels of resolution in molecular simulations. *The Journal of Chemical Physics*, *132*(11), 114101. Advance online publication. DOI: 10.1063/1.3357982

Possenti, E., Colombo, C., Realini, M., Song, C. L., & Kazarian, S. G. (2021). Time-resolved ATR–FTIR spectroscopy and macro ATR–FTIR spectroscopic imaging of inorganic treatments for stone conservation. *Analytical Chemistry*, *93*(44), 14635–14642. DOI: 10.1021/acs.analchem.1c02392

Potestio, R., Fritsch, S., Español, P., Delgado-Buscalioni, R., Kremer, K., Everaers, R., & Donadio, D. (2013). Hamiltonian adaptive resolution simulation for molecular liquids. *Physical Review Letters*, *110*(10), 108301. Advance online publication. DOI: 10.1103/PhysRevLett.110.108301

Praprotnik, M., Delle Site, L., & Kremer, K. (2005). Adaptive resolution molecular-dynamics simulation: Changing the degrees of freedom on the fly. *The Journal of Chemical Physics*, *123*(22), 224106. Advance online publication. DOI: 10.1063/1.2132286

Praprotnik, M., Matysiak, S., Site, L. D., Kremer, K., & Clementi, C. (2008). Adaptive resolution simulation of liquid water. *Journal of Physics Condensed Matter*, *21*(49), 499801–499801. DOI: 10.1088/0953-8984/21/49/499801

Praprotnik, M., Site, L. D., & Kremer, K. (2008). Multiscale simulation of soft matter: From scale bridging to adaptive resolution. *Annual Review of Physical Chemistry*, *59*(1), 545–571. DOI: 10.1146/annurev.physchem.59.032607.093707

Prudhomme, S., Ben Dhia, H., Bauman, P. T., Elkhodja, N., & Oden, J. T. (2008). Computational analysis of modeling error for the coupling of particle and continuum models by the Arlequin method. *Computer Methods in Applied Mechanics and Engineering*, *197*(41–42), 3399–3409. DOI: 10.1016/j.cma.2008.03.014

Pütz, M., & Nielaba, P. (2015). Effects of temperature on spinodal decomposition and domain growth of liquid-vapor systems with smoothed particle hydrodynamics. *Physical Review E: Statistical, Nonlinear, and Soft Matter Physics*, *91*(3), 032303. DOI: 10.1103/PhysRevE.91.032303

Qi, Y., Wang, L.-P., Guo, Z., & Chen, S. (2023). Consistent lifting relations for the initialization of total-energy double-distribution-function kinetic models. *Physical Review. E*, *108*(6), 065301. DOI: 10.1103/PhysRevE.108.065301

Rajapriyadharshini, J. R. (2022). An improved smoothed particle hydrodynamics approach using new inverse kernel function. *Journal of Ocean Engineering and Science*, *7*(4), 327–336. DOI: 10.1016/j.joes.2021.09.004

Ramachandran, P., Bhosale, A., Puri, K., Negi, P., Muta, A., Dinesh, A., Menon, D., Govind, R., Sanka, S., Sebastian, A. S., Sen, A., Kaushik, R., Kumar, A., Kurapati, V., Patil, M., Tavker, D., Pandey, P., Kaushik, C., Dutt, A., & Agarwal, A. (2021). PySPH: A Python-based framework for smoothed particle hydrodynamics. [TOMS]. *ACM Transactions on Mathematical Software*, *47*(4), 1–38. DOI: 10.1145/3460773

Rappe, A. K., & Goddard, W. A.III. (1991). Charge equilibration for molecular dynamics simulations. *Journal of Physical Chemistry*, *95*(8), 3358–3363. DOI: 10.1021/j100161a070

Rudd, R. E., & Broughton, J. Q. (1998). Coarse-grained molecular dynamics and the atomic limit of finite elements. *Physical Review B: Condensed Matter*, *58*(10), R5893–R5896. DOI: 10.1103/PhysRevB.58.R5893

Sadjina, S., Kyllingstad, L. T., Skjong, S., & Pedersen, E. (2017). Energy conservation and power bonds in co-simulations: Non-iterative adaptive step size control and error estimation. *Engineering with Computers*, *33*(3), 607–620. DOI: 10.1007/s00366-016-0492-8

Schütt, K. T., Sauceda, H. E., Kindermans, P. J., Tkatchenko, A., & Müller, K. R. (2018). SchNet - A deep learning architecture for molecules and materials. *The Journal of Chemical Physics*, *148*(24), 241722. Advance online publication. DOI: 10.1063/1.5019779

Senftle, T. P., Hong, S., Islam, M. M., Kylasa, S. B., Zheng, Y., Shin, Y. K., Junkermeier, C., Engel-Herbert, R., Janik, M. J., Aktulga, H. M., Verstraelen, T., Grama, A., & Van Duin, A. C. T. (2016). *The ReaxFF reactive force-field: Development, applications and future directions. In npj Computational Materials* (Vol. 2). Nature Publishing Group., DOI: 10.1038/npjcompumats.2015.11

Senftle, T. P., Hong, S., Islam, M. M., Kylasa, S. B., Zheng, Y., Shin, Y. K., Junkermeier, C., Engel-Herbert, R., Janik, M. J., Aktulga, H. M., Verstraelen, T., Grama, A., & van Duin, A. C. T. (2016a). The ReaxFF reactive force-field: Development, applications and future directions. *npj Computational Materials*, *2*(1), 1–14. DOI: 10.1038/npjcompumats.2015.11

Senftle, T. P., Hong, S., Islam, M. M., Kylasa, S. B., Zheng, Y., Shin, Y. K., Junkermeier, C., Engel-Herbert, R., Janik, M. J., Aktulga, H. M., Verstraelen, T., Grama, A., & van Duin, A. C. T. (2016b). The ReaxFF reactive force-field: Development, applications and future directions. *npj Computational Materials*, *2*(1), 1–14. DOI: 10.1038/npjcompumats.2015.11

Sengul, M. Y., Song, Y., Nayir, N., Gao, Y., Hung, Y., Dasgupta, T., & van Duin, A. C. T. (2021). INDEEDopt: A deep learning-based ReaxFF parameterization framework. *npj Computational Materials*, *7*(1), 68. DOI: 10.1038/s41524-021-00534-4

Shapeev, A. V. (2011). Consistent energy-based atomistic/continuum coupling for two-body potentials in one and two dimensions. *Multiscale Modeling & Simulation*, *9*(3), 905–932. DOI: 10.1137/100792421

Sheidaei, M., Zhu, J., & Agardh, S. (2024). Assessing the Effect of Specimen Preparation Methods on DSR Test Results of Bitumen Using Factorial Design Analysis. *Materials (Basel)*, *17*(20), 5117. DOI: 10.3390/ma17205117

Shilkrot, L. E., Curtin, W. A., & Miller, R. E. (2002). A coupled atomistic/continuum model of defects in solids. *Journal of the Mechanics and Physics of Solids*, *50*(10), 2085–2106. DOI: 10.1016/S0022-5096(02)00017-0

Smith, E. R., & Theodorakis, P. E. (2024a). Multiscale simulation of fluids: Coupling molecular and continuum. *Physical Chemistry Chemical Physics*, *26*(2), 724–744. DOI: 10.1039/D3CP03579D

Smith, E. R., & Theodorakis, P. E. (2024b). Multiscale simulation of fluids: Coupling molecular and continuum. *Physical Chemistry Chemical Physics*, *26*(2), 724–744. DOI: 10.1039/D3CP03579D

Sockwell, K. C., Bochev, P., Peterson, K., & Kuberry, P. (2023). Interface flux recovery framework for constructing partitioned heterogeneous time-integration methods. *Numerical Methods for Partial Differential Equations*, *39*(5), 3572–3593. DOI: 10.1002/num.23015

Sohag, M., Chausalkar, A., Li, L., & Yang, X. (2022). Numerical study of drop spread and rebound on heated surfaces with consideration of high pressure. *Physics of Fluids*, *34*(11), 113319. DOI: 10.1063/5.0124794

Stillinger, F. H., & Weber, T. A. (1985). Computer simulation of local order in condensed phases of silicon. *Physical Review B: Condensed Matter*, *31*(8), 5262–5271. DOI: 10.1103/PhysRevB.31.5262

Su, J. T., & Goddard, W. A.III. (2007). Excited electron dynamics modeling of warm dense matter. *Physical Review Letters*, *99*(18), 185003. DOI: 10.1103/PhysRevLett.99.185003

Subedi, K. K., Kong, S.-C., & Kweon, C.-B. M. (2022). Numerical study of consecutive drop/wall impacts using smoothed particle hydrodynamics. *International Journal of Multiphase Flow*, *151*, 104060. DOI: 10.1016/j.ijmultiphaseflow.2022.104060

Sun, J., He, Y., Tao, W., Yin, X., & Wang, H. (2012). Roughness effect on flow and thermal boundaries in microchannel/nanochannel flow using molecular dynamics-continuum hybrid simulation. *International Journal for Numerical Methods in Engineering*, *89*(1), 2–19. DOI: 10.1002/nme.3229

Tang, X., Wu, D., Wang, Z., Haidn, O., & Hu, X. (2023). An explicit multi-time stepping algorithm for multi-time scale coupling problems in SPH. ArXiv Preprint ArXiv:2309.04010.

Tao, K., Zhou, X., & Ren, H. (2021). A Local Semi-Fixed Ghost Particles Boundary Method for WCSPH. *Journal of Marine Science and Engineering*, *9*(4), 416. DOI: 10.3390/jmse9040416

Techawinyutham, L., Techawinyutham, W., Laohapanich, C., Insawang, K., Sanjay, M. R., & Siengchin, S. (2024). Possibility of infrared (IR) thermography camera to investigate properties of injected plastic product and optimisation injection process parameter using a design of experiment. *Advances in Materials and Processing Technologies*, *10*(3), 1501–1522.

Tersoff, J. (1988). New empirical approach for the structure and energy of covalent systems. *Physical Review B: Condensed Matter*, *37*(12), 6991–7000. DOI: 10.1103/PhysRevB.37.6991

Tersoff, J. (1989). Modeling solid-state chemistry: Interatomic potentials for multicomponent systems. *Physical Review B: Condensed Matter*, *39*(8), 5566–5568. DOI: 10.1103/PhysRevB.39.5566

Unke, O. T., Chmiela, S., Sauceda, H. E., Gastegger, M., Poltavsky, I., Schütt, K. T., Tkatchenko, A., & Müller, K. R. (2021). Machine Learning Force Fields. In Chemical Reviews (Vol. 121, Issue 16, pp. 10142–10186). American Chemical Society. DOI: 10.1021/acs.chemrev.0c01111

Unke, O. T., Chmiela, S., Sauceda, H. E., Gastegger, M., Poltavsky, I., Schütt, K. T., Tkatchenko, A., & Müller, K.-R. (2021). Machine learning force fields. *Chemical Reviews*, *121*(16), 10142–10186. DOI: 10.1021/acs.chemrev.0c01111

Van Duin, A. C. T., Dasgupta, S., Lorant, F., & Goddard, W. A. (2001). ReaxFF: A reactive force field for hydrocarbons. *The Journal of Physical Chemistry A*, *105*(41), 9396–9409. DOI: 10.1021/jp004368u

van Duin, A. C. T., Dasgupta, S., Lorant, F., & Goddard, W. A. (2001). ReaxFF: A reactive force field for hydrocarbons. *The Journal of Physical Chemistry A*, *105*(41), 9396–9409. DOI: 10.1021/jp004368u

Violeau, D., & Rogers, B. D. (2016). Smoothed particle hydrodynamics (SPH) for free-surface flows: Past, present and future. *Journal of Hydraulic Research, 54*(1), 1–26. DOI: 10.1080/00221686.2015.1119209

Violi, A. (2004). Modeling of soot particle inception in aromatic and aliphatic premixed flames. *Combustion and Flame, 139*(4), 279–287. DOI: 10.1016/j.combustflame.2004.08.013

Vogiatzis, G. G., & Theodorou, D. N. (2018). Multiscale molecular simulations of polymer-matrix nanocomposites: Or what molecular simulations have taught us about the fascinating nanoworld. *Archives of Computational Methods in Engineering, 25*(3), 591–645. DOI: 10.1007/s11831-016-9207-y

Vreven, T., & Morokuma, K. (2006). Hybrid methods: Oniom (qm: mm) and qm/mm. *Annual Reports in Computational Chemistry, 2*, 35–51. DOI: 10.1016/S1574-1400(06)02003-2

Wagner, G. J., Karpov, E. G., & Liu, W. K. (2004). Molecular dynamics boundary conditions for regular crystal lattices. *Computer Methods in Applied Mechanics and Engineering, 193*(17–20), 1579–1601. DOI: 10.1016/j.cma.2003.12.012

Wagner, G. J., & Liu, W. K. (2003). Coupling of atomistic and continuum simulations using a bridging scale decomposition. *Journal of Computational Physics, 190*(1), 249–274. DOI: 10.1016/S0021-9991(03)00273-0

Wang, D., Qiang, H., & Shi, C. (2020). A multiphase SPH framework for solving the evaporation and combustion process of droplets. *International Journal of Numerical Methods for Heat & Fluid Flow, 30*(3), 1547–1575. DOI: 10.1108/HFF-08-2019-0666

Wang, S., Zhao, L., & Liu, Y. (2023). A concurrent multiscale method based on smoothed molecular dynamics for large-scale parallel computation at finite temperature. *Computer Methods in Applied Mechanics and Engineering, 406*, 115898. DOI: 10.1016/j.cma.2023.115898

Wei, X., Li, Z., & Zhao, G. (2024). A review of multiscale numerical modeling of rock mechanics and rock engineering. *Deep Underground Science and Engineering.*

Weinan, E., & Engquist, B. (2003). The heterognous multiscale methods. *Communications in Mathematical Sciences, 1*(1), 87–132. DOI: 10.4310/CMS.2003.v1.n1.a8

Werder, T., Walther, J. H., & Koumoutsakos, P. (2005a). Hybrid atomistic-continuum method for the simulation of dense fluid flows. *Journal of Computational Physics, 205*(1), 373–390. DOI: 10.1016/j.jcp.2004.11.019

Xiao, S. P., & Belytschko, T. (2004). A bridging domain method for coupling continua with molecular dynamics. *Computer Methods in Applied Mechanics and Engineering, 193*(17–20), 1645–1669. DOI: 10.1016/j.cma.2003.12.053

Xiao, X., Ma, X., Kari, T., Xu, Q., & Fan, M. (2023). Numerical simulation of droplet evaporation based on the smoothed particle hydrodynamics method. *Thermal Science, 27*(5 Part A), 3745–3756. DOI: 10.2298/TSCI230311101X

Yan, L., Duan, X., Liu, B., & Xu, J. (2018). Gaussian processes and polynomial chaos expansion for regression problem: Linkage via the RKHS and comparison via the KL divergence. *Entropy (Basel, Switzerland), 20*(3), 191. DOI: 10.3390/e20030191

Yang, X., & Kong, S.-C. (2017). Smoothed particle hydrodynamics method for evaporating multiphase flows. *Physical Review. E, 96*(3), 033309. DOI: 10.1103/PhysRevE.96.033309

Yang, X., & Kong, S.-C. (2020). Smoothed particle hydrodynamics modeling of fuel drop impact on a heated surface at atmospheric and elevated pressures. *Physical Review. E, 102*(3), 033313. DOI: 10.1103/PhysRevE.102.033313

Yang, X., Ray, M., Kong, S.-C., & Kweon, C.-B. M. (2019). SPH simulation of fuel drop impact on heated surfaces. *Proceedings of the Combustion Institute, 37*(3), 3279–3286. DOI: 10.1016/j.proci.2018.07.078

Yasuda, S., & Yamamoto, R. (2008). A model for hybrid simulations of molecular dynamics and computational fluid dynamics. *Physics of Fluids, 20*(11), 113101. Advance online publication. DOI: 10.1063/1.3003218

Yasuda, S., & Yamamoto, R. (2010). Multiscale modeling and simulation for polymer melt flows between parallel plates. *Physical Review E: Statistical, Nonlinear, and Soft Matter Physics, 81*(3), 036308. Advance online publication. DOI: 10.1103/PhysRevE.81.036308

Ye, T., Pan, D., Huang, C., & Liu, M. (2019). Smoothed particle hydrodynamics (SPH) for complex fluid flows: Recent developments in methodology and applications. *Physics of Fluids, 31*(1), 011301. DOI: 10.1063/1.5068697

Ye, T., Pan, D., Huang, C., & Liu, M. (2019). Smoothed particle hydrodynamics (SPH) for complex fluid flows: Recent developments in methodology and applications. *Physics of Fluids, 31*(1), 011301. DOI: 10.1063/1.5068697

Yu, J., Sinnott, S. B., & Phillpot, S. R. (2007). Charge optimized many-body potential for the Si/ SiO 2 system. *Physical Review B: Condensed Matter and Materials Physics, 75*(8), 085311. DOI: 10.1103/PhysRevB.75.085311

Zhang, C., Zhu, Y., Wu, D., Adams, N. A., & Hu, X. (2022). Smoothed particle hydrodynamics: Methodology development and recent achievement. *Journal of Hydrodynamics, 34*(5), 767–805. DOI: 10.1007/s42241-022-0052-1

Zhang, W., Starr, F. W., Beers, K. L., & Douglas, J. F. (2022). Reactive Molecular Dynamics Simulations of the Depolymerization of Polyethylene Using Graphene-Oxide-Supported Platinum Nanoparticles. *The Journal of Physical Chemistry A, 126*(20), 3167–3173. DOI: 10.1021/acs.jpca.2c01167

Chapter 9

Green Synthesis, Characterization, and Applications of Tin Nanoparticles

Aditya Shrivastav
https://orcid.org/0009-0000-8461-8171
Ajeenkya D.Y. Patil University, India

Susanta Das
https://orcid.org/0000-0002-9314-3988
Ajeenkya D.Y. Patil University, India

ABSTRACT

Nanotechnology has evolved as a popular, central and core part of research and development activities, in the current time of advanced science. Metal & metal oxides have a good amount of contribution in experiments & research. Specifically, from metal oxides, Tin oxide (SnO2) has gained interest among researchers due to its wide range of applications. Research on tin nanoparticles (SnNPs) has helped researchers discover the therapeutic properties of SnNPs in recent years. There are many methods for the synthesis of nanoparticles of different size & shapes. There are generally three methods to synthesize NPs - physical, chemical & biological. In particular, plant mediated green (biological) method uses different parts of plants. Detailed Analysis have been done based on the experiments, revolving around research on Sn and SnO2 NPs, which further helped in discovering various properties of green or biologically synthesized nanoparticles. The green synthesis of SnO2 \ NPs through various media, their characteristics, & ample uses are covered in this review.

1. INTRODUCTION

Nanotechnology, is an intriguing concept, which revolves around activities & experiments reducing large and bulky materials to tiny ones with size varying between one to one hundred nanometers. Nanotechnology covers diverse areas under science, for example, physics, chemistry, human biology, genetics,

DOI: 10.4018/979-8-3373-6058-4.ch009

energy sciences, food industry, electronics, aviation, medicine or any other field of scientific research. Nanostructures are very relevant in today's world, especially for the modern science, due to which they turned up to be creative part of research. It has become a significant area of focus for most of the curious researchers. The reduction in size of materials into nanoparticles increases the surface to volume ratio which gives rise to unique physical, biological, chemical, electrical, and optical properties. This is the reason for the various applications of nanoparticles. These tiny particles are light in weight which make them useful, they also have properties like higher screening, better electrical conductivity, more strength, more durability, and better reactivity with other significant traits. The tin nanoparticles are generally considered light in weight, even though the original or principal material has relatively high density, the individual nanoparticles are light in weight due to their nanoscale dimensions (Gari et al., 2023).

Nanoparticles (NPs) can be categorized based on size, shape and morphology, for example, the most common ones are metal based NPs, carbon-based NPs, ceramic NPs, polymeric NPs, semiconductor NPs, lipid-based NPs, etc. Different nanoparticles have different applications, like Silver NPs are popular for their anti-microbial effect, Nickel NPs are the foundation of fuel cells, nanowires, textiles and coatings. Moreover, the Gold NPs help in treatment of cancer and cardiac illness, Copper Oxide NPs have antibiotic, antifungal and antimicrobial properties, similarly Zinc Oxide NPs, Titanium Oxide NPs, Cerium Oxide NPs are fantastic as UV absorber. Generally, in today's time, metal oxide nanoparticles are equally important and used, they have been one of the most trending topics for research and development in the study of nanoparticles because of their high resistance to bacteria, fungus and other infection-causing microbes. Moreover, metal oxide nanoparticles synthesized through green methods show amazing results and they are economically and environmentally favourable to produce. However, among the metal oxide nanoparticles, Tin Oxide, a group-14 metal oxide which has commendable chemical, physical, optical and medicinal properties, these NPs have attracted the researchers owing to their remarkably perceptible magnetic, photo-catalytic, anti-bacterial, anti-fungal properties. Nanostructured SnO_2 particles possess exceedingly large potential in environment remediation because of their reasonably designed structure and flexible band energy gap ranging from 3.6 to 3.8 eV. SnO_2 NPs are commonly found or used in in chemiresistant gas sensors and can detect various gases like nitrous oxide, nitrogen dioxide, hydrogen disulphide and carbon monoxide owing to its low manufacture cost, rapid response and high sensitivity (Din et al., 2022).

SnO_2 nanostructures are further manipulated and used as nanowires, quantum wires, nanorods, hollow spheres, nanofilms, hollow nanotubes, core-shells etc. The ability of SnO_2 NPs to adapt themselves to different ways of doping make them probable and more approachable materials for electronic storage device making and also energy generation. Taking another example of biologically synthesized nanoparticles, Tin Oxide nanoparticles produced from the leaf extracts of *Calotropis gigantea* show the photocatalytic activity by degrading methylene orange. This innovative use of natural plant extracts showcases the potential of green chemistry while providing a sustainable way to produce environmentally beneficial nanoparticles. The properties of SnO_2 NPs are appealing and peculiar because their nano-scaled size makes them acceptable for grey water treatment, pollutant sensing and restoration of environment. One interesting aspect about SnO_2 NPs concerning the high surface-to-volume ratio with small particle size is its mobility, while the low atom count on edge and vertex gives them effective adsorption property. Their high thermal resistivity up to around 500 ^{0}C and their strong physical and chemical stability help in manufacture of dye-sensitizing solar cells, lithium-ion batteries and photoelectrochemical cells. Many applications, such as gas sensors, transparent conducting electrodes, catalyst support, and Li-ion battery anode materials, have been investigated for SnO_2 which acts as an n-type semiconductor. They are considered crucial for identifying gases, for example, nitrogen dioxide (NO_2), which is naturally hazardous and can lead to respiratory and environmental problems. Therefore, SnO_2 gas sensors in particular have received a lot of attention. Structured and measurable efforts are being made to create nanostructured SnO_2 materials for enhanced selectivity and sensitivity in detecting the gas, as research has shown that modifications in structured way can significantly improve the performance of SnO_2 based sensors. In terms of applications, as we know, Nanoscience and nanoparticles are used and their concept is applied in many fields including medicine, especially herbal medicine. The medicine obtained using plant parts such as seeds, roots, leaves and flowers for therapeutic purposes is called herbal medicine, botanical medicine, or phytomedicine. Since research has shown its advantages in both ways, treating and preventing diseases, it has been used for centuries and is still being studied and implemented. As an illustration, consider the herb *Ziziphora clinopodioides Lam.* which is used as traditional Iranian medicine. This plant is mostly used to treat colds, digestive problems, blood disorders, and inflammation. Moreover, it has antifungal, antibacterial, anti-inflammatory and antioxidant properties (Gebreslassie & Gebretnsae,2021).

The chemical and physical synthesis methods like chemical precipitation, hydrothermal, microemulsions, sol-gel methods were also implemented for synthesis of SnO_2 NPs. However, toxic effects, expenses and high energy and pressure requirements lead to limited use of chemical and physical methods. Hence, green synthesis methods grabbed more importance as they were known to be cost effective and environment friendly. Green Synthesis use plant extracts, biomolecules, agriculture products etc. as reducing and capping agents. There are two methods of preparing NPs, bottom-up and top-down approach, which are applied to all kind of ways for preparation of NPs, physical method, chemical method, and biological method. Bottom-up method, also known as constructive method, includes the manipulation or transformation of bulk materials from atoms to SnO_2 NPs. For example, sol-gel method, spinning method, chemical vapor deposition (CVD), pyrolysis and biosynthesis are some of the bottom-up approaches. These are easy and efficient to implement. Top-down, also known as destructive method, is the method in which SnO_2 NPs are made from reduction of bulk material or particles, they are reduced to nano-scale particles. For example, mechanical milling, laser ablation, sputtering and thermal decomposition, nanolithography. In this method, the particle size cannot be controlled, which is one of the main disadvantages. Hence, it is not used frequently. On the other hand, green or biological methods of synthesis of SnO_2 NPs are more suitable, eco-friendly and cost effective as compared to physical and chemical methods. Many researchers have suggested the use of other biological sources like bacteria, vitamins, proteins, and other metabolites for green synthesis of SnO_2 NPs (Sagadevan et al, 2021).

2. GREEN SYNTHESIS OF Sn NPs

2.1 Synthesis of Sn NPs from Plant Extract

There are plentiful of renewable resources like plants which make Green Synthesis cost effective, eco-friendly and non-toxic in nature. The vitamins, proteins, polysaccharide, amino acids, enzymes and organic constituents like phenolic acids, citrates, flavonoids, alkaloids, tannins, carbonyls, amines, and amides are the key active components responsible for capping, reduction, and stabilization (Gebreslassie & Gebretnsae,2021).

2.1.1 Leaves

The leaves extract of *Aquilaria malaccensis* plant, also called agarwood, in solution with tin(IV) chloride pentahydrate at room temperature supported with eco-friendly, non-toxic, and rapid green method gives SnO_2 NPs, which are high in quality. These NPs display immense ability in catalysis and optoelectronic devices. Another example is using *Daphne Mucronata* leaves extract as reducing and capping agent for SnO_2 NPs fabrication. The SnO_2 NPs produced using water extracts from *Limonia acidissima* and biologically derived leaves were found to have anti-cancer ability. Similarly, there are many other ways to produce SnO_2 NPs through different leaves extract such as simple, environment-friendly, and low-cost process using guava leaf extract (*Psidium guajava*). Moreover, the synthesis of SnO_2 NPs from the leaf extract of the *S. rebaudiana* plant is clearly simple. *Ficus carica* can also be used synthesize SnO NPs using water and tin (II) chloride solution at around 75 to 80 ^{0}C (Jaji et al., 2020).

2.1.2 Fruits

In this method, researchers used *Cyphomandra betacea* fruit extract in methanol to biosynthesize SnO_2 NPs, a simple biological method. SnO_2 NPs were also synthesized from *Averrhoa bilimbi* fruit essence, which is considered to be cost effective and environment-friendly method, being a one-step process. Even jujube fruit was used in consideration of it being a non-toxic renewable reducing agent and excellent stabilizer for producing SnO_2 NPs (Gebreslassie & Gebretnsae, 2021).

2.1.3 Vegetable

Researchers tried the idea of biosynthesis of $Sn(OH)_2$ using the water extract of fresh cauliflower (*Brassica oleracea L. var botrytis*), from which SnO_2 NPs were produced which were rutile tetragonal phase having quasi-spherical and spherical morphologies for two samples. Their size range was between 3.62 and 6.34 nm (Jaji et al., 2020).

2.1.4 Seeds

Water taken out from Soaked Bengal gram beans (*Cicer arietinum L.*) extract is used in synthesizing Nickel-SnO_2 nano material doping. This is again an environment- friendly and economical biological synthesis method. There are immense applications of nanostructured materials doped with transition metals and the advantages of this method or production are considered to be fruitful, serving versatile applications (Gebreslassie & Gebretnsae,2021).

2.1.5 Peels

C. aurantifolia peels extract was used at different concentrations as reducing agent to synthesize SnO_2 NPs. Through this method, highly pure SnO_2 NPs with great photocatalytic performance was synthesized. Different concentrations of *Lycopersicon esculentum* peel extract were also used in research, which produced SnO_2 NPs being crystalline and purely tetragonal crystal structure (it had a bandgap of around 3.3 eV) (Jaji et al., 2020).

2.1.6 Flower

Saraca indica flower was a key component for biological synthesis of spherical tetragonal rutile SnO_2 NPs of average particles size ranging from 2.1 nm to 4.1 nm. These NPs showed antibacterial activity against *GNB E. coli*. (Garrafa-Galvez et al., 2019).

2.1.7 Roots

Researchers developed a biological synthesization method for producing materials that decompose toxic substances such as dyes through environmental synthesis. In this method, SnO_2 NPs were made using the water extract of the root cortex of *Catunaregamspinosa* (Gebreslassie & Gebretnsae, 2021).

2.1.8 Bacterial Sources

Research has proved that Tin NPs can also be produced from Bacterial and fungal extracts. The metabolites for example, bacterial proteins and biomolecules present in the bacterial biomass act as strong stabilizing and reducing agent for the green synthesis of SnO_2 NPs. Although much bacterial and fungal species are reported to act as reducing and capping agents in the SnO_2 NPs synthesis, yet till date only one research study has been reported in which SnO_2 NPs produced with bacterium *Erwinia herbicola* (Sagadevan et al. 2021).

2.2 Synthesis of Sn NPs from Aqueous Extract

2.2.1 About the Plant

Among the *Asteraceae* family, *calendula officinalis* is a well-known medicinal plant. It has been approved by the European Medicines Agency as a herbal remedy since 2008. Throughout the history, traditional medicine has used this plant to treat a wide range of illnesses. It is a useful treatment for pain management, inflammation, perspiration, wound healing, and minor burns. The presence of numerous chemicals, including phenolics, flavonoids, saponins, carotenes, and mucilage, gives the plant its ability to treat a variety of ailments. *Calendula officinalis* extract has been used in recent years to help prevent cancer in a number of forms. Nanoparticles have gained popularity in medical applications, particularly metal nanoparticles and metal oxides. By releasing reactive oxygen species (ROS), which are harmful to cancer cells, nanoparticles combat the disease. The methods for creating nanoparticles used in cancer treatment and diagnosis will change drastically as our knowledge of scalable nanoparticle production advances and as we create multifunctional nanoparticles (Zhai et al., 2022).

2.2.2 Preparation and Extraction of Aqueous Extract

For the aqueous extract of the leaves of *C. officinalis*, researchers took 15 g of the dried plant leaves and mixed it with 200 mL of the boiling distilled water, then they put it aside for three hours. After that the extract was filtered and evaporated under low pressure. The concentrated crude extract was put in an oven at 50 ^{0}C for 24 hours. The result was a powder in brown colour (Zhai et al., 2022).

2.2.3 Green Synthesis

The tin nanoparticles were synthesized using a green method with reference to previous experiments and studies. 150 millilitres of a 0.1 M Stannous Chloride Dihydrate solution were mixed with 25 millilitres of *Calendula officinalis* extract, which was prepared by dissolving two grams in fifteen millilitres of deionized water. Next, for 55 minutes, the mixture was heated to 90 ^{0}C and refluxed. Pale yellow instead of yellow-green was the colour shift that signalled the creation of tin nanoparticles during process. After being centrifuged for 15 minutes at 12,000 rpm, the resultant tin nanoparticles were cleaned three times using water and ethanol. Using a 50 ^{0}C oven, the finished product was dried. For chemical analysis and biological activity testing, the yellow powder that had formed was kept in a vial (Zhai et al., 2022).

2.3 Synthesis of Sn NPs by Averrhoa Bilimbi Fruit Extract

2.3.1 Plant Extract

Averrhoa bilimbi is a plant of the kingdom Plantae, which belongs to family *Oxalidaceae* and *genus Averrhoa* and species *bilimbi*. Fresh and natural fruits were taken in a fresh sterilized bag at their perfect blooming phase from Alappuzha district Kerala. In the lab, they were washed with tap water and distilled water to clean up the external junk parts. Then they were chopped into pieces and were put in water bath, around 50g of pieces in 100ml distilled water. Finally, the extract was cleaned with a muslin cloth and a Whatmann filter paper (Sunny & Kumar, 2019).

2.3.2 Synthesis of Tin Oxide Nanoparticles

Around 90 ml of distilled water was used to dilute 10 mL of A. bilimbi extract. Moreover, a magnetic stirrer was used to mix 0.1 M of tin chloride (Sigma Aldrich) with an equal volume of A. bilimbi for four hours. After that, it was centrifuged at 5000 rpm for 15 minutes and then washed with ethanol followed by washing it three times with distilled water., by which a precipitate was obtained and it was dehydrated inside an oven for one hour at 80 ^{0}C. Finally, the powder was calcinated for about two hours at 400 ^{0}C (Sunny & Kumar, 2019).

2.4 Synthesis of Sn NPs from Z. clinopodioides

2.4.1 Preparation and Extraction of Z. clinopodioides

Z. clinopodioides was identified and in the spring of 2019, the plant was taken from a site in Kermanshah, Iran, that was 1400 meters above sea level. The plant's leaf powder, weighing 100 grams, was combined with 1000 milliliters of distilled water and shaken for a full day to create the extract. The mixture was then filtered, and a rotary evaporator set at around 60 ^{0}C was used to concentrate the liquid. The extract was then preserved for later use by freezedrying it for 24 hours at a temperature between -50 to -40 ^{0}C (Gebreslassie & Gebretnsae, 2021).

2.4.2 Preparation and Synthesis of Sn NPs

With a few modifications, methods from the past were used to biosynthesize Sn NPs (tin nanoparticles). 90 ml of a 3 mM Stannous Chloride solution was combined with 10 ml of a plant extract solution. An hour was taken for heating this mixture up to 80 ^{0}C. Tin oxide nanoparticle formation was indicated by the change in colour from light green to deep yellow. The nanoparticles were washed with water thrice before being centrifuged. The resultant yellow powder was put away in a vial to be examined and tested further in terms of its biological effects (Gebreslassie & Gebretnsae, 2021).

2.5 Preparation and Extraction of Aqueous Extract

After being gathered in Bozhou City, China, the Foeniculum (F. vulgare) leaves were shaded from the sun. Following the drying process, the leaves were ground into a powder, and 15 grams of the powder were soaked for three days in 200 mL of distilled water Following the solution's filtering, the liquid was bain-marie evaporated to between 40 to 50 ^{0}C. After that, the leftover solid material was combined to create a 10 mg/mL solution in DMEM medium, which was used in the experiments (Shu et al., 2023).

A modified method was used to biosynthesize tin nanoparticles (Sn NPs). 90 mL of a 0.1 M Stannous Chloride Dihydrate solution was mixed with 10 mL of plant extract, which was earlier made by combining 1 gram of the plant with 100 mL of water. Next, the mixture was heated for 60 minutes at a higher temperature, around 80 ^{0}C. A colour change from yellow-green to pale yellow depicted the formation of tin nanoparticles. In preparation for additional biological and chemical testing, the Sn NPs underwent three times of washing with ethanol and water, centrifugation, and drying in an oven at 50 ^{0}C (Shu et al., 2023).

2.6 Preparation of Sn NPs from the Leaves of Ceropegia jainii

2.6.1 Preparation of the Leaves Extract of Plant Ceropegia jainii:

The plant assisted nanoparticles was synthesized using the leaves of *Ceropegia jainii*. This plant was collected from the Rajapur region of the Konkan, Maharashtra, India. 100 g of the *Ceropegia jainii* leaves powder placed in the round bottom flask and 250 mL methanol were added to it and reflux this mixture for 2 hours. Then cool at the general room temperature, filtered it through the Whatman filter paper. The filtrate was evaporated in Rota Evaporator to get powder. This powder stored in glass bottle and used for the study (Nayak et al., 2020).

2.6.2 Synthesis of the Tin Nanoparticles Using Leaves Extract

A conical flask was filled with 100 mL of a tin chloride solution (0.1 mol). The flask was filled with 5 mL of benzyl alcohol and 1 gram of Ceropegia jainii plant extract. The mixture was then cooled to room temperature after being heated to 100 ^{0}C for 18 hours. Then, for making the solution basic, drop by drop was added sodium hydroxide solution. Furthermore, the mixture was centrifuged at 3000 rpm. The uppermost layer was gathered into a silica crucible and subjected to 6 hours at 200 ^{0}C in a muffle furnace to produce fine nanoparticles (Nayak et al., 2020).

3. CHARACTERIZATION AND RESULTS

Based on the Tin or Tin oxide nanoparticles produced from various sources, it is possible to characterize them, as in analyse it to understand their shape, size, structure, features or properties and applications. This can be done by many methods, like FI-TR spectroscopy, X-ray diffraction, SEM and TEM, energy-dispersive X-ray analysis, XRD analysis, etc (Garrafa-Galvez et al., 2019).

3.1 Chemical Characterization of Sn NPs Produced from C. officinalis

In the characterization and examination of nanoparticles, FT-IR spectroscopy is stated to be one of the most widely used techniques for the characterization of green synthesized metallic nanoparticles (Zhai et al., 2022). The peaks associated with Metal-Oxygen bonds have a range of 400 to 700 cm^{-1}. Following peaks above 1000 cm^{-1}, these relate to the functional groups of organic compounds acting as capping and reducing agents in the synthesis of metallic nanoparticles. The FT-IR spectra of Sn NPs exhibit peaks at the wavenumbers of 430, 597 and 661 cm^{-1} which correspond to the Sn-O bond. The EDS analysis of metallic nanoparticles is another qualitative method for the evaluation of characterization of metallic nanoparticles. The method gives the elemental analysis of the nanoparticles through some kind of signals. In such way, signals were recorded for the green synthesis of tin nanoparticles. It also showed signals for carbon and oxygen, revealing their presence at the surface of Sn NPs. The FE-SEM technique is used to examine morphology and particle size, which are two essential characteristics of metallic nanoparticles. The FE-SEM results depicted spherical morphology of Sn NPs with particle size ranging from 21.63 nm to 64.07 nm. XRD method, a different method, is one of the most popular methods to examine the crystallinity of the metallic nanoparticles. The result or XRD pattern show that Sn NPs have small crystal size and well crystallizing ability. Using Scherer's equation, the crystal size of Sn NPs was found to be around 18 nm (Sudhaparimala, 2014).

3.2 Chemical Characterization of Tin Oxide NPs Produced from Averrhoa bilimbi

X-ray diffraction was used to analyse the crystal structure of the biosynthesized tin oxide nanoparticles. Scherrer's formula was used to determine the nanoparticles' size, which resulted to be above 2.5 nm. 400-4500 cm^{-1} was the range of the nanoparticles FTIR spectrum, which were derived from A. bilimbi. Using the scanning electron microscopy (SEM), the nanoparticles' shape was identified as spherical and the presence of tin (Sn) and oxygen (O) was verified by the EDS (Energy-dispersive spectroscopy) spectrum. The nanoparticles' size revealed by the TEM (Transmission Electron Microscopy) images was around 3.08 nm, which corresponded with the XRD data (Sunny & Kumar, 2019).

3.3 Chemical Characterization of Sn NPs Produced from F. vulgare

From the results of this analysis surface plasmon resonance (SPR) bands at about 219 nm and 456 nm in the UV–vis spectrum confirmed the formation of tin nanoparticles (Sn NPs). The formation of Sn NPs associated with Sn–O bonds was confirmed by the observation of peaks at 441 cm^{-1}, 520 cm^{-1}, and 609 cm^{-1}in the FT-IR spectrum. The Sn NPs were found to have a Tin Oxide crystal structure, with a calculated crystal size of around 8.40 nm to 8.50 nm. Images taken from scanning electron microscopy (SEM) showed that the Sn NPs were spherical and the size of the nanoparticles were ranging from 26.45 nm to 38.53 nm. The presence of carbon, oxygen, and tin in the Sn NPs was verified through EDX Analysis (Shu et al., 2023).

3.4 Chemical Characterization of Sn NPs Produced from Ceropegia jainii

The XRD analysis revealed peaks at specific planes such as 110, 200, 111, and others, which confirmed the formation of tin oxide nanoparticles with a cassiterite structure. The UV analysis revealed that the nanoparticles successfully formed as depicted by their absorption of light at 330 nm. The well-formed nanoparticles, which range in size from 100 nm to 150 nm, were revealed by scanning electron microscopy (SEM). A key peak was observed at 540 cm^{-1} in the FTIR spectra, which revealed the formation of tin oxide nanoparticles (Nayak et al., 2020).

There are many methods to characterize and assess the Tin or Tin Oxide nanoparticles formed or produced. In similar ways, other Sn NPs were characterized and assessed in different experimental procedures. They could also be assessed in methods like phytochemical assessment. They could be assessed or experimented upon to check antibacterial activity and antioxidant activity.

4. THEORETICAL AND COMPUTATIONAL INSIGHTS OF SnO_2 NANOPARTICLES

Tin dioxide (SnO_2) is a model wide-bandgap semiconductor with a dominant crystal structure of tetragonal (space group P42/mnm) rutile, with each Sn atom coordinated with six oxygen atoms, and each oxygen atom coordinated with three Sn atoms. It has intrinsic n-type conductivity, which is primarily due to intrinsic point defects, including oxygen vacancies and tin interstitials, which provide shallow donor levels immediately below the conduction band minimum. SnO_2 has a direct band gap of 3.63.8 e V, which allows it to be exceptionally optically transparent in the visible range and, as such, is very easy to use in optoelectronic, photocatalytic, and gas-sensing applications (Tameh et al., 2025; Karmaoui et al., 2018; Rahimi et al., 2024; Mashkovtsev et al., 2024; Chellamuthu et al., 2025). Once the particle size enters the nanoscale regime, the effect of quantum confinement becomes significant and can modulate the electronic and optical properties of the particle to a discrete energy state and a blue shift in the absorption edge is observed. At the same time, the augmentation of the ratio of the surface to volume boosts the defect density, reactivity at the surface, and charge carrier dynamics. DFT has become essential in understanding such quantum phenomena, allowing the accurate determination of the electronic band structure, defect formation energies, and charge density distributions, and therefore provides atomistic insights into the size-dependence of the change of optoelectronic and catalytic properties in SnO_2 (Fatimah et al., 2022; Tasisa et al., 2024; Ajp, 2024; Tameh et al., 2025; Rahimi et al., 2024; Soumane et al., 2025).

Green synthesis of SnO_2 nanoparticles takes advantage of the reducing and stabilizing ability of the phytochemical constituents found in plant extracts and include flavonoids, tannin, terpenoids, phenolic acid, and polyhydroxylated aromatics. These biomolecules work in concert to reduce, via the electron donation mechanism, to eliminate the growth of nuclei of snap-on nanoparticle, to tertiary SnO_2 or SnO, and to cap the growing nuclei with the functional groups of the biomolecule's hydroxyl, methoxy, carbonyl, and carboxyl groups, which complex the nanoparticle surface and prevent agglomeration. It is a biomolecular mediation that determines the kinetics of nanoparticle nucleation and growth as well as the crystallographic orientation, the distribution of defects, and the ultimate morphology of the particles (Matussin et al., 2020; Tasisa et al., 2024; Borah et al., 2024; Gebreslassie & Gebretnsae, 2021; Narasaiah et al., 2022). Computational chemistry has played a pivotal role in uncovering such atomistic processes: DFT and molecular dynamics (MD) simulations of their energy landscapes of Sn-phytochemical intermediates, model electron transfer processes in reduction, and computation of adsorption energies which govern the stability of ligand-surface interactions. MD trajectories also reveal dynamics like the exchange of ligands, aggregation, and stabilization by solvents, which can serve as computational validation of the entire structure, which in turn can be used to select plant materials rationally to obtain desired nanostructure shapes (Saif et al., 2023).

Theoretically, the morphology and stability of SnO_2 nanoparticles can be explained by the DFT-based surface energy calculations and Wulff constructions, which assume the equilibrium shapes are established by comparing the relative energy of the facets to thermodynamic stability (Jaszkaniec et al., 2021; Ringe et al., 2013; Laranjeira et al., 2024). Such computational findings combine the principles of crystal chemistry to the actual syntheses by determining the prevailing growth faces and their reactivity in the conditions of green syntheses. Atypically, at the atomic level, modeling has shown that lattice parameters, c/a ratios, and surface terminations can have a direct impact on the electronic density of states, as well as on local chemical potentials, which can be used to control catalytic and adsorption behavior. The narrowing of the band gap as particle size decreases due to quantum confinement has been verified by DFT calculations, again as experimentally shown by UV-Vis data. In addition, surface reconstructions of oxygen vacancies or undercoordinated tin atoms create mid-gap states which promote surface charge exchange and catalytic activity, particularly on high-index planes and the planes which are more reactive because of unsaturated coordination sites (Soumane et al., 2025; Fatimah et al., 2022).

Computational and experimental characterization synergy provides a holistic insight into structure to property relationship in SnO_2. The lattice constants and vibrational frequencies calculated by DFT are close to both the X-ray diffraction (XRD) and FTIR data, which prove the purity of the phase and the functionalization of the surface. Time-dependent DFT (TD-DFT) is capable of useful prediction on the electronic transitions and useful assignments of UV-Vis absorption bands and confirms blue shifts achieved through confinement effects. Theoretical validation such as this is used to refine the empirical interpretation of spectroscopic signatures, such as phytochemical bonding motifs or defect-induced modes in FTIR spectra (Tasisa et al., 2024; Karmaoui et al., 2018). This combination of in silico and experimental data has now become the standard of nanomaterials research in the modern generation, such that the observed phenomena are not being rationalized phenomenologically but on first principles.

SnO_2 nanoparticles are tunable electronic nanoparticles with versatile nanoparticles that are employed in numerous fields, in particular, gas sensing, photocatalysis and in biomedical applications. Adsorption energetics, charge transfer pathways, and defect-mediated activation sites in gas-sensing systems can be explained by DFT-based modeling, sensitivity and selectivity of the systems to analytes, such as NO_2, NH_3, and H_2S, predicted (Tameh et al., 2025; Chellamuthu et al., 2025). Ab initio and TD-DFT simulations can also be used to elucidate the photocatalytic activity by examining the dynamics of the charge carriers induced by light, surface traps and recombination dynamics, and the relationship between the electronic excitation processes and the efficiencies of energy loss at organic pollutants degradation is directly correlated (Tasisa et al., 2024). In addition, the bio-nano interface of computational modeling has aided in the development of the knowledge regarding the antibacterial and anticancer activity as well as contributed to the understanding of the mechanism of the formation of the reactive oxygen species (ROS), the disruption of the membrane and the denaturation of the protein. Such atomistic analyses reveal that the surface chemistry of nanoparticles and states of defects and capping-layers characterize nanoparticle biocompatibility and cytotoxicity (Takhar and Singh, 2025). The discovery process has remained accelerated with new computational techniques such as machine learning (ML) and high-throughput screening, where big datasets of DFT are used to predict band gaps, adsorption energies and catalytic descriptors quickly, allowing the design of functional SnO_2 nanostructures to be designed using data (Masood et al., 2019).

In spite of these major developments, it has been observed that there are a number of problems that must be overcome in obtaining reproducible and mechanistically defined green synthesis of SnO_2 nanoparticles. Fluctuating levels of phytochemicals in plant extracts, complicated interactions of phytochemicals, and uncontrolled capping layer structure are factors that cause unevenness in particle size, morphology, and surface chemistry (Borah et al., 2024; Narasaiah et al., 2022). The stochastic nucleation pathways of natural extracts due to its inherent combinatorial complexity make it difficult to predictively control the properties of nanoparticles in nature. Complex computational models offer a positive path to overcome these problems by connecting the molecular reaction processes at the molecular level to the macroscopic synthesis products. First-principles calculations explain the thermodynamics of individual phytochemical reducers, which ligands are best to use in electron transfer and stabilizing the surface. Multiscale simulations Multiscale simulations consist of a DFT-MD exchange or coarse-grained model and are able to trace energy landscapes between starting reduction and eventual particle morphology. Optimization of synthesis aided with ML on the basis of experimental-computational databases can predict the effects of extract composition and reaction parameters on the properties of nanoparticles to provide rational approaches toward the standardization of green synthesis. In addition, capping layer dynamics simulations illuminate the strength of ligand binding, desorption energy, and solvent interaction informing capping layer strategies to improve colloidal stability, biocompatibility, and functional performance (Masood et al., 2019; Sharma et al., 2025).

The combination of theory, and computational models and the experimental synthesis of nanomaterials therefore represents a paradigm shift, basing the design of SnO_2 nanomaterials on prediction and systematic rational design, rather than on empirical trial and error design approaches. With quantum mechanical modelling, atomistic simulations, and machine learning analytics coming to a common language, researchers now have a chance to explore the intricate interaction between crystal structure, surface chemistry, and functional performance at a level never before witnessed. This intersection creates a rational and sustainable nanomaterials paradigm: theoretically foresighted experimental synthesis facilitating increased reproducibility, scalability, and environmental friendliness in diverse applications, including catalysis and sensing to biomedicine. Computation and green chemistry serve as an ever-changing synergy, which will re-establish the design principles of oxide nanostructures turning SnO_2 into a model of the next generation of sustainable and smart material systems (Masood et al., 2019; Sharma et al., 2025; Borah et al., 2024; Narasaiah et al., 2022).

5. APPLICATIONS OF Sn NPs

There has been a high demand for alternative energy sources and highly efficient energy storage methods for a sustainable future. This is happening due to the global warming, faster environment deterioration, faster fossil fuel depletion, rapid water contamination, etc. These all things have been increasing due to accelerated industrial revolution, development and urbanization. Basically, in this situation, it is not practically possible to stop industrial revolution or urbanisation, but to reduce its effects or to protect environment with even small efforts would be far better than completely letting everything get destroyed. For high energy storage and efficiency, modern technology is one of the big sources that could help it. Specifically, nanotechnology has a deep impact and good scope for the future. Many metal NPs have been accessible and helpful, in the same way SnO_2 has immense applications (Din et al., 2022). Research depicts the antimicrobial, antioxidant, anticancer, and excellent photocatalytic degradation properties of the Tin Oxide nanoparticles produced through green synthesis (Gebreslassie & Gebretnsae,2021).

For anti-bacterial, there have been studies and experiments on efficiency of Tin Oxide nanoparticles over various kinds of bacteria. Gram-negative and gram-positive bacteria are both vulnerable to these green-synthesised Tin Oxide nanoparticles' defences. A lot of plant-based Tin Oxide nanoparticles have shown antibacterial activity against different and various strains of bacteria. Varying concentrations of nanoparticle samples were applied to *Escherichia coli* and *S. aureus* bacterial samples, by using the agar method antibacterial inhibition zones were created. *E. coli* displayed the strongest antibacterial activity among them (Sagadevan et al, 2021). Anti-fungal and Antioxidant: Scientists tried to study the anti-fungal activity and efficiency of doped and undoped NPs against A. niger, A. flavus, and C. albicans fungi specimens. The green synthesized Tin Oxide NPs displayed anti-bacterial activity. The further studies revealed that Asperjillus nigar had a significant activity as compared to *T. viridea* fungi species. The plant extracts have alkaloids, cyclopeptides, flavonoids, phenols, tannins and minerals such as Cu, Fe, Mn, Zn, Se which give the therapeutic and antioxidant property to the NPs (Sudhaparimala, 2014). For Anti-cancerous, fabricated green synthesized NPs show cytotoxic effect on the breast cancer MCF-7 in comparison to human lung fibroblast (WI38) and human amnion (WISH) cell lines. Moreover, green synthesized undoped Tin Oxide and Co-doped Tin Oxide NPs exhibit the anticancer activity. They showed distinct characteristics in vitro anticancer activity and in vivo antitumor propensity against mammary gland breast cancer (MCF-7) cell lines and Ehrlich ascites tumor respectively, which exhibited their potential applications in Oncology field (Sunny & Kumar, 2019).

Physiochemical applications:

1. Gas sensing: Ni-doped Tin Oxide sensors, in the research, showed good gas sensing against many flammables, poisonous, toxic and corrosive gases like hydrogen sulphide, liquefied petroleum gas, ammonia and nitrogen dioxide at 200 ^{0}C which is optimal operating temperature for NO_2 gas. The studies depict that the Ni-doped Tin Oxide films can be used to monitor the NO_2 emissions (Jaji et al., 2020).
2. Adsorption of Cadmium cation (2+): Studies and research confirmed that the green fabricated Tin Oxide NPs have great ability to be used as heavy metal ion like cadmium cation (2+) (Hamedani & Farzaneh, 2006).
3. Mercury cation (2+) electrochemical sensor: The biosynthesized Tin Oxide NPs could be used as electrode modifier for electrochemical detection of Mercury cation (2+) in various water samples (Jaji et al., 2020).

6. BIBLIOMETRIC ANALYSIS

The bibliometric analysis performed by VOSviewer based on the data available on Lens.org provides an in-depth representation of the structural and temporal development of the field of Tin nanoparticles and the fields that it represents. Figure 1 network clustering map may be considered a comprehensive reflection of the intellectual framework of the area of the study of tin nanoparticles. In the case of VOSViewer, several distinct clusters are found each consisting of terms that are closely interrelated and represent significant areas of the theme and research interest. The most prominent cluster, often addressing the terms of such basic research as the production and physicochemical properties of tin-based nanomaterials, includes the keywords of tin oxide nanoparticles, nanostructures, synthesis, sol-gel method, and characterization. The focus of this cluster is the creation of high-precision synthesis approaches that aim at controlling particle size, morphology, and crystalline structure all of which are important factors in influencing their optical, electrical, and catalytic properties.

A second significant cluster is defined by such application-oriented terms as electrocatalysis, lithium-ion batteries, anode materials, energy storage, gas sensors, and photocatalysis. This group of nanoparticles is the technological integration of tin nanoparticles in highly complex functional systems, and it is closely related to environmental remediation and green energy. The fact that this application cluster is related to the synthesis centred cluster implies that the development of surface engineering and synthesis has a direct impact on the performance of application. Another important cluster consists of such keywords as nanocomposites, graphene, carbon nanotubes, hybrid materials, polymer composites, and mechanical properties. This demonstrates that it is of high essence to integrate tin nanoparticles in matrix materials in order to enhance their conductivity, stability and multifunctionality. Smaller but significant groups of the field shown by co-occurrence lines include smaller but significant groups such as biological and medical applications, antibacterial activity, biosensors, toxicity and drug delivery.

Figure 1. Network Clustering of Tin Nanoparticles (Source: Lens Database; created using VOSviewer) (Lens, 2025; Van Eck & Waltman, 2009).

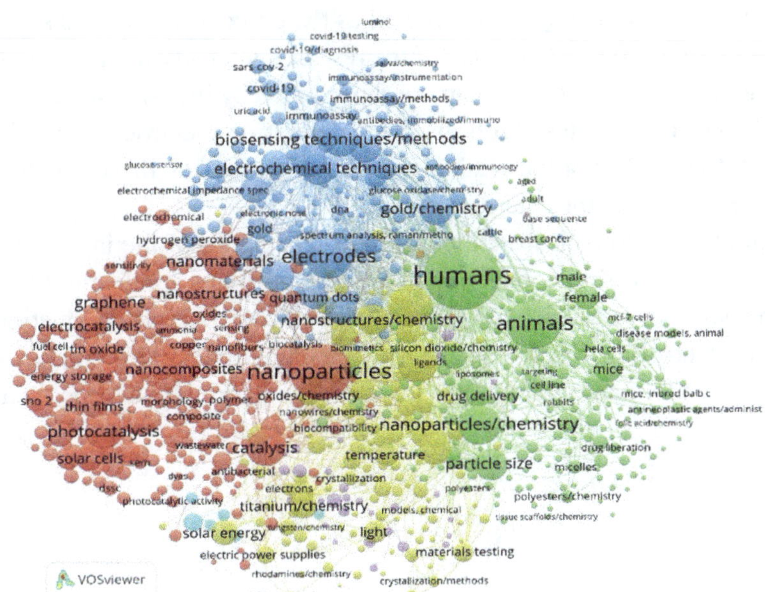

The interdisciplinary nature of the research on tin nanoparticles is reflected by the close-knit and high number of connections between groups. As an illustration, dual-purpose materials under investigation, both in environmental and health-related applications are indicated in the intersection between the terms photocatalysis and antibacterial. All in all, the network clustering shows that the study of tin nanoparticles is not tied to one field of study but, instead, it involves biotechnology, environmental engineering, chemistry, and materials science with common approaches and applications.

The change in the research subjects of the tin nanoparticle is demonstrated in Figure 2 with a visual representation in terms of time. The prevailing issues of interest, during the early study of the field, between the 1990s and early 2000s, included: synthesis, nanostructure, morphology, crystallinity, thin films, and sol-gel process. The chemical and physical properties of tin and its oxide forms are understood at the nanoscale and this is the primary area of research in this early phase. These initial efforts formed a basis of subsequent application-based research. It can be seen in the timeline that the subject matter began to diversify rather swiftly as early as 2010. The appearance of such terms as gas sensors, photocatalysis, and electrocatalysis, energy storage, nanocomposites, and SnO_2 anodes indicated the transition to the production of basic materials to their practical applications. This shift is connected to the global focus on nanomaterials as a sustainable environment and renewable energy. It is interesting to note that advancements in battery materials, especially the use of tin oxide and Tin sulfide nanoparticles as anode material in lithium-ion and sodium-ion batteries are associated with the peak in the number of papers following 2015. The timeline in Figure 2 shows in the recent years (2020-2025) of new areas of interest such as green synthesis, biomedical applications, toxicity assessment, biosensors, and environmental remediation. These articles reveal that the use of sustainable, biocompatible and multipurpose means of making and utilizing tin nanoparticles is gaining increasing significance. The continuous historical evolution of keywords and the interrelated development of the keywords over time depict a coherent and evolutional development of the material development to the practical applications.

Following the frequency of the key words as well as the co-occurrence intensity, the density visualization provided in Figure 3 shows a heat-map of the research intensity in the topic of tin nanoparticles. The brightest and most dense areas of the map are represented by high-impact terms such as "tin oxide nanoparticles," SnO_2, nanocomposites, synthesis, characterization, electrocatalysis, lithium-ion batteries, and "photocatalytic activity. Most of the academic work is clustered on these overriding research fronts, all of which are encompassed by these terms.

Light zone around SnO_2 and nanostructures is the active centre of the importance of tin oxide as a versatile nanomaterial due to its semiconducting properties and many applications in energy and sensors. The proximity of the word's synthesis, hydrothermal, sol-gel, and chemical reduction implies the diversity of the methods of controlling the size and shape of nanoparticles. This emphasis demonstrates that manufacturing and structural control continues to be the primary effects of performance optimization advancements.

Figure 2. Overlay Visualization of Tin Nanoparticles (Source: Lens Database; created using VOSviewer) (Lens, 2025; Van Eck & Waltman, 2009).

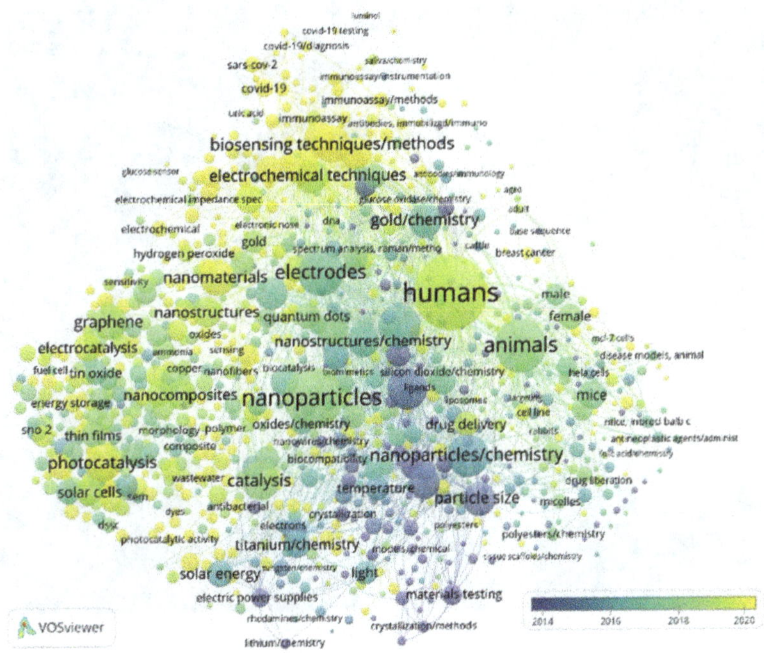

These applications, which include: electrocatalysis, gas sensing, supercapacitors and photocatalysis are emphasized in high density regions nearby. The applications of Tin nanoparticles are catalytic reactions, environmental remediation, and energy devices; all of them are closely related concepts that imply cross-functional research. New terms such as antibacterial activity, biosensors, drug delivery, nanotoxicology, and biomedical applications come into medium-density areas, which means that more attention is drawn to the biological activities and potential therapeutic applications of tin nanoparticles.

Figure 3. Density plot of Tin Nanoparticles (Source: Lens Database; created using VOSviewer) (Lens, 2025; Van Eck & Waltman, 2009).

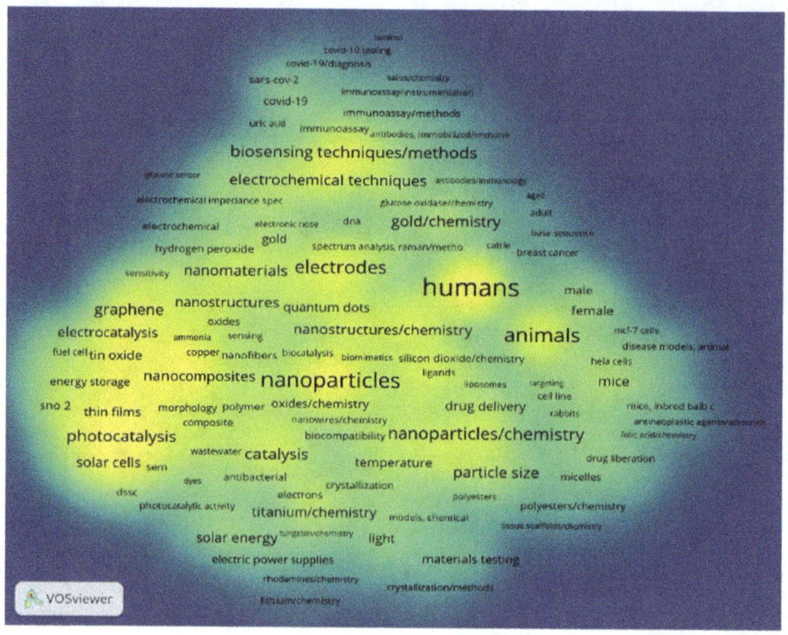

The shift of the gradient between high and low density is a graphic representation of the development of the well-established themes towards emerging, exploratory directions. This stratified density pattern brings out the dynamic nature of the field where newer and multidisciplinary uses are slowly catching up with the fundamental fields, which are constantly developing. Consequently, the general density distribution provides a clear image of those areas with the highest concentration of academic interest and also those areas where opportunities of further research may be mostly expected to increase.

The publication trend as shown in Figure 4 presents a temporal summary of academic productivity of tin nanoparticles between 1991 and 2025. The earliest articles to deal chiefly with the basic synthesis of the chemical, thin-film formation, and early optical characterisation of tin-based substances date to the early 1990s, when interest in the field started. This period was dominated by largely exploratory research and limited by the ability of equipment needed to conduct the research due to the wider infancy of nanotechnology. The rising tendency can be traced back to the beginning of the 2000s, which coincided with the rapid development of electron microscopy, spectroscopy, and the surface analysis method resulting in the possibility of studying nanoparticles on an atomic scale. The growing awareness of the semiconducting and catalytic properties of tin oxide (SnO_2) nanoparticles is a reason that causes a growth in the rate of publications between 2005 and 2010. The practical stage of the discipline commenced with the investigations of the scientists on their potential in gas sensors, photocatalytic systems, and electrochemical devices.

Figure 4. Publication Trend on Tin Nanoparticles from 1991-2025 (Source: Lens Database) (Lens, 2025).

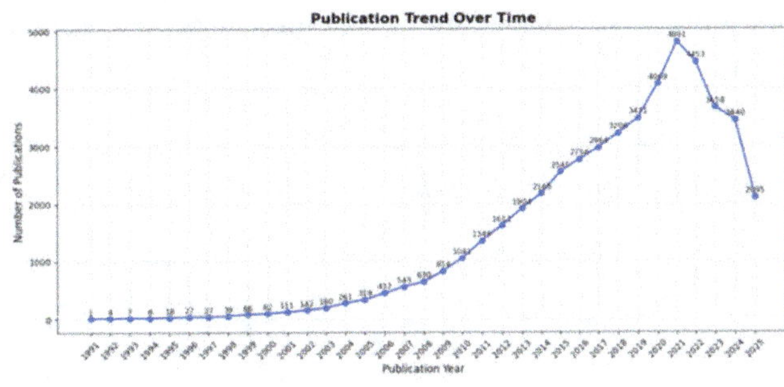

Following 2010, this curve steepens and it is associated with the global attention on energy conversion and storage technologies. In this decade, there was the appearance of such high-frequency terms as energy storage, electrocatalysis, nanocomposites, performance optimization which signified the hegemony of the research about using tin-based anode materials in the lithium-ion batteries, supercapacitors, and fuel cells. The exponential rate of growth in the period of 2015-2020 means interdisciplinary work and diversification of research in the applications of biomedical analysis, environmental recovery, and photocatalysis.

Publications still increase following the year 2020, and they are expected to perform higher than ever before, in 2024. This recent growth is due to the integration of eco-friendly materials engineering, hybrid nanocomposites, and the green synthesis. Also, the trend line reveals that the interest will keep growing at least until 2025, which promotes the stability and maturity of the field. The recent discovery of tin nanoparticles has become one of the most important nanomaterials in academic and industrial research activities, as the trend in the long term, which indicates the development of nanotechnology in the world over time since the time of discovery of the nanomaterial to the innovation of its application.

The word cloud in Figure 5 indicates the key and frequent research problems in the literature on tin nanoparticles qualitatively and insightfully. The practical and theoretical nature of the field is demonstrated by the large, uninhibited words such as Tin oxide (SnO_2) synthesis, nanocomposite, electrocatalysis, lithium-ion batteries, performance, and photocatalysis. Due to its outstanding stability, broad band gap, and ability to sense as well as energy devices, the most studied type of tin at the nanoscale is the tin oxide.

Closely related terms, including synthesis, characterization, morphology and crystallinity, reinforce the fact that the objective remains to control surface and structural properties through a set of diverse fabrication methods, including sol-gel, hydrothermal, chemical vapor deposition and green synthesis approaches. Meanwhile, the names nanocomposite and hybrid material are often used, indicating that the incorporation of tin nanoparticles into carbonaceous or polymeric matrices such as graphene, carbon nanotubes or conducting polymers to enhance conductivity, stability, and catalytic activity is gaining more and more popularity. It is also interesting to observe that new terminologies such as the biomedical applications, toxicity, antibacterial, biosensor and drug delivery suggest that the research in life science and medical engineering is significantly diversified in addition to the conventional material science. This shift is an expression of a global effort to evaluate the biocompatibility of nanomaterials and potential benefits to health. At the same time, the concepts of sustainability, such as green synthesis, eco-friendly, bio-reduction, and the use of plant extract, signify the increase in the adoption of the ecologically-friendly synthesis procedures. The size of words and their relative positions to each other in the cloud are used to demonstrate the complexity of the interactions between the synthesis, structure, and performance. The image in general demonstrates how the study of the tin nanoparticles is transforming to be not merely materials-focused but sustainable and multipurpose based ones due to environmental awareness and practical need.

Figure 5. Word Cloud on Tin Nanoparticles (Source: Lens Database) (Lens, 2025)

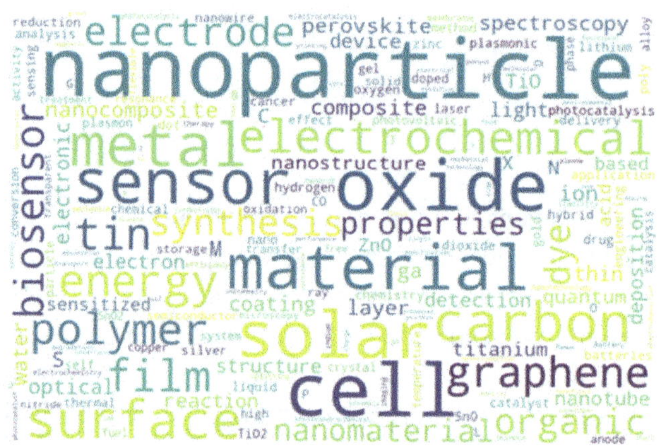

Figure 6 provides a geographical distribution of the contribution of the global community to the output of research on tin nanoparticles. The map is clear in its demonstration of how the spread of the scientific activity is distributed around the world yet in some countries it is strongly concentrated. The most prolific contributor is China which constitutes a large portion of international articles. This leadership is attributed to the massive investments by the country in materials science, nanotechnology, and renewable energy research supported by government financing initiatives, as well as large research consortia. Chinese research usually focuses on energy storage, catalytic activity, and large-scale production of SnO_2 nanoparticles.

In the second position is the United States which is often at the front in the field of basic research, theoretical modelling, and advanced characterization of the nanostructures of tin. American researchers have made an enormous contribution to understanding electronic structure, surface modification and integrating devices. It also appears to have key centres in Germany and India, with a contribution being made in Germany through high-tech materials design, engineering of nanocomposites and thin films and in India at least by green synthesis and photo-catalytical uses.

Figure 6. World Map on Contribution by Countries in Publication (Source: Lens Database) (Lens, 2025).

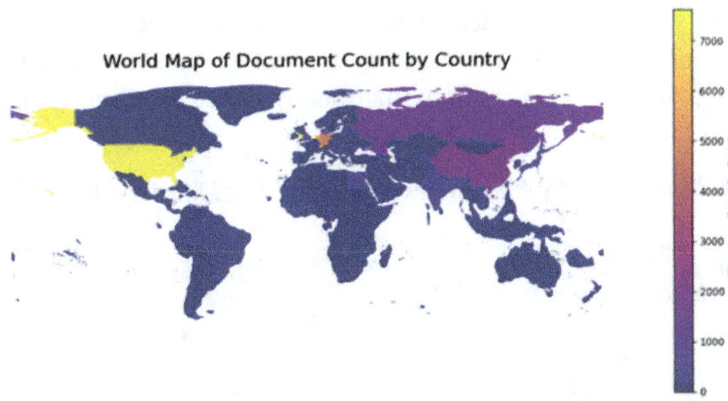

The engagement of other nations in the research regularly is perceived to be a well-connected international community like South Korea, Japan, Iran, Brazil and United Kingdom. The international nature of nanotechnology study is evidenced by the fact that there is an intercontinental communication network particularly between Asia and Europe. The high- and middle-income economies have access to tin nanoparticles due to their inexpensive production and a large variety of applications that are becoming increasingly popular with the inclusion of underdeveloped countries. The reason why tin nanoparticles are considered to be strategic as a research resource of a common type is its wide distribution around the world. Since cross-border cooperation yields an opportunity to exchange ideas, methods, and knowledge, the convergence of knowledge in numerous locations accelerates innovation as well. To sum up, Figure 6 shows us the international scope, diversity, and interconnectivity of the scientific community to develop tin nanoparticle technology.

The publishing organizations and journals categorize research output into the tree map in Figure 7 and provides knowledge regarding the pattern of diffusion as well as concentration of the research activity among the academic publishers. The publications of the research on tin nanoparticles are done through Elsevier, Springer, MDPI, Wiley, Taylor and Francis, and the Royal Society of chemistry (RSC), which are represented by the biggest blocks in the picture. These are key in the distribution of contents in the field of material science as evidenced by the number of expert journals available, diversified in nature to address applied chemistry, nanotechnology and catalysis. Synthesis, characterization, structure and electrochemical behaviour Studies of tin nanoparticles are reviewed in Elsevier many high impact journals, including Applied Surface Science, Materials Letters and Journal of Alloys and Compounds. Likewise, the articles about the subject of theoretical modeling, the generation of nanocomposites, and the process of photocatalysis, are offered by Springer and Wiley by concentrating on the interdisciplinary articles. The MDPI salience is a sign of the open-access movement where the aim is to reach the research outputs on tin nanoparticles more accessible and citable. The variety of smaller rectangles denotes the level of the publication diffusion of the map (not only by up-and-coming publishers but also by specialized journals, dedicated to specific applications, e.g., environmental nanotechnology or biomedical engineering). The relative discrepancies between the segments show what disciplines receive, on the average, more or less editorial coverage; e.g. one of the publishers may concentrate on energy and materials research, others on chemical or environmental applications. The dynamic and interdisciplinary nature of the field with the enhanced scholarly communication ecosystem is disclosed with this distribution. It appears that the research on nanoparticles consisting of tin is of interest to a rather wide audience in the academic community, with chemists, physicists, materials scientists, and engineers representing the most varied range of channels to be involved in. By so doing, the tree map indicates maturity and diversity of the field, existence of strong, multi-channel, and internationally linked research dissemination.

Figure 7. Tree Map on Publication by Publishers (Source: Lens Database) (Lens, 2025).

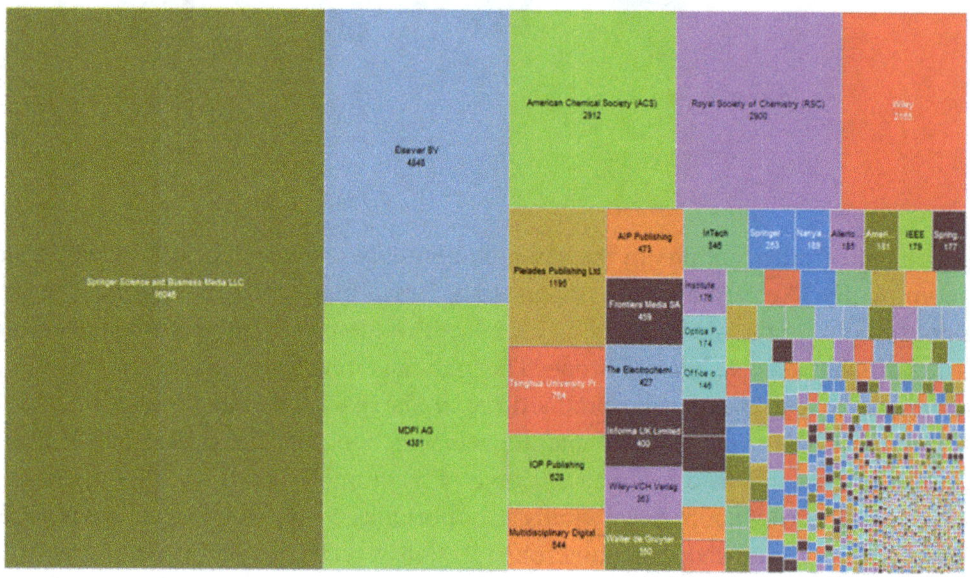

7. CONCLUSION

The many green synthesis methods for producing Tin Oxide nanoparticles were the main topic of this review article. The objective of the research and study was to gain knowledge and analyse the biological and green techniques or methods of producing Tin Oxide nanoparticles and a wide range of potential uses for Tin Oxide nanoparticles, such as physiochemical, photocatalytic, and antimicrobial applications. Furthermore, the article described the characteristics of Tin Oxide nanoparticles and their various uses, as well as highlighting various plant-based green synthesis methods. In short, Tin Nanoparticles were synthesised biologically, because green synthesis is more eco-friendly, economical or cost-friendly and does not have big requirements as compared to physical and chemical synthesis method, applied with topdown or bottom-up approach. After synthesising the nanoparticles, they were characterized using X-Ray diffraction method, FT-IR spectroscopy, SEM, TEM, EDX, UV analysis, etc. Then, after getting the results, the characteristics were displayed and the nanoparticles were also assessed for their properties and applications.

REFERENCES

Ajp. (2024). Structural, morphological and spectroscopic characterization of chemically synthesized SnO2 nanoparticles. *Asian Journal of Physics*. https://asianjournalofphysics.com/structural-morphological-and-spectroscopic-characterization-of-chemically/

Borah, D., Saikia, P., Rout, J., Gogoi, D., Ghosh, N. N., & Bhattacharjee, C. R. (2024). Sustainable green synthesis of SnO2 quantum dots: A stable, phase-pure and highly efficient photocatalyst for degradation of toxic dyes. *Materials Today Sustainability*, *26*, 100770. DOI: 10.1016/j.mtsust.2024.100770

Chellamuthu, P., Savarimuthu, K., Alsath, M. G. N., R, K., T, Y., Bajaj, M., & Shabaz, M. (2025). Fabrication and comprehensive experimental evaluation of surfactant-activated PEDOT:PSS/SnO2 thin films deposited via spin coating for advanced sensing applications. *Scientific Reports*, *15*(1), 30628. DOI: 10.1038/s41598-025-12499-1

Din, S. U., Kiani, S. H., Haq, S., Ahmad, P., Khandaker, M. U., Faruque, M. R. I., Idris, A. M., & Sayyed, M. I. (2022). Bio-Synthesized tin oxide nanoparticles: Structural, optical, and biological studies. *Crystals*, *12*(5), 614. DOI: 10.3390/cryst12050614

Fatimah, I., Purwiandono, G., Jauhari, M. H., Maharani, A. P., Sagadevan, S., Oh, W., & Doong, R. (2022). Synthesis and control of the morphology of SnO2 nanoparticles via various concentrations of Tinospora cordifolia stem extract and reduction methods. *Arabian Journal of Chemistry*, *15*(4), 103738. DOI: 10.1016/j.arabjc.2022.103738

Gari, M. S., Narasaiah, B. P., Pandurengan, A., Mandal, B. K., & Natarajan, M. (2023). Synthesis and Antioxidant Activity of some novel 4H-Chromene Derivatives Catalysed by Biogenic Tin Oxide Nanoparticles. *Biointerface Research in Applied Chemistry*, *13*(6), 521. DOI: 10.33263/BRIAC136.521

Garrafa-Galvez, H., Nava, O., Soto-Robles, C., Vilchis-Nestor, A., Castro-Beltrán, A., & Luque, P. (2019). Green synthesis of SnO2 nanoparticle using Lycopersicon esculentum peel extract. *Journal of Molecular Structure*, *1197*, 354–360. DOI: 10.1016/j.molstruc.2019.07.052

Gebreslassie, Y. T., & Gebretnsae, H. G. (2021). Green and Cost-Effective Synthesis of Tin Oxide Nanoparticles: A review on the synthesis methodologies, mechanism of formation, and their potential applications. *Nanoscale Research Letters*, *16*(1), 97. Advance online publication. DOI: 10.1186/s11671-021-03555-6

Hamedani, N. F., & Farzaneh, F. (2006). Synthesis of ZnO Nanocrystals with Hexagonal (Wurtzite) Structure in Water Using Microwave Irradiation. *Journal of Science*, *17*(3), 231–234. https://journals.ut.ac.ir/article_31762_e1fc4d21c2aa63931c4f50ef9758cd52.pdf

Jaji, N., Lee, H. L., Hussin, M. H., Akil, H. M., Zakaria, M. R., & Othman, M. B. H. (2020). Advanced nickel nanoparticles technology: From synthesis to applications. *Nanotechnology Reviews*, *9*(1), 1456–1480. DOI: 10.1515/ntrev-2020-0109

Karmaoui, M., Jorge, A. B., McMillan, P. F., Aliev, A. E., Pullar, R. C., Labrincha, J. A., & Tobaldi, D. M. (2018). One-Step synthesis, structure, and band gap properties of SNO2 nanoparticles made by a low temperature nonaqueous Sol–Gel technique. *ACS Omega*, *3*(10), 13227–13238. DOI: 10.1021/acsomega.8b02122

Laranjeira, J. A., Azevedo, S. A., Martins, N. F., La Porta, F. A., Longo, E., & Sambrano, J. R. (2024). Insights into the role of surface properties on the optical, electronic and nanoparticles morphology of scheelite BaMoO4. *Surfaces and Interfaces*, *46*, 103894. DOI: 10.1016/j.surfin.2024.103894

Lens, (2025). *Search results for "Tin Nanoparticles"*. The Lens. Retrieved November 07, 2025, from https://www.lens.org/lens/search/scholar/list?q=tin%20nanoparticles

Mashkovtsev, M. A., Kosykh, A. S., Ishchenko, A. V., Chukin, A. V., Kukharenko, A. I., Troshin, P. A., & Zhidkov, I. S. (2024). Unraveling oxygen vacancies effect on chemical composition, electronic structure and optical properties of EU doped SNO2. *Nanomaterials (Basel, Switzerland)*, *14*(20), 1675. DOI: 10.3390/nano14201675

Masood, H., Toe, C. Y., Teoh, W. Y., Sethu, V., & Amal, R. (2019). Machine learning for Accelerated discovery of solar photocatalysts. *ACS Catalysis*, *9*(12), 11774–11787. DOI: 10.1021/acscatal.9b02531

Matussin, S., Harunsani, M. H., Tan, A. L., & Khan, M. M. (2020). Plant-Extract-Mediated SnO2 Nanoparticles: Synthesis and Applications. *ACS Sustainable Chemistry & Engineering*, *8*(8), 3040–3054. DOI: 10.1021/acssuschemeng.9b06398

Narasaiah, B. P., Banoth, P., Sohan, A., Mandal, B. K., Dominguez, A. G. B., De Los Santos Valladares, L., & Kollu, P. (2022). Green biosynthesis of tin oxide nanomaterials mediated by Agro-Waste cotton boll peel extracts for the remediation of environmental pollutant dyes. *ACS Omega*, *7*(18), 15423–15438. DOI: 10.1021/acsomega.1c07099

Nayak, S. S., Wadhawa, G. C., Shivankar, V. S., Patil, D. D., Sonawale, M. C., & Mirgane, N. A. (2020). Tin oxide plant assisted nanoparticle catalyzed green synthesis of imidazole derivatives. *Materials Today: Proceedings*, *37*, 2490–2494. DOI: 10.1016/j.matpr.2020.08.301

Ringe, E., Van Duyne, R. P., & Marks, L. D. (2013). Kinetic and thermodynamic modified wulff constructions for twinned nanoparticles. *The Journal of Physical Chemistry. C, Nanomaterials and Interfaces*, *117*(31), 15859–15870. DOI: 10.1021/jp401566m

Sagadevan, S., Lett, J. A., Fatimah, I., Lokanathan, Y., Léonard, E., Oh, W. C., Hossain, M. M., & Johan, M. R. (2021). Current trends in the green syntheses of tin oxide nanoparticles and their biomedical applications. *Materials Research Express*, *8*(8), 082001. DOI: 10.1088/2053-1591/ac187e

Saif, N., Tariq, A., Intisar, A., Batool, Z., Arshad, M., Rehman, A. U., & Kousar, R. (2023). Green synthesis of SnO2 nanorods and Mo– SnO2 nanocomposite for efficient sunlight driven organic dye photodegradation in aqueous solution. *Materials Chemistry and Physics*, *305*, 127920. DOI: 10.1016/j.matchemphys.2023.127920

Sharma, S., Shukla, S. K., Govender, K. K., & Govender, P. P. (2025). Unveiling the multifunctionality of iron oxide nanoparticle: A synergistic experimental and computational investigation. *Chemical Physics Impact*, *10*, 100845. DOI: 10.1016/j.chphi.2025.100845

Shu, M., Mahdavi, B., Balčiūnaitienė, A., Goorani, S., & Mahdavi, A. A. (2023). Novel green synthesis of tin nanoparticles by medicinal plant: Chemical characterization and determination of cytotoxicity, cutaneous wound healing and antioxidant properties. *Micro & Nano Letters*, *18*(2), e12157. Advance online publication. DOI: 10.1049/mna2.12157

Soumane, M., Bouaaliouat, O., Fazouan, N., Ydir, B., Dloo, A., Celzard, A., Fierro, V., Prades, J. D., & Lahlou, H. (2025). Experimental and DFT simulation of SnO2 monotube-based gas sensing for hazardous VOCs. *Surfaces and Interfaces*, *60*, 105971. DOI: 10.1016/j.surfin.2025.105971

Sudhaparimala, S. (2014). Green Synthesis of tin based Nano Medicine: Assessment of microstructure and surface property. *American Journal of Nanoscience and Nanotechnology*, *2*(4), 75. DOI: 10.11648/j.nano.20140204.13

Sunny, N. E., & Kumar, S. V. (2019). Biogenesis, Characterization and Bioefficacy of Tin Oxide Nanoparticles from Averrhoa Bilimbi Fruit Extract. [IJRTE]. *International Journal of Recent Technology and Engineering*, *8*(4), 10309–10315. DOI: 10.35940/ijrte.D4523.118419

Takhar, V., & Singh, S. (2025). Nanomaterials ROS: A comprehensive review for environmental applications. *Environmental Science. Nano*, *12*(5), 2516–2550. DOI: 10.1039/D5EN00049A

Tameh, M. S., Gladfelter, W. L., & Goodpaster, J. D. (2025). Unraveling surface chemistry of SNO2 through formation of charged oxygen species and oxygen vacancies. *International Journal of Quantum Chemistry*, *125*(4), e70017. Advance online publication. DOI: 10.1002/qua.70017

Tasisa, Y. E., Sarma, T. K., Sahu, T. K., & Krishnaraj, R. (2024). Phytosynthesis and characterization of tin-oxide nanoparticles (SnO2-NPs) from Croton macrostachyus leaf extract and its application under visible light photocatalytic activities. *Scientific Reports*, *14*(1), 10780. DOI: 10.1038/s41598-024-60633-2

Van Eck, N. J., & Waltman, L. (2009). Software survey: VOSviewer, a computer program for bibliometric mapping. *Scientometrics*, *84*(2), 523–538. DOI: 10.1007/s11192-009-0146-3

Zhai, C., Shi, C., Hu, Y., Xu, Z., & Wang, R. (2022). Anti-breast carcinoma effects of green synthesized tin nanoparticles from Calendula officinalis leaf aqueous extract inhibits MCF7, Hs 319.T, and MCF10 cells proliferation. *Journal of Experimental Nanoscience*, *17*(1), 351–361. DOI: 10.1080/17458080.2022.2076836

Chapter 10
Predictive Analytics in Phytochemistry:
Modeling Extraction Yields From Opuntia ficus-indica via Solvent Polarity and Regional Variation

Sara Razzak
Faculty of Sciences, Ibn Tofail University, Kenitra, Morocco

Yassine Mouniane
https://orcid.org/0000-0002-6037-5301
Faculty of Sciences, Ibn Tofail University, Kenitra, Morocco

Imane Aitouhanni
https://orcid.org/0009-0004-1400-8027
ENSIAS, SSLAB, Mohammed V University, Rabat, Morocco

Chaima Sabri
Faculty of Sciences, Ibn Tofail University, Kenitra, Morocco

Malak Zirari
Faculty of Sciences, Ibn Tofail University, Kenitra, Morocco

Hiba Benchehida
Faculty of Sciences, Ibn Tofail University, Kenitra, Morocco

Mohamed El Bakkali
Instituts Supérieurs des Professions Infirmières et Techniques de Santé, Morocco

Youness Taboz
Faculty of Sciences, Ibn Tofail University, Kenitra, Morocco

ABSTRACT

This chapter presents a data-driven framework for predicting the extraction yield of bioactive compounds from Opuntia ficus-indica cladodes using machine learning techniques. Plant materials were collected from three geographically distinct Moroccan provinces and subjected to chemical extraction using solvents of varying polarity. A Random Forest regression model was developed using solvent type, province of origin, and quantified phytochemical content as input variables. The model demonstrated high predictive accuracy ($R^2 \approx 0.99$), with solvent type emerging as the most influential factor, followed by flavonoid and flavonol concentrations. The findings underscore the potential of predictive modeling to optimize experimental design, reduce resource consumption, and guide solvent and region selection

DOI: 10.4018/979-8-3373-6058-4.ch010

in phytochemical research. This chapter provides insights into how supervised learning algorithms can be effectively applied to natural product data to support green extraction processes and enhance reproducibility in phytochemical workflows.

1. INTRODUCTION

The improvement of extraction methods is an important step in natural products and phytochemistry research (Salem et al., 2020; Heinrich et al., 2022). Yield from extraction is an important measure of process efficiency that can be influenced by several variables: solvent, source of the plant or bioactive component (Khaw et al., 2017; Chaves et al., 2020; da Silva et al., 2022). Until recently, optimal extraction variables based on yield were primarily explored through experimental or classical approaches which require substantial resources to perform, and yield cannot always be generalized or replicated (Franceschini & Macchietto, 2008; Naviglio et al., 2019; Reckling et al., 2021; Plathottam et al., 2023). With the development of machine learning-based algorithms, predictive models facilitate prediction of yield by learning complex and interesting patterns from empirical data (Sarker, 2021; Richard-Bollans et al., 2023; Varghese et al., 2025).

Extraction yield is dependent on many variables but it is also an indication of nutritional value (Carmen et al., 2023; Shi et al., 2023; Caminiti et al., 2024). Differences in plant origin, climate, soil composition, and farming practices among provinces can produce significant variation in both qualitative (how many meta-compounds such as polyphenols, flavonoids, or tannins) and quantitative aspects of biosynthesis of plant secondary metabolites (Loveland & Webb, 2003; Jeffery et al., 2011; Sun, & Fernie, 2024; Zimba et al., 2025). The affinity of a solvent relates to its intrinsic chemical properties that dictate union, dissolution and extraction of certain classes of compounds, arguably sociality (Tshilande et al., 2024; Ulfa et al., 2025). Therefore building a combined geospatial and chemical predictive model can allow researchers to better examine extraction yields while increasing understanding of the extraction process (Liu et al., 2024; Niazi & Mariam, 2025).

Opuntia ficus-indica, commonly known as prickly pear cactus, is a resilient and multifunctional succulent species predominantly found in arid and semi-arid ecosystems (Rodrigues et al., 2023; Slot & Fort, 2024). Renowned for its adaptability to harsh climatic conditions, it plays a significant agro-ecological role in soil conservation, water retention, and food security (Acharya et al., 2019; Jorge et al., 2023). Besides its ecological benefits, *O. ficus-indica* has gained popularity for its bioactive compound contents, including polyphenols, flavonoids, tannins, and important vitamins (Tahir et al., 2019; Barba et al., 2022). These ingredients can offer antioxidant, anti-inflammatory and antimicrobial properties, making this cactus a potential candidate for pharmaceutical, cosmetic, and functional food industries (Zeghbib et al., 2024; Liu et al., 2025).

While there is growing academic and industry interest in its phytochemical profile, there is still a significant knowledge gap in its variability across environmental and geographical contexts (Macías et al., 2007; Amon et al., 2022; Ogwu et al., 2025). The soil, climate, elevation, and harvest outcomes can impact the metabolism of the plant (Pant et al., 2021; Kumar et al., 2025), but characterization of these impacts is limited, if not inconsistent (Hoffmann, 2010; Pérez-Lucas et al., 2024). There is also the challenge that any traditional methods used to study the variability are typically very laborious, time-consuming, and not reproducible. Therefore, the present study introduces an integrated, data-driven strategy using machine learning algorithms specifically Random Forest regression - to model and predict the extraction yield of bioactive compounds isolated from *O. ficus-indica*. The methods of chemical analysis, in conjunction with AI-based modelling- predictive modelling, provide a significant contribution to our scientific understanding of phytochemical dynamics from O. figus-indica across region and solvents through the integration of these two methodological approaches. In the context of enhancing scientific understanding of phytochemical variability this also offers a reproducible, scalable framework for enhancing extraction methodologies. The study outlines significant methodological approaches that can assist in setting the groundwork for future valorization approaches of *O. ficus-indica* specifically as well as similar underutilized medicinal phytochemical medicinal plants.

In this research, we describe the use of supervised learning to predict the extraction yield of bioactive compounds from plant samples based on the separate categorical factors of province and solvent and the five continuous variables that represent each of the concentrations of the phytochemicals. The justification is that these categorical factors account for biological and procedural variability to a level that sufficient to model extraction behaviour. It is also not unreasonable to want to focus on extraction yield as our target variable because our goal is to generate information that is useful for selection of future extraction workflows by solvents and region of origin. Interpretable models also allow you to consider the proportional contribution of each input, which gives you base quantitative information to optimize the process.

This work exemplifies the application of machine learning, effectively merging empirical knowledge to computational inference, which can produce robust and reproducible predictions based on modeling approaches. Another broader contribution of the work is that the model may be generalizable across other natural product systems to aid designing and rationalizing experiments to enhance growth in the acceleration of discovery in phytochemistry and green extraction technologies.

2. BACKGROUND STUDY

Extraction yield refers to the percentage of compounds extracted and ultimately obtained from the plant material during the chemical treatment. It is an important measure for quantifying and evaluating the efficacy and efficiency of extraction protocols in phytochemical research. Yield is influenced by a number of factors that include the polarity and type of solvent used, the molecular complexity of the plant matrix, extraction time, temperature and agitation. Extraction yield tends to be higher when all of the variables have been optimized. Therefore, extraction yield is both a product and reference point for experimental design. In predictive modeling, yield can be used as a dependent variable that encompasses all the contributions from the experimental variables to model or predict expected performance under different conditions.

2.1. Solvent Type

Solvents are important in determining the types and amounts of phytochemicals found in plant materials. The solubility of phytoconstituents is largely determined by the polarity of the solvent. Polar solvents, for example, methanol, allows extraction of polar phytoconstituents such as polyphenols and tannins. Non-polar solvents, for example, dichloromethane, provide extraction opportunities for non-polar or moderately polar compounds (Royani et al., 2025). The chemical affinity between solvent and target metabolite provides the basis of extraction capacity and determines the hormones and compounds that are qualitative and quantitative identifiable in the extract (López-Yerena et al., 2021). Keeping in mind how compound polarity and characteristics relate to solvent selection is important not only modifications of methodology, but at the planning level, chemical solvent choice determines the default extent and outcome of what metabolites are extracted, thereby impacting all biological analyses that are dependent on that extract including quantifying total extract method vs. polyethylene glycol extracted, extract per gram of fresh instead of dry plant material (Pateiro et al., 2021; Mukherjee et al., 2024).

2.2. Phytochemical Composition

Phytochemicals are biologically active compounds that are responsible for the therapeutic potential and chemical diversity of plant extracts (Chintada & Golla, 2025). The phytochemicals most frequently studied are polyphenols, flavonoids, flavonols, and tannins, and there are many structural variations and biological activities (Heim et al., 2002). Phytochemicals will not have the same solubility profile or behave similarly when in solution with solvents, which makes concentration as a measure of extraction success (Mastellone et al., 2022). For this research, at least because these phytochemicals were quantified as input to the predictive models, and also because they are chemical descriptors of the functional value of the extracts, I felt their effects would add rich, chemically meaningful content to the models. The phytochemical data added meaningful chemical information and would help improve yield predictions.

2.3. Geographic Origin (province)

The province of origin of the plant material can affect its phytochemistry due to environmental and agricultural differences. Soil chemistry, altitude, rainfall, temperature, and how the crop was cultivated all effect plant production and therefore the secondary metabolites synthesized (Figueiredo et al., 2008; Ahmed et al., 2019; Palit & Mandal, 2021; Qaderi et al., 2023). Consequently, the same species under cultivation in two different provinces may have different polyphenol, flavonoid, and tannin content (Xu et al., 2020). The geographic phytochemistry difference adds a spatial component to the analysis, giving models the opportunity to learn geographic patterns that contribute to yield variability. Encoding province of origin will allow the model to consider environmental variance when making predictions in this study (Zonyane et al., 2019; Shinyuy et al., 2025).

2.4. Machine Learning in Phytochemistry

Machine learning is a rapidly growing methodology in phytochemistry as it can analyze multivariate types of data and find associations where traditional statistical approaches would be less successful (Zhang et al., 2024). Algorithmic approaches, like Random Forests, are advantageous in particular due to their capacity to manage categorical and numeric data, being less susceptible to overfitting, and producing accompanying metrics such as feature importance indices that are simple to interpret (Genuer et al., 2017). In predictive modeling, machine learning allows phytochemists to simulate and evaluate the effects of different experimental variables on outputs, like extract yield, allowing for hypothesis testing and experimental design. Incorporation of machine learning is a major step toward something we call "data-driven discovery," as well as rationalizing design in phytochemical protocols (Bacong & Juanico, 2021; Yang et al., 2024).

3. METHODOLOGY

This section presents the complete methodological pipeline followed to construct, clean, and analyze the dataset and to develop a predictive model for extraction yield. The approach is organized into clearly defined steps, each with a specific objective and contribution to the overall predictive framework.

3.1. Plant Material

Depending on several topographical and climatic conditions three ecologically distinct regions of Morocco were chosen to identify mature cladodes of *Opuntia ficus-indica* OFI, Imzouren in the Al Hoceima province on the northern Mediterranean coastline (Tangier-Tetouan-Al Hoceima region); Oulad Boubker a rural area of Driouch province (northeastern area of the Oriental region), and Skoura (located 40 km from Ouarzazate) located in southern Souss-Massa region. After harvesting cladode, a standard preparation was followed to remove spines and thoroughly rinsed each cladode with distilled water, then surface sanitized using a 10% ethanol solution, cut and dehydrated at 60 °C in a ventilated oven for 18 to 20 hours. Once dehydrated, the samples were ground in a laboratory grinder to obtain powdered samples. The powder was preserved in opaque containers at room temperature and away from light to avoid degradation from light.

3.2. Data Collection

The dataset was based on a series of extractions performed on plant material collected from three Moroccan provinces (Oulad Boubker, Imzouren, and Skoura). Each sample was treated with one of three solvents (dichloromenthane, chloroform, or methanol), selected for their respective polarities, and the extraction conditions were standardized across all samples to ensure comparability (Razzak et al., 2024a). We quantified several biochemical parameters for each sample that was treated, including the concentration of total polyphenols, flavonoids, flavones and flavonols, condensed and hydrolysable tannins and hydrolysable tannins (Razzak et al., 2024b), which were used as measurements to model the extraction yield (i.e ., the response variable).

3.3. Data Preprocessing and Cleaning

The original dataset in Excel had some structural limitations that demanded a clean, systematic analysis. Specifically, solvent labels were indicated only at the start of each block of rows, simply indicating that a solvent label applied to all rows with a solvent label until a new solvent label was shown in at the start of a new block of rows. We performed a forward-fill operation to take the last known solvent label and repeat it for all requisite rows. Thus, each entry in the dataset can now be explicitly labelled by the solvent in the dataset. We then encoded categorical variables (province and solvent) numerically, through label encoding, to preserve the non-ordinal variable categorization and allow for numerical processing by machine learning algorithms with the continuous variables as their original scale, and standardization was unnecessary due to tree-type models like the Random Forest. All missing values were visually inspected and programatically validated and confirmed all were addressed appropriately. The cleaned dataset was then ready for supervised machine learning approaches.

3.4. Model Training and Evaluation

After preprocessing, the dataset was divided into training and testing subsets with an 80/20 ratio in order to assess model performance. A Random Forest regression model was trained on the training subset. Random Forest is an ensemble learning technique that builds multiple decision trees and averages their predictions, allowing for the identification of non-linear relationships and reduced overfitting. Since the dataset was small, the model was trained using default hyperparameters. We evaluated the model's performance on three criteria, the coefficient of determination (R^2), which indicates the extent to which the model captures variability; mean absolute error (MAE) reflects average prediction error; and root mean square error (RMSE), which captures the larger errors. We chose Random Forest because of its strong capacity to manage mixed data types (categorical and continuous modeling), non-linear relationships, complex feature interactions, and the ability to provide interpretative metrics of feature importance. This is different from linear regression models, which require an additive relationship among features; and is also in contrast to other predictive analysis methods, such as support vector machines or neural networks, which require tuning hyper parameters and larger sample sizes. Random Forest often performs well using moderate sample sizes to predict, and protects against overfitting based on its use of ensemble average. These features lend themselves to chemometric and phytochemical modeling.

3.5. Interpretability and Visualization

To better understand how the model made its predictions, we derived feature importances that ranked the contribution of each variable to the predictions. The feature importances were then shown visually via plots bar plots for ease of interpretation. We then created a scatter plot of predicted versus actual yields, where the predicted yield from the model were plotted along the y-axis against the actual or true yield values were plotted along x-axis, so that the visual could show how accurate the model's predictions were and where were veering off error wise. A series of additional plots were created to further visualize the data - including yield distributions via provinces and solvents, and a correlation heatmap of all the variables, presented to contextualize the data and show any possible patterns. All of the plots were included in the analysis and become the basis of the further discussion below in the next chapter section.

The methodology used in this analysis was proposed framework that examined various aspects of data pre-processing, model training, evaluation and interpretation all within a supervised learning context. In general, directionally we combined categorical and continuous feature types to produce an establish predictive model train capable learning to predict extraction yield, accurately based off the features input to the model. Given the structured nature of the data and the relatively small sample size, Random Forest Regression was selected for its robustness, ability to capture non-linear relationships, and inherent importance metrics.

During preprocessing, we encoded categorical variables using label encoding so that solvent types and provinces of origin were given numeric identities without assigning any unintended false ordinal relationship. Continuous features with respect to their phytochemical composition were kept in their original scale to preserve chemical interpretability. The data was then randomly assigned into training and test datasets with an 80/20 split for model validation and the ability for generalizability. The models were generated multiple times with a Random Forest model having 100 decision trees. The hyperparameters were set to default options due to the small size of the dataset. The decision tree models were developed using the feature matrix that had been preprocessed and with extraction yield as the dependent variable.

Each model was assessed using three common metrics of evaluation (coefficient of determination (R^2), mean absolute error (MAE), and root mean square error (RMSE). There metrics all gave a more rounded 'view' about the accuracy, bias, and precision of the models. To improve the explainability of the model, I determined the feature importances and visualized them, providing an insight into the variables relative contributions to yield prediction. Additional diagnostic plotting included a predicted vs actual scatter plot and residual analysis to examine the consistency and stability of predictions. The process described above provides a transparent and replicable example of using machine learning methods for predictive purposes with phytochemical data.

4. RESULTS AND VISUAL ANALYSIS

4.1. Model Performance

The evaluation of the Random Forest model's performance showed good predictive ability. The coefficient of determination (R^2) of the model is around 0.99, meaning that almost all variation in extraction yield can be accounted for by the selected variables as features. Accordingly, this result is consistent with the low mean absolute error (MAE) and root mean square error (RMSE), both of which indicate very small differences between predicted and actual values, and in combination provide confirming support that the model can be established as accurate and precise for predicting extraction yield. In order to evaluate the robustness of the model further, we applied a 5-fold cross-validation strategy. The mean cross-validated R^2 was also high (0.95 ± 0.03), which supports the stability and generalizability of the model. This framework also addressed concerns over possible overfitting, especially when the sample size is not large.

4.2. Actual vs. Predicted Yield

The scatter plot displaying the predicted yield as compared to the actual yield (Figure 1) is a very compelling aspect of the performance assessment of the model. Most of the data points are very near the diagonal red line that indicates perfect prediction, suggesting a very high relationship between the model output values and values measured from an experiment. Further, the distribution of points clustered around the reference line gives additional confidence to the very low prediction error and the outperformance of the model after being trained. This enables the conclusion that the model was able to generalize very well to the new, and previously unseen, test data, and overall met the expectations for the purpose.

Figure 1. Predicted vs. Actual Yield scatter plot.

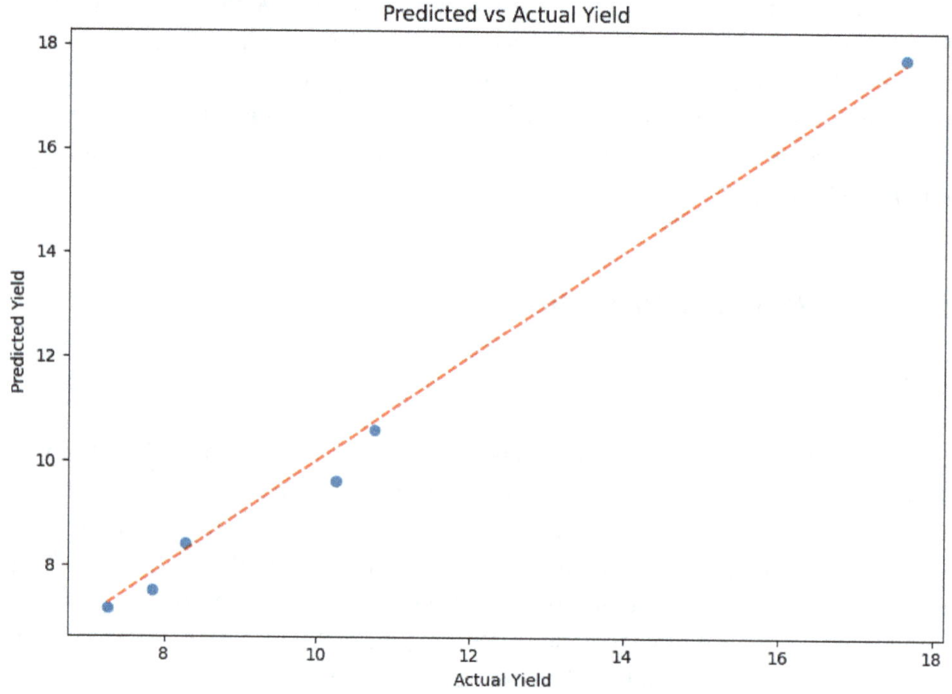

The scatter plot of predicted yields versus observed yields gives an visual representation of the model performing exceptionally well. The majority of data point are grouped around the diagonal, red line representing perfect prediction, indicating that the model outputs are well-matched with experimental data. The tight grouping of points around the red, hope line confirms that the model has low prediction error and that the model was trained reliably.

4.3. Yield Distribution by Province

The yield distribution across the provinces of Oulad Boubker, Imzouren, and Skoura suggests that geographical origin can introduce systematic differences in yield variability (Figure 2). Although the medians were similar across provinces, there was slight variation in interquartile range which suggests that some provinces may naturally produce a greater or varying yield. These differences between provinces could be attributed to environmental variables, such as soil type, climate, or altitude, that may impact the phytochemicals characterizing plant samples. Although the province variable was not the highest ranked predictor, it does add an ecological context to the data and increases the ecological validity of the model.

Figure 2. Yield distribution across different provinces.

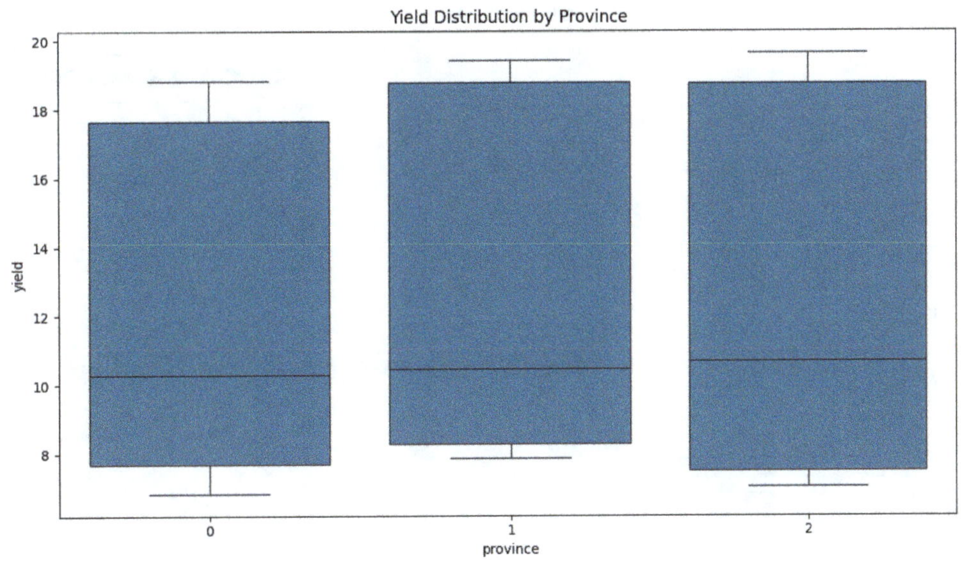

The analysis of yield distributions in the three provinces of Oulad Boubker, Imzouren and Skoura suggests that geographic provenance is a source of variation in extraction yields. While median values were similar across the three sites, the interquartile ranges were slightly different suggesting that certain provinces may have inherent higher and/or more variable yields than others. Differences observed with geographic provenance could be related to environmental changes of factors such as soil type, climate, or elevation that can affect phytochemical profiles in plant samples.

4.4. Yield Distribution by Solvent

Yield distributions stratified by solvent type demonstrate a clear and consistent pattern (Figure 3): methanol consistently yields greater yields than chloroform and dichloromethane. This is expected because methanol is a very polar solvent, clearly able to solubilize a wider range of phytochemicals than chloroform or dichloromethane, especially many polar and semi-polar compounds. It is evident from the boxplot that methanol not only yields extract but does so at lower variability, indicating methanol is consistently a higher-yielding and more efficient solvent for the plant-based extractions studied. As such, highlighting the choice of solvent, is an important consideration for the overall design of extraction experiment.

Figure 3. Yield distribution by solvent type.

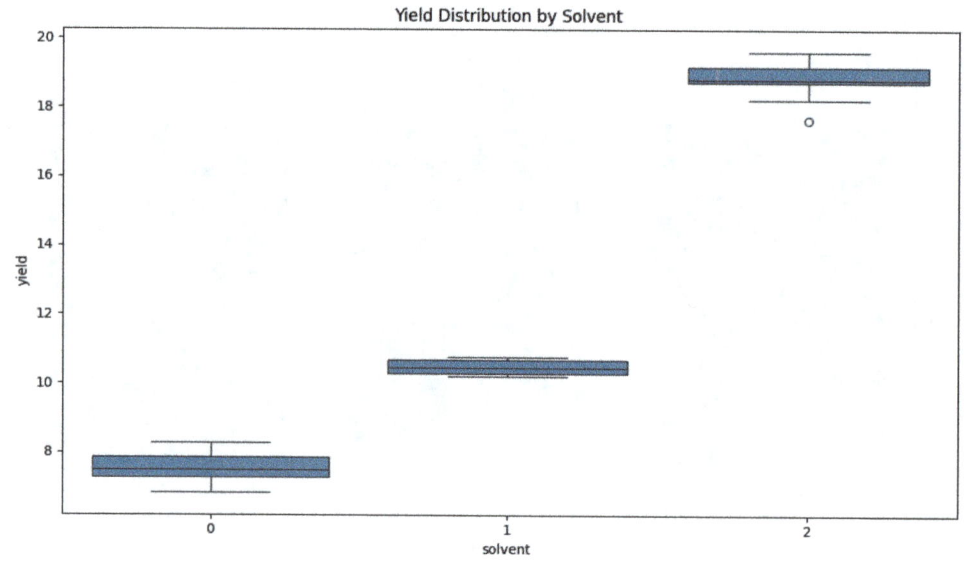

As shown in Figure 3, it is apparent that methanol gave higher yields than chloroform or dichloromethane. This is what we expected, since methanol has a high polarity which allows it to solubilize a greater diversity of amphiphilic phytochemicals. The boxplot indicates that methanol was not only able to yield more extract on average, but it was also less variable, meaning that it is a better solvent for conducting plant extractions in this instance.

4.5. Correlation Analysis

The correlation matrix (Figure 4) provides additional statistical detail regarding the linear relationships of the variables in the dataset. Yield had strong positive correlations with flavonoid content, flavonol concentration, and total polyphenol content, all of which are as expected based on chemical theory. The type of solvent also had relatively high correlation with yield, which corroborated its fundamental importance in the extraction process. Conversely, province showed weaker correlations, which were consistent with its lower feature importance. The utility of the correlation matrix is to support the logical coherence of the dataset as well as to confirm that the model is utilizing real domain-relevant relationships.

Figure 4. Correlation heatmap of all model variables.

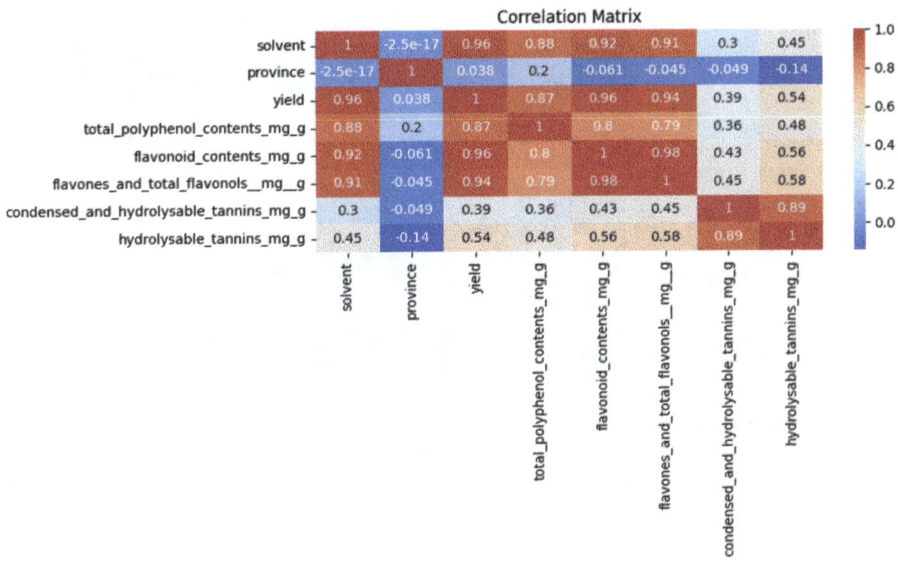

The correlation matrix illustrated the linear relationships between different variables. Yield had strong positive correlations with flavonoid content, flavonol concentration, and totals of polyphenols, supporting the chemical rationale for the reliance on these features. Solvent type also showed a strong correlation with yield, further confirming its importance in the extraction process. The variable province, on the other hand, had weaker correlations, which was consistent with its lower feature importance score. Once again, this matrix serves to confirm the logical coherence of the dataset and the model's ability to take advantage of true relationships.

4.6. Feature Importance Interpretation

The feature importance analysis based on the Random Forest model provided an interpretable account of which variables most strongly contributed to the yield predictions (Figure 5). The solvent type was the most significant variable in this breakdown, receiving considerably more weight than any other feature. The flavonol and flavonoid content were also determined to be significant contributors to the model output within the other variables considered. The minor weight of province signifies that geographic origin may demonstrate some variability, but it wasn't a significant influence on extraction performance in this specific dataset. Overall, the results validate your understanding for the context, but provided a useful quantification of each feature in this way. The visual breakdown allows you be more targeted to the features that matter the most as you think about potential experimental program designs in the future.

Figure 5. Feature importance ranking from the Random Forest model.

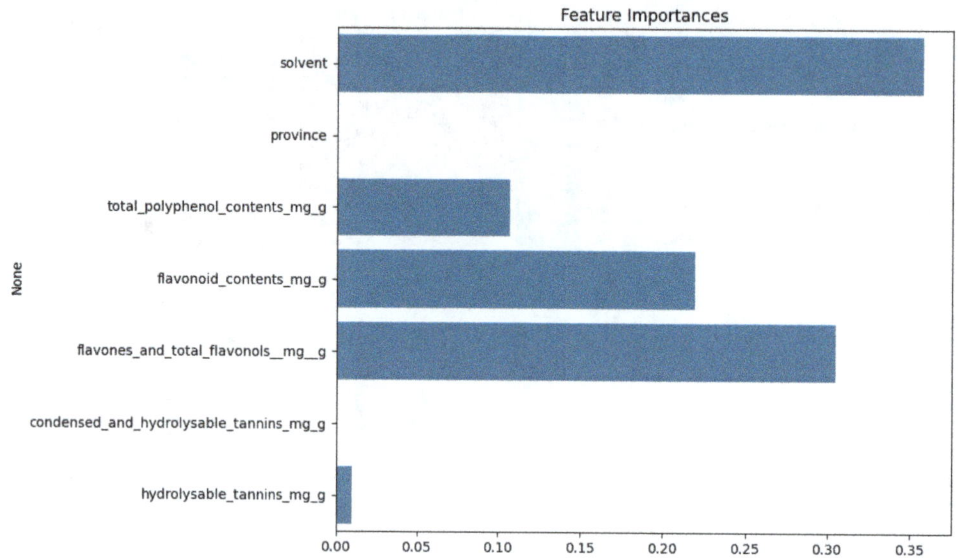

The Random Forest model's feature importance assessment gave a clear breakdown of the variables acknowledging how paranoid determinations may generate predictive consequences. Solvent was the leading variable and substantially far more significant than any other feature. Flavonol and flavonoid are important as the next two features following solvent, fraying toward a minimally significant level with the subsequent feature, province. Since province has a relatively low weight, having province would suggest province of origin is not a significant variable that contributes to the variability introduced by sourcing flips. Bearing these meanings in mind is congruent with domain knowledge while generating a quantitative output from the data reporting the significance of each feature; this output is meaningful.

5. DISCUSSION

5.1. Summary of Key Findings

The present study verifies the ability of machine learning, especially in the form of Random Forest regression, to make accurate predictions of extraction yield from a limited number of chemical and contextual predictors. Of the 11 predictors evaluated, the type of solvent had the largest impact. This is consistent with chemical principles that polarity drives extractive efficacy. Concentration of flavonoids and flavonols also had significant impacts, while province was the least influential but served to support the model with some environmental context.

5.2. Scientific Interpretation

The strong relationship between solvent and yield and the corresponding rankings of feature importance, confirm the foundational importance of solvent polarity in extraction science. Methanol yielded the highest yield because of its high polarity, confirming experience-based knowledge. Similarly, strong correlation between yield and associated phytochemical concentrations also supports the use of these metrics as sensible predictive input. This also follows intuitively, that a higher abundance of the precursor should lead to a higher respective extract recovery.

In addition to bulk phytochemical concentration, additional molecular-level descriptors (e.g., octanol-water partition coefficient (logP), molar volume, polar surface area, and number of hydrogen-bond donors and acceptors) could be offered in a more mechanistic representation of solubility and extraction behavior. These descriptors could also be calculated for the dominant compounds of *O. ficus-indica* with cheminformatics tools, which would directly relate molecular structure with extraction efficiency. The feature options of computational chemistry could be incorporate models' interpretability by indicating predictive performance to molecular interactions.

It should also be noted that feature importance in Random Forest is a measure of statistical association not causal mechanisms and as such, the associations established in this study should be understood as empical relationships that inform, but are not prescriptive to, mechanistic-inferrred hypotheses for testable experiments.

5.3. Theoretical Basis of Solvation and Extraction Efficiency

The efficiency of extraction among the solvents can be described through solvation theory's mating between a quantitative understanding of solvent–solute interactions. As suggested by Hildebrand's solubility parameter (δ), best results occur when solutes dissolve in solvents with like cohesive energy densities, which is based on the adage that "like dissolves like". Hansen then expanded this concept into three factors within the solubility parameter - a dispersion (δD), polar (δP), and hydrogen-bonding parameter (δH) - that defines the appropriate solubility space in a multidimensional scale. The COSMO-RS (Conductor-like Screening Model for Real Solvents) model extends these models even more, allowing for quantum chemical calculation of solute–solvent interactions. Here, our Random Forest model implicitly learns the nonlinear relationships between the extraction yield, the solvent polarity and molecular properties of phytochemicals and environmental factors. Incorporating solubility parameters determined experimentally, or derived from computational predictions could potentially allow the parameters to make predictions that are physically interpretable and grounded in molecular thermodynamics.

5.4. Practical Application and Industrial Perspectives

The model can substantively be used as a decision-support system to improve extraction protocols as long as there is limited research expenditure and/ or research capacity. Researchers can input knowns related to chemical composition and categorical parameters and obtain predicted yield outcomes prior to conducting wet-lab experiments. This can save time and money, while improving the planning process and likelihood of generating a hypothesis. From the research standpoint, the framework illustrates the possibilities of extending modeling techniques similar format to larger and more diverse datasets in the future (Bibri & Krogstie, 2017; Shen, 2018). Additional variables such as temperature, pH, extraction time, and agitation method would hugely benefit numbers of different target variables and models capturing ranges and behaviors into a more predictive and even adaptable tool (Baş & Boyacı, 2007; Mao et al., 2023).

The predictive model, based on phytochemical extraction provides illustrated possibilities for industry applications in cosmetics, pharmaceuticals, agri-food, and nutritional supplements, etc., (Musa & Elnour, 2024; Roselli et al., 2024). The combination of artificial intelligence and random forest modeling in this study enables very accurate predictions of extraction yields while also reducing the effort required in developing new natural extracts (Wang et al., 2021). Manufacturers can therefore concentrate efforts with three or four tests instead of sufficient experimental test work reducing variables substantially, the most obvious benefit. With the ability to make predictions based on easily measured characteristics (solvent type, phytochemical profile, etc.), it can be possible to select optimal variables upstream in order to maximize yield or target specific families of bioactive compounds for their desired applications (Sharma et al., 2019).

Additionally, the system could also be applied to normalize extracts for quality control and regulatory purposes, especially in the pharmaceutical and cosmetic related markets (Chemat et al., 2012; Lores et al., 2016; de Hond et al., 2022). Hence, this model could be utilized as part of supportive predictive quality control systems regarding formulated products. Ultimately, this type of model could lead to more sustainable and responsible practice to ensure environmental stewardship by avoiding waste when processing plant materials and solvents in adherence to principles of green chemistry (Chen et al., 2020). It would be possible to further develop the model, utilizing all relevant environmental and process variables, and assist the development of intelligent decision-support tools for process selection that could be applied to other plants species or other areas within the world (Bhardwaj et al., 2022; Pantic et al., 2022; Zhou et al., 2024).

5.5. Limitations and Future Work

While we have observed promising results, there are limitations to note. One, the sample size was limited (27 samples), which may impact the generalizability of results. While Random Forest can handle small datasets, a larger sample size could have allowed us to utilize more advanced modeling approaches (i.e., deep learning) or more reliable cross-validation. Secondly, environmental variables known to impact plant growth (for example, seasonality or nutrient content) and could potentially add significant value if included were excluded from our study. Future work should assess the generalizability of this model in terms of species, solvents, and geography, as well as predicting on unused extraction conditions. By enlarging the data's feature space and the scale of the data, we would could enhance the width and applicability of predictive models like this. Even though random forests help alleviate the overfitting through averaging out of the ensemble, the root causes of the small sample size is still a limiting factor. For example, while we consider cross-validation to assess a "wider range" of model robustness, the bigger picture moving forward should be about the collection of more data to be used as an independent validation step to assess generalizability. In addition, adding more environmental metadata including, but not limited to, soil nutrient profiles, temperature, precipitation, and altitude, would allow for breaking-up the currently categorical province variable into a set of potentially quantitative new predictors that would lend not only predictive power but also improve upon the interpretability of the environmental variables that inform phytochemical yield.

6. CONCLUSION

Through this study, we illustrated some possibilities and the promise of algorithms for integrating machine learning in the study of phytochemistry, as we were able to optimize our prediction model based on a small, but chemically diverse data set and a unique emphasis on solvent selection, geographic origin and phytochemical profile, to realize excellent models with good predictability and interpretability. Our findings suggest that solvent polarity was the main driving factor towards extraction yield, and we demonstrated the importance of some phytochemical variables such as concentration of flavonoides included in our variable list as contributors; both in the context of limitations of these variables, and their predictive capabilities. The methodological framework - including data scrubbing, feature coding, and rapt model interpretation - was sufficient to shed light on the complexity of the experimental conditions and draw predictive inferences from the experiments into one coherent framework. The use of visualization analytics improved our capacity to validate the model's internal logic and communicate its findings better. Potentially, this study also offers researchers a low cost and scalable model, for predicting yield possibilities of extracts that does not require the labour-intensive approach of trial and error, and therefore streamlining our processes and enhancing the identification of discovery. Going forward, we will work on applying this model to a larger array of plants, as well as different solvents, extraction techniques and other factors (e.g., environmental, temporal) to further generalize the model across cannabis varieties. By merging our design experience with computer-aided inference, this project and related research build on the increasing drive for digitization, taking advantage of data convergence to approach the study of natural products in a new way.

REFERENCES

Acharya, P., Biradar, C., Louhaichi, M., Ghosh, S., Hassan, S., Moyo, H., & Sarker, A. (2019). Finding a suitable niche for cultivating cactus pear (*Opuntia ficus-indica*) as an integrated crop in resilient dryland agroecosystems of India. *Sustainability (Basel)*, *11*(21), 5897. DOI: 10.3390/su11215897

Ahmed, S., Griffin, T. S., Kraner, D., Schaffner, M. K., Sharma, D., Hazel, M., Leitch, A. R., Orians, C. M., Han, W., Stepp, J. R., Robbat, A., Matyas, C., Long, C., Xue, D., Houser, R. F., & Cash, S. B. (2019). Environmental factors variably impact tea secondary metabolites in the context of climate change. *Frontiers in Plant Science*, *10*, 939. DOI: 10.3389/fpls.2019.00939

Amon, D. J., Gollner, S., Morato, T., Smith, C. R., Chen, C., Christiansen, S., Currie, B., Drazen, J. C., Fukushima, T., Gianni, M., Gjerde, K. M., Gooday, A. J., Grillo, G. G., Haeckel, M., Joyini, T., Ju, S.-J., Levin, L. A., Metaxas, A., Mianowicz, K., & Pickens, C. (2022). Assessment of scientific gaps related to the effective environmental management of deep-seabed mining. *Marine Policy*, *138*, 105006. DOI: 10.1016/j.marpol.2022.105006

Bacong, J. R. C., & Juanico, D. E. O. (2021). Predictive Chromatography of Leaf Extracts Through Encoded Environmental Forcing on Phytochemical Synthesis. *Frontiers in Plant Science*, *12*, 613507. DOI: 10.3389/fpls.2021.613507

Barba, F. J., Garcia, C., Fessard, A., Munekata, P. E. S., Lorenzo, J. M., Aboudia, A., Ouadia, A., & Remize, F. (2022). *Opuntia ficus-indica* edible parts: A food and nutritional security perspective. *Food Reviews International*, *38*(5), 930–952. DOI: 10.1080/87559129.2020.1756844

Baş, D., & Boyacı, İ. H. (2007). Modeling and optimization I: Usability of response surface methodology. *Journal of Food Engineering*, *78*(3), 836–845. DOI: 10.1016/j.jfoodeng.2005.11.024

Bhardwaj, A., Kishore, S., & Pandey, D. K. (2022). Artificial intelligence in biological sciences. *Life (Basel, Switzerland)*, *12*(9), 1430. DOI: 10.3390/life12091430

Bibri, S. E., & Krogstie, J. (2017). The core enabling technologies of big data analytics and context-aware computing for smart sustainable cities: A review and synthesis. *Journal of Big Data*, *4*(1), 38. DOI: 10.1186/s40537-017-0091-6

Caminiti, R., Serra, M., Nucera, S., Ruga, S., Oppedisano, F., Scarano, F., Macrì, R., Muscoli, C., Palma, E., Musolino, V., Statti, G., Mollace, V., & Maiuolo, J. (2024). Antioxidant activity and seasonal variations in the composition of insoluble fiber from the cladodes of *Opuntia ficus-indica* (L.) Miller: Development of new extraction procedures to improve fiber yield. *Plants*, *13*(4), 544. DOI: 10.3390/plants13040544

Carmen, F., Frances, C., & Barthe, L. (2023). Trends on valorization of pitaya fruit biomass through value-added and green extraction technology – A critical review of advancements and processes. *Trends in Food Science & Technology*, *138*, 339–354. DOI: 10.1016/j.tifs.2023.06.014

Chaves, J. O., de Souza, M. C., da Silva, L. C., Lachos-Perez, D., Torres-Mayanga, P. C., Machado, A. P. F., Forster-Carneiro, T., Vázquez-Espinosa, M., González-de-Peredo, A. V., Barbero, G. F., & Rostagno, M. A. (2020). Extraction of flavonoids from natural sources using modern techniques. *Frontiers in Chemistry*, *8*, 507887. DOI: 10.3389/fchem.2020.507887

Chemat, F., Vian, M. A., & Cravotto, G. (2012). Green extraction of natural products: Concept and principles. *International Journal of Molecular Sciences*, *13*(7), 8615–8627. DOI: 10.3390/ijms13078615

Chen, T.-L., Kim, H., Pan, S.-Y., Tseng, P.-C., Lin, Y.-P., & Chiang, P.-C. (2020). Implementation of green chemistry principles in circular economy system towards sustainable development goals: Challenges and perspectives. *The Science of the Total Environment*, *716*, 136998. DOI: 10.1016/j.scitotenv.2020.136998

Chintada, V., & Golla, N. (2025). Exploring the therapeutic potential of bioactive compounds from plant sources. In Devi, J. (Ed.), *Biotechnological Intervention in Production of Bioactive Compounds* (pp. 229–247). Springer., DOI: 10.1007/978-3-031-76859-0_15

da Silva, R.F., & Carneiro, C.N., do C. de Sousa, C.B., Gomez, F.J.V., Espino, M., Boiteux, J., Fernández, M. de los Á., Silva, M.F., & de S. Dias, F. (2022). Sustainable extraction of bioactive compounds procedures in medicinal plants based on the principles of green analytical chemistry: A review. *Microchemical Journal*, *175*, 107184. DOI: 10.1016/j.microc.2022.107184

de Hond, A. A. H., Leeuwenberg, A. M., Hooft, L., Kant, I. M. J., Nijman, S. W. J., van Os, H. J. A., Aardoom, J. J., Debray, T. P. A., Schuit, E., van Smeden, M., Reitsma, J. B., Steyerberg, E. W., Chavannes, N. H., & Moons, K. G. M. (2022). Guidelines and quality criteria for artificial intelligence-based prediction models in healthcare: A scoping review. npj. *NPJ Digital Medicine*, *5*(1), 2. DOI: 10.1038/s41746-021-00549-7

Figueiredo, A. C., Barroso, J. G., Pedro, L. G., & Scheffer, J. J. C. (2008). Factors affecting secondary metabolite production in plants: Volatile components and essential oils. *Flavour and Fragrance Journal*, *23*(4), 213–226. DOI: 10.1002/ffj.1875

Franceschini, G., & Macchietto, S. (2008). Model-based design of experiments for parameter precision: State of the art. *Chemical Engineering Science*, *63*(19), 4846–4872. DOI: 10.1016/j.ces.2007.11.034

Genuer, R., Poggi, J.-M., Tuleau-Malot, C., & Villa-Vialaneix, N. (2017). Random Forests for Big Data. *Big Data Research*, *9*, 28–46. DOI: 10.1016/j.bdr.2017.07.003

Heim, K. E., Tagliaferro, A. R., & Bobilya, D. J. (2002). Flavonoid antioxidants: Chemistry, metabolism and structure-activity relationships. *The Journal of Nutritional Biochemistry*, *13*(10), 572–584. DOI: 10.1016/S0955-2863(02)00208-5

Heinrich, M., Jalil, B., Abdel-Tawab, M., Echeverria, J., Kulić, Ž., McGaw, L. J., Pezzuto, J. M., Potterat, O., & Wang, J.-B. (2022). Best practice in the chemical characterisation of extracts used in pharmacological and toxicological research—The ConPhyMP—Guidelines12. *Frontiers in Pharmacology*, *13*, 953205. DOI: 10.3389/fphar.2022.953205

Hoffmann, I. (2010). Climate change and the characterization, breeding and conservation of animal genetic resources. *Animal Genetics*, *41*(s1, Suppl. 1), 32–46. DOI: 10.1111/j.1365-2052.2010.02043.x

Jeffery, S., Verheijen, F. G. A., van der Velde, M., & Bastos, A. C. (2011). A quantitative review of the effects of biochar application to soils on crop productivity using meta-analysis. *Agriculture, Ecosystems & Environment, 144*(1), 175–187. DOI: 10.1016/j.agee.2011.08.015

Jorge, A. O. S., Costa, A. S. G., & Oliveira, M. B. P. P. (2023). Adapting to climate change with Opuntia. *Plants, 12*(16), 2907. DOI: 10.3390/plants12162907

Khaw, K.-Y., Parat, M.-O., Shaw, P. N., & Falconer, J. R. (2017). Solvent supercritical fluid technologies to extract bioactive compounds from natural sources: A review. *Molecules (Basel, Switzerland), 22*(7), 1186. DOI: 10.3390/molecules22071186

Kumar, N., Kaur, B., Sharma, D., Korpole, S., Shukla, S., Bhardwaj, P., Patel, M. K., Acharya, S., Khatri, M., & Saxena, S. (2025). Impact of altitudinal variations on plant growth dynamics, nutritional composition, and free living rhizospheric N_2 fixing bacterial community of Eruca sativa. *Scientific Reports, 15*(1), 13839. Advance online publication. DOI: 10.1038/s41598-025-98242-2

Liu, X., An, H., Cai, W., & Shao, X. (2024). Deep learning in spectral analysis: Modeling and imaging. *Trends in Analytical Chemistry, 172*, 117612. DOI: 10.1016/j.trac.2024.117612

Liu, X., Xing, Y., Liu, G., Bao, D., Hu, W., Bi, H., & Wang, M. (2025). Extraction, purification, structural features, biological activities, and applications of polysaccharides from *Opuntia ficus-indica* (L.) Mill. (cactus): A review. *Frontiers in Pharmacology, 16*, 1566000. Advance online publication. DOI: 10.3389/fphar.2025.1566000

López-Yerena, A., Domínguez-López, I., Vallverdú-Queralt, A., Pérez, M., Jáuregui, O., Escribano-Ferrer, E., & Lamuela-Raventós, R. M. (2021). Metabolomics technologies for the identification and quantification of dietary phenolic compound metabolites: An overview. *Antioxidants, 10*(6), 846. DOI: 10.3390/antiox10060846

Lores, M., Llompart, M., Alvarez-Rivera, G., Guerra, E., Vila, M., Celeiro, M., Lamas, J. P., & Garcia-Jares, C. (2016). Positive lists of cosmetic ingredients: Analytical methodology for regulatory and safety controls – A review. *Analytica Chimica Acta, 915*, 1–26. DOI: 10.1016/j.aca.2016.02.033

Loveland, P., & Webb, J. (2003). Is there a critical level of organic matter in the agricultural soils of temperate regions: A review. *Soil & Tillage Research, 70*(1), 1–18. DOI: 10.1016/S0167-1987(02)00139-3

Macías, F. A., Galindo, J. L. G., & Galindo, J. C. G. (2007). Evolution and current status of ecological phytochemistry. *Phytochemistry, 68*(22–24), 2917–2936. DOI: 10.1016/j.phytochem.2007.10.010

Mao, Y., Robinson, J. P., & Binner, E. R. (2023). Current status of microwave-assisted extraction of pectin. *Chemical Engineering Journal, 473*, 145261. DOI: 10.1016/j.cej.2023.145261

Mastellone, G., Marengo, A., Sgorbini, B., Rubiolo, P., & Cagliero, C. (2022). Development of a dispersive solid-liquid microextraction method using natural eutectic solvents for a greener extraction of phytochemicals from fiber-type Cannabis sp. Industrial Crops and Products, 187(Part B), 115476. https://doi.org/DOI: 10.1016/j.indcrop.2022.115476

Mukherjee, C., Suryawanshi, P. G., Kalita, M. C., Deka, D., Aranda, D. A. G., & Goud, V. V. (2024). Polarity-wise successive solvent extraction of Scenedesmus obliquus biomass and characterization of the crude extracts for broad-spectrum antibacterial activity. *Biomass Conversion and Biorefinery*, *14*(2), 2467–2483. DOI: 10.1007/s13399-022-02432-1

Musa, K. H., & Elnour, A. A. M. (2024). Advances and future perspectives in biotechnological and bioconversional of dates byproducts. *Journal of Agriculture and Food Research*, *16*, 101145. DOI: 10.1016/j.jafr.2024.101145

Naviglio, D., Scarano, P., Ciaravolo, M., & Gallo, M. (2019). Rapid solid-liquid dynamic extraction (RSLDE): A powerful and greener alternative to the latest solid-liquid extraction techniques. *Foods*, *8*(7), 245. DOI: 10.3390/foods8070245

Niazi, S. K., & Mariam, Z. (2025). Artificial intelligence in drug development: Reshaping the therapeutic landscape. *Therapeutic Advances in Drug Safety*, *16*, 20420986251321704. Advance online publication. DOI: 10.1177/20420986251321704

Ogwu, M. C., Izah, S. C., & Joshua, M. T. (2025). Ecological and environmental determinants of phytochemical variability in forest trees. *Phytochemistry Reviews*, *24*(6), 5109–5137. Advance online publication. DOI: 10.1007/s11101-025-10066-0

Palit, P., & Mandal, S. C. (2021). Climate change, geographical location, and other allied triggering factors modulate the standardization and characterization of traditional medicinal plants: A challenge and prospect for phyto-drug development. In S.C. Mandal, R. Chakraborty & S. Sen (Eds.), Evidence Based Validation of Traditional Medicines (pp. [numéros de pages si connus]). Springer, Singapore. https://doi.org/DOI: 10.1007/978-981-15-8127-4_18

Pant, P., Pandey, S., & Dall'Acqua, S. (2021). The influence of environmental conditions on secondary metabolites in medicinal plants: A literature review. *Chemistry & Biodiversity*, *18*(11), e2100345. DOI: 10.1002/cbdv.202100345

Pantic, I., Paunovic, J., Pejic, S., Drakulic, D., Todorovic, A., Stankovic, S., Vucevic, D., Cumic, J., & Radosavljevic, T. (2022). Artificial intelligence approaches to the biochemistry of oxidative stress: Current state of the art. *Chemico-Biological Interactions*, *358*, 109888. DOI: 10.1016/j.cbi.2022.109888

Pateiro, M., Gómez-Salazar, J. A., Jaime-Patlán, M., Sosa-Morales, M. E., & Lorenzo, J. M. (2021). Plant extracts obtained with green solvents as natural antioxidants in fresh meat products. *Antioxidants*, *10*(2), 181. DOI: 10.3390/antiox10020181

Pérez-Lucas, G., Navarro, G., & Navarro, S. (2024). Adapting agriculture and pesticide use in Mediterranean regions under climate change scenarios: A comprehensive review. *European Journal of Agronomy*, *161*, 127337. DOI: 10.1016/j.eja.2024.127337

Plathottam, S. J., Rzonca, A., Lakhnori, R., & Iloeje, C. O. (2023). A review of artificial intelligence applications in manufacturing operations. *Journal of Advanced Manufacturing and Processing*, *5*(3), e10159. DOI: 10.1002/amp2.10159

Qaderi, M. M., Martel, A. B., & Strugnell, C. A. (2023). Environmental factors regulate plant secondary metabolites. *Plants*, *12*(3), 447. DOI: 10.3390/plants12030447

Razzak, S., Aouji, M., Sabri, C., Benchehida, H., Taibi, M., & Taboz, Y. (2024b). Effect of solvent polarity on phytochemical analysis of different Moroccan sites of *Opuntia ficus-indica* (L.) Mill. pads. *International Journal of Secondary Metabolite*, *11*(1), 93–104. DOI: 10.21448/ijsm.1354740

Razzak, S., Aouji, M., Zirari, M., Benchehida, H., Taibi, M., Bengueddour, R., Wondmie, G. F., Ibenmoussa, S., Bin Jardan, Y. A., & Taboz, Y. (2024a). Nutritional composition, functional and chemical characterization of Moroccan *Opuntia ficus-indica* cladode powder. *International Journal of Food Properties*, *27*(1), 1167–1179. DOI: 10.1080/10942912.2024.2382877

Reckling, M., Ahrends, H., Chen, T.-W., Eugster, W., Hadasch, S., Knapp, S., Laidig, F., Linstädter, A., Macholdt, J., Piepho, H.-P., Schiffers, K., & Friedrich, T. (2021). Methods of yield stability analysis in long-term field experiments: A review. *Agronomy for Sustainable Development*, *41*(2), 27. DOI: 10.1007/s13593-021-00681-4

Richard-Bollans, A., Aitken, C., Antonelli, A., Bitencourt, C., Goyder, D., Lucas, E., Ondo, I., Pérez-Escobar, O. A., Pironon, S., Richardson, J. E., Russell, D., Silvestro, D., Wright, C. W., & Howes, M.-J. R. (2023). Machine learning enhances prediction of plants as potential sources of antimalarials. *Frontiers in Plant Science*, *14*, 1173328. DOI: 10.3389/fpls.2023.1173328

Rodrigues, C., Paula, C. D., Lahbouki, S., Meddich, A., Outzourhit, A., Rashad, M., Pari, L., Coelhoso, I., Fernando, A. L., & Souza, V. G. L. (2023). Opuntia spp.: An overview of the bioactive profile and food applications of this versatile crop adapted to arid lands. *Foods*, *12*(7), 1465. DOI: 10.3390/foods12071465

Roselli, V., Pugliese, G., Leuci, R., Brunetti, L., Gambacorta, L., Tufarelli, V., & Piemontese, L. (2024). Green methods to recover bioactive compounds from food industry waste: A sustainable practice from the perspective of the circular economy. *Molecules (Basel, Switzerland)*, *29*(11), 2682. DOI: 10.3390/molecules29112682

Royani, A., Mubarak, N. M., Hanafi, M., Verma, C., Lotulung, P. D. N., Prastya, M. E., Alfantazi, A., & Manaf, A. (2025). Effect of solvent polarity on yield extract, antioxidant and antibacterial activities of phytochemicals from Andrographis paniculata leaves. *Indian Chemical Engineer*, *67*(1), 1–15. DOI: 10.1080/00194506.2024.2409260

Salem, M. A., Perez de Souza, L., Serag, A., Fernie, A. R., Farag, M. A., Ezzat, S. M., & Alseekh, S. (2020). Metabolomics in the context of plant natural products research: From sample preparation to metabolite analysis. *Metabolites*, *10*(1), 37. DOI: 10.3390/metabo10010037

Sarker, I. H. (2021). Machine learning: Algorithms, real-world applications and research directions. *SN Computer Science*, *2*(3), 160. DOI: 10.1007/s42979-021-00592-x

Sharma, M., Koul, A., Sharma, D., Kaul, S., Swamy, M. K., & Dhar, M. K. (2019). Metabolic engineering strategies for enhancing the production of bio-active compounds from medicinal plants. In Akhtar, M., & Swamy, M. (Eds.), *Natural bio-active compounds* (pp. 287–316). Springer., DOI: 10.1007/978-981-13-7438-8_12

Shen, C. (2018). A transdisciplinary review of deep learning research and its relevance for water resources scientists. *Water Resources Research*, *54*(11), 8558–8593. DOI: 10.1029/2018WR022643

Shi, T., Cao, J., Cao, J., Zhu, F., Cao, F., & Su, E. (2023). Almond (Amygdalus communis L.) kernel protein: A review on the extraction, functional properties and nutritional value. *Food Research International, 167*, 112721. DOI: 10.1016/j.foodres.2023.112721

Shinyuy, L. M., Loe, G. E., Jansen, O., Ledoux, A., Palmaerts, B., Mamede, L., Boussif, N., Bonnet, O., Enone, B. S., Noukimi, S. F., Nchang, A. S., Demeyer, K., Robert, A., Ghogomu, S. M., Souopgui, J., Hallot, E., & Frederich, M. (2025). Exploring the Phytochemical Diversity and Anti-Plasmodial Potential of Artemisia annua and Artemisia afra from Different Geographical Locations in Cameroon. *Molecules (Basel, Switzerland), 30*(3), 596. DOI: 10.3390/molecules30030596

Slot, L. E., & Fort, F. (2024). Drivers of adoption of sustainable prickly pear (*Opuntia ficus-indica*) innovations and conservation agriculture by smallholder farmers in Morocco. *Agronomy (Basel), 14*(12), 3014. DOI: 10.3390/agronomy14123014

Sun, Y., & Fernie, A. R. (2024). Plant secondary metabolism in a fluctuating world: Climate change perspectives. *Trends in Plant Science, 29*(5), 560–571. DOI: 10.1016/j.tplants.2023.11.008

Tahir, H. E., Xiaobo, Z., Komla, M. G., & Mariod, A. A. (2019). Nopal cactus (*Opuntia ficus-indica* (L.) Mill) as a source of bioactive compounds. In A. Mariod (Ed.), Wild fruits: Composition, nutritional value and products (pp. [non précisé]). Springer. https://doi.org/DOI: 10.1007/978-3-030-31885-7_26

Tshilande, N., Mammino, L., & Bilonda, M. K. (2024). The study of molecules and processes in solution: An overview of questions, approaches and applications. *Computation (Basel, Switzerland), 12*(4), 78. DOI: 10.3390/computation12040078

Ulfa, D. M., Bayu, A., Rahmawati, S. I., Ahmadi, P., Putra, M. Y., Karnjanakom, S., Guan, G., & Mun'im, A. (2025). New horizon in selective tocols extraction from deodorizer distillates under mild conditions by using deep eutectic solvents. *Molecules (Basel, Switzerland), 30*(6), 1217. DOI: 10.3390/molecules30061217

Varghese, R., Shringi, H., Efferth, T., & Ramamoorthy, S. (2025). Artificial intelligence driven approaches in phytochemical research: Trends and prospects. *Phytochemistry Reviews, 2025*(5), 10096. Advance online publication. DOI: 10.1007/s11101-025-10096-8

Wang, J., Sun, X., Cheng, Q., & Cui, Q. (2021). An innovative random forest-based nonlinear ensemble paradigm of improved feature extraction and deep learning for carbon price forecasting. *The Science of the Total Environment, 762*, 143099. DOI: 10.1016/j.scitotenv.2020.143099

Xu, Z., Meenu, M., Chen, P., & Xu, B. (2020). Comparative study on phytochemical profiles and antioxidant capacities of chestnuts produced in different geographic area in China. *Antioxidants, 9*(3), 190. DOI: 10.3390/antiox9030190

Yang, S., Li, Z., Li, W., Li, C., Yang, X., Zhao, Y., & Liu, R. (2024). Unraveling the variations and relationships between phytochemical constituents and anti-inflammatory potentials of three Zanthoxylum species through metabolomics and explainable machine learning. *Industrial Crops and Products, 212*, 118281. DOI: 10.1016/j.indcrop.2024.118281

Zeghbib, W., Boudjouan, F., Carneiro, J., Oliveira, A. L. S., Sousa, S. F., Pintado, M. E., Ourabah, A., Vasconcelos, V., & Lopes, G. (2024). LC-ESI-UHR-QqTOF-MS/MS profiling and anti-inflammatory potential of the cultivated *Opuntia ficus-indica* (L.) Mill. and the wild *Opuntia stricta* (Haw.) Haw. fruits from the Algerian region. *Food Chemistry, 460*(Part 1), 140414. DOI: 10.1016/j.foodchem.2024.140414

Zhang, Y., Bao, X., Zhu, Y., Dai, Z., Shen, Q., & Xue, Y. (2024). Advances in machine learning screening of food bioactive compounds. *Trends in Food Science & Technology*, *150*, 104578. DOI: 10.1016/j.tifs.2024.104578

Zhou, Z., Xu, T., & Zhang, X. (2024). Empowerment of AI algorithms in biochemical sensors. *Trends in Analytical Chemistry*, *173*, 117613. DOI: 10.1016/j.trac.2024.117613

Zimba, S., Dougill, A., Chanza, C., Boesch, C., & Kepinski, S. (2025). Gender differential in choices of crop variety traits and climate-smart cropping systems: Insights from sorghum and millet farmers in drought-prone areas of Malawi. *Plants, People, Planet*, *7*(3), 689–703. DOI: 10.1002/ppp3.10467

Zonyane, S., Chen, L., Xu, M.-J., Gong, Z.-N., Xu, S., & Makunga, N. P. (2019). Geographic-based metabolomic variation and toxicity analysis of Sutherlandia frutescens L. R.Br. – An emerging medicinal crop in South Africa. *Industrial Crops and Products*, *133*, 414–423. DOI: 10.1016/j.indcrop.2019.03.010

Chapter 11
Modeling of Adsorption by Artificial Neural Networks:
Predicting the Performance of Adsorbent Materials

Soukayna Maitouf
https://orcid.org/0009-0000-4619-7863
LGP, Faculty of Sciences Agadir. Eco-Process, Optimization, and Decision Support, Jules Verne University of Picardie, France

Karim Benhabib
Eco-Process, Optimization, and Decision Support, Jules Verne University of Picardie, France

Toufik Saadi
Eco-Process, Optimization, and Decision Support, Jules Verne University of Picardie, France

Ahmed Bachar
Faculty of Applied Sciences of Ait-Melloul, Ibn Zohr University, Agadir, Morocco

Assia Mabrouk
https://orcid.org/0000-0001-6399-0644
Faculty of Applied Sciences of Ait-Melloul, Ibn Zohr University, Agadir, Morocco

Nadia Faska
https://orcid.org/0000-0001-6203-6904
Faculty of Applied Sciences of Ait-Melloul, Ibn Zohr University, Agadir, Morocco

ABSTRACT

The combination of computational power and theoretical approaches has enabled the solution of complex problems in chemistry. In wastewater treatment, it is challenging to model nonlinear and complex techniques using classical methods, as these models cannot understand the relationships between the different parameters that affect the process. For instance, the adsorption process is a commonly used process for removing contaminants from wastewater. The adsorption technique requires two elements: the adsorbent material and the pollutant; however, the complexity of molecule-surface interactions makes the simulation challenging. The utilization of artificial neural networks in the adsorption process is crucial. Considering that ANN provides close prediction results for adsorption performance and materials efficiency, which helps to design and improve the next generation of adsorbent materials.

DOI: 10.4018/979-8-3373-6058-4.ch011

INTRODUCTION

In chemistry, various fields focus on research on subatomic-scale phenomena that the human eye cannot perceive. Therefore, computational chemistry and molecular modeling are the most effective solutions for working under these constraints. On the other hand, modeling and computational approaches play a crucial role in minimizing pollution problems, in the same vein as green chemistry (Boikos et al., 2025). Based on these methods, the use of chemicals will be reduced, and researchers will only use them in important steps, such as the validation stage. This would better protect both researchers and the environment. In addition, time and cost savings are achieved, as the technology can absorb massive amounts of data and solve complex problems in a short time. In addition, replacing parts of experimental studies with theoretical studies reduces the cost of materials and chemicals (Mohsenzadeh et al., 2024).

Today, artificial intelligence (AI) tools have been combined with wastewater treatment processes as an effective method for modeling and optimizing them. Various AI models have been tested for various water purification processes, proving AI's ability to understand various complex phenomena and produce results that closely resemble experimental results (Maitouf et al., 2025). Adsorption is a common technique used to eliminate contaminants from water and wastewater. They offer advantages such as simple implementation, eco-friendliness, and high efficiency (Dbik et al., 2022) (El Messaoudi et al., 2023),(El Khomri et al., 2022). This technique is based on the application of an adsorbent material with several properties that enable it to adsorb the target pollutant. The performance of the adsorbent determines the efficiency of the adsorption process. According to the literature, factors such as pH, temperature, initial concentration, and dosage are widely encountered as parameters that influence adsorption. Therefore, understanding the impact of these factors provides a clear vision for further studies on the behavior of adsorption techniques. The adsorption technique is a physicochemical process that is divided into two main types. The first chemical adsorption and the second physical adsorption (Mei et al., 2025).

1. THEORETICAL FOUNDATIONS

The machine learning discipline focuses on the development and study of statistical algorithms. The specific characteristic of these algorithms is their ability to understand and learn from huge amounts of data and then make generalizations for unseen data. In addition, the power of machine learning is evident in its mechanisms. A simple example that explains its potential is the problem of solving a linear function with an intercept. The classical programming system takes the slope (a) and the y-intercept (b) as constants, and the variable is x. However, in the machine learning domain, the fixed element is x, which represents the input, and the other elements are the variables, which are optimized to determine the best model.

In general, machine learning presents various models, including artificial neural networks, adaptive neuro-fuzzy inference systems, support vector machines, K-nearest neighbors, decision trees, and random forests. These models are distinguished based on the types of supervision. Nevertheless, there are two types of ML. The first is supervised learning, and the second is unsupervised learning. The difference between them is that in unsupervised learning, we have no truth about what the data are known, as the true value of the labels is unknown (Deshmukh et al., 2025). Figure 1 presents the basic types of machine learning.

Figure 1. Basic types of Machine Learning.

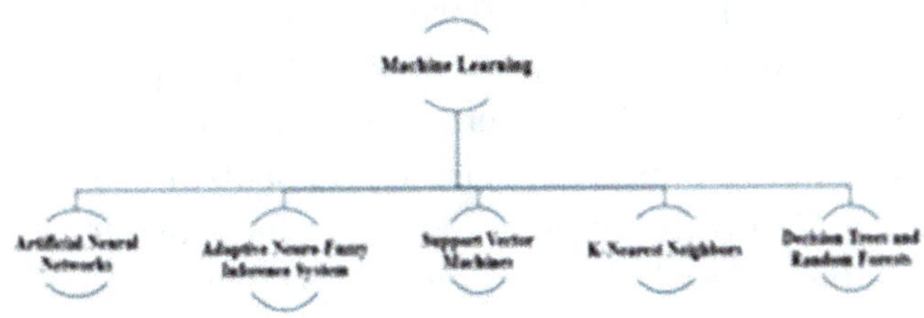

2. OVERVIEW OF NEURAL NETWORK ARCHITECTURES

2.1 Feedforward Neural Networks (FNN)

A feedforward neural network (FNN) is an architecture formed by three layers. The first layer is the input layer. The second layer is a hidden layer that contains several layers. Finally, the output layer is used. In this architecture, the data follows only one direction, which starts from the inputs and ends in the outputs (B. Yang et al., 2024). Figure 2 presents the Feedforward Neural Networks structure. In the study by (Xu & Zhang, 2022), a virtual injection enhancement front-neural network (VIP-FNN) was developed to solve the X-to-Y function representing the measurement of isotherm adsorption curve parameters. The role of the front-neural network is to identify relationships between variables. Furthermore, in the study (Cagcag Yolcu et al., 2021), a front-neural network model was applied to approximate ammonium adsorption by identifying variables' nonlinear relationships.

Figure 2. Feedforward Neural Networks

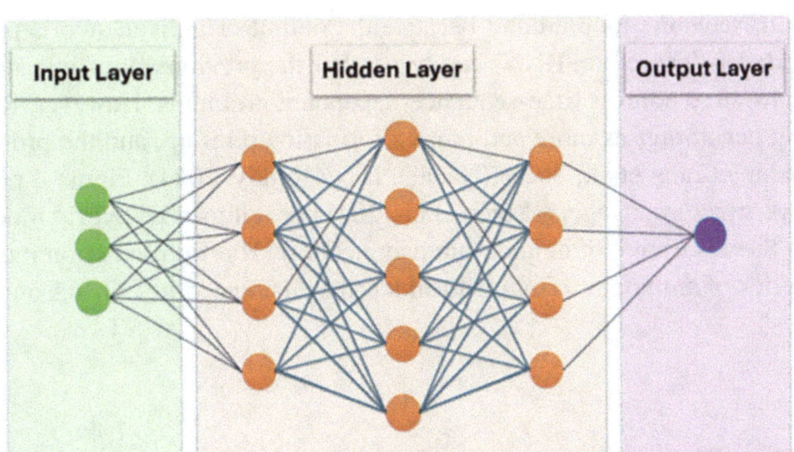

2.2 Convolutional Neural Networks (CNN)

The convolutional neural network could be defined as part of a neural network or a section of layers. The CNN architecture has two layers, the input layer and an output layer, and is characterized by a convolutional layer and a pooling layer. Moreover, each layer has a filter. The filters perform pattern recognition in the CNN. Therefore, it is considered the most popular model for analyzing images (Zeng et al., 2024),(Zheng & Luo, 2024). Figure 3 presents the Convolutional Neural Network structure. Under varying surface roughness at the molecular level, (Li et al., 2022) applied a deep learning convolutional neural network to predict the adsorption density distribution of atomic and organic liquids. In addition, in a study by (Karmakar et al., 2023), the adsorption properties of gases were predicted using CNN models. In another study of (Cheng et al., 2023), the content of rare earth metals adsorbed by ions was quantified using a convolutional neural network method.

Figure 3. Convolutional Neural Networks

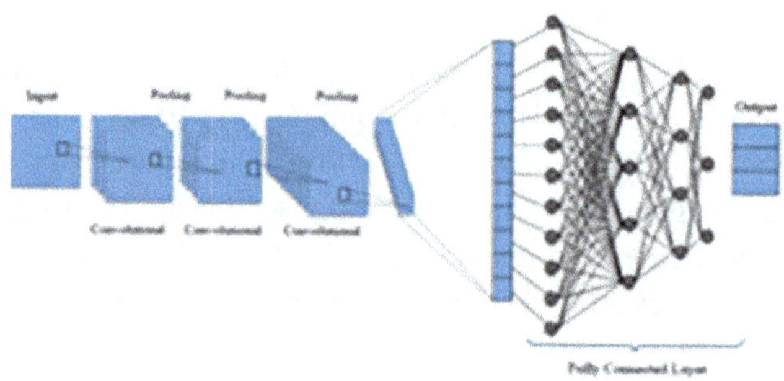

2.3 Recurrent Neural Networks (RNN)

Recurrent Neural Networks loop around because the output of a given step is provided alongside the input in the next step. Therefore, RNNs can remember the previous steps in a sequence. An RNN algorithm was employed to address time-sequence-dependent problems. However, RNNs become less effective at learning new things as more and more information piles up, and the problem of exploding gradients and vanishing (Tang et al., 2025),(Nadakuditi & Reddy, 2024). Figure 4 presents the Recurrent Neural Network structure. (Cagcag Yolcu et al., 2021) develloped an hybrid model of FF-NN and ER-NNs to predict the adsorption of ammonium using zeolite. Hybrid prediction models were used to estimate the adsorption of ammonium from landfill leachate using zeolite in batch and column systems.

Figure 4. Recurrent Neural Networks

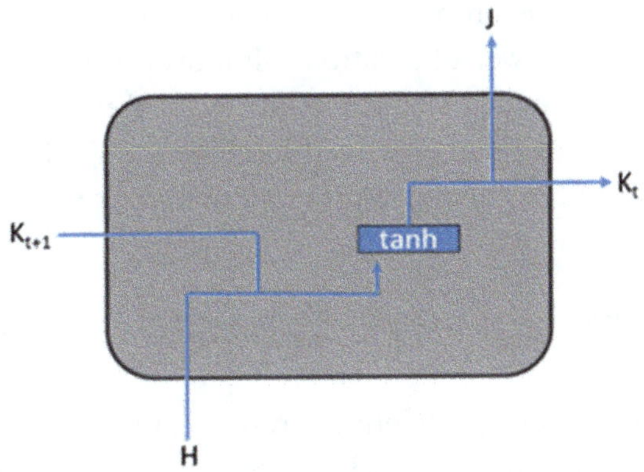

2.4 Long Short-Term Memory (LSTM)

Long Short-Term Memory is an RNN algorithm that allows an artificial neural network (ANN) to remember the information that it needs but also to forget useless the information. To overcome the problems of ordinary RNNs, the *LSTM* model is trained using temporal backpropagation. Therefore, in the LSTM architecture, regular neurons are replaced with more intelligent and performant storage units. This algorithm has been successfully applied to research related to video and audio data processing and speech recognition (Wu et al., 2024). Figure 5 presents the Long Short-Term Memory Network structure. To predict the loading capacity of the drying layer, (Yusoff et al., 2023) proposed combining two models, long-term memory and particle swarm optimization algorithms (LSTM+PSO) as an effective approach. Furthermore, the kinetics of total chromium removal capacity [Cr(T)] and hexavalent chromium removal capacity [Cr(VI)] were predicted in studies (Cruz-Victoria et al., 2024) using LSTM-type neural networks and LSTM-type bidirectional neural networks.

Figure 5. Long Short-Term Memory

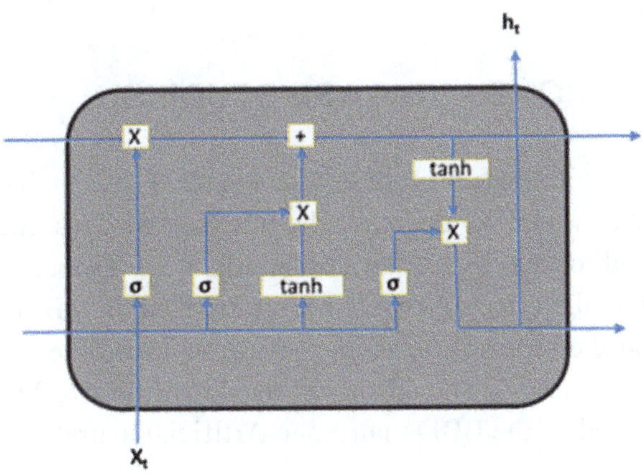

2.5 Artificial Neural Networks (ANNs)

Artificial intelligence has become a strong science, as it offers several advantages over traditional methods. It offers a new, rapid, and powerful solution. Moreover, it is not limited to a special domain or field but is considered a widely applied science. ANN is a subpart of the machine learning field, which is a subpart of the principal domain of Artificial Intelligence. ANN is considered a simulation of the neurons of the human brain and can solve complex and nonlinear processes (Hidayat et al., 2025a).

3. BRIDGING DATA-DRIVEN AND PHYSICS-BASED MODELS

There are two important approaches to modeling physical and chemical processes. The first is the explanatory physical and chemical approach, and the second is the predictive statistical approach. Each has its own characteristics and advantages. The first is based on understanding the fundamental laws of physics and chemistry: it seeks to describe the actual mechanisms that govern the behavior of a process (Sellaoui et al., 2024). These models are interpretable; each parameter has a precise physical meaning. However, the limitations of these methods lie in their need to simplify assumptions and their ineffectiveness in complex or nonlinear systems (Ben et al., 2021). Statistical or artificial intelligence methods, such as artificial neural networks, aim for predictive performance more than explanation. They can learn directly from experimental data. A neural network can understand the nonlinear relationships and interactions between variables. However, the problem is that the internal weights of the network do not have a physical meaning and therefore cannot be explained physically (Allahkarami et al., 2024).

Isotherm models, including Langmuir, Freundlich, Temkin, and kinetic models the widely applied classical models for describing adsorption, degradation, and mass transfer phenomena. They are based on equations derived from physicochemical principles. These models are interpretable and allow the obtained parameters to be linked to concrete physical properties. However, they often assume idealized conditions and can fail when the system is nonlinear or influenced by several coupled factors (Y. Wang et al., 2024),(Khan et al., 2024). Artificial neural networks, on the other hand, constitute an empirical and predictive approach, capable of modeling complex systems without a priori assumptions about the functional form of the phenomenon. They can simultaneously integrate several variables and learn to predict experimental results with very high accuracy. However, unlike physicochemical models, ANNs are often perceived as black boxes, which limits their explanatory power (Y. Wang et al., 2024).

4. PRACTICAL APPLICATIONS AND CASE STUDIES

Dyes are the most common pollutants present in wastewater. They can be easily observed since they change the transparent color of water. To decrease the intensity of dyes, the adsorption process is the most common process applied to treat this water contaminant. In (Deepa, 2025) box–behnken design (BBD) was applied to evaluate the effect of dye concentration (mg/L), Biochar weight Proportion (g), and time of contact (min) on the removal efficiency of Methylene blue (MB). The coded and actual levels of variables are given in Table 1. The adsorbent used to eliminate Methylene blue was biochar–polymer composite beads. Based on the BBD matrix, an Artificial neural network was applied to model the sorption of Methylene blue using biochar–polymer composite beads. The experiments, the predicted and experimental removal percentage of MB, are shown in Table 2.

The ANN models with three layers were developed (input layer, a hidden layer, and an output layer). The input layer was the BBD controlled parameters (dye concentration, Biochar weight Proportion, and time of contact). Therefore, the model input layer had three. Moreover, the output layer had one neuron that represented the removal efficiency of Methylene blue. The number of neurons in the hidden layer was optimized to be between 1 and 10 neurons. ANN has a tangent sigmoid transfer function (tansig) at the hidden layer and a linear transfer function (purelin) at the output layer. The algorithm chosen for training was the Levenberg-Marquardt (LM) backpropagation algorithm, as it is fast and efficient. In addition, the dataset was split into 70% for network training, 15% for network testing, and 15% for model validation.

Table 1. The experimental factors, their actual and coded values

Factors	Description	Low (-1)	High (+1)
A	Conc. of Dye (mg/L)	2	20
B	Biochar weight Proportion (g)	0.1	0.5
C	Contact Time (min)	2	60

Table 2. Experimental and predicted values of MB removal by biochar–polymer composite beads

Run	Conc. of Dye (mg/L)	Biochar weight Proportion (g)	Contact Time (min)	% Removal	
				Actual	ANN
1	20	0.5	30	70.0	70.01
2	2	0.5	30	95.0	95.03
3	2	0.2	60	92.5	92.51
4	10	0.5	2	66.0	66.02
5	10	0.2	30	82.0	82
6	20	0.2	60	62.5	62.5
7	20	0.1	30	36.0	35.99
8	10	0.5	60	85.0	85.02
9	10	0.1	60	53.0	49.63
10	10	0.1	2	35.0	34.99
11	20	0.2	2	53.5	61.28
12	10	0.2	30	82.0	82.00
13	2	0.1	30	65.0	65.02
14	2	0.2	2	72.0	72
15	10	0.2	30	82.0	82

The neurons of the hidden layer are the important parameter that affects the model's efficiency. For the architecture optimization, 1 to 10 hidden-neurons were applied in the hidden layer. The number of neuron test was low to avoid the overfitting problem. Eight neurons in the hidden layer were found to be the best, and the configuration of the neural network was 3-8-1, presented in Figure 6.

The performance comparison of the ANN model, including error analysis, is illustrated in Figs. 7,8 and 9. The correlation coefficients (R) for all the data points are nearly 1, which reveals proper performance of the designed ANN in the modeling of the process. R^2 values for all model was 0.99263. As the number of samples was small, the testing and validation was donne using 2 random samples for each one, which is not sufficient to consider the model as a general model. However, with small data, the model is considered a local model. Many studies applied a small amount of data from the RSM model to train the ANN model, and get performance predicted results (Yousefi et al., 2024),(Hidayat et al., 2025b),(Ni et al., 2024).

The training of the networks was stopped after 5 epochs with a low MSE value of 30.2775. Figure 8 shows the MSE versus the number of epochs for optimal ANN models. The error histogram in Figure 9 shows the verification network performance. The graph shows that most of the bars are centered around zero, particularly between -0.259 and 0.309, indicating a small difference between the predicted and actual values for most of the data. Two isolated values, (~-7.5) and (~+3.1), were observed, which may suggest some outliers. Overall, the fact that most errors are close to zero suggests that the network generalizes well to the entire dataset; there is no significant bias in the predictions. Figure 10 presents a comparison between the real and predicted values. It is clearly observed that the ANN predictions are very close to experimental values.

Figure 6. Topology (3-8-1) of ANN architecture.

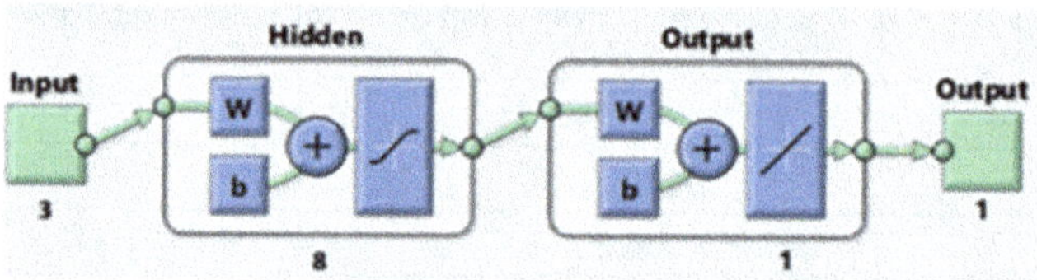

Figure 7. The ANN model performance of adsorption methylene dye

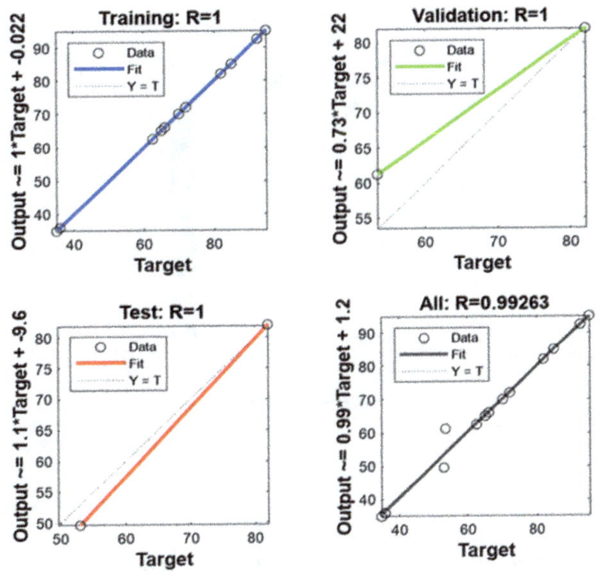

Figure 8. The training, validation and test performance

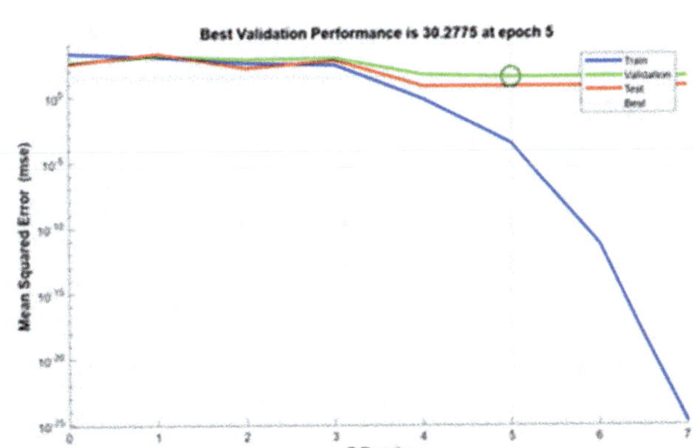

Figure 9. The error histogram of the network

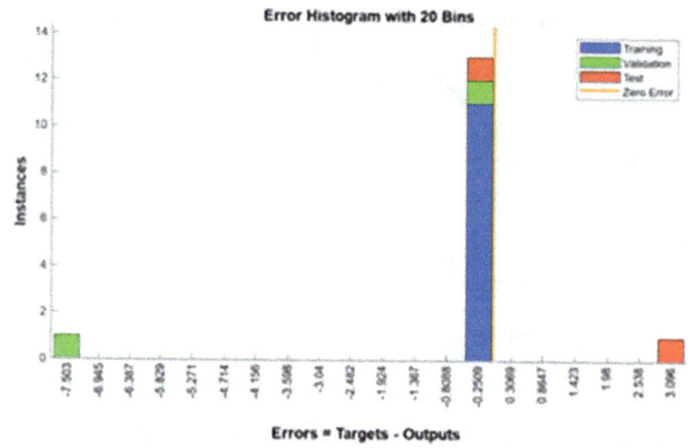

Figure 10. Experimental vs ANN-predicted output removal of MB

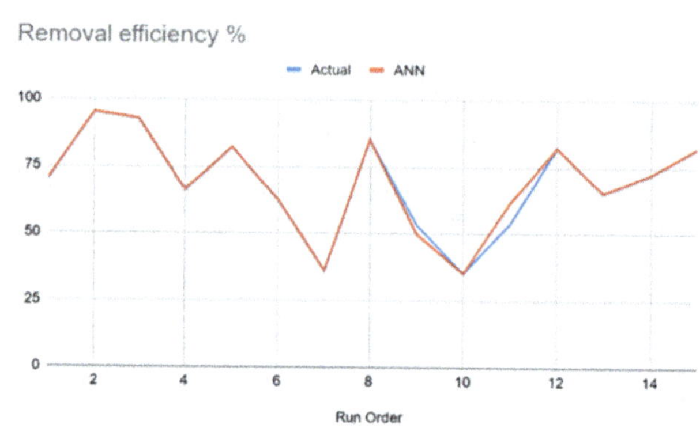

5. ANN MODEL DEVELOPMENT FOR ADSORPTION

5.1 Data Collection and Preprocessing

Generally, data are collected using instruments that allow the response to be measured. However, one must always be sure that the measurements are correct and precise, as a wrong measurement can throw off the whole study. Therefore, it is recommended to repeat each measurement several times to achieve higher precision. After collecting data, a preprocessing step is important before using them for training. In this stage, we ensured that no data were missing and that the value in each column had the same unit. Moreover, to ensure the high quality of the data and increase the model performance, normalization of the datasets is crucial (Canay & Kocabıçak, 2023),(M. Wang et al., 2025). Normalization reduces data noise and redundancy (Rasul et al., 2024). Additionally, it improves the data quality and equalizes the contribution of each feature. In most cases, the columns of a dataset matrix have very different ranges; therefore, the model pays more attention to large values, considering them more important than others in other columns. Normalization scales or transforms these values before they are used in subsequent steps (Ounoughi & Ben Yahia, 2024).

5.2 Selection of Input and Output Variables

The input variables selected for training a network are the same variables used to control the process or phenomenon studied. The output variable represents the response of the study. In the adsorption technique, the inputs are usually the temperature, initial concentration, adsorbent dose, and pH. The output was the removal percentage. Furthermore, Elemental composition, porosity, and surface area have been used as input parameters to create general ANN model that have in consideration the intern propriety of the adsorbent. Sensitivity analysis can also be a useful technique for selecting inputs, as it helps to identify the input variables that have a higher effect on the model response. (Yousefi et al., 2024)

The material fingerprints concept has a crucial role in connecting propriety to structure. Molecular fingerprinting (MF) is a technique that decomposes molecules into substructures and converts these substructures into digital representations. That can be used directly as ANN inputs (Ni et al., 2024). These fingerprints are created by dividing the molecule's structure into specific substructures or parts, and then encoding the presence or absence of these parts in binary form (Elsenety, 2024). Designed for filtering big data, molecular fingerprinting (MF) involves a rule for converting molecules into a series of bits. It uses an array containing "1s" or "0s" to describe the presence or absence of substructures. This binary vector is used as a descriptor, capturing structural features, and serves as input for a machine learning model (Yuan et al., 2022).

5.3 Choosing Optimal Topology and Learning Algorithms

The optimal ANN topology provides the model with greater accuracy. The algorithm can better understand the problem and predict results that are very close to the experimental findings. The optimal topology is when the model has not too many hidden neurons that cause overfitting or a low number of hidden neurons that cause underfitting. The best topology should have a high R^2 value and low RMSE and MSE values.

5.3.1 Feedforward Backpropagation Networks (FBNNs)

Feedforward backpropagation networks are models with two training algorithms, a feed-forward neural network (FNN) that ensures the stable direction of the model, which starts from inputs and finishes in outputs. Backpropagation (BP) returns a portion of the total error to each neuron, proportional to its influence on the final output (de León-Delgado et al., 2025). The prediction errors were reduced after each epoch; therefore, the loss function was also minimized. This is attended due to the gradient descent method used by the backpropagation algorithm, which updates weights and biases automatically until the minimum error is obtained (Nguyen & Truong, 2024),(Chinatamby & Jewaratnam, 2023).

5.3.2 Radial Basis Function Networks (RBFNs)

Radial Basis Function Networks are one of the most commonly used architectures in adsorption modeling. It contains an input layer, an RBF layer, and an output layer, and uses radial basis functions as activation functions. The RBFN is characterized by its ability to approximate the model by exploiting the linear combinations of radial basis functions. RBF networks provide a nonlinear mapping from the input to the output space, making it a very strong algorithm for supervised learning problems. The RBFN training process involves two important steps. The first step is to find the appropriate centers and width of the RBFN. However, the second step is the determination of the optimal output weights (Dai et al., 2024),(Chang et al., 2025).

5.3.3 Hybrid ANN Models

The hybrid ANN model includes a combination of an ANN model with optimization algorithms such as Particle Swarm Optimization (PSO) and genetic algorithm (GA). The PSO algorithm optimizes the ANN parameters by updating the velocities and positions of the particles until the best individual and global best solutions are found (Cu et al., 2025). The GA applied random initialization, crossover, selection, and mutation processes. It produces several independent solutions (Şen, 2024). Therefore, it is important to note that this algorithm has been used for two different optimization objectives. The first objective was to optimize the ANN architecture by determining the best number of hidden layers, hidden neurons, weights, biases, epochs, learning rates, adaptive activation functions, and training algorithms.

In (Jiao et al., 2023), modified pomelo peels were used to adsorb Cu (II) ions (Jiao et al., 2023). The process was optimized using an artificial neural network (ANN) model. A PSO algorithm was incorporated to improve the model accuracy by determining the best weights and biases. The model architecture was 4-9-1. Moreover, the results show that $RMSE_{ANN-PSO} = 0.0080$ which was lower than the $RMSE_{ANN} = 0.0207$. In another study (Tripathi et al., 2022), the GA algorithm was applied to optimize the ANN model of As (III) adsorption by adopting an iron-impregnated fungal biofilter. The best architecture found to be 4-10-1 wih a high R^2 of 0.9997 and MSE_{training} of 0023 an MSE_{test} of 0.031. In (Baskaran et al., 2024), an ANN-PSO model was implemented to optimize the removal of chemical oxygen demand (COD) and chromium adsorption on palm husk powder. The best architecture was 3-10-1, with a coefficient of correlation of 0.99998 and an MSE of 0.0172.

The second objective of a hybrid ANN is to optimize the operational conditions to achieve optimal or targeted results. (Wu et al., 2021) applied both GA and PSO algorithms to optimize the adsorption parameters of As (III) removal by an iron-impregnated fungal biofilter (Wu et al., 2021). The parameters used were As (III) Concentration, Temperature, Contact time, and pH. Under the optimal conditions of 10.78 mg/L, 25.66 °C, 46.41 min, and pH=7.58, the experimental percentage removal of As (III) was 84.64% for the GA algorithm. Additionally, for the PSO algorithm, the optimal conditions were 18.03 mg/L, 24.68 °C, 46.46 min, and pH=8. The experimental percentage removal of As (III) was 81.31%. In an equivalent study, (Prabhu et al., 2021) used alginate-immobilized Sargassum sp. as Cr (VI) adsorbent. Moreover, the ANN-GA algorithm was applied to determine the optimal values of the operating conditions, namely, Cr concentration, bed height, and flow rate. The removal percentage of Cr (VI) attained is 90.79% under condition 25 mg/L, 11.97 cm 5.29 mL/min. In another study, (Hou et al., 2020) explored the optimal operating parameter value by hybrid ANN-GA and ANN-PSO. The initial concentration, pH, contact time, and dosage were studied. The experimental results showed that the removal percentages attained by the GA and PSO algorithms were 92.03% and 88.83%, respectively. The optimal conditions for the ANN-GA model were 299.22 mg/L, 13.19 min, 17.55 mg, pH=6.1. However, for the ANN-PSO model, optimal conditions were 371.23 mg/L, 13.95 min, 17.2 mg and pH=5.

5.4 Training methods, Error Functions, and validation Techniques

5.4.1 Training Methods

The Levenberg–Marquardt (ML):

The Levenberg–Marquardt algorithm combines the advantages of Gaussian-Newtonian regression and stepwise regression, and is known for its ability to achieve rapid local convergence and improve performance. It is characterized by its rapid convergence, stability, reliability, and ability to enhance algorithm learning by reducing overall error through weight and bias adjustments (Kattimani et al., 2025),(M. Zhou et al., 2024).

Gradient Descent (GD):

The gradient descent approach is a traditional and mathematically accurate method used in loss reduction optimization problems (Chen & Chen, 2025). Gradient descent algorithms start with an initial point and then gradually move towards a local or global optimal value by ensuring the negative regression direction (D. Yang et al., 2025). This type of iteration enables the GD algorithm to efficiently find a minimum value (Ji et al., 2024). In artificial neural networks, the stepwise regression (GD) method facilitates rapid and accurate model training (X. Zhou et al., 2025).

Bayesian Regularization (BR)

Bayesian Regularization (BR) is a training method that incorporates prior probabilities of weights while minimizing the loss function (L. Wang et al., 2024). In this model, system biases and weights are treated as random variables with prior distributions. BR can be applied to avoid overfitting problems and improve generalization. This is achieved by adding penalty factors to the cost function, which discourages large values of biases and weights. (Sabir et al., 2025).

5.4.2 Error Functions

Squared Error (MSE) by definition is the average squared difference between the predicted and actual values; lower values indicate a better fit (Mehedi Shamrat et al., 2024). The MSE mathematical notation is expressed in Eq 1.

$$\mathrm{MSE} = \frac{\sum_{i=1}^{N}(y_i - \hat{y}_i)^2}{N} \quad (1)$$

The Root Mean Squared Error (RMSE) was used as an indicator to test the accuracy of the predictive models. It measures the average deviation between the model's predicted and real values (Huang et al., 2024). The RMSE mathematical expression is presented in Eq 2.

$$\text{RMSE} = \sqrt{\frac{1}{n}\sum_{i=1}^{n}(y_i - \hat{y}_i)^2} \tag{2}$$

Mean Absolute Error (MAE) is one of the regression model evaluation metrics that calculates the average absolute difference between predicted and real values (Harrou et al., 2024). The MAE mathematical expression is shown in Eq 3.

$$\text{MAE} = \frac{1}{n}\sum_{i=1}^{n}|y_i - \hat{y}_i| \tag{3}$$

The coefficient of determination (R^2) indicates the proportion of variance in the observed data explained by the model; values closer to 1 represent higher accuracy (Kaak et al., 2024). The R^2 mathematical notation is shown in Eq 4.

$$R^2 = 1 - \frac{\sum_{i=1}^{n}(y_i - \hat{y}_i)^2}{\sum_{i=1}^{n}(y_i - \hat{y}_i)^2} \tag{4}$$

5.4.3 Validation Techniques

K-Fold Cross-Validation

K-fold cross-validation is used for selecting and conserving error estimates of the model. The basic principle of this technique is to divide the data into two classes: training sets and test sets. The data is divided into K equal parts, and for K times, the model is trained and tested. One fold is used as the cross-validation set in each iteration while the other K-1 folds are applied to the model training (Bajpai et al., 2024). Then, the average of the error measures from all K different runs on the model is taken. The final model is the one with the lowest average error score (Rasul et al., 2024).

Leave-One-Out (LOO)

In the case of small sample sizes, the Leave-One-Out technique is the most applicable method for evaluating the generalizability of predictive models. Leave-One-Out Cross-Validation is a specific form of multifold (k-fold) cross-validation, where one data point is left out from the training step and applied for model validation, while the remaining data is used for the model training (Janga & Rao, 2025). This technique is advantageous because it can accurately evaluate the model on all available data points, thus minimizing overfitting (Ali Üncü et al., 2025).

7 FUTURE PERSPECTIVES

The combination of artificial neural networks and Density functional theory (DFT) or molecular simulation is paving the way for a new era of rational adsorbent material design, where machine learning models guide experiments and dramatically reduce the development time for new, high-performance materials. Density functional theory calculations provide precise atomic-level information about the interactions between the adsorbent and the adsorbate while molecular simulation describes the dynamic behavior of the system at the nanoscale. However, these methods are often computationally expensive

(Amaka et al., 2024),(Bouzid et al., 2025). ANNs can replace or accelerate some of these computational steps by learning the complex structure-property relationships from a limited number of initial simulations. Furthermore, ANNs can be used in the computer-aided design of novel adsorbent materials. By learning from experimental and theoretical databases, they can select promising materials and even optimize surface textures, porosity, and functionality to achieve specific selectivity or adsorption capacity.

6. CONCLUSION

Modeling adsorption using artificial neural networks (ANNs) has emerged as a powerful and flexible approach to predicting the performance of adsorbent materials. Unlike traditional isothermal and kinetic models, which rely on simplified assumptions, ANNs capture the nonlinearity and complexity of the interactions between operating parameters and adsorption capacity. Thanks to their ability to learn from experimental data, they provide accurate predictions, even under conditions where experimental models fail. However, the predictive power of ANNs is accompanied by a lack of physical explainability. This chapter explores a case study of the adsorption process. The results show high model performance with a coefficient of determination (R^2) of 0.9926. However, the predictive power of artificial neural networks is accompanied by a lack of physical explanation. To overcome this limitation, the combined use of ANNs and physical and chemical models allows for a blend of predictive power and mechanistic understanding. Neural networks can also be used to optimize adsorption conditions and select the most efficient materials before experiments, reducing development costs and time. Thus, neural networks are no longer just simple predictive tools but have become effective instruments for accelerating the discovery and improvement of advanced materials for water and environmental remediation.

ACKNOWLEDGEMENT

This work was supported by the National Center for Scientific and Technical Research (CNRST), which funded this research through the PhD-Associate Scholarship- PASS Program.

REFERENCES

Ali Üncü, Y., Danışman, T., & Özdoğan, H. (2025). Predicting (n,3n) nuclear reaction cross-sections using XGBoost and Leave-One-Out Cross-Validation. *Applied Radiation and Isotopes*, *219*(January), 111714. DOI: 10.1016/j.apradiso.2025.111714 PMID: 39947034

Allahkarami, E., Allahkarami, E., Heydari, M., Azadmehr, A., & Maghsoudi, A. (2024). Assessment of chromite ore wastes for methylene blue adsorption: Isotherm, kinetic, thermodynamic studies, ANN, and statistical physics modeling. *Chemosphere*, *358*(March), 142098. DOI: 10.1016/j.chemosphere.2024.142098 PMID: 38677606

Amaka, N., Sunday, C., Chukwudi, P., Bartholomew, N., Dagdag, O., Okechukwu, J., Thakur, A., Chikaodili, V., & Dominic, O. (2024). Physisorption or chemisorption : Insight from AI computing model based on DFT, MC / MD-simulation for prediction of MOF-based inhibitor adsorption on Cu in brine solution. *Computational & Theoretical Chemistry*, *1238*(June), 114730. DOI: 10.1016/j.comptc.2024.114730

Bajpai, A., Chaurasia, D., & Tiwari, N. (2024). A novel methodology for anomaly detection in smart home networks via Fractional Stochastic Gradient Descent. *Computers and Electrical Engineering, 119*(PB), 109604. DOI: 10.1016/j.compeleceng.2024.109604

Baskaran, D., Natarajan, A. P. A., & Byun, H.-S. (2024). Potential of palm husk powder in the treatment of real tannery wastewater: Performance comparison between response surface methodology and artificial neural network-particle swarm optimization. *Journal of Water Process Engineering*, *67*(August), 106101. DOI: 10.1016/j.jwpe.2024.106101

Ben, M., Wazzan, N., & Ben, A. (2021). Physico-chemical interpretations of the adsorption isotherms of D e p e A sensitizers with pyridyl group on TiO 2 for dye sensitized solar cells using statistical physics and density functional theory. *Journal of Materials Research and Technology*, *15*, 369–383. DOI: 10.1016/j.jmrt.2021.08.017

Boikos, C., Ioannidis, G., Rapkos, N., Tsegas, G., Katsis, P., & Ntziachristos, L. (2025). Estimating daily road traffic pollution in Hong Kong using CFD modelling: Validation and application. *Building and Environment, 267*(PA), 112168. DOI: 10.1016/j.buildenv.2024.112168

Bouzid, T., Grich, A., Naboulsi, A., Yazid, H., Elbasraoui, A., El Himri, M., Regti, A., & El Haddad, M. (2025). Investigating the surface structure effects of dual polymer coatings on eriochrome black T adsorption : A DFT and artificial intelligence approach based on ANN modeling. *Journal of Molecular Structure*, *1343*(May), 142726. DOI: 10.1016/j.molstruc.2025.142726

Cagcag Yolcu, O., Aydın Temel, F., & Kuleyin, A. (2021). New hybrid predictive modeling principles for ammonium adsorption: The combination of Response Surface Methodology with feed-forward and Elman-Recurrent Neural Networks. *Journal of Cleaner Production, 311*(January), 127688. Advance online publication. DOI: 10.1016/j.jclepro.2021.127688

Canay, Ö., & Kocabıçak, Ü. (2023). An innovative data collection method to eliminate the preprocessing phase in web usage mining. *Engineering Science and Technology, an International Journal, 40*, 101360. DOI: 10.1016/j.jestch.2023.101360

Chang, L., Yang, W., Khabaz, M. K., Eftekhari, S. A., Alkhalifah, T., & Khairy, Y. (2025). Introducing an integrated self-organizing map and radial basis function network for accurate prediction of water-Fe3O4 nanofluid viscosity versus solid volume fraction and temperature. *Renewable and Sustainable Energy Reviews, 209*(November 2024), 115119. DOI: 10.1016/j.rser.2024.115119

Chen, Z., & Chen, K. (2025). A new gradient descent divergence cleaning method with optimized high-order low-dissipation TENO schemes for ideal magnetohydrodynamic simulations. *Computer Physics Communications, 315*(May), 109744. DOI: 10.1016/j.cpc.2025.109744

Cheng, G., Ban, Y., Deng, X., Li, H., Zhang, H., Li, G., Liao, L., & Khan, R. (2023). Research on quantitative inversion of ion adsorption type rare earth ore based on convolutional neural network. *Frontiers in Earth Science (Lausanne), 10*(January), 1–14. DOI: 10.3389/feart.2022.1086325

Chinatamby, P., & Jewaratnam, J. (2023). A performance comparison study on PM2.5 prediction at industrial areas using different training algorithms of feedforward-backpropagation neural network (FBNN). *Chemosphere, 317*, 137788. DOI: 10.1016/j.chemosphere.2023.137788 PMID: 36642141

Cruz-Victoria, J. C., Netzahuatl-Muñoz, A. R., & Cristiani-Urbina, E. (2024). Long Short-Term Memory and Bidirectional Long Short-Term Memory Modeling and Prediction of Hexavalent and Total Chromium Removal Capacity Kinetics of Cupressus lusitanica Bark. *Sustainability (Basel), 16*(7), 2874. Advance online publication. DOI: 10.3390/su16072874

Cu, W., Fang, J., Guo, X., Chen, K., Zheng, N., Xiao, B., & Wei, J. (2025). Prediction and analysis of bed-to-tube heat transfer in fluidized bed heat exchangers based on ANN-PSO hybrid approach and CFD simulation. *International Journal of Heat and Mass Transfer, 252*(June), 127490. DOI: 10.1016/j.ijheatmasstransfer.2025.127490

Dai, Y., Wu, Q., & Zhang, Y. (2024). Generalized sparse radial basis function networks for multi-classification problems. *Applied Soft Computing, 154*(December 2022), 111361. DOI: 10.1016/j.asoc.2024.111361

Dbik, A., El Messaoudi, N., Bentahar, S., El Khomri, M., Lacherai, A., & Faska, N. (2022). Optimization of methylene blue adsorption on agricultural solid waste using box–behnken design (BBD) combined with response surface methodology (RSM) modeling. *Biointerface Research in Applied Chemistry, 12*(4), 4567–4583. DOI: 10.33263/BRIAC124.45674583

de León-Delgado, H., Praga-Alejo, R. J., & González-González, D. S. (2025). Backpropagation neural network model with statistical inference in manufacturing processes. *Journal of Industrial Information Integration, 44*(January), 100783. Advance online publication. DOI: 10.1016/j.jii.2025.100783

Deepa, A. (2025). Biochar-infused polyethersulfone beads for dye adsorption: RSM-BBD optimization coupled with isotherm and kinetic insights. *Materials Today. Communications*, *48*(June), 113652. DOI: 10.1016/j.mtcomm.2025.113652

Deshmukh, M. T., Wankhede, P. R., Chakole, N., Kale, P. D., Jadhav, M. R., Kulkarni, M. B., & Bhaiyya, M. (2025). Towards intelligent food safety: Machine learning approaches for aflatoxin detection and risk prediction. *Trends in Food Science & Technology*, *161*(April), 105055. DOI: 10.1016/j.tifs.2025.105055

El Khomri, M., El Messaoudi, N., Dbik, A., Bentahar, S., Lacherai, A., Faska, N., & Jada, A. (2022). Regeneration of argan nutshell and almond shell using HNO3 for their reusability to remove cationic dye from aqueous solution. *Chemical Engineering Communications*, *209*(10), 1304–1315. DOI: 10.1080/00986445.2021.1963960

El Messaoudi, N., El Mouden, A., Fernine, Y., El Khomri, M., Bouich, A., Faska, N., Ciğeroğlu, Z., Américo-Pinheiro, J. H. P., Jada, A., & Lacherai, A. (2023). Green synthesis of Ag2O nanoparticles using Punica granatum leaf extract for sulfamethoxazole antibiotic adsorption: Characterization, experimental study, modeling, and DFT calculation. *Environmental Science and Pollution Research International*, *30*(34), 81352–81369. DOI: 10.1007/s11356-022-21554-7 PMID: 35729389

Elsenety, M. M. (2024). Design and predict the potential of imidazole-based organic dyes in dye-sensitized solar cells using fingerprint machine learning and supported by a web application. *Scientific Reports*, *14*(1), 1–11. DOI: 10.1038/s41598-024-76739-6 PMID: 39489728

Harrou, F., Dairi, A., Dorbane, A., & Sun, Y. (2024). Enhancing wind power prediction with self-attentive variational autoencoders: A comparative study. *Results in Engineering*, *23*(May), 102504. DOI: 10.1016/j.rineng.2024.102504

Hidayat, E., Hannani Hamid, N. A., Mohamad Sarbani, N. M., Samitsu, S., Aoyagi, M., Harada, H., & Safari, M. A. M. (2025a). Optimization and comparative modelling of RSM and ANN for the adsorptive removal of Remazol Brilliant Blue R dye using spent coffee ground biochar. *Chemosphere*, *389*(June), 144709. DOI: 10.1016/j.chemosphere.2025.144709 PMID: 41033296

Hidayat, E., Hannani Hamid, N. A., Mohamad Sarbani, N. M., Samitsu, S., Aoyagi, M., Harada, H., & Safari, M. A. M. (2025b). Optimization and comparative modelling of RSM and ANN for the adsorptive removal of Remazol Brilliant Blue R dye using spent coffee ground biochar. *Chemosphere*, *389*(September), 144709. DOI: 10.1016/j.chemosphere.2025.144709 PMID: 41033296

Hou, Y., Qi, J., Hu, J., Xiang, Y., Xin, L., & Wei, X. (2020). Mesoporous Mn-Doped Fe Nanoparticle-Modified Reduced Graphene Oxide for Ethyl Violet Elimination: Modeling and Optimization Using Artificial Intelligence. *Processes (Basel, Switzerland)*, *8*(4), 488. DOI: 10.3390/pr8040488

Huang, M., Zhang, X. S., Bhatti, U. A., Wu, Y., Zhang, Y., & Yasin Ghadi, Y. (2024). An interpretable approach using hybrid graph networks and explainable AI for intelligent diagnosis recommendations in chronic disease care. *Biomedical Signal Processing and Control*, *91*, 105913. DOI: 10.1016/j.bspc.2023.105913

Janga, K., & Rao, G. (2025). Bioresource Technology Reports Effect of residence time and temperature on pyrolysis of Water Hyacinth Leaf Biomass (WHLB): Experiments and modelling through ML and validation by Leave One Out Cross Validation (LOOCV). *Bioresource Technology Reports*, *32*(August), 102369. DOI: 10.1016/j.biteb.2025.102369

Ji, J. Y., Tan, Z., Zeng, S., & Wong, M. L. (2024). An ε-constrained multiobjective differential evolution with adaptive gradient-based repair method for real-world constrained optimization problems. *Applied Soft Computing*, *152*, 111202. DOI: 10.1016/j.asoc.2023.111202

Jiao, M., Jacquemin, J., Zhang, R., Zhao, N., & Liu, H. (2023). The Prediction of Cu(II) Adsorption Capacity of Modified Pomelo Peels Using the PSO-ANN Model. *Molecules (Basel, Switzerland)*, *28*(19), 6957. DOI: 10.3390/molecules28196957 PMID: 37836799

Kaak, A. R. S., Çelebioğlu, K., Bozkuş, Z., Ulucak, O., & Aylı, E. (2024). A novel CFD-ANN approach for plunger valve optimization: Cost-effective performance enhancement. *Flow Measurement and Instrumentation*, *97*, 102589. DOI: 10.1016/j.flowmeasinst.2024.102589

Karmakar, T., Finney, A. R., Salvalaglio, M., Yazaydin, A. O., & Perego, C. (2023). Non-Equilibrium Modeling of Concentration-Driven processes with Constant Chemical Potential Molecular Dynamics Simulations. *Accounts of Chemical Research*, *56*(10), 1156–1167. DOI: 10.1021/acs.accounts.2c00811 PMID: 37120847

Kattimani, P., Prasad, K. V., Anwar, T., Jena, S. P., Shankhyan, A., & Naveen Kumar, R. (2025). Application of Levenberg-Marquardt artificial neural network to study nanoparticle aggregation phenomena in stagnation point flow towards an off-centered rotating disk. *International Journal of Thermofluids*, *27*, 101276. DOI: 10.1016/j.ijft.2025.101276

Khan, A., Al-hazmi, H. E., Agustiono, T., Majtacz, J., Piechota, G., Kumar, G., Ezzati, P., Saeb, R., Rabiee, N., Karimi-maleh, H., Claudio, E., & Jacek, M. (2024). *Hydrochar as a bio-based adsorbent for heavy metals removal : A review of production processes, adsorption mechanisms, kinetic models, regeneration and reusability. 945*(April). DOI: 10.1016/j.scitotenv.2024.173972

Li, G., Guo, Y., Mabuchi, T., Surblys, D., Ohara, T., & Tokumasu, T. (2022). Prediction of the adsorption properties of liquid at solid surfaces with molecular scale surface roughness via encoding-decoding convolutional neural networks. *Journal of Molecular Liquids*, *349*, 118489. DOI: 10.1016/j.molliq.2022.118489

Maitouf, S., Bachar, A., Mabrouk, A., & Faska, N. (2025). Wastewater Treatment and Recycling Using Adsorption Enhanced by Artificial Neural Networks. In *Computational Intelligence and Optimization Methods for Sustainable Water Management* (pp. 441–476). IGI Global Scientific Publishing., DOI: 10.4018/979-8-3373-2700-6.ch013

Mehedi Shamrat, F. M. J., Shakil, R., Sharmin, , Hoque ovy, N., Akter, B., Ahmed, M. Z., Ahmed, K., Bui, F. M., & Moni, M. A. (2024). An advanced deep neural network for fundus image analysis and enhancing diabetic retinopathy detection. *Healthcare Analytics*, *5*, 100303. DOI: 10.1016/j.health.2024.100303

Mei, Y., Zhuang, S., & Wang, J. (2025). Adsorption of heavy metals by biochar in aqueous solution: A review. *The Science of the Total Environment*, *968*(February), 178898. DOI: 10.1016/j.scitotenv.2025.178898 PMID: 39986038

Mohsenzadeh, E., Ratautaite, V., Brazys, E., Ramanavicius, S., Zukauskas, S., Plausinaitis, D., & Ramanavicius, A. (2024). Application of computational methods in the design of molecularly imprinted polymers (review). *TrAC Trends in Analytical Chemistry, 171*(August 2023), 117480. DOI: 10.1016/j.trac.2023.117480

Nadakuditi, A., & Reddy, V. (2024). An optimized framework for metal ion analysis in automobile industrial sewage. *Journal of the Indian Chemical Society*, *101*(8), 101200. DOI: 10.1016/j.jics.2024.101200

Nguyen, H. H., & Truong, V. H. (2024). Machine Learning-based prediction of seismic lateral deflection of steel trusses using nonlinear time-history analysis. *Structures (Oxford)*, *69*, 107369. DOI: 10.1016/j.istruc.2024.107369

Ni, J., Li, J., Li, S., Zheng, H., Ming, Z., Li, L., Li, H., Zhang, S., Zhao, Y., Liang, H., & Qiao, Z. (2024). Molecular fingerprint and machine learning enhance high-performance MOFs for mustard gas removal. *iScience*, *27*(6), 110042. Advance online publication. DOI: 10.1016/j.isci.2024.110042 PMID: 38883811

Ounoughi, C., & Ben Yahia, S. (2024). Sequence to sequence hybrid Bi-LSTM model for traffic speed prediction. *Expert Systems with Applications*, *236*, 121325. DOI: 10.1016/j.eswa.2023.121325

Prabhu, A. A., Chityala, S., Jayachandran, D., Deshavath, N. N., & Veeranki, V. D. (2021). A two step optimization approach for maximizing biosorption of hexavalent chromium ions (Cr (VI)) using alginate immobilized Sargassum sp in a packed bed column. *Separation Science and Technology*, *56*(1), 90–106. DOI: 10.1080/01496395.2019.1708933

Rasul, R. A., Saha, P., Bala, D., Karim, S. M. R. U., Abdullah, M. I., & Saha, B. (2024). An evaluation of machine learning approaches for early diagnosis of autism spectrum disorder. *Healthcare Analytics*, *5*, 100293. DOI: 10.1016/j.health.2023.100293

Sabir, Z., Ismail, T., Sleem, H., Umar, M., & Salahshour, S. (2025). A radial basis Bayesian regularization neural network process for the malaria disease model. *Knowledge-Based Systems, 322*(October 2024), 113722. DOI: 10.1016/j.knosys.2025.113722

Sellaoui, L., Nour, S., Dhaouadi, F., & Bonilla-petriciolet, A. (2024). Physicochemical study of the adsorption of chlorophenols on activated carbon using an advanced double layer model. *Journal of Molecular Liquids, 414*(PA), 126098. DOI: 10.1016/j.molliq.2024.126098

Şen, M. (2024). Optimization of performance and emission of a diesel engine fueled with isopropyl alcohol Blends: A comparative ANN-GA and RSM-HCO application. *Engineering Science and Technology, an International Journal*, *55*(June), 101733. Advance online publication. DOI: 10.1016/j.jestch.2024.101733

Tang, S., Duan, Z., Tian, Z., Du, W., & Qian, F. (2025). A fine-tuned RNN model for accurately predicting the spatial distribution of parameters in light hydrocarbon cracking tubular reactor. *Chemical Engineering Journal, 504*(December 2024), 158521. DOI: 10.1016/j.cej.2024.158521

Tripathi, A., Ranjan, M. R., Verma, D. K., Singh, Y., Shukla, S. K., Rajput, V. D., Minkina, T., Mishra, P. K., & Garg, M. C. (2022). ANN-GA based biosorption of As(III) from water through chemo-tailored and iron impregnated fungal biofilter system. *Scientific Reports*, *12*(1), 12414. DOI: 10.1038/s41598-022-14802-w PMID: 35858932

Wang, L., Huangfu, Z., Li, R., Wen, X., Sun, Y., & Chen, Y. (2024). Iterative learning control with parameter estimation for non-repetitive time-varying systems. *Journal of the Franklin Institute, 361*(3), 1455–1466. DOI: 10.1016/j.jfranklin.2024.01.011

Wang, M., Zhang, G., Peng, W., Gao, Y., He, Z., Wang, Y., Liu, W., Liu, J., & Chen, Y. (2025). A novel framework for enhancing postpartum hemorrhage prediction: Balanced data preprocessing meets neural network modeling. *Biomedical Signal Processing and Control, 110*(May), 108073. Advance online publication. DOI: 10.1016/j.bspc.2025.108073

Wang, Y., Wang, C., Huang, X., Zhang, Q., Wang, T., & Guo, X. (2024). Guideline for modeling solid-liquid adsorption : Kinetics, isotherm, fixed bed, and thermodynamics. *Chemosphere, 349*(September 2023), 140736. DOI: 10.1016/j.chemosphere.2023.140736

Wu, X., Du, Z., Ma, R., Zhang, X., Yang, D., Liu, H., & Zhang, Y. (2024). Qualitative and quantitative studies of phthalates in extra virgin olive oil (EVOO) by surface-enhanced Raman spectroscopy (SERS) combined with long short term memory (LSTM) neural network. *Food Chemistry, 433*(August 2023). DOI: 10.1016/j.foodchem.2023.137300

Wu, X., Hu, J., Wang, X., Xin, L., Li, C., & Wei, X. (2021). Effective removal of arsenide from aqueous solutions using mesoporous CoFe2O4/graphene oxide nanocomposites assisted by artificial intelligence. *Carbon Letters, 31*(6), 1163–1178. DOI: 10.1007/s42823-021-00239-0

Xu, C., & Zhang, Y. (2022). Estimating adsorption isotherm parameters in chromatography via a virtual injection promoting double feed-forward neural network. *Journal of Inverse and Ill-Posed Problems, 30*(5), 693–712. DOI: 10.1515/jiip-2020-0121

Yang, B., Liang, B., Qian, Y., Zheng, R., Su, S., Guo, Z., & Jiang, L. (2024). Parameter identification of PEMFC via feedforward neural network-pelican optimization algorithm. *Applied Energy, 361*(February), 122857. DOI: 10.1016/j.apenergy.2024.122857

Yang, D., He, Q., Wang, H., Gao, Y., Zhang, S., Yao, G., & Zhou, M. (2025). High-dimensional railway vertical alignment optimization using hybrid differential evolution and gradient descent. *Automation in Construction, 179*(December 2024), 106413. DOI: 10.1016/j.autcon.2025.106413

Yousefi, M., Akbari, H., Adibzadeh, A., Mohammadi, A. A., Baziar, M., Abbasi Farajzadeh, M., & Akbari, H. (2024). Adsorption of diazinon from aqueous solution using metal organic framework and functionalized graphene: Comparison of BBD, ANN models. *Chemosphere, 351*(December 2023), 141222. DOI: 10.1016/j.chemosphere.2024.141222

Yuan, X., Li, L., Shi, Z., Liang, H., Li, S., & Qiao, Z. (2022). Molecular-fingerprint machine-learning-assisted design and prediction for high-performance MOFs for capture of NMHCs from air. *Advanced Powder Materials, 1*(3), 100026. DOI: 10.1016/j.apmate.2021.12.002

Yusoff, M., Sallehud-din, M. T. M., Tahir, N., Yaacob, W. F. W., Maarof, N. N. N. A., Zain, J. M., Azmi, P. A. R., & Karunakumar, C. (2023). Drier Bed Adsorption Predictive Model with Enhancement of Long Short-Term Memory and Particle Swarm Optimization. *International Journal of Advanced Computer Science and Applications, 14*(12), 67–74. DOI: 10.14569/IJACSA.2023.0141206

Zeng, S., Zhang, Z., Cheng, X., Cai, X., Cao, M., & Guo, W. (2024). Prediction of soluble solids content using near-infrared spectra and optical properties of intact apple and pulp applying PLSR and CNN. *Spectrochimica Acta Part A: Molecular and Biomolecular Spectroscopy, 304*(July 2023), 123402. DOI: 10.1016/j.saa.2023.123402

Zheng, M., & Luo, X. (2024). Joint estimation of State of Charge (SOC) and State of Health (SOH) for lithium ion batteries using Support Vector Machine (SVM), Convolutional Neural Network (CNN) and Long Sort Term Memory Network (LSTM) models. *International Journal of Electrochemical Science, 19*(9), 100747. DOI: 10.1016/j.ijoes.2024.100747

Zhou, M., Yu, B., Shi, J., Zhou, W., Fang, C., Gong, Y., Zuo, C., & Zhang, J. (2024). Improved phase-generated carrier demodulation technique based on LM algorithm with low harmonic distortion and high stability. *Optical Fiber Technology, 87*, 103735. DOI: 10.1016/j.yofte.2024.103735

Zhou, X., Zhao, C., Huang, Y., Zhou, C., & Ye, J. (2025). Improved fractional-order gradient descent method based on multilayer perceptron. *Neural Networks, 183*(March 2024), 106970. DOI: 10.1016/j.neunet.2024.106970

Chapter 12
Artificial Intelligence-Enhanced Detection of Sesame Oil Adulteration Using FTIR, d13C Analysis, and Symbolic Regression

Widad Ezhani
Faculty of Sciences, Ibn Tofail University, Kenitra, Morocco

Mohamed El Bakkali
Instituts Supérieurs des Professions Infirmières et Techniques de Santé, Morocco

Said Bouchefra
Higher Institute of Nursing and Health Techniques, Morocco

Wafaa Benkayba
https://orcid.org/0000-0001-5196-1678
Chemistry, Laboratory of Materials, Nanotechnology, and Environment, Faculty of Sciences, University of Mohammed V, Rabat, Morocco

Yassine Mouniane
https://orcid.org/0000-0002-6037-5301
Faculty of Sciences, Ibn Tofail University, Kenitra, Morocco

Mohamed Rektouti
https://orcid.org/0009-0004-5560-4925
Faculty of Sciences, Ibn Tofail University, Kenitra, Morocco

Brahim El Ouardi
https://orcid.org/0009-0000-4559-6141
Faculty of Sciences, Ibn Tofail University, Kenitra, Morocco

Imane Aitouhanni
https://orcid.org/0009-0004-1400-8027
ENSIAS, SSLAB, Mohammed V University, Rabat, Morocco

Nour Eddine Amenzou
https://orcid.org/0000-0002-6601-0151
DEC, Division of Water and Climate, Direction of Studies and Scientific Research, National Center for Nuclear Energy, Sciences and Techniques, Kenitra, Morocco

Yassine Kadmi
LASIRE, Equipe Physico-Chimie de l'Environnement, CNRS UMR 8516, Université Lille, France

DOI: 10.4018/979-8-3373-6058-4.ch012

Amar Habsaoui
Faculty of Sciences, Ibn Tofail University, Kenitra, Morocco

ABSTRACT

The adulteration of sesame oil poses significant challenges for public health, food safety, and environmental sustainability, requiring advanced analytical and decision-making tools for detection. This study presents an integrated framework combining Fourier Transform Infrared (FTIR) spectroscopy, $\delta^{13}C$ isotope ratio mass spectrometry, and artificial intelligence within an environmental engineering perspective. FTIR spectra were processed using an unsupervised deep autoencoder, producing a Spectral Deviation Index (SDI) with excellent discriminatory power (AUC = 1.00). In parallel, a Box-Behnken design quantified $\delta^{13}C$ isotopic responses, while machine learning models, including Random Forest ($R^2 = 0.93$) and symbolic regression, accurately predicted adulteration patterns even at low levels. Symbolic regression provided interpretable equations supporting traceability and quantitative decision-making. The hybrid methodology is rapid, non-destructive, and automatable, offering a scalable pathway for sustainable food quality control and environmental monitoring.

INTRODUCTION

Edible oils have been an integral part of human diets all across the globe for centuries. At present, there are many different types of oils available to purchase. These include not only the more common, inexpensive types of cooking oils, such as canola or peanut or soybean, but also higher-end specialty oils such as linseed and sesame and olive oils that have unique characteristics that make them valuable (Aghili 2022). Among these cooking oils, sesame oil is noted to have one of the highest percentages of total oil (Aghili 2022). Sesame seed oils are also high in a number of bioactive components that are believed to contribute to good health, including oil, protein, fatty acids, lignans and vitamin E (Morris et al. 2021).

Sesame oil (Sesamum indicum), is high in bioactive substances called bioactive compounds including sesamin, sesamolin, phytosterols and tocopherols; they are all known to be antibacterial, anticancer, antioxidant and antihypertensive agents; they also help to lower cholesterol levels (Firouz et al., 2021). Sesame oil is used primarily as a flavouring in many food products and is also used in pharmaceutical products, cosmetics and shortening/margarine (Wu et al., 2021). Because it is a premium oil compared to many other vegetable oils, sesame oil also tends to have a higher price point which makes it a prime target for financially motivated frauds (Pan et al., 2021). Adulteration of sesame is typically done via adding cheaper oils to the sesame oil which poses significant risk to consumers' health and creates unbalanced markets. As such it highlights that there is a need to have trustworthy and effective mechanisms to prevent the different forms of economic fraud that lead to adulteration of items or develop efficient tools to better monitor the marketplaces (Li et al., 2023).

Over the years, several different kinds of analytical techniques have become available to help solve this problem: gas chromatography (GC), gas chromatography/mass spectrometry (GC-MS), high performance liquid chromatography (HPLC), and liquid chromatography/mass spectrometry (LC-MS) are all techniques that scientists have successfully employed to identify counterfeit culinary oils in history (Green et al., 2020); while, at the same time, variaspectroscopic methods, including nuclear magnetic resonance (NMR), near infrared spectroscopy (NIR), Fourier transform infrared spectroscopy (FT-IR), Raman and UV-visible spectroscopy, have also been employed for their rapidity as well as for their ability to fingerprint samples (Jamwal et al., 2021). However, as a result of their classical nature, these techniques usually do not work well with complex mixtures, especially those created by poorly defined combinations of different types of oils being used in an adulteration (Martinez-Castillo et al., 2020).

There has been recent interest in strategies applying artificial intelligence and chemometrics to enhance the effectiveness of fraud detection. Methods such as Principal Component Analysis (PCA), Linear Discriminant Analysis (LDA), Support Vector machines (SVM) and Artificial Neural Networks (ANN) have been successfully used to classify and quantify samples as either authentic or adulterated (Rasekh et al., 2021). The obvious disadvantage of these approaches is that training the models usually requires a large amount of training samples, and this can be costly and time-consuming. Meanwhile, measurement of $\delta^{13}C$ isotope analysis has been shown to be a useful technique in food fraud detection. On the other hand, using $\delta^{13}C$ isotope analysis can differentiate vegetable oils from C3 plants versus C4 plants because they each have their own carbon isotope signature (Zhang et al., 2017). Isotope analysis differs from conventional spectroscopy-based methods because it also gives us an idea of where the oils were produced, which is a useful identity marker regardless of where they are grown. Despite all the improvements that have been made in the past, the problem of detecting multiple types of adulteration still presents an enormous challenge. Current methods for detecting fraud based on GC-MS or LC-MS have high costs, require a large amount of organic reagents, and require a long and tedious analytical process. Conversely, using methods such as FT-IR or Raman may be limited because they can have many complicated overlaps that complicate the interpretation of the data, particularly when comparing similar classes of oils and oils with closely related chemical structures (Yuan et al., 2020).

A new method is proposed combining FTIR (Fourier Transform Infrared), isotopic $\delta^{13}C$ spectroscopy (an analysis of stable isotopes), and advanced machine learning techniques to identify and analyse fraudulent sesame oils. Our original method splits the use of FTIR into two steps: first, extracting the FTIR spectra using Deep Autoencoder (DAE); second, using a symbolic regression model to predict isotopic values allowing for an explicit interpretation of these values based on the type of adulterant oil, its dilution level (or %), and geographical origin. Moreover, supervised model algorithms such as Random Forest Analysis, Support Vector Machines (SVMs), Extreme Gradient Boosting (XGBoost), and Multilayer Perceptron neural networks (MLP) have been created to offer a hierarchical structure for enhanced fraud detection. One goal of our study is providing easier, faster and more clearly interpretable results than had been previously available, therefore complying with the requirements of existing traceability objects and the level of objectivity required within Quality Control (QC) processes for all stages of the supply chain. While conventional models tend to be built around PCA, LDA and/or SVM together with other methods of analysis such as chemometrics on FTIR data, the approach presented herein is different – it utilizes a unique hybrid of Deep Autoencoder-based Spectral Deviation Index (SDI), δ13C Isotope Modelling, and Explainable Artificial Intelligence (XAI) modelling methodologies. In contrast to traditional multivariate statistical modelling, which is based on linear regression analysis of data through the combination of a pre-established separation boundary and linear projections for classifying oils according to their respective qualities, the SDI method utilizes unsupervised learning via its design to extract and analyze data patterns in order to establish and then differentiate qualitative hierarchies in pure sesame oil. The SDI methodology is able to discern subtle differences in spectral properties of both pure and diluted sesame oils – prior to establishing the classifications for those classes of oils. The application of symbolic regression provides a means of establishing a mathematical relationship between the isotopic signatures and the dilution rate and type of adulterant. Symbolic regression addresses a key limitation of black box techniques when used to detect fraud for compliance with regulations and meet quality control standards. Thus, the above-mentioned factors together with the development of combined sensitivity and interpretability of results mark a significant improvement over current methods used to detect article adulteration utilizing FTIR spectroscopy and isotope analyses.

METHODOLOGY

FTIR Spectroscopy Analysis

Samples and Oil Extraction

Certified markets that specialize in pure seeds in Morocco and Egypt were used to collect sesame seeds for this research project (Table 1). The seeds were sorted to remove damaged seeds and sieved to remove any foreign matter and other contaminants that might affect the oil extraction process. An automatic oil press system was used to extract the oil from the sesame seeds in the laboratory using machinery designed for automated oil extraction. For each set of sesame seeds (totaling 100 g), the quantity of oil extracted from the seeds was recorded as the extracted yield.

Table 1. Comparative yields of sesame oil extraction from Egypt and Morocco

Sample	Oil Mass (g)	Cake Mass (g)	Total Mass (g)	Oil Yield (%)	Cake Yield (%)
Egypt Oil	36.909	58.632	95.541	36.909	58.632
Morocco Oil	38.5936	58.961	97.554	38.5936	58.961

Two sets of oil samples were prepared for analysis. The first set consisted of authentic oils: sesame oils of Moroccan and Egyptian origin, pure table oil (a cheaper oil available on the Moroccan market), and pure paraffin oil. The second set consisted of mixed oils: several dilutions were prepared by blending the Moroccan and Egyptian sesame oils with table oil and paraffin oil. Adulteration mixtures were created at concentrations ranging from 10% to 90% sesame oil. This preparation aimed to simulate adulteration scenarios for isotopic analysis (Figure 1).

Figure 1. Preparation of sesame oil mixtures with table and paraffin oils for isotopic analysis.

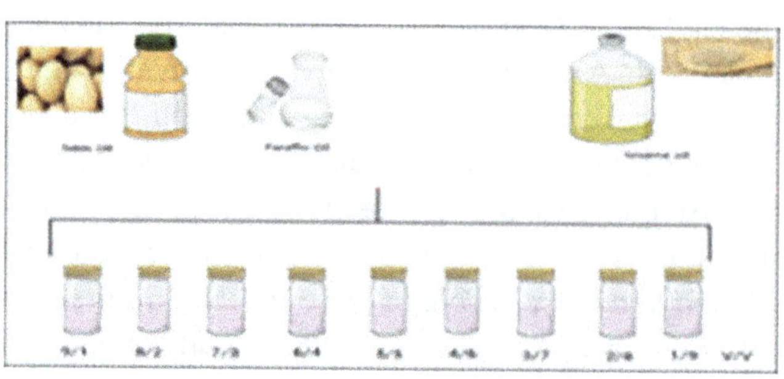

Using an isotope ratio mass spectrometer attached to an elemental analyzer (EA-IRMS) that uses stable isotopes of carbon-13, we characterized oil by providing an assessment of sesame oil authenticity (total carbon content and nullifying the $\delta^{13}C‰$) based on amounts of carbon-13 which suggest adulterants. Additional analyses included Fourier Transform Infrared (FTIR) analysis on each sample. For FTIR spectroscopy, spectral data were captured on all samples using a Spectrum Two FT-IR spectrometer (Perkin-Elmer, USA) with an Attenuated Total Reflectance (ATR) crystal and data were captured in the spectral range of 650-4000 cm^{-1} and as 32 co-additions at a spectral resolution of 4 cm^{-1}. Adulterated oil sample were not allowed to settle before spectral acquisition and were homogenized by shaking. Approximately 10 ml of oil was placed on the ATR surface, and the air background spectrum was subtracted from each spectrum collected. For carbon isotope analysis, 0.15 mg of each oil sample was placed in tin capsules and analyzed using IRMS (Isochrom-EA, VG ISOTECH, UK). The carbon-containing compounds were oxidized at 1030 °C in a ceramic reactor to produce CO_2 gas, which was then analyzed. Carbon dioxide gas standards were injected before and after the sample of interest to allow for accurate calculation of $\delta^{13}C$ values, which are expressed in per mille (‰) relative to the Pee Dee Belemnite (PDB) marine carbonate standard. The analysis was carried out using helium as a carrier gas throughout. Finally, an agreed coding scheme was produced to make it much easier to identify the various oil samples being analysed (Table 2) - the system describes the sample's composition and shows the actual percentages of pure sesame oil and table oils and also shows the blends of these oils as well as the sources of each oil, whether Moroccan or Egyptian. This coding system is very important in ensuring we are interpreting results correctly.

Table 2. Coding and Meaning of Oil Samples

Code	Meaning
Egypt 100%	100% pure sesame oil from Egypt, without any mixture.
Egypt 1/9 Table Oil	Egyptian sesame oil diluted to 10% with 90% table oil.
Egypt 2/8 Table Oil	Egyptian sesame oil diluted to 20% with 80% table oil.
Egypt 3/7 Table Oil	30% pure sesame oil mixed with 70% table oil.
Morocco 100%	100% pure sesame oil from Morocco, without any mixture.
Morocco 1/9 Table Oil	Moroccan sesame oil diluted to 10% with 90% table oil.
Paraffin 100%	Pure paraffin oil.
Table Oil	Table oil used as a diluent.
Egypt	Refers to Egypt (origin of sesame oil).
Morocco	Refers to Morocco (origin of sesame oil).

Intelligent Approach to Fraud Detection by FTIR Spectroscopy: Innovative Integration of Deep Autoencoder and Supervised Learning for Sesame Oil Analysis.

The study developed a systematic method for identifying adulteration of sesame oil from FTIR (Fourier Transform Infrared) spectroscopy. Initially, FTIR spectra were taken in the mid-infrared range (4000 – 400 cm^{-1}), and then pre-processed through min-max normalisation and centering and reducing, to make spectra from each sample comparable to one another prior to dimension reduction using a UMAP (Uniform Manifold Approximation and Projection) algorithm that projects the sample FTIR spectra into a 2D space, revealing clusters (grouped) of samples by adulterating agent type. After clustering was completed, an unsupervised deep autoencoder was trained on 100% pure samples of sesame oil to generate an autoencoder characteristic compressed representation (signature) for future reference and comparison. The Autoencoder Model was developed in Python 3.10 using TensorFlow/Keras for the model architecture, pandas for parsing and managing the data and Scikit-learn for scaling and validating the model. The developed architecture consists of a three-layer top-down encoder, with an equivalent mirrored layer, using a Mean Squared Error (MSE) Cost Function and the Adam Optimiser. The model, trained over 200 epochs, converged to a minimum reconstruction error (final loss = 5.48×10^{-5}), enabling the calculation of the Spectral Deviation Index (SDI) defined as the mean squared error between the original spectrum x and its reconstruction $\widehat{x}$ (Goodfellow et al., 2016):

$$SDI(x) = \frac{1}{n}\sum_{i=1}^{n}(x_i - \widehat{x}_i)^2$$

These SDI scores were interpreted as continuous indicators of fraud, then used as input variables in a supervised model calibrated by ROC curve. The probabilistic classification model was trained using Logistic Regression from scikit-learn, with k-fold cross-validation (k = 5), and evaluated according to the area under the curve (AUC) (Fawcett, 2006). The optimal decision threshold was determined automatically by maximizing the Youden index (J = Sensitivity + Specificity - 1), resulting in a threshold SDI* = 0.844. Finally, a second multi-class classification model was trained on the estimated chemical proportions (kerosene, table, pure) to assign each sample a nature of fraud (three-output softmax model), validated by consistency between actual and predicted labels (Goodfellow et al., 2016). All processing, calculations and visualizations were performed in Python 3.10, using the pandas, numpy, matplotlib, umap-learn, tensorflow, scikit-learn and seaborn libraries.

Isotopic Analysis of Paraffin Oil and Table Oil Mixtures

The experimental analysis was conducted using a Box-Behnken design to model the δ13C isotopic response as a function of three factors: country of origin (Morocco or Egypt), type of blended oil (table oil or kerosene), and the percentage of dilution applied to pure sesame oil. The experimental design was constructed to ensure statistical balance in sampling the levels of each factor, while limiting the total number of experiments, in keeping with the quadratic nature of the model envisaged. The data were processed in the Python environment (version 3.10), using the pandas' library for data management, numpy for numerical operations, statsmodels for fitting generalized linear models and ANOVA, and matplotlib for visualization (Harris et al., 2020). The regression model used is a linear model with quadratic effects and interactions, whose general equation is written as follows:

$$Y = \beta_0 + \beta_1 x_1 + \beta_2 x_2 + \beta_3 x_3 + \beta_{12} x_1 x_2 + \beta_{13} x_1 x_3 + \beta_{23} x_2 x_3 + \beta_{33} x_3^2 + \varepsilon$$

Where Y is the dependent variable representing the isotopic ratio δ13C, x1 the country of origin (binary variable), x2 the type of oil (binary variable), x3 the dilution rate expressed as a percentage, and ε the residual error assumed to follow a centered reduced normal distribution. The coefficients βi were estimated using the ordinary least squares (OLS) method, which minimizes the sum of the squares of the deviations between the observed and predicted values (Montgomery, 2017). Model fit was assessed using the R^2 determination index, the Fisher statistic, the AIC (Akaike Information Criterion) and the AIC (Akaike Information Criterion) (Akaike, 1974), du BIC (Bayesian Information Criterion) (Schwarz, 1978), as well as tests on residuals including omnibus, Jarque-Bera, skewness and kurtosis. Analysis of variance (ANOVA) was used to assess the statistical significance of main effects, cross-interactions and quadratic terms, with significance thresholds set at $\alpha=0.05$. This rigorous statistical approach made it possible to identify parameters with a significant influence on the isotopic behavior of the system, complementing the artificial intelligence-based approaches developed in other parts of the work.

This study combines continuous regression, binary classification, symbolic modeling and hybrid machine learning techniques to detect and quantify fraud in sesame oils based on their isotopic signature. All processing was implemented in Python 3.10 using the pandas, numpy and other libraries (Harris et al., 2020) and openpyxl for tabular management of experimental data, matplotlib and seaborn for scientific visualization, scikit-learn (Pedregosa et al., 2011) for standard algorithms (SVM, Random Forest, MLP, VotingClassifier, StackingRegressor), xgboost for boosting models, tensorflow.keras for deep neural networks, and pysr for symbolic regression based on genetic evolution. The data used consist of simulated samples of sesame oils diluted with table oil or kerosene at different rates (0% to 90%), each characterized by an experimental δ13C isotope ratio, a dilution variable expressed as a percentage, and a categorical encoding of the type of adulterant oil (Chollet, 2015; Cranmer et al., 2020). The tasks were subdivided into regression and classification according to the nature of the target variable. In regression, the aim is to estimate δ13C as a continuous variable; in classification, to automatically detect the presence of fraud based on binary labeling. The data were segmented into 70% training, 20% validation and 10% testing. Nine artificial intelligence models were applied to the regression task: SVM (Support Vector Machine) Random Forest, XGBoost, MLP, Voting-RF-XGB, Voting-RF-XGB-SVM, Stacking-RF-XGB-MLP, GA-ANN et RSM-ANN (Breiman, 2001; Chen & Guestrin, 2016; Goodfellow et al., 2016). For each model, training was performed on the training set with parameter optimization on validation, then frozen for evaluation on the test set. The regression metrics used were the coefficient of determination R^2, the mean absolute error (MAE) and the root mean square error (RMSE), defined by Montgomery and Runger (Montgomery & Runger, 2010):

$$RMSE = \sqrt{\frac{1}{n}\sum_{i=1}^{n}(y_i - \widehat{y}_i)^2}$$

For binary classification tasks, the same architectures were used in classificatory configuration, with outputs evaluated in terms of precision, recall, F1-score and AUC, calculated separately for the two classes (fraud vs. non-fraud). Simulated fraud detection was carried out by generating a set of blends of sesame oils with kerosene and table oil at increasing rates and predicting isotopic ratios via a Random Forest regression model, then evaluating whether the predicted values fell outside the natural range of pure oils, producing results organized in a synthetic table. In parallel, a symbolic regression was conducted to extract an analytical equation of the form (Koza, 1993; Schmidt & Lipson, 2009):

$$\delta^{13}C = f(Dilution(\%), Oil\ type)$$

By combining the following elementary transformations:

$$f(x) = log(1 + x), f(x) = x, f(x) = x^2, f(x) = exp(x)$$

To facilitate understanding of the data sets involved in this research, there were three categories of data. The experimental data set contained FTIR spectra and $\delta^{13}C$ isotopic measurements for pure and adulterated (or intentionally mixed) sesame oil samples. The modeled data set contained data from machine learning models (both supervised and unsupervised) that predict isotopic ratios, SDI scores, and classification probabilities. Finally, the design-based data set comes from the design of the Box–Behnken experiment, whereby the effect of dilution level, oil type, and geographic region was examined in a controlled manner. Distinguishing between these types of data sets makes it easier to interpret the results and emphasizes the different ways in which experimental measurement, statistical modeling, and AI all provide complementary information. The goal was to develop a simple but effective mathematical model based on symbolic algorithmic operators to predict the value of δ13C as a function of C using a multi-objective cost function (Goldberg, 1989) to ensure that the model was optimal with little variance between training data and independent reference (validation) data. Both calibration and validation methods were performed by graphically comparing simulated isotopic variation within a known range of values, and through visualizing classification performance across various data set divides (training data, validation data, test data) using confusion matrices. Overall, the results from fitting all regressive models were summarized in a performance summary table, and the binary classifier classification results from all models were grouped by classification performance into one summary table. All Python scripts created to automate the training of each model, the evaluation and simulation of predicted values, the exportation of results to Excel, and the generation of scientific publications, were pre-created and therefore ready to execute at any time.

RESULTS

In this study, FTIR spectroscopic analysis of pure and adulterated sesame oils revealed subtle but consistent chemical variations, exploitable for fraud detection. Figure 2 shows the complete spectra of the different samples, with the characteristic bands of the hydroxyl group (around 3470 cm^{-1}), methylenes (CH_2, CH_3 around 2924-2854 cm^{-1}), and double bonds (around 1740 cm^{-1}) annotated. These variations indicate the impact of adulteration even at low concentration. To simplify the multi-dimensional visualization of the spectra, a UMAP reduction was applied (Figure 3). This two-dimensional projection reveals clear groupings of samples according to their type: pure, adulterated with table oil or kerosene.

This clear clustering shows that spectroscopic signals are sufficient to discriminate chemical profiles according to the nature of the adulterant, providing a solid basis for supervised learning and the development of automated suspicion indices.

Figure 2. FTIR spectra of pure and adulterated sesame oils (paraffin/table oil).

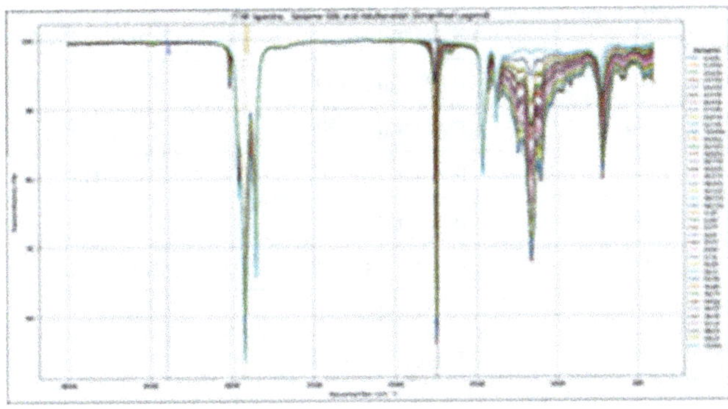

Figure 3. Projection UMAP des spectres FTIR pour la détection de fraude dans les huiles de sésame.

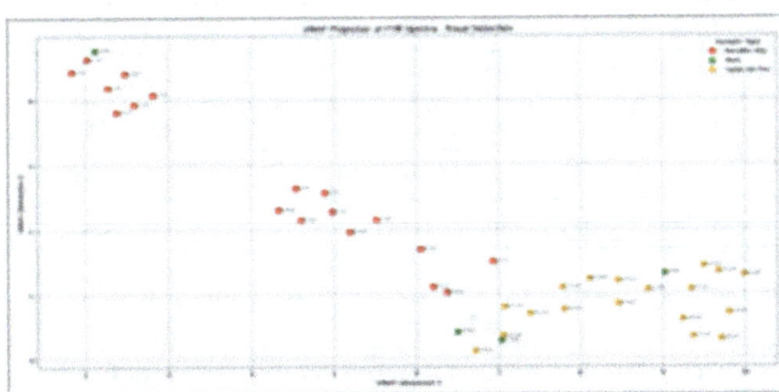

A multi-class supervised artificial intelligence classifier model has been developed to automatically assign each sample a type of fraud (pure oil, kerosene or table oil), based on the chemical proportions measured. This table summarizes the proportions of kerosene, pure oil and table oil detected in each sample, in relation to the actual labels and the model's predictions. There is very good consistency between the expected values (True Label column) and those automatically predicted (Predicted Label), particularly for mixtures dominated by kerosene (e.g. M 4/6 p, E 3/7 P) or table oil (E 8/2 HT, M 7/3 HT), demonstrating the model's robustness in finely classifying adulteration types. Isolated errors exist (e.g. HT 100%) where pure oil is confused with table oil, indicating possible spectral overlap at very low SDI (Table 3).

Table 3. Classification of sesame samples by nature of fraud (A supervised artificial intelligence multi-class classifier model).

Paraffine	Pure	Table Oil	Sample	True Label	Predicted Label
0,12	**0,52**	**0,36**	**E 100%**	**Pure**	**Pure**
0	0	1	E 1/9 Table Oil	Table Oil	Table Oil
0,01	0,01	0,98	E 2/8 Table Oil	Table Oil	Table Oil
0	0	1	E 3/7 Table Oil	Table Oil	Table Oil
0	0	1	E 4/6 Table Oil	Table Oil	Table Oil
0	0,01	0,99	E 5/5 Table Oil	Table Oil	Table Oil
0,02	0,04	0,94	E 6/4 Table Oil	Table Oil	Table Oil
0	0,01	0,99	E 7/3 Table Oil	Table Oil	Table Oil
0	0,01	0,99	E 8/2 Table Oil	Table Oil	Table Oil
0,16	0,34	0,5	E 9/1 Table Oil	Table Oil	Table Oil
0,02	0	0,98	Table Oil 100%	Pure	Table Oil
0,06	0,68	0,26	M 100%	Pure	Pure
0	0	1	M 1/9 Table Oil	Table Oil	Table Oil
0	0,01	0,99	M 2/8 Table Oil	Table Oil	Table Oil
0	0	1	M 3/7 Table Oil	Table Oil	Table Oil
0	0	1	M4/6 Table Oil	Table Oil	Table Oil
0,01	0,02	0,97	M 5/5 Table Oil	Table Oil	Table Oil
0	0,01	0,99	M 6/4 Table Oil	Table Oil	Table Oil
0,01	0	0,99	M 7/3 Table Oil	Table Oil	Table Oil
0,04	0,11	0,85	M 8/2 Table Oil	Table Oil	Table Oil
0,05	0,21	0,74	M 9/1 Table Oil	Table Oil	Table Oil
0,85	0,15	0	E 1/9 Paraffine	Paraffine	Paraffine
0,99	0,01	0	E 2/8 Paraffine	Paraffine	Paraffine
1	0	0	E 3/7 Paraffine	Paraffine	Paraffine
0,96	0,04	0	E 4/6 Paraffine	Paraffine	Paraffine
1	0	0	E 5/5 Paraffine	Paraffine	Paraffine
1	0	0	E 6/4 Paraffine	Paraffine	Paraffine
0,99	0,01	0	E 7/3 Paraffine	Paraffine	Paraffine
0,97	0	0,03	E 8/2 Paraffine	Paraffine	Paraffine
0,87	0,1	0,03	E 9/1 Paraffine	Paraffine	Paraffine
0,96	0,04	0	M 1/9 Paraffine	Paraffine	Paraffine
0,99	0,01	0	M 2/8 Paraffine	Paraffine	Paraffine
0,99	0,01	0	M 3/7 Paraffine	Paraffine	Paraffine
0,89	0,09	0,02	M 4/6 Paraffine	Paraffine	Paraffine
1	0	0	M 5/5 Paraffine	Paraffine	Paraffine
0,87	0,08	0,05	M 6/4 Paraffine	Paraffine	Paraffine
0,99	0,01	0	M 7/3 Paraffine	Paraffine	Paraffine

continued on following page

Table 3. Continued

Paraffine	Pure	Table Oil	Sample	True Label	Predicted Label
0,12	**0,52**	**0,36**	**E 100%**	**Pure**	**Pure**
0,97	0,02	0,01	M 8/2 Paraffine	Paraffine	Paraffine
0,95	0,04	0,01	M 9/1 Paraffine	Paraffine	Paraffine
0,39	0,61	0	Paraffine 100%	Pure	Pure

To characterize the spectral signature of pure sesame oils, an unsupervised deep autoencoder was trained exclusively on the normalized FTIR spectra of 100% authentic samples. This training, conducted over 200 epochs, demonstrated excellent convergence, with the loss function decreasing from 1.0048 to just 5.48×10^{-5}, indicating an almost perfect reconstruction of the original spectral profiles. The Compressive Models serve as a digital fingerprint for reference oils and contain their spectral signature within a latent space. Once the Compressive Model is stabilized, it is used to reconstruct each sample's spectrum and determine the difference between the original and reconstructed sample spectrum. A continuous metric called the Spectral Deviation Index (SDI) represents the normalized error for the reconstructed sample spectrum compared to the original sample spectrum and is used to calculate the Spectral Deviation Indices (SDIs) for each of the samples. A horizontal representation of each of the SDI scores sorted by sample is shown in Figure 4. The SDIs show a trend of increasing deviation, with pure oils clustered to the left (SDIs near zero) and the highly adulterated mixtures further from pure oils based on their composition. This visualization demonstrates that the SDI has the ability to identify the spectral distance of samples from pure oils and is useful for detecting adulteration using an unsupervised approach. The pattern described above also reinforces the application of SDI as an input variable in the development of supervised prediction models, specifically for establishing the lower limit of detection. Thus, the SDI serves as the basis for the next step in the development process: utilizing the statistical optimization of the classification threshold via the ROC curve and establishing the Minimum Detectable Adulteration (MDA) expressed in percent dilution.

Figure 4. Comparative visualization of SDI versus pure sesame oil (reference).

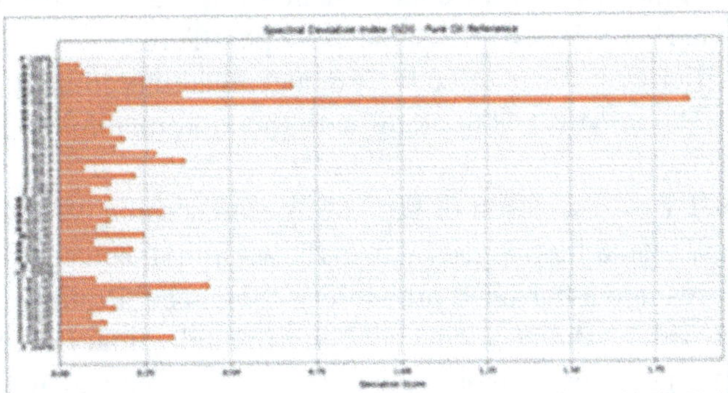

By creating a supervised model using only the spectral deviation index (SDI), and then using the receiver operating characteristic (ROC) method of calibrating it, I was able to achieve an AUC (area under the curve) of 1.00 (See Figure 5). This indicates that my model is able to effectively separate between purely true (no pronounced spectral deviations) and purely fraudulent (clearly pronounced spectral deviation) samples without any overlap or confusion. Therefore, the probability threshold (i.e., threshold for classifying a sample as fraudulent) calculated automatically is determined to be 0.844, making it so that any sample with a probability higher than this threshold, by the SDI model, is classified as fraudulent. The AUC calibration provides a balanced trade-off between sensitivity (ability to detect minor levels of fraud) and specificity (ability to minimize false positive rates), allowing for reliable identification of potential fraudulent samples based on the initial observable spectral deviations.

Figure 5. ROC Curve of the fraud detection model based on the SDI (Balanced Model).

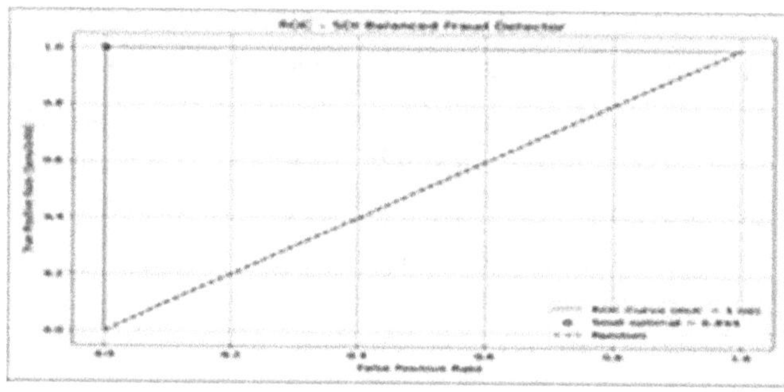

To improve reliability and transparency when detecting fraud associated with sesame oil, a supervised classification method was developed that utilizes the Spectral Deviation Index (SDI) derived from an autoencoder trained using pure oils. A probabilistic classifier was further refined to discriminate between pure and diluted samples of sesame oil diluted using table oil or paraffin, with an additional focus on the use of a significantly lower dilution range.

An ROC curve was analyzed to select the optimal threshold for mathematic decision making; with a SDI of ≥ 0.844, which corresponds with the greatest balance of sensitivity and specificity (100% sensitivity and the absence of false positives for pure oils) and AUC of 1.00. The classifier was able to detect contamination, with an accuracy of over 90%, for any sample containing a minimum of 10% contamination from non-sesame, whether the contaminant was a vegetable oil (E 1/9 HT = 10% sesame; 90% table) or non-vegetable (E 1/9 P = 10% sesame; 90% paraffin).

Additionally, the Classifier was able to effectively detect all adulteration of pure sesame at a dilution rate as low as 90% (E 9/1 HT or M 9/1 HT) along with the aforementioned figures, all of which had high sensitivity.

Only one sample (M 9/1 P) – composed of 90% Moroccan sesame + 10% paraffin – scores at the threshold limit (0.8441) and would be flagged in an early warning mechanism ("Early Warning"). The conversion of raw SDI scores into a balanced probability (Balanced_Prob) and then into final classification (Balanced_Pred) provides a direct, robust, and interpretable binary output for operators.

This system enables automated, explainable, and scientifically grounded monitoring of mixtures, with operational thresholds expressible concretely as a percentage of fraud starting from 10%, thus meeting regulatory and industrial requirements for food safety (Table 4).

Table 4. Balanced fraud predictions based on spectral deviation index (SDI) with optimized threshold calibration.

Sample	SDI	IsFraud	Fraud Prob	Prediction	Pred Prob	Pred Optimal	Balanced Prob	Balanced Pred
E 100%	0,0001	0	0,8568	0	0,8863	0	0,7942	0
E 1/9 TO	0,3324	1	0,8846	0	0,9083	1	0,9610	1
E 2/8 TO	0,1134	1	0,8668	0	0,8943	1	0,8790	1
E 3/7 TO	0,1375	1	0,8689	0	0,8959	1	0,8925	1
E 4/6 TO	0,0933	1	0,8651	0	0,8929	1	0,8665	1
E 5/5 TO	0,1632	1	0,8711	0	0,8976	1	0,9055	1
E 6/4 TO	0,1336	1	0,8686	0	0,8957	1	0,8904	1
E 7/3 TO	0,2644	1	0,8793	0	0,9041	1	0,9440	1
E 8/2 TO	0,4362	1	0,8923	0	0,9143	1	0,9777	1
E 9/1 TO	0,1051	1	0,8661	0	0,8937	1	0,8739	1
TO 100%	8,62E-05	0	0,8568	0	0,8863	0	0,7942	0
M 100%	7,321E-06	0	0,8568	0	0,8863	0	0,7941	0
M 1/9 TO	0,1394	1	0,8691	0	0,8961	1	0,8935	1
M 2/8 TO	0,2136	1	0,8752	0	0,9009	1	0,9269	1
M 3/7 TO	0,1001	1	0,8657	0	0,8934	1	0,8708	1
M4/6 TO	0,2466	1	0,8779	0	0,9030	1	0,9385	1
M 5/5 TO	0,1079	1	0,8664	0	0,8939	1	0,8756	1
M 6/4 TO	0,14929	1	0,8699	0	0,8967	1	0,8986	1
M 7/3 TO	0,3043	1	0,8825	0	0,9066	1	0,9546	1
M 8/2 TO	0,1310	1	0,8684	0	0,8955	1	0,8890	1
M 9/1 TO	0,1523	1	0,8702	0	0,8969	1	0,9002	1
E 1/9 PO	0,0913	1	0,8649	0	0,8928	1	0,8652	1
E 2/8 PO	0,15105	1	0,8701	0	0,8968	1	0,8995	1
E 3/7 PO	0,2229	1	0,8760	0	0,9015	1	0,9304	1
E 4/6 PO	0,0728	1	0,8633	0	0,8915	1	0,8527	1
E 5/5 PO	0,3673	1	0,8872	0	0,9104	1	0,9676	1
E 6/4 PO	0,2842	1	0,8809	0	0,9054	1	0,9495	1
E 7/3 PO	0,1674	1	0,8714	0	0,8979	1	0,9075	1
E 8/2 PO	0,1918	1	0,8735	0	0,8995	1	0,9183	1
E 9/1 PO	0,1478	1	0,8698	0	0,8966	1	0,8979	1
M 1/9 PO	0,1291	1	0,8682	0	0,8954	1	0,8880	1
M 2/8 PO	0,1534	1	0,8703	0	0,8970	1	0,9007	1
M 3/7 PO	0,1669	1	0,8714	0	0,8979	1	0,9073	1
M 4/6 PO	1,8516	1	0,9597	0	0,9673	1	0,9999	1
M 5/5 PO	0,3587	1	0,8866	0	0,9099	1	0,9661	1
M 6/4 PO	0,6819	1	0,9087	0	0,9272	1	0,9942	1
M 7/3 PO	0,2510	1	0,8783	0	0,9033	1	0,9399	1

continued on following page

Table 4. Continued

Sample	SDI	IsFraud	Fraud Prob	Prediction	Pred Prob	Pred Optimal	Balanced Prob	Balanced Pred
M 8/2 PO	0,0753	1	0,8635	0	0,8917	1	0,8545	1
M 9/1 PO	0,0608	1	0,8622	0	0,8907	1	0,8441	0
PO 100%	2,95E^{-06}	0	0,8568	0	0,8863	0	0,7941	0

P: Paraffin Oil; TO: Table Oil; M: Moroccan Sesame Oil; E: Egyptian Sesame Oil; SDI: Spectral Deviation Index, representing the deviation of the sample from the pure reference; IsFraud: 1 = known fraud (true label), 0 = non-fraudulent; Fraud_Prob: fraud probability predicted by the SDI-based model; Prediction: raw model prediction (based on a 0.5 threshold); Pred_Prob: probability output by the classifier; Pred_Optimal: prediction based on the optimal ROC threshold, set here at ~0.844; Balanced_Prob: recalibrated probability for the balanced model; Balanced_Pred: final classification using the balanced model.

The results of the Box-Behnken experimental design are presented in Table 5, which compiles the observed isotopic values ($\delta^{13}C$), the model-predicted values, and the residuals, considering different dilution levels, oil types, and geographic origins. The measured and predicted values of the model are very close together and show only a small difference from each other (the average difference is less than two-tenths), indicating that this model is a good fit to the actual data, as demonstrated by a high overall coefficient of determination (R^2) of 0.9591 and adjusted R^2 value of 0.9516, combined with a root mean square error (RMSE) of 0.2433. Analysis of variance (ANOVA) for the model is provided in Table 6, which looks at the significant main effects, cross-interactions, and quadratic terms of the model. The results of ANOVA show that the "oil type (Paraffin)" factor is highly significant ($p < 10^{-21}$) and plays a major role in changing the isotopic ratios, while the interaction of oil type and dilution is also significant ($p < 10^{-14}$). On the other hand, the country of origin factor and dilution by itself, as well as the interaction between these two factors, do not contribute to the relativistic response of the isotopes. The quadratic term for dilution reflects a marginal level of significance ($p = 0.0776$), suggesting that once dilution reaches a certain level, there may be curvature in the isotopic response.

The full set of estimated ordinary least squares (OLS) coefficients, their standard errors of estimate, t-statistics, p-values, and corresponding 95% confidence intervals for the coefficients can be found in Table 6 and Table 7. The dilution variable has a negative correlation with the response variable ($\beta = -0.0199$, $p = 0.002$), meaning that as you increase mixing percentage (i.e., the dilution factor), $\delta^{13}C$ decreases significantly. Additionally, the interaction of dilution and oil type enhances this trend ($\beta = 0.0402$, $p < 0.001$).

Global performance indicators confirm the model's quality and robustness: F-statistic = 128.9, overall p-value < 10^{-20}, AIC = 14.45, BIC = 26.27, Durbin-Watson statistic = 1.714, skewness = 1.878, kurtosis = 11.418, and the Jarque-Bera normality test = 141.621 with a p-value < 10^{-30}, indicating an asymmetric residual distribution. These results highlight the relevance of the Box-Behnken model in providing a detailed statistical understanding of the factors influencing the isotopic signature of sesame oils.

Table 5. Box-Behnken Design with Observed, Predicted and Residual Values.

Country	Dilution (%)	Isotopic ratio	Oil type	Predicted	Residuals
Morocco	0	-28,92	Paraffine	-28,79	-0,13
Morocco	10	-28,64	Paraffine	-28,56	-0,08
Morocco	20	-28,25	Paraffine	-28,34	0,09
Morocco	30	-28,18	Paraffine	-28,15	-0,02
Morocco	40	-27,91	Paraffine	-27,98	0,07
Morocco	50	-27,71	Paraffine	-27,83	0,12
Morocco	60	-27,42	Paraffine	-27,71	0,28
Morocco	70	-27,86	Paraffine	-27,60	-0,26
Morocco	80	-27,51	Paraffine	-27,52	0,01
Morocco	90	-27,31	Paraffine	-27,45	0,14
Egypt	0	-29,14	Paraffine	-28,64	-0,49
Egypt	10	-28,38	Paraffine	-28,45	0,07
Egypt	20	-27,18	Paraffine	-28,28	1,10
Egypt	30	-27,99	Paraffine	-28,13	0,14
Egypt	40	-28,38	Paraffine	-28,00	-0,40
Egypt	50	-28,29	Paraffine	-27,89	-0,39
Egypt	60	-28,03	Paraffine	-27,81	-0,22
Egypt	70	-27,91	Paraffine	-27,74	-0,17
Egypt	80	-27,71	Paraffine	-27,69	-0,012
Egypt	90	-27,54	Paraffine	-27,67	0,13
Morocco	0	-28,92	Oil table	-28,97	0,05
Morocco	10	-29,03	Oil table	-29,15	0,12
Morocco	20	-29,36	Oil table	-29,34	-0,02
Morocco	30	-29,55	Oil table	-29,55	-0,001
Morocco	40	-29,89	Oil table	-29,78	-0,12
Morocco	50	-30,07	Oil table	-30,04	-0,03
Morocco	60	-30,4	Oil table	-30,31	-0,08
Morocco	70	-30,65	Oil table	-30,61	-0,04
Morocco	80	-31,06	Oil table	-30,93	-0,13
Morocco	90	-31,24	Oil table	-31,26	0,02
Egypt	0	-29,14	Oil table	-28,83	-0,31
Egypt	10	-28,84	Oil table	-29,04	0,20
Egypt	20	-29,19	Oil table	-29,27	0,08
Egypt	30	-29,45	Oil table	-29,52	0,07
Egypt	40	-29,75	Oil table	-29,79	0,05
Egypt	50	-30,14	Oil table	-30,09	-0,04
Egypt	60	-30,44	Oil table	-30,41	-0,03

continued on following page

Table 5. Continued

Country	Dilution (%)	Isotopic ratio	Oil type	Predicted	Residuals
Egypt	70	-30,79	Oil table	-30,75	-0,04
Egypt	80	-31,09	Oil table	-31,10	0,014
Egypt	90	-31,26	Oil table	-31,48	0,22

R^2: 0.9591 ; R^2 ajusté: 0.9516 ; RMSE: 0.2433

Table 6. Analysis of Variance (ANOVA) for Main Effects, Interactions and Quadratic Terms.

Source	Sum of squares	df	F	p-value	Sig.
Morocco vs Egypt	0,01444	1	0,201189	0,656695	Non Significatif
Paraffine	40	1	557,3095	3,05E-22	Significatif
Dilution	0,012210766	1	0,170129	0,682664	Non Significatif
Country × Dilution	0,134814848	1	1,87834	0,179774	Non Significatif
Oil Type × Dilution	13,34037121	1	185,8679	4,12E-15	Significatif
Dilution^2	0,238000189	1	3,315994	0,077687	Limite
Résidu	2,368522538	33			

Table 7. Detailed results of the OLS regression model based on the Box-Behnken design: estimated coefficients, 95% confidence intervals and overall performance indicators.

Parameter	Coefficient	Standard error	t	p-value	CI 95% Min	CI 95% Max
Intercept	-28,8349	0,153	-188,14	0	-29,147	-28,523
Morocco	-0,1439	0,157	-0,914	0,367	-0,464	0,176
Paraffine	0,1905	0,157	1,21	0,235	-0,13	0,511
Dilution	-0,0199	0,006	-3,407	0,002	-0,032	-0,008
Country x Dilution	0,004	0,003	1,371	0,18	-0,002	0,01
Oil type x Dilution	0,0402	0,003	13,633	0	0,034	0,046
Dilution^2	-0,0001	0,0000583	-1,821	0,078	0	0,0000124

Overall performance indicators for the OLS model derived from the Box-Behnken design: Statistic F= 128.9; overall p-value (F)= 1.88e-21; AIC= 14.45; BIC= 26.27; Durbin-Watson= 1.714; Skewness= 1.878; Kurtosis= 11.418; Jarque-Bera (JB)= 141.621; p-value JB= 1.77e-31.

Figure 6.a depicts how the Isotope Ratio continues to decrease as dilution increases, and how two different types of adulterating oils effect this decrease in Sesame Oil from Morocco and Egypt. Sesame Oil mixed with Table Oil significantly reduces the isotope ratio more than when mixed with Paraffin at dilutions above 20%, confirming that the model can detect complex adulteration profiles. An artificial intelligence advancements framework has been developed to investigate the practical application of Isotope Ratio measurements as fraud detection in diluted Sesame Oil. The basis of this framework is a Random Forest regression model trained to predict the Isotope Ratio based on both dilution percentage and Oil Type, achieving a root mean square error (RMSE) of 2.3769, which represents an excellent level of predictive accuracy. Simultaneously, a Random Forest classifier was developed that identified fraudulent samples with a detection accuracy of 90.91%. The classifier performance is demonstrated in figure 6.b, showing 10 true positives and only 1 false negative, with no false positives. Thus, the robustness of the model is demonstrated when differentiating pure oils from adulterated mixtures. Detection thresholds were quantified by performing inverse simulation using the trained models over a range of dilution levels.

Figure 6. a: Variation in isotope ratio with dilution; b. confusion matrix: Fraud detection.

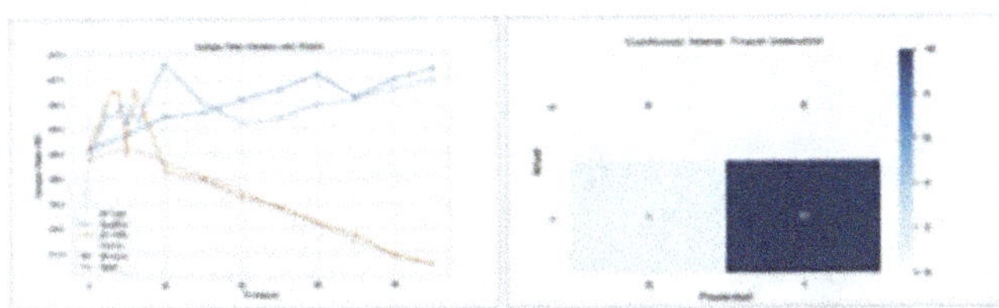

In Table 8, results from paraffin show that the first detection of fraud occurred at a dilution of 5%. At this point, the predicted IR value will begin to stabilize at approximately -28.2‰. In addition, results from table oils show that 5-50% dilutions indicated active fraud detection and only at 20% dilution was an outlier detected with an altered -29.15‰, but not associated with any fraud. Thus, modelled isotopic ratios through machine learning algorithms provide a reliable "fingerprint" for the identification of fraudulently counterfeit sesame oil products.

Table 8. Fraud detection simulation for kerosene and table oils (Random Forest regression model).

Oil type				
Parraffine Oil			**Table oil**	
Dilution	**Predicted Isotopic Ratio**	**Fraud Detected**	**Predicted Isotopic Ratio**	**Fraud Detected**
0	-28,89345433	No	-28,91736305	No
5	-28,24078062	Yes	-28,18914655	Yes
10	-28,44993762	Yes	-28,62117578	Yes
15	-28,20694059	Yes	-28,26791417	Yes
20	-28,23020628	Yes	-29,15589441	No
25	-28,23006845	Yes	-29,28524779	Yes
30	-28,22469903	Yes	-29,47469782	Yes
35	-28,1808721	Yes	-29,53959869	Yes
40	-28,1757035	Yes	-29,77679833	Yes
45	-28,15575702	Yes	-29,8551625	Yes
50	-28,21906369	Yes	-30,06663417	Yes

Analysis of the simulated values against those calculated using the symbolic equation indicates significant predictive ability by machine learning that confirms the usefulness of the resulting numerical predictions for analytical quality control and food physicochemical modeling. A summary of the results for numerical validation of the symbolically derived prediction against the Random Forest simulated values can be found in Table 8. The differences between the two analyses were very small and support the conclusion that the symbolic equation captures the essential dynamics of the isotopic signatures regardless of the oil type. The similarity between the numerical predictions of Random Forest and the symbolically derived equation provides excellent value for providing clarity and improved quality assurance.

Figure 7. Correlation between isotope ratios measured and predicted by the symbolic model.

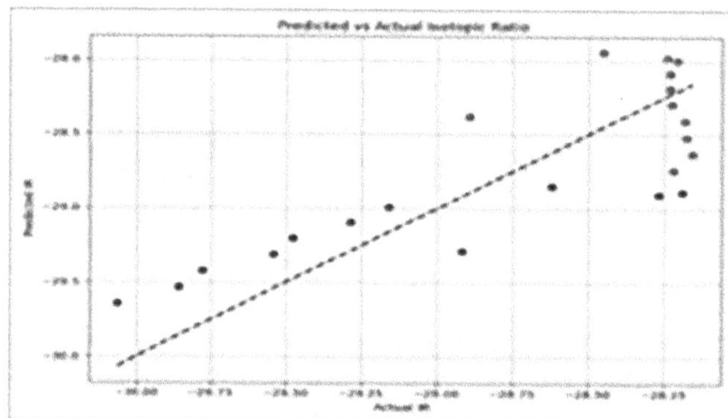

Random Forest regressor was first introduced into the exploratory analysis since it has been shown to be very robust, with low sensitivity to outliers, and can effectively model moderate levels of non-linearity without excessive hyperparameterization. Random Forest provides a balanced combination of both the quality of predictive accuracy and stability, producing a coefficient of determination (R^2=0.9334) and root mean square error (RMSE=0.3935‰) on the testing dataset, confirming its capability to generalize the correlations between dilution and isotopic change. While advanced hybrid models, such as Voting-RF-XGB-SVM and Stacking-RF-XGB-MLP, yield OSRMSE® values slightly higher than 0.97, (R^2 >0.97) these models are built from complex associations between base models; thus, they require greater amounts of cross-validation, longer periods for model fitting and less interpretability (Table 9). Because of these advantages, we chose Random Forest to perform our exploratory analyses; it is a variable importance-sensitive approach providing both partial interpretability, good quality of predictive accuracy and added robustness to "small" isotopic datasets. Using this approach enabled us to demonstrate a viable proof of concept before progressing to using more complex hybrid structures during later stages of the analysis.

Table 9. Comparative performance of regression models for the prediction of δ13C isotope ratio in sesame oils at different dilution stages.

Model	Train R^2	Train MAE	Train RMSE	Validation R^2	Validation MAE	Validation RMSE	Test R^2	Test MAE	Test RMSE
SVM	0,47	0,56	1,69	0,88	0,21	0,29	0,83	0,44	0,62
Random Forest	0,88	0,36	0,79	0,87	0,27	0,31	0,93	0,26	0,39
XGBoost	1	0,002	0,002	0,84	0,28	0,34	0,96	0,22	0,27

continued on following page

Table 9. Continued

Model	Train R^2	Train MAE	Train RMSE	Validation R^2	Validation MAE	Validation RMSE	Test R^2	Test MAE	Test RMSE
MLP	0,73	0,92	1,20	-1,47	1,16	1,38	0,39	1,08	1,18
Voting-RF-XGB	0,97	0,18	0,39	0,91	0,23	0,26	0,97	0,18	0,23
Voting-RF-XGB-SVM	0,89	0,28	0,77	0,92	0,22	0,24	0,98	0,14	0,17
Stacking-RF-XGB-MLP	0,97	0,26	0,36	0,77	0,36	0,41	0,93	0,35	0,37
RSM-ANN	-10,67	5,68	7,96	-173,72	9,28	11,65	-25,42	6,56	7,83
GA-ANN	0,85	0,52	0,88	0,53	0,54	0,60	0,77	0,56	0,72

DISCUSSION

The results obtained in this study demonstrate the feasibility and relevance of a combined spectroscopic, isotopic and algorithmic approach for the early, automatic and interpretable detection of fraud in sesame oils. Combining the analytical techniques with Artificial Intelligence algorithms creates a powerful combination to eliminate limitations of previous methods and allow for sensitivity, accuracy, traceability and applicability. The results of SEMS-FTIR analysis demonstrated inherent differences in the banding structure associated with the OH, CH_2, and C=O functional groups, indicating changes to the chemistry due to adulteration. In particular, the sensitivity to band frequencies from 2924-2854 cm^{-1} (asymmetric and symmetric C-H stretches) and at 1740 cm^{-1} (C=O groups) corroborates that even the smallest amounts of structural change are able to be detected in the early stages of a fraudulent sample, consistent with previous work (Rohman & Che Man, 2011; Rohman & Man, 2012). The stability in the spectral signatures, even though samples came from a wide variety of geographic locations, indicates that the lipid matrix of the sesame oil has a robust signature inherent to it. This aspect, combined with the fast response time of FTIR and limited sample preparation procedures, provides a basis for both Industrial and Regulatory uses of this technology as demonstrated in previous studies (Setyaningrum et al., 2013). FTIR has outperformed other methods of detecting fraud in other edible oils, specifically Olive Oil and Coconut Oil, by providing good discrimination between pure and adulterated samples in the form of forensic classification (Lumakso et al., 2015; Rohman et al., 2019).

Analysis of FTIR spectra by UMAP projection reveals a clear separation of samples according to their chemical nature, confirming that these signatures contain rich discriminating information. This approach is in line with observations made with PCA, SIMCA or SVM, which show a comparable ability to group samples according to their composition (Upadhyay et al., 2018; Bodner & Scampicchio, 2020). This exploitable structure justifies the integration of dimensional reduction as a rapid visualization and diagnostic tool.

The supervised multi-class classifier developed here demonstrates a robust ability to identify the nature of the adulterant. However, some confusions in weakly adulterated samples suggest a partial overlap of spectral signatures, typical of systems with low spectral differentiation (Ma et al., 2025). The integration of adaptive thresholds could improve accuracy.

The unsupervised deep autoencoder approach proved effective in encoding the spectroscopic profiles of pure oils, generating a reliable latent fingerprint of the FTIR spectra. Calculation of the SDI as the root-mean-square deviation between the original spectrum and its reconstruction provides a robust distinction between genuine oils and adulterated blends, in agreement with other studies (Gyftokostas et al., 2021). The SDI provides an interpretable metric for quantifying the spectral conformity of a sample, facilitating its integration into automated quality control devices.

Although CV s may be useful in comparison of data and sampling point comparison, the SDI indicators have a clear separation between pure and defrauded oils and this supports the idea of it possibly being used as a semi-quantitative indicator. In addition, the integration of SDI into OCR can establish an advisable threshold of SDI which can detect defrauded oils at levels lower than when measured by other thresholds from different spectral applications (Balan et al., 2020; Jamwal et al., 2020).

Isotopic work includes the relevance of δ13C, as an isotopic tracer of fraud, that is agnostic towards geographical differences and has predictable yiEld dilution patterns (Zhang et al., 2016). Taking this as your baseline for irrelevant dilution analysis, it is easy to develop a Random Forest model that can predict the degree of dilution > $R^2 \geq 0.93$ (Karami et al., 2020; Rasekh et al., 2021). Creation of an explicit symbolic model follows normative requirements for a symbol or verbal model that enhances its application in regulatory (Aghili et al., 2022). Although more complex hybrid models exhibit higher overall accuracy (>0.97), their low interpretability limits them from being used in the industry (Khorramifar et al., 2022). A step-wise approach, which starts by applying a robust and interpretable model as Random Forest before progressing towards a more complex architecture, avoids uncertainty and promotes a more reliable and traceable method of implementation for agri-food quality assurance (Raschka & Mirjalili, 2019).

The present study demonstrates an integrated approach for sesame oil fraud detection, that combines FTIR spectroscopy, isotope analysis and artificial intelligence, while providing traceability and interpretability to meet industry and regulatory standards. The methodology developed in this work has some important methodological strengths. The pairwise combination of FTIR, $\delta^{13}C$ isotopes and artificial intelligence algorithms and explainable models represents a novel and strong analytic approach capable of detecting adulteration from 5 - 10% dilution. The use of a variational autoencoder, which does not require prior annotations to create a spectral deviation index based on ROC calibration, can be used to develop an approach to making general-purpose detection of non-compliance and illicitly produced oils automated, sensitive and interpretable in real time. Alternatively, the use of the random forest algorithm provides an ideal balance between significant predictive power and insights into the algorithm model. Similarly, the symbolic regression output equation presents to the reader with an easy-to-understand format, thus providing an additional means by which the reader may find potential practical applications in both embedded or regulatory tools. Overall, this integrated solution is designed for non-destructive testing and demonstrates speed and the potential for automation, thus offering increased industrial and/or quality assurance value. However, this technology is hindered by the current understanding of the limitations that exist with class imbalance during binary classification processes that hinder a pure oil detection process; therefore, it is very likely that these limitations will lead to inconsistencies in non-compliance verdicts in cases where the underlying distribution of the sample set is not symmetrical. The presence of a threshold response of the isotopic response to extreme dilution indicates that further refinement of the model is required to fully identify nonlinear dynamics. Finally, the lack of validation of the solution on commercial samples has prevented extrapolation of real-world scenarios, particularly in terms of evaluating the combined frauds/degradations.

Furthermore, while geographic factors have been shown to have little to no effect, additional testing on a wider range of countries will help ensure that this model is robust enough for the multi-national nature of fraud and other frauds. The extensive character of the dataset should lead to a cautious approach when making generalizations to other products and fraud types, however, the current accuracy demonstrates that this model has considerable potential for targeted use. While machine learning architectures were evaluated for predicting fraudulent behaviour across the dataset, a smaller subset of representative models was used to reduce duplication and increase interpretability. Each model was evaluated to identify similar predictive capabilities, and the goal was to select a model that balances all three aspects of predictive performance, that is, accuracy, robustness and explainability. Of these, Random Forests were chosen as the reference model because of their stability and partial interpretability. Symbolic Regression offers an analytical approach that will be relevant to regulatory requirements. Thus, while the proposed model has performed well experimentally, consideration must be given to how the proposed model will perform in real-world cases. At present, validation has been done using mixtures of known adulterant type and dilution rates that were prepared in a laboratory. In the industrial and/or regulatory setting oils may be much more complicated due to having multiple adulterants present, going through oxidative degradation, having batch-to-batch variability etc. Therefore, additional validation of this method using market samples and inter-laboratory testing is still necessary to adequately evaluate its robustness and transferability. The non-destructive characteristics, rapid turn-around time and automated decision thresholds make this method amenable to routine screening, after which only suspected samples will need to be further validated by additional confirmatory analyses.

CONCLUSION

The proposed methodology includes the combination of three fraud analysis techniques through deep learning algorithms utilizing both autoencoders and probabilistic methods. The means used to identify adulteration include FTIR spectral analysis as well as measurement of $\delta^{13}C$ ratios. The development of automated systems that can generate $\delta^{13}C$ values based on a type of oil and a dilution factor are highly accurate (AUC=1.00). The suitability of Random Forest models for describing data is evident in the results obtained through this study; each model produced highly reliable predictive equations that offer advantages of operational clarity and increased flexibility. Examples of opportunities where this method could be effectively utilized would include: 1) industry, through improved quality control of their products; 2) traceability within the agricultural production chain; and 3) real-time monitoring of high-value food oils, providing an answer to some ethical concerns about transparency in food production.

REFERENCES

Aghili, N. S., Rasekh, M., Karami, H., Azizi, V., & Gancarz, M. (2022). Detection of fraud in sesame oil with the help of artificial intelligence combined with chemometrics methods and chemical compounds characterization by gas chromatography–mass spectrometry. *Lebensmittel-Wissenschaft + Technologie, 167*, 113863. DOI: 10.1016/j.lwt.2022.113863

Akaike, H. (1974). A new look at the statistical model identification. *IEEE Transactions on Automatic Control, 19*(6), 716–723. DOI: 10.1109/TAC.1974.1100705

Balan, B., Dhaulaniya, A. S., Jamwal, R., Amit, A., Sodhi, K. K., Kelly, S., Cannavan, A., & Singh, D. K. (2020). Application of attenuated total reflectance-Fourier transform infrared (ATR-FTIR) spectroscopy coupled with chemometrics for detection and quantification of formalin in cow milk. *Vibrational Spectroscopy, 107*, 103033. DOI: 10.1016/j.vibspec.2020.103033

Bodner, M., & Scampicchio, M. (2020). Analytical authentication of butter using Fourier transform infrared spectroscopy coupled with chemometrics. In Rapid Analytical Methods in Quality Control and Authenticity Analysis on Different Foods (p. 81).

Breiman, L. (2001). Random forests. *Machine Learning, 45*(1), 5–32. DOI: 10.1023/A:1010933404324

Chen, T., & Guestrin, C. (2016). XGBoost: A scalable tree boosting system. In *Proceedings of the 22nd ACM SIGKDD International Conference on Knowledge Discovery and Data Mining* (pp. 785–794). https://doi.org/DOI: 10.1145/2939672.2939785

Chollet, F. (2015). Keras [Computer software]. GitHub. https://github.com/fchollet/keras (Accessed May 25, 2025)

Cranmer, M., Sanchez-Gonzalez, A., Battaglia, P., Xu, R., Cranmer, K., Spergel, D., & Ho, S. (2020). Discovering symbolic models from deep learning with inductive biases. *Advances in Neural Information Processing Systems, 33*, 17429–17442.

Fawcett, T. (2006). An introduction to ROC analysis. *Pattern Recognition Letters, 27*(8), 861–874. DOI: 10.1016/j.patrec.2005.10.010

Firouz, M. S., Rashvand, M., & Omid, M. (2021). Rapid identification and quantification of sesame oils adulteration using low frequency dielectric spectroscopy combined with chemometrics. *Lebensmittel-Wissenschaft + Technologie, 140*, 110736. DOI: 10.1016/j.lwt.2020.110736

Goldberg, D. E. (1989). *Genetic algorithms in search, optimization and machine learning*. Addison-Wesley Publishing Company.

Goodfellow, I., Bengio, Y., & Courville, A. (2016). *Deep learning* (Vol. 1). MIT Press.

Green, H. S., Li, X., De Pra, M., Lovejoy, K. S., Steiner, F., Acworth, I. N., & Wang, S. C. (2020). A rapid method for the detection of extra virgin olive oil adulteration using UHPLC-CAD profiling of triacylglycerols and PCA. *Food Control, 107*, 106773. DOI: 10.1016/j.foodcont.2019.106773

Gyftokostas, N., Nanou, E., Stefas, D., Kokkinos, V., Bouras, C., & Couris, S. (2021). Classification of Greek olive oils from different regions by machine learning-aided laser-induced breakdown spectroscopy and absorption spectroscopy. *Molecules (Basel, Switzerland), 26*(5), 1241. DOI: 10.3390/molecules26051241 PMID: 33669128

Harris, C. R., Millman, K. J., van der Walt, S. J., Gommers, R., Virtanen, P., Cournapeau, D., Wieser, E., Taylor, J., Berg, S., Smith, N. J., Kern, R., Picus, M., Hoyer, S., van Kerkwijk, M. H., Brett, M., Haldane, A., Fernández del Río, J., Wiebe, M., Peterson, P., & Oliphant, T. E. (2020). Array programming with NumPy. *Nature, 585*(7825), 357–362. DOI: 10.1038/s41586-020-2649-2 PMID: 32939066

Jamwal, R., Amit, , Kumari, S., Balan, B., Kelly, S., Cannavan, A., & Singh, D. K. (2021). Rapid and non-destructive approach for the detection of fried mustard oil adulteration in pure mustard oil via ATR-FTIR spectroscopy-chemometrics. *Spectrochimica Acta. Part A: Molecular and Biomolecular Spectroscopy, 244*, 118822. DOI: 10.1016/j.saa.2020.118822 PMID: 32829154

Jamwal, R., Kumari, S., Dhaulaniya, A. S., Balan, B., Kelly, S., Cannavan, A., & Singh, D. K. (2020). Utilizing ATR-FTIR spectroscopy combined with multivariate chemometric modelling for the swift detection of mustard oil adulteration in virgin coconut oil. *Vibrational Spectroscopy, 109*, 103066. DOI: 10.1016/j.vibspec.2020.103066

Karami, H., Rasekh, M., & Mirzaee–Ghaleh, E. (2020). Comparison of chemometrics and AOCS official methods for predicting the shelf life of edible oil. *Chemometrics and Intelligent Laboratory Systems, 206*, 104165. DOI: 10.1016/j.chemolab.2020.104165

Khorramifar, A., Karami, H., Wilson, A. D., Sayyah, A. H. A., Shuba, A., & Lozano, J. (2022). Grape cultivar identification and classification by machine olfaction analysis of leaf volatiles. *Chemosensors (Basel, Switzerland), 10*(4), 125. DOI: 10.3390/chemosensors10040125

Koza, J. R. (1993). *On the programming of computers by means of natural selection*. MIT Press.

Li, X., Wang, D., Ma, F., Yu, L., Mao, J., Zhang, W., Jiang, J., Zhang, L., & Li, P. (2023). Rapid detection of sesame oil multiple adulteration using a portable Raman spectrometer. Food Chemistry, 405(Part B), 134884. https://doi.org/DOI: 10.1016/j.foodchem.2022.134884

Lumakso, F. A., Riyanto, S., Ahmad, S. A., Rosman, A. S., Yusoff, F. M., & Rohman, A. (2015). Application of chemometrics in combination with Fourier Transform Mid Infrared spectroscopy for authentication of avocado oil. *Journal of Food and Pharmaceutical Sciences, 2*, 12–17.

Ma, H., Guo, J., Liu, G., Xie, D., Zhang, B., Li, X., Zhang, Q., Cao, Q., Li, X., Ma, F., Li, Y., Wan, G., Li, Y., Wu, D., Ma, P., Guo, M., & Yin, J. (2025). Raman spectroscopy coupled with chemometrics for identification of adulteration and fraud in muscle foods: A review. *Critical Reviews in Food Science and Nutrition, 65*(11), 2008–2030. DOI: 10.1080/10408398.2024.2329956 PMID: 38523442

Martinez-Castillo, C., Astray, G., Mejuto, J. C., & Simal-Gandara, J. (2020). Random forest, artificial neural network, and support vector machine models for honey classification. *eFood, 1*(1), 69–76. DOI: 10.2991/efood.k.191004.001

Montgomery, D. C. (2017). *Design and analysis of experiments*. John Wiley & Sons.

Montgomery, D. C., & Runger, G. C. (2010). *Applied statistics and probability for engineers*. John Wiley & Sons.

Morris, J. B., Wang, M. L., & Tonnis, B. D. (2021). Variability for oil, protein, lignan, tocopherol, and fatty acid concentrations in eight sesame (Sesamum indicum L.) genotypes. *Industrial Crops and Products, 164*, 113355. DOI: 10.1016/j.indcrop.2021.113355

Pan, Z., Li, R. H., Cui, Y. Y., Wu, X. J., Zhang, Y. Y., & Wang, Y. T. (2021). A simple and quick method to detect adulterated sesame oil using 3D fluorescence spectra. *Spectrochimica Acta. Part A: Molecular and Biomolecular Spectroscopy, 245*, 118948. DOI: 10.1016/j.saa.2020.118948 PMID: 32980759

Pedregosa, F., Varoquaux, G., Gramfort, A., Michel, V., Thirion, B., Grisel, O., Blondel, M., Prettenhofer, P., Weiss, R., & Dubourg, V. (2011). Scikit-learn: Machine learning in Python. *Journal of Machine Learning Research, 12*, 2825–2830.

Raschka, S., & Mirjalili, V. (2019). *Python machine learning: Machine learning and deep learning with Python, scikit-learn, and TensorFlow 2*. Packt Publishing Ltd.

Rasekh, M., Karami, H., Wilson, A. D., & Gancarz, M. (2021). Classification and identification of essential oils from herbs and fruits based on a MOS electronic-nose technology. *Chemosensors (Basel, Switzerland), 9*(6), 142. DOI: 10.3390/chemosensors9060142

Rohman, A., & Che Man, Y. B. (2011). The use of Fourier transform mid infrared (FT-MIR) spectroscopy for detection and quantification of adulteration in virgin coconut oil. *Food Chemistry, 129*(2), 583–588. DOI: 10.1016/j.foodchem.2011.04.070 PMID: 30634271

Rohman, A., Che Man, Y. B., & Ali, M. E. (2019). The authentication of virgin coconut oil from grape seed oil and soybean oil using FTIR spectroscopy and chemometrics. *International Journal of Applied Pharmaceutics, 11*(2), 259–263. DOI: 10.22159/ijap.2019v11i2.31758

Rohman, A., & Man, C. (2012). Y. B. (2012). Quantification and classification of corn and sunflower oils as adulterants in olive oil using chemometrics and FTIR spectra. *TheScientificWorldJournal, 2012*, 250795. Advance online publication. DOI: 10.1100/2012/250795 PMID: 22448127

Schmidt, M., & Lipson, H. (2009). Distilling free-form natural laws from experimental data. *Science, 324*(5923), 81–85. DOI: 10.1126/science.1165893 PMID: 19342586

Schwarz, G. (1978). Estimating the dimension of a model. *Annals of Statistics, 6*(2), 461–464. DOI: 10.1214/aos/1176344136

Setyaningrum, D., Riyanto, S., & Rohman, A. (2013). Analysis of corn and soybean oils in red fruit oil using FTIR spectroscopy in combination with partial least square. *International Food Research Journal, 20*, 1977.

Upadhyay, N., Jaiswal, P., & Jha, S. N. (2018). Application of attenuated total reflectance Fourier Transform Infrared spectroscopy (ATR–FTIR) in MIR range coupled with chemometrics for detection of pig body fat in pure ghee (heat clarified milk fat). *Journal of Molecular Structure, 1153*, 275–281. DOI: 10.1016/j.molstruc.2017.09.116

Wu, X., Zhao, Z., Tian, R., Niu, Y., Gao, S., & Liu, H. (2021). Total synchronous fluorescence spectroscopy coupled with deep learning to rapidly identify the authenticity of sesame oil. *Spectrochimica Acta. Part A: Molecular and Biomolecular Spectroscopy*, *244*, 118841. DOI: 10.1016/j.saa.2020.118841 PMID: 32871392

Yuan, Z., Zhang, L., Wang, D., Jiang, J., Harrington, P. de B., Mao, J., Zhang, Q., & Li, P. (2020). Detection of flaxseed oil multiple adulteration by near-infrared spectroscopy and nonlinear one class partial least squares discriminant analysis. *Lebensmittel-Wissenschaft + Technologie*, *125*, 109247. DOI: 10.1016/j.lwt.2020.109247

Zhang, L., Huang, X., Li, P., Na, W., Jiang, J., Mao, J., Ding, X., & Zhang, Q. (2017). Multivariate adulteration detection for sesame oil. *Chemometrics and Intelligent Laboratory Systems*, *161*, 147–150. DOI: 10.1016/j.chemolab.2016.11.009

Zhang, L., Shuai, Q., Li, P., Zhang, Q., Ma, F., Zhang, W., & Ding, X. (2016). Ion mobility spectrometry fingerprints: A rapid detection technology for adulteration of sesame oil. *Food Chemistry*, *192*, 60–66. DOI: 10.1016/j.foodchem.2015.06.096 PMID: 26304320

Compilation of References

Abd El Qader, M., Venkat, R., Kumar, R., Hartmann, T., Ginobbi, P., Newman, N., & Singh, R. (2013). Structural, electrical, and thermoelectric properties of CrSi 2 thin films. *Thin Solid Films*, *545*, 100–105. DOI: 10.1016/j.tsf.2013.07.040

Abdullah, A. S., Wang, Y., Menger, M. F., Sami, S., & Head-Gordon, T. (2025). Improved treatment of 1–4 interactions in force fields for molecular dynamics simulations. *Journal of Chemical Theory and Computation*, *21*(22), 11669–11678. DOI: 10.1021/acs.jctc.5c01458 PMID: 41197020

Abraham, R. J., Byrne, J. J., Griffiths, L., & Perez, M. (2006). 1H chemical shifts in NMR: Part 23, the effect of dimethyl sulphoxide versus chloroform solvent on 1H chemical shifts. *Magnetic Resonance in Chemistry*, *44*(5), 491–509. DOI: 10.1002/mrc.1747 PMID: 16395732

Acharya, P., Biradar, C., Louhaichi, M., Ghosh, S., Hassan, S., Moyo, H., & Sarker, A. (2019). Finding a suitable niche for cultivating cactus pear (*Opuntia ficus-indica*) as an integrated crop in resilient dryland agroecosystems of India. *Sustainability (Basel)*, *11*(21), 5897. DOI: 10.3390/su11215897

Acosta, M., Novak, N., Rojas, V., Patel, S., Vaish, R., Koruza, J., Rossetti, G. A.Jr, & Rödel, J. (2017). BaTiO3-based piezoelectrics: Fundamentals, current status, and perspectives. *Applied Physics Reviews*, *4*(4), 041305. Advance online publication. DOI: 10.1063/1.4990046

Agarwal, A., Wang, H., Schütte, C., & Site, L. D. (2014). Chemical potential of liquids and mixtures via adaptive resolution simulation. *The Journal of Chemical Physics*, *141*(3), 034102. Advance online publication. DOI: 10.1063/1.4886807

Agarwal, M., Pasupathy, P., Wu, X., Recchia, S. S., & Pelegri, A. A. (2024). Multiscale computational and artificial intelligence models of linear and nonlinear composites: A review. *Small Science*, *4*(5), 2300185. DOI: 10.1002/smsc.202300185

Ager, J. W., Veirs, D. K., & Rosenblatt, G. M. (1991). Spatially resolved Raman studies of diamond films grown by chemical vapor deposition. *Physical Review B: Condensed Matter*, *43*(8), 6491–6499. DOI: 10.1103/PhysRevB.43.6491

Ageyeva, T., Horváth, S., & Kovács, J. G. (2019). In-mold sensors for injection molding: On the way to industry 4.0. *Sensors (Basel)*, *19*(16), 3551. DOI: 10.3390/s19163551

Aghili, N. S., Rasekh, M., Karami, H., Azizi, V., & Gancarz, M. (2022). Detection of fraud in sesame oil with the help of artificial intelligence combined with chemometrics methods and chemical compounds characterization by gas chromatography–mass spectrometry. *Lebensmittel-Wissenschaft + Technologie*, *167*, 113863. DOI: 10.1016/j.lwt.2022.113863

Agu, P. C., Afiukwa, C. A., Orji, O. U., Ezeh, E. M., Ofoke, I. H., Ogbu, C. O., Ugwuja, E. I., & Aja, P. M. (2023). Molecular docking as a tool for the discovery of molecular targets of nutraceuticals in diseases management. *Scientific Reports*, *13*(1), 13398. DOI: 10.1038/s41598-023-40160-2 PMID: 37592012

Ahmed, I. (2021). Synthesis, characterization, ferroelectric, and piezoelectric properties of (1 – x)BaTiO 3 – x (BaNi 0.5 Nb 0.5 O 3) perovskite ceramics. *Journal of Materials Science Materials in Electronics*, *32*(8), 10769–10777. DOI: 10.1007/s10854-021-05734-8

Ahmed, S. J., Kivinen, J., Zaporzan, B., Curiel, L., Pichardo, S., & Rubel, O. (2013). BerryPI : A software for studying polarization of crystalline solids with WIEN2k density functional all-electron package. *Computer Physics Communications*, *184*(3), 647–651. DOI: 10.1016/j.cpc.2012.10.028

Ahmed, S., Griffin, T. S., Kraner, D., Schaffner, M. K., Sharma, D., Hazel, M., Leitch, A. R., Orians, C. M., Han, W., Stepp, J. R., Robbat, A., Matyas, C., Long, C., Xue, D., Houser, R. F., & Cash, S. B. (2019). Environmental factors variably impact tea secondary metabolites in the context of climate change. *Frontiers in Plant Science*, *10*, 939. DOI: 10.3389/fpls.2019.00939

Aittokallio, T., & Fang, E. F. (2024). Editorial overview–artificial intelligence methodologies in structural biology: Bridging the gap to medical applications. *Current Opinion in Structural Biology*, *87*, 102862. DOI: 10.1016/j.sbi.2024.102862 PMID: 38870638

Ajp. (2024). Structural, morphological and spectroscopic characterization of chemically synthesized SnO2 nanoparticles. *Asian Journal of Physics*. https://asianjournalofphysics.com/structural-morphological-and-spectroscopic-characterization-of-chemically/

Akaike, H. (1974). A new look at the statistical model identification. *IEEE Transactions on Automatic Control*, *19*(6), 716–723. DOI: 10.1109/TAC.1974.1100705

Aktulga, H. M., Pandit, S. A., Van Duin, A. C. T., & Grama, A. Y. (2012a). Reactive molecular dynamics: Numerical methods and algorithmic techniques. *SIAM Journal on Scientific Computing*, *34*(1), C1–C23. DOI: 10.1137/100808599

Alexiadis, A., Ghraybeh, S., & Qiao, G. (2018). Natural convection and solidification of phase-change materials in circular pipes: A SPH approach. *Computational Materials Science*, *150*, 475–483. DOI: 10.1016/j.commatsci.2018.04.037

Ali Üncü, Y., Danışman, T., & Özdoğan, H. (2025). Predicting (n,3n) nuclear reaction cross-sections using XGBoost and Leave-One-Out Cross-Validation. *Applied Radiation and Isotopes*, *219*(January), 111714. DOI: 10.1016/j.apradiso.2025.111714 PMID: 39947034

Ali, M. K. M., Mohsen, A. A., & Allam, N. K. (2026). Lead-free perovskite materials for optoelectronic and solar energy applications. *Solar Energy Materials and Solar Cells*, *295*, 114025. DOI: 10.1016/j.solmat.2025.114025

Allahkarami, E., Allahkarami, E., Heydari, M., Azadmehr, A., & Maghsoudi, A. (2024). Assessment of chromite ore wastes for methylene blue adsorption: Isotherm, kinetic, thermodynamic studies, ANN, and statistical physics modeling. *Chemosphere*, *358*(March), 142098. DOI: 10.1016/j.chemosphere.2024.142098 PMID: 38677606

Alshomrany, A. S., Bousbih, R., Sajid, M., Jamil, M., Munir, J., Bani-Fwaz, M. Z., Jabir, M. S., Majdi, H., Assem, E. E., Shaban, M., & Nazar, M. (2025). DFT-Based Tailoring of the Thermoelectric and Photovoltaic Response of the Halide Double Perovskite Cs2TlYF6 (Y = Ag, Co). *Journal of Inorganic and Organometallic Polymers and Materials*, *35*(3), 1591–1602. DOI: 10.1007/s10904-024-03187-0

Amaka, N., Sunday, C., Chukwudi, P., Bartholomew, N., Dagdag, O., Okechukwu, J., Thakur, A., Chikaodili, V., & Dominic, O. (2024). Physisorption or chemisorption : Insight from AI computing model based on DFT, MC / MD-simulation for prediction of MOF-based inhibitor adsorption on Cu in brine solution. *Computational & Theoretical Chemistry*, *1238*(June), 114730. DOI: 10.1016/j.comptc.2024.114730

Amicarelli, A., Manenti, S., Albano, R., Agate, G., Paggi, M., Longoni, L., Mirauda, D., Ziane, L., Viccione, G., Todeschini, S., Sole, A., Baldini, L. M., Brambilla, D., Papini, M., Khellaf, M. C., Tagliafierro, B., Sarno, L., & Pirovano, G. (2020). SPHERA v. 9.0. 0: A Computational Fluid Dynamics research code, based on the Smoothed Particle Hydrodynamics mesh-less method. *Computer Physics Communications*, *250*, 107157. DOI: 10.1016/j.cpc.2020.107157

Amin, B., Khenata, R., Bouhemadou, A., Ahmad, I., & Maqbool, M. (2012). Opto-electronic response of spinels MgAl2O4 and MgGa2O4 through modified Becke-Johnson exchange potential. *Physica B, Condensed Matter*, *407*(13), 2588–2592. DOI: 10.1016/j.physb.2012.03.075

Amon, D. J., Gollner, S., Morato, T., Smith, C. R., Chen, C., Christiansen, S., Currie, B., Drazen, J. C., Fukushima, T., Gianni, M., Gjerde, K. M., Gooday, A. J., Grillo, G. G., Haeckel, M., Joyini, T., Ju, S.-J., Levin, L. A., Metaxas, A., Mianowicz, K., & Pickens, C. (2022). Assessment of scientific gaps related to the effective environmental management of deep-seabed mining. *Marine Policy*, *138*, 105006. DOI: 10.1016/j.marpol.2022.105006

Anderson, O. L. (1963). A simplified method for calculating the debye temperature from elastic constants. *Journal of Physics and Chemistry of Solids*, *24*(7), 909–917. DOI: 10.1016/0022-3697(63)90067-2

Apon, I. A., Kriaa, K., Rahman, M. A., Hossain, M. A., Maatki, C., Assadi, A. A., & Elboughdiri, N. (2025). Rb 2 BX 6 double perovskites: unlocking 22% efficiency through structural, electronic, mechanical, and optical insights. DOI: 10.1039/D5RA03981A

Arul, N. S., & Nithya, V. D. (Eds.). (2020). *Revolution of Perovskite : Synthesis, Properties and Applications*. Springer Singapore., DOI: 10.1007/978-981-15-1267-4

Ashcroft, N., & Mermin, N. D. (1976). Electron Dynamics in Solids. *Solid State Physics*, *116*(117). https://www.scirp.org/reference/referencespapers?referenceid=1476522

Aslam, N., Pfender, M., Neumann, P., Reuter, R., Zappe, A., De Oliveira, F. F., Denisenko, A., Sumiya, H., Onoda, S., Isoya, J., & Wrachtrup, J. (2017). Nanoscale nuclear magnetic resonance with chemical resolution. *Science*, *357*(6346), 67–71. DOI: 10.1126/science.aam8697 PMID: 28572453

ASTM G173–03. (2012). Standard Tables for Reference Solar Spectral Irradiances: Direct Normal and Hemispherical on 37° Tilted Surface. In G173–03. https://www.astm.org/g0173-23.html

Autschbach, J., & Zheng, S. (2008). Analyzing Pt chemical shifts calculated from relativistic density functional theory using localized orbitals: The role of Pt 5d lone pairs. *Magnetic Resonance in Chemistry*, *46*(S1), S45–S55. DOI: 10.1002/mrc.2289 PMID: 18855339

Autschbach, J., Zheng, S., & Schurko, R. W. (2010). Analysis of electric field gradient tensors at quadrupolar nuclei in common structural motifs. *Concepts in Magnetic Resonance. Part A, Bridging Education and Research., 36A*(2), 84–126. DOI: 10.1002/cmr.a.20155

Ayyaz, A., Murtaza, G., Shafiq, M., Qasim Shah, M., Sfina, N., & Ali, S. (2023). Exploring structural, thermodynamic, elastic, electro-optic, and thermoelectric characteristics of double perovskites Rb2X-InBr6 (X = Na, K) for photovoltaic applications: A DFT approach. *Solar Energy, 265*, 112131. DOI: 10.1016/j.solener.2023.112131

Bacong, J. R. C., & Juanico, D. E. O. (2021). Predictive Chromatography of Leaf Extracts Through Encoded Environmental Forcing on Phytochemical Synthesis. *Frontiers in Plant Science, 12*, 613507. DOI: 10.3389/fpls.2021.613507

Bae, B.-H., Choi, J., Seok, C., & Park, H. (2025). Physics-inspired accuracy estimator for model-docked ligand complexes. *Journal of Chemical Theory and Computation, 21*(4), 2140–2152. DOI: 10.1021/acs.jctc.4c01373 PMID: 39928910

Baeurle, S. A., Usami, T., & Gusev, A. A. (2006). A new multiscale modeling approach for the prediction of mechanical properties of polymer-based nanomaterials. *Polymer, 47*(26), 8604–8617. DOI: 10.1016/j.polymer.2006.10.017

Bagheri, M., Mohammadi, M., & Riazi, M. (2024). A review of smoothed particle hydrodynamics. *Computational Particle Mechanics, 11*(3), 1163–1219. DOI: 10.1007/s40571-023-00679-7

Bai, D. Z., Zhu, J.-G., Yu, W., & Bain, J. A. (2004). Micromagnetic simulation of effect of stress-induced anisotropy in soft magnetic thin films. *Journal of Applied Physics, 95*(11), 6864–6866. DOI: 10.1063/1.1667445

Bajpai, A., Chaurasia, D., & Tiwari, N. (2024). A novel methodology for anomaly detection in smart home networks via Fractional Stochastic Gradient Descent. *Computers and Electrical Engineering, 119*(PB), 109604. DOI: 10.1016/j.compeleceng.2024.109604

Baker, C. M., Anisimov, V. M., & MacKerell, A. D.Jr. (2010). Development of CHARMM Polarizable Force Field for nucleic acid bases based on the classical Drude Oscillator model. *The Journal of Physical Chemistry B, 115*(3), 580–596. DOI: 10.1021/jp1092338 PMID: 21166469

Balaji, S., Djaoued, Y., & Robichaud, J. (2006). Phonon confinement studies in nanocrystalline anatase-TiO 2 thin films by micro Raman spectroscopy. *Journal of Raman Spectroscopy : JRS, 37*(12), 1416–1422. DOI: 10.1002/jrs.1566

Balan, B., Dhaulaniya, A. S., Jamwal, R., Amit, A., Sodhi, K. K., Kelly, S., Cannavan, A., & Singh, D. K. (2020). Application of attenuated total reflectance-Fourier transform infrared (ATR-FTIR) spectroscopy coupled with chemometrics for detection and quantification of formalin in cow milk. *Vibrational Spectroscopy, 107*, 103033. DOI: 10.1016/j.vibspec.2020.103033

Balıkçı, E., Marques, A.-S. M. C., Bauer, L. G., Seupel, R., Bennett, J., Raux, B., Buchan, K., Simelis, K., Singh, U., Rogers, C., Ward, J., Cheng, C., Szommer, T., Schützenhofer, K., Elkins, J. M., Sloman, D. L., Ahel, I., Fedorov, O., Brennan, P. E., & Huber, K. V. M. (2024). Unexpected noncovalent off-target activity of clinical BTK inhibitors leads to discovery of a dual NUDT5/14 antagonist. *Journal of Medicinal Chemistry*, *67*(9), 7245–7259. DOI: 10.1021/acs.jmedchem.4c00072 PMID: 38635563

Banerjee, S., Das, H. S., & Maity, H. (2024). *An introduction to optoelectronics*. LAP Lambert Academic Publishing.

Bao, W., Liu, D., & Duan, Y. (2018). A first-principles prediction of anisotropic elasticity and thermal properties of potential superhard WB3. *Ceramics International*, *44*(12), 14053–14062. DOI: 10.1016/j.ceramint.2018.05.002

Barba, F. J., Garcia, C., Fessard, A., Munekata, P. E. S., Lorenzo, J. M., Aboudia, A., Ouadia, A., & Remize, F. (2022). *Opuntia ficus-indica* edible parts: A food and nutritional security perspective. *Food Reviews International*, *38*(5), 930–952. DOI: 10.1080/87559129.2020.1756844

Barman, A., & Sinha, J. (2018). *Spin Dynamics and Damping in Ferromagnetic Thin Films and Nanostructures*. Springer International Publishing., DOI: 10.1007/978-3-319-66296-1

Bartel, C. J., Clary, J. M., Sutton, C., Vigil-Fowler, D., Goldsmith, B. R., Holder, A. M., & Musgrave, C. B. (2020). Inorganic Halide Double Perovskites with Optoelectronic Properties Modulated by Sublattice Mixing. *Journal of the American Chemical Society*, *142*(11), 5135–5145. DOI: 10.1021/jacs.9b12440

Baş, D., & Boyacı, İ. H. (2007). Modeling and optimization I: Usability of response surface methodology. *Journal of Food Engineering*, *78*(3), 836–845. DOI: 10.1016/j.jfoodeng.2005.11.024

Bas, J. A., Calero, J. A., & Dougan, M. J. (2003). Sintered soft magnetic materials. Properties and applications. *Journal of Magnetism and Magnetic Materials*, *254–255*(255), 391–398. DOI: 10.1016/S0304-8853(02)00934-4

Baskaran, D., Natarajan, A. P. A., & Byun, H.-S. (2024). Potential of palm husk powder in the treatment of real tannery wastewater: Performance comparison between response surface methodology and artificial neural network-particle swarm optimization. *Journal of Water Process Engineering*, *67*(August), 106101. DOI: 10.1016/j.jwpe.2024.106101

Baskes, M. I. (1987). Application of the embedded-atom method to covalent materials: A semiempirical potential for silicon. *Physical Review Letters*, *59*(23), 2666–2669. DOI: 10.1103/PhysRevLett.59.2666

Bechmann, R. (1956). Elastic, Piezoelectric, and Dielectric Constants of Polarized Barium Titanate Ceramics and Some Applications of the Piezoelectric Equations. *The Journal of the Acoustical Society of America*, *28*(3), 347–350. DOI: 10.1121/1.1908324

Bechstedt, F., Seino, K., Hahn, P. H., & Schmidt, W. G. (2005). Quasiparticle bands and optical spectra of highly ionic crystals: AlN and NaCl. *Physical Review B: Condensed Matter and Materials Physics*, *72*(24), 245114. DOI: 10.1103/PhysRevB.72.245114

Becke, A. D. (1993). Density-functional thermochemistry. III. The role of exact exchange. *The Journal of Chemical Physics*, *98*(7), 5648–5652. DOI: 10.1063/1.464913

Becke, A. D., & Roussel, M. R. (1989). Exchange holes in inhomogeneous systems: A coordinate-space model. *Physical Review A: General Physics*, *39*(8), 3761–3767. DOI: 10.1103/PhysRevA.39.3761

BECQUEREL. E. (1839). On Electric Effects under the Influence of Solar Radiation. Compt. Rend., 9, 561. https://cir.nii.ac.jp/crid/1570009750806226944.bib?lang=ja

Becquerel, E. (1839). On electron effects under the influence of solar radiation. *Comptes Rendus Hebdomadaires des Séances de l'Académie des Sciences*, *9*(9), 561–567.

Behler, J., & Parrinello, M. (2007). Generalized neural-network representation of high-dimensional potential-energy surfaces. *Physical Review Letters*, *98*(14), 146401. DOI: 10.1103/PhysRevLett.98.146401

Bekker, V., Seemann, K., & Leiste, H. (2006). Development and optimisation of thin soft ferromagnetic Fe–Co–Ta–N and Fe–Co–Al–N films with in-plane uniaxial anisotropy for HF applications. *Journal of Magnetism and Magnetic Materials*, *296*(1), 37–45. DOI: 10.1016/j.jmmm.2005.01.025

Belytschko, T., & Xiao, S. P. (2003). Coupling Methods for Continuum Model with Molecular Model. *International Journal for Multiscale Computational Engineering*, *1*(1), 12. DOI: 10.1615/IntJMultCompEng.v1.i1.100

Ben Dhia, H., & Rateau, G. (2005a). The Arlequin method as a flexible engineering design tool. *International Journal for Numerical Methods in Engineering*, *62*(11), 1442–1462. DOI: 10.1002/nme.1229

Ben Youssef, J., Vukadinovic, N., Billet, D., & Labrune, M. (2004). Thickness-dependent magnetic excitations in Permalloy films with nonuniform magnetization. *Physical Review B: Condensed Matter and Materials Physics*, *69*(17), 174402. DOI: 10.1103/PhysRevB.69.174402

Benassi, E. (2016). Benchmarking of density functionals for a soft but accurate prediction and assignment of 1H and 13C NMR chemical shifts in organic and biological molecules. *Journal of Computational Chemistry*, *38*(2), 87–92. DOI: 10.1002/jcc.24521 PMID: 27796077

Ben, M., Wazzan, N., & Ben, A. (2021). Physico-chemical interpretations of the adsorption isotherms of D e p e A sensitizers with pyridyl group on TiO 2 for dye sensitized solar cells using statistical physics and density functional theory. *Journal of Materials Research and Technology*, *15*, 369–383. DOI: 10.1016/j.jmrt.2021.08.017

Beran, G. J. O. (2016). Modeling Polymorphic Molecular Crystals with Electronic Structure Theory. *Chemical Reviews*, *116*(9), 5567–5613. DOI: 10.1021/acs.chemrev.5b00648 PMID: 27008426

Béranger, G., Duffaut, F., Morlet, J., & Tiers, J.-F. (1996). A Hundred Years After the Discovery of Invar... the Iron-nickel Alloys. Londres.

Berlincourt, D., & Jaffe, H. (1958). Elastic and Piezoelectric Coefficients of Single-Crystal Barium Titanate. *Physical Review*, *111*(1), 143–148. DOI: 10.1103/PhysRev.111.143

Bertotti, G. (1998). *Hysteresis in magnetism: for physicists, materials scientists, and engineers*. Academic press.

Best Research-Cell Efficiency Chart | Photovoltaic Research | NREL. (n.d.). Retrieved November 3, 2025, from https://www.nrel.gov/pv/cell-efficiency

Beveridge, J., Söderström, M., Prieto-Díaz, R., Gutierrez-de-Teran, H., Odell, L. R., Hallberg, M., Larhed, M., & Gising, J. (2024). Benzylhydroxamic acids as inhibitors of insulin regulated aminopeptidase (IRAP). *European Journal of Medicinal Chemistry Reports*, *12*, 100215. DOI: 10.1016/j.ejmcr.2024.100215

Bhalla, A. S., Guo, R., & Roy, R. (2000). The perovskite structure - A review of its role in ceramic science and technology. *Materials Research Innovations*, *4*(1), 3–26. DOI: 10.1007/s100190000062

Bhardwaj, A., Kishore, S., & Pandey, D. K. (2022). Artificial intelligence in biological sciences. *Life (Basel, Switzerland)*, *12*(9), 1430. DOI: 10.3390/life12091430

Bian, X., Litvinov, S., Qian, R., Ellero, M., & Adams, N. A. (2012). Multiscale modeling of particle in suspension with smoothed dissipative particle dynamics. *Physics of Fluids*, *24*(1), 012002. DOI: 10.1063/1.3676244

Bibri, S. E., & Krogstie, J. (2017). The core enabling technologies of big data analytics and context-aware computing for smart sustainable cities: A review and synthesis. *Journal of Big Data*, *4*(1), 38. DOI: 10.1186/s40537-017-0091-6

Biega, R. I., Filip, M. R., Leppert, L., & Neaton, J. B. (2021). Chemically Localized Resonant Excitons in Silver-Pnictogen Halide Double Perovskites. *The Journal of Physical Chemistry Letters*, *12*(8), 2057–2063. DOI: 10.1021/acs.jpclett.0c03579

Bilotta, G., Hérault, A., Cappello, A., Ganci, G., & Del Negro, C. (2016). GPUSPH: a Smoothed Particle Hydrodynamics model for the thermal and rheological evolution of lava flows.

Birch, F. (1947). Finite elastic strain of cubic crystals. *Physical Review*, *71*(11), 809–824. DOI: 10.1103/PhysRev.71.809

Blaha, P., Schwarz, K., Madsen, G. K. H., Kvasnicka, D., & Luitz, J. (2001). wien2k. An Augmented Plane Wave+ Local Orbitals Program for Calculating Crystal Properties, 60(1), 155–169.

Blaha, P., Schwarz, K., Tran, F., Laskowski, R., Madsen, G. K. H., & Marks, L. D. (2020). WIEN2k : An APW+lo program for calculating the properties of solids. *The Journal of Chemical Physics*, *152*(7), 074101. Advance online publication. DOI: 10.1063/1.5143061

Bloch, F. (1930). Zur theorie des ferromagnetismus. *The European Physical Journal A*, *61*(3), 206–219. DOI: 10.1007/BF01339661

Blum, V., Asahi, R., Autschbach, J., Bannwarth, C., Bihlmayer, G., Blügel, S., Burns, L. A., Crawford, T. D., Dawson, W., De Jong, W. A., Draxl, C., Filippi, C., Genovese, L., Giannozzi, P., Govind, N., Hammes-Schiffer, S., Hammond, J. R., Hourahine, B., Jain, A., & Windus, T. L. (2024). Roadmap on methods and software for electronic structure-based simulations in chemistry and materials. *Electronic Structure*, *6*(4), 042501. DOI: 10.1088/2516-1075/ad48ec

Bocklage, L., Scholtyssek, J. M., Merkt, U., & Meier, G. (2007). Spin polarization of Ni2MnIn and Ni80Fe20 determined by point-contact Andreev spectroscopy. Journal of Applied Physics, 101(9), 09J512. DOI: 10.1063/1.2710238

Bocklage, L., Krüger, B., Matsuyama, T., Bolte, M., Merkt, U., Pfannkuche, D., & Meier, G. (2009). Dependence of Magnetic Domain-Wall Motion on a Fast Changing Current. *Physical Review Letters, 103*(19), 197204. DOI: 10.1103/PhysRevLett.103.197204

Bodner, M., & Scampicchio, M. (2020). Analytical authentication of butter using Fourier transform infrared spectroscopy coupled with chemometrics. In Rapid Analytical Methods in Quality Control and Authenticity Analysis on Different Foods (p. 81).

Boereboom, J. M., Fleurat-Lessard, P., & Bulo, R. E. (2018). Explicit solvation matters: Performance of QM/MM solvation models in nucleophilic addition. *Journal of Chemical Theory and Computation, 14*(4), 1841–1852. DOI: 10.1021/acs.jctc.7b01206 PMID: 29438621

Bogaerts, A., Yusupov, M., Van der Paal, J., Verlackt, C. C. W., & Neyts, E. C. (2014). Reactive molecular dynamics simulations for a better insight in plasma medicine. *Plasma Processes and Polymers, 11*(12), 1156–1168. DOI: 10.1002/ppap.201400084

Boikos, C., Ioannidis, G., Rapkos, N., Tsegas, G., Katsis, P., & Ntziachristos, L. (2025). Estimating daily road traffic pollution in Hong Kong using CFD modelling: Validation and application. *Building and Environment, 267*(PA), 112168. DOI: 10.1016/j.buildenv.2024.112168

Bonin, R., Schneider, M. L., Silva, T. J., & Nibarger, J. P. (2005). Dependence of magnetization dynamics on magnetostriction in NiFe alloys. *Journal of Applied Physics, 98*(12), 123904. Advance online publication. DOI: 10.1063/1.2143121

Boochani, A., Rasoolian, G., Kafi, F., Aminian, A., & Elahi, S. M. (2019). Electronic, optical, vibrational and EFG properties of tetragonal BaTiO3 under pressure : By first-principles study. *Zhongguo Wuli Xuekan, 59*, 357–371. DOI: 10.1016/j.cjph.2019.02.023

Borah, D., Saikia, P., Rout, J., Gogoi, D., Ghosh, N. N., & Bhattacharjee, C. R. (2024). Sustainable green synthesis of SnO2 quantum dots: A stable, phase-pure and highly efficient photocatalyst for degradation of toxic dyes. *Materials Today Sustainability, 26*, 100770. DOI: 10.1016/j.mtsust.2024.100770

Borg, M. K., Lockerby, D. A., & Reese, J. M. (2015). A hybrid molecular-continuum method for unsteady compressible multiscale flows. *Journal of Fluid Mechanics, 768*, 388–414. DOI: 10.1017/jfm.2015.83

Born, M., Huang, K., & Lax, M. (1954). Dynamical Theory of Crystal Lattices. Clarendon Press, Oxford, 141. DOI: 10.1119/1.1934059

Born, M., & Oppenheimer, R. (1927). Zur Quantentheorie der Molekeln. *Annalen der Physik, 389*(20), 457–484. DOI: 10.1002/andp.19273892002

Bose, B. K., & Wang, F. (Fred). (2019). ENERGY, ENVIRONMENT, POWER ELECTRONICS, RENEWABLE ENERGY SYSTEMS, AND SMART GRID. In Power Electronics in Renewable Energy Systems and Smart Grid (pp. 1–83). Wiley. DOI: 10.1002/9781119515661.ch1

Boubaia, A., Assali, A., Berrah, S., Bennacer, H., Zerifi, I., & Boukortt, A. (2021). Band gap and emission wavelength tuning of Sr-doped BaTiO3 (BST) perovskites for high-efficiency visible-light emitters and solar cells. *Materials Science in Semiconductor Processing, 130*, 105837. DOI: 10.1016/j.mssp.2021.105837

Boulanger, E., & Harvey, J. N. (2018). QM/MM methods for free energies and photochemistry. *Current Opinion in Structural Biology*, *49*, 72–76. DOI: 10.1016/j.sbi.2018.01.003 PMID: 29414514

Bouzid, T., Grich, A., Naboulsi, A., Yazid, H., Elbasraoui, A., El Himri, M., Regti, A., & El Haddad, M. (2025). Investigating the surface structure effects of dual polymer coatings on eriochrome black T adsorption : A DFT and artificial intelligence approach based on ANN modeling. *Journal of Molecular Structure*, *1343*(May), 142726. DOI: 10.1016/j.molstruc.2025.142726

Bozorth, R. M., & Ferromagnetism, D. (1951). Van Nostrand Company. Inc., New York, 867.

Breiman, L. (2001). Random forests. *Machine Learning*, *45*(1), 5–32. DOI: 10.1023/A:1010933404324

Brini, E., Algaer, E. A., Ganguly, P., Li, C., Rodríguez-Ropero, F., & Van Der Vegt, N. F. A. (2013). Systematic coarse-graining methods for soft matter simulations-a review. In Soft Matter (Vol. 9, Issue 7, pp. 2108–2119). Royal Society of Chemistry. DOI: 10.1039/C2SM27201F

Broughton, J. Q., Abraham, F. F., Bernstein, N., & Kaxiras, E. (n.d.). Concurrent coupling of length scales: Methodology and application.

Bruno, P., & Renard, J.-P. (1989). Magnetic surface anisotropy of transition metal ultrathin films. *Applied Physics. A, Solids and Surfaces*, *49*(5), 499–506. DOI: 10.1007/BF00617016

Burgelman, M., Nollet, P., & Degrave, S. (2000). Modelling polycrystalline semiconductor solar cells. *Thin Solid Films*, *361–362*, 527–532. DOI: 10.1016/S0040-6090(99)00825-1

Butler, K. T., Frost, J. M., & Walsh, A. (2015). Band alignment of the hybrid halide perovskites CH 3 NH 3 PbCl 3, CH 3 NH 3 PbBr 3 and CH 3 NH 3 PbI 3. *Materials Horizons*, *2*(2), 228–231. DOI: 10.1039/C4MH00174E

Buznikov, N. A., & Rozanov, K. N. (2005). The effect of stripe domain structure on dynamic permeability of thin ferromagnetic films with out-of-plane uniaxial anisotropy. *Journal of Magnetism and Magnetic Materials*, *285*(3), 314–326. DOI: 10.1016/j.jmmm.2004.07.050

Cagcag Yolcu, O., Aydın Temel, F., & Kuleyin, A. (2021). New hybrid predictive modeling principles for ammonium adsorption: The combination of Response Surface Methodology with feed-forward and Elman-Recurrent Neural Networks. *Journal of Cleaner Production*, *311*(January), 127688. Advance online publication. DOI: 10.1016/j.jclepro.2021.127688

Camarero, J., Sort, J., Hoffmann, A., García-Martín, J. M., Dieny, B., Miranda, R., & Nogués, J. (2005). Origin of the Asymmetric Magnetization Reversal Behavior in Exchange-Biased Systems: Competing Anisotropies. *Physical Review Letters*, *95*(5), 057204. DOI: 10.1103/PhysRevLett.95.057204

Caminiti, R., Serra, M., Nucera, S., Ruga, S., Oppedisano, F., Scarano, F., Macrì, R., Muscoli, C., Palma, E., Musolino, V., Statti, G., Mollace, V., & Maiuolo, J. (2024). Antioxidant activity and seasonal variations in the composition of insoluble fiber from the cladodes of *Opuntia ficus-indica* (L.) Miller: Development of new extraction procedures to improve fiber yield. *Plants*, *13*(4), 544. DOI: 10.3390/plants13040544

Canay, Ö., & Kocabıçak, Ü. (2023). An innovative data collection method to eliminate the preprocessing phase in web usage mining. *Engineering Science and Technology, an International Journal, 40*, 101360. DOI: 10.1016/j.jestch.2023.101360

Cancès, E., Mennucci, B., & Tomasi, J. (1997). A new integral equation formalism for the polarizable continuum model: Theoretical background and applications to isotropic and anisotropic dielectrics. *The Journal of Chemical Physics, 107*(8), 3032–3041. DOI: 10.1063/1.474659

Capone, M., Romanelli, M., Castaldo, D., Parolin, G., Bello, A., Gil, G., & Vanzan, M. (2024). A vision for the future of multiscale modeling. *ACS Physical Chemistry Au, 4*(3), 202–225. DOI: 10.1021/acsphyschemau.3c00080

Carlsson, J., & Luttens, A. (2024). Structure-based virtual screening of vast chemical space as a starting point for drug discovery. *Current Opinion in Structural Biology, 87*, 102829. DOI: 10.1016/j.sbi.2024.102829 PMID: 38848655

Carmen, F., Frances, C., & Barthe, L. (2023). Trends on valorization of pitaya fruit biomass through value-added and green extraction technology – A critical review of advancements and processes. *Trends in Food Science & Technology, 138*, 339–354. DOI: 10.1016/j.tifs.2023.06.014

Car, R., & Parrinello, M. (1985). Unified Approach for Molecular Dynamics and Density-Functional Theory. *Physical Review Letters, 55*(22), 2471–2474. DOI: 10.1103/PhysRevLett.55.2471 PMID: 10032153

Carreras-Puigvert, J., & Spjuth, O. (2024). Artificial intelligence for high content imaging in drug discovery. *Current Opinion in Structural Biology, 87*, 102842. DOI: 10.1016/j.sbi.2024.102842 PMID: 38797109

Carvalho, A. P., Santos, S. M., Pérez-Sánchez, G., Gouveia, J. D., Gomes, J. R. B., & Jorge, M. (2022). Sticky-MARTINI as a reactive coarse-grained model for molecular dynamics simulations of silica polymerization. *npj Computational Materials, 8*(1), 49. DOI: 10.1038/s41524-022-00722-w

Casabianca, L. B. (2020). Calculating nuclear magnetic resonance chemical shifts in solvated systems. *Magnetic Resonance in Chemistry, 58*(7), 611–624. DOI: 10.1002/mrc.4994 PMID: 31916612

Castro, A. C., Balcells, D., Repisky, M., Helgaker, T., & Cascella, M. (2020). First-Principles Calculation of 1H NMR chemical shifts of complex metal polyhydrides: The essential inclusion of relativity and dynamics. *Inorganic Chemistry, 59*(23), 17509–17518. DOI: 10.1021/acs.inorgchem.0c02753 PMID: 33226791

Cercos-Pita, J. L. (2015). AQUAgpusph, a new free 3D SPH solver accelerated with OpenCL. *Computer Physics Communications, 192*, 295–312. DOI: 10.1016/j.cpc.2015.01.026

Ceriotti, M., Fang, W., Kusalik, P. G., McKenzie, R. H., Michaelides, A., Morales, M. A., & Markland, T. E. (2016). Nuclear quantum effects in water and aqueous Systems: Experiment, theory, and current challenges. *Chemical Reviews, 116*(13), 7529–7550. DOI: 10.1021/acs.chemrev.5b00674 PMID: 27049513

Chakraborty, A., Liton, M. N. H., Sarker, M. S. I., Rahman, M. M., & Khan, M. K. R. (2023). A comprehensive DFT evaluation of catalytic and optoelectronic properties of BaTiO3 polymorphs. *Physica B, Condensed Matter, 648*, 414418. DOI: 10.1016/j.physb.2022.414418

Chamoin, L., Prudhomme, S., Ben Dhia, H., & Oden, T. (2010). Ghost forces and spurious effects in atomic-to-continuum coupling methods by the Arlequin approach. *International Journal for Numerical Methods in Engineering*, *83*(8-9), 1081–1113. DOI: 10.1002/nme.2879

Chandra, S., Wang, Z., Tao, X., Chen, O., Luo, X., Ji, R.-R., & Bortsov, A. V. (2020). Computer-aided discovery of a new Nav1.7 inhibitor for treatment of pain and itch. *Anesthesiology*, *133*(3), 611–627. DOI: 10.1097/ALN.0000000000003427 PMID: 32788559

Chang, L., Yang, W., Khabaz, M. K., Eftekhari, S. A., Alkhalifah, T., & Khairy, Y. (2025). Introducing an integrated self-organizing map and radial basis function network for accurate prediction of water-Fe3O4 nanofluid viscosity versus solid volume fraction and temperature. *Renewable and Sustainable Energy Reviews, 209*(November 2024), 115119. DOI: 10.1016/j.rser.2024.115119

Charisiadis, P., Kontogianni, V., Tsiafoulis, C., Tzakos, A., Siskos, M., & Gerothanassis, I. (2014). 1H-NMR as a structural and analytical tool of intra- and intermolecular hydrogen bonds of Phenol-Containing natural products and model compounds. *Molecules (Basel, Switzerland)*, *19*(9), 13643–13682. DOI: 10.3390/molecules190913643 PMID: 25185070

Charles, A. N., & Daivis, P. (2011). A three dimensional smooth particle hydrodynamics model of the nanoscale condensation of water. *19th International Congress on Modelling and Simulation*, 516–522.

Charpentier, T., Menziani, M. C., & Pedone, A. (2013). Computational simulations of solid state NMR spectra: A new era in structure determination of oxide glasses. *RSC Advances*, *3*(27), 10550. DOI: 10.1039/c3ra40627j

Chaves, J. O., de Souza, M. C., da Silva, L. C., Lachos-Perez, D., Torres-Mayanga, P. C., Machado, A. P. F., Forster-Carneiro, T., Vázquez-Espinosa, M., González-de-Peredo, A. V., Barbero, G. F., & Rostagno, M. A. (2020). Extraction of flavonoids from natural sources using modern techniques. *Frontiers in Chemistry*, *8*, 507887. DOI: 10.3389/fchem.2020.507887

Chellamuthu, P., Savarimuthu, K., Alsath, M. G. N., R, K., T, Y., Bajaj, M., & Shabaz, M. (2025). Fabrication and comprehensive experimental evaluation of surfactant-activated PEDOT:PSS/SnO2 thin films deposited via spin coating for advanced sensing applications. *Scientific Reports*, *15*(1), 30628. DOI: 10.1038/s41598-025-12499-1

Chemat, F., Vian, M. A., & Cravotto, G. (2012). Green extraction of natural products: Concept and principles. *International Journal of Molecular Sciences*, *13*(7), 8615–8627. DOI: 10.3390/ijms13078615

Chen, C.-W. (2013). *Magnetism and metallurgy of soft magnetic materials*. Courier Corporation.

Cheng, G., Ban, Y., Deng, X., Li, H., Zhang, H., Li, G., Liao, L., & Khan, R. (2023). Research on quantitative inversion of ion adsorption type rare earth ore based on convolutional neural network. *Frontiers in Earth Science (Lausanne)*, *10*(January), 1–14. DOI: 10.3389/feart.2022.1086325

Chen, L., Yang, W., Fu, H., Liu, W., Shao, G., Tang, B., & Zheng, J. (2021). Mn2+-doped Cs2NaInCl6 double perovskites and their photoluminescence properties. *Journal of Materials Science*, *56*(13), 8048–8059. DOI: 10.1007/s10853-021-05822-4

Chen, S., & Izgorodina, E. I. (2017). Prediction of 1H NMR chemical shifts for clusters of imidazolium-based ionic liquids. *Physical Chemistry Chemical Physics, 19*(26), 17411–17425. DOI: 10.1039/C7CP02951A PMID: 28650016

Chen, T., & Guestrin, C. (2016). XGBoost: A scalable tree boosting system. In *Proceedings of the 22nd ACM SIGKDD International Conference on Knowledge Discovery and Data Mining* (pp. 785–794). https://doi.org/DOI: 10.1145/2939672.2939785

Chen, T.-L., Kim, H., Pan, S.-Y., Tseng, P.-C., Lin, Y.-P., & Chiang, P.-C. (2020). Implementation of green chemistry principles in circular economy system towards sustainable development goals: Challenges and perspectives. *The Science of the Total Environment, 716*, 136998. DOI: 10.1016/j.scitotenv.2020.136998

Chen, Y., Lin, H., Yang, H., Tan, R., Bian, Y., Fu, T., Li, W., Wu, L., Pei, Y., & Sun, H. (2017). Discovery of new acetylcholinesterase and butyrylcholinesterase inhibitors through structure-based virtual screening. *RSC Advances, 7*(6), 3429–3438. DOI: 10.1039/C6RA25887E

Chen, Z., & Chen, K. (2025). A new gradient descent divergence cleaning method with optimized high-order low-dissipation TENO schemes for ideal magnetohydrodynamic simulations. *Computer Physics Communications, 315*(May), 109744. DOI: 10.1016/j.cpc.2025.109744

Cherrad Ounissa. (2023). *Élaboration sous vide et investigations de propriétés physiques de films minces ferromagnétiques de Permalloy*. Dépôt Institutionnel de l'Université Ferhat ABBAS.

Chinatamby, P., & Jewaratnam, J. (2023). A performance comparison study on PM2.5 prediction at industrial areas using different training algorithms of feedforward-backpropagation neural network (FBNN). *Chemosphere, 317*, 137788. DOI: 10.1016/j.chemosphere.2023.137788 PMID: 36642141

Chintada, V., & Golla, N. (2025). Exploring the therapeutic potential of bioactive compounds from plant sources. In Devi, J. (Ed.), *Biotechnological Intervention in Production of Bioactive Compounds* (pp. 229–247). Springer., DOI: 10.1007/978-3-031-76859-0_15

Choe, H., Yoon, H., Kim, I., & Kim, S. Y. (2025). Machine Learning Unlocks New Directions in Halide Perovskite Research. *ChemElectroChem*. Advance online publication. DOI: 10.1002/celc.202500282

Chollet, F. (2015). Keras [Computer software]. GitHub. https://github.com/fchollet/keras (Accessed May 25, 2025)

Cohen, A. J., Mori-Sánchez, P., & Yang, W. (2012). Challenges for density Functional Theory. *Chemical Reviews, 112*(1), 289–320. DOI: 10.1021/cr200107z PMID: 22191548

Cohen, R. D., Wood, J. S., Lam, Y., Buevich, A. V., Sherer, E. C., Reibarkh, M., Williamson, R. T., & Martin, G. E. (2023). DELTA50: A highly accurate database of experimental 1H and 13C NMR chemical shifts applied to DFT benchmarking. *Molecules (Basel, Switzerland), 28*(6), 449. DOI: 10.3390/molecules28062449 PMID: 36985422

Coller, D. H., Vicente, B. C., & Scott, S. L. (2016). Rapid extraction of quantitative kinetic information from variable-temperature reaction profiles. *Chemical Engineering Journal, 303*, 182–193. DOI: 10.1016/j.cej.2016.05.136

Collins, K. D. (2012). Why continuum electrostatics theories cannot explain biological structure, polyelectrolytes or ionic strength effects in ion–protein interactions. *Biophysical Chemistry*, *167*, 43–59. DOI: 10.1016/j.bpc.2012.04.002 PMID: 22608112

Cordeiro, D. S., Wang, J., Wang, Y., Liu, X., & Wang, X. (2025). Academic Editor: M. Natália Prediction and Screening of Lead-Free Double Perovskite Photovoltaic Materials Based on Machine Learning. DOI: 10.3390/molecules30112378

Couderchon, G. (1994). Alliages fer-nickel et fer-cobalt Propriétés magnétiques. Techniques de l'ingénieur. Matériaux Fonctionnels, 2(D2130).

Couderchon, G. (1998). *Alliages magnétiques doux* (Ingénieur, T., Ed.).

Couderchon, G., & Tiers, J. F. (1982). Some aspects of magnetic properties of Ni-Fe and Co-Fe alloys. *Journal of Magnetism and Magnetic Materials*, *26*(1–3), 196–214. DOI: 10.1016/0304-8853(82)90152-4

Counil, G., Kim, J.-V., Devolder, T., Chappert, C., Shigeto, K., & Otani, Y. (2004). Spin wave contributions to the high-frequency magnetic response of thin films obtained with inductive methods. *Journal of Applied Physics*, *95*(10), 5646–5652. DOI: 10.1063/1.1697641

Cranmer, M., Sanchez-Gonzalez, A., Battaglia, P., Xu, R., Cranmer, K., Spergel, D., & Ho, S. (2020). Discovering symbolic models from deep learning with inductive biases. *Advances in Neural Information Processing Systems*, *33*, 17429–17442.

Crespo, A. J. C., Domínguez, J. M., Rogers, B. D., Gómez-Gesteira, M., Longshaw, S., Canelas, R., Vacondio, R., Barreiro, A., & García-Feal, O. (2015). DualSPHysics: Open-source parallel CFD solver based on Smoothed Particle Hydrodynamics (SPH). *Computer Physics Communications*, *187*, 204–216. DOI: 10.1016/j.cpc.2014.10.004

Cruz-Victoria, J. C., Netzahuatl-Muñoz, A. R., & Cristiani-Urbina, E. (2024). Long Short-Term Memory and Bidirectional Long Short-Term Memory Modeling and Prediction of Hexavalent and Total Chromium Removal Capacity Kinetics of Cupressus lusitanica Bark. *Sustainability (Basel)*, *16*(7), 2874. Advance online publication. DOI: 10.3390/su16072874

Cui, Q., & Karplus, M. (2000). Molecular properties from combined QM/MM methods. 2. Chemical shifts in large molecules. *The Journal of Physical Chemistry B*, *104*(15), 3721–3743. DOI: 10.1021/jp994154g

Cui, X., Bakkar, A., & Habashi, W. G. (2019). A multiphase SPH framework for supercooled large droplets dynamics. *International Journal of Numerical Methods for Heat & Fluid Flow*, *29*(7), 2434–2449. DOI: 10.1108/HFF-10-2018-0547

Cullity, B. D., & Graham, C. D. (2011). *Introduction to magnetic materials*. John Wiley & Sons.

Cu, W., Fang, J., Guo, X., Chen, K., Zheng, N., Xiao, B., & Wei, J. (2025). Prediction and analysis of bed-to-tube heat transfer in fluidized bed heat exchangers based on ANN-PSO hybrid approach and CFD simulation. *International Journal of Heat and Mass Transfer*, *252*(June), 127490. DOI: 10.1016/j.ijheatmasstransfer.2025.127490 da Rocha, M. N., de Sousa, D. S., da Silva Mendes, F. R., dos Santos, H. S., Marinho, G. S., Marinho, M. M., & Marinho, E. S. (2025). Ligand and structure-based virtual screening approaches in drug discovery: Minireview. *Molecular Diversity, 29*(3), 2799–2809. DOI: 10.1007/s11030-024-10979-6 PMID: 39223358

Da Silva, H. C., Hernandes, I. S., & De Almeida, W. B. (2025). Modeling solvent effects in quantum chemical calculation of relative energies and NMR chemical shifts for azithromycin. *The Journal of Physical Chemistry A, 129*(9), 2200–2216. DOI: 10.1021/acs.jpca.4c08015 PMID: 39985808

da Silva, R.F., & Carneiro, C.N., do C. de Sousa, C.B., Gomez, F.J.V., Espino, M., Boiteux, J., Fernández, M. de los Á., Silva, M.F., & de S. Dias, F. (2022). Sustainable extraction of bioactive compounds procedures in medicinal plants based on the principles of green analytical chemistry: A review. *Microchemical Journal, 175*, 107184. DOI: 10.1016/j.microc.2022.107184

Dai, Y., Wu, Q., & Zhang, Y. (2024). Generalized sparse radial basis function networks for multi-classification problems. *Applied Soft Computing, 154*(December 2022), 111361. DOI: 10.1016/j.asoc.2024.111361

Dai, G., Wang, S., Huang, G., Chen, G., Lu, B., Li, D., Tao, T., Yao, Y., Liang, B., & Lu, S. (2020). Direct and indirect measurement of large electrocaloric effect in barium strontium titanate ceramics. *International Journal of Applied Ceramic Technology, 17*(3), 1354–1361. DOI: 10.1111/ijac.13384

Darhi, Z., Malki, S., Abbadi, H., El Farh, L., Guesmi, I., & Challioui, A. (2023). Ab-initio calculation of the structural, electronic, mechanical, optical, and thermoelectric properties of orthorhombic ZnAs compound. *Physica B, Condensed Matter, 654*, 414722. DOI: 10.1016/j.physb.2023.414722

Das, H. S., Basak, A., & Maity, S. (2024a). Materials Science and nanotechnology. In *Advances in chemical and materials engineering book series* (pp. 175–206). DOI: 10.4018/979-8-3693-3398-3.ch008

Das, H. S., Mishra, S., & Roymahapatra, G. (2024b). Advanced Nano-Structured materials for energy storage devices. In *Advances in chemical and materials engineering book series* (pp. 1–34). DOI: 10.4018/979-8-3693-5320-2.ch001

Das, S., & Merz, K. M.Jr. (2025). Exploring the frontiers of computational NMR: Methods, applications, and challenges. *Chemical Reviews, 125*(19), 9256–9295. DOI: 10.1021/acs.chemrev.5c00259 PMID: 40923454

Datta, A. K., & Rahman, Md. S., Mia, N., & Ahmed, S. (2024). Exploration of Structural, Electronic and Optical Properties of Be-doped $BaTiO_3$ Perovskite : First Principles DFT Calculations. 2024 3rd International Conference on Advancement in Electrical and Electronic Engineering (ICAEEE), 1-6. DOI: 10.1109/ICAEEE62219.2024.10561704

Daw, M. S., & Baskes, M. I. (1984). Embedded-atom method: Derivation and application to impurities, surfaces, and other defects in metals. *Physical Review B: Condensed Matter, 29*(12), 6443–6453. DOI: 10.1103/PhysRevB.29.6443

Dbik, A., El Messaoudi, N., Bentahar, S., El Khomri, M., Lacherai, A., & Faska, N. (2022). Optimization of methylene blue adsorption on agricultural solid waste using box–behnken design (BBD) combined with response surface methodology (RSM) modeling. *Biointerface Research in Applied Chemistry, 12*(4), 4567–4583. DOI: 10.33263/BRIAC124.45674583

De Gortari, I., Portella, G., Salvatella, X., Bajaj, V. S., Van Der Wel, P. C. A., Yates, J. R., Segall, M. D., Pickard, C. J., Payne, M. C., & Vendruscolo, M. (2010). Time averaging of NMR chemical shifts in the MLF peptide in the solid state. *Journal of the American Chemical Society*, *132*(17), 5993–6000. DOI: 10.1021/ja9062629 PMID: 20387894

de Hond, A. A. H., Leeuwenberg, A. M., Hooft, L., Kant, I. M. J., Nijman, S. W. J., van Os, H. J. A., Aardoom, J. J., Debray, T. P. A., Schuit, E., van Smeden, M., Reitsma, J. B., Steyerberg, E. W., Chavannes, N. H., & Moons, K. G. M. (2022). Guidelines and quality criteria for artificial intelligence-based prediction models in healthcare: A scoping review. npj. *NPJ Digital Medicine*, *5*(1), 2. DOI: 10.1038/s41746-021-00549-7

de León-Delgado, H., Praga-Alejo, R. J., & González-González, D. S. (2025). Backpropagation neural network model with statistical inference in manufacturing processes. *Journal of Industrial Information Integration*, *44*(January), 100783. Advance online publication. DOI: 10.1016/j.jii.2025.100783

de Moraes, A., Bohn, F., Gamino, M., da Silva, E. F., de Oliveira, D. F., Lopes, C., Ferreira, A., Vaz, F., & Correa, M. A. (2025). Asymmetrical magnetoimpedance on Permalloy/Ag multilayer for high-frequency sensor applications. *Surface and Coatings Technology*, *495*, 131580. DOI: 10.1016/j.surfcoat.2024.131580

De Oliveira, M. T., Alves, J. M. A., Braga, A. C., Wilson, D. J. D., & Barboza, C. A. (2021). Do Double-Hybrid Exchange–Correlation functionals provide accurate chemical shifts? A benchmark assessment for proton NMR. *Journal of Chemical Theory and Computation*, *17*(11), 6876–6885. DOI: 10.1021/acs.jctc.1c00604 PMID: 34637284

De Souza, L. A., Da Silva, H. C., & De Almeida, W. B. (2018). Structural Determination of Antioxidant and Anticancer Flavonoid Rutin in Solution through DFT Calculations of 1H NMR Chemical Shifts. *ChemistryOpen*, *7*(11), 902–913. DOI: 10.1002/open.201800209 PMID: 30460171

De Vivo, M., Masetti, M., Bottegoni, G., & Cavalli, A. (2016). Role of molecular dynamics and related methods in drug discovery. *Journal of Medicinal Chemistry*, *59*(9), 4035–4061. DOI: 10.1021/acs.jmedchem.5b01684 PMID: 26807648

Deepa, A. (2025). Biochar-infused polyethersulfone beads for dye adsorption: RSM-BBD optimization coupled with isotherm and kinetic insights. *Materials Today. Communications*, *48*(June), 113652. DOI: 10.1016/j.mtcomm.2025.113652

Delgado-Buscalioni, R., Kremer, K., & Praprotnik, M. (2008). Concurrent triple-scale simulation of molecular liquids. *The Journal of Chemical Physics*, *128*(11), 114110. Advance online publication. DOI: 10.1063/1.2890729

Delle Site, L., Praprotnik, M., Bell, J. B., & Klein, R. (2020). Particle–continuum coupling and its scaling regimes: Theory and applications. *Advanced Theory and Simulations*, *3*(5), 1900232. DOI: 10.1002/adts.201900232

Demokritov, S. O., Hillebrands, B., & Slavin, A. N. (2001). Brillouin light scattering studies of confined spin waves: Linear and nonlinear confinement. *Physics Reports*, *348*(6), 441–489. DOI: 10.1016/S0370-1573(00)00116-2

Dempsey, J. L., Li, F., O'Mullane, A. P., & Unwin, P. R. (2025). Welcome to ACS Electrochemistry. *ACS Electrochemistry*, *1*(1), 1–2. DOI: 10.1021/acselectrochem.4c00205

Deng, B., & Shi, Y. (2016). A reactive coarse-grained model for polydisperse polymers. *Polymer*, *98*, 88–99. DOI: 10.1016/j.polymer.2016.06.018

Deng, Z., Wei, F., Sun, S., Kieslich, G., Cheetham, A. K., & Bristowe, P. D. (2016). Exploring the properties of lead-free hybrid double perovskites using a combined computational-experimental approach. *Journal of Materials Chemistry. A, Materials for Energy and Sustainability*, *4*(31), 12025–12029. DOI: 10.1039/C6TA05817E

Deringer, V. L., Caro, M. A., & Csányi, G. (2019). Machine learning interatomic potentials as emerging tools for materials science. *Advanced Materials*, *31*(46), 1902765. DOI: 10.1002/adma.201902765

Deshmukh, M. T., Wankhede, P. R., Chakole, N., Kale, P. D., Jadhav, M. R., Kulkarni, M. B., & Bhaiyya, M. (2025). Towards intelligent food safety: Machine learning approaches for aflatoxin detection and risk prediction. *Trends in Food Science & Technology*, *161*(April), 105055. DOI: 10.1016/j.tifs.2025.105055

Dib, M., El-Yakoubi, E. Y., ElHarfaoui, M., & Oublal, M. (2025). The study analyzes the impacts of pressure the permeability of permalloy (Ni81Fe19)1−x(Al2O3)x thin films. *Applied Physics. A, Materials Science & Processing*, *131*(9), 707. DOI: 10.1007/s00339-025-08844-w

Din, S. U., Kiani, S. H., Haq, S., Ahmad, P., Khandaker, M. U., Faruque, M. R. I., Idris, A. M., & Sayyed, M. I. (2022). Bio-Synthesized tin oxide nanoparticles: Structural, optical, and biological studies. *Crystals*, *12*(5), 614. DOI: 10.3390/cryst12050614

Dirac, P. A. M. (1930). Note on Exchange Phenomena in the Thomas Atom. *Mathematical Proceedings of the Cambridge Philosophical Society*, *26*(3), 376–385. DOI: 10.1017/S0305004100016108

Dolhen, M., Mahajan, A., Pinho, R., Costa, M. E., Trolliard, G., & Vilarinho, P. M. (2015). Sodium potassium niobate ($K_{0.5}$ $Na_{0.5}$ NbO_3, KNN) thick films by electrophoretic deposition. *RSC Advances*, *5*(6), 4698–4706. DOI: 10.1039/C4RA11058G

Dong, C., Guan, X., Wang, Z., Zhao, H., Kuai, Y., & Gao, S. (2023). The effects of cation and halide anion on the stability, electronic and optical properties of double perovskite Cs2NaMX6 (M = In, Tl, Sb, bi; X = Cl, Br, I). 220, 112058. DOI: 10.1016/j.commatsci.2023.112058

Dong, X., Hao, G., & Yu, R. (2022). Two-dimensional smoothed particle hydrodynamics (SPH) simulation of multiphase melting flows and associated interface behavior. *Engineering Applications of Computational Fluid Mechanics*, *16*(1), 588–629. DOI: 10.1080/19942060.2022.2026820

Dračínský, M., Čechová, L., Hodgkinson, P., Procházková, E., & Janeba, Z. (2015). Resonance-assisted stabilisation of hydrogen bonds probed by NMR spectroscopy and path integral molecular dynamics. *Chemical Communications*, *51*(73), 13986–13989. DOI: 10.1039/C5CC05199A PMID: 26245988

Dračínský, M., & Hodgkinson, P. (2013). A molecular dynamics study of the effects of fast molecular motions on solid-state NMR parameters. *CrystEngComm*, *15*(43), 8705–8712. DOI: 10.1039/c3ce40612a

Dračínský, M., Möller, H. M., & Exner, T. E. (2013). Conformational sampling by AB initio molecular dynamics simulations improves NMR chemical shift predictions. *Journal of Chemical Theory and Computation*, *9*(8), 3806–3815. DOI: 10.1021/ct400282h PMID: 26584127

Dračínský, M., Unzueta, P., & Beran, G. J. O. (2019). Improving the accuracy of solid-state nuclear magnetic resonance chemical shift prediction with a simple molecular correction. *Physical Chemistry Chemical Physics*, *21*(27), 14992–15000. DOI: 10.1039/C9CP01666J PMID: 31237586

Duan, J., Hu, S., Wang, P., Zhang, D., & Jiang, Y. (2024). Topological quantum materials for spintronics. *MetalMat*, *1*(2), e24. Advance online publication. DOI: 10.1002/metm.24

Du, K., Meng, W., Wang, X., Yan, Y., & Mitzi, D. B. (2017). Bandgap Engineering of Lead-Free Double Perovskite Cs 2 AgBiBr 6 through Trivalent Metal Alloying. *Angewandte Chemie*, *129*(28), 8270–8274. DOI: 10.1002/ange.201703970

Eglitis, R. I., Piskunov, S., Heifets, E., Kotomin, E. A., & Borstel, G. (2004). Ab initio study of the SrTiO3, BaTiO3 and PbTiO3 (001) surfaces. *Ceramics International*, *30*(7), 1989–1992. DOI: 10.1016/j.ceramint.2003.12.176

Ehrenreich, H., & Cohen, M. H. (1959). Self-consistent field approach to the many-electron problem. *Physical Review*, *115*(4), 786–790. DOI: 10.1103/PhysRev.115.786

El Khomri, M., El Messaoudi, N., Dbik, A., Bentahar, S., Lacherai, A., Faska, N., & Jada, A. (2022). Regeneration of argan nutshell and almond shell using HNO3 for their reusability to remove cationic dye from aqueous solution. *Chemical Engineering Communications*, *209*(10), 1304–1315. DOI: 10.1080/00986445.2021.1963960

El Messaoudi, N., El Mouden, A., Fernine, Y., El Khomri, M., Bouich, A., Faska, N., Ciğeroğlu, Z., Américo-Pinheiro, J. H. P., Jada, A., & Lacherai, A. (2023). Green synthesis of Ag2O nanoparticles using Punica granatum leaf extract for sulfamethoxazole antibiotic adsorption: Characterization, experimental study, modeling, and DFT calculation. *Environmental Science and Pollution Research International*, *30*(34), 81352–81369. DOI: 10.1007/s11356-022-21554-7 PMID: 35729389

Elgammal, W. E., Shaban, S. S., Eliwa, E. M., Halawa, A. H., Abd El-Gilil, S. M., Hassan, R. A., Abdou, A. M., Elhagali, G. A. M., & Reheim, M. A. M. A. (2024). Thiazolation of phenylthiosemicarbazone to access new thiazoles: Anticancer activity and molecular docking. *Future Medicinal Chemistry*, *16*(12), 1219–1237. DOI: 10.1080/17568919.2024.2342668 PMID: 38989988

Ellero, M., Kröger, M., & Hess, S. (2002). Viscoelastic flows studied by smoothed particle dynamics. *Journal of Non-Newtonian Fluid Mechanics*, *105*(1), 35–51. DOI: 10.1016/S0377-0257(02)00059-9

Ellis, E. A. I., Chmielus, M., Han, S., & Baker, S. P. (2020). Effect of sputter pressure on microstructure and properties of β-Ta thin films. *Acta Materialia*, *183*, 504–513. DOI: 10.1016/j.actamat.2019.10.056

Elsenety, M. M. (2024). Design and predict the potential of imidazole-based organic dyes in dye-sensitized solar cells using fingerprint machine learning and supported by a web application. *Scientific Reports*, *14*(1), 1–11. DOI: 10.1038/s41598-024-76739-6 PMID: 39489728

Emadi, F., Nemati, A., Hinterstein, M., & Adabifiroozjaei, E. (2019). Microstructural, optical, and electrical characteristics of Ni/C doped BST thin films. *Ceramics International*, *45*(5), 5503–5510. DOI: 10.1016/j.ceramint.2018.12.007

Engel, E., & Vosko, S. H. (1993). Exact exchange-only potentials and the virial relation as microscopic criteria for generalized gradient approximations. *Physical Review B: Condensed Matter*, *47*(20), 13164–13174. DOI: 10.1103/PhysRevB.47.13164

Eurofins Pharma Discovery Services. (2025). SafetyScreen44™ Panel. Retrieved 11 August 2025 from https://cdnmedia.eurofins.com/corporate-eurofins/media/safetyscreen44_epdsfl420june16.pdf?utm_source=chatgpt.com

European Medicines Agency. (2020). ICH M9 guideline on biopharmaceutics classification system-based biowaivers - questions and answers. Retrieved 11 August 2025 from https://www.ema.europa.eu/en/documents/other/ich-m9-biopharmaceutics-classification-system-based-biowaivers-step-5-questions-and-answers_en.pdf?utm_source=chatgpt.com

Exner, T. E., Frank, A., Onila, I., & Möller, H. M. (2012). Toward the quantum chemical calculation of NMR chemical shifts of proteins. 3. Conformational sampling and explicit solvents model. *Journal of Chemical Theory and Computation*, *8*(11), 4818–4827. DOI: 10.1021/ct300701m PMID: 26605634

Fan, B., Jiang, Q., Dai, X., Xu, X., Wu, Y., & Peng, Q. (2026). Coordinating energy-level alignment and morphology in perovskite–organic heterostructures for efficient and stable solar cells. *Materials Science and Engineering R Reports*, *168*, 101165. DOI: 10.1016/j.mser.2025.101165

Fan, B., Yang, G., Li, M.-Y., Liu, P., Qiu, S., Shen, M., Liu, H., Zhang, G., Wang, Q., & Jiang, S. (2018). Lead-free Ba(1-x)SrxTiO3 ceramics for room-temperature pyroelectric energy conversion. *Ceramics International*, *44*(7), 8270–8276. DOI: 10.1016/j.ceramint.2018.02.009

Fan, B., Zeng, F., Chen, C., Yang, Y. C., Yang, P. Y., & Pan, F. (2009). Influence of strain and grain boundary variations on magnetism of Cr-doped AlN films. *Journal of Applied Physics*, *106*(7), 073907. Advance online publication. DOI: 10.1063/1.3234377

Fang, J., Owens, R. G., Tacher, L., & Parriaux, A. (2006). A numerical study of the SPH method for simulating transient viscoelastic free surface flows. *Journal of Non-Newtonian Fluid Mechanics*, *139*(1–2), 68–84. DOI: 10.1016/j.jnnfm.2006.07.004

Farah, K., Karimi-Varzaneh, H. A., Müller-Plathe, F., & Böhm, M. C. (2010). Reactive molecular dynamics with material-specific coarse-grained potentials: Growth of polystyrene chains from styrene monomers. *The Journal of Physical Chemistry B*, *114*(43), 13656–13666. DOI: 10.1021/jp107001e

Farah, K., Müller-Plathe, F., & Böhm, M. C. (2012). Classical reactive molecular dynamics implementations: State of the art. *ChemPhysChem*, *13*(5), 1127–1151. DOI: 10.1002/cphc.201100681

Faska, N., Auhmani, A., & Esseffar, M. (2004). Theoretical solvent effects on 13C NMR shielding of some aromatic hydrocarbons. Linear regression analysis. *Journal of Molecular Structure THEOCHEM*, *709*(1–3), 171–181. DOI: 10.1016/j.theochem.2004.05.038

Faska, N., Auhmani, A., & Esseffar, M. (2007). Solvent effects on 13C and 1H NMR shielding of cyclic ketones: An experimental and theoretical study. *Journal of Molecular Structure THEOCHEM*, *811*(1–3), 203–213. DOI: 10.1016/j.theochem.2006.12.057

Faska, N., Auhmani, A., Esseffar, M., & Abboud, J. L. (2011). Computational (solute–solvent cluster + PCM) study of medium effects on the experimental 13C and 1H NMR chemical shifts of lactones and lactams. *Journal of Physical Organic Chemistry*, *24*(12), 1209–1221. DOI: 10.1002/poc.1849

Fatehi, R., & Manzari, M. T. (2011). Error estimation in smoothed particle hydrodynamics and a new scheme for second derivatives. *Computers & Mathematics with Applications (Oxford, England)*, *61*(2), 482–498. DOI: 10.1016/j.camwa.2010.11.028

Fatimah, I., Purwiandono, G., Jauhari, M. H., Maharani, A. P., Sagadevan, S., Oh, W., & Doong, R. (2022). Synthesis and control of the morphology of SnO2 nanoparticles via various concentrations of Tinospora cordifolia stem extract and reduction methods. *Arabian Journal of Chemistry*, *15*(4), 103738. DOI: 10.1016/j.arabjc.2022.103738

Fawcett, T. (2006). An introduction to ROC analysis. *Pattern Recognition Letters*, *27*(8), 861–874. DOI: 10.1016/j.patrec.2005.10.010

Fergen, I., Seemann, K., Weth, A., & Schüppen, A. (2002). Soft ferromagnetic thin films for high frequency applications. *Journal of Magnetism and Magnetic Materials*, *242–245*, 146–151. DOI: 10.1016/S0304-8853(01)01185-4

Fermi, E. (1928). Eine statistische Methode zur Bestimmung einiger Eigenschaften des Atoms und ihre Anwendung auf die Theorie des periodischen Systems der Elemente. *The European Physical Journal A*, *48*(1–2), 73–79. DOI: 10.1007/BF01351576

Ferrari, B. S., Manica, M., Giro, R., Laino, T., & Steiner, M. B. (2024). Predicting polymerization reactions via transfer learning using chemical language models. *npj Computational Materials*, *10*(1), 119. DOI: 10.1038/s41524-024-01304-8

Feynman, R. P. (1939). Forces in Molecules. *Physical Review*, *56*(4), 340–343. DOI: 10.1103/PhysRev.56.340

Figueiredo, A. C., Barroso, J. G., Pedro, L. G., & Scheffer, J. J. C. (2008). Factors affecting secondary metabolite production in plants: Volatile components and essential oils. *Flavour and Fragrance Journal*, *23*(4), 213–226. DOI: 10.1002/ffj.1875

Figueirinhas, J. L., & Dong, R. Y. (2014). NMR of bent-core nematogens: A mini-review. *Magnetic Resonance in Chemistry*, *52*(10), 614–624. DOI: 10.1002/mrc.4097 PMID: 24975632

Filandr, F., Sarpe, V., Raval, S., Crowder, D. A., Khan, M. F., Douglas, P., Coales, S., Viner, R., Syed, A., Tainer, J. A., Lees-Miller, S. P., & Schriemer, D. C. (2024). Automating data analysis for hydrogen/deuterium exchange mass spectrometry using data-independent acquisition methodology. *Nature Communications*, *15*(1), 2200. DOI: 10.1038/s41467-024-46610-3 PMID: 38467655

Filip, M. R., Hillman, S., Haghighirad, A. A., Snaith, H. J., & Giustino, F. (2016). Band Gaps of the Lead-Free Halide Double Perovskites Cs2BiAgCl6 and Cs2BiAgBr6 from Theory and Experiment. *The Journal of Physical Chemistry Letters*, *7*(13), 2579–2585. DOI: 10.1021/acs.jpclett.6b01041

Firouz, M. S., Rashvand, M., & Omid, M. (2021). Rapid identification and quantification of sesame oils adulteration using low frequency dielectric spectroscopy combined with chemometrics. *Lebensmittel-Wissenschaft + Technologie*, *140*, 110736. DOI: 10.1016/j.lwt.2020.110736

Fish, J. (2006). Bridging the scales in nano engineering and science. *Journal of Nanoparticle Research*, *8*(5), 577–594. DOI: 10.1007/s11051-006-9090-9

Fish, J., & Chen, W. (2004). Discrete-to-continuum bridging based on multigrid principles. *Computer Methods in Applied Mechanics and Engineering*, *193*(17–20), 1693–1711. DOI: 10.1016/j.cma.2003.12.022

Fnidiki, A., Lemarchand, D., & Ben Youssef, J. (2010). Structural and magnetic properties of granular (Ni0.84Fe0.16)54(alumina)46 sputtered thin films. *Physica B, Condensed Matter*, *405*(12), 2632–2639. DOI: 10.1016/j.physb.2010.03.037

Fock, V. (1930). Näherungsmethode zur Lösung des quantenmechanischen Mehrkörperproblems. *The European Physical Journal A*, *61*(1–2), 126–148. DOI: 10.1007/BF01340294

Fogarty, J. C., Aktulga, H. M., Grama, A. Y., Van Duin, A. C. T., & Pandit, S. A. (2010). A reactive molecular dynamics simulation of the silica-water interface. *The Journal of Chemical Physics*, *132*(17), 174704. DOI: 10.1063/1.3407433

Franceschini, G., & Macchietto, S. (2008). Model-based design of experiments for parameter precision: State of the art. *Chemical Engineering Science*, *63*(19), 4846–4872. DOI: 10.1016/j.ces.2007.11.034

Frenkel, D., & Smit, B. (1996). Understanding molecular simulation: From algorithms to applications. In *Understanding molecular simulation: From algorithms to applications*. Academic Press, Incorporated., DOI: 10.1063/1.881812

Freysoldt, C., Grabowski, B., Hickel, T., Neugebauer, J., Kresse, G., Janotti, A., & Van De Walle, C. G. (2014). First-principles calculations for point defects in solids. *Reviews of Modern Physics*, *86*(1), 253–305. DOI: 10.1103/RevModPhys.86.253

Fritsch, S., Poblete, S., Junghans, C., Ciccotti, G., Delle Site, L., & Kremer, K. (2012). Adaptive resolution molecular dynamics simulation through coupling to an internal particle reservoir. *Physical Review Letters*, *108*(17), 170602. Advance online publication. DOI: 10.1103/PhysRevLett.108.170602

Fuchs, S. L., Meier, C., Wall, W. A., & Cyron, C. J. (2021a). An SPH framework for fluid–solid and contact interaction problems including thermo-mechanical coupling and reversible phase transitions. *Advanced Modeling and Simulation in Engineering Sciences*, *8*(1), 15. DOI: 10.1186/s40323-021-00200-w

Fu, H., Li, D., Peng, F., Gao, T., & Cheng, X. (2008). Ab initio calculations of elastic constants and thermodynamic properties of NiAl under high pressures. *Computational Materials Science*, *44*(2), 774–778. DOI: 10.1016/j.commatsci.2008.05.026

Gahlawat, S., Hopmann, K. H., & Castro, A. C. (2024). Advancing 19F NMR Prediction of Metal-Fluoride Complexes in Solution: Insights from Ab Initio Molecular Dynamics. *The Journal of Physical Chemistry A, 128*(49), 10498–10506. DOI: 10.1021/acs.jpca.4c05408 PMID: 39626023

Gaillac, R., Pullumbi, P., & Coudert, F.-X. (2016). ELATE: An open-source online application for analysis and visualization of elastic tensors. *Journal of Physics Condensed Matter, 28*(27), 275201. DOI: 10.1088/0953-8984/28/27/275201

Gan, J. L., Kumar, D., Chen, C., Taylor, B. C., Jagger, B. R., Amaro, R. E., & Lee, C. T. (2022). Benchmarking ensemble docking methods in D3R Grand Challenge 4. *Journal of Computer-Aided Molecular Design, 36*(2), 87–99. DOI: 10.1007/s10822-021-00433-2 PMID: 35199221

Gari, M. S., Narasaiah, B. P., Pandurengan, A., Mandal, B. K., & Natarajan, M. (2023). Synthesis and Antioxidant Activity of some novel 4H-Chromene Derivatives Catalysed by Biogenic Tin Oxide Nanoparticles. *Biointerface Research in Applied Chemistry, 13*(6), 521. DOI: 10.33263/BRIAC136.521

Garrafa-Galvez, H., Nava, O., Soto-Robles, C., Vilchis-Nestor, A., Castro-Beltrán, A., & Luque, P. (2019). Green synthesis of SnO2 nanoparticle using Lycopersicon esculentum peel extract. *Journal of Molecular Structure, 1197*, 354–360. DOI: 10.1016/j.molstruc.2019.07.052

Gartner, T. E.III, & Jayaraman, A. (2019). Modeling and simulations of polymers: A roadmap. *Macromolecules, 52*(3), 755–786. DOI: 10.1021/acs.macromol.8b01836

Garza, A. J., & Scuseria, G. E. (2016). Predicting Band Gaps with Hybrid Density Functionals. *The Journal of Physical Chemistry Letters, 7*(20), 4165–4170. DOI: 10.1021/acs.jpclett.6b01807

Gautam, S., Pathak, S., & Dubey, S. H. (2024). The role of molecular docking in modern drug discovery and development: A comprehensive review. *Journal of Drug Discovery and Health Sciences, 1*(03), 129–137. DOI: 10.21590/jddhs.01.03.02

Gear, C. W., Li, J., & Kevrekidis, I. G. (2003). The gap-tooth method in particle simulations. *Physics Letters. [Part A], 316*(3–4), 190–195. DOI: 10.1016/j.physleta.2003.07.004

Gebreslassie, Y. T., & Gebretnsae, H. G. (2021). Green and Cost-Effective Synthesis of Tin Oxide Nanoparticles: A review on the synthesis methodologies, mechanism of formation, and their potential applications. *Nanoscale Research Letters, 16*(1), 97. Advance online publication. DOI: 10.1186/s11671-021-03555-6

Gentile, F., Yaacoub, J. C., Gleave, J., Fernandez, M., Ton, A.-T., Ban, F., Stern, A., & Cherkasov, A. (2022). Artificial intelligence–enabled virtual screening of ultralarge chemical libraries with deep docking. *Nature Protocols, 17*(3), 672–697. DOI: 10.1038/s41596-021-00659-2 PMID: 35121854

Genuer, R., Poggi, J.-M., Tuleau-Malot, C., & Villa-Vialaneix, N. (2017). Random Forests for Big Data. *Big Data Research, 9*, 28–46. DOI: 10.1016/j.bdr.2017.07.003

Gerrits, M. L. (2006). Schneider, and TJ Silva. *Journal of Applied Physics, 99*, 023901. DOI: 10.1063/1.2159076

Ge, S., Yao, D., Yamaguchi, M., Yang, X., Zuo, H., Ishii, T., Zhou, D., & Li, F. (2007). Microstructure and magnetism of FeCo–SiO 2 nano-granular films for high frequency application. *Journal of Physics. D, Applied Physics*, *40*(12), 3660–3664. DOI: 10.1088/0022-3727/40/12/016

Getzlaff, M. (2007). *Fundamentals of magnetism*. Springer Science & Business Media.

Gholami, A., Klein, R., & Delle Site, L. (2022). Simulation of a particle domain in a continuum, fluctuating hydrodynamics reservoir. *Physical Review Letters*, *129*(23), 230603. DOI: 10.1103/PhysRevLett.129.230603

Ghoniem, N. M., Busso, E. P., Kioussis, N., & Huang, H. (2003). Multiscale modelling of nanomechanics and micromechanics: An overview. *Philosophical Magazine*, *83*(31–34), 3475–3528. DOI: 10.1080/14786430310001607388

Ghosh, S., Shankar, H., & Kar, P. (2022). Recent developments of lead-free halide double perovskites: A new superstar in the optoelectronic field. *Materials Advances*, *3*(9), 3742–3765. DOI: 10.1039/D2MA00071G

Gilbert, T. L. (1955). A Lagrangian formulation of the gyromagnetic equation of the magnetization field. *Physical Review*, *100*, 1243.

Gilbert, T. L. (2004). Classics in Magnetics A Phenomenological Theory of Damping in Ferromagnetic Materials. *IEEE Transactions on Magnetics*, *40*(6), 3443–3449. DOI: 10.1109/TMAG.2004.836740

Gilmore, K., Stiles, M. D., Seib, J., Steiauf, D., & Fähnle, M. (2010). Anisotropic damping of the magnetization dynamics in Ni, Co, and Fe. *Physical Review B: Condensed Matter and Materials Physics*, *81*(17), 174414. DOI: 10.1103/PhysRevB.81.174414

Gingold, R. A., & Monaghan, J. J. (1977). Smoothed particle hydrodynamics: Theory and application to non-spherical stars. *Monthly Notices of the Royal Astronomical Society*, *181*(3), 375–389. DOI: 10.1093/mnras/181.3.375

Giri, A., Park, G., & Jeong, U. (2023). Layer-Structured Anisotropic Metal Chalcogenides: Recent advances in synthesis, modulation, and applications. *Chemical Reviews*, *123*(7), 3329–3442. DOI: 10.1021/acs.chemrev.2c00455 PMID: 36719999

Gkekas, I., Katsamakas, S., Mylonas, S., Fotopoulou, T., Magoulas, G. E., Tenchiu, A. C., Dimitriou, M., Axenopoulos, A., Rossopoulou, N., Kostova, S., Wanker, E. E., Katsila, T., Papahatjis, D., Gorgoulis, V. G., Koufaki, M., Karakasiliotis, I., Calogeropoulou, T., Daras, P., & Petrakis, S. (2024). AI promoted virtual screening, structure-based hit optimization, and synthesis of novel COVID-19 S-RBD domain inhibitors. *Journal of Chemical Information and Modeling*, *64*(22), 8562–8585. DOI: 10.1021/acs.jcim.4c01110 PMID: 39535926

Glazer, A. M. (1972). The classification of tilted octahedra in perovskites. *Acta Crystallographica. Section B, Structural Crystallography and Crystal Chemistry*, *28*(11), 3384–3392. DOI: 10.1107/S0567740872007976

Glazer, A. M. (2002). **Perovskites modern and ancient**. By Roger H. Mitchell.Thunder Bay, Ontario : AlmazPress, 2002. Price USD 70.00. ISBN0-9689411-0-9. *Acta Crystallographica. Section B, Structural Science*, *58*(6), 1075–1075. DOI: 10.1107/S0108768102020220

Goldberg, D. E. (1989). *Genetic algorithms in search, optimization and machine learning*. Addison-Wesley Publishing Company.

Goldey, M., Dutoi, A., & Head-Gordon, M. (2013). Attenuated second-order Møller–Plesset perturbation theory: Performance within the aug-cc-pVTZ basis. *Physical Chemistry Chemical Physics*, *15*(38), 15869. DOI: 10.1039/c3cp51826d PMID: 23942866

Goldman, S., & Celinski, Z. (2023). Magnetic properties of (Ni0.81Fe0.19/SiO2)n multilayers for high frequency on-wafer inductor applications. *Journal of Magnetism and Magnetic Materials*, *569*, 170440. DOI: 10.1016/j.jmmm.2023.170440

Goldschmidt, V. M. (1926). The Laws of Crystallochemistry. *Naturwissenschaften*, *14*(21), 477–485. DOI: 10.1007/BF01507527

Goloviznina, K., Gong, Z., & Padua, A. H. (2021). The CL&Pol polarizable force field for the simulation of ionic liquids and eutectic solvents. *Wiley Interdisciplinary Reviews. Computational Molecular Science*, *12*(3), e1572. Advance online publication. DOI: 10.1002/wcms.1572

Goodfellow, I., Bengio, Y., & Courville, A. (2016). *Deep learning* (Vol. 1). MIT Press.

Gräfenstein, J., & Cremer, D. (2009). The self-interaction error and the description of non-dynamic electron correlation in density functional theory. *Theoretical Chemistry Accounts*, *123*(3–4), 171–182. DOI: 10.1007/s00214-009-0545-9

Gray, M. B., Hariyani, S., Strom, T. A., Majher, J. D., Brgoch, J., & Woodward, P. M. (2020). High-efficiency blue photoluminescence in the Cs2NaInCl6:Sb3+ double perovskite phosphor. *Journal of Materials Chemistry. C, Materials for Optical and Electronic Devices*, *8*(20), 6797–6803. DOI: 10.1039/D0TC01037E

Green, H. S., Li, X., De Pra, M., Lovejoy, K. S., Steiner, F., Acworth, I. N., & Wang, S. C. (2020). A rapid method for the detection of extra virgin olive oil adulteration using UHPLC-CAD profiling of triacylglycerols and PCA. *Food Control*, *107*, 106773. DOI: 10.1016/j.foodcont.2019.106773

Green, M. A. (1981). Solar cell fill factors: General graph and empirical expressions. *Solid-State Electronics*, *24*(8), 788–789. DOI: 10.1016/0038-1101(81)90062-9

Green, M. A. (2001). Third generation photovoltaics: Ultra-high conversion efficiency at low cost. *Progress in Photovoltaics: Research and Applications*, *9*(2), 123–135. DOI: 10.1002/pip.360

Green, M. A., Ho-Baillie, A., & Snaith, H. J. (2014). The emergence of perovskite solar cells. *Nature Photonics*, *8*(7), 506–514. DOI: 10.1038/nphoton.2014.134

Grimblat, N., & Sarotti, A. M. (2016). Computational Chemistry to the Rescue: Modern Toolboxes for the assignment of complex molecules by GIAO NMR Calculations. *Chemistry (Weinheim an der Bergstrasse, Germany)*, *22*(35), 12246–12261. DOI: 10.1002/chem.201601150 PMID: 27405775

Grimme, S., Antony, J., Ehrlich, S., & Krieg, H. (2010). A consistent and accurate ab initio parametrization of density functional dispersion correction (DFT-D) for the 94 elements H-Pu. *The Journal of Chemical Physics, 132*(15), 154104. DOI: 10.1063/1.3382344 PMID: 20423165

Guan, Y., Shree Sowndarya, S. V., Gallegos, L. C., St. John, P. C., & Paton, R. S. (2021). Real-time prediction of ^{1}H and ^{13}C chemical shifts with DFT accuracy using a 3D graph neural network. *Chemical Science (Cambridge), 12*(36), 12012–12026. DOI: 10.1039/D1SC03343C PMID: 34667567

Guechi Nacir. (2017). *Élaboration et caractérisations structurale, électrique et magnétique des couches minces de FexNi100-x déposées sur Si (100).* Dépôt Institutionnel de l'Université Ferhat ABBAS.

Guedes, I. A., Pereira, F. S. S., & Dardenne, L. E. (2018). Empirical scoring functions for structure-based virtual screening: Applications, critical aspects, and challenges. *Frontiers in Pharmacology, 9*, 1089. DOI: 10.3389/fphar.2018.01089 PMID: 30319422

Gurevich, A. G., & Melkov, G. A. (2020). *Magnetization oscillations and waves.* CRC press. DOI: 10.1201/9780138748487

Gyftokostas, N., Nanou, E., Stefas, D., Kokkinos, V., Bouras, C., & Couris, S. (2021). Classification of Greek olive oils from different regions by machine learning-aided laser-induced breakdown spectroscopy and absorption spectroscopy. *Molecules (Basel, Switzerland), 26*(5), 1241. DOI: 10.3390/molecules26051241 PMID: 33669128

Haddadi-Asl, V., Abdollahi, E., Rostamipoor, H., Shahrasari, A., & Narjili, S. (2025). Multiscale simulation of gas-phase propylene polymerization initiated by $MgCl_2$-supported Ziegler-Natta catalysts: A reactive molecular dynamics and smoothed particle hydrodynamics approach. *Journal of Polymer Research, 32*(7), 1–21.

Hadjiconstantinou, N. G. (1999). Hybrid Atomistic-Continuum Formulations and the Moving Contact-Line Problem. In Journal of Computational Physics (Vol. 154). http://www.idealibrary.comon

Hamedani, N. F., & Farzaneh, F. (2006). Synthesis of ZnO Nanocrystals with Hexagonal (Wurtzite) Structure in Water Using Microwave Irradiation. *Journal of Science, 17*(3), 231–234. https://journals.ut.ac.ir/article_31762_e1fc4d21c2aa63931c4f50ef9758cd52.pdf

Hamideddine, I., Tahiri, N., Bounagui, O. E., & Ez-Zahraouy, H. (2022). Ab initio study of structural and optical properties of the halide perovskite KBX3 compound. *Journal of the Korean Ceramic Society, 59*(3), 350–358. DOI: 10.1007/s43207-021-00178-6

Hamler, A., Goričan, V., Šuštaršič, B., & Sirc, A. (2006). The use of soft magnetic composite materials in synchronous electric motor. *Journal of Magnetism and Magnetic Materials, 304*(2), e816–e819. DOI: 10.1016/j.jmmm.2006.03.003

Hamuro, Y. (2024). Interpretation of hydrogen/deuterium exchange mass spectrometry. *Journal of the American Society for Mass Spectrometry, 35*(5), 819–828. DOI: 10.1021/jasms.4c00044 PMID: 38639434

Han, X., & Guo, K. (2024). Pharmacophore-based virtual screening of commercial databases against β-secretase 1 for drug development against Alzheimer's disease. *Frontiers in Chemistry, 12*, 1412349. DOI: 10.3389/fchem.2024.1412349 PMID: 39045333

Hao, F., Stoumpos, C. C., Cao, D. H., Chang, R. P. H., & Kanatzidis, M. G. (2014). Lead-free solid-state organic–inorganic halide perovskite solar cells. Nature Photonics 2014 8:6, 8(6), 489–494. DOI: 10.1038/nphoton.2014.82

Hao, J., Li, W., Zhai, J., & Chen, H. (2019). Progress in high-strain perovskite piezoelectric ceramics. *Materials Science and Engineering R Reports*, *135*, 1–57. DOI: 10.1016/j.mser.2018.08.001

Harris, C. R., Millman, K. J., van der Walt, S. J., Gommers, R., Virtanen, P., Cournapeau, D., Wieser, E., Taylor, J., Berg, S., Smith, N. J., Kern, R., Picus, M., Hoyer, S., van Kerkwijk, M. H., Brett, M., Haldane, A., Fernández del Río, J., Wiebe, M., Peterson, P., & Oliphant, T. E. (2020). Array programming with NumPy. *Nature*, *585*(7825), 357–362. DOI: 10.1038/s41586-020-2649-2 PMID: 32939066

Harrou, F., Dairi, A., Dorbane, A., & Sun, Y. (2024). Enhancing wind power prediction with self-attentive variational autoencoders: A comparative study. *Results in Engineering*, *23*(May), 102504. DOI: 10.1016/j.rineng.2024.102504

Hartree, D. R. (1928). The Wave Mechanics of an Atom with a Non-Coulomb Central Field. Part I. Theory and Methods. *Mathematical Proceedings of the Cambridge Philosophical Society*, *24*(1), 89–110. DOI: 10.1017/S0305004100011919

Hasan, Z., Rahman, M. A., Das, D. K., & Rouf, H. K. (s. d.). OPEN Influence of Ca doping in structural, electronic, optical and mechanical properties of B a1−xCaxTiO3 perovskite from first-principles investigation. Scientific Reports.

Hasan, M., & Akther Hossain, A. K. M. (2021a). Structural, electronic and optical properties of strontium and nickel co-doped BaTiO3 : A DFT based study. *Computational Condensed Matter*, *28*, e00578. DOI: 10.1016/j.cocom.2021.e00578

Hattabi, I., Abdiche, A., Naqib, S. H., & Khenata, R. (2019). First-principles calculations of elastic and thermodynamic properties under hydrostatic pressure of cubic InNxP1-x ternary alloys. *Zhongguo Wuli Xuekan*, *59*, 449–464. DOI: 10.1016/j.cjph.2019.04.011

Hatzikraniotis, E. S., Polymeris, G., & Kyratsi, T. (2024). Thermal Conductivity in Thermoelectric Materials. In R. Vargas-Bernal & R. Palma Guerrero (Éds.), Novel Applications of Piezoelectric and Thermoelectric Materials. IntechOpen. DOI: 10.5772/intechopen.106168

Heim, K. E., Tagliaferro, A. R., & Bobilya, D. J. (2002). Flavonoid antioxidants: Chemistry, metabolism and structure-activity relationships. *The Journal of Nutritional Biochemistry*, *13*(10), 572–584. DOI: 10.1016/S0955-2863(02)00208-5

Heinrich, M., Jalil, B., Abdel-Tawab, M., Echeverria, J., Kulić, Ž., McGaw, L. J., Pezzuto, J. M., Potterat, O., & Wang, J.-B. (2022). Best practice in the chemical characterisation of extracts used in pharmacological and toxicological research—The ConPhyMP—Guidelines12. *Frontiers in Pharmacology*, *13*, 953205. DOI: 10.3389/fphar.2022.953205

Hermann, J., Lorusso, A., Perrone, A., Strafella, F., Dutouquet, C., & Torralba, B. (2015). Simulation of emission spectra from nonuniform reactive laser-induced plasmas. *Physical Review E: Statistical, Nonlinear, and Soft Matter Physics*, *92*(5), 053103. DOI: 10.1103/PhysRevE.92.053103

Hernandes, I. S., Da Silva, H. C., Santos, H. F. D., Ávila, E. P., De Almeida, M. V., & De Almeida, W. B. (2023). Quantum chemical investigation of predominant conformation of the antibiotic azithromycin in water and DMSO solutions: Thermodynamic and NMR analysis. *Royal Society Open Science*, *10*(10), 230409. DOI: 10.1098/rsos.230409 PMID: 37830015

Herrera-Arozamena, C., Estrada-Valencia, M., García-Díez, G., Pérez, C., León, R., Infantes, L., Morales-García, J. A., Pérez-Castillo, A., del Sastre, E., López, M. G., & Rodríguez-Franco, M. I. (2024). Discovery of a potent melatonin-based inhibitor of quinone reductase-2 with neuroprotective and neurogenic properties. *European Journal of Medicinal Chemistry*, *277*, 116763. DOI: 10.1016/j.ejmech.2024.116763 PMID: 39146834

He, X., Wang, B., & Merz, K. M.Jr. (2009). Protein NMR chemical shift calculations based on the automated fragmentation QM/MM approach. *The Journal of Physical Chemistry B*, *113*(31), 10380–10388. DOI: 10.1021/jp901992p PMID: 19575540

Heyd, J., Scuseria, G. E., & Ernzerhof, M. (2003). Hybrid functionals based on a screened Coulomb potential. *The Journal of Chemical Physics*, *118*(18), 8207–8215. DOI: 10.1063/1.1564060

Hidayat, E., Hannani Hamid, N. A., Mohamad Sarbani, N. M., Samitsu, S., Aoyagi, M., Harada, H., & Safari, M. A. M. (2025a). Optimization and comparative modelling of RSM and ANN for the adsorptive removal of Remazol Brilliant Blue R dye using spent coffee ground biochar. *Chemosphere*, *389*(June), 144709. DOI: 10.1016/j.chemosphere.2025.144709 PMID: 41033296

Hoffer, J. G., Geiger, B. C., & Kern, R. (2022). Gaussian process surrogates for modeling uncertainties in a use case of forging superalloys. *Applied Sciences (Basel, Switzerland)*, *12*(3), 1089. DOI: 10.3390/app12031089

Hoffmann, H. (1979). Magnetic properties of thin ferromagnetic films in relation to their structure. *Thin Solid Films*, *58*(2), 223–233. DOI: 10.1016/0040-6090(79)90241-4

Hoffmann, I. (2010). Climate change and the characterization, breeding and conservation of animal genetic resources. *Animal Genetics*, *41*(s1, Suppl. 1), 32–46. DOI: 10.1111/j.1365-2052.2010.02043.x

Hohenberg, P., & Kohn, W. (1964). Inhomogeneous Electron Gas. *Physical Review*, *136*(3B), B864–B871. DOI: 10.1103/PhysRev.136.B864

Hollingsworth, S. A., & Dror, R. O. (2018). Molecular dynamics simulation for all. *Neuron*, *99*(6), 1129–1143. DOI: 10.1016/j.neuron.2018.08.011 PMID: 30236283

Hou, Y., Qi, J., Hu, J., Xiang, Y., Xin, L., & Wei, X. (2020). Mesoporous Mn-Doped Fe Nanoparticle-Modified Reduced Graphene Oxide for Ethyl Violet Elimination: Modeling and Optimization Using Artificial Intelligence. *Processes (Basel, Switzerland)*, *8*(4), 488. DOI: 10.3390/pr8040488

Huang, M., Zhang, X. S., Bhatti, U. A., Wu, Y., Zhang, Y., & Yasin Ghadi, Y. (2024). An interpretable approach using hybrid graph networks and explainable AI for intelligent diagnosis recommendations in chronic disease care. *Biomedical Signal Processing and Control*, *91*, 105913. DOI: 10.1016/j.bspc.2023.105913

Hubin, P. O., Jacquemin, D., Leherte, L., & Vercauteren, D. P. (2016). Parameterization of the ReaxFF reactive force field for a proline-catalyzed aldol reaction. *Journal of Computational Chemistry*, *37*(29), 2564–2572. DOI: 10.1002/jcc.24481

Huffer, J. M. (1977). Special aspects of primary care. *Postgraduate Medicine*, *62*(5), 223–228. DOI: 10.1080/00325481.1977.11714687

Hui, R., & O'sullivan, M. (2023). Fundamentals of optical devices. In *Fiber-Optic Measurement Techniques* (pp. 1–135). Elsevier., DOI: 10.1016/B978-0-323-90957-0.00002-3

Huitema, L., Reveyrand, T., Mattei, J.-L., Arnaud, E., Decroze, C., & Monediere, T. (2013). Frequency Tunable Antenna Using a Magneto-Dielectric Material for DVB-H Application. *IEEE Transactions on Antennas and Propagation*, *61*(9), 4456–4466. DOI: 10.1109/TAP.2013.2269474

Hu, J., Fan, H., Wu, S., Tang, L., Qin, L., & Luo, W. (2022). Characterization of temperature dependence of dielectric, elastic and piezoelectric properties of BaTiO3 ceramics. *Ceramics International*, *48*(18), 25741–25746. DOI: 10.1016/j.ceramint.2022.05.245

Hung, C., Chih, C., Huang, K., Xue, Y., Chu, H., Tseng, C., Li, C., Chen, J., Chen, B., Yang, S., Cheng, Y., & Chou, P. (2025). Perovskite-Coupled NIR Organic Hybrid Solar Cells Achieving an 84.2% Fill Factor and a 25.2% Efficiency: A Comprehensive Mechanistic Exploration. *Angewandte Chemie International Edition*, *64*(21), e202501375. Advance online publication. DOI: 10.1002/anie.202501375

Hussain, S., Aftab, S., Muneer, S., Rehman, J. U., Tahir, M. B., Tirth, V., & Algahtani, A. (2025). Computational predictions of physical properties of perovskite CsGeR3 (R = F, Cl, Br, and I) materials for optoelectronic applications. *Inorganic Chemistry Communications*, *173*, 113811. DOI: 10.1016/j.inoche.2024.113811

Hu, X. Y., & Adams, N. A. (2006). A multi-phase SPH method for macroscopic and mesoscopic flows. *Journal of Computational Physics*, *213*(2), 844–861. DOI: 10.1016/j.jcp.2005.09.001

Hwang, I., Jeong, I., Lee, J., Ko, M. J., & Yong, K. (2015). Enhancing Stability of Perovskite Solar Cells to Moisture by the Facile Hydrophobic Passivation. *ACS Applied Materials & Interfaces*, *7*(31), 17330–17336. DOI: 10.1021/acsami.5b04490

Ibrahim, R. S., Mahrous, R. S. R., Fathy, H. M., Omar, A. A., & Abu El-Khair, R. M. (2020). Anticoagulant activity screening of an in-house database of natural compounds for discovering novel selective factor Xa inhibitors; a combined in silico and in vitro approach. *Medicinal Chemistry Research*, *29*(4), 707–726. DOI: 10.1007/s00044-020-02516-5

Iron, M. A. (2017). Evaluation of the Factors Impacting the Accuracy of 13C NMR Chemical Shift Predictions using Density Functional Theory: The Advantage of Long-Range Corrected Functionals. *Journal of Chemical Theory and Computation*, *13*(11), 5798–5819. DOI: 10.1021/acs.jctc.7b00772 PMID: 29016125

Islam, M. A., Khan, S., Mim, J. J., Rahman, S. M. M., Patwary, M. I., Islam, M. S., & Hossain, N. (2025). Recent advances of 2D materials in semiconductor application: A review. *Advanced Sensor and Energy Materials*, *4*(4), 100161. DOI: 10.1016/j.asems.2025.100161

Iype, E., Hütter, M., Jansen, A. P. J., Nedea, S. V., & Rindt, C. C. M. (2013). Parameterization of a reactive force field using a Monte Carlo algorithm. *Journal of Computational Chemistry*, *34*(13), 1143–1154. DOI: 10.1002/jcc.23246

Jain, M., Chelikowsky, J. R., & Louie, S. G. (2011). Reliability of hybrid functionals in predicting band gaps. *Physical Review Letters*, *107*(21), 216806. DOI: 10.1103/PhysRevLett.107.216806 PMID: 22181911

Jaji, N., Lee, H. L., Hussin, M. H., Akil, H. M., Zakaria, M. R., & Othman, M. B. H. (2020). Advanced nickel nanoparticles technology: From synthesis to applications. *Nanotechnology Reviews*, *9*(1), 1456–1480. DOI: 10.1515/ntrev-2020-0109

Jamal, M., Bilal, M., Ahmad, I., & Jalali-Asadabadi, S. (2018). IRelast package. *Journal of Alloys and Compounds*, *735*, 569–579. DOI: 10.1016/j.jallcom.2017.10.139

Jamaluddin, A., Suwarni, , Supriyanto, A., & Iriani, Y. (2016). Properties of strontium doped barium titanate powder prepared by solid state reaction. *Journal of Physics: Conference Series*, *776*, 012052. DOI: 10.1088/1742-6596/776/1/012052

Jameel, M. H., Rehman, A., Bin Roslan, M. S., & Agam, M. A. B. (2023). To investigate the structural, electronic, optical and magnetic properties of Sr-doped $KNbO_3$ for perovskite solar cell applications : A DFT study. *Physica Scripta*, *98*(5), 055802. DOI: 10.1088/1402-4896/acc6fb

Jamwal, R., Amit, , Kumari, S., Balan, B., Kelly, S., Cannavan, A., & Singh, D. K. (2021). Rapid and non-destructive approach for the detection of fried mustard oil adulteration in pure mustard oil via ATR-FTIR spectroscopy-chemometrics. *Spectrochimica Acta. Part A: Molecular and Biomolecular Spectroscopy*, *244*, 118822. DOI: 10.1016/j.saa.2020.118822 PMID: 32829154

Jamwal, R., Kumari, S., Dhaulaniya, A. S., Balan, B., Kelly, S., Cannavan, A., & Singh, D. K. (2020). Utilizing ATR-FTIR spectroscopy combined with multivariate chemometric modelling for the swift detection of mustard oil adulteration in virgin coconut oil. *Vibrational Spectroscopy*, *109*, 103066. DOI: 10.1016/j.vibspec.2020.103066

Janesko, B. G., Henderson, T. M., & Scuseria, G. E. (2009). Screened hybrid density functionals for solid-state chemistry and physics. *Physical Chemistry Chemical Physics*, *11*(3), 443–454. DOI: 10.1039/B812838C

Janga, K., & Rao, G. (2025). Bioresource Technology Reports Effect of residence time and temperature on pyrolysis of Water Hyacinth Leaf Biomass (WHLB): Experiments and modelling through ML and validation by Leave One Out Cross Validation (LOOCV). *Bioresource Technology Reports*, *32*(August), 102369. DOI: 10.1016/j.biteb.2025.102369

Jang, C., Yadav, D. K., Subedi, L., Venkatesan, R., Venkanna, A., Afzal, S., Lee, E., Yoo, J., Ji, E., Kim, S. Y., & Kim, M. (2018). Identification of novel acetylcholinesterase inhibitors designed by pharmacophore-based virtual screening, molecular docking and bioassay. *Scientific Reports*, *8*(1), 14921. DOI: 10.1038/s41598-018-33354-6 PMID: 30297729

Jayalakshmi, D. S., & Nivetha, B. (2023). First-principles calculations for comparative band structure study of SrTiO3 perovskite on bulk and layered phases for efficient optoelectronic conversion. *Computational & Theoretical Chemistry*, *1220*, 114006. DOI: 10.1016/j.comptc.2022.114006

Jedema, F. J., Filip, A. T., & van Wees, B. J. (2001). Electrical spin injection and accumulation at room temperature in an all-metal mesoscopic spin valve. *Nature*, *410*(6826), 345–348. DOI: 10.1038/35066533

Jeevaraj, M., Sivaganesh, D., Saravanakumar, S., Bahadur, S. A., Sudhahar, S., & Kumar, M. K. (2024). Extrinsic electronic states to tune the luminescence and bonding nature of Cs2NaInCl6 double perovskite. Materials Chemistry and Physics, 311(June 2023), 128569. DOI: 10.1016/j.matchemphys.2023.128569

Jeffery, S., Verheijen, F. G. A., van der Velde, M., & Bastos, A. C. (2011). A quantitative review of the effects of biochar application to soils on crop productivity using meta-analysis. *Agriculture, Ecosystems & Environment*, *144*(1), 175–187. DOI: 10.1016/j.agee.2011.08.015

Jiang, S., & Webb, M. (2024). Physics-Guided Neural Networks for Transferable Prediction of Polymer Properties.

Jiang, C., Xue, D., & Sui, W. (2011). Broadband microwave absorption in [NiFe/FeMn]n exchange-coupled multilayer films. *Thin Solid Films*, *519*(8), 2527–2530. DOI: 10.1016/j.tsf.2010.11.030

Jiang, D., Tan, V. G. W., Gong, Y., Shao, H., Mu, X., Luo, Z., & He, S. (2025). Semiconducting covalent organic frameworks. *Chemical Reviews*, *125*(13), 6203–6308. DOI: 10.1021/acs.chemrev.4c00950 PMID: 40366230

Jiang, X., Qin, S., Meng, L., He, G., Zhang, J., Wang, Y., Zhu, Y., Zou, T., Gong, Y., Chen, Z., Sun, G., Liu, M., Li, X., Lang, F., & Li, Y. (2024). Isomeric diammonium passivation for perovskite–organic tandem solar cells. *Nature*, *635*(8040), 860–866. DOI: 10.1038/s41586-024-08160-y

Jiao, M., Jacquemin, J., Zhang, R., Zhao, N., & Liu, H. (2023). The Prediction of Cu(II) Adsorption Capacity of Modified Pomelo Peels Using the PSO-ANN Model. *Molecules (Basel, Switzerland)*, *28*(19), 6957. DOI: 10.3390/molecules28196957 PMID: 37836799

Jia, Z., Zhao, M., Chen, Q., Tian, Y., Liu, L., Zhang, F., Zhang, D., Ji, Y., Camargo, B., Ye, K., Sun, R., Wang, Z., & Jiang, Y. (2025). Spintronic Devices upon 2D Magnetic Materials and Heterojunctions. *ACS Nano*, *19*(10), 9452–9483. DOI: 10.1021/acsnano.4c14168 PMID: 40053908

Ji, J. Y., Tan, Z., Zeng, S., & Wong, M. L. (2024). An ε-constrained multiobjective differential evolution with adaptive gradient-based repair method for real-world constrained optimization problems. *Applied Soft Computing*, *152*, 111202. DOI: 10.1016/j.asoc.2023.111202

Jiles, D. (2015). *Introduction to magnetism and magnetic materials*. CRC press. DOI: 10.1201/b18948

Jiles, D. C. (2003). Recent advances and future directions in magnetic materials. *Acta Materialia*, *51*(19), 5907–5939. DOI: 10.1016/j.actamat.2003.08.011

Jiles, D. C. (2017). *Introduction to the electronic properties of materials*. CRC Press.

Johnsson, M., & Lemmens, P. (2007). Crystallography and Chemistry of Perovskites. In H. Kronmüller & S. Parkin (Éds.), Handbook of Magnetism and Advanced Magnetic Materials (1re éd.). Wiley. DOI: 10.1002/9780470022184.hmm411

Jorge, A. O. S., Costa, A. S. G., & Oliveira, M. B. P. P. (2023). Adapting to climate change with Opuntia. *Plants*, *12*(16), 2907. DOI: 10.3390/plants12162907

Jorn, R., & Voth, G. A. (2012). Mesoscale simulation of proton transport in proton exchange membranes. *The Journal of Physical Chemistry. C, Nanomaterials and Interfaces*, *116*(19), 10476–10489. DOI: 10.1021/jp300040w

Jose, K. V. J., & Raghavachari, K. (2017). Fragment-Based approach for the evaluation of NMR chemical shifts for large biomolecules incorporating the effects of the solvent environment. *Journal of Chemical Theory and Computation*, *13*(3), 1147–1158. DOI: 10.1021/acs.jctc.6b00922 PMID: 28194972

Jug, A., Bratkovič, T., & Ilaš, J. (2024). Biolayer interferometry and its applications in drug discovery and development. *Trends in Analytical Chemistry*, *176*, 117741. DOI: 10.1016/j.trac.2024.117741

Junghans, C., Praprotnik, M., & Kremer, K. (2007). Transport properties controlled by a thermostat: An extended dissipative particle dynamics thermostat. *Soft Matter*, *4*(1), 156–161. DOI: 10.1039/B713568H

K., P. M., B., B., T.P., V., T.V., R. M., & M., V. G. B. (2024). Experimental and first-principles investigation of structural, electronic, optical, elastic and ferroelectric properties of BaTiO 3. Physica B: Condensed Matter, 674, 415548. DOI: 10.1016/j.physb.2023.415548

Kaak, A. R. S., Çelebioğlu, K., Bozkuş, Z., Ulucak, O., & Aylı, E. (2024). A novel CFD-ANN approach for plunger valve optimization: Cost-effective performance enhancement. *Flow Measurement and Instrumentation*, *97*, 102589. DOI: 10.1016/j.flowmeasinst.2024.102589

Kacimi-Naciri, H., Rguiti, M., Mabrouk, A., Ben Achour, M. A., Lorgouilloux, Y., Courtois, C., Faska, N., & Bachar, A. (2024). Perovskite Progress : The Role of Sr-Doping in Advancing BaTiO3-Based Technologies. In Amrousse, R., Bachar, A., Mabrouk, A., & Azat, S. (Eds.), (pp. 201–246). Advances in Chemical and Materials Engineering. IGI Global., DOI: 10.4018/979-8-3373-0669-8.ch006

Kacimi-Naciri, H., Rguiti, M., Mabrouk, A., Courtois, C., Ben Achour, M. A., Lorgouilloux, Y., Amrousse, R., Faska, N., & Bachar, A. (2025). DFT-based and experimental study on Sr-doped BaTiO3 : Impacts on piezoelectric and ferroelectric performance. *Ceramics International*, *51*(17), 23801–23813. DOI: 10.1016/j.ceramint.2025.03.069

Kalarickal, S. S., Krivosik, P., Das, J., Kim, K. S., & Patton, C. E. (2008). Microwave damping in polycrystalline Fe-Ti-N films: Physical mechanisms and correlations with composition and structure. *Physical Review B: Condensed Matter and Materials Physics*, *77*(5), 054427. DOI: 10.1103/PhysRevB.77.054427

Kalateh, F., & Koosheh, A. (2020). Simulation of cavitating fluid–Structure interaction using SPH–FE method. *Mathematics and Computers in Simulation*, *173*, 51–70. DOI: 10.1016/j.matcom.2020.01.019

Kale, A. J., Chaurasiya, R., & Dixit, A. (2022). Lead-Free Cs 2 BB′X 6 (B: Ag/Au/Cu, B′: Bi/Sb/Tl, and X: Br/Cl/I) Double Perovskites and Their Potential in Energy Conversion Applications. *ACS Applied Energy Materials*, *5*(9), 10427–10445. DOI: 10.1021/acsaem.2c00672

Kaplanskiy, M. V., Shitov, D. A., Tolstoy, P. M., & Tupikina, E. Y. (2025). Deconstructing 1H NMR chemical shifts in strong hydrogen bonds: A computational investigation of solvation, dynamics, and nuclear delocalization effects. *Journal of Chemical Information and Modeling*, *65*(10), 5019–5034. DOI: 10.1021/acs.jcim.5c00566 PMID: 40350596

Karami, H., Rasekh, M., & Mirzaee–Ghaleh, E. (2020). Comparison of chemometrics and AOCS official methods for predicting the shelf life of edible oil. *Chemometrics and Intelligent Laboratory Systems, 206*, 104165. DOI: 10.1016/j.chemolab.2020.104165

Kariminejad, M., Tormey, D., Huq, S., Morrison, J., & McAfee, M. (2021). Ultrasound sensors for process monitoring in injection moulding. *Sensors (Basel), 21*(15), 5193. DOI: 10.3390/s21155193

Karimi, S., & Nakshatrala, K. B. (2015). A monolithic multi-time-step computational framework for first-order transient systems with disparate scales. *Computer Methods in Applied Mechanics and Engineering, 283*, 419–453. DOI: 10.1016/j.cma.2014.10.003

Karimi-Varzaneh, H. A., Van Der Vegt, N. F. A., Müller-Plathe, F., & Carbone, P. (2012). How good are coarse-grained polymer models? A comparison for atactic polystyrene. In ChemPhysChem (Vol. 13, Issue 15, pp. 3428–3439). Wiley-VCH Verlag. DOI: 10.1002/cphc.201200111

Karmakar, T., Finney, A. R., Salvalaglio, M., Yazaydin, A. O., & Perego, C. (2023). Non-Equilibrium Modeling of Concentration-Driven processes with Constant Chemical Potential Molecular Dynamics Simulations. *Accounts of Chemical Research, 56*(10), 1156–1167. DOI: 10.1021/acs.accounts.2c00811 PMID: 37120847

Karmaoui, M., Jorge, A. B., McMillan, P. F., Aliev, A. E., Pullar, R. C., Labrincha, J. A., & Tobaldi, D. M. (2018). One-Step synthesis, structure, and band gap properties of SNO2 nanoparticles made by a low temperature nonaqueous Sol–Gel technique. *ACS Omega, 3*(10), 13227–13238. DOI: 10.1021/acsomega.8b02122

Kattimani, P., Prasad, K. V., Anwar, T., Jena, S. P., Shankhyan, A., & Naveen Kumar, R. (2025). Application of Levenberg-Marquardt artificial neural network to study nanoparticle aggregation phenomena in stagnation point flow towards an off-centered rotating disk. *International Journal of Thermofluids, 27*, 101276. DOI: 10.1016/j.ijft.2025.101276

Kawabata, K., Mashimo, K., & Takimiya, K. (2024). Core- versus End-Alkylation: Tailoring Solid-State Structures and Properties of Near-Infrared-Absorbing Organic Semiconductors Based on Naphthodithiophenediones. *Chemistry of Materials, 36*(24), 11920–11933. DOI: 10.1021/acs.chemmater.4c02436

Kawagoe, Y., & Okabe, T. (2023). Evaluations of atomic-resolution strain fields using molecular dynamics simulations combined with corrected smoothed particle hydrodynamics. *Computational Materials Science, 228*, 112333. DOI: 10.1016/j.commatsci.2023.112333

Kayal, R., Baldinelli, L., Harden, I., Neese, F., & Bistoni, G. (2024). Understanding and quantifying the impact of solute–solvent van der Waals interactions on the selectivity of asymmetric catalytic transformations. *Chemical Science (Cambridge), 16*(6), 2700–2709. DOI: 10.1039/D4SC04329D PMID: 39802687

Khalal, A., Khatib, D., & Jannot, B. (1999). Elastic and piezoelectric properties of BaTiO3 at room temperature. *Physica B, Condensed Matter, 271*(1), 343–347. DOI: 10.1016/S0921-4526(99)00202-1

Khan, A., Al-hazmi, H. E., Agustiono, T., Majtacz, J., Piechota, G., Kumar, G., Ezzati, P., Saeb, R., Rabiee, N., Karimi-maleh, H., Claudio, E., & Jacek, M. (2024). *Hydrochar as a bio-based adsorbent for heavy metals removal : A review of production processes, adsorption mechanisms, kinetic models, regeneration and reusability. 945*(April). DOI: 10.1016/j.scitotenv.2024.173972

Khandelwal, P., Ahlawat, N., Kundu, R. S., Sangwan, K. M., & Bajia, S. (2019). Influence of Sr-doping on structural and ferroelectric properties of BaTiO3 ceramics. 030599. DOI: 10.1063/1.5113438

Khan, S. Y., Rauf, S., Liu, S., Chen, W., Shen, Y., & Kumar, M. (2025). Revolutionizing the solar photovoltaic efficiency: A comprehensive review on the cutting-edge thermal management methods for advanced and conventional solar photovoltaics. *Energy & Environmental Science*, *18*(3), 1130–1175. DOI: 10.1039/D4EE03525A

Kharazmi, A., Student, M., & Kamali, R. (n.d.). A Continuum: Molecular Dynamics Hybrid Method for Micro- and Nano-Fluid Flow. http://www.asme.or

Khaw, K.-Y., Parat, M.-O., Shaw, P. N., & Falconer, J. R. (2017). Solvent supercritical fluid technologies to extract bioactive compounds from natural sources: A review. *Molecules (Basel, Switzerland)*, *22*(7), 1186. DOI: 10.3390/molecules22071186

Khorramifar, A., Karami, H., Wilson, A. D., Sayyah, A. H. A., Shuba, A., & Lozano, J. (2022). Grape cultivar identification and classification by machine olfaction analysis of leaf volatiles. *Chemosensors (Basel, Switzerland)*, *10*(4), 125. DOI: 10.3390/chemosensors10040125

Khristenko, U., Constantinescu, A., Le Tallec, P., & Wohlmuth, B. (2022). Statistically equivalent surrogate material models: Impact of random imperfections on the elasto-plastic response. *Computer Methods in Applied Mechanics and Engineering*, *402*, 115278. DOI: 10.1016/j.cma.2022.115278

Kibbou, M., Haman, Z., Essaoudi, I., & Ainane, A. (2023). Designing new halide double perovskite materials Rb2AgGaX6 (X: Br, Cl) with direct band gaps and high power conversion efficiency. *Journal of Solid State Chemistry*, *317*, 123698. DOI: 10.1016/j.jssc.2022.123698

Kim, J., & Suh, Y. J. (2017). Temperature- and pressure-dependent elastic properties, thermal expansion ratios, and minimum thermal conductivities of ZrC, ZrN, and Zr(C0.5N0.5). *Ceramics International*, *43*(15), 12968–12974. DOI: 10.1016/j.ceramint.2017.06.195

Kinani, M. A., Lekdadri, A., Mir, Y., & Zazoui, M. (2022). First-principles study of double perovskites Cs2Sn(Br1−xIx)6 for the design of high-efficiency thin-film photovoltaics. Computational Condensed Matter, 30(September 2021), e00634. DOI: 10.1016/j.cocom.2021.e00634

Kipgen, L., Fulara, H., Raju, M., & Chaudhary, S. (2012). In-plane magnetic anisotropy and coercive field dependence upon thickness of CoFeB. *Journal of Magnetism and Magnetic Materials*, *324*(19), 3118–3121. DOI: 10.1016/j.jmmm.2012.05.012

Kirschner, K. N., Yongye, A. B., Tschampel, S. M., González-Outeiriño, J., Daniels, C. R., Foley, B. L., & Woods, R. J. (2007). GLYCAM06: A generalizable biomolecular force field. Carbohydrates. *Journal of Computational Chemistry*, *29*(4), 622–655. DOI: 10.1002/jcc.20820 PMID: 17849372

Kittel, C. (1948). On the Theory of Ferromagnetic Resonance Absorption. *Physical Review*, *73*(2), 155–161. DOI: 10.1103/PhysRev.73.155

Kjær, H., & Sauer, S. P. A. (2011). Pople Style Basis Sets for the Calculation of NMR Spin–Spin Coupling Constants: The 6-31G-J and 6-311G-J Basis Sets. *Journal of Chemical Theory and Computation*, *7*(12), 4070–4076. DOI: 10.1021/ct200546q PMID: 26598352

Klamt, A., Moya, C., & Palomar, J. (2015). A Comprehensive Comparison of the IEFPCM and SS(V)PE Continuum Solvation Methods with the COSMO Approach. *Journal of Chemical Theory and Computation, 11*(9), 4220–4225. DOI: 10.1021/acs.jctc.5b00601 PMID: 26575917

Klamt, A., & Schüürmann, G. (1993). COSMO: A new approach to dielectric screening in solvents with explicit expressions for the screening energy and its gradient. *Journal of the Chemical Society, Perkin Transactions 2: Physical Organic Chemistry*, (5), 799–805. DOI: 10.1039/P29930000799

Koch, W., & Holthausen, M. C. (2001). A Chemist's Guide to Density Functional Theory (1re éd.). Wiley. DOI: 10.1002/3527600043

Koes, D. R., & Vries, J. K. (2017). Evaluating amber force fields using computed NMR chemical shifts. *Proteins, 85*(10), 1944–1956. DOI: 10.1002/prot.25350 PMID: 28688107

Kohn, W. (1999). Nobel Lecture: Electronic structure of matter wave functions and density functionals. *Reviews of Modern Physics, 71*(5), 1253–1266. DOI: 10.1103/RevModPhys.71.1253

Kohn, W., & Sham, L. J. (1965). Self-Consistent Equations Including Exchange and Correlation Effects. *Physical Review, 140*(4A), A1133–A1138. DOI: 10.1103/PhysRev.140.A1133

Kojima, A., Teshima, K., Shirai, Y., & Miyasaka, T. (2009). Organometal halide perovskites as visible-light sensitizers for photovoltaic cells. *Journal of the American Chemical Society, 131*(17), 6050–6051. DOI: 10.1021/ja809598r

Kokalj, A. (1999). XCrySDen A new program for displaying crystalline structures and electron densities. *Journal of Molecular Graphics & Modelling, 17*(3-4), 176–179. DOI: 10.1016/S1093-3263(99)00028-5

Koohkan, R., Sharafi, S., Shokrollahi, H., & Janghorban, K. (2008). Preparation of nanocrystalline Fe–Ni powders by mechanical alloying used in soft magnetic composites. *Journal of Magnetism and Magnetic Materials, 320*(6), 1089–1094. DOI: 10.1016/j.jmmm.2007.10.033

Korlepara, D. B., C. S, V., Srivastava, R., Pal, P. K., Raza, S. H., Kumar, V., . . . Priyakumar, U. D. (. (2024). PLAS-20k: Extended dataset of protein–ligand affinities from MD simulations for machine learning applications. *Scientific Data, 11*(1), 180. DOI: 10.1038/s41597-023-02872-y PMID: 38336857

Kovács, D. P., Moore, J. H., Browning, N. J., Batatia, I., Horton, J. T., Pu, Y., Kapil, V., Witt, W. C., Magdău, I., Cole, D. J., & Csányi, G. (2025). MACE-OFF: Short-Range Transferable Machine Learning Force Fields for Organic Molecules. *Journal of the American Chemical Society, 147*(21), 17598–17611. DOI: 10.1021/jacs.4c07099 PMID: 40387214

Koza, J. R. (1993). *On the programming of computers by means of natural selection.* MIT Press.

Kremer, K., & Müller-Plathe, F. (2002). Multiscale simulation in polymer science. *Molecular Simulation, 28*(8–9), 729–750. DOI: 10.1080/0892702021000002458

Kreminska, J. (2025). *Thickness Dependent Fermiology in 2D materials.* DOI: 10.33612/diss.1328567056

Krukau, A. V., Vydrov, O. A., Izmaylov, A. F., & Scuseria, G. E. (2006). Influence of the exchange screening parameter on the performance of screened hybrid functionals. *The Journal of Chemical Physics, 125*(22), 224106. Advance online publication. DOI: 10.1063/1.2404663

Kuanr, B. K., Camley, R. E., & Celinski, Z. (2003). Exchange bias of NiO/NiFe: Linewidth broadening and anomalous spin-wave damping. *Journal of Applied Physics*, *93*(10), 7723–7725. DOI: 10.1063/1.1557964

Kuanr, B. K., Camley, R. E., & Celinski, Z. (2005). Extrinsic contribution to Gilbert damping in sputtered NiFe films by ferromagnetic resonance. *Journal of Magnetism and Magnetic Materials*, *286*(SPEC. ISS.), 276–281. DOI: 10.1016/j.jmmm.2004.09.080

Kube, C. M. (2016). Elastic anisotropy of crystals. *AIP Advances*, *6*(9), 095209. DOI: 10.1063/1.4962996

Kumagai, S., Sato, M., Ma, C., Nakai, Y., Kameda, T., Saito, Y., Watanabe, A., Watanabe, C., Teramae, N., & Yoshioka, T. (2022). A comprehensive study into the thermo-oxidative degradation of sulfur-based engineering plastics. *Journal of Analytical and Applied Pyrolysis*, *168*, 105754. DOI: 10.1016/j.jaap.2022.105754

Kumar, N., Guin, S. N., Manna, K., Shekhar, C., & Felser, C. (2020). Topological Quantum Materials from the Viewpoint of Chemistry. *Chemical Reviews*, *121*(5), 2780–2815. DOI: 10.1021/acs.chemrev.0c00732 PMID: 33151662

Kumar, N., Kaur, B., Sharma, D., Korpole, S., Shukla, S., Bhardwaj, P., Patel, M. K., Acharya, S., Khatri, M., & Saxena, S. (2025). Impact of altitudinal variations on plant growth dynamics, nutritional composition, and free living rhizospheric N_2 fixing bacterial community of Eruca sativa. *Scientific Reports*, *15*(1), 13839. Advance online publication. DOI: 10.1038/s41598-025-98242-2

Kussmann, J., & Ochsenfeld, C. (2007). Linear-scaling method for calculating nuclear magnetic resonance chemical shifts using gauge-including atomic orbitals within Hartree-Fock and density-functional theory. *The Journal of Chemical Physics*, *127*(5), 054103. DOI: 10.1063/1.2749509 PMID: 17688330

Lafiosca, P., Rossi, F., Egidi, F., Giovannini, T., & Cappelli, C. (2023). Multiscale Frozen Density Embedding/Molecular Mechanics approach for simulating magnetic response properties of solvated systems. *Journal of Chemical Theory and Computation*, *20*(1), 266–279. DOI: 10.1021/acs.jctc.3c00850 PMID: 38109486

Lamy, Y., & Viala, B. (2006). NiMn, IrMn, and NiO Exchange Coupled CoFe Multilayers for Microwave Applications. *IEEE Transactions on Magnetics*, *42*(10), 3332–3334. DOI: 10.1109/TMAG.2006.878871

Landau, L., & Lifshitz, E. (1935). On the theory of the dispersion of magnetic permeability in ferromagnetic bodies. *Phys. Z. Sowjetunion*, *8*(153), 101–114.

Lanka, G., Banerjee, S., Adhikari, N., & Ghosh, B. (2025). Fragment-based discovery of new potential DNMT1 inhibitors integrating multiple pharmacophore modeling, 3D-QSAR, virtual screening, molecular docking, ADME, and molecular dynamics simulation approaches. *Molecular Diversity*, *29*(1), 117–137. DOI: 10.1007/s11030-024-10837-5 PMID: 38637479

Laranjeira, J. A., Azevedo, S. A., Martins, N. F., La Porta, F. A., Longo, E., & Sambrano, J. R. (2024). Insights into the role of surface properties on the optical, electronic and nanoparticles morphology of scheelite BaMoO4. *Surfaces and Interfaces*, *46*, 103894. DOI: 10.1016/j.surfin.2024.103894

Larsson, H. R., & Hartke, B. (2013). Fitting reactive force fields using genetic algorithms. Comput. Method Mater. Sci, 13(1).

Laun, J., & Bredow, T. (2021). BSSE-corrected consistent Gaussian basis sets of triple-zeta valence with polarization quality of the sixth period for solid-state calculations. *Journal of Computational Chemistry*, *42*(15), 1064–1072. DOI: 10.1002/jcc.26521 PMID: 33792062

Ledbetter, H., & Migliori, A. (2006). A general elastic-anisotropy measure. *Journal of Applied Physics*, *100*(6), 063516. DOI: 10.1063/1.2338835

Ledieu, M., Schoenstein, F., Le Gallou, J.-H., Valls, O., Queste, S., Duverger, F., & Acher, O. (2003). Microwave permeability spectra of ferromagnetic thin films over a wide range of temperatures. *Journal of Applied Physics*, *93*(10), 7202–7204. DOI: 10.1063/1.1555902

Lemkul, J. A., Huang, J., Roux, B., & MacKerell, A. D.Jr. (2016). An empirical polarizable force field based on the classical Drude oscillator model: Development history and recent applications. *Chemical Reviews*, *116*(9), 4983–5013. DOI: 10.1021/acs.chemrev.5b00505 PMID: 26815602

Leng, M., Chen, Z., Yang, Y., Li, Z., Zeng, K., Li, K., Niu, G., He, Y., Zhou, Q., & Tang, J. (2016). Lead-Free, Blue Emitting Bismuth Halide Perovskite Quantum Dots. *Angewandte Chemie International Edition*, *55*(48), 15012–15016. DOI: 10.1002/anie.201608160

Lens, (2025). *Search results for "Tin Nanoparticles"*. The Lens. Retrieved November 07, 2025, from https://www.lens.org/lens/search/scholar/list?q=tin%20nanoparticles

Lenz, K., Wende, H., Kuch, W., Baberschke, K., Nagy, K., & Jánossy, A. (2006). Two-magnon scattering and viscous Gilbert damping in ultrathin ferromagnets. *Physical Review B: Condensed Matter and Materials Physics*, *73*(14), 144424. DOI: 10.1103/PhysRevB.73.144424

Li, X., Wang, D., Ma, F., Yu, L., Mao, J., Zhang, W., Jiang, J., Zhang, L., & Li, P. (2023). Rapid detection of sesame oil multiple adulteration using a portable Raman spectrometer. Food Chemistry, 405(Part B), 134884. https://doi.org/DOI: 10.1016/j.foodchem.2022.134884

Liang, Y. Y., Khalid, K., Le, H. V., Teo, H. M. V., Raitelaitis, M., Gerault, M.-A., Lee, J. J. H., Lyu, J., Chan, A., Jeyasekharan, A. D., Tam, W. L., Nordlund, P., & Prabhu, N. (2025). MS CETSA deep functional proteomics uncovers DNA repair programs leading to gemcitabine resistance. *Nature Communications*, *16*(1), 4234. DOI: 10.1038/s41467-025-59505-8 PMID: 40335468

Li, C., Soh, K. C. K., & Wu, P. (2004). Formability of ABO3 perovskites. *Journal of Alloys and Compounds*, *372*(1–2), 40–48. DOI: 10.1016/j.jallcom.2003.10.017

Li, G., Guo, Y., Mabuchi, T., Surblys, D., Ohara, T., & Tokumasu, T. (2022). Prediction of the adsorption properties of liquid at solid surfaces with molecular scale surface roughness via encoding-decoding convolutional neural networks. *Journal of Molecular Liquids*, *349*, 118489. DOI: 10.1016/j.molliq.2022.118489

Li, H., Li, F., Shen, Z., Han, S.-T., Chen, J., Dong, C., Chen, C., Zhou, Y., & Wang, M. (2021). Photoferroelectric perovskite solar cells: Principles, advances and insights. *Nano Today*, *37*, 101062. DOI: 10.1016/j.nantod.2020.101062

Li, H., Wang, L., Xu, L., Li, A., Mao, P., Wu, Q., & Xie, Z. (2021). First-principles study on the structural, elastic, piezoelectric and electronic properties of (BaTiO3, LiTaO3)-modified KNbO3. *Materials Today. Communications*, *26*, 102092. DOI: 10.1016/j.mtcomm.2021.102092

Li, J., Liang, J., Wang, Z., Ptaszek, A. L., Liu, X., Ganoe, B., Head-Gordon, M., & Head-Gordon, T. (2024). Highly Accurate Prediction of NMR Chemical Shifts from Low-Level Quantum Mechanics Calculations Using Machine Learning. *Journal of Chemical Theory and Computation*, *20*(5), 2152–2166. DOI: 10.1021/acs.jctc.3c01256 PMID: 38331423

Li, L., Yang, X., Sohag, M. M. A., Wang, X., & Liu, Q. (2022). SPH-ASR study of drop impact on a heated surface with consideration of inclined angle and evaporation. *Engineering Analysis with Boundary Elements*, *141*, 235–249. DOI: 10.1016/j.enganabound.2022.05.016

Lind, S. J., Rogers, B. D., & Stansby, P. K. (2020). Review of smoothed particle hydrodynamics: Towards converged Lagrangian flow modelling. *Proceedings - Royal Society. Mathematical, Physical and Engineering Sciences*, *476*(2241), 20190801. DOI: 10.1098/rspa.2019.0801

Li, Q., Wu, K., Zhu, H., Yang, Y., He, S., & Lian, T. (2024). Charge Transfer from Quantum-Confined 0D, 1D, and 2D Nanocrystals. *Chemical Reviews*, *124*(9), 5695–5763. DOI: 10.1021/acs.chemrev.3c00742 PMID: 38629390

Li, S., Liu, S., Yuan, D., Liu, R., Hu, L., & Zhu, X. (2024). Discovery of quinazoline-benzothiazole derivatives as novel potent protease-activated receptor 4 antagonists with improved pharmacokinetics and low bleeding liability. *European Journal of Medicinal Chemistry*, *280*, 116980. DOI: 10.1016/j.ejmech.2024.116980 PMID: 39442337

Liu, B., & Zhang, W. (2023). Research Progress of Topological Quantum Materials: From First-Order to Higher-Order. *Symmetry*, *15*(9), 1651. DOI: 10.3390/sym15091651

Liu, F., Sanchez, D. M., Kulik, H. J., & Martínez, T. J. (2018). Exploiting graphical processing units to enable quantum chemistry calculation of large solvated molecules with conductor-like polarizable continuum models. *International Journal of Quantum Chemistry*, *119*(1), e25760. Advance online publication. DOI: 10.1002/qua.25760

Liu, G. R., Wang, G. Y., Peng, Q., & De, S. (2017). A micro-macro coupling approach of MD-SPH method for reactive energetic materials. *AIP Conference Proceedings*, *1793*(1), 050006. DOI: 10.1063/1.4971540

Liu, J., Chen, S., Nie, X., & Robbins, M. O. (2007). A continuum-atomistic simulation of heat transfer in micro- and nano-flows. *Journal of Computational Physics*, *227*(1), 279–291. DOI: 10.1016/j.jcp.2007.07.014

Liu, M. B., & Liu, G. (2010). Smoothed particle hydrodynamics (SPH): An overview and recent developments. *Archives of Computational Methods in Engineering*, *17*(1), 25–76. DOI: 10.1007/s11831-010-9040-7

Liu, M., Du, S., Wang, F., Adam, R., Li, Q., Ma, X., Guo, X., Chen, X., Yu, J., Song, Y., Xu, J., Li, S., & Cao, D. (2021). Influence of surface pinning in the domain on the magnetization dynamics in permalloy striped domain films. *Journal of Alloys and Compounds*, *869*, 159327. DOI: 10.1016/j.jallcom.2021.159327

Liu, M., Mirza, A., McAndrew, P. C., Thapaliya, A., Pierrat, O. A., Stubbs, M., Hahner, T., Chessum, N. E. A., Innocenti, P., Caldwell, J., Cheeseman, M. D., Bellenie, B. R., van Montfort, R. L. M., Newton, G. K., Burke, R., Collins, I., & Hoelder, S. (2023). Determination of ligand-binding affinity (Kd) using transverse relaxation rate (R_2) in the ligand-observed ^{1}H NMR experiment and applications to fragment-based drug discovery. *Journal of Medicinal Chemistry, 66*(15), 10617–10627. DOI: 10.1021/acs.jmedchem.3c00758 PMID: 37467168

Liu, S., Savage, J., & Voth, G. A. (2015). Mesoscale study of proton transport in proton exchange membranes: Role of morphology. *The Journal of Physical Chemistry. C, Nanomaterials and Interfaces, 119*(4), 1753–1762. DOI: 10.1021/jp511830f

Liu, X., An, H., Cai, W., & Shao, X. (2024). Deep learning in spectral analysis: Modeling and imaging. *Trends in Analytical Chemistry, 172*, 117612. DOI: 10.1016/j.trac.2024.117612

Liu, X., Cai, J., Zhang, J., Zhao, P., Wang, G., Bi, S., & Hou, Z. (2025). Surface-state-constrained topological insulator Bi2Te3 nanorods for electromagnetic wave trapping and conversion into electricity. *Journal of Materials Science and Technology, 244*, 149–155. DOI: 10.1016/j.jmst.2025.03.102

Liu, X., Xing, Y., Liu, G., Bao, D., Hu, W., Bi, H., & Wang, M. (2025). Extraction, purification, structural features, biological activities, and applications of polysaccharides from *Opuntia ficus-indica* (L.) Mill. (cactus): A review. *Frontiers in Pharmacology, 16*, 1566000. Advance online publication. DOI: 10.3389/fphar.2025.1566000

Liu, Z., Li, S., Pan, X., & Fang, H. (2022). Mechanism study on spreading dynamics of nanofluids droplet coupled with thermal evaporation. *International Journal of Heat and Mass Transfer, 183*, 122172. DOI: 10.1016/j.ijheatmasstransfer.2021.122172

Li, Y., Lin, H., Yang, H., Yuan, Y., Zou, R., Zhou, G., Zhang, L., & Zheng, H. (2024). Synergistic application of molecular docking and machine learning for improved binding pose. *National Science Open, 3*(2). Advance online publication. DOI: 10.1360/nso/20230058

Li, Z., Yu, J., Hao, S., & Janolin, P.-E. (2022). Enhancing properties of lead-free ferroelectric BaTiO3 through doping. *Journal of the European Ceramic Society, 42*(12), 4693–4701. DOI: 10.1016/j.jeurceramsoc.2022.05.023

Lodewyk, M. W., Siebert, M. R., & Tantillo, D. J. (2011). Computational prediction of ^{1}H and ^{13}C chemical shifts: A useful tool for natural product, mechanistic, and synthetic organic chemistry. *Chemical Reviews, 112*(3), 1839–1862. DOI: 10.1021/cr200106v PMID: 22091891

Lomas, J. S. (2015). 1H NMR spectra of alcohols in hydrogen bonding solvents: DFT/GIAO calculations of chemical shifts. *Magnetic Resonance in Chemistry, 54*(1), 28–38. DOI: 10.1002/mrc.4312 PMID: 26256675

López-Yerena, A., Domínguez-López, I., Vallverdú-Queralt, A., Pérez, M., Jáuregui, O., Escribano-Ferrer, E., & Lamuela-Raventós, R. M. (2021). Metabolomics technologies for the identification and quantification of dietary phenolic compound metabolites: An overview. *Antioxidants, 10*(6), 846. DOI: 10.3390/antiox10060846

Lordan, D., Wei, G., McCloskey, P., O'Mathuna, C., & Masood, A. (2021). Origin of perpendicular magnetic anisotropy in amorphous thin films. *Scientific Reports*, *11*(1), 3734. DOI: 10.1038/s41598-020-78950-7

Lores, M., Llompart, M., Alvarez-Rivera, G., Guerra, E., Vila, M., Celeiro, M., Lamas, J. P., & Garcia-Jares, C. (2016). Positive lists of cosmetic ingredients: Analytical methodology for regulatory and safety controls – A review. *Analytica Chimica Acta*, *915*, 1–26. DOI: 10.1016/j.aca.2016.02.033

Loveland, P., & Webb, J. (2003). Is there a critical level of organic matter in the agricultural soils of temperate regions: A review. *Soil & Tillage Research*, *70*(1), 1–18. DOI: 10.1016/S0167-1987(02)00139-3

Lu, A. H., Salabas, E. L., & Schüth, F. (2007). Magnetic nanoparticles: Synthesis, protection, functionalization, and application. *Angewandte Chemie International Edition*, *46*(8), 1222–1244. DOI: 10.1002/anie.200602866

Lucy, L. B. (1977). A numerical approach to the testing of the fission hypothesis. *The Astronomical Journal*, *82*(12), 1013–1024. DOI: 10.1086/112164

Ludwig, A., Tewes, M., Glasmachers, S., Löhndorf, M., & Quandt, E. (2002). High-frequency magnetoelastic materials for remote-interrogated stress sensors. *Journal of Magnetism and Magnetic Materials*, *242–245*(PART II), 1126–1131. DOI: 10.1016/S0304-8853(01)00979-9

Lumakso, F. A., Riyanto, S., Ahmad, S. A., Rosman, A. S., Yusoff, F. M., & Rohman, A. (2015). Application of chemometrics in combination with Fourier Transform Mid Infrared spectroscopy for authentication of avocado oil. *Journal of Food and Pharmaceutical Sciences*, *2*, 12–17.

Luo, W., Yu, L., Lu, S., Yu, Y., Bai, Y., Wang, S., Li, D., Han, Z., Zheng, Y., Li, F., Zhao, J., & He, L. (2025). Design and screening of novel molecular compounds targeting lactate dehydrogenase of Babesia microti. *Parasites & Vectors*, *18*(1), 69. DOI: 10.1186/s13071-024-06623-9 PMID: 39988659

Luttens, A., Cabeza de Vaca, I., Sparring, L., Brea, J., Martínez, A. L., Kahlous, N. A., Radchenko, D. S., Moroz, Y. S., Loza, M. I., Norinder, U., & Carlsson, J. (2025). Rapid traversal of vast chemical space using machine learning-guided docking screens. *Nature Computational Science*, *5*(4), 301–312. DOI: 10.1038/s43588-025-00777-x PMID: 40082701

Luttens, A., Vo, D. D., Scaletti, E. R., Wiita, E., Almlöf, I., Wallner, O., Davies, J., Košenina, S., Meng, L., Long, M., Mortusewicz, O., Masuyer, G., Ballante, F., Michel, M., Homan, E., Scobie, M., Kalderén, C., Warpman Berglund, U., Tarnovskiy, A. V., & Carlsson, J. (2025). Virtual fragment screening for DNA repair inhibitors in vast chemical space. *Nature Communications*, *16*(1), 1741. DOI: 10.1038/s41467-025-56893-9 PMID: 39966348

Lu, W., Zhang, J., Huang, W., Zhang, Z., Jia, X., Wang, Z., Shi, L., Li, C., Wolynes, P. G., & Zheng, S. (2024). DynamicBind: Predicting ligand-specific protein–ligand complex structure with a deep equivariant generative model. *Nature Communications*, *15*(1), 1071. DOI: 10.1038/s41467-024-45461-2 PMID: 38316797

Lu, X., Wang, X., Li, Q., Huang, X., Han, S., & Wang, G. (2015). A ReaxFF-based molecular dynamics study of the pyrolysis mechanism of polyimide. *Polymer Degradation & Stability*, *114*, 72–80. DOI: 10.1016/j.polymdegradstab.2015.02.004

Lu, X., Zhao, Y., & Dennis, D. J. C. (2018). Flow measurements in microporous media using micro-particle image velocimetry. *Physical Review Fluids*, *3*(10), 104202. DOI: 10.1103/PhysRevFluids.3.104202

Lv, K., Li, Z., Huang, X., Cheng, Z., Wang, Z., & Zhao, H. (2025). $CsPbBr_3$ Perovskite-Based Heterostructures in Photocatalysis : Mechanisms, Stability, and Multifunctional Performance. *Advanced Science (Weinheim, Baden-Wurttemberg, Germany)*, *12*(34), e07747. DOI: 10.1002/advs.202507747

Machado, R., Sepliarsky, M., & Stachiotti, M. G. (2012). First-Principles Calculations of Structural Properties of $NaNbO_3$. *Ferroelectrics*, *427*(1), 98–104. DOI: 10.1080/00150193.2012.674419

Macías, F. A., Galindo, J. L. G., & Galindo, J. C. G. (2007). Evolution and current status of ecological phytochemistry. *Phytochemistry*, *68*(22–24), 2917–2936. DOI: 10.1016/j.phytochem.2007.10.010

Madsen, G. K. H., & Singh, D. J. (2006a). BoltzTraP. A code for calculating band-structure dependent quantities. *Computer Physics Communications*, *175*(1), 67–71. DOI: 10.1016/j.cpc.2006.03.007

Ma, H., Guo, J., Liu, G., Xie, D., Zhang, B., Li, X., Zhang, Q., Cao, Q., Li, X., Ma, F., Li, Y., Wan, G., Li, Y., Wu, D., Ma, P., Guo, M., & Yin, J. (2025). Raman spectroscopy coupled with chemometrics for identification of adulteration and fraud in muscle foods: A review. *Critical Reviews in Food Science and Nutrition*, *65*(11), 2008–2030. DOI: 10.1080/10408398.2024.2329956 PMID: 38523442

Mahedi, A. H., Uddin, S., Yasin, M., Khan, H., Tarekuzzaman, M., Alsalmi, O., Rasheduzzaman, M., Mostafa, S. M. G., Hossen, M. M., & Hasan, M. Z. (2025). Multiscale analysis of Rb 2 NaInI 6 : from electronic structure to device performance for next-generation perovskite solar cells. DOI: 10.1039/D5RA04739K

Mahraj, I., Oublal, M., & Ptok, A. (2025). Theoretical investigation of the photovoltaic properties of MgSnN2 for multi-junction solar cells. *Physica B, Condensed Matter*, *713*, 417261. DOI: 10.1016/j.physb.2025.417261

Maitouf, S., Bachar, A., Mabrouk, A., & Faska, N. (2025). Wastewater Treatment and Recycling Using Adsorption Enhanced by Artificial Neural Networks. In *Computational Intelligence and Optimization Methods for Sustainable Water Management* (pp. 441–476). IGI Global Scientific Publishing., DOI: 10.4018/979-8-3373-2700-6.ch013

Maksymov, I. S., & Kostylev, M. (2015). Broadband stripline ferromagnetic resonance spectroscopy of ferromagnetic films, multilayers and nanostructures. *Physica E, Low-Dimensional Systems and Nanostructures*, *69*, 253–293. DOI: 10.1016/j.physe.2014.12.027

Ma, L., Tang, J., Chen, F., Liu, Q., Huang, J., Liu, X., Zhou, Z., & Yi, W. (2024). Structure-based screening, optimization and biological evaluation of novel chrysin-based derivatives as selective PPARγ modulators for the treatment of T2DM and hepatic steatosis. *European Journal of Medicinal Chemistry*, *276*, 116728. DOI: 10.1016/j.ejmech.2024.116728 PMID: 39089002

Malki, S., & El Farh, L. (2020). First-Principles Investigation on Thermoelectric Properties of $$VSb_{2}$$ Material. *International Journal of Thermophysics*, *41*(5), 58. DOI: 10.1007/s10765-020-02630-x

Manago, T., Yamanoi, K., Kasai, S., & Mitani, S. (2015). Damping factor estimation using spin wave attenuation in permalloy film. *Journal of Applied Physics, 117*(17), 17D121. Advance online publication. DOI: 10.1063/1.4914538

Manfredi, P., & Trinchero, R. (2024). Nonparametric formulation of polynomial chaos expansion based on least-square support-vector machines. *Engineering Applications of Artificial Intelligence, 133*, 108182. DOI: 10.1016/j.engappai.2024.108182

Manni, G. L., Galván, I. F., Alavi, A., Aleotti, F., Aquilante, F., Autschbach, J., Avagliano, D., Baiardi, A., Bao, J. J., Battaglia, S., Birnoschi, L., Blanco-González, A., Bokarev, S. I., Broer, R., Cacciari, R., Calio, P. B., Carlson, R. K., Couto, R. C., Cerdán, L., & Lindh, R. (2023). The OpenMoLCAS Web: A Community-Driven Approach to Advancing Computational Chemistry. *Journal of Chemical Theory and Computation, 19*(20), 6933–6991. DOI: 10.1021/acs.jctc.3c00182 PMID: 37216210

Mao, P., Bi, W., Lv, J., Zhang, Z., Wang, B., & Zhong, Y. (2025). Decoding Buried Interfaces in Perovskite Solar Cells: Core Issues, Strategic Engineering, and Prospects for High-Efficiency Stable Devices. *Advanced Science (Weinheim, Baden-Wurttemberg, Germany), 12*(40), e12523. Advance online publication. DOI: 10.1002/advs.202512523

Mao, Q., Feng, M., Jiang, X. Z., Ren, Y., Luo, K. H., & van Duin, A. C. T. (2023). Classical and reactive molecular dynamics: Principles and applications in combustion and energy systems. *Progress in Energy and Combustion Science, 97*, 101084. DOI: 10.1016/j.pecs.2023.101084

Mao, Y., Robinson, J. P., & Binner, E. R. (2023). Current status of microwave-assisted extraction of pectin. *Chemical Engineering Journal, 473*, 145261. DOI: 10.1016/j.cej.2023.145261

Marcarino, M. O., Cicetti, S., Zanardi, M. M., & Sarotti, A. M. (2021). A critical review on the use of DP4+ in the structural elucidation of natural products: The good, the bad and the ugly. A practical guide. *Natural Product Reports, 39*(1), 58–76. DOI: 10.1039/D1NP00030F PMID: 34212963

Marenich, A. V., Cramer, C. J., & Truhlar, D. G. (2009). Universal solvation model based on solute electron density and on a continuum model of the solvent defined by the bulk dielectric constant and atomic surface tensions. *The Journal of Physical Chemistry B, 113*(18), 6378–6396. DOI: 10.1021/jp810292n PMID: 19366259

Mari, S. H., Varras, P. C., Atia-Tul-Wahab, N., Choudhary, I. M., Siskos, M. G., & Gerothanassis, I. P. (2019). Solvent-Dependent structures of natural products based on the combined use of DFT calculations and 1H-NMR chemical shifts. *Molecules (Basel, Switzerland), 24*(12), 2290. DOI: 10.3390/molecules24122290 PMID: 31226776

Martinez-Castillo, C., Astray, G., Mejuto, J. C., & Simal-Gandara, J. (2020). Random forest, artificial neural network, and support vector machine models for honey classification. *eFood, 1*(1), 69–76. DOI: 10.2991/efood.k.191004.001

Marx, D., & Hutter, J. (2009). AB Initio Molecular Dynamics. In Cambridge University Press eBooks. DOI: 10.1017/CBO9780511609633

Marzo, A., Ferrada, P., Beiza, F., Alonso-Montesinos, J., Ballestrín, J., & Román, R. (2016). Comparison of atacama desert solar spectra vs. ASTM G173-03 reference spectra for solar energy application. International Energy Society. Conference Proceedings.

Mashkovtsev, M. A., Kosykh, A. S., Ishchenko, A. V., Chukin, A. V., Kukharenko, A. I., Troshin, P. A., & Zhidkov, I. S. (2024). Unraveling oxygen vacancies effect on chemical composition, electronic structure and optical properties of EU doped SNO2. *Nanomaterials (Basel, Switzerland)*, *14*(20), 1675. DOI: 10.3390/nano14201675

Masood, H., Toe, C. Y., Teoh, W. Y., Sethu, V., & Amal, R. (2019). Machine learning for Accelerated discovery of solar photocatalysts. *ACS Catalysis*, *9*(12), 11774–11787. DOI: 10.1021/acscatal.9b02531

Mastellone, G., Marengo, A., Sgorbini, B., Rubiolo, P., & Cagliero, C. (2022). Development of a dispersive solid-liquid microextraction method using natural eutectic solvents for a greener extraction of phytochemicals from fiber-type Cannabis sp. Industrial Crops and Products, 187(Part B), 115476. https://doi.org/DOI: 10.1016/j.indcrop.2022.115476

Matussin, S., Harunsani, M. H., Tan, A. L., & Khan, M. M. (2020). Plant-Extract-Mediated SnO2 Nanoparticles: Synthesis and Applications. *ACS Sustainable Chemistry & Engineering*, *8*(8), 3040–3054. DOI: 10.1021/acssuschemeng.9b06398

Matveeva, R., Folkestad, S. D., & Høyvik, I. (2023). Particle-Breaking Hartree–Fock Theory for open Molecular Systems. *The Journal of Physical Chemistry A*, *127*(5), 1329–1341. DOI: 10.1021/acs.jpca.2c07686 PMID: 36720055

Mavko, G., Mukerji, T., & Dvorkin, J. (2020). Elasticity and Hooke's Law. The Rock Physics Handbook, 37–120. DOI: 10.1017/9781108333016.003

Mazaleyrat, F. (2006). *Introduction à la physique des matériaux électromagnétiques*. Inst. Univ. Form. Des Maîtres Crétl.

Mazzei, L., Ranieri, S., Silvestri, D., Greene-Cramer, R., Cioffi, C., Montelione, G. T., & Ciurli, S. (2024). An isothermal calorimetry assay for determining steady state kinetic and Ensitrelvir inhibition parameters for SARS-CoV-2 3CL-protease. *Scientific Reports*, *14*(1), 32175. DOI: 10.1038/s41598-024-81990-y PMID: 39741150

McClure, E. T., Ball, M. R., Windl, W., & Woodward, P. M. (2016). Cs2AgBiX6 (X = Br, Cl): New Visible Light Absorbing, Lead-Free Halide Perovskite Semiconductors. *Chemistry of Materials*, *28*(5), 1348–1354. DOI: 10.1021/acs.chemmater.5b04231

McCord, J., Mattheis, R., & Elefant, D. (2004). Dynamic magnetic anisotropy at the onset of exchange bias: The <math display="inline"> <mrow> <mi mathvariant="normal">Ni</mi> <mi mathvariant="normal">Fe</mi> <mo>/</mo> <mi mathvariant="normal">Ir</mi> <mi mathvariant="normal">Mn</mi> </mrow> </math> ferromagnet/antiferromagnet system. *Physical Review. B*, *70*(9), 094420. DOI: 10.1103/PhysRevB.70.094420

McMichael, R. D., Lee, C. G., Bonevich, J. E., Chen, P. J., Miller, W., & Egelhoff, W. F.Jr. (2000). Strong anisotropy in thin magnetic films deposited on obliquely sputtered Ta underlayers. *Journal of Applied Physics*, *88*(9), 5296–5299. DOI: 10.1063/1.1323436

McNutt, A. T., Francoeur, P., Aggarwal, R., Masuda, T., Meli, R., Ragoza, M., Sunseri, J., & Koes, D. R. (2021). GNINA 1.0: Molecular docking with deep learning. *Journal of Cheminformatics*, *13*(1), 43. DOI: 10.1186/s13321-021-00522-2 PMID: 34108002

McNutt, A. T., Li, Y., Meli, R., Aggarwal, R., & Koes, D. R. (2025). GNINA 1.3: The next increment in molecular docking with deep learning. *Journal of Cheminformatics*, *17*(1), 28. DOI: 10.1186/s13321-025-00973-x PMID: 40025560

Mehedi Shamrat, F. M. J., Shakil, R., Sharmin, , Hoque ovy, N., Akter, B., Ahmed, M. Z., Ahmed, K., Bui, F. M., & Moni, M. A. (2024). An advanced deep neural network for fundus image analysis and enhancing diabetic retinopathy detection. *Healthcare Analytics*, *5*, 100303. DOI: 10.1016/j.health.2024.100303

Mei, Y., Zhuang, S., & Wang, J. (2025). Adsorption of heavy metals by biochar in aqueous solution: A review. *The Science of the Total Environment*, *968*(February), 178898. DOI: 10.1016/j.scitotenv.2025.178898 PMID: 39986038

Mennucci, B. (2012). Polarizable continuum model. *Wiley Interdisciplinary Reviews. Computational Molecular Science*, *2*(3), 386–404. DOI: 10.1002/wcms.1086

Merrill, A. T., & Tantillo, D. J. (2019). Solvent optimization and conformational flexibility effects on 1H and 13C NMR scaling factors. *Magnetic Resonance in Chemistry*, *58*(6), 576–583. DOI: 10.1002/mrc.4986 PMID: 31883397

Miah, M. H., Khandaker, M. U., Rahman, M. B., Nur-E-Alam, M., & Islam, M. A. (2024). Band gap tuning of perovskite solar cells for enhancing the efficiency and stability: Issues and prospects. *RSC Advances*, *14*(23), 15876–15906. DOI: 10.1039/D4RA01640H

Miller, R. E., & Tadmor, E. B. (2009). A unified framework and performance benchmark of fourteen multiscale atomistic/continuum coupling methods. *Modelling and Simulation in Materials Science and Engineering*, *17*(5), 053001. Advance online publication. DOI: 10.1088/0965-0393/17/5/053001

Mills, D. L., & Arias, R. (2006). The damping of spin motions in ultrathin films: Is the Landau–Lifschitz–Gilbert phenomenology applicable? *Physica B, Condensed Matter*, *384*(1–2), 147–151. DOI: 10.1016/j.physb.2006.05.209

Milne, B. F. (2022). Scoring functions, ScotChem protein–ligand docking course documentation. Retrieved 12 August 2025 from https://arts.st-andrews.ac.uk/scotchem/pldc/scoring_functions.html#:~:text=%5C%5BE_%7Bbind%7D%20%3D%20%5Csum_%7Bi%3D1%7D,6%7D%20%2B%20%5Cfrac%7Bq_iq_j%7D%7B%5Cepsilon%20r_%7Bij%7D%7D%20%5Cright

Miran, H. A. J., Abdullah, Z. N., Altarawneh, M., Rahman, M. M., Al-Bayati, A. T., & Salman, E. M.-T. (2023). First-Principles Analysis of Cr-Doped SrTiO3 Perovskite as Optoelectronic Materials. *Iranian Journal of Materials Science and Engineering*, *20*(1).

Mirzaei, S., Ivanov, M. V., & Timerghazin, Q. K. (2019). Improving performance of the SMD solvation model: BoNdi Radii improve predicted aqueous solvation free energies of ions and PKA values of thiols. *The Journal of Physical Chemistry A*, *123*(44), 9498–9504. DOI: 10.1021/acs.jpca.9b02340 PMID: 31318553

Mishin, Y. (2021). Machine-learning interatomic potentials for materials science. *Acta Materialia, 214*, 116980. DOI: 10.1016/j.actamat.2021.116980

Mishra, R., Jain, S., & Prasad, C. D. (2015). A review on piezoelectric material as a source of generating electricity and its possibility to fabricate devices for daily uses of army personnel. International Journal of Systems. *Control and Communications*, *6*(3), 212. DOI: 10.1504/IJSCC.2015.068908

Miura, K., Kasai, S., Kobayashi, K., & Ono, T. (2006). Non Local Spin Detection in Ferromagnet/Superconductor/Ferromagnet Spin-Valve Device with Double-Tunnel Junctions. *Japanese Journal of Applied Physics*, *45*(4R), 2888. DOI: 10.1143/JJAP.45.2888

Mocci, F., & Laaksonen, A. (2015). Combining MD simulations and NMR spectroscopy for molecular insight and methodological synergy: the integrated MD-NMR method. In Nuclear magnetic resonance (pp. 592–616). DOI: 10.1039/9781782622758-00592

Mohsenzadeh, E., Ratautaite, V., Brazys, E., Ramanavicius, S., Zukauskas, S., Plausinaitis, D., & Ramanavicius, A. (2024). Application of computational methods in the design of molecularly imprinted polymers (review). *TrAC Trends in Analytical Chemistry, 171*(August 2023), 117480. DOI: 10.1016/j.trac.2023.117480

Moliton, A. (2007). *Basic electromagnetism and materials*. Springer.

Monaghan, J. J. (2005). Smoothed particle hydrodynamics. In Reports on Progress in Physics (Vol. 68, Issue 8, pp. 1703–1759). DOI: 10.1088/0034-4885/68/8/R01

Monaghan, J. J. (1992). Smoothed particle hydrodynamics. *Annual Review of Astronomy and Astrophysics*, *30*(1), 543–574. DOI: 10.1146/annurev.aa.30.090192.002551

Monaghan, J. J. (2011). Smoothed particle hydrodynamics and its diverse applications. *Annual Review of Fluid Mechanics*, *44*(1), 323–346. DOI: 10.1146/annurev-fluid-120710-101220

Montgomery, D. C. (2017). *Design and analysis of experiments*. John Wiley & Sons.

Montgomery, D. C., & Runger, G. C. (2010). *Applied statistics and probability for engineers*. John Wiley & Sons.

Mopoung, K., Dávid, A., Liu, X., Fahlman, M., Buyanova, I. A., Chen, W. M., & Puttisong, Y. (2024). Spin Centers in Vanadium-Doped Cs2NaInCl6 Halide Double Perovskites. *ACS Materials Letters*, *6*(2), 566–571. DOI: 10.1021/acsmaterialslett.3c01590

Morab, S., Sundaram, M. M., & Pivrikas, A. (2023). Review on Charge Carrier Transport in Inorganic and Organic Semiconductors. *Coatings*, *13*(9), 1657. DOI: 10.3390/coatings13091657

Morales-Acevedo, A. (2023). Fundamentals of solar cell physics revisited: Common pitfalls when reporting calculated and measured photocurrent density, open-circuit voltage, and efficiency of solar cells. *Solar Energy*, *262*, 111774. DOI: 10.1016/j.solener.2023.05.051

Morales-Bayuelo, A., & Sánchez-Márquez, J. (2020). Discerning the thermal cyclotrimerizations of fluoro- and chloroacetylenes through ELF, NBO descriptors and QTAIM analysis: Pseudodiradical character. *Heliyon*, *6*(7), e04441. DOI: 10.1016/j.heliyon.2020.e04441 PMID: 32715128

Moreira, R. L., & Dias, A. (2007). Comment on "Prediction of lattice constant in cubic perovskites". *Journal of Physics and Chemistry of Solids*, *68*(8), 1617–1622. DOI: 10.1016/j.jpcs.2007.03.050

Morris, J. P., Fox, P. J., & Zhu, Y. (1997). Modeling Low Reynolds Number Incompressible Flows Using SPH. In JOURNAL OF COMPUTATIONAL PHYSICS (Vol. 136). DOI: 10.1006/jcph.1997.5776

Morris, J. B., Wang, M. L., & Tonnis, B. D. (2021). Variability for oil, protein, lignan, tocopherol, and fatty acid concentrations in eight sesame (Sesamum indicum L.) genotypes. *Industrial Crops and Products*, *164*, 113355. DOI: 10.1016/j.indcrop.2021.113355

Mortier, W. J., Ghosh, S. K., & Shankar, S. (1986). Electronegativity-equalization method for the calculation of atomic charges in molecules. *Journal of the American Chemical Society*, *108*(15), 4315–4320. DOI: 10.1021/ja00275a013

Moshawih, S., Bu, Z. H., Goh, H. P., Kifli, N., Lee, L. H., Goh, K. W., & Ming, L. C. (2024). Consensus holistic virtual screening for drug discovery: A novel machine learning model approach. *Journal of Cheminformatics*, *16*(1), 62. DOI: 10.1186/s13321-024-00855-8 PMID: 38807196

Mubarak, A. A. (2016). The first-principle study of the electronic, optical and thermoelectric properties of XTiO3 (X = Ca, Sr and Ba) compounds. *International Journal of Modern Physics B*, *30*(20), 1650141. DOI: 10.1142/S0217979216501411

Mu, J., Liu, H., Zhang, J., Luo, R., & Chen, H. (2021). Recent force field strategies for intrinsically disordered proteins. *Journal of Chemical Information and Modeling*, *61*(3), 1037–1047. DOI: 10.1021/acs.jcim.0c01175 PMID: 33591749

Mukherjee, C., Suryawanshi, P. G., Kalita, M. C., Deka, D., Aranda, D. A. G., & Goud, V. V. (2024). Polarity-wise successive solvent extraction of Scenedesmus obliquus biomass and characterization of the crude extracts for broad-spectrum antibacterial activity. *Biomass Conversion and Biorefinery*, *14*(2), 2467–2483. DOI: 10.1007/s13399-022-02432-1

Mukherji, D., Marques, C. M., & Kremer, K. (2014). Polymer collapse in miscible good solvents is a generic phenomenon driven by preferential adsorption. *Nature Communications*, *5*(1), 4882. Advance online publication. DOI: 10.1038/ncomms5882

Mukherji, D., Marques, C. M., & Kremer, K. (2018). Collapse in two good solvents, swelling in two poor solvents: Defying the laws of polymer solubility? *Journal of Physics Condensed Matter*, *30*(2), 024002. Advance online publication. DOI: 10.1088/1361-648X/aa9c77

Müller-Plathe, F. (n.d.). Coarse-Graining in Polymer Simulation: From the Atomistic to the Mesoscopic Scale and Back.

Murnaghan, F. D. (1944). The Compressibility of Media under Extreme Pressures. *Proceedings of the National Academy of Sciences of the United States of America*, *30*(9), 244–247. DOI: 10.1073/pnas.30.9.244

Mursaleen, I., Ain, Q., Kumar, A., Kanabar, B., Rajiv, A., Thatoi, D. N., Ojha, M. K., Karamti, H., Munir, J., & Oza, A. D. (2025). Exploring the effect of pressure on Rb2CsGaCl6 perovskite halide: A pathway for enhanced optoelectronic performance. *Inorganic Chemistry Communications*, *179*, 114723. DOI: 10.1016/j.inoche.2025.114723

Musa, K. H., & Elnour, A. A. M. (2024). Advances and future perspectives in biotechnological and bioconversional of dates byproducts. *Journal of Agriculture and Food Research, 16*, 101145. DOI: 10.1016/j.jafr.2024.101145

Nadakuditi, A., & Reddy, V. (2024). An optimized framework for metal ion analysis in automobile industrial sewage. *Journal of the Indian Chemical Society, 101*(8), 101200. DOI: 10.1016/j.jics.2024.101200

Nadh, A. G., Kunhikrishnan, M. J., Ravi, V., Ramakrishnan, K., Rehman, N., Adithya, K. S. B., Revikumar, A., Sudhakaran, P. R., & Raju, R. (2025). Convolidine as potent BACE1 inhibitor for Alzheimer's disease; in-silico coupled with in vitro assessment. *Journal of Computer-Aided Molecular Design, 39*(1), 13. DOI: 10.1007/s10822-025-00592-6 PMID: 40208466

Nakamura, K., Higuchi, S., & Ohnuma, T. (2018). Density Functional Perturbation Theory to Predict Piezoelectric Properties. In Bakırtaş, İ. (Ed.), *Perturbation Methods with Applications in Science and Engineering*. InTech., DOI: 10.5772/intechopen.76827

Nakamura, K., Tokiwa, T., & Kawamura, Y. (2002). Domain structures in KNbO3 crystals and their piezoelectric properties. *Journal of Applied Physics, 91*(11), 9272–9276. DOI: 10.1063/1.1476078

Narasaiah, B. P., Banoth, P., Sohan, A., Mandal, B. K., Dominguez, A. G. B., De Los Santos Valladares, L., & Kollu, P. (2022). Green biosynthesis of tin oxide nanomaterials mediated by Agro-Waste cotton boll peel extracts for the remediation of environmental pollutant dyes. *ACS Omega, 7*(18), 15423–15438. DOI: 10.1021/acsomega.1c07099

Naviglio, D., Scarano, P., Ciaravolo, M., & Gallo, M. (2019). Rapid solid-liquid dynamic extraction (RSLDE): A powerful and greener alternative to the latest solid-liquid extraction techniques. *Foods, 8*(7), 245. DOI: 10.3390/foods8070245

Nayak, S. S., Wadhawa, G. C., Shivankar, V. S., Patil, D. D., Sonawale, M. C., & Mirgane, N. A. (2020). Tin oxide plant assisted nanoparticle catalyzed green synthesis of imidazole derivatives. *Materials Today: Proceedings, 37*, 2490–2494. DOI: 10.1016/j.matpr.2020.08.301

Nayir, N., Van Duin, A. C. T., & Erkoc, S. (2019). Development of the ReaxFF Reactive Force Field for Inherent Point Defects in the Si/Silica System. *The Journal of Physical Chemistry A, 123*(19), 4303–4313. DOI: 10.1021/acs.jpca.9b01481

Nelson, J. (2003). The Physics of Solar Cells. The Physics of Solar Cells. DOI: 10.1142/p276

Nepravishta, R., Ramírez-Cárdenas, J., Rocha, G., Walpole, S., Hicks, T., Monaco, S., Muñoz-García, J. C., & Angulo, J. (2024). Fast quantitative validation of 3D models of low-affinity protein–ligand complexes by STD NMR spectroscopy. *Journal of Medicinal Chemistry, 67*(12), 10025–10034. DOI: 10.1021/acs.jmedchem.4c00204 PMID: 38848103

Neves, M. A. D., Ni, T. T., Mackeigan, D. T., Shoara, A. A., Lei, X., Slavkovic, S., Yu, S.-Y., Stratton, T. W., Gallant, R. C., Zhang, D., Xu, X. R., Fernandes, C., Zhu, G., Hu, X., Chazot, N., Donaldson, L. W., Johnson, P. E., Connelly, K., Rand, M., & Ni, H. (2024). Salvianolic acid B inhibits thrombosis and directly blocks the thrombin catalytic site. *Research and Practice in Thrombosis and Haemostasis, 8*(4), 102443. DOI: 10.1016/j.rpth.2024.102443 PMID: 38993621

Nguyen, H. H., & Truong, V. H. (2024). Machine Learning-based prediction of seismic lateral deflection of steel trusses using nonlinear time-history analysis. *Structures (Oxford)*, *69*, 107369. DOI: 10.1016/j.istruc.2024.107369

Niazi, S. K., & Mariam, Z. (2025). Artificial intelligence in drug development: Reshaping the therapeutic landscape. *Therapeutic Advances in Drug Safety*, *16*, 20420986251321704. Advance online publication. DOI: 10.1177/20420986251321704

Ni, J., Li, J., Li, S., Zheng, H., Ming, Z., Li, L., Li, H., Zhang, S., Zhao, Y., Liang, H., & Qiao, Z. (2024). Molecular fingerprint and machine learning enhance high-performance MOFs for mustard gas removal. *iScience*, *27*(6), 110042. Advance online publication. DOI: 10.1016/j.isci.2024.110042 PMID: 38883811

Niklasson, A. M. (2008). Extended born-oppenheimer molecular dynamics. *Physical Review Letters*, *100*(12), 123004. DOI: 10.1103/PhysRevLett.100.123004 PMID: 18517861

Ning, W., & Gao, F. (2019). Structural and Functional Diversity in Lead-Free Halide Perovskite Materials. *Advanced Materials*, *31*(22), 1900326. DOI: 10.1002/adma.201900326

Noel, N. K., Stranks, S. D., Abate, A., Wehrenfennig, C., Guarnera, S., Haghighirad, A. A., Sadhanala, A., Eperon, G. E., Pathak, S. K., Johnston, M. B., Petrozza, A., Herz, L. M., & Snaith, H. J. (2014). Lead-free organic-inorganic tin halide perovskites for photovoltaic applications. *Energy & Environmental Science*, *7*(9), 3061–3068. DOI: 10.1039/C4EE01076K

Noor, F., Junaid, M., Almalki, A. H., Almaghrabi, M., Ghazanfar, S., & Tahir ul Qamar, M. (2024). Deep learning pipeline for accelerating virtual screening in drug discovery. *Scientific Reports*, *14*(1), 28321. DOI: 10.1038/s41598-024-79799-w PMID: 39550439

Noor, N. A., Iqbal, M. W., Zelai, T., Mahmood, A., Shaikh, H. M., Ramay, S. M., & Al-Masry, W. (2021). Analysis of direct band gap A2ScInI6 (A=Rb, Cs) double perovskite halides using DFT approach for renewable energy devices. *Journal of Materials Research and Technology*, *13*, 2491–2500. DOI: 10.1016/j.jmrt.2021.05.080

Nyden, M. R., Stoliarov, S. I., Westmoreland, P. R., Guo, Z. X., & Jee, C. (2004). Applications of reactive molecular dynamics to the study of the thermal decomposition of polymers and nanoscale structures. *Materials Science and Engineering A*, *365*(1–2), 114–121. DOI: 10.1016/j.msea.2003.09.060

O, S. T., & Thompson, P. A. (1995). Molecular dynamics-continuum hybrid computations: A tool for studying complex fluid flows (Vol. 52, Issue 6).

O'Shea, J. P., Augustijns, P., Brandl, M., Brayden, D. J., Brouwers, J., Griffin, B. T., Holm, R., Jacobsen, A.-C., Lennernäs, H., Vinarov, Z., & O'Driscoll, C. M. (2022). Best practices in current models mimicking drug permeability in the gastrointestinal tract - An UNGAP review. *European Journal of Pharmaceutical Sciences*, *170*, 106098. DOI: 10.1016/j.ejps.2021.106098 PMID: 34954051

Odian, G. (n.d.). *PRINCIPLES OF POLYMERIZATION* (4th ed.). DOI: 10.1002/047147875X

Ogwu, M. C., Izah, S. C., & Joshua, M. T. (2025). Ecological and environmental determinants of phytochemical variability in forest trees. *Phytochemistry Reviews*, *24*(6), 5109–5137. Advance online publication. DOI: 10.1007/s11101-025-10066-0

Olamit, J., & Liu, K. (2007). Rotational hysteresis of the exchange anisotropy direction in Co/FeMn thin films. *Journal of Applied Physics*, *101*(9), 09E508. Advance online publication. DOI: 10.1063/1.2694378

Olatomiwa, A., Adam, T., Edet, C., Adewale, A., Chik, A., Mohammed, M., Gopinath, S. C., & Hashim, U. (2023). Recent advances in density functional theory approach for optoelectronics properties of graphene. *Heliyon*, *9*(3), e14279. DOI: 10.1016/j.heliyon.2023.e14279 PMID: 36950613

Omran, H. A., Majed, A. A., Hussein, K., Abid, D. S., Abdel-Maksoud, M. A., Elwahsh, A., Aufy, M., & Kotob, M. H. (2024). Anticancer activity, DFT and molecular docking study of new bisthiazolidine amide. *Results in Chemistry*, *12*, 101835. DOI: 10.1016/j.rechem.2024.101835

Oublal, M., Dib, M., & El-yakoubi, E. Y. (2025). First-principles study of optoelectronic properties of new lead-free double perovskites Cs2NaInX6 (X=Br, I) for next generation thin-film photovoltaics. *Micro and Nanostructures*, *209*, 208435. DOI: 10.1016/j.micrna.2025.208435

Oublal, M., Dib, M., Khalal, A., Chibane, F., Es-Salhi, E.-S., Nafidi, A., & El-Yakoubi, E. Y. (2025). Prediction of optoelectronic properties and efficiency of halide double perovskites Rb2NaInX6 (X=Br, I) for solar cell applications: A density functional theory approach. *Materials Science in Semiconductor Processing*, *200*, 109917. DOI: 10.1016/j.mssp.2025.109917

Ounoughi, C., & Ben Yahia, S. (2024). Sequence to sequence hybrid Bi-LSTM model for traffic speed prediction. *Expert Systems with Applications*, *236*, 121325. DOI: 10.1016/j.eswa.2023.121325

Padding, J. T., & Briels, W. J. (2011). Systematic coarse-graining of the dynamics of entangled polymer melts: The road from chemistry to rheology. In Journal of Physics Condensed Matter (Vol. 23, Issue 23). DOI: 10.1088/0953-8984/23/23/233101

Paggi, J. M., Pandit, A., & Dror, R. O. (2024). The art and science of molecular docking. *Annual Review of Biochemistry*, *93*(1), 389–410. DOI: 10.1146/annurev-biochem-030222-120000 PMID: 38594926

Paier, J., Marsman, M., & Kresse, G. (2007). Why does the B3LYP hybrid functional fail for metals? *The Journal of Chemical Physics*, *127*(2), 024103. DOI: 10.1063/1.2747249

Pakhomov, A. S., Skirdkov, P. N., Yurlov, V. V., Kichin, G. A., Hai, N. T., Wu, J. C., Huang, J. C. A., Su, S. H., Chernov, A. I., & Zvezdin, K. A. (2025). Temperature resolved spin-pumping in permalloy-bismuth selenide heterostructure. *Journal of Applied Physics*, *137*(20), 203903. Advance online publication. DOI: 10.1063/5.0259659

Palit, P., & Mandal, S. C. (2021). Climate change, geographical location, and other allied triggering factors modulate the standardization and characterization of traditional medicinal plants: A challenge and prospect for phyto-drug development. In S.C. Mandal, R. Chakraborty & S. Sen (Eds.), Evidence Based Validation of Traditional Medicines (pp. [numéros de pages si connus]). Springer, Singapore. https://doi.org/DOI: 10.1007/978-981-15-8127-4_18

Panda, P. K., & Sahoo, B. (2015). PZT to Lead Free Piezo Ceramics : A Review. *Ferroelectrics*, *474*(1), 128–143. DOI: 10.1080/00150193.2015.997146

Pandiarajan, I., Walunj, S. B., Banerjee, N., Rout, J., Srivastava, S., Patankar, S., & Kaledhonkar, S. (2024). Application of biolayer interferometry for the analysis of ribosome-protein interactions. *Frontiers in Molecular Biosciences, 11*, 1398964. DOI: 10.3389/fmolb.2024.1398964 PMID: 39148630

Pan, L., Xie, H., Cheng, X., Zhao, C., Feng, H., Cao, D., Wang, J., & Liu, Q. (2018). Tuning the ferromagnetic resonance frequency of soft magnetic film by patterned permalloy micro-stripes with stripe-domain. *Journal of Magnetism and Magnetic Materials, 457*, 46–51. DOI: 10.1016/j.jmmm.2018.02.060

Pannico, M., Mensitieri, G., & Musto, P. (2024). In-situ FTIR spectroscopy of epoxy resin degradation: Kinetics and mechanisms. *Frontiers in Chemistry, 12*, 1476965. DOI: 10.3389/fchem.2024.1476965

Panomsuwan, G., & Manuspiya, H. (2018). Structural and dielectric properties of sol–gel derived $Ba_{1-x}Sr_xTiO_3$ ($0 \leq x \leq 0.5$) ceramics for energy storage applications. *Materials Research Express, 6*(2), 026310. DOI: 10.1088/2053-1591/aaedf2

Pantic, I., Paunovic, J., Pejic, S., Drakulic, D., Todorovic, A., Stankovic, S., Vucevic, D., Cumic, J., & Radosavljevic, T. (2022). Artificial intelligence approaches to the biochemistry of oxidative stress: Current state of the art. *Chemico-Biological Interactions, 358*, 109888. DOI: 10.1016/j.cbi.2022.109888

Pant, P., Pandey, S., & Dall'Acqua, S. (2021). The influence of environmental conditions on secondary metabolites in medicinal plants: A literature review. *Chemistry & Biodiversity, 18*(11), e2100345. DOI: 10.1002/cbdv.202100345

Pan, Z., Li, R. H., Cui, Y. Y., Wu, X. J., Zhang, Y. Y., & Wang, Y. T. (2021). A simple and quick method to detect adulterated sesame oil using 3D fluorescence spectra. *Spectrochimica Acta. Part A: Molecular and Biomolecular Spectroscopy, 245*, 118948. DOI: 10.1016/j.saa.2020.118948 PMID: 32980759

Papon, P. (2021). AIE (Agence internationale de l'énergie), Net Zero by 2050: A Roadmap for the Global Energy Sector, Paris: AIE / OCDE, mai 2021, 224 p. Futuribles, N° 444(5), VI–VI. DOI: 10.3917/futur.444.0113f

Park, H. S., Hwang, J. S., Choi, W. Y., Shim, D. S., Na, K. W., & Choi, S. O. (2004). Development of micro-fluxgate sensors with electroplated magnetic cores for electronic compass. *Sensors and Actuators. A, Physical, 114*(2–3), 224–229. DOI: 10.1016/j.sna.2003.11.037

Park, H. S., Karpov, E. G., Wing, K. L., & Klein, P. A. (2005). The bridging scale for two-dimensional atomistic/continuum coupling. *Philosophical Magazine, 85*(1), 79–113. DOI: 10.1080/14786430412331300163

Park, J. H., Lee, J. W., Choi, H. J., Jang, W. G., Kim, T. S., Suh, D. S., Jeong, H. Y., Chang, S. Y., Roh, J. C., Yoo, C. S., Kim, K. H., Park, C., & Suh, S. J. (2019). Electromagnetic interference shielding effectiveness of sputtered NiFe/Cu multi-layer thin film at high frequencies. *Thin Solid Films, 677*, 130–136. DOI: 10.1016/j.tsf.2019.03.015

Parmar, V. B., Raval, D., Gupta, S. K., Gajjar, P. N., & Vora, A. M. (2023). BaTiO3 perovskite for optoelectronics application : A DFT study. Materials Today: Proceedings, S2214785323005059. DOI: 10.1016/j.matpr.2023.01.410

Paruzzo, F. M., Hofstetter, A., Musil, F., De, S., Ceriotti, M., & Emsley, L. (2018). Chemical shifts in molecular solids by machine learning. *Nature Communications*, *9*(1), 4501. DOI: 10.1038/s41467-018-06972-x PMID: 30374021

Pascual-Borràs, M., López, X., Rodríguez-Fortea, A., Errington, R. J., & Poblet, J. M. (2014). 17O NMR chemical shifts in oxometalates: From the simplest monometallic species to mixed-metal polyoxometalates. *Chemical Science (Cambridge)*, *5*(5), 2031. DOI: 10.1039/c4sc00083h

Pateiro, M., Gómez-Salazar, J. A., Jaime-Patlán, M., Sosa-Morales, M. E., & Lorenzo, J. M. (2021). Plant extracts obtained with green solvents as natural antioxidants in fresh meat products. *Antioxidants*, *10*(2), 181. DOI: 10.3390/antiox10020181

Patil, S. P., Heider, Y., Hernandez Padilla, C. A., Cruz-Chú, E. R., & Markert, B. (2016). A comparative molecular dynamics-phase-field modeling approach to brittle fracture. *Computer Methods in Applied Mechanics and Engineering*, *312*, 117–129. DOI: 10.1016/j.cma.2016.04.005

Patil, S. U., Shah, S. P., Olaya, M., Deshpande, P. P., Maiaru, M., & Odegard, G. M. (2021). Reactive molecular dynamics simulation of epoxy for the full cross-linking process. *ACS Applied Polymer Materials*, *3*(11), 5788–5797. DOI: 10.1021/acsapm.1c01024

Patrahau, B., Piejko, M., Mayer, R. J., Antheaume, C., Sangchai, T., Ragazzon, G., Jayachandran, A., Devaux, E., Genet, C., Moran, J., & Ebbesen, T. W. (2024). Direct observation of polaritonic chemistry by nuclear magnetic resonance spectroscopy. *Angewandte Chemie International Edition*, *63*(23), e202401368. DOI: 10.1002/anie.202401368 PMID: 38584127

Pedregosa, F., Varoquaux, G., Gramfort, A., Michel, V., Thirion, B., Grisel, O., Blondel, M., Prettenhofer, P., Weiss, R., & Dubourg, V. (2011). Scikit-learn: Machine learning in Python. *Journal of Machine Learning Research*, *12*, 2825–2830.

Peng, B., Zhang, W. L., Xie, Q. Y., Zhang, W. X., & Jiang, H. C. (2013). Effect of sputtering pressure on microstructure and magnetic properties of amorphous FeCoSiB films. *Journal of Non-Crystalline Solids*, *365*(1), 59–62. DOI: 10.1016/j.jnoncrysol.2013.01.028

Peng, W., Wang, L., Li, Y., Du, Y., He, Z., Wang, C., Zhao, Y., Jiang, Z., Zhou, Z., & Liu, M. (2022). Photoelectron-induced quantitative regulation of ferromagnetism in Permalloy at room temperature for photovoltaic flexible spintronics. *Journal of Alloys and Compounds*, *910*, 164903. DOI: 10.1016/j.jallcom.2022.164903

Perdew, J. P., Burke, K., & Ernzerhof, M. (1996). Generalized Gradient Approximation Made Simple. *Physical Review Letters*, *77*(18), 3865–3868. DOI: 10.1103/PhysRevLett.77.3865

Perdew, J. P., Chevary, J. A., Vosko, S. H., Jackson, K. A., Pederson, M. R., Singh, D. J., & Fiolhais, C. (1993). Erratum: Atoms, molecules, solids, and surfaces: Applications of the generalized gradient approximation for exchange and correlation. *Physical Review B: Condensed Matter*, *48*(7), 4978. DOI: 10.1103/PhysRevB.48.4978.2

Perdew, J. P., Ruzsinszky, A., Csonka, I., Vydrov, O. A., Scuseria, G. E., Constantin, L. A., Zhou, X., & Burke, K. (2008). Restoring the Density-Gradient Expansion for Exchange in Solids and Surfaces. *Physical Review Letters*, *100*(13), 136406. DOI: 10.1103/PhysRevLett.100.136406

Pérez-Lucas, G., Navarro, G., & Navarro, S. (2024). Adapting agriculture and pesticide use in Mediterranean regions under climate change scenarios: A comprehensive review. *European Journal of Agronomy*, *161*, 127337. DOI: 10.1016/j.eja.2024.127337

Peter, C., & Kremer, K. (2009b). Multiscale simulation of soft matter systems - From the atomistic to the coarse-grained level and back. In Soft Matter (Vol. 5, Issue 22, pp. 4357–4366). DOI: 10.1039/b912027k

Peter, C., & Kremer, K. (2009a). Multiscale simulation of soft matter systems. *Faraday Discussions*, *144*, 9–24. DOI: 10.1039/B919800H

Petrauskas, V., Kazlauskas, E., Gedgaudas, M., Baranauskienė, L., Zubrienė, A., & Matulis, D. (2024). Thermal shift assay for protein–ligand dissociation constant determination. *Trends in Analytical Chemistry*, *170*, 117417. DOI: 10.1016/j.trac.2023.117417

Petsev, N. D., Leal, L. G., & Shell, M. S. (2017). Coupling discrete and continuum concentration particle models for multiscale and hybrid molecular-continuum simulations. *The Journal of Chemical Physics*, *147*(23), 234112. DOI: 10.1063/1.5001703

Pfeifer, F., & Radeloff, C. (1980). Soft magnetic Ni-Fe and Co-Fe alloys - some physical and metallurgical aspects. *Journal of Magnetism and Magnetic Materials*, *19*(1–3), 190–207. DOI: 10.1016/0304-8853(80)90592-2

Pham, L. V. D., Hinsche, N. F., & Mertig, I. (2025). Quantum spin Hall phase in the truncated trihexagonal lattice: a topological archimedean structure. *2D Materials, 13*(1), 015021. DOI: 10.1088/2053-1583/ae23e1

Pham, P., Nguyen, V. T. D., Cho, K. H., & Hy, T.-S. (2025). DrugPipe: Generative artificial intelligence-assisted virtual screening pipeline for generalizable and efficient drug repurposing. *Biology Methods and Protocols*, *10*(1), bpaf038. Advance online publication. DOI: 10.1093/biomethods/bpaf038 PMID: 40486179

Phuoc, N. N., Chai, G., & Ong, C. K. (2012). Temperature-dependent dynamic magnetization of FeCoHf thin films fabricated by oblique deposition. *Journal of Applied Physics*, *112*(8), 083925. Advance online publication. DOI: 10.1063/1.4763361

Phuoc, N. N., Chapon, P., Acher, O., & Ong, C. K. (2013). Large magneto-elastic anisotropy enhancement with temperature in composition-graded FeCoTa thin films. *Journal of Applied Physics*, *114*(15), 153903. Advance online publication. DOI: 10.1063/1.4825225

Phuoc, N. N., & Ong, C. K. (2011). Influence of ferromagnetic thickness on dynamic anisotropy in exchange-biased MnIr/FeCo multilayered thin films. *Physica B, Condensed Matter*, *406*(18), 3514–3518. DOI: 10.1016/j.physb.2011.05.063

Phuoc, N. N., & Ong, C. K. (2013). Anomalous Temperature Dependence of Magnetic Anisotropy in Gradient-Composition Sputterred Thin Films. *Advanced Materials*, *25*(7), 980–984. DOI: 10.1002/adma.201203995

Phuoc, N. N., & Ong, C. K. (2014). Gradient-Composition Sputtering: An Approach to Fabricate Magnetic Thin Films With Magnetic Anisotropy Increased With Temperature. *IEEE Transactions on Magnetics*, *50*(7), 1–6. DOI: 10.1109/TMAG.2013.2296936

Phuoc, N. N., Xu, F., Ma, Y., & Ong, C. K. (2009). Permalloy–FeMn exchange-biased multilayers grown on flexible substrates for microwave applications. *Journal of Magnetism and Magnetic Materials*, *321*(17), 2685–2690. DOI: 10.1016/j.jmmm.2009.03.073

Picado, A., Chaikuad, A., Wells, C. I., Shrestha, S., Zuercher, W. J., Pickett, J. E., Kwarcinski, F. E., Sinha, P., de Silva, C. S., Zutshi, R., Liu, S., Kannan, N., Knapp, S., Drewry, D. H., & Willson, T. M. (2020). A chemical probe for dark kinase STK17B derives its potency and high selectivity through a unique P-loop conformation. *Journal of Medicinal Chemistry*, *63*(23), 14626–14646. DOI: 10.1021/acs.jmedchem.0c01174 PMID: 33215924

Pierens, G. K. (2014). 1H and 13C NMR scaling factors for the calculation of chemical shifts in commonly used solvents using density functional theory. *Journal of Computational Chemistry*, *35*(18), 1388–1394. DOI: 10.1002/jcc.23638 PMID: 24854878

Pierens, G. K., Venkatachalam, T. K., & Reutens, D. C. (2017). NMR and DFT investigations of structure of colchicine in various solvents including density functional theory calculations. *Scientific Reports*, *7*(1), 5605. DOI: 10.1038/s41598-017-06005-5 PMID: 28717218

Plathottam, S. J., Rzonca, A., Lakhnori, R., & Iloeje, C. O. (2023). A review of artificial intelligence applications in manufacturing operations. *Journal of Advanced Manufacturing and Processing*, *5*(3), e10159. DOI: 10.1002/amp2.10159

Platt, C. L., Berkowitz, A. E., Smith, D. J., & McCartney, M. R. (2000). Correlation of coercivity and microstructure of thin CoFe films. *Journal of Applied Physics*, *88*(4), 2058–2062. DOI: 10.1063/1.1305833

Poblete, S., Praprotnik, M., Kremer, K., & Delle Site, L. (2010). Coupling different levels of resolution in molecular simulations. *The Journal of Chemical Physics*, *132*(11), 114101. Advance online publication. DOI: 10.1063/1.3357982

Polo, V., Kraka, E., & Cremer, D. (2002). Electron correlation and the self-interaction error of density functional theory. *Molecular Physics*, *100*(11), 1771–1790. DOI: 10.1080/00268970110111788

Ponder, J. W., Wu, C., Ren, P., Pande, V. S., Chodera, J. D., Schnieders, M. J., Haque, I., Mobley, D. L., Lambrecht, D. S., DiStasio, R. A.Jr, Head-Gordon, M., Clark, G. N. I., Johnson, M. E., & Head-Gordon, T. (2010). Current status of the AMOEBA polarizable force field. *The Journal of Physical Chemistry B*, *114*(8), 2549–2564. DOI: 10.1021/jp910674d PMID: 20136072

Popov, K. I., Wellnitz, J., Maxfield, T., & Tropsha, A. (2024). Hit discovery using docking ENriched by GEnerative modeling (HIDDEN GEM): A novel computational workflow for accelerated virtual screening of ultra-large chemical libraries. *Molecular Informatics*, *43*(1), e202300207. DOI: 10.1002/minf.202300207 PMID: 37802967

Possenti, E., Colombo, C., Realini, M., Song, C. L., & Kazarian, S. G. (2021). Time-resolved ATR–FTIR spectroscopy and macro ATR–FTIR spectroscopic imaging of inorganic treatments for stone conservation. *Analytical Chemistry*, *93*(44), 14635–14642. DOI: 10.1021/acs.analchem.1c02392

Potestio, R., Fritsch, S., Español, P., Delgado-Buscalioni, R., Kremer, K., Everaers, R., & Donadio, D. (2013). Hamiltonian adaptive resolution simulation for molecular liquids. *Physical Review Letters, 110*(10), 108301. Advance online publication. DOI: 10.1103/PhysRevLett.110.108301

Prabhu, A. A., Chityala, S., Jayachandran, D., Deshavath, N. N., & Veeranki, V. D. (2021). A two step optimization approach for maximizing biosorption of hexavalent chromium ions (Cr (VI)) using alginate immobilized Sargassum sp in a packed bed column. *Separation Science and Technology, 56*(1), 90–106. DOI: 10.1080/01496395.2019.1708933

Praprotnik, M., Delle Site, L., & Kremer, K. (2005). Adaptive resolution molecular-dynamics simulation: Changing the degrees of freedom on the fly. *The Journal of Chemical Physics, 123*(22), 224106. Advance online publication. DOI: 10.1063/1.2132286

Praprotnik, M., Matysiak, S., Site, L. D., Kremer, K., & Clementi, C. (2008). Adaptive resolution simulation of liquid water. *Journal of Physics Condensed Matter, 21*(49), 499801–499801. DOI: 10.1088/0953-8984/21/49/499801

Praprotnik, M., Site, L. D., & Kremer, K. (2008). Multiscale simulation of soft matter: From scale bridging to adaptive resolution. *Annual Review of Physical Chemistry, 59*(1), 545–571. DOI: 10.1146/annurev.physchem.59.032607.093707

Priya, P., & Stonier, A. A. (2025). Emerging innovations in solar photovoltaic (PV) technologies: The perovskite solar cells and more. *Energy Reports, 14*, 216–242. DOI: 10.1016/j.egyr.2025.06.003

Provorse, M. R., Peev, T., Xiong, C., & Isborn, C. M. (2016). Convergence of excitation energies in mixed quantum and classical solvent: Comparison of continuum and point charge models. *The Journal of Physical Chemistry B, 120*(47), 12148–12159. DOI: 10.1021/acs.jpcb.6b09176 PMID: 27797196

Prudhomme, S., Ben Dhia, H., Bauman, P. T., Elkhodja, N., & Oden, J. T. (2008). Computational analysis of modeling error for the coupling of particle and continuum models by the Arlequin method. *Computer Methods in Applied Mechanics and Engineering, 197*(41–42), 3399–3409. DOI: 10.1016/j.cma.2008.03.014

Pruitt, S. R., Brorsen, K. R., & Gordon, M. S. (2015). Ab initio investigation of the aqueous solvation of the nitrate ion. *Physical Chemistry Chemical Physics, 17*(40), 27027–27034. DOI: 10.1039/C5CP04445F PMID: 26412597

Punyawatthananukool, S., Matsuura, R., Wongchang, T., Katsurada, N., Tsuruyama, T., Tajima, M., Enomoto, Y., Kitamura, T., Kawashima, M., Toi, M., Yamanoi, K., Hamanishi, J., Hisamori, S., Obama, K., Charoensawan, V., Thumkeo, D., & Narumiya, S. (2024). Prostaglandin E2-EP2/EP4 signaling induces immunosuppression in human cancer by impairing bioenergetics and ribosome biogenesis in immune cells. *Nature Communications, 15*(1), 9464. DOI: 10.1038/s41467-024-53706-3 PMID: 39487111

Pütz, M., & Nielaba, P. (2015). Effects of temperature on spinodal decomposition and domain growth of liquid-vapor systems with smoothed particle hydrodynamics. *Physical Review E: Statistical, Nonlinear, and Soft Matter Physics, 91*(3), 032303. DOI: 10.1103/PhysRevE.91.032303

Qaderi, M. M., Martel, A. B., & Strugnell, C. A. (2023). Environmental factors regulate plant secondary metabolites. *Plants, 12*(3), 447. DOI: 10.3390/plants12030447

Qiao, J., Pizzi, G., & Marzari, N. (2023). Automated mixing of maximally localized Wannier functions into target manifolds. *npj Computational Materials*, *9*(1), 206. Advance online publication. DOI: 10.1038/s41524-023-01147-9 PMID: 38666055

Qin, M., Gild, J., Hu, C., Wang, H., Hoque, M. S., Braun, J. L., Harrington, T. J., Hopkins, P. E., Vecchio, K. S., & Luo, J. (2020). Dual-Phase High-Entropy Ultra-High Temperature Ceramics. *Journal of the European Ceramic Society*, *40*(15), 5037–5050. DOI: 10.1016/j.jeurceramsoc.2020.05.040

Qi, X., Ren, P., Tong, X., Wang, X., & Zhuo, F. (2025). Enhanced piezoelectric properties of KNN-based ceramics by synergistic modulation of phase constitution, grain size and domain configurations. *Journal of the European Ceramic Society*, *45*(1), 116874. DOI: 10.1016/j.jeurceramsoc.2024.116874

Qi, Y., Wang, L.-P., Guo, Z., & Chen, S. (2023). Consistent lifting relations for the initialization of total-energy double-distribution-function kinetic models. *Physical Review. E*, *108*(6), 065301. DOI: 10.1103/PhysRevE.108.065301

Quan, Y., Ren, W., Niu, G., Wang, L., Zhao, J., Zhang, N., Liu, M., Ye, Z., Liu, L., & Karaki, T. (2018). Large Piezoelectric Strain with Superior Thermal Stability and Excellent Fatigue Resistance of Lead-Free Potassium Sodium Niobate-Based Grain Orientation-Controlled Ceramics. *ACS Applied Materials & Interfaces*, *10*(12), 10220–10226. DOI: 10.1021/acsami.8b01554

Radziuk, D., & Möhwald, H. (2016). Ultrasonically treated liquid interfaces for progress in cleaning and separation processes. *Physical Chemistry Chemical Physics*, *18*(1), 21–46. DOI: 10.1039/C5CP05142H

Rahman, B. M. F., Divan, R., Rosenmann, D., Wang, T., Peng, Y., & Wang, G. (2015). Application of sub-micrometer patterned permalloy thin film in tunable radio frequency inductors. Journal of Applied Physics, 117(17), 17C121. DOI: 10.1063/1.4918766

Raingeval, C., Cala, O., Brion, B., Le Borgne, M., Hubbard, R. E., & Krimm, I. (2019). ^{1}D NMR WaterLOGSY as an efficient method for fragment-based lead discovery. *Journal of Enzyme Inhibition and Medicinal Chemistry*, *34*(1), 1218–1225. DOI: 10.1080/14756366.2019.1636235 PMID: 31286785

Rajapriyadharshini, J. R. (2022). An improved smoothed particle hydrodynamics approach using new inverse kernel function. *Journal of Ocean Engineering and Science*, *7*(4), 327–336. DOI: 10.1016/j.joes.2021.09.004

Ramachandran, P., Bhosale, A., Puri, K., Negi, P., Muta, A., Dinesh, A., Menon, D., Govind, R., Sanka, S., Sebastian, A. S., Sen, A., Kaushik, R., Kumar, A., Kurapati, V., Patil, M., Tavker, D., Pandey, P., Kaushik, C., Dutt, A., & Agarwal, A. (2021). PySPH: A Python-based framework for smoothed particle hydrodynamics. [TOMS]. *ACM Transactions on Mathematical Software*, *47*(4), 1–38. DOI: 10.1145/3460773

Ramos, S. A., Mueller, L. J., & Beran, G. J. O. (2024). The interplay of density functional selection and crystal structure for accurate NMR chemical shift predictions. *Faraday Discussions*, *255*(0), 119–142. DOI: 10.1039/D4FD00072B PMID: 39258864

Ranchal, R., Aroca, C., & López, E. (2008). Domain walls and exchange-interaction in Permalloy/Gd films. *New Journal of Physics*, *10*(1), 013013. DOI: 10.1088/1367-2630/10/1/013013

Ranganathan, S. I., & Ostoja-Starzewski, M. (2008). Universal elastic anisotropy index. *Physical Review Letters, 101*(5), 055504. DOI: 10.1103/PhysRevLett.101.055504

Rantschler, J. O., & Alexander, C.Jr. (2003). Ripple field effect on high-frequency measurements of FeTiN films. *Journal of Applied Physics, 93*(10), 6665–6667. DOI: 10.1063/1.1556097

Rappe, A. K., & Goddard, W. A.III. (1991). Charge equilibration for molecular dynamics simulations. *Journal of Physical Chemistry, 95*(8), 3358–3363. DOI: 10.1021/j100161a070

Raschka, S., & Mirjalili, V. (2019). *Python machine learning: Machine learning and deep learning with Python, scikit-learn, and TensorFlow 2*. Packt Publishing Ltd.

Rasekh, M., Karami, H., Wilson, A. D., & Gancarz, M. (2021). Classification and identification of essential oils from herbs and fruits based on a MOS electronic-nose technology. *Chemosensors (Basel, Switzerland), 9*(6), 142. DOI: 10.3390/chemosensors9060142

Rasul, R. A., Saha, P., Bala, D., Karim, S. M. R. U., Abdullah, M. I., & Saha, B. (2024). An evaluation of machine learning approaches for early diagnosis of autism spectrum disorder. *Healthcare Analytics, 5*, 100293. DOI: 10.1016/j.health.2023.100293

Razzak, S., Aouji, M., Sabri, C., Benchehida, H., Taibi, M., & Taboz, Y. (2024b). Effect of solvent polarity on phytochemical analysis of different Moroccan sites of *Opuntia ficus-indica* (L.) Mill. pads. *International Journal of Secondary Metabolite, 11*(1), 93–104. DOI: 10.21448/ijsm.1354740

Razzak, S., Aouji, M., Zirari, M., Benchehida, H., Taibi, M., Bengueddour, R., Wondmie, G. F., Ibenmoussa, S., Bin Jardan, Y. A., & Taboz, Y. (2024a). Nutritional composition, functional and chemical characterization of Moroccan *Opuntia ficus-indica* cladode powder. *International Journal of Food Properties, 27*(1), 1167–1179. DOI: 10.1080/10942912.2024.2382877

Reckling, M., Ahrends, H., Chen, T.-W., Eugster, W., Hadasch, S., Knapp, S., Laidig, F., Linstädter, A., Macholdt, J., Piepho, H.-P., Schiffers, K., & Friedrich, T. (2021). Methods of yield stability analysis in long-term field experiments: A review. *Agronomy for Sustainable Development, 41*(2), 27. DOI: 10.1007/s13593-021-00681-4

Richard-Bollans, A., Aitken, C., Antonelli, A., Bitencourt, C., Goyder, D., Lucas, E., Ondo, I., Pérez-Escobar, O. A., Pironon, S., Richardson, J. E., Russell, D., Silvestro, D., Wright, C. W., & Howes, M.-J. R. (2023). Machine learning enhances prediction of plants as potential sources of antimalarials. *Frontiers in Plant Science, 14*, 1173328. DOI: 10.3389/fpls.2023.1173328

Richter, H., Wang, Z. P., & Ley, L. (1981). The one phonon Raman spectrum in microcrystalline silicon. *Solid State Communications, 39*(5), 625–629. DOI: 10.1016/0038-1098(81)90337-9

Ringe, E., Van Duyne, R. P., & Marks, L. D. (2013). Kinetic and thermodynamic modified wulff constructions for twinned nanoparticles. *The Journal of Physical Chemistry. C, Nanomaterials and Interfaces, 117*(31), 15859–15870. DOI: 10.1021/jp401566m

Ripka, P. (1992). Review of fluxgate sensors. *Sensors and Actuators. A, Physical, 33*(3), 129–141. DOI: 10.1016/0924-4247(92)80159-Z

Ripka, P. (2003). Advances in fluxgate sensors. *Sensors and Actuators. A, Physical, 106*(1–3), 8–14. DOI: 10.1016/S0924-4247(03)00094-3

Ripka, P. (2008). Sensors based on bulk soft magnetic materials: Advances and challenges. *Journal of Magnetism and Magnetic Materials, 320*(20), 2466–2473. DOI: 10.1016/j.jmmm.2008.04.079

Robert, P. (1999). *Matériaux de l'électronique* (Vol. 2). EPFL Press.

Rodrigues, C., Paula, C. D., Lahbouki, S., Meddich, A., Outzourhit, A., Rashad, M., Pari, L., Coelhoso, I., Fernando, A. L., & Souza, V. G. L. (2023). Opuntia spp.: An overview of the bioactive profile and food applications of this versatile crop adapted to arid lands. *Foods, 12*(7), 1465. DOI: 10.3390/foods12071465

Rohman, A., & Che Man, Y. B. (2011). The use of Fourier transform mid infrared (FT-MIR) spectroscopy for detection and quantification of adulteration in virgin coconut oil. *Food Chemistry, 129*(2), 583–588. DOI: 10.1016/j.foodchem.2011.04.070 PMID: 30634271

Rohman, A., Che Man, Y. B., & Ali, M. E. (2019). The authentication of virgin coconut oil from grape seed oil and soybean oil using FTIR spectroscopy and chemometrics. *International Journal of Applied Pharmaceutics, 11*(2), 259–263. DOI: 10.22159/ijap.2019v11i2.31758

Rohman, A., & Man, C. (2012). Y. B. (2012). Quantification and classification of corn and sunflower oils as adulterants in olive oil using chemometrics and FTIR spectra. *TheScientificWorldJournal, 2012*, 250795. Advance online publication. DOI: 10.1100/2012/250795 PMID: 22448127

Roknuzzaman, M., Ostrikov, K., Chandula Wasalathilake, K., Yan, C., Wang, H., & Tesfamichael, T. (2018). Insight into lead-free organic-inorganic hybrid perovskites for photovoltaics and optoelectronics: A first-principles study. *Organic Electronics, 59*, 99–106. DOI: 10.1016/j.orgel.2018.04.051

Roselli, V., Pugliese, G., Leuci, R., Brunetti, L., Gambacorta, L., Tufarelli, V., & Piemontese, L. (2024). Green methods to recover bioactive compounds from food industry waste: A sustainable practice from the perspective of the circular economy. *Molecules (Basel, Switzerland), 29*(11), 2682. DOI: 10.3390/molecules29112682

Royani, A., Mubarak, N. M., Hanafi, M., Verma, C., Lotulung, P. D. N., Prastya, M. E., Alfantazi, A., & Manaf, A. (2025). Effect of solvent polarity on yield extract, antioxidant and antibacterial activities of phytochemicals from Andrographis paniculata leaves. *Indian Chemical Engineer, 67*(1), 1–15. DOI: 10.1080/00194506.2024.2409260

Rudd, R. E., & Broughton, J. Q. (1998). Coarse-grained molecular dynamics and the atomic limit of finite elements. *Physical Review B: Condensed Matter, 58*(10), R5893–R5896. DOI: 10.1103/PhysRevB.58.R5893

Rudeichuk, T., Olekšáková, D., Maciaszek, R., Matysiak, W., & Kollár, P. (2024). Exploring the Impact of Different Milling Parameters of Fe/SiO2 Composites on Their Structural and Magnetic Properties. *Materials (Basel), 17*(4), 862. DOI: 10.3390/ma17040862

Rychnovsky, S. D. (2006). Predicting NMR spectra by computational methods: Structure revision of hexacyclinol. *Organic Letters, 8*(13), 2895–2898. DOI: 10.1021/ol0611346 PMID: 16774284

Sabir, Z., Ismail, T., Sleem, H., Umar, M., & Salahshour, S. (2025). A radial basis Bayesian regularization neural network process for the malaria disease model. *Knowledge-Based Systems, 322*(October 2024), 113722. DOI: 10.1016/j.knosys.2025.113722

Sadjina, S., Kyllingstad, L. T., Skjong, S., & Pedersen, E. (2017). Energy conservation and power bonds in co-simulations: Non-iterative adaptive step size control and error estimation. *Engineering with Computers, 33*(3), 607–620. DOI: 10.1007/s00366-016-0492-8

Saeed, Y., Amin, B., Khalil, H., Rehman, F., Ali, H., Khan, M. I., Mahmood, A., & Shafiq, M. (2020). Cs2NaGaBr6: A new lead-free and direct band gap halide double perovskite. *RSC Advances, 10*(30), 17444–17451. DOI: 10.1039/D0RA01764G

Sagadevan, S., Lett, J. A., Fatimah, I., Lokanathan, Y., Léonard, E., Oh, W. C., Hossain, M. M., & Johan, M. R. (2021). Current trends in the green syntheses of tin oxide nanoparticles and their biomedical applications. *Materials Research Express, 8*(8), 082001. DOI: 10.1088/2053-1591/ac187e

Sagredo, F., Gant, S. E., Ohad, G., Haber, J. B., Filip, M. R., Kronik, L., & Neaton, J. B. (2024). Electronic structure and optical properties of halide double perovskites from a Wannier-localized optimally-tuned screened range-separated hybrid functional. *Physical Review Materials, 8*(10), 105401. DOI: 10.1103/PhysRevMaterials.8.105401

Sahakyan, H. (2021). Improving virtual screening results with MM/GBSA and MM/PBSA rescoring. *Journal of Computer-Aided Molecular Design, 35*(6), 731–736. DOI: 10.1007/s10822-021-00389-3 PMID: 33983518

Saif, N., Tariq, A., Intisar, A., Batool, Z., Arshad, M., Rehman, A. U., & Kousar, R. (2023). Green synthesis of SnO2 nanorods and Mo– SnO2 nanocomposite for efficient sunlight driven organic dye photodegradation in aqueous solution. *Materials Chemistry and Physics, 305*, 127920. DOI: 10.1016/j.matchemphys.2023.127920

Saito, Y., Takao, H., Tani, T., Nonoyama, T., Takatori, K., Homma, T., Nagaya, T., & Nakamura, M. (2004). Lead-free piezoceramics. *Nature, 432*(7013), 84–87. DOI: 10.1038/nature03028

Sakhya, A. P., Maibam, J., Saha, S., Chanda, S., Dutta, A., Sharma, B. I., Thapa, R. K., & Sinha, T. P. (2015). Electronic structure and elastic properties of ATiO3 (A = Ba, Sr, Ca) perovskites : A first principles study. *APPL PHYS*, •••, 53.

Salahun, E., Tanné, G., Quéffélec, P., Lefloc'h, M., Adenot, A., & Acher, O. (2001). Application of ferromagnetic composite in different planar tunable microwave devices. *Microwave and Optical Technology Letters, 30*(4), 272–276. DOI: 10.1002/mop.1288

Salem, M. A., Perez de Souza, L., Serag, A., Fernie, A. R., Farag, M. A., Ezzat, S. M., & Alseekh, S. (2020). Metabolomics in the context of plant natural products research: From sample preparation to metabolite analysis. *Metabolites, 10*(1), 37. DOI: 10.3390/metabo10010037

Sarker, I. H. (2021). Machine learning: Algorithms, real-world applications and research directions. *SN Computer Science, 2*(3), 160. DOI: 10.1007/s42979-021-00592-x

Sarotti, A. M., & Pellegrinet, S. C. (2009). A multi-standard approach for GIAO ^{13}C NMR calculations. *The Journal of Organic Chemistry, 74*(19), 7254–7260. DOI: 10.1021/jo901234h PMID: 19725561

Sarotti, A. M., & Pellegrinet, S. C. (2012). Application of the multi-standard methodology for calculating 1H NMR chemical shifts. *The Journal of Organic Chemistry, 77*(14), 6059–6065. DOI: 10.1021/jo3008447 PMID: 22713105

Scardino, V., Galarce, M. J., Mignone, M. E., & Cavasotto, C. N. (2025). Enhancing the reliability of integrated consensus strategies to boost docking-based screening campaigns using publicly available docking programs. *Molecular Informatics, 44*(5-6), e2445. DOI: 10.1002/minf.2445 PMID: 40537017

Schleder, G. R., Padilha, A. C. M., Acosta, C. M., Costa, M., & Fazzio, A. (2019). From DFT to machine learning : Recent approaches to materials science–a review. *JPhys Materials, 2*(3), 032001. DOI: 10.1088/2515-7639/ab084b

Schmidt, M., & Lipson, H. (2009). Distilling free-form natural laws from experimental data. *Science, 324*(5923), 81–85. DOI: 10.1126/science.1165893 PMID: 19342586

Schmidt-Mende, L., Fakharuddin, A., & Kraner, S. (2022). Organic and Hybrid Solar Cells: An Introduction. Organic and Hybrid Solar Cells: An Introduction, 2nd Edition, 1–348. https://doi.org/DOI: 10.1515/9783110736939/XML

Schmidt-Mende, L., Kraner, S., & Fakharuddin, A. (2022). *Organic and hybrid solar cells*. Walter de Gruyter GmbH & Co KG. DOI: 10.1515/9783110736939

Schütt, K. T., Sauceda, H. E., Kindermans, P. J., Tkatchenko, A., & Müller, K. R. (2018). SchNet - A deep learning architecture for molecules and materials. *The Journal of Chemical Physics, 148*(24), 241722. Advance online publication. DOI: 10.1063/1.5019779

Schwarz, G. (1978). Estimating the dimension of a model. *Annals of Statistics, 6*(2), 461–464. DOI: 10.1214/aos/1176344136

Schwarz, K. (2003). DFT calculations of solids with LAPW and WIEN2k. *Journal of Solid State Chemistry, 176*(2), 319–328. DOI: 10.1016/S0022-4596(03)00213-5

Seemann, K., Leiste, H., & Bekker, V. (2006). A new generation of CMOS-compatible high frequency micro-inductors with ferromagnetic cores: Theory, fabrication and characterisation. *Journal of Magnetism and Magnetic Materials, 302*(2), 321–326. DOI: 10.1016/j.jmmm.2005.05.042

Sellaoui, L., Nour, S., Dhaouadi, F., & Bonilla-petriciolet, A. (2024). Physicochemical study of the adsorption of chlorophenols on activated carbon using an advanced double layer model. *Journal of Molecular Liquids, 414*(PA), 126098. DOI: 10.1016/j.molliq.2024.126098

Sendner, M., Pabitra Nayak, ab K., Egger, D. A., Beck, S., Christian Mü ller, ab, Bernd Epding, abe, Wolfgang Kowalsky, ae, Leeor Kronik, abe, Snaith, H. J., & Pucci abf, A. (2016). Optical phonons in methylammonium lead halide perovskites and implications for charge transport †. 3, 613. DOI: 10.1039/C6MH00275G

Senftle, T. P., Hong, S., Islam, M. M., Kylasa, S. B., Zheng, Y., Shin, Y. K., Junkermeier, C., Engel-Herbert, R., Janik, M. J., Aktulga, H. M., Verstraelen, T., Grama, A., & Van Duin, A. C. T. (2016). *The ReaxFF reactive force-field: Development, applications and future directions. In npj Computational Materials* (Vol. 2). Nature Publishing Group., DOI: 10.1038/npjcompumats.2015.11

Sengul, M. Y., Song, Y., Nayir, N., Gao, Y., Hung, Y., Dasgupta, T., & van Duin, A. C. T. (2021). INDEEDopt: A deep learning-based ReaxFF parameterization framework. *npj Computational Materials*, *7*(1), 68. DOI: 10.1038/s41524-021-00534-4

Şen, M. (2024). Optimization of performance and emission of a diesel engine fueled with isopropyl alcohol Blends: A comparative ANN-GA and RSM-HCO application. *Engineering Science and Technology, an International Journal*, *55*(June), 101733. Advance online publication. DOI: 10.1016/j.jestch.2024.101733

Senn, H. M., & Thiel, W. (2009). QM/MM methods for biomolecular systems. *Angewandte Chemie International Edition*, *48*(7), 1198–1229. DOI: 10.1002/anie.200802019 PMID: 19173328

Serrano, J., Manjón, F. J., Romero, A. H., Widulle, F., Lauck, R., & Cardona, M. (2003). Dispersive Phonon Linewidths: The <math display="inline"> <msub> <mi mathvariant="normal">E</mi> <mn>2</mn> </msub> </math> Phonons of ZnO. *Physical Review Letters*, *90*(5), 055510. DOI: 10.1103/PhysRevLett.90.055510

Sethio, D., Bopp, P., & Hermansson, K. (2025). ^{7}Li NMR chemical shifts in Battery-Relevant electrolytes: Ligand-Induced shifts cancel in both experiment and calculation. *ChemSusChem*, *18*(21), e202500798. DOI: 10.1002/cssc.202500798 PMID: 40971230

Setyaningrum, D., Riyanto, S., & Rohman, A. (2013). Analysis of corn and soybean oils in red fruit oil using FTIR spectroscopy in combination with partial least square. *International Food Research Journal*, *20*, 1977.

Shahbazi, M., & Wang, H. (2016). Progress in research on the stability of organometal perovskite solar cells. *Solar Energy*, *123*, 74–87. DOI: 10.1016/j.solener.2015.11.008

Shakeel, S., Song, P., Shah, S. H., Zada, Z., Huang, T., Laref, A., Hakimi, N., & Faizan, M. (2024). DFT analysis of elastic and optoelectronic properties of Cs2NaXCl6 (X=In, La, Sc, Y) double perovskite compounds. *Materials Chemistry and Physics*, *324*, 129683. DOI: 10.1016/j.matchemphys.2024.129683

Shao, B., Song, X., Zhu, H., Bioud, Y. A., Wu, W., Abulikemu, M., Saiari, H., Aqeel, S., Gereige, I., Mohammed, O. F., & Bakr, O. M. (2025). Single-crystal perovskites for photovoltaic and high-energy detection applications. *Chemical Society Reviews*, *54*(21), 9939–9977. DOI: 10.1039/D5CS00625B

Shapeev, A. V. (2011). Consistent energy-based atomistic/continuum coupling for two-body potentials in one and two dimensions. *Multiscale Modeling & Simulation*, *9*(3), 905–932. DOI: 10.1137/100792421

Sharma, A., Tyagi, V. V., Chen, C. R., & Buddhi, D. (2009). Review on thermal energy storage with phase change materials and applications. *Renewable & Sustainable Energy Reviews*, *13*(2), 318–345. DOI: 10.1016/j.rser.2007.10.005

Sharma, G., Kumar, N., Sharma, C. S., Alqahtani, T., Tiruneh, Y. K., Sultana, S., Rolim Silva, G. V., de Lima Menezes, G., Zaki, M. E. A., & Nobre Oliveira, J. I. (2025). Identification of promising SARS-CoV-2 main protease inhibitor through molecular docking, dynamics simulation, and ADMET analysis. *Scientific Reports*, *15*(1), 2830. DOI: 10.1038/s41598-025-86016-9 PMID: 39843610

Sharma, M., Koul, A., Sharma, D., Kaul, S., Swamy, M. K., & Dhar, M. K. (2019). Metabolic engineering strategies for enhancing the production of bio-active compounds from medicinal plants. In Akhtar, M., & Swamy, M. (Eds.), *Natural bio-active compounds* (pp. 287–316). Springer., DOI: 10.1007/978-981-13-7438-8_12

Sharma, S., & Khan, S. (2023). Enhancement of dielectric and ferroelectric properties of (Sr, Ni) Co-doped $BaTiO_3$ ceramics. *Ferroelectrics*, *617*(1), 18–24. DOI: 10.1080/00150193.2023.2271131

Sharma, S., Shukla, S. K., Govender, K. K., & Govender, P. P. (2025). Unveiling the multifunctionality of iron oxide nanoparticle: A synergistic experimental and computational investigation. *Chemical Physics Impact*, *10*, 100845. DOI: 10.1016/j.chphi.2025.100845

Shaukat, H., Ali, A., Bibi, S., Mehmood, S., Altabey, W. A., Noori, M., & Kouritem, S. A. (2023). Piezo-electric materials : Advanced applications in electro-chemical processes. *Energy Reports*, *9*, 4306–4324. DOI: 10.1016/j.egyr.2023.03.077

Sheidaei, M., Zhu, J., & Agardh, S. (2024). Assessing the Effect of Specimen Preparation Methods on DSR Test Results of Bitumen Using Factorial Design Analysis. *Materials (Basel)*, *17*(20), 5117. DOI: 10.3390/ma17205117

Shen, C. (2018). A transdisciplinary review of deep learning research and its relevance for water resources scientists. *Water Resources Research*, *54*(11), 8558–8593. DOI: 10.1029/2018WR022643

Shenderovich, I. G. (2023). Experimentally established 15N NMR absolute shielding scale for theoretical calculations. *The Journal of Physical Chemistry A*, *127*(26), 5547–5555. DOI: 10.1021/acs.jpca.3c02350 PMID: 37344362

Shilkrot, L. E., Curtin, W. A., & Miller, R. E. (2002). A coupled atomistic/continuum model of defects in solids. *Journal of the Mechanics and Physics of Solids*, *50*(10), 2085–2106. DOI: 10.1016/S0022-5096(02)00017-0

Shinyuy, L. M., Loe, G. E., Jansen, O., Ledoux, A., Palmaerts, B., Mamede, L., Boussif, N., Bonnet, O., Enone, B. S., Noukimi, S. F., Nchang, A. S., Demeyer, K., Robert, A., Ghogomu, S. M., Souopgui, J., Hallot, E., & Frederich, M. (2025). Exploring the Phytochemical Diversity and Anti-Plasmodial Potential of Artemisia annua and Artemisia afra from Different Geographical Locations in Cameroon. *Molecules (Basel, Switzerland)*, *30*(3), 596. DOI: 10.3390/molecules30030596

Shi, T., Cao, J., Cao, J., Zhu, F., Cao, F., & Su, E. (2023). Almond (Amygdalus communis L.) kernel protein: A review on the extraction, functional properties and nutritional value. *Food Research International*, *167*, 112721. DOI: 10.1016/j.foodres.2023.112721

Shi, Y., Xia, Z., Zhang, J., Best, R., Wu, C., Ponder, J. W., & Ren, P. (2013). Polarizable Atomic Multipole-Based AMOEBA force field for proteins. *Journal of Chemical Theory and Computation*, *9*(9), 4046–4063. DOI: 10.1021/ct4003702 PMID: 24163642

Shockley, W., & Queisser, H. J. (1961). Detailed balance limit of efficiency of p-n junction solar cells. *Journal of Applied Physics*, *32*(3), 510–519. DOI: 10.1063/1.1736034

Shokrollahi, H., & Janghorban, K. (2006). The effect of compaction parameters and particle size on magnetic properties of iron-based alloys used in soft magnetic composites. *Materials Science and Engineering B*, *134*(1), 41–43. DOI: 10.1016/j.mseb.2006.07.015

Shu, J., Wang, Y., Guo, W., Liu, T., Cai, S., Shi, T., & Hu, W. (2024). Carbenoid-involved reactions integrated with scaffold-based screening generates a Nav1.7 inhibitor. *Communications Chemistry*, *7*(1), 135. DOI: 10.1038/s42004-024-01213-3 PMID: 38866907

Shu, M., Mahdavi, B., Balčiūnaitienė, A., Goorani, S., & Mahdavi, A. A. (2023). Novel green synthesis of tin nanoparticles by medicinal plant: Chemical characterization and determination of cytotoxicity, cutaneous wound healing and antioxidant properties. *Micro & Nano Letters*, *18*(2), e12157. Advance online publication. DOI: 10.1049/mna2.12157

Siekmann, J., Kulkarni, A., Akel, S., Klingebiel, B., Saliba, M., Rau, U., & Kirchartz, T. (2023). Characterizing the Influence of Charge Extraction Layers on the Performance of Triple-Cation Perovskite Solar Cells. *Advanced Energy Materials*, *13*(32), 2300448. Advance online publication. DOI: 10.1002/aenm.202300448

Silva, N. M., Deglmann, P., & Pliego, J. R.Jr. (2016). CMIRS Solvation Model for Methanol: Parametrization, Testing, and Comparison with SMD, SM8, and COSMO-RS. *The Journal of Physical Chemistry B*, *120*(49), 12660–12668. DOI: 10.1021/acs.jpcb.6b10249 PMID: 27973833

Siskos, M., Choudhary, M., & Gerothanassis, I. (2017). Hydrogen Atomic Positions of O–H···O Hydrogen Bonds in Solution and in the Solid State: The Synergy of Quantum Chemical Calculations with 1H-NMR Chemical Shifts and X-ray Diffraction Methods. *Molecules (Basel, Switzerland)*, *22*(3), 415. DOI: 10.3390/molecules22030415 PMID: 28272366

Slater, J. C. (1930). Atomic Shielding Constants. *Physical Review*, *36*(1), 57–64. DOI: 10.1103/PhysRev.36.57

Slavney, A. H., Hu, T., Lindenberg, A. M., & Karunadasa, H. I. (2016). A Bismuth-Halide Double Perovskite with Long Carrier Recombination Lifetime for Photovoltaic Applications. *Journal of the American Chemical Society*, *138*(7), 2138–2141. DOI: 10.1021/jacs.5b13294

Slavney, A. H., Smaha, R. W., Smith, I. C., Jaffe, A., Umeyama, D., & Karunadasa, H. I. (2017). Chemical Approaches to Addressing the Instability and Toxicity of Lead–Halide Perovskite Absorbers. *Inorganic Chemistry*, *56*(1), 46–55. DOI: 10.1021/acs.inorgchem.6b01336

Słoczyńska, K., Gunia-Krzyżak, A., Koczurkiewicz, P., Wójcik-Pszczoła, K., Żelaszczyk, D., Popiół, J., & Pękala, E. (2019). Metabolic stability and its role in the discovery of new chemical entities. *Acta Pharmaceutica (Zagreb, Croatia)*, *69*(3), 345–361. DOI: 10.2478/acph-2019-0024 PMID: 31259741

Slot, L. E., & Fort, F. (2024). Drivers of adoption of sustainable prickly pear (*Opuntia ficus-indica*) innovations and conservation agriculture by smallholder farmers in Morocco. *Agronomy (Basel)*, *14*(12), 3014. DOI: 10.3390/agronomy14123014

Smith, E. R., & Theodorakis, P. E. (2024a). Multiscale simulation of fluids: Coupling molecular and continuum. *Physical Chemistry Chemical Physics*, *26*(2), 724–744. DOI: 10.1039/D3CP03579D

Smith, J. S., Isayev, O., & Roitberg, A. E. (2017). ANI-1: An extensible neural network potential with DFT accuracy at force field computational cost. *Chemical Science (Cambridge)*, *8*(4), 3192–3203. DOI: 10.1039/C6SC05720A PMID: 28507695

Smith, S. G., & Goodman, J. M. (2010). Assigning stereochemistry to single diastereoisomers by GIAO NMR calculation: The DP4 probability. *Journal of the American Chemical Society*, *132*(37), 12946–12959. DOI: 10.1021/ja105035r PMID: 20795713

Sobha, A., Ganapathy, A., Mohan, S., Madhusoodanan, N., Babysulochana, A. D., Alaganandan, K., & Somappa, S. B. (2024). Novel small molecule-based acetylcholinesterase (AChE) inhibitors: From biological perspective to recent developments. *European Journal of Medicinal Chemistry Reports*, *12*, 100237. DOI: 10.1016/j.ejmcr.2024.100237

Sockwell, K. C., Bochev, P., Peterson, K., & Kuberry, P. (2023). Interface flux recovery framework for constructing partitioned heterogeneous time-integration methods. *Numerical Methods for Partial Differential Equations*, *39*(5), 3572–3593. DOI: 10.1002/num.23015

Sodhi, J. K., & Benet, L. Z. (2021). Successful and unsuccessful prediction of human hepatic clearance for lead optimization. *Journal of Medicinal Chemistry*, *64*(7), 3546–3559. DOI: 10.1021/acs.jmedchem.0c01930 PMID: 33765384

Sohag, M., Chausalkar, A., Li, L., & Yang, X. (2022). Numerical study of drop spread and rebound on heated surfaces with consideration of high pressure. *Physics of Fluids*, *34*(11), 113319. DOI: 10.1063/5.0124794

Sohn, J., Han, S. H., Yamaguchi, M., & Lim, S. H. (2006). Tunable electromagnetic noise suppressor integrated with a magnetic thin film. *Applied Physics Letters*, *89*(10), 103501. DOI: 10.1063/1.2335389

Sonia, P., Patel, R. K., Kumar, P., Prakash, C., & Agrawal, D. K. (2012). Low temperature synthesis and dielectric, ferroelectric and piezoelectric study of microwave sintered BaTiO3 ceramics. *Ceramics International*, *38*(2), 1585–1589. DOI: 10.1016/j.ceramint.2011.09.045

Soumane, M., Bouaaliouat, O., Fazouan, N., Ydir, B., Dloo, A., Celzard, A., Fierro, V., Prades, J. D., & Lahlou, H. (2025). Experimental and DFT simulation of SnO2 monotube-based gas sensing for hazardous VOCs. *Surfaces and Interfaces*, *60*, 105971. DOI: 10.1016/j.surfin.2025.105971

Spenato, D., Pogossian, S. P., Dekadjevi, D. T., & Ben Youssef, J. (2007). Study of dynamic properties and magnetic anisotropies of NiFe/MnPt in the critical thickness range. *Journal of Physics. D, Applied Physics*, *40*(11), 3306–3313. DOI: 10.1088/0022-3727/40/11/008

Srinivas, K., Manivel Raja, M., Sridhara Rao, D. V., & Kamat, S. V. (2014). Effect of sputtering pressure and power on composition, surface roughness, microstructure and magnetic properties of as-deposited Co2FeSi thin films. *Thin Solid Films*, *558*, 349–355. DOI: 10.1016/j.tsf.2014.02.052

St John, P. C., Phillips, C., Kemper, T. W., Wilson, A. N., Guan, Y., Crowley, M. F., Nimlos, M. R., & Larsen, R. E. (2019). Message-passing neural networks for high-throughput polymer screening. *The Journal of Chemical Physics*, *150*(23), 234111. DOI: 10.1063/1.5099132 PMID: 31228909

Stahn, M., Ehlert, S., & Grimme, S. (2023). Extended Conductor-like Polarizable Continuum Solvation Model (CPCM-X) for semiempirical methods. *The Journal of Physical Chemistry A*, *127*(33), 7036–7043. DOI: 10.1021/acs.jpca.3c04382 PMID: 37567769

Steinbeck, C., Krause, S., & Kuhn, S. (2003). NMRShiftDBConstructing a Free Chemical Information System with Open-Source Components. *Journal of Chemical Information and Computer Sciences*, *43*(6), 1733–1739. DOI: 10.1021/ci0341363 PMID: 14632418

Steiner, M., Holzknecht, T., Schauperl, M., & Podewitz, M. (2021). Quantum chemical microsolvation by automated water placement. *Molecules (Basel, Switzerland)*, *26*(6), 1793. DOI: 10.3390/molecules26061793 PMID: 33806731

Stillinger, F. H., & Weber, T. A. (1985). Computer simulation of local order in condensed phases of silicon. *Physical Review B: Condensed Matter*, *31*(8), 5262–5271. DOI: 10.1103/PhysRevB.31.5262

Subedi, K. K., Kong, S.-C., & Kweon, C.-B. M. (2022). Numerical study of consecutive drop/wall impacts using smoothed particle hydrodynamics. *International Journal of Multiphase Flow*, *151*, 104060. DOI: 10.1016/j.ijmultiphaseflow.2022.104060

Sudhaparimala, S. (2014). Green Synthesis of tin based Nano Medicine: Assessment of microstructure and surface property. *American Journal of Nanoscience and Nanotechnology*, *2*(4), 75. DOI: 10.11648/j.nano.20140204.13

Su, J. T., & Goddard, W. A.III. (2007). Excited electron dynamics modeling of warm dense matter. *Physical Review Letters*, *99*(18), 185003. DOI: 10.1103/PhysRevLett.99.185003

Sun, J., He, Y., Tao, W., Yin, X., & Wang, H. (2012). Roughness effect on flow and thermal boundaries in microchannel/nanochannel flow using molecular dynamics-continuum hybrid simulation. *International Journal for Numerical Methods in Engineering*, *89*(1), 2–19. DOI: 10.1002/nme.3229

Sunny, N. E., & Kumar, S. V. (2019). Biogenesis, Characterization and Bioefficacy of Tin Oxide Nanoparticles from Averrhoa Bilimbi Fruit Extract. [IJRTE]. *International Journal of Recent Technology and Engineering*, *8*(4), 10309–10315. DOI: 10.35940/ijrte.D4523.118419

Sun, P.-P., Li, Q.-S., Yang, L.-N., & Li, Z.-S. (2016). Theoretical insights into a potential lead-free hybrid perovskite: Substituting Pb2+ with Ge2+. *Nanoscale*, *8*(3), 1503–1512. DOI: 10.1039/C5NR05337D

Sun, S., Hartono, N. T. P., Ren, Z. D., Oviedo, F., Buscemi, A. M., Layurova, M., Chen, D. X., Ogunfunmi, T., Thapa, J., Ramasamy, S., Settens, C., DeCost, B. L., Kusne, A. G., Liu, Z., Tian, S. I. P., Peters, I. M., Correa-Baena, J.-P., & Buonassisi, T. (2019). Accelerated Development of Perovskite-Inspired Materials via High-Throughput Synthesis and Machine-Learning Diagnosis. *Joule*, *3*(6), 1437–1451. DOI: 10.1016/j.joule.2019.05.014

Sun, Y., & Fernie, A. R. (2024). Plant secondary metabolism in a fluctuating world: Climate change perspectives. *Trends in Plant Science*, *29*(5), 560–571. DOI: 10.1016/j.tplants.2023.11.008

Swails, J., Zhu, T., He, X., & Case, D. A. (2015). AFNMR: Automated fragmentation quantum mechanical calculation of NMR chemical shifts for biomolecules. *Journal of Biomolecular NMR, 63*(2), 125–139. DOI: 10.1007/s10858-015-9970-3 PMID: 26232926

Szántó, J. K., Dietschreit, J. C. B., Shein, M., Schütz, A. K., & Ochsenfeld, C. (2024). Systematic QM/MM study for predicting 31P NMR chemical shifts of adenosine nucleotides in solution and stages of ATP hydrolysis in a protein environment. *Journal of Chemical Theory and Computation*, *20*(6), 2433–2444. DOI: 10.1021/acs.jctc.3c01280 PMID: 38497488

Sze, S. M., & Ng, K. K. (2006). LEDs and lasers. *Physics of Semiconductor Devices*, *3*, 601–657.

Sze, S. M., & Ng, K. K. (2006). Physics of Semiconductor Devices. In *Physics of Semiconductor Devices*. Wiley., DOI: 10.1002/0470068329

Tahir, H. E., Xiaobo, Z., Komla, M. G., & Mariod, A. A. (2019). Nopal cactus (*Opuntia ficus-indica* (L.) Mill) as a source of bioactive compounds. In A. Mariod (Ed.), Wild fruits: Composition, nutritional value and products (pp. [non précisé]). Springer. https://doi.org/DOI: 10.1007/978-3-030-31885-7_26

Takeda, S., & Naoe, M. (2018). Size optimization for complex permeability measurement of magnetic thin films using a short-circuited microstrip line up to 30 GHz. *Journal of Magnetism and Magnetic Materials*, *449*, 530–537. DOI: 10.1016/j.jmmm.2017.10.025

Takhar, V., & Singh, S. (2025). Nanomaterials ROS: A comprehensive review for environmental applications. *Environmental Science. Nano*, *12*(5), 2516–2550. DOI: 10.1039/D5EN00049A

Tameh, M. S., Gladfelter, W. L., & Goodpaster, J. D. (2025). Unraveling surface chemistry of SNO2 through formation of charged oxygen species and oxygen vacancies. *International Journal of Quantum Chemistry*, *125*(4), e70017. Advance online publication. DOI: 10.1002/qua.70017

Tang, S., Duan, Z., Tian, Z., Du, W., & Qian, F. (2025). A fine-tuned RNN model for accurately predicting the spatial distribution of parameters in light hydrocarbon cracking tubular reactor. *Chemical Engineering Journal, 504*(December 2024), 158521. DOI: 10.1016/j.cej.2024.158521

Tang, X., Wu, D., Wang, Z., Haidn, O., & Hu, X. (2023). An explicit multi-time stepping algorithm for multi-time scale coupling problems in SPH. ArXiv Preprint ArXiv:2309.04010.

Tang, G., Ghosez, P., & Hong, J. (2021). Band-Edge Orbital Engineering of Perovskite Semiconductors for Optoelectronic Applications. *The Journal of Physical Chemistry Letters*, *12*(17), 4227–4239. DOI: 10.1021/acs.jpclett.0c03816

Tang, T. Y., & Tang, Y. L. (2023). Physical and optoelectronic properties of double halide perovskites A2CuSbX6 (A = Cs, Rb, K; X = Cl, Br, I) based on first principles calculations. *Chemical Physics*, *570*, 111897. DOI: 10.1016/j.chemphys.2023.111897

Tang, T. Y., Zhao, X. H., Hu, D. Y., Liang, Q. Q., Wei, X. N., & Tang, Y. L. (2022). An ab-initio investigation of novel double halide perovskite Cs2InCoX6(X=F, Cl, Br) materials with direct band structure and broadband light absorption. *Materials Science in Semiconductor Processing*, *152*, 107047. DOI: 10.1016/j.mssp.2022.107047

Tang, T., Hu, D., Zhao, X., Li, L., & Tang, Y. (2022). First-principles study on the mechanical, electronic and optical properties of double halide perovskite Cs2TlSbX6 (X = Cl, Br, I). *Physica Scripta*, *97*(12), 125821. DOI: 10.1088/1402-4896/ac9ff7

Tao, K., Zhou, X., & Ren, H. (2021). A Local Semi-Fixed Ghost Particles Boundary Method for WCSPH. *Journal of Marine Science and Engineering*, *9*(4), 416. DOI: 10.3390/jmse9040416

Tariq, S., Ahmed, A., Saad, S., & Tariq, S. (2015). Structural, electronic and elastic properties of the cubic CaTiO3 under pressure : A DFT study. *AIP Advances*, *5*(7), 077111. DOI: 10.1063/1.4926437

Tasisa, Y. E., Sarma, T. K., Sahu, T. K., & Krishnaraj, R. (2024). Phytosynthesis and characterization of tin-oxide nanoparticles (SnO2-NPs) from Croton macrostachyus leaf extract and its application under visible light photocatalytic activities. *Scientific Reports*, *14*(1), 10780. DOI: 10.1038/s41598-024-60633-2

Techawinyutham, L., Techawinyutham, W., Laohapanich, C., Insawang, K., Sanjay, M. R., & Siengchin, S. (2024). Possibility of infrared (IR) thermography camera to investigate properties of injected plastic product and optimisation injection process parameter using a design of experiment. *Advances in Materials and Processing Technologies*, *10*(3), 1501–1522.

Terentjev, A. V., Constantin, L. A., & Pitarke, J. M. (2018). Dispersion-corrected PBEsol exchange-correlation functional. *Physical Review. B*, *98*(21), 214108. DOI: 10.1103/PhysRevB.98.214108

Tersoff, J. (1988). New empirical approach for the structure and energy of covalent systems. *Physical Review B: Condensed Matter*, *37*(12), 6991–7000. DOI: 10.1103/PhysRevB.37.6991

Tersoff, J. (1989). Modeling solid-state chemistry: Interatomic potentials for multicomponent systems. *Physical Review B: Condensed Matter*, *39*(8), 5566–5568. DOI: 10.1103/PhysRevB.39.5566

Teshima, M., Monobe, K., Okubo, S., & Aoki, S. (2024). Discovery of antibacterial compounds with potential multipharmacology against Staphylococcus mur ligase family members by in silico structure-based drug screening. *Molecules (Basel, Switzerland)*, *29*(16), 3792. DOI: 10.3390/molecules29163792 PMID: 39202871

The fundamental equations of quantum mechanics. (1925). Proceedings of the Royal Society of London. Series A, Containing Papers of a Mathematical and Physical Character, 109(752), 642-653. DOI: 10.1098/rspa.1925.0150

Thi, M. P., Marchet, P., Levassort, F., & Blantignie, C. (2014). Matériaux piézoélectriques sans plomb pour transducteurs acoustiques.

Tihtih, M., Ibrahim, J. E. F. M., Basyooni, M. A., En-nadir, R., Belaid, W., Abdelfattah, M. M., Hussainova, I., Pszota, G., & Kocserha, I. (2023). Enhanced optical and thermal conductivity properties of barium titanate ceramic via strontium doping for thermo-optical applications. *Optical and Quantum Electronics*, *55*(3), 226. DOI: 10.1007/s11082-022-04516-8

Tihtih, M., Ibrahim, J. E. F. M., Basyooni, M. A., Kurovics, E., Belaid, W., Hussainova, I., & Kocserha, I. (2023). Role of A-site (Sr), B-site (Y), and A, B sites (Sr, Y) substitution in lead-free BaTiO3 ceramic compounds : Structural, optical, microstructure, mechanical, and thermal conductivity properties. *Ceramics International*, *49*(2), 1947–1959. DOI: 10.1016/j.ceramint.2022.09.160

Timoshenko, S. P., & Goodier, J. N. (n.d.). Theory of Elasticity THIRD EDITION PROFESSOR EMERITUS OF. *Gongcheng Lixue*.

Tokay, O., & Yazıcı, M. (2022). A review of potassium sodium niobate and bismuth sodium titanate based lead free piezoceramics. *Materials Today. Communications*, *31*, 103358. DOI: 10.1016/j.mtcomm.2022.103358

Tolba, S. A., Gameel, K. M., Ali, B. A., Almossalami, H. A., & Allam, N. K. (2018). The DFT+U : Approaches, Accuracy, and Applications. In Yang, G. (Ed.), *Density Functional Calculations Recent Progresses of Theory and Application*. InTech., DOI: 10.5772/intechopen.72020

Tomasi, J., Mennucci, B., & Cammi, R. (2005). Quantum mechanical continuum solvation models. *Chemical Reviews*, *105*(8), 2999–3094. DOI: 10.1021/cr9904009 PMID: 16092826

Trainor, K., Palumbo, J. A., MacKenzie, D. W. S., & Meiering, E. M. (2019). Temperature dependence of NMR chemical shifts: Tracking and statistical analysis. *Protein Science : a Publication of the Protein Society*, *29*(1), 306–314. DOI: 10.1002/pro.3785 PMID: 31730280

Tran, F., & Blaha, P. (2009). Accurate band gaps of semiconductors and insulators with a semilocal exchange-correlation potential. *Physical Review Letters*, *102*(22), 5–8. DOI: 10.1103/PhysRevLett.102.226401

Trihalide S B, P., Christian, Y., & Mølle, K. N. (1959). Matematisk-fysiske Meddelelser udgivet af Det Kongelige Danske Videnskabernes Selskab THE STRUCTUR E OF PEROVSKITE-LIKE CAESIUM Introductio n. Mat. Fys. Medd. Dan. Vid. Selsk, 32(2).

Tripathi, A., Ranjan, M. R., Verma, D. K., Singh, Y., Shukla, S. K., Rajput, V. D., Minkina, T., Mishra, P. K., & Garg, M. C. (2022). ANN-GA based biosorption of As(III) from water through chemo-tailored and iron impregnated fungal biofilter system. *Scientific Reports*, *12*(1), 12414. DOI: 10.1038/s41598-022-14802-w PMID: 35858932

Truflandier, L. A., & Autschbach, J. (2010). Probing the Solvent Shell with 195Pt Chemical Shifts: Density Functional Theory Molecular Dynamics Study of PtII and PtIV Anionic Complexes in Aqueous Solution. *Journal of the American Chemical Society*, *132*(10), 3472–3483. DOI: 10.1021/ja9096863 PMID: 20166712

Tserkovnyak, Y., Brataas, A., & Bauer, G. E. W. (2002). Enhanced Gilbert Damping in Thin Ferromagnetic Films. *Physical Review Letters*, *88*(11), 117601. DOI: 10.1103/PhysRevLett.88.117601

Tshilande, N., Mammino, L., & Bilonda, M. K. (2024). The study of molecules and processes in solution: An overview of questions, approaches and applications. *Computation (Basel, Switzerland)*, *12*(4), 78. DOI: 10.3390/computation12040078

Tuckerman, M. E. (2002). Ab initiomolecular dynamics: Basic concepts, current trends and novel applications. *Journal of Physics Condensed Matter*, *14*(50), R1297–R1355. DOI: 10.1088/0953-8984/14/50/202

Ulfa, D. M., Bayu, A., Rahmawati, S. I., Ahmadi, P., Putra, M. Y., Karnjanakom, S., Guan, G., & Mun'im, A. (2025). New horizon in selective tocols extraction from deodorizer distillates under mild conditions by using deep eutectic solvents. *Molecules (Basel, Switzerland)*, *30*(6), 1217. DOI: 10.3390/molecules30061217

Unke, O. T., Chmiela, S., Sauceda, H. E., Gastegger, M., Poltavsky, I., Schutt, K. T., Tkatchenko, A., & Muller, K. R. (2021). Machine learning force fields. *Chemical Reviews*, *121*(16), 10142–10186. DOI: 10.1021/acs.chemrev.0c01111 PMID: 33705118

Ünlü, F., Deo, M., Mathur, S., Kirchartz, T., & Kulkarni, A. (2022). Bismuth-based halide perovskite and perovskite-inspired light absorbing materials for photovoltaics. *Journal of Physics. D, Applied Physics*, *55*(11), 113002. DOI: 10.1088/1361-6463/ac3033

Upadhyay, N., Jaiswal, P., & Jha, S. N. (2018). Application of attenuated total reflectance Fourier Transform Infrared spectroscopy (ATR–FTIR) in MIR range coupled with chemometrics for detection of pig body fat in pure ghee (heat clarified milk fat). *Journal of Molecular Structure*, *1153*, 275–281. DOI: 10.1016/j.molstruc.2017.09.116

Upadhyay, V., Lucas, A., Patrick, C., & Mallela, K. M. G. (2024). Isothermal titration calorimetry and surface plasmon resonance methods to probe protein–protein interactions. *Methods (San Diego, Calif.)*, *225*, 52–61. DOI: 10.1016/j.ymeth.2024.03.007 PMID: 38492901

Urban, A., Reese, M., Mrovec, M., Elsässer, C., & Meyer, B. (2011). Parameterization of tight-binding models from density functional theory calculations. *Physical Review B: Condensed Matter and Materials Physics*, *84*(15), 155119. Advance online publication. DOI: 10.1103/PhysRevB.84.155119

Usman, M., & Yan, Q. (2020). Recent Advancements in Crystalline Pb-Free Halide Double Perovskites. *Crystals*, *10*(2), 62. DOI: 10.3390/cryst10020062

Van Duin, A. C. T., Dasgupta, S., Lorant, F., & Goddard, W. A. (2001). ReaxFF: A reactive force field for hydrocarbons. *The Journal of Physical Chemistry A*, *105*(41), 9396–9409. DOI: 10.1021/jp004368u

Van Eck, N. J., & Waltman, L. (2009). Software survey: VOSviewer, a computer program for bibliometric mapping. *Scientometrics*, *84*(2), 523–538. DOI: 10.1007/s11192-009-0146-3

van Staa, A., Wulfhorst, J., Vogel, A., Merkt, U., & Meier, G. (2008). Spin precession in lateral all-metal spin valves: Experimental observation and theoretical description. *Physical Review B: Condensed Matter and Materials Physics*, *77*(21), 214416. DOI: 10.1103/PhysRevB.77.214416

Van Veen, F., Kölling, S., De Wit, S. R., Metsch, R., Rosenbach, D., Li, C., & Brinkman, A. (2025). Observation of the surface hybridization gap in the electrical transport properties of the ultrathin topological insulator (Bi1−xSbx)2Te3. *Physical Review. B. Physical Review. B*, *112*(4). Advance online publication. DOI: 10.1103/d5xc-bvmx

Van Vranken, J. G., Li, J., Mintseris, J., Wei, T.-Y., Sniezek, C. M., Gadzuk-Shea, M., Gygi, S. P., & Schweppe, D. K. (2024). Large-scale characterization of drug mechanism of action using proteome-wide thermal shift assays. *eLife*, *13*, RP95595. Advance online publication. DOI: 10.7554/eLife.95595 PMID: 39526730

Varghese, R., Shringi, H., Efferth, T., & Ramamoorthy, S. (2025). Artificial intelligence driven approaches in phytochemical research: Trends and prospects. *Phytochemistry Reviews*, *2025*(5), 10096. Advance online publication. DOI: 10.1007/s11101-025-10096-8

Vícha, J., Novotný, J., Straka, M., Repisky, M., Ruud, K., Komorovsky, S., & Marek, R. (2015). Structure, solvent, and relativistic effects on the NMR chemical shifts in square-planar transition-metal complexes: Assessment of DFT approaches. *Physical Chemistry Chemical Physics*, *17*(38), 24944–24955. DOI: 10.1039/C5CP04214C PMID: 26344822

Victora, A., Möller, H. M., & Exner, T. E. (2014). Accurate ab initio prediction of NMR chemical shifts of nucleic acids and nucleic acids/protein complexes. *Nucleic Acids Research*, *42*(22), e173. DOI: 10.1093/nar/gku1006 PMID: 25404135

Violeau, D., & Rogers, B. D. (2016). Smoothed particle hydrodynamics (SPH) for free-surface flows: Past, present and future. *Journal of Hydraulic Research*, *54*(1), 1–26. DOI: 10.1080/00221686.2015.1119209

Violi, A. (2004). Modeling of soot particle inception in aromatic and aliphatic premixed flames. *Combustion and Flame*, *139*(4), 279–287. DOI: 10.1016/j.combustflame.2004.08.013

Vittorio, S., Lunghini, F., Morerio, P., Gadioli, D., Orlandini, S., Silva, P., Jan Martinovic, , Pedretti, A., Bonanni, D., Del Bue, A., Palermo, G., Vistoli, G., & Beccari, A. R. (2024). Addressing docking pose selection with structure-based deep learning: Recent advances, challenges and opportunities. *Computational and Structural Biotechnology Journal*, *23*, 2141–2151. DOI: 10.1016/j.csbj.2024.05.024 PMID: 38827235

Vogel, A., Wulfhorst, J., & Meier, G. (2009). Enhanced spin injection and detection in spin valves with integrated tunnel barriers. *Applied Physics Letters*, *94*(12), 122510. DOI: 10.1063/1.3109787

Vogiatzis, G. G., & Theodorou, D. N. (2018). Multiscale molecular simulations of polymer-matrix nanocomposites: Or what molecular simulations have taught us about the fascinating nanoworld. *Archives of Computational Methods in Engineering*, *25*(3), 591–645. DOI: 10.1007/s11831-016-9207-y

Volonakis, G., Haghighirad, A. A., Milot, R. L., Sio, W. H., Filip, M. R., Wenger, B., Johnston, M. B., Herz, L. M., Snaith, H. J., & Giustino, F. (2017). Cs 2 InAgCl 6 : A New Lead-Free Halide Double Perovskite with Direct Band Gap. *The Journal of Physical Chemistry Letters*, *8*(4), 772–778. DOI: 10.1021/acs.jpclett.6b02682

Vreven, T., & Morokuma, K. (2006). Hybrid methods: Oniom (qm: mm) and qm/mm. *Annual Reports in Computational Chemistry*, *2*, 35–51. DOI: 10.1016/S1574-1400(06)02003-2

Wagner, G. J., Karpov, E. G., & Liu, W. K. (2004). Molecular dynamics boundary conditions for regular crystal lattices. *Computer Methods in Applied Mechanics and Engineering*, *193*(17–20), 1579–1601. DOI: 10.1016/j.cma.2003.12.012

Wagner, G. J., & Liu, W. K. (2003). Coupling of atomistic and continuum simulations using a bridging scale decomposition. *Journal of Computational Physics*, *190*(1), 249–274. DOI: 10.1016/S0021-9991(03)00273-0

Walker, C. T. (1967). Solid State Physics over 13 Years: Introduction to Solid State Physics. Charles Kittel. Wiley, New York, ed. 3, 1966. 662 pp., illus. $12.50. Science, 155(3765), 991. DOI: 10.1126/science.155.3765.991.b

Wallach, I., Bernard, D., Nguyen, K., Ho, G., Morrison, A., Stecula, A., Rosnik, A., O'Sullivan, A. M., Davtyan, A., Samudio, B., Thomas, B., Worley, B., Butler, B., Laggner, C., Thayer, D., Moharreri, E., Friedland, G., Truong, H., van den Bedem, H., & Heifets, A. (2024). AI is a viable alternative to high throughput screening: A 318-target study. *Scientific Reports*, *14*(1), 7526. DOI: 10.1038/s41598-024-54655-z PMID: 38565852

Wang, Y., Wang, C., Huang, X., Zhang, Q., Wang, T., & Guo, X. (2024). Guideline for modeling solid-liquid adsorption : Kinetics, isotherm, fixed bed, and thermodynamics. *Chemosphere, 349*(September 2023), 140736. DOI: 10.1016/j.chemosphere.2023.140736

Wang, D., Qiang, H., & Shi, C. (2020). A multiphase SPH framework for solving the evaporation and combustion process of droplets. *International Journal of Numerical Methods for Heat & Fluid Flow*, *30*(3), 1547–1575. DOI: 10.1108/HFF-08-2019-0666

Wang, D., Wang, G., Lu, Z., Al-Jlaihawi, Z., & Feteira, A. (2020). Crystal Structure, Phase Transitions and Photoferroelectric Properties of KNbO3-Based Lead-Free Ferroelectric Ceramics : A Brief Review. *Frontiers in Materials*, *7*, 91. DOI: 10.3389/fmats.2020.00091

Wang, H., Vlcek, L., Neuefeind, J. C., Page, K., Irle, S., Simonson, J. M., & Stack, A. G. (2018). Decoding oxyanion aqueous solvation structure: A potassium nitrate example at saturation. *The Journal of Physical Chemistry B*, *122*(30), 7584–7589. DOI: 10.1021/acs.jpcb.8b05895 PMID: 29991255

Wang, H., Zhang, Y., Yang, D., Liu, J., Huang, W., Dai, B., Ren, Y., Li, X., & Chen, M. (2025). Preparation and properties of high magnetic permeability [Fe20Ni80-Al2O3]n magnetic thin films. *Applied Physics. A, Materials Science & Processing*, *131*(4), 254. DOI: 10.1007/s00339-025-08416-y

Wang, J., Liu, Y., Li, W., & Gao, G. (2018). Prediction of 1H NMR chemical shifts for ionic liquids: Strategy and application of a relative reference standard. *RSC Advances*, *8*(50), 28604–28612. DOI: 10.1039/C8RA04822C PMID: 35548389

Wang, J., Sun, X., Cheng, Q., & Cui, Q. (2021). An innovative random forest-based nonlinear ensemble paradigm of improved feature extraction and deep learning for carbon price forecasting. *The Science of the Total Environment*, *762*, 143099. DOI: 10.1016/j.scitotenv.2020.143099

Wang, L., Huangfu, Z., Li, R., Wen, X., Sun, Y., & Chen, Y. (2024). Iterative learning control with parameter estimation for non-repetitive time-varying systems. *Journal of the Franklin Institute*, *361*(3), 1455–1466. DOI: 10.1016/j.jfranklin.2024.01.011

Wang, M., Zhang, G., Peng, W., Gao, Y., He, Z., Wang, Y., Liu, W., Liu, J., & Chen, Y. (2025). A novel framework for enhancing postpartum hemorrhage prediction: Balanced data preprocessing meets neural network modeling. *Biomedical Signal Processing and Control*, *110*(May), 108073. Advance online publication. DOI: 10.1016/j.bspc.2025.108073

Wang, R., Wang, J., Tan, S., Duan, Y., Wang, Z.-K., & Yang, Y. (2019). Opportunities and Challenges of Lead-Free Perovskite Optoelectronic Devices. *Trends in Chemistry*, *1*(4), 368–379. DOI: 10.1016/j.trechm.2019.04.004

Wang, R., & Wang, S. (2001). How does consensus scoring work for virtual library screening? an idealized computer experiment. *Journal of Chemical Information and Computer Sciences*, *41*(5), 1422–1426. DOI: 10.1021/ci010025x PMID: 11604043

Wang, S., Zhao, L., & Liu, Y. (2023). A concurrent multiscale method based on smoothed molecular dynamics for large-scale parallel computation at finite temperature. *Computer Methods in Applied Mechanics and Engineering*, *406*, 115898. DOI: 10.1016/j.cma.2023.115898

Wang, T., Peng, Y., Jiang, W., Huang, Y. M., Rahman, B. M. F., Divan, R., Rosenmann, D., & Wang, G. (2016). Integrating nanopatterned ferromagnetic and ferroelectric thin films for electrically tunable RF applications. *IEEE Transactions on Microwave Theory and Techniques*, *65*(2), 504–512. DOI: 10.1109/TMTT.2016.2616869

Wang, X., Zhang, Y., Yang, Y., Wu, X., Fan, H., & Qiao, Y. (2017). Identification of berberine as a direct thrombin inhibitor from traditional Chinese medicine through structural, functional and binding studies. *Scientific Reports*, *7*(1), 44040. DOI: 10.1038/srep44040 PMID: 28276481

Wang, Y., Verma, P., Zhang, L., Li, Y., Liu, Z., Truhlar, D. G., & He, X. (2020). M06-SX screened-exchange density functional for chemistry and solid-state physics. *Proceedings of the National Academy of Sciences of the United States of America*, *117*(5), 2294–2301. DOI: 10.1073/pnas.1913699117 PMID: 31953258

Webber, K. G., Vögler, M., Khansur, N. H., Kaeswurm, B., Daniels, J. E., & Schader, F. H. (2017). Review of the mechanical and fracture behavior of perovskite lead-free ferroelectrics for actuator applications. *Smart Materials and Structures*, *26*(6), 063001. DOI: 10.1088/1361-665X/aa590c

Weber, O. J. (2018). Structural Chemistry of Hybrid Halide Perovskites for Thin Film Photovoltaics.

Wei, F., Deng, Z., Sun, S., Xie, F., Kieslich, G., Evans, D. M., Carpenter, M. A., Bristowe, P. D., & Cheetham, A. K. (2016). The synthesis, structure and electronic properties of a lead-free hybrid inorganic-organic double perovskite (MA)2KBiCl6 (MA = methylammonium). *Materials Horizons*, *3*(4), 328–332. DOI: 10.1039/C6MH00053C

Weinan, E., & Engquist, B. (2003). The heterognous multiscale methods. *Communications in Mathematical Sciences*, *1*(1), 87–132. DOI: 10.4310/CMS.2003.v1.n1.a8

Wei, X., Li, Z., & Zhao, G. (2024). A review of multiscale numerical modeling of rock mechanics and rock engineering. *Deep Underground Science and Engineering*.

Werder, T., Walther, J. H., & Koumoutsakos, P. (2005a). Hybrid atomistic-continuum method for the simulation of dense fluid flows. *Journal of Computational Physics*, *205*(1), 373–390. DOI: 10.1016/j.jcp.2004.11.019

Westermayr, J., & Marquetand, P. (2020). Machine learning for electronically excited states of molecules. *Chemical Reviews*, *121*(16), 9873–9926. DOI: 10.1021/acs.chemrev.0c00749 PMID: 33211478

Wiitala, K. W., Al-Rashid, Z. F., Dvornikovs, V., Hoye, T. R., & Cramer, C. J. (2007). Evaluation of various DFT protocols for computing 1H and 13C chemical shifts to distinguish stereoisomers: Diastereomeric 2-, 3-, and 4-methylcyclohexanols as a test set. *Journal of Physical Organic Chemistry*, *20*(5), 345–354. DOI: 10.1002/poc.1151

Wilhelm, J., Golze, D., Talirz, L., Hutter, J., & Pignedoli, C. A. (2017). Toward GW calculations on thousands of atoms. *The Journal of Physical Chemistry Letters*, *9*(2), 306–312. DOI: 10.1021/acs.jpclett.7b02740 PMID: 29280376

Willoughby, P. H., Jansma, M. J., & Hoye, T. R. (2014). A guide to small-molecule structure assignment through computation of (1H and 13C) NMR chemical shifts. *Nature Protocols*, *9*(3), 643–660. DOI: 10.1038/nprot.2014.042 PMID: 24556787

Winiewska-Szajewska, M., & Poznański, J. (2025). Differential scanning fluorimetry followed by microscale thermophoresis and/or isothermal titration calorimetry as an efficient tool for ligand screening. *Biophysical Reviews*, *17*(1), 199–223. DOI: 10.1007/s12551-025-01280-3 PMID: 40060009

Wong, K. M., Irfan, M., Mahmood, A., & Murtaza, G. (2017). First principles study of the structural and optoelectronic properties of the A2InSbO6 (A = Ca, Sr, Ba) compounds. *Optik (Stuttgart)*, *130*, 517–524. DOI: 10.1016/j.ijleo.2016.10.139

Wu, X., Du, Z., Ma, R., Zhang, X., Yang, D., Liu, H., & Zhang, Y. (2024). Qualitative and quantitative studies of phthalates in extra virgin olive oil (EVOO) by surface-enhanced Raman spectroscopy (SERS) combined with long short term memory (LSTM) neural network. *Food Chemistry, 433*(August 2023). DOI: 10.1016/j.foodchem.2023.137300

Wu, X., Hu, J., Wang, X., Xin, L., Li, C., & Wei, X. (2021). Effective removal of arsenide from aqueous solutions using mesoporous CoFe2O4/graphene oxide nanocomposites assisted by artificial intelligence. *Carbon Letters*, *31*(6), 1163–1178. DOI: 10.1007/s42823-021-00239-0

Wu, X., Zhao, Z., Tian, R., Niu, Y., Gao, S., & Liu, H. (2021). Total synchronous fluorescence spectroscopy coupled with deep learning to rapidly identify the authenticity of sesame oil. *Spectrochimica Acta. Part A: Molecular and Biomolecular Spectroscopy*, *244*, 118841. DOI: 10.1016/j.saa.2020.118841 PMID: 32871392

Wu, Z., & Cohen, R. E. (2006). More accurate generalized gradient approximation for solids. *Physical Review B: Condensed Matter and Materials Physics*, *73*(23), 235116. DOI: 10.1103/PhysRevB.73.235116

Xiao, S. P., & Belytschko, T. (2004). A bridging domain method for coupling continua with molecular dynamics. *Computer Methods in Applied Mechanics and Engineering*, *193*(17–20), 1645–1669. DOI: 10.1016/j.cma.2003.12.053

Xiao, W., Hu, Y., Yan, K., Tang, L., Chen, X., D'Agosta, R., & Yang, K. (2025). Phonon-limited carrier mobility modeling of two-dimensional semiconductors based on first principles. *Journal of Physics Condensed Matter*, *37*(26), 263001. DOI: 10.1088/1361-648X/ade18e PMID: 40472870

Xiao, X., Ma, X., Kari, T., Xu, Q., & Fan, M. (2023). Numerical simulation of droplet evaporation based on the smoothed particle hydrodynamics method. *Thermal Science*, *27*(5 Part A), 3745–3756. DOI: 10.2298/TSCI230311101X

Xin, D., Sader, C. A., Chaudhary, O., Jones, P., Wagner, K., Tautermann, C. S., Yang, Z., Busacca, C. A., Saraceno, R. A., Fandrick, K. R., Gonnella, N. C., Horspool, K., Hansen, G., & Senanayake, C. H. (2017). Development of a 13C NMR chemical shift prediction procedure using B3LYP/CC-PVDZ and empirically derived systematic error correction terms: A computational small molecule structure elucidation method. *The Journal of Organic Chemistry*, *82*(10), 5135–5145. DOI: 10.1021/acs.joc.7b00321 PMID: 28398046

Xu, W., Hou, X., Lu, T., Jiao, Z., Xu, X., & Dai, J. (2024). Research on the High Permeability Permalloy and its Application to the Hydrogen Maser. 2024 European Frequency and Time Forum (EFTF), 300–303. DOI: 10.1109/EFTF61992.2024.10722257

Xu, C., & Zhang, Y. (2022). Estimating adsorption isotherm parameters in chromatography via a virtual injection promoting double feed-forward neural network. *Journal of Inverse and Ill-Posed Problems*, *30*(5), 693–712. DOI: 10.1515/jiip-2020-0121

Xu, D., Zhang, Q., Huo, X., Wang, Y., & Yang, M. (2023). Advances in data-assisted high-throughput computations for material design. *Materials Genome Engineering Advances*, *1*(1), e11. Advance online publication. DOI: 10.1002/mgea.11

Xu, X., Jin, L., Wen, T., Liao, Y., Tang, X., Zhang, H., & Zhong, Z. (2021). Effects of substrate annealing on uniaxial magnetic anisotropy and ferromagnetic resonance frequency of Ni80Fe20 films deposited on self-organized periodically rippled sapphire substrates. *Vacuum*, *186*, 110047. DOI: 10.1016/j.vacuum.2021.110047

Xu, Z., Meenu, M., Chen, P., & Xu, B. (2020). Comparative study on phytochemical profiles and antioxidant capacities of chestnuts produced in different geographic area in China. *Antioxidants*, *9*(3), 190. DOI: 10.3390/antiox9030190

Yamaguchi, M., Suezawa, K., Takahashi, Y., Arai, K. I., Kikuchi, S., Shimada, Y., Tanabe, S., & Ito, K. (2000). Magnetic thin-film inductors for RF-integrated circuits. *Journal of Magnetism and Magnetic Materials*, *215–216*, 807–810. DOI: 10.1016/S0304-8853(00)00293-6

Yanai, T., Koda, K., Kaji, J., Aramaki, H., Eguchi, K., Takashima, K., Nakano, M., & Fukunaga, H. (2018). Electroplated Fe-Co-Ni films prepared in ammonium-chloride-based plating baths. *AIP Advances*, *8*(5), 056127. DOI: 10.1063/1.5007782

Yang, D., He, Q., Wang, H., Gao, Y., Zhang, S., Yao, G., & Zhou, M. (2025). High-dimensional railway vertical alignment optimization using hybrid differential evolution and gradient descent. *Automation in Construction, 179*(December 2024), 106413. DOI: 10.1016/j.autcon.2025.106413

Yang, B., Liang, B., Qian, Y., Zheng, R., Su, S., Guo, Z., & Jiang, L. (2024). Parameter identification of PEMFC via feedforward neural network-pelican optimization algorithm. *Applied Energy*, *361*(February), 122857. DOI: 10.1016/j.apenergy.2024.122857

Yang, F., Lin, S., Yang, L., Liao, J., Chen, Y., & Wang, C.-Z. (2017). First-principles investigation of metal-doped cubic BaTiO 3. *Materials Research Bulletin*, *96*, 372–378. DOI: 10.1016/j.materresbull.2017.03.023

Yang, P. Y., Song, C., Fan, B., Zeng, F., & Pan, F. (2009). The role of rotatable anisotropy in the asymmetric magnetization reversal of exchange biased NiO/Ni bilayers. *Journal of Applied Physics*, *106*(1), 013902. Advance online publication. DOI: 10.1063/1.3157176

Yang, S., Li, Z., Li, W., Li, C., Yang, X., Zhao, Y., & Liu, R. (2024). Unraveling the variations and relationships between phytochemical constituents and anti-inflammatory potentials of three Zanthoxylum species through metabolomics and explainable machine learning. *Industrial Crops and Products*, *212*, 118281. DOI: 10.1016/j.indcrop.2024.118281

Yang, T., Kimura, T., & Otani, Y. (2008). Giant spin-accumulation signal and pure spin-current-induced reversible magnetization switching. *Nature Physics*, *4*(11), 851–854. DOI: 10.1038/nphys1095

Yang, X., & Kong, S.-C. (2017). Smoothed particle hydrodynamics method for evaporating multiphase flows. *Physical Review. E*, *96*(3), 033309. DOI: 10.1103/PhysRevE.96.033309

Yang, X., & Kong, S.-C. (2020). Smoothed particle hydrodynamics modeling of fuel drop impact on a heated surface at atmospheric and elevated pressures. *Physical Review. E*, *102*(3), 033313. DOI: 10.1103/PhysRevE.102.033313

Yang, X., Ray, M., Kong, S.-C., & Kweon, C.-B. M. (2019). SPH simulation of fuel drop impact on heated surfaces. *Proceedings of the Combustion Institute*, *37*(3), 3279–3286. DOI: 10.1016/j.proci.2018.07.078

Yang, X., Wei, J. Q., Li, X. H., Gong, L. Q., Wang, T., & Li, F. S. (2012). Thickness dependence of microwave magnetic properties in electrodeposited FeCo soft magnetic films with in-plane anisotropy. *Physica B, Condensed Matter*, *407*(3), 555–559. DOI: 10.1016/j.physb.2011.11.049

Yang, Y., Ma, L., Gan, G. Y., Wang, W., & Tang, B. Y. (2019). Investigation of thermodynamic properties of high entropy (TaNbHfTiZr)C and (TaNbHfTiZr)N. *Journal of Alloys and Compounds*, *788*, 1076–1083. DOI: 10.1016/j.jallcom.2019.02.254

Yang, Z., Chakraborty, M., & White, A. D. (2021). Predicting chemical shifts with graph neural networks. *Chemical Science (Cambridge)*, *12*(32), 10802–10809. DOI: 10.1039/D1SC01895G PMID: 34476061

Yan, L., Duan, X., Liu, B., & Xu, J. (2018). Gaussian processes and polynomial chaos expansion for regression problem: Linkage via the RKHS and comparison via the KL divergence. *Entropy (Basel, Switzerland)*, *20*(3), 191. DOI: 10.3390/e20030191

Yaseen, H., Ain, Q., Murtaza, H., Munir, J., Ahmed, A. A. A., Qaid, S. M. H., & Aldwayyan, A. S. (2025). The Influence of Multiple Potentials on the Bandgap Improvement and the Physical Properties of Cs2InSbF6 Fluoroperovskite: A DFT Study. *Journal of Inorganic and Organometallic Polymers and Materials*, *35*(12), 1–14. DOI: 10.1007/s10904-025-03903-4

Yasuda, S., & Yamamoto, R. (2008). A model for hybrid simulations of molecular dynamics and computational fluid dynamics. *Physics of Fluids*, *20*(11), 113101. Advance online publication. DOI: 10.1063/1.3003218

Yasuda, S., & Yamamoto, R. (2010). Multiscale modeling and simulation for polymer melt flows between parallel plates. *Physical Review E: Statistical, Nonlinear, and Soft Matter Physics*, *81*(3), 036308. Advance online publication. DOI: 10.1103/PhysRevE.81.036308

Ye, T., Pan, D., Huang, C., & Liu, M. (2019). Smoothed particle hydrodynamics (SPH) for complex fluid flows: Recent developments in methodology and applications. *Physics of Fluids*, *31*(1), 011301. DOI: 10.1063/1.5068697

Yi, J.-C., Yang, Z.-Y., Zhao, W.-T., Yang, Z.-J., Zhang, X.-C., Wu, C.-K., Lu, A.-P., & Cao, D.-S. (2024). ChemMORT: An automatic ADMET optimization platform using deep learning and multiobjective particle swarm optimization. *Briefings in Bioinformatics*, *25*(2), bbae008. Advance online publication. DOI: 10.1093/bib/bbae008 PMID: 38385872

Yingngam, B. (2025a). AI-Driven Drug Discovery and Repurposing for Neurodegenerative Disorders. In Machine Learning for Neurodegenerative Disorders: Advancements and Applications (pp. 80-124). CRC Press. DOI: 10.1201/9781032661025-4

Yingngam, B. (2024a). AI Models for Biopharmaceutical Property Prediction. In Khanna, A., Barachi, M. E., Jain, S., Kumar, M., & Nayyar, A. (Eds.), *Artificial Intelligence and Machine Learning in Drug Design and Development* (pp. 407–449). Wiley., DOI: 10.1002/9781394234196.ch13

Yingngam, B. (2024b). Introduction to Alzheimer's Disease, Biomarkers, and the AI Revolution. In Lilhore, U. K., Anand, A., Kumar, A., Yadav, S. P., & Vyas, N. (Eds.), *AI-Driven Alzheimer's Disease Detection and Prediction* (pp. 1–23). IGI Global Scientific Publishing., DOI: 10.4018/979-8-3693-3605-2.ch001

Yingngam, B. (2025b). AI in Predictive Toxicology. In Shaik, A. (Ed.), *AI-Powered Advances in Pharmacology* (pp. 79–134). IGI Global Scientific Publishing., DOI: 10.4018/979-8-3693-3212-2.ch005

Yingngam, B., Navabhatra, A., & Sillapapibool, P. (2024). AI-Driven Decision-Making Applications in Pharmaceutical Sciences. In Nguyen, T. V. T., & Vo, N. T. M. (Eds.), *Using Traditional Design Methods to Enhance AI-Driven Decision Making* (pp. 1–63). IGI Global Scientific Publishing., DOI: 10.4018/979-8-3693-0639-0.ch001

Yingngam, B., & Sethabouppha, B. (2024). Advanced AI Applications for Drug Discovery. In Shah, I. A., & Sial, Q. (Eds.), *Advances in Computational Intelligence for the Healthcare Industry 4.0* (pp. 42–86). IGI Global Scientific Publishing., DOI: 10.4018/979-8-3693-2333-5.ch003

You, J., Li, G., Zhang, S., Zhang, X., Luo, J., Rao, M., & Peng, Z. (2021). Synthesis, characterization and thermodynamic properties of KNbO3. *Journal of Alloys and Compounds*, *882*, 160641. DOI: 10.1016/j.jallcom.2021.160641

Yousefi, M., Akbari, H., Adibzadeh, A., Mohammadi, A. A., Baziar, M., Abbasi Farajzadeh, M., & Akbari, H. (2024). Adsorption of diazinon from aqueous solution using metal organic framework and functionalized graphene: Comparison of BBD, ANN models. *Chemosphere*, *351*(December 2023), 141222. DOI: 10.1016/j.chemosphere.2024.141222

Yuan, X., Li, L., Shi, Z., Liang, H., Li, S., & Qiao, Z. (2022). Molecular-fingerprint machine-learning-assisted design and prediction for high-performance MOFs for capture of NMHCs from air. *Advanced Powder Materials*, *1*(3), 100026. DOI: 10.1016/j.apmate.2021.12.002

Yuan, Z., Zhang, L., Wang, D., Jiang, J., Harrington, P. de B., Mao, J., Zhang, Q., & Li, P. (2020). Detection of flaxseed oil multiple adulteration by near-infrared spectroscopy and nonlinear one class partial least squares discriminant analysis. *Lebensmittel-Wissenschaft + Technologie*, *125*, 109247. DOI: 10.1016/j.lwt.2020.109247

Yu, J., Sinnott, S. B., & Phillpot, S. R. (2007). Charge optimized many-body potential for the Si/ SiO 2 system. *Physical Review B: Condensed Matter and Materials Physics*, *75*(8), 085311. DOI: 10.1103/PhysRevB.75.085311

Yu, L., & Zunger, A. (2012). Identification of Potential Photovoltaic Absorbers Based on First-Principles Spectroscopic Screening of Materials. *Physical Review Letters*, *108*(6), 068701. DOI: 10.1103/PhysRevLett.108.068701

Yusoff, M., Sallehud-din, M. T. M., Tahir, N., Yaacob, W. F. W., Maarof, N. N. N. A., Zain, J. M., Azmi, P. A. R., & Karunakumar, C. (2023). Drier Bed Adsorption Predictive Model with Enhancement of Long Short-Term Memory and Particle Swarm Optimization. *International Journal of Advanced Computer Science and Applications*, *14*(12), 67–74. DOI: 10.14569/IJACSA.2023.0141206

Zafar, Z., Yi, S., Li, J., Li, C., Zhu, Y., Zada, A., Yao, W., Liu, Z., & Yue, X. (2021). Recent development in defects Engineered Photocatalysts: An overview of the experimental and theoretical strategies. *Energy & Environmental Materials*, *5*(1), 68–114. DOI: 10.1002/eem2.12171

Zakeri, Kh., Lindner, J., Barsukov, I., Meckenstock, R., Farle, M., von Hörsten, U., Wende, H., Keune, W., Rocker, J., Kalarickal, S. S., Lenz, K., Kuch, W., Baberschke, K., & Frait, Z. (2007). Spin dynamics in ferromagnets: Gilbert damping and two-magnon scattering. *Physical Review B: Condensed Matter and Materials Physics*, *76*(10), 104416. DOI: 10.1103/PhysRevB.76.104416

Zangwill, A. (2013). Hartree and Thomas: the forefathers of density functional theory. Archive for History of Exact Sciences 2013 67:3, 67(3), 331–348. DOI: 10.1007/s00407-013-0114-4

Zeghbib, W., Boudjouan, F., Carneiro, J., Oliveira, A. L. S., Sousa, S. F., Pintado, M. E., Ourabah, A., Vasconcelos, V., & Lopes, G. (2024). LC-ESI-UHR-QqTOF-MS/MS profiling and anti-inflammatory potential of the cultivated *Opuntia ficus-indica* (L.) Mill. and the wild *Opuntia stricta* (Haw.) Haw. fruits from the Algerian region. *Food Chemistry*, *460*(Part 1), 140414. DOI: 10.1016/j.foodchem.2024.140414

Zeng, S., Zhang, Z., Cheng, X., Cai, X., Cao, M., & Guo, W. (2024). Prediction of soluble solids content using near-infrared spectra and optical properties of intact apple and pulp applying PLSR and CNN. *Spectrochimica Acta Part A: Molecular and Biomolecular Spectroscopy*, *304*(July 2023), 123402. DOI: 10.1016/j.saa.2023.123402

Zezyulina, P. A., Iakubov, I. T., Lagarkov, A. N., Maklakov, S. A., Maklakov, S. S., Naboko, A. S., Osipov, A. V., Petrov, D. A., Rozanov, K. N., & Ryzhikov, I. A. (2016). Effect of Perpendicular Anisotropy and Eddy Currents on the Microwave Performance of Single-Layer and Multi-Layer Permalloy Films. *IEEE Magnetics Letters*, *7*, 1–4. DOI: 10.1109/LMAG.2016.2597230

Zhai, C., Shi, C., Hu, Y., Xu, Z., & Wang, R. (2022). Anti-breast carcinoma effects of green synthesized tin nanoparticles from Calendula officinalis leaf aqueous extract inhibits MCF7, Hs 319.T, and MCF10 cells proliferation. *Journal of Experimental Nanoscience*, *17*(1), 351–361. DOI: 10.1080/17458080.2022.2076836

Zhang, Z., Fung, V., & Hu, G. (2025). Pushing the boundary of the stability and band gap Pareto front by going towards high-entropy perovskites † Lead-free Cs 2 BX 6 (B = Zr 4+, Sn 4+, Te 4+, Hf 4+, Re 4+, Os 4+, Ir 4+, and Pt 4+ and X = Cl. 17, 11376. DOI: 10.1039/D4NR05013D

Zhang, Z., Rao, J., Zhong, J., Bai, W., Wang, D., Ning, S., . . . Hua, W. (2025). Fitness aligned structural modeling enables scalable virtual screening with AuroBind. arXiv preprint arXiv:2508.02137. https://doi.org//arXiv.2508.02137DOI: 10.48550

Zhang, C., Zhu, Y., Wu, D., Adams, N. A., & Hu, X. (2022). Smoothed particle hydrodynamics: Methodology development and recent achievement. *Journal of Hydrodynamics*, *34*(5), 767–805. DOI: 10.1007/s42241-022-0052-1

Zhang, H., Li, Y.-Y., Yang, M.-Y., Zhang, B., Yang, G., Wang, S.-G., & Wang, K.-Y. (2015). Tuning the magnetic anisotropy of CoFeB grown on flexible substrates. *Chinese Physics B*, *24*(7), 077501. DOI: 10.1088/1674-1056/24/7/077501

Zhang, L., Huang, X., Li, P., Na, W., Jiang, J., Mao, J., Ding, X., & Zhang, Q. (2017). Multivariate adulteration detection for sesame oil. *Chemometrics and Intelligent Laboratory Systems*, *161*, 147–150. DOI: 10.1016/j.chemolab.2016.11.009

Zhang, L., Shuai, Q., Li, P., Zhang, Q., Ma, F., Zhang, W., & Ding, X. (2016). Ion mobility spectrometry fingerprints: A rapid detection technology for adulteration of sesame oil. *Food Chemistry*, *192*, 60–66. DOI: 10.1016/j.foodchem.2015.06.096 PMID: 26304320

Zhang, L., Wang, Y., Zheng, C., Zhou, Z., & Chen, Z. (2024). Cellular thermal shift assay: An approach to identify and assess protein target engagement. *Expert Review of Proteomics*, *21*(9-10), 387–400. DOI: 10.1080/14789450.2024.2406785 PMID: 39317941

Zhang, L., Xie, X., Li, D., Yuan, Y., Xue, X., Li, Q., Xu, J., Wang, H., Hu, F., & Zhang, X. (2021). Investigation on lead-free Mn-doped Cs2NaInCl6 double perovskite phosphors and their optical properties. *Optical Materials*, *122*, 111802. DOI: 10.1016/j.optmat.2021.111802

Zhang, N., Zheng, T., & Wu, J. (2020). Lead-Free (K,Na)NbO_3-Based Materials : Preparation Techniques and Piezoelectricity. *ACS Omega*, *5*(7), 3099–3107. DOI: 10.1021/acsomega.9b03658

Zhang, P., Yang, J., & Wei, S.-H. (2018). Manipulation of cation combinations and configurations of halide double perovskites for solar cell absorbers. *Journal of Materials Chemistry. A, Materials for Energy and Sustainability*, *6*(4), 1809–1815. DOI: 10.1039/C7TA09713A

Zhang, W. F., He, Y. L., Zhang, M. S., Yin, Z., & Chen, Q. (2000). Raman scattering study on anatase TiO 2 nanocrystals. *Journal of Physics. D, Applied Physics*, *33*(8), 912–916. DOI: 10.1088/0022-3727/33/8/305

Zhang, W., Starr, F. W., Beers, K. L., & Douglas, J. F. (2022). Reactive Molecular Dynamics Simulations of the Depolymerization of Polyethylene Using Graphene-Oxide-Supported Platinum Nanoparticles. *The Journal of Physical Chemistry A*, *126*(20), 3167–3173. DOI: 10.1021/acs.jpca.2c01167

Zhang, W.-X., Huang, J., Tian, X.-Y., Liu, Y.-H., Jia, M.-Q., Wang, W., Jin, C.-Y., Song, J., & Zhang, S.-Y. (2023). A review of progress in o-aminobenzamide-based HDAC inhibitors with dual targeting capabilities for cancer therapy. *European Journal of Medicinal Chemistry, 259*, 115673. DOI: 10.1016/j.ejmech.2023.115673 PMID: 37487305

Zhang, Y., Bao, X., Zhu, Y., Dai, Z., Shen, Q., & Xue, Y. (2024). Advances in machine learning screening of food bioactive compounds. *Trends in Food Science & Technology*, *150*, 104578. DOI: 10.1016/j.tifs.2024.104578

Zhao, X.-G., Fu, Y., Yang, D., Xu, Q., Zhang, L., Yang, J.-H., Yu, L., Wei, S.-H., & Zhang, L. (2017). Design of Lead-Free Inorganic Halide Perovskites for Solar Cells via Cation-Transmutation. *Journal of the American Chemical Society*, *139*(7), 2630–2638. DOI: 10.1021/jacs.6b09645

Zhao, Y., & Truhlar, D. G. (2007). The M06 suite of density functionals for main group thermochemistry, thermochemical kinetics, noncovalent interactions, excited states, and transition elements: Two new functionals and systematic testing of four M06-class functionals and 12 other functionals. *Theoretical Chemistry Accounts*, *120*(1–3), 215–241. DOI: 10.1007/s00214-007-0310-x

Zhao, Z.-J., Liu, S., Zha, S., Cheng, D., Studt, F., Henkelman, G., & Gong, J. (2020). Publisher Correction: Theory-guided design of catalytic materials using scaling relationships and reactivity descriptors. *Nature Reviews. Materials*, *7*(8), 671–671. DOI: 10.1038/s41578-020-0192-2

Zheng, L., Zou, Y., Xie, T., Wu, X., Tan, Y., Mei, S., Geng, Y., Chen, S., Xu, S., & Niu, M.-M. (2023). Discovery of a dual tubulin and neuropilin-1 (NRP1) inhibitor with potent in vivo anti-tumor activity via pharmacophore-based docking screening, structure optimization, and biological evaluation. *Journal of Medicinal Chemistry*, *66*(23), 16187–16200. DOI: 10.1021/acs.jmedchem.3c01572 PMID: 38093696

Zheng, M., & Luo, X. (2024). Joint estimation of State of Charge (SOC) and State of Health (SOH) for lithium ion batteries using Support Vector Machine (SVM), Convolutional Neural Network (CNN) and Long Sort Term Memory Network (LSTM) models. *International Journal of Electrochemical Science*, *19*(9), 100747. DOI: 10.1016/j.ijoes.2024.100747

Zhou, X., Zhao, C., Huang, Y., Zhou, C., & Ye, J. (2025). Improved fractional-order gradient descent method based on multilayer perceptron. *Neural Networks, 183*(March 2024), 106970. DOI: 10.1016/j.neunet.2024.106970

Zhou, A. (2012). Hohenberg–Kohn theorem for Coulomb type systems and its generalization. *Journal of Mathematical Chemistry*, *50*(10), 2746–2754. DOI: 10.1007/s10910-012-0061-3

Zhou, A. (2019). A mathematical aspect of Hohenberg-Kohn theorem. *Science China. Mathematics*, *62*(1), 63–68. DOI: 10.1007/s11425-018-9337-2

Zhou, C., Li, T., Wei, X., & Yan, B. (2020). Effect of the sputtering power on the structure, morphology and magnetic properties of Fe films. *Metals*, *10*(7), 896. DOI: 10.3390/met10070896

Zhou, G., Rusnac, D.-V., Park, H., Canzani, D., Nguyen, H. M., Stewart, L., Bush, M. F., Nguyen, P. T., Wulff, H., Yarov-Yarovoy, V., Zheng, N., & DiMaio, F. (2024). An artificial intelligence accelerated virtual screening platform for drug discovery. *Nature Communications*, *15*(1), 7761. DOI: 10.1038/s41467-024-52061-7 PMID: 39237523

Zhou, M., Yu, B., Shi, J., Zhou, W., Fang, C., Gong, Y., Zuo, C., & Zhang, J. (2024). Improved phase-generated carrier demodulation technique based on LM algorithm with low harmonic distortion and high stability. *Optical Fiber Technology*, *87*, 103735. DOI: 10.1016/j.yofte.2024.103735

Zhou, X., Xu, D., Yu, T., Duan, X., Zhao, H., & Zhao, W. (2023). Effects of Mn doping and sintering condition on the microstructure, dielectric, and energy storage properties of Ba0.8Sr0.2TiO3 ceramics. *Journal of Materials Science Materials in Electronics*, *34*(20), 1511. DOI: 10.1007/s10854-023-10918-5

Zhou, Z., Xu, T., & Zhang, X. (2024). Empowerment of AI algorithms in biochemical sensors. *Trends in Analytical Chemistry*, *173*, 117613. DOI: 10.1016/j.trac.2024.117613

Zhu, M., Qu, J., & Deng, Q. (2024). Identification of potential inhibitors against Staphylococcus aureus shikimate dehydrogenase through virtual screening and susceptibility test. *Journal of Enzyme Inhibition and Medicinal Chemistry*, *39*(1), 2301768. DOI: 10.1080/14756366.2024.2301768 PMID: 38234148

Zhu, T., Zhang, J. Z. H., & He, X. (2013). Automated Fragmentation QM/MM Calculation of Amide Proton Chemical Shifts in Proteins with Explicit Solvent Model. *Journal of Chemical Theory and Computation*, *9*(4), 2104–2114. DOI: 10.1021/ct300999w PMID: 26583557

Zimba, S., Dougill, A., Chanza, C., Boesch, C., & Kepinski, S. (2025). Gender differential in choices of crop variety traits and climate-smart cropping systems: Insights from sorghum and millet farmers in drought-prone areas of Malawi. *Plants, People, Planet*, *7*(3), 689–703. DOI: 10.1002/ppp3.10467

Zonyane, S., Chen, L., Xu, M.-J., Gong, Z.-N., Xu, S., & Makunga, N. P. (2019). Geographic-based metabolomic variation and toxicity analysis of Sutherlandia frutescens L. R.Br. – An emerging medicinal crop in South Africa. *Industrial Crops and Products*, *133*, 414–423. DOI: 10.1016/j.indcrop.2019.03.010

Zunger, A., & Freeman, A. J. (1977). Ground-state electronic properties of diamond in the local-density formalism. *Physical Review. B, Solid State*, *15*(10), 5049–5065. DOI: 10.1103/PhysRevB.15.5049

About the Contributors

Nadia Faska holds a Ph.D. in Physical Chemistry from Cadi Ayyad University, Marrakech, Morocco, and a European Master's degree in Theoretical Chemistry and Computational Modeling from the Autonomous University of Madrid, Spain. She is currently an Associate Professor (MCH) at the Faculty of Applied Sciences, Ibn Zohr University, Ait-Melloul, where she conducts research in computational chemistry, material science, and environmental applications. With a career spanning over two decades, Pr. FASKA has made significant contributions to the fields of molecular modeling, solvent effects on NMR spectroscopy, and the development of advanced materials for energy and environmental remediation. Her work combines theoretical approaches (DFT, Ab-initio) with experimental techniques to study intermolecular interactions, corrosion inhibition, and piezoelectric materials. She has published extensively in peer-reviewed journals. Pr. FASKA is an active member of the scientific community, serving as a reviewer for leading chemistry and engineering journals and participating in international conferences as a speaker and committee member. Her interdisciplinary research bridges chemistry, materials science, and artificial intelligence, with a focus on sustainable solutions for industrial and environmental challenges.

***

Noureddine Amenzou is a research Scientist since 2006 at Centre National de l'Energie, des Sciences et des Techniques Nucléaires (CNESTEN). He was the coordinator of several research projects in CNESTEN, supervisor of many national and international trainees and lecturer in many conferences. He has accomplished more than 10 expert missions for IAEA in Africa and Asia in the field Use of Isotope and Radiation Methods in Soil and Water Management and Crop Nutrition. He is the local organizer of two CRP projects meetings and IAEA Regional training courses related work.

Susanta Das is a seasoned educator with a proven record of teaching and mentoring diverse students, working at multiple institutions, and providing service to the various initiatives of organizations. He received his Ph.D. and M.A. degrees from Western Michigan University (WMU), USA, and M.Sc. degree from Banaras Hindu University (BHU), India all in Physics. He continued his research as a Marie-Curie post-doctoral fellow at Stockholm University, Stockholm, Sweden on beam diagnostics for the DESIREE (Double ElectroStatic Ion Ring ExpEriment) in the project DITANET (Diagnostic Techniques for particle Accelerators – a Marie-Curie initial training NETwork), at the Indian Institute of Science Education and Research-Kolkata, India on high-pressure physics as a Project Scientist-B, and at the University of Electro-Communications, Tokyo, Japan on ion-surface interactions as a Post-doctoral fellow. Throughout his career, Dr. Das worked and collaborated with many researchers, Ph.D., master, and visiting students from different countries. He further visited many countries to discuss research and an international conference participant (UK, Italy, Germany, Greece, Belgium, Bulgaria, Romania, Brazil etc.). He worked at several institutions before assuming his current role at Ajeenkya D Y Patil University. He received the Marie-Curie post-doctoral fellowship in Sweden, Science Academies' Summer Research Fellowship in India, Gowen Frostic Doctoral Fellowship, Department Graduate Research and Creative Scholar Award by WMU, and Leo R. Parpart Doctoral Fellowship by Dept. of Physics, WMU, among many others, throughout his academic journey. He co-authored several research articles in WoS/Scopus indexed international journals, conference proceedings, and scholarly book chapters and co-edited several books published by renowned international publishing houses (IGI-Global, Springer Nature, Nova Science Publishers, Wiley, Scrivener-Wiley, Notion Press). At present, Dr. Das continues to delve into cutting-edge fields, with a keen interest in nanotechnology, UN SDGs, and data science. His multifaceted contributions, spanning teaching, research, and administrative leadership, showcase a dedicated professional who is instrumental in advancing the vision and mission of the institutions he serves.

Hicham Kacimi-Naciri is a PhD student in materials science specializing in the modeling of piezoelectric materials for energy harvesting transducers and sensor applications using the Wien2k package, followed by synthesis.

Aditya Shrivastav is an aspiring technologist currently pursuing a Bachelor of Technology (B. Tech.) in Computer Science and Engineering at Ajeenkya D. Y. Patil University. He holds solid programming expertise in C, C++, and Python, alongside strong command over data and business tools such as Microsoft Excel and Tally software. Aditya is a member of Team Vac-Vectoor Hyperloop and participated with the team in the Global Hyperloop Competition organized by IIT Madras, contributing to cutting-edge transportation solutions. He actively engages in hackathons and various technical events to expand his skills and contribute to innovative problem-solving. He has authored and co-authored book chapters in the domains of solid waste management, bioplastics, nanotechnology applications, and artificial intelligence, showcasing his multidisciplinary approach and research acumen. Aditya is also working on an innovative project on predictive e-waste management. The project is targeted toward IT departments, schools, and colleges, aiming to support sustainable practices and resource optimization. In parallel with his academic journey, Aditya is serving as a faculty intern at Disha Computer Institute, where he teaches C, C++, basic Python, basic Java, Excel formulas, and Tally software. His dedication and teaching proficiency earned him a Letter of Appreciation and the opportunity to conduct a workshop on Advanced Excel at G. H. Raisoni College of Arts, Commerce and Science. With a strong foundation in technology, a passion for sustainability, and hands-on experience in both teaching and project development, Aditya Shrivastav is committed to driving impact through research, education, and innovation.

Pichit Sudta is a faculty member at the Faculty of Science and Technology, Phetchaburi Rajabhat University, Thailand. His research focuses on natural product chemistry, phytochemical analysis, and drug discovery from medicinal plants. Dr. Sudta has extensive experience in the isolation, characterization, and biological evaluation of bioactive compounds with potential pharmaceutical and therapeutic applications. His work aims to bridge traditional herbal knowledge with modern scientific approaches for the development of new drugs and health-promoting products. He has published in peer-reviewed journals and actively contributes to advancing natural product research and innovation in Thailand.

Charuwan Thanawiroon is an Assistant Professor in the Faculty of Pharmacy, Ubon Ratchathani University, Thailand. She received her PhD in Medicinal and Natural Product Chemistry from the University of Iowa, USA. Her academic and research interests focus on the analysis of synthetic drugs, medicinal compounds, and pharmaceutical care. With extensive expertise in pharmaceutical chemistry, Dr. Thanawiroon contributes significantly to both research and education, establishing herself as an important figure in the advancement of pharmaceutical sciences in Thailand.

Bancha Yingngam is an Associate Professor in the Faculty of Pharmacy, Ubon Ratchathani University, Thailand, specializing in Pharmaceutical Sciences and Technology. He earned both his pharmacy degree and PhD from Ubon Ratchathani University, and subsequently undertook a postdoctoral fellowship at Karl-Franzens-Universität Graz, Austria (2013–2021). Dr. Yingngam has published more than 70 scholarly works and serves as a reviewer for over 100 Scopus-indexed journals. His academic achievements include an h-index of 20 (Scopus) and recognition on Stanford University and Elsevier's World Top 2% Scientists list (2025) in the field of pharmacology and pharmacy, highlighting his substantial impact in pharmaceutical research.

Index

A

B

C

D

E

F

G

H

L

M

N

O

P

Q

R

S

T

V

W

www.ingramcontent.com/pod-product-compliance
Lightning Source LLC
LaVergne TN
LVHW060253011026
849161LV00011B/2710
9798337360584